CONTROL SYSTEMS-I

(With MATLAB Programs)

By :

Syed Hasan Saeed
Prof. & HOD
Deptt. of Electronics and Communication Engg.
Integral University, Lucknow

S.K. KATARIA & SONS®
Publisher of Engineering Books
4885/109, Prakash Mahal, Dr. Subhash Bhargav Lane,
Opposite Delhi Medical Association, Daryaganj, New Delhi–110002
Phone: +91-11-23243489, 43551243; Mobile: +91-9871775858
e-mail: katariabook@gmail.com, katariabooks@yahoo.com
Website: www.skkatariaandsons.com
(e-books, subscriptions and bulk buying also available on our website)

Published by:

S.K. Kataria & Sons®
4885/109, Prakash Mahal, Dr. Subhash Bhargav Lane,
Opposite Delhi Medical Association, Daryaganj, New Delhi–110002
Phone: +91-11-23243489, 43551243; Mobile: +91-9871775858
e-mail: katariabook@gmail.com, katariabooks@yahoo.com

First Edition: August, 2002; Eighth Edition: August, 2015, Reprint: 2023

Price: ₹ 595/-

Laser Typesetting:
Plus Computers, Delhi

Printed at:
New A.S. Printing Press, Delhi

PREFACE TO THE EIGHTH EDITION

It is a matter of great pleasure to present the eighth edition of the book *Control Systems-I*. Most of the chapters have been reinforced with additional materials where-ever considered necessary. Some more solved examples are added. Few examples are solved with MATLAB 6.1. I hope that this will be useful to the students.

The author is grateful to Dr. Mairajuddin Mufti, Asstt. Prof. NIT Srinagar, Mr. Afzal Ahmad & Dr. N.R. Kidwai for their valuable suggestions and feedback for improvement in the book.

The author express thanks to everybody who are directly or indirectly involved during the preparation of eighth edition.

Suggestion for the further improvement in the book will be gratefully acknowledged.

August, 2015 **SYED HASAN SAEED**
Lucknow

PREFACE TO THE FIRST EDITION

One of the primary purposes of this book is to provide the text to the students to understand the basic principles of Automatic control system. The material presented has been classtested in the present form. The presented text covers the course content of the subject control system of Indian universities. An effort is made to present the subject in simple language and a lot of suitable examples are given in each chapter which will help the students to understand the concept well. The salient feature of this book is the inclusion of objective type multiple choice questions given in the last. The illustration of principles and concepts has been done by means of solved examples. Step by step procedure for solving the problems are provided. The author hopes that this book will helps the students for whom it is written.

It is now my duty to make acknowledgements. First I would like to express my sincere appreciation to Prof. S.W. Akhtar, Prof. S.M. Iqbal and Prof. M. M Hasan, for their cooperation. I am thankful to my brother Syed Ishrat Saeed for his cooperation during this project. I am also thankful to Mr. Ateequr Rehman, Mr. Munawwar Alam and Mr. Ahsan Beg for their cooperation and suggestions during writing this book.

Last but not least, I would like to thank Mr. Sanjeev Kataria for bringing out this book in a nice manner.

Hope this book will prove to be useful for students. Further suggestions are solicited to increase the utility of this book for students who have been ultimate cause for writing this book.

January, 2006 **SYED HASAN SAEED**
Lucknow

ACKNOWLEDGEMENT

I am thankful to my brothers Er. S. Ahsan Saeed and S. Husain Saeed for their encouragement and support.

I am grateful to my colleagues Mr. Narendra Kumar Prajapati, Mr. Nirmesh K. Awasthi and Mr. Abdul Waheed Khan for their suggestions and help during revision process of this edition. I am also thankful to Mr. Manoj Tyagi of plus computers and his staff, Delhi for typing this manuscript in scheduled time.

Thanks are also due to Mr. Sanjeev K. Kataria for his great persuation and personal interest in publishing this edition.

January, 2006,
Lucknow **SYED HASAN SAEED**

To

My Parents

CONTENTS

Properties of Laplace Transforms

1.	$£\pm\left[\frac{d}{dt} f(t)\right] = sF(s) - f(0\pm)$
2.	$£\pm\left[\frac{d^n}{dt^n} f(t)\right] = s^n F(s) - \sum_{K=1}^{n} s^{n-K^{(K-1)}} f(0\pm)$
	where $\overset{K-1}{f(t)} = \frac{d^{K-1}}{dt^{K-1}} f(t)$
3.	$£\pm\left[\int f(t)\,dt\right] = \frac{F(s)}{s} + \frac{1}{s}\left[\int f(t)\,dt\right]_{t=0\pm}$
4.	$£\left[\int_0^t f(t)\,dt\right] = \frac{F(s)}{s}$
5.	$£\left[e^{-at} f(t)\right] = F(s+a)$
6.	$£\left[t\, f(t)\right] = \frac{-dF(s)}{ds}$
7.	$£\left[t^n f(t)\right] = (-1)^n \frac{d^n}{ds^n} F(s), \quad n = 1, 2, 3, 4 \ldots.$
8.	$£\left[\frac{1}{t} f(t)\right] = \int_s^{\infty} F(s)\,ds$ if $\lim_{t\to 0} \frac{1}{t} f(t)$ exists
9.	$£\left[f\left(\frac{t}{a}\right)\right] = aF(as)$
10.	$£\left[f_1(t) * f_2(t)\right] = F_1(s) \cdot F_2(s)$ (complex multiplication)
11.	Intial value Theorem $\lim_{t\to 0} f(t) = \lim_{s\to\infty} sF(s)$
12.	Final value theorem $\lim_{t\to\infty} sF(s) = \lim_{s\to 0} sF(s)$

Laplace Transform Pairs

S. No.	$f(t)$	$F(s)$
1.	Unit Impulse	1
2.	Unit Step	$1/s$
3.	t	$1/s^2$
4.	$\dfrac{t^{n-1}}{\lfloor n-1}, \quad n=1,2,3,\ldots$	$1/s^n$
5.	e^{-at}	$\dfrac{1}{s+a}$
6.	$t\,e^{-at}$	$\dfrac{1}{(s+a)^2}$
7.	$\dfrac{t^{n-1}e^{-at}}{\lfloor n-1}, \quad n=1,2,3,\ldots$	$\dfrac{1}{(s+a)^n}$
8.	$\dfrac{e^{-at}-e^{-bt}}{b-a}$	$\dfrac{1}{(s+a)(s+b)}$
9.	$\sin\omega t$	$\dfrac{\omega}{s^2+\omega^2}$
10.	$\cos\omega t$	$\dfrac{s}{s^2+\omega^2}$
11.	$e^{-at}\sin\omega t$	$\dfrac{\omega}{(s+a)^2+\omega^2}$
12.	$e^{-at}\cos\omega t$	$\dfrac{s+a}{(s+a)^2+\omega^2}$
13.	$\dfrac{\omega_n}{\sqrt{1-\xi^2}}\,e^{-\xi\omega_n t}\sin\omega_n\sqrt{1-\xi^2}\,t$	$\dfrac{{\omega_n}^2}{s^2+2\xi\omega_n s+{\omega_n}^2}$
14.	$1-\dfrac{1}{\sqrt{1-\xi^2}}e^{-\xi\omega_n t}\sin\left(\omega_n\sqrt{1-\xi^2}\,t+\tan^{-1}\dfrac{\sqrt{1-\xi^2}}{\xi}\right)$	$\dfrac{{\omega_n}^2}{s(s^2+2\xi\omega_n s+{\omega_n}^2)}$
15.	$-\dfrac{1}{\sqrt{1-\xi^2}}\,e^{-\xi\omega_n t}\sin\left(\omega_n\sqrt{1-\xi^2}\,t-\tan^{-1}\dfrac{\sqrt{1-\xi^2}}{\xi}\right)$	$\dfrac{s}{s^2+2\xi\omega_n s+{\omega_n}^2}$
16.	$\dfrac{1}{ab}\left[1+\dfrac{1}{a-b}\left(be^{-at}-ae^{-bt}\right)\right]$	$\dfrac{1}{s\,(s+a)\,(s+b)}$

1 INPUT-OUTPUT RELATIONSHIP

1.1. INTRODUCTION

The control system is very important for all engineers. The first significant control device was James Watt's flyball governer. This was invented in 1767 to keep the speed of the engine constant by regulating the supply of the steam to the engine.

In control system the behaviour of the system is described by the differential equations. Minorsky, in 1922 showed that how to determine the stability from the differential equations describing the systems. The differential equations may be ordinary differential equations or the difference equations. The control system can be classified as open loop control system and closed loop control systems.

1.2. OPEN LOOP CONTROL SYSTEM

The open loop control system is also known as control system without feedback or non-feedback control systems. In open loop systems the control action is independent of the desired output. In this system the output is not compared with the reference input.

The component of the open loop systems are controller and controlled process. The controller may be amplifier, filter, etc., depends upon the system. An input is applied to the controller and the output of the controller gives to the controlled process and we get the output (desired).

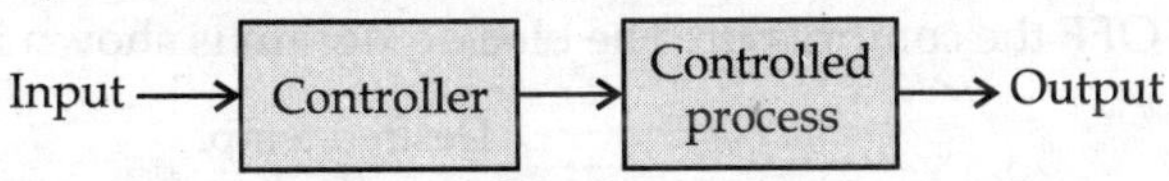

Fig. 1.1.

Examples :

1. Automatic washing machine is the example of the open loop systems. In the machine the operating time is set manually. After the completion of set time the machine will stops, with the result we may or may not get the desired (output) amount of cleanliness of washed cloths because there is no feedback is provided to the machine for desired output.
2. Immersion rod is another example of open loop system. The rod heats the water but how much heating is required is not sense by the rod because of no feedback to the rod.
3. A field control d.c. motor is the example of open loop system.

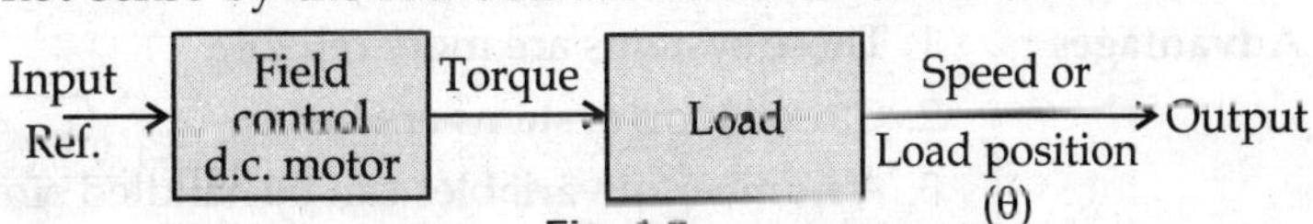

Fig. 1.2.

4. For automatic control of traffic the lamps of three different colours (red, yellow and green) are used. The time for each lamp is fixed. The operation of each lamp does not depends upon the density of the traffic but depends upon the fixed time. Thus, we can say that the control system which operates on the time basis is open loop system.

Advantages :
1. Open loop control systems are simple.
2. Open loop control systems are economical.
3. Less maintenance is required and not difficult.
4. Proper calibration is not a problem.

Disadvantages :
1. Open loop systems are inaccurate.
2. These are not reliable.
3. These are slow.
4. Optimization is not possible.

1.3. CLOSED LOOP CONTROL SYSTEM

Closed loop control systems are also known as feedback control systems. In closed loop control systems the control action is dependent on the desired output. If any system having one or more feedback paths forming a closed loop system.

In closed loop systems the output is compared with the reference input and error signal is produced. The error signal is fed to the controller to reduce the error and desired output is obtained.

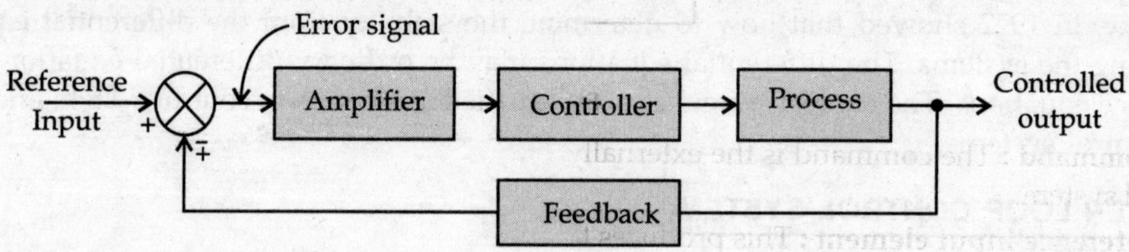

Fig. 1.3.

Example : In a room we need to regulate the temperature and humidity for comfortable living. Air conditioners are provided with thermostat. By measuring the actual room temperature and compared it with desired temperature, an error signal is produced, the thermostat turns ON the compressor or OFF the compressor. The block diagram is shown in Fig. 1.4.

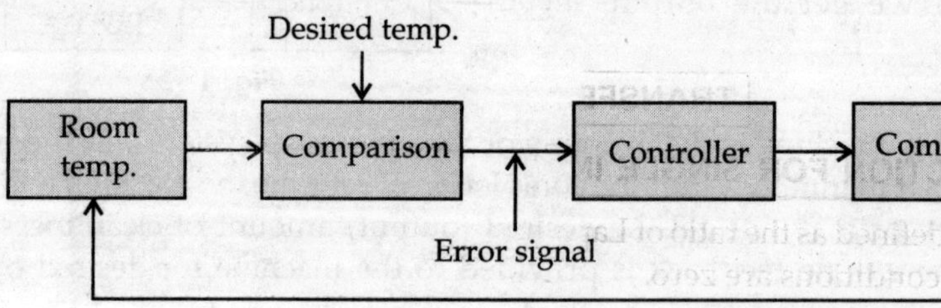

Fig. 1.4.

Advantages :
1. These systems are more reliable.
2. Closed loop systems are faster.
3. A number of variables can be handled simultaneously.
4. Optimization is possible.

Disadvantages :
1. Closed loop systems are expensive
2. Maintenance difficult.
3. Complicated installation.

1.4. COMPARISON BETWEEN OPEN LOOP AND CLOSED LOOP

Table 1.1

S.No.	Open loop system	Closed loop system
1.	These are not reliable	These are reliable
2.	It is easier to build	It is difficult to build
3.	If calibration is good, they perform accurately	They are accurate because of feedback
4.	Open loop systems are generally more stable	These are less stable
5.	Optimization is not possible	Optimization is possible

1.5. ELEMENTS OR COMPONENTS OF CLOSED LOOP SYSTEMS

The various components of closed loop system are shown in Fig. 1.5

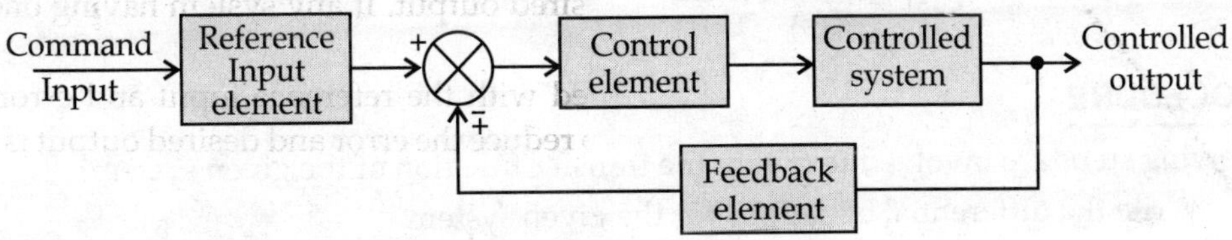

Fig. 1.5.

Command : The command is the externally produced input and independent of the feedback control system.

Reference input element : This produces the standard signals proportional to the command.

Error Detector : The error detector receives the measured signal and compare it with reference input. The difference of two signals produces the error signal.

Control element : This regulates the output according to the signal obtained from error detector.

Controlled system : This represents what we are controlling by the feedback loop.

Feedback element : This element fed back the output to the error detector for comparison with the reference input.

TRANSFER FUNCTION

1.6. TRANSFER FUNCTION FOR SINGLE INPUT SINGLE OUTPUT SYSTEM

The transfer function is defined as the ratio of Laplace transform of the output to the Laplace transform of input with all initial conditions are zero.

Consider a linear system having input $r(t)$ & $c(t)$ is the output of the system, the input-output relation can be described by the following n^{th}-order differential equation :

$$a_n \frac{d^n C(t)}{dt^n} + a_{n-1}\frac{d^{n-1}c(t)}{dt^{n-1}} + ... + a_1 \frac{dc(t)}{dt} + a_0 c(t) = b_m \frac{d^m r(t)}{dt^m} + b_{m-1}\frac{d^{m-1}r(t)}{dt^{m-1}} + ... + b_1 \frac{dr(t)}{dt} + b_0 r(t) \quad ...(1.1)$$

Where 'a' & 'b' are constants.

Take the laplace transform of equation (1.1)

$$(a_n s^n + a_{n-1}s^{n-1} + ... + a_1 s + a_0)\, C(s) = (b_m s^m + b_{m-1}s^{m-1} + ... + b_1 s + b_0)R(s) \quad ...(1.2)$$

We can define the transfer function as

$$G(s) = \frac{C(s)}{R(s)} = \frac{(b_m s^m + b_{m-1} s^{m-1} + ... + b_1 s + b_0)}{(a_n s^n + a_{n-1} s^{n-1} + ... + a_1 s + a_0)} \quad ...(1.3)$$

In equation (1.3), if the order of the denominator polynomial is greater than the order of the numerator polynomial then the transfer function is said to be STRICTLY PROPER. If the order of both polynomials are same, then the transfer function is PROPER. The transfer function is said to be IMPROPER, if the order of numerator polynomial is greater than the order of denominator polynomial.

Consider the block diagram of open loop control system (Fig. 1.6) where $R(s)$ & $C(s)$ are the laplace transform of input and output respectively, then the transfer function $G(s)$ can be expressed as

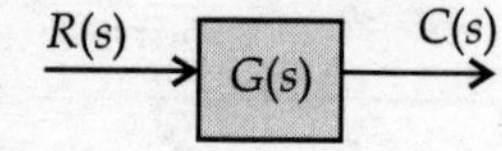

Fig. 1.6

$$G(s) = \frac{C(s)}{R(s)} = \frac{£c(t)}{£r(t)} \quad ...(1.4)$$

1.7. PROCEDURE

The following steps are involved to obtain the transfer function of the given system.

Step 1 : Write the differential equations for the given system.

Step 2 : Take the laplace transform of the equations obtained in step 1, with assumption; all initial conditions are zero.

Step 3 : Take the ratio of transformed output to input.

Step 4 : The ratio of transformed output to the input, obtained in step 3 is the required transfer function of the given system.

1.8. CHARACTERISTIC EQUATION OF A TRANSFER FUNCTION

The characteristic equation of a linear system can be obtained by equating the denominator polynomial of the transfer function to zero. Thus, the characteristic equation of the transfer function of equation 1.3 will be

$$a_n s^n + a_{n-1} s^{n-1} + ... + a_1 s + a_0 = 0 \quad ...(1.5)$$

1.9. POLES AND ZEROS OF A TRANSFER FUNCTION

Consider the equation 1.3, the numerator and denominator can be factored in m and n terms respectively, then the equation 1.3 can be expressed as

$$\frac{C(s)}{R(s)} = G(s) = \frac{K(s+Z_1)(s+Z_2)(as^2+bs+c)}{(s+p_1)(s+p_2)(As^2+Bs+C)} \quad ...(1.6)$$

where $K = \frac{b_m}{a_n}$ is known as the gain factor, s is the complex frequency

POLES : The poles of $G(s)$ are those values of 's' which make $G(s)$ tend to infinity. For example, in equation 1.6 we have poles at $s = -p_1$, $s = -p_2$ and a pair of poles at

$$s = \frac{-B \pm \sqrt{B^2 - 4AC}}{2A} \quad ...(1.7)$$

ZEROS : The zeros of $G(s)$ are those values of 's' which make $G(s)$ tend to zero. For example in eq. 1.6 we have zeros at $s_1 = -Z_1$, $s_2 = -Z_2$ and a pair of zeros at

$$s = \frac{-b \pm \sqrt{b^2 - 4ac}}{2a} \quad ...(1.8)$$

If either poles or zeros coincide, then such type of poles or zeros are called multiple poles or multiple zeros, otherwise they are known as simple poles or simple zeros. Multiple poles are due to the repetitive factor in denominator and multiple zeros are due to the repetitive factor in numerator of a transfer function.

For example, consider the transfer function

$$G(s) = \frac{50(s+3)}{s(s+2)(s+4)^2} \quad ...(1.9)$$

The above transfer function having the simple poles at $s = 0$, $s = -2$, multiple poles at $s = -4$, *i.e.*, the pole of order 2 at $s = -4$ and simple zero at $s = -3$.

The above mentioned poles and zeros are of finite values. If we consider the entire 's' plane including infinity then two cases arises.

1. If the no. of zeros are less than the no of poles *i.e.*, $Z < P$ then the value of the transfer function becomes zero for $s \to \infty$. Hence we can say that there are zeros at infinity and the order of such zeros is $P - Z$. For example, in equation 1.9 there are four finite poles at $s = 0, -2, -4$ and -4; there is one finite zero at $s = -3$ but there are three zeros at infinity $P-Z = 4-1 = 3$. Therefore, the function has a total of four poles and four zeros in the entire s-plane including infinity.
2. If the no. of poles are less than the no. of zeros $P < Z$ then the value of the transfer function becomes infinity for $s \to \infty$. Hence, we can say there are poles at infinity ($s \to \infty$) and the order of the poles will be Z-P.

Therefore, in addition to finite poles and zeros, if we consider poles and zeros at infinity, then for a rational function the total number of zeros will be equal to the total number of poles.

The pole is represented by 'X' and zero by 'O'. These symbols are used to locate the poles and zeros on s-plane. The pole-zero plot of eq^n 1.9 is shown in Fig. 1.7.

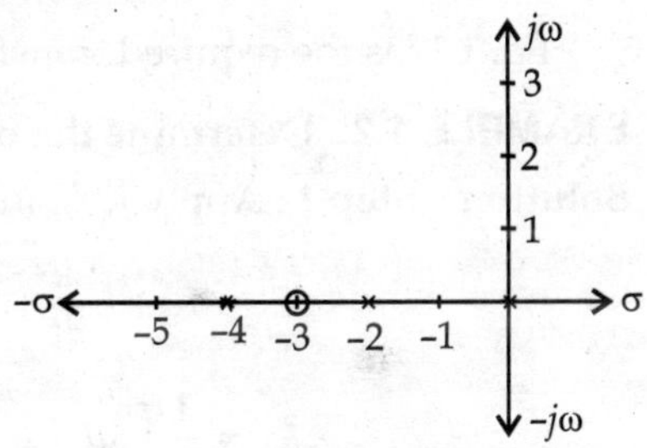

Fig. 1.7

1.10. IMPULSE RESPONSE

From equation 1.4

$$C(s) = R(s)\,G(s) \quad ...(1.10)$$

Suppose, a system is subjected to a unit impulse, then the output will be

$$C(s) = G(s) \quad ...(1.11)$$

because the Laplace transform of unit impulse function is unity.

The inverse laplace of eq 1.11 will be

$$C(t) = g(t) \quad ...(1.12)$$

where $g(t)$ is the unit impulse response of a system.

Therefore inverse laplace of $G(s)$

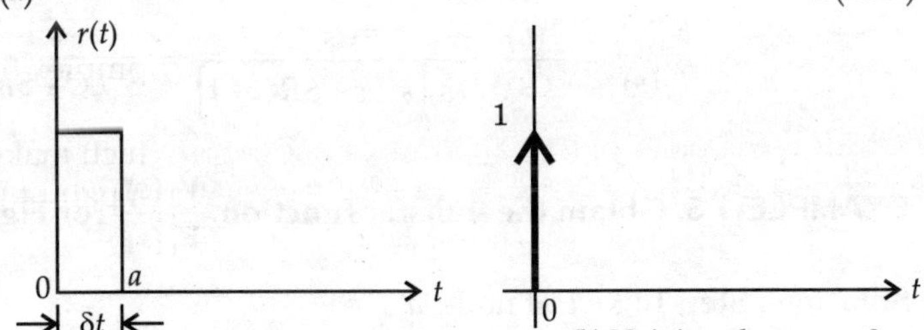

(a) Impulse function

(b) Unit impulse as $a \to 0$

Fig. 1.8.

is called impulse response or the transfer function of a system is the laplace transform of its impulse response.

Practically, it is not possible to generate a true impulse. A pulse with less duration than the time constant of the system can be considered as an impulse and denoted by $\delta(t)$.

EXAMPLE 1.1. Find the transfer function of the given network

Solution : Step 1 : Apply *KVL* in mesh (1)

$$V_i = Ri + L\frac{di}{dt} \quad ...(1.13)$$

Apply *KVL* in mesh (2)

$$V_0 = L\frac{di}{dt} \quad ...(1.14)$$

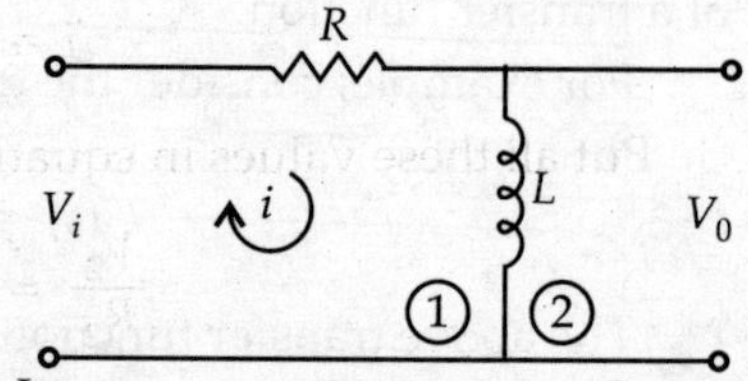

Fig. 1.9

Step 2 : Take Laplace transform of equation 1.13 and 1.14 with assumption that all initial conditions are zero.

$$V_i(s) = RI(s) + sLI(s) \quad ...(1.15)$$

$$V_0(s) = sLI(s) \quad ...(1.16)$$

Step 3 : Calculation of transfer function

$$\frac{V_0(s)}{V_i(s)} = \frac{sLI(s)}{(R+sL)I(s)} = \frac{sL}{R+sL} \quad ...(1.17)$$

Eq. 1.17 is the required transfer function.

EXAMPLE 1.2. Determine the transfer function of the electrical network shown in Fig. 1.10.

Solution : Step 1 : Apply *KVL* in both meshes

$$E_i = Ri + L\frac{di}{dt} + \frac{1}{C}\int idt \quad ...(1.18)$$

$$E_0 = \frac{1}{C}\int idt \quad ...(1.19)$$

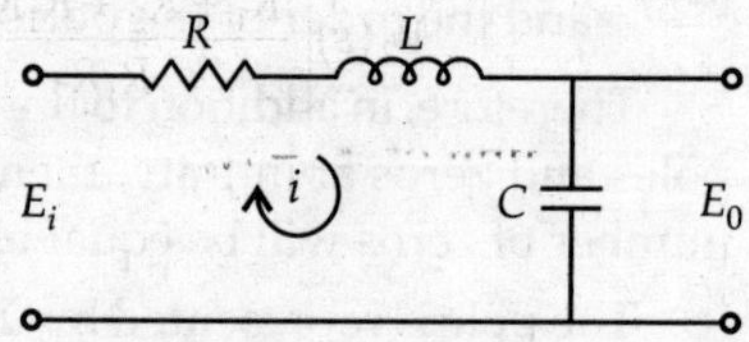

Fig. 1.10.

Step 2 : Take Laplace transform of equation 1.18 and 1.19

$$E_i(s) = RI(s) + sLI(s) + \frac{1}{Cs}I(s) = I(s)\left[R + sL + \frac{1}{Cs}\right] = I(s)\left[\frac{RCs + s^2LC + 1}{Cs}\right] \quad ...(1.20)$$

$$E_0(s) = \frac{1}{Cs}I(s) \quad ...(1.21)$$

Step 3 : Determination of transfer function

$$\frac{E_0(s)}{E_i(s)} = \frac{I(s)}{Cs} \cdot \frac{Cs}{I(s)\left[s^2Lc + SRC + 1\right]} = \frac{1}{S^2LC + SRC + 1} \quad \textbf{Ans.} \quad ...(1.22)$$

EXAMPLE 1.3. Obtain the transfer function $\frac{V_2(s)}{V_1(s)}$ for Fig. 1.11.

Solution : Step 1 : *KCL* at node '*a*'

$$i = i_1 + i_2 \quad ...(1.23)$$

$$i_1 = \frac{V_1 - V_2}{R_1}$$

$$i_2 = C\frac{d}{dt}(V_1 - V_2)$$

$$i = i_3 = \frac{V_2}{R_2}$$

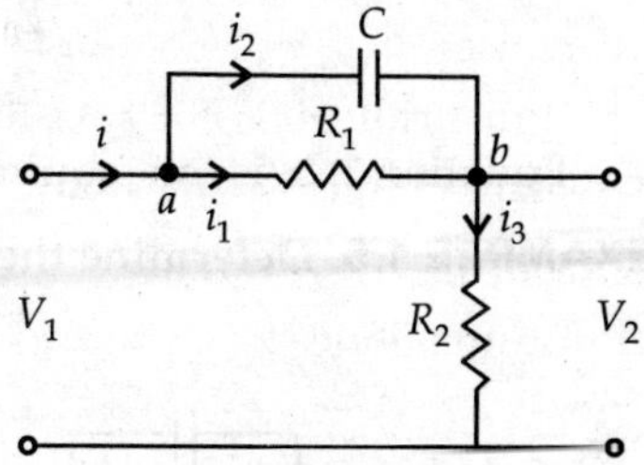

Fig. 1.11.

Put all these values in equation 1.23

$$\frac{V_2}{R_2} = \frac{V_1 - V_2}{R_1} + C\frac{d}{dt}(V_1 - V_2) \qquad \text{...(1.24)}$$

Step 2 : Take Laplace transform of equation 1.24

$$\frac{V_2(s)}{R_2} = \frac{1}{R_1}V_1(s) - \frac{1}{R_1}V_2(s) + Cs\,V_1(s) - Cs\,V_2(s)$$

$$\frac{V_2(s)}{R_2} + \frac{1}{R_1}V_2(S) + Cs\,V_2(s) = \frac{1}{R_1}V_1(s) + Cs\,V_1(s)$$

$$V_2(s)\left[\frac{1}{R_1} + \frac{1}{R_2} + Cs\right] = V_1(s)\left[\frac{1}{R_1} + Cs\right]$$

Step 3 : Determination of transfer function

$$V_2(s)\left[\frac{R_1 + R_2 + R_1R_2Cs}{R_1R_2}\right] = V_1(s)\left[\frac{1 + R_1Cs}{R_1}\right]$$

$$\frac{V_2(s)}{V_1(s)} = \frac{R_2 + R_1R_2Cs}{R_1 + R_2 + R_1R_2Cs} \quad \textbf{Ans.} \qquad \text{...(1.25)}$$

EXAMPLE 1.4. Find the transfer function of lag network shown in Fig. 1.12.

Solution : Step 1 : Apply *KVL* in both meshes

$$e_i(t) = R_1 i(t) + R_2 i(t) + \frac{1}{C}\int i(t)\,dt \qquad \text{...(1.26)}$$

$$e_0(t) = R_2 i(t) + \frac{1}{C}\int i(t)\,dt \qquad \text{...(1.27)}$$

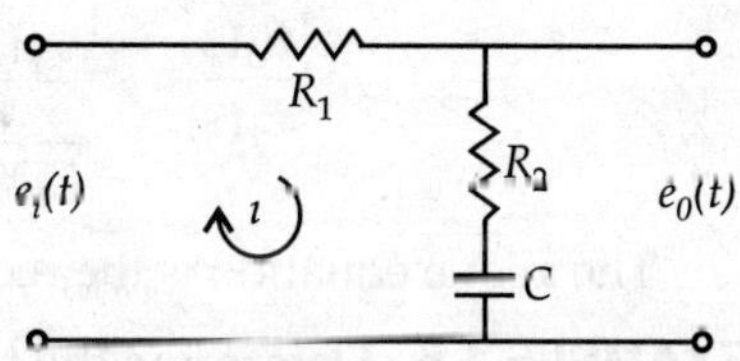

Fig. 1.12

Step 2 : Laplace transform of equation (1.26) and (1.27)

$$E_i(s) = \left[R_1 + R_2 + \frac{1}{Cs}\right]I(s) \quad \text{and} \quad E_0(s) = \left[R_2 + \frac{1}{Cs}\right]I(s)$$

Step 3 : Calculation of transfer function

$$E_0(s) = \frac{\left[R_2 + \frac{1}{Cs}\right]I(s)}{\left[\frac{R_1Cs + R_2Cs + 1}{Cs}\right]I(s)}$$

$$\frac{E_0(s)}{E_i(s)} = \frac{1+R_2Cs}{1+R_1Cs+R_2Cs} \quad ...(1.28)$$

Equation 1.28 is the required transfer function.

EXAMPLE 1.5. Determine the transfer function of Fig. 1.13.

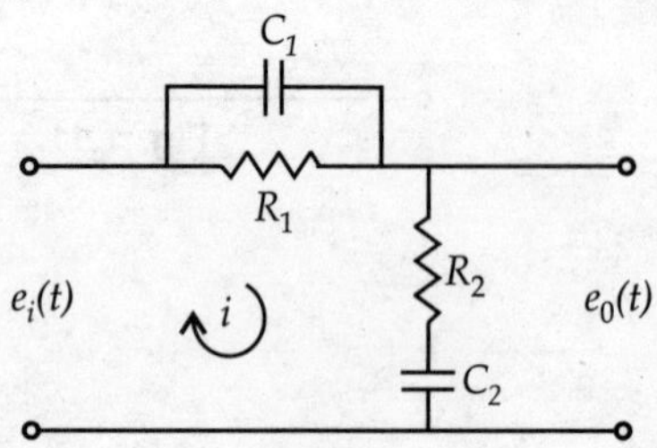

Fig. 1.13.

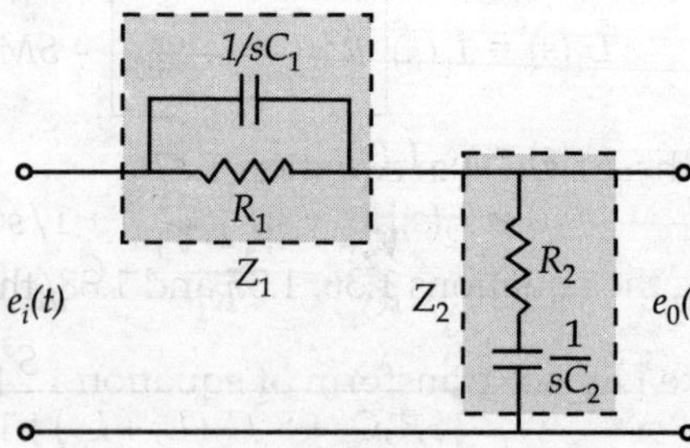

Fig. 1.14.

Solution : Step 1 : calculation *of* Z_1 :

$$Z_1 = \frac{R_1\frac{1}{sC_1}}{R_1+\frac{1}{sC_1}} = \frac{R_1}{R_1C_1s+1} \quad ...(1.29)$$

Step 2 : Calculation of Z_2 :

$$Z_2 = R_2+\frac{1}{sC_2} = \frac{R_2C_2S+1}{sC_2} \quad ...(1.30)$$

Step 3 : Calculation of transfer function in terms of Z_1 and Z_2

$$\frac{E_0(s)}{E_i(s)} = \frac{Z_2(s)}{Z_1(s)+Z_2(s)} \quad ...(1.31)$$

Step 4 : Calculation of transfer function in terms of R_1, R_2, C_1 and C_2. Put the values of Z_1 (s) and Z_2(s) from equation 1.29 and 1.30 in equation 1.31

$$\frac{E_0(s)}{E_i(s)} = \frac{(1+R_2C_2S)/Sc_2}{\frac{R_1}{C_1R_1S+1}+\frac{R_2C_2S+1}{SC_2}} = \frac{(1+R_1C_1S)(1+R_2C_2S)}{(1+R_1C_1S)(1+R_2C_2S)+R_1C_2S} \quad ...(1.32)$$

The above equation is the required transfer function of the given circuit.

EXAMPLE 1.6. Determine the transfer function of given transformer coupled circuit (Fig. 1.15).

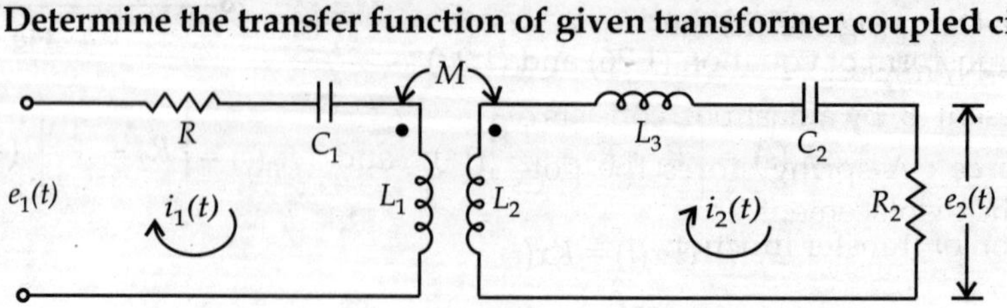

Fig. 1.15.

Solution : Apply *KVL* in both meshes

$$e_i(t) = Ri_1(t)+\frac{1}{C_1}\int i_1\,dt+L_1\frac{di_1(t)}{dt}-M\frac{di_2(t)}{dt} \quad ...(1.33)$$

$$e_2(t) = R_2 i_2(t) \quad ...(1.34)$$

$$0 = R_2 i_2(t) + (L_3 + L_2)\frac{d}{dt}i_2(t) + \frac{1}{C_2}\int i_2(t)dt - M\frac{d}{dt}i_1(t) \quad ...(1.35)$$

Take Laplace transform of equations 1.33, 1.34 and 1.35

$$E_1(s) = I_1(s)\left[R + \frac{1}{C_1 S} + SL_1\right] - SMI_2(s) \quad ...(1.36)$$

$$E_2(s) = R_2 I_2(s) \quad ...(1.37)$$

$$0 = I_2(s)[R_2 + s(L_2 + L_3) + 1/sC_2] - s\,MI_1(s) \quad ...(1.38)$$

Solving the equations 1.36, 1.37 and 1.38, the required transfer function.

$$G(s) = \frac{S^3 R_2 C_1 C_2 M}{\left[SR_2C_2 + S^2C_2(L_2 + L_3) + 1\right]\left[S^2L_1C_1 + SC_1R_1 + 1\right] - M^2S^4C_1C_2} \quad \textbf{Ans.}$$

MECHANICAL SYSTEM

1.11. TRANSLATIONAL SYSTEMS

The motion takes place along a straight line is known as translational motion. There are three types of forces that resists motion.

1. Intertia Force : Consider a body of mass 'M' & acceleration 'a', then according to Newton's second law of motion the inertia force will be equal to the product of mass 'M' & acceleration 'a'.

$$F_M(t) = Ma(t) \quad ...(1.39)$$

In terms of velocity the equation (1.39) becomes

$$F_M(t) = M\frac{dv(t)}{dt} \quad ...(1.40)$$

In terms of displacement the equation (1.39) can be expressed as

$$F_M(t) = M\frac{d^2x(t)}{dt^2} \quad ...(1.41)$$

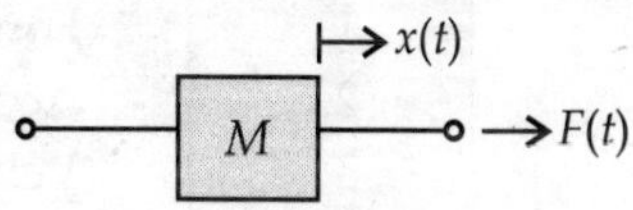

Fig. 1.16.

2. Damping Force : For viscous friction we assume that the damping force is proportional to the velocity.

$$F_D(t) = Bv(t) = B\frac{d}{dt}x(t) \quad ...(1.42)$$

where, B = Damping coefficient

unit of B = N/m/sec.

We can represent 'B' by a dashpot, consists of piston and cylinder.

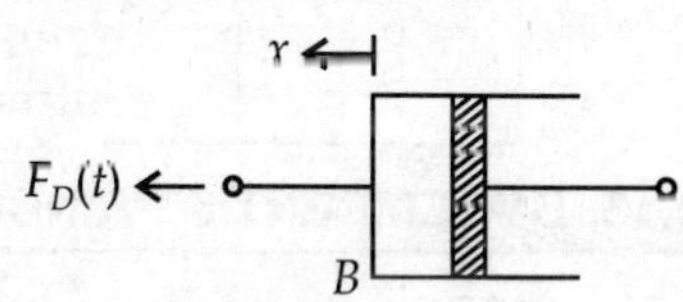

Fig. 1.17.

3. Spring Force : A spring stores the potential energy. The restoring force of a spring is proportional to the displacement.

$$F_K(t) \propto x(t) = Kx(t)$$

$$F_K(t) = K\int v(t)dt \quad ...(1.43)$$

where 'K' = spring constant or stiffness

unit of $K = N/m$

The stiffness of a spring can be defined as restoring force per unit displacement.

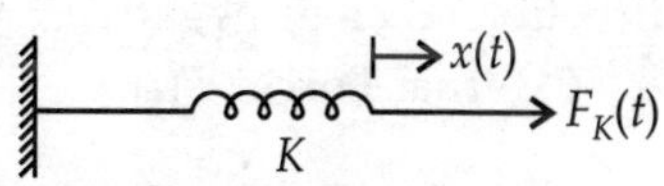

Fig. 1.18.

1.12. ROTATIONAL SYSTEM

The rotational motion of a body can be defined as the motion of a body about a fixed axis. There are three types of torques resists the rotational motion.

1. Inertia Torque : Inertia (J) is the property of an element that stores the kinetic energy of rotational motion. The inertia torque T_I is the product of moment of inertia J and angular acceleration $\alpha(t)$

$$T_I(t) = J\,\alpha(t)$$

$$T_I(t) = J\frac{d}{dt}\omega(t) = J\frac{d^2}{dt^2}\theta(t) \qquad \text{...(1.44)}$$

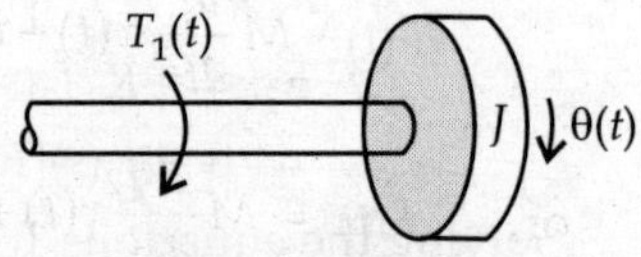

Fig. 1.19.

where $\omega(t)$ = angular velocity

$\theta(t)$ = angular displacement

unit of Torque = $N\text{-}m$

2. Damping Torque : The damping torque $T_D(t)$ is the product of damping coefficient B and angular velocity ω.

$$T_D(t) = Bw(t) = B\frac{d}{dt}\theta(t) \qquad \text{...(1.45)}$$

3. Spring Torque : Spring torque $T_\theta(t)$ is the product of torsional stiffness and angular displacement.

$$T_\theta(t) = K\theta(t)$$

unit of K = $N\text{-}m/rad$. ...(1.46)

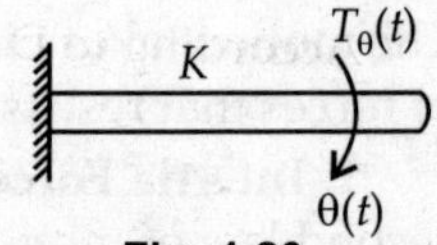

Fig. 1.20.

If we compare the equations

We get an analogous system. The analogous quantities are tabulated as follows :

Table 1.2.

S.No.	Translational	Rotational
1.	Force, F	Torque, T
2.	Acceleration, a	Angular acceleration, α
3.	Velocity, v	Angular velocity, ω
4.	Displacement, x	Angular displacement, θ
5.	Mass, M	Moment of inertia, J
6.	Damping coefficient B	Rotational damping coeff., B
7.	Stiffness	Torsional stiffness

1.13. D'ALEMBERT'S PRINCIPLE

This principle states that "for any body, the algebraic sum of externally applied forces and the forces resisting motion in any given direction is zero".

D'Alembert principle is useful in writing the equation of motion of mechanical system. Consider, a system shown in Fig. 1.21, consisting of a mass M, spring and dashpot.

First choose a reference direction. All the forces in the direction of reference direction considered as positive and the forces opposite to the reference direction taken as negative.

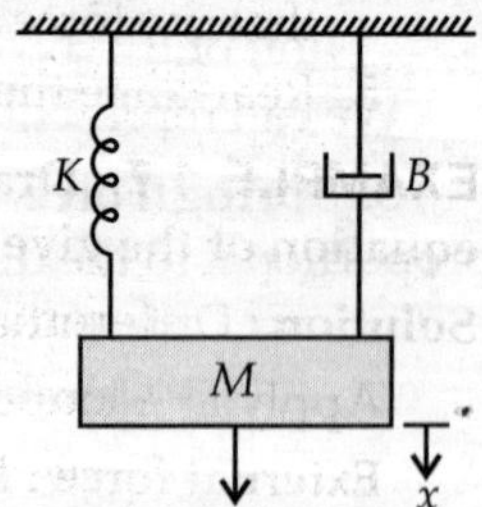

Fig. 1.21.

External Force : $F(t)$

Resisting Forces : (a) Inertia force $F_m(t) = -M\frac{d^2}{dt^2}x(t)$

(b) Damping force $F_D(t) = -B\frac{d}{dt}x(t)$

(c) Spring force $F_K(t) = -Kx(t)$

According to D'Alembert's principle

$$F(t) + F_m(t) + F_D(t) + F_K(t) = 0$$

$$F(t) - M\frac{d^2}{dt^2}x(t) - B\frac{d}{dt}x(t) - Kx(t) = 0$$

or, $$F(t) = M\frac{d^2}{dt^2}x(t) + B\frac{d}{dt}x(t) + Kx(t) \quad ...(1.47)$$

Consider a rotational system shown in Fig. 1.22.

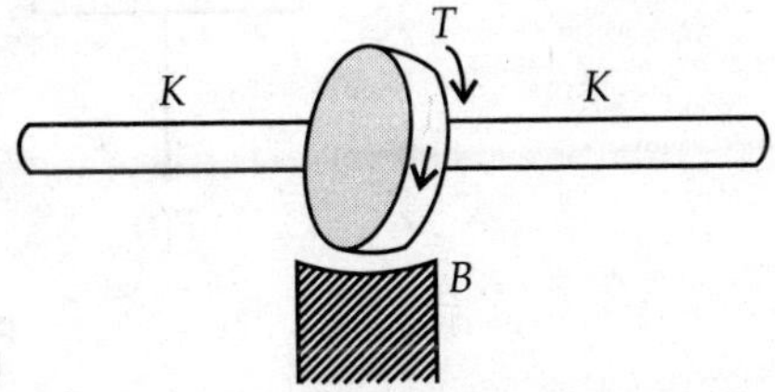

Fig. 1.22.

External Torque : *T (t)*

Resisting Torque : (a) Inertia torque $T_I(t) = -J\frac{d\omega(t)}{dt}$

(b) Damping torque $T_D(t) = -B\frac{d}{dt}\theta(t)$

(c) Spring torque $T_K(t) = -K\theta$

According to D'Alembert principle

$$T(t) + T_I + T_D + T_K = 0$$

$$T(t) - J\frac{d}{dt}\omega(t) - B\frac{d}{dt}\theta(t) - K\theta(t) = 0$$

or, $$T(t) = J\frac{d}{dt}\omega(t) + B\frac{d}{dt}\theta(t) + K\theta(t) \quad ...(1.48)$$

So, D'Alembert principle for rotational motion is :

"For any body, the algebaric sum of externally applied torques and the torques resisting rotation about any axis is zero".

1.14. PROCEDURE OF WRITING THE MODELS OF MECHANICAL SYSTEM

1. Assume system is in equilibrium.
2. Assume that the system is given same arbitrary displacement if no. of distributing forces are present.
3. Draw the free body diagram of forces exerted on each mass in the system.
4. Apply Newton's law of motion to each diagram, using the convention that any force acting in the direction of assume displacement is positive.
5. Rearrange the equations in suitable form to be solved by any means.

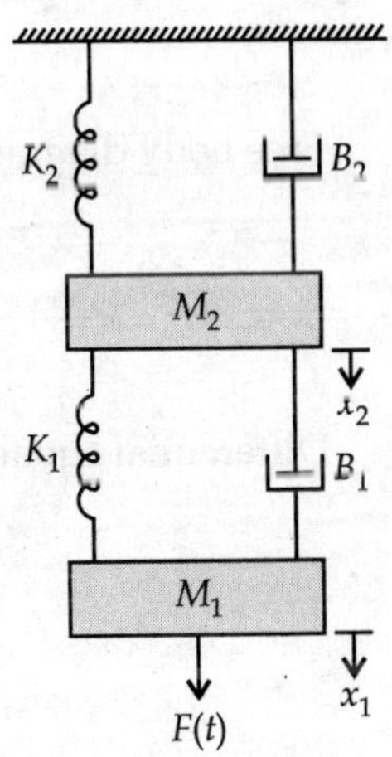

Fig. 1.23

EXAMPLE 1.7. Draw the free body diagram and write the diffrential equation of the given system shown in Fig. 1.23.

Solution : Differential equation for M_1 :

Apply D'Alembert principle

External force : $F(t)$

Resisting forces :

1. Inertia force $\quad F_M = -M_1\frac{d^2}{dt^2}x_1$

2. Damping force $\quad F_D = -B_1 \frac{d}{dt}(x_1 - x_2)$

3. Spring force $\quad F_K = K_1(x_1 - x_2)$

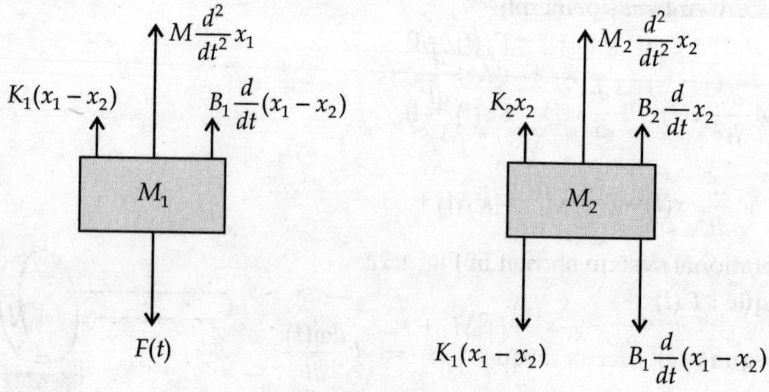

Fig. 1.24. Free body diagram

$$\therefore \qquad F(t) = M_1\frac{d^2}{dt^2}x_1 + B_1\frac{d}{dt}(x_1 - x_2) + K_1(x_1 - x_2) \qquad ...(1.49)$$

Similarly for mass M_2 :

$$K(x_1 - x_2) + B_1\frac{d}{dt}(x_1 - x_2) = M_2\frac{d^2x_2}{dt^2} + B_2\frac{dx_2}{dt} + K_2x_2 \qquad ...(1.50)$$

Equation (1.49) and (1.50) are the required equations.

EXAMPLE 1.8. Write the differential equations describing the dynamics of the system shown in Fig. 1.25 and find the ratio $\frac{X_2(s)}{F(s)}$.

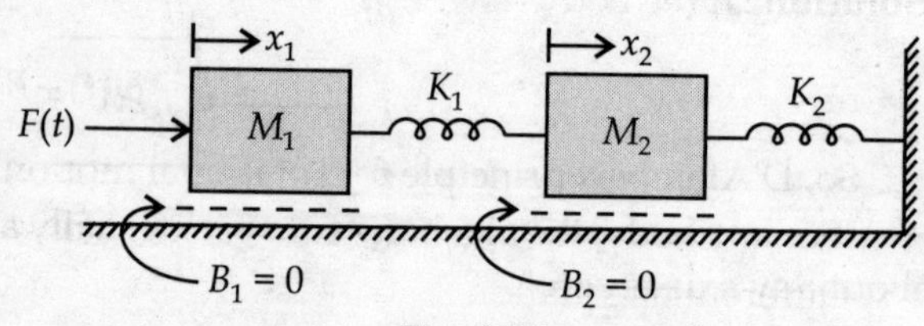

Fig. 1.25

Solution : Free body diagram For mass M_1 :

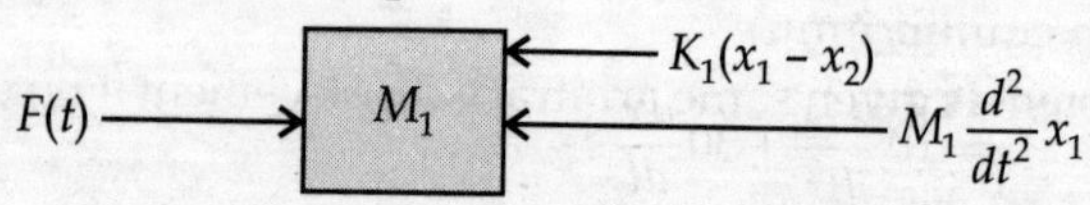

Fig. 1.26 (a)

Free body diagram For mass M_2 :

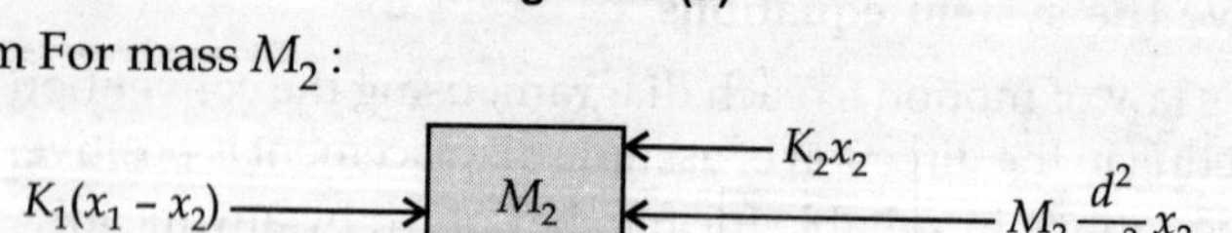

Fig. 1.26 (b)

Differential equation for mass M_1

$$F(t) = M_1\frac{d^2}{dt^2}x_1 + K_1(x_1 - x_2) \qquad ...(1.51)$$

$$K_1(x_1 - x_2) = K_2x_2 + M_2\frac{d^2}{dt^2}x_2 \qquad ...(1.52)$$

Take Laplace transform of equation (1.51), assume initial conditions zero.

$$F(t) = M_1 \frac{d^2}{dt^2} x_1 + K_1 x_1 - K_1 x_2$$

$$F(s) = M_1 S^2 X_1(s) + K_1 X_1(s) - K_1 X_2(s) \quad ...(1.53)$$

laplace transform of equation (1.52)

$$K_1X_1 - K_1X_2 = K_2X_2 + M_2 \frac{d^2}{dt^2} X_2$$

$$K_1X_1(s) - K_1 X_2(s) = K_2X_2(s) + M_2 s^2 X_2(s)$$

$$X_1(s) = \frac{X_2(s)}{K_1} [s^2M_2 + K_1 + K_2] \quad ...(1.54)$$

Put the value of $X_1(s)$ in equation (1.53)

$$F(s) = \frac{X_2(s)}{K_1} [s^2M_2 + K_1 + K_2] [s^2M_1 + K_1] - K_1X_2(s)$$

$$\frac{X_2(s)}{F(s)} = \frac{K_1}{(S^2M_2 + K_1 + K_2)(K_1 + s^2M_1) - K_1^2} \quad \textbf{Ans.}$$

EXAMPLE 1.9. A mass spring system under equilibrium condition is shown in Fig. 1.27. Drive the system equation where $M = 10$ kg. $B = 30$ N/m/sec. and $K = 20$ N/m.

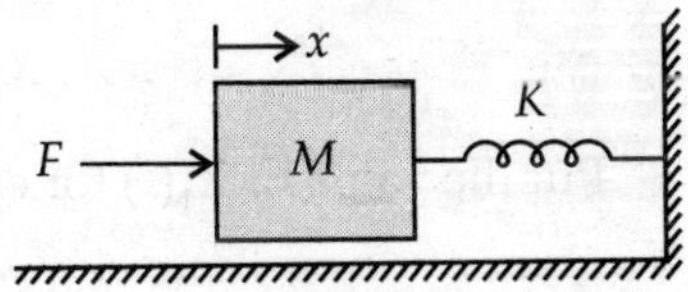

Fig. 1.27

Solution : Free body diag.

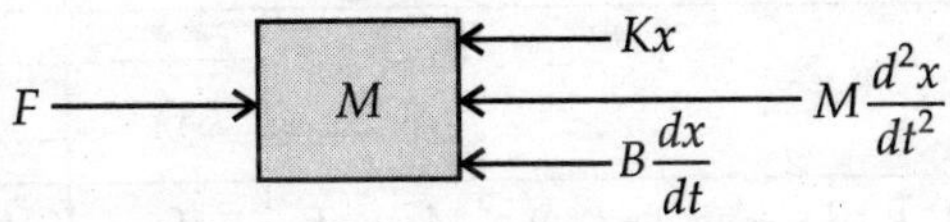

Fig. 1.28

Differential equation

$$F = M\frac{d^2x}{dt^2} + B\frac{dx}{dt} + Kx$$

Put the given values

$$F = 10\frac{d^2x}{dt^2} + 30\frac{dx}{dt} + 20x \quad \textbf{Ans.} \quad ....(1.55)$$

EXAMPLE 1.10. Drive the system equations and find the value of $X_2(s)/F(s)$ for the system shown in Fig. 1.29. *(R.M.L., University, Faizabad, 2003)*

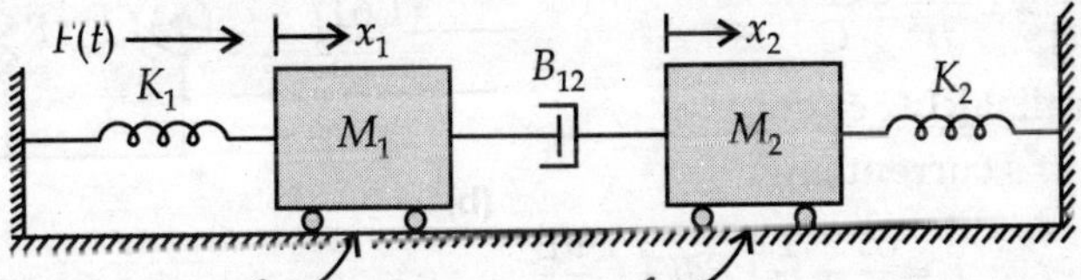

Fig. 1.29.

Solution : Free body diagram for mass M_1:

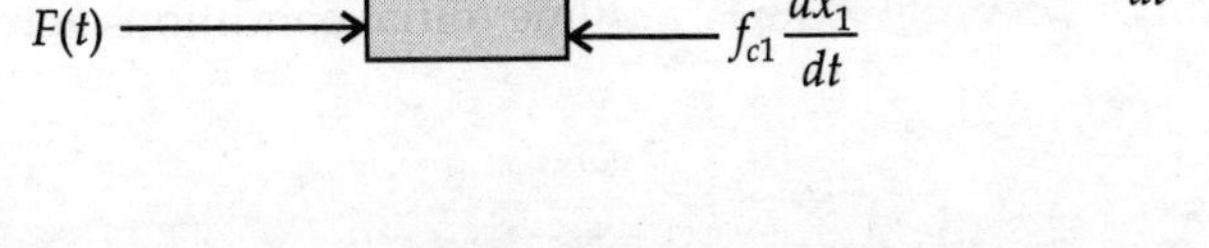

Free body diagram for mass M_2:

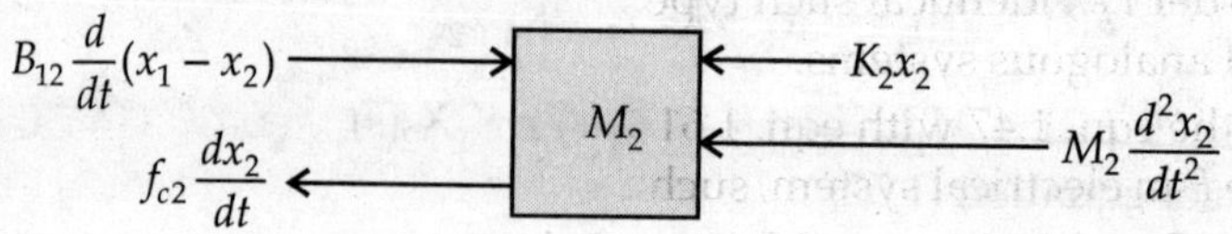

System equation for mass M_1:

$$F(t) = M_1\frac{d^2x_1}{dt^2} + fc_1\frac{dx_1}{dt} + B_{12}\frac{d}{dt}(x_1 - x_2) + K_1x_1 \qquad .....(1.56)$$

System equation for mass M_2:

$$B_{12}\frac{d}{dt}(x_1 - x_2) = M_2\frac{d^2x_2}{dt^2} + fc_2\frac{dx_2}{dt} + K_2x_2 \qquad .....(1.57)$$

Laplace transform of equation 1.56

$$F(s) = X_1(s)[s^2M_1 + sB_{12} + sfc_1 + K_1] - B_{12}\,sX_2(s) \qquad .....(1.58)$$

Laplace transform of equation 1.57

$$B_{12}sX_1(s) = (s^2M_2 + B_{12}s + sfc_2 + K_2)\,X_2(s)$$

$$\therefore \quad X_1(s) = \frac{X_2(s)\left(s^2M_2 + B_{12}s + sfc_2 + K_2\right)}{B_{12}s} \qquad .....(1.59)$$

Put the value of $X_1(s)$ for equation (1.59) in equation (1.58)

$$F(s) = \frac{X_2(s)\left[S^2M_2 + B_{12}S + sfc_2 + K_2\right]\left[S^2M_1 + SB_{12} + sfc_1 + K_1\right]}{B_{12}S} - B_{12}SX_2(s)$$

$$\therefore \quad \frac{X_2(s)}{F(s)} = \frac{SB_{12}}{(s^2M_1 + sB_{12} + sfc_1 + K_1)\left(s^2M_2 + sB_{12} + sfc_2 + K_2\right) - s^2{B_{12}}^2} \qquad \textbf{Ans.}$$

1.15. ANALOGOUS SYSTEM

Consider a series *RLC* circuit

Apply Lirchhoff's voltage law

$$E = Ri + L\frac{di}{dt} + \frac{1}{C}\int idt \qquad .....(1.60)$$

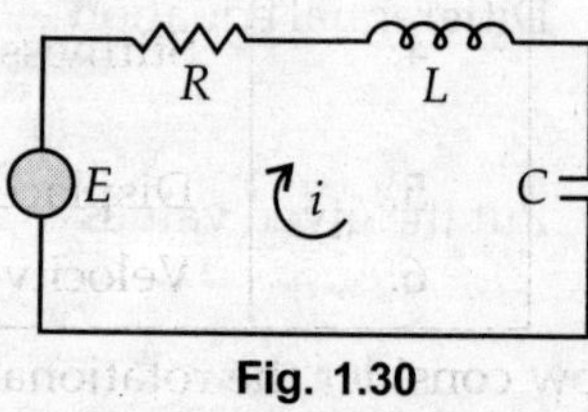

Fig. 1.30

In terms of charge equation 1.60 becomes

$$E = R\frac{dq}{dt} + L\frac{d^2q}{dt^2} + \frac{1}{C}q \qquad ....(1.61)$$

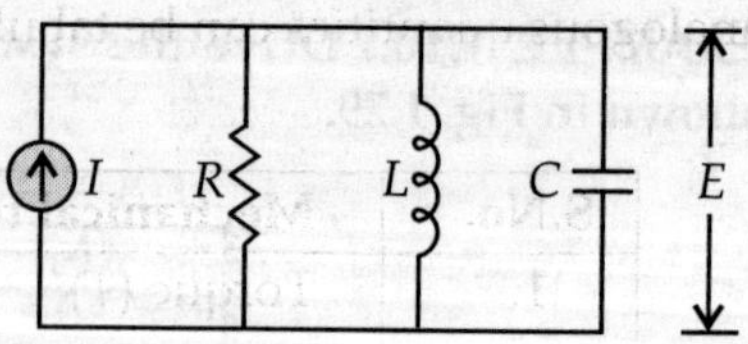

Fig. 1.31.

Now consider a parallal RLC circuit

Now apply Kirchhoff's current law

$$I = \frac{E}{R} + \frac{1}{L}\int Edt + C\frac{dE}{dt} \qquad ....(1.62)$$

In terms of magnetic flux linkage, the equation 1.62 becomes

Since $\phi = \int Edt$

$$I = \frac{1}{R}\left(\frac{d\phi}{dt}\right) + \frac{1}{L}\phi + C\frac{d^2\phi}{dt^2} \qquad ....(1.63)$$

Now, compare the equation (1.47) with the equation (1.61), both equations are differential equations of same order *i.e.,* identical such type of systems whose differential equations are in the same form are called analogous systems.

On comparison the eqn. 1.47 with eqn. 1.61 we can see that in mechanical system the force F is analogous to voltage E in electrical system, such type of analogy is known as force-voltage analogy.

From equation 1.47 and 1.61, mass (M) is analogous to inductance (L), coeff. of viscous friction (B) is analogous to resistance (R), spring stiffness (K) is analogous to $\frac{1}{C}$ and so on.

Table 1.3.

S.No.	Mechanical translational system	Electrical system
1.	Force (F)	Voltage E
2.	Mass (M)	Inductance (L)
3.	Stiffness (K) (Elastance, $1/K$)	Reciprocal of capacitance ($^1/_C$) Capacitance (C)
4.	Damping coeff. (B)	Resistance R
5.	Displacement (x)	Charge (q)

Now compare the equation (1.47) with equation (1.63), since the force F is analogous to the current source, such type of analogy is known as force current analogy. The analogous quantities are tabulated as.

Table 1.4.

S.No.	Mechanical translational system	Electrical system
1.	Force (F)	Current (I)
2.	Mass (M)	Capacitance (c)
3.	Damping coeff. (B)	Reciprocal of resistance ($1/R$) *i.e.,* conductance G
4.	Stiffness (K) (Elastance, $1/K$)	Reciprocal of Inductance ($1/L$) Inductance (L)
5.	Displacement (x)	Flux linkage (ϕ)
6.	Velocity ($\dot{x}$)	Voltage (E)

Now consider the rotational system and compare the equation 1.48 with 1.61. On comparison the analogous quantities can be tabulated as

Table 1.5. Torque-voltage analogy

S.No.	Mechanical rotational system	Electrical system
1.	Torque (T)	Voltage (E)
2.	Moment of Inertia (J)	Inductance (L)
3.	Damping coeff. (B)	Resistance (R)
4.	Stiffness (K) (Elastance, $1/K$)	Reciprocal of capacitance ($^1/_C$) capacitance (C)
5.	Angular displacement (θ)	Charge (q)
6.	Angular Velocity (w)	Current (i)

On comparison the equation 1.48 with equation 1.63 we get Torque current analogy and can be tabulated as

Table 1.6

S.No.	Mechanical rotational system	Electrical system
1.	Torque (*T*)	Current (*I*)
2.	Moment of Inertia (*J*)	Capacitance (*c*)
3.	Damping coeff. (*B*)	Reciprocal of Resistance ($^1/_R$) *i.e.,* conductance (*G*)
4.	Stiffness (*K*) (Elastance, $1/_K$)	Reciprocal of Inductance ($^1/_L$) Inductance (*L*)
5.	Angular displacement (θ)	Flux linkage (ϕ)

Following Cheng*, the rule for drawing the analogous circuit can be stated as

(*a*) For force-voltage analogy : "Each junction in the mechanical system corresponds to a closed loop which consists of electrical excitation sources and passive elements analogous to the mechanical driving sources and passive elements connected to the junction. All points on a rigid mass are considered as the junction."

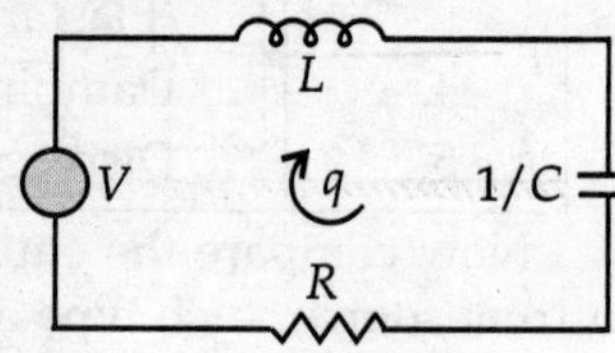

Fig. 1.32

(*b*) For force-current analogy : "Each junction in the mechanical system corresponds to a node which joins electrical excitation sources and passive elements analogous to the mechanical driving sources and passive elements connected to the junction. All points on a rigid mass are considered as the same junction and one terminal of the capacitance analogous to a mass is always connected to the ground."

The reason for connecting one terminal of capacitance analogous to a mass is always connected to the ground is that the vetocity (or displacement) of a mass is always referred to the earth.

Consider the system shown in Fig. 1.21, corresponding to *x* in mechanical system, we will have one loop in *f-v* analogous electrical circuit. The circuit consists of a voltage source, inductance *L*(*M*), resistance *R*(*B*), the reciprocal of capacitance (*c*) and charge *q*(*x*).

The equation for the mechanical system shown in Fig. 1.21 is

$$f = M\frac{d^2x}{dt^2} + B\frac{dx}{dt} + Kx \qquad \text{....(1.64)}$$

For the electrical circuit the equation will be

$$V = L\frac{d^2q}{dt^2} + R\frac{dq}{dt} + \frac{1}{C}q \qquad \text{....(1.65)}$$

Since, q = $\int idt$ the equation (1.65) will becomes

$$V = L\frac{di}{dt} + Ri + \frac{1}{C}\int idt \qquad \text{.....(1.66)}$$

The equation (1.66) is the voltage equation for series *RLC* circuit.

Series *RLC* circuit, is the analogous electrical circuit of the mechanical system shown in Fig. 1.21.

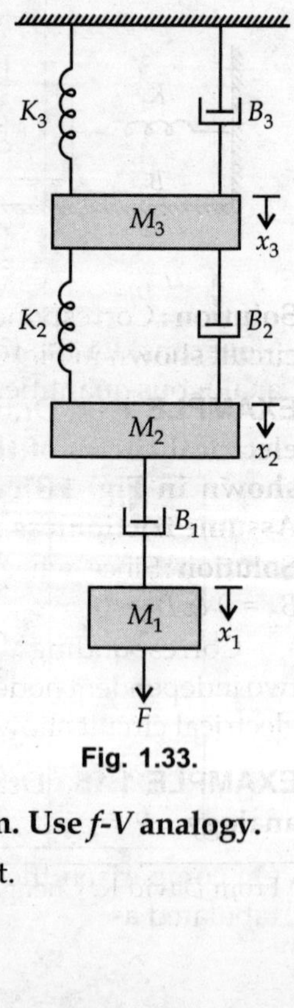

Fig. 1.33.

EXAMPLE 1.11. Draw the analogous electrical network of the given system. Use *f-V* analogy.

Solution : Now corresponding x_1, x_2, x_3 we have three loops in electrical circuit.

First loop consists of a voltage source $V(f)$, Inductance $L_1(M_1)$ resistance R_1 (B_1). The resistance R_1 is common in between first and second loop.

Second loop consists of inductance $L_2(M_2)$, resistance $R_2(B_2)$ and $\frac{1}{C_2}(K_2)$. Resistance R_2 and $\frac{1}{C_2}$ are common in loop second and third loop.

Similarly, third loop consists of Inductance $L_3(M_3)$, $\frac{1}{C_3}$ (K_3) and R_3 (B_3).

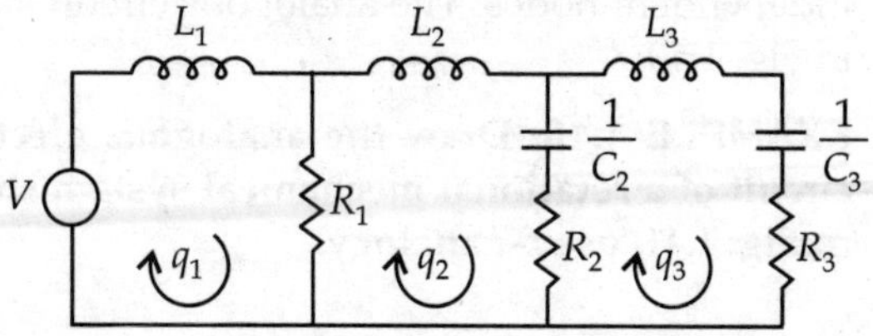

Fig. 1.34. Analogous electrical circuit of Fig. 1.33.

q_1,q_2 and q_3 are analogous quantities of x_1,x_2 and x_3 respectively.

EXAMPLE 1.12. Draw the electrical analogous circuit of the system shown in Fig. 1.35.

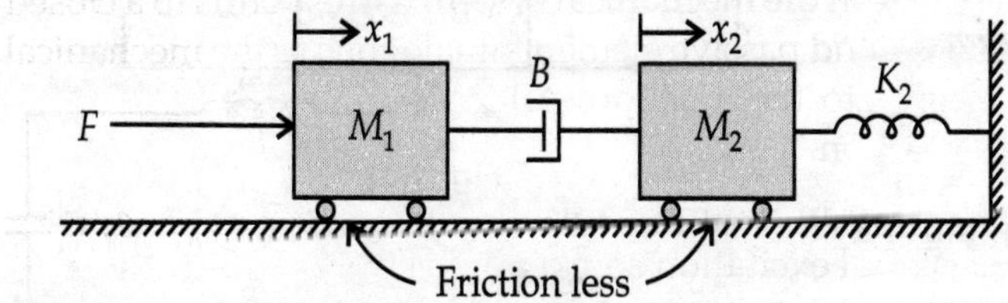

Fig. 1.35.

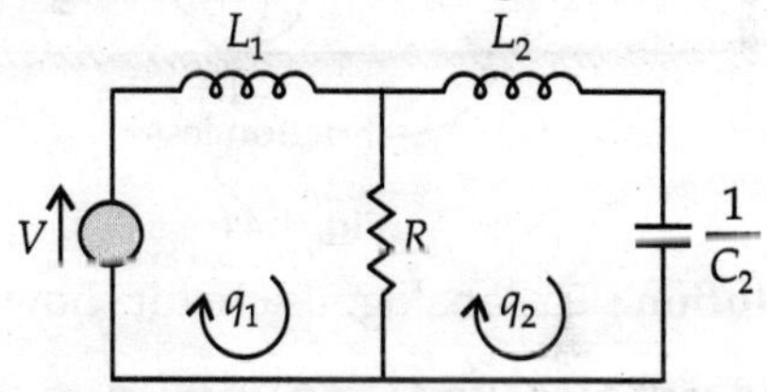

Fig. 1.36.

Solution : By inspection and with the help of table we can draw the electrical circuit shown in Fig. 1.36.

EXAMPLE. 1.13. Draw the analogous electrical circuit of the system shown in Fig. 1.37. Use *f*-V analogy.

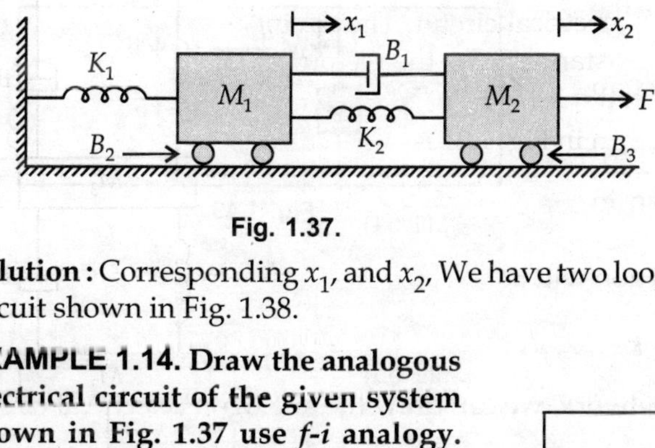

Fig. 1.37.

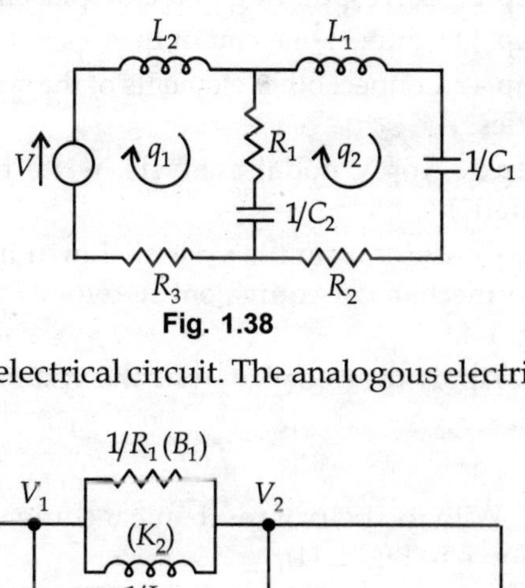

Fig. 1.38

Solution : Corresponding x_1, and x_2, We have two loops in electrical circuit. The analogous electrical circuit shown in Fig. 1.38.

EXAMPLE 1.14. Draw the analogous electrical circuit of the given system shown in Fig. 1.37 use *f-i* analogy. Assume frictionless wheels.

Solution : Since, wheels are frictionless, $B_2 = 0$ & $B_3 = 0$

Corresponding x_1 and x_2, we have two independent nodes. The analogous electrical circuit shown in Fig. 1.39.

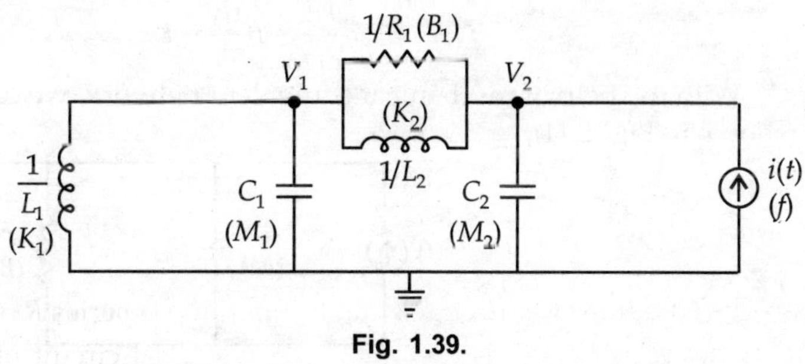

Fig. 1.39.

EXAMPLE 1.15. Draw the analogous electrical circuit of the system shown in Fig. 1.35 use *f-i* analogy.

* From David K. Cheng, Analysis of linear systems., Addison-Wesley

Solution : Corresponding x_1 and x_2 we have two independent nodes. The analogous circuit shown in Fig. 1.40.

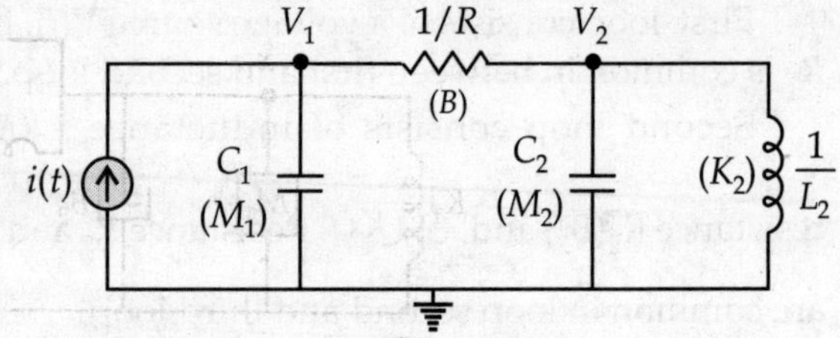

Fig. 1.40.

EXAMPLE 1.16. Draw the analogous electrical circuit of a rotational mechanical system shown in Fig. 1.41. use *f–i* analogy.

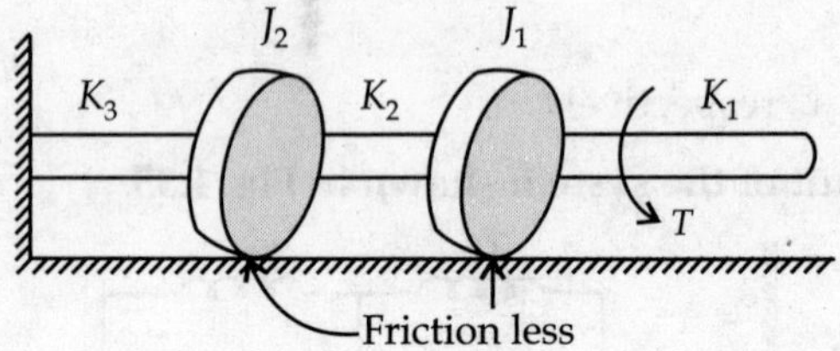

Fig. 1.41.

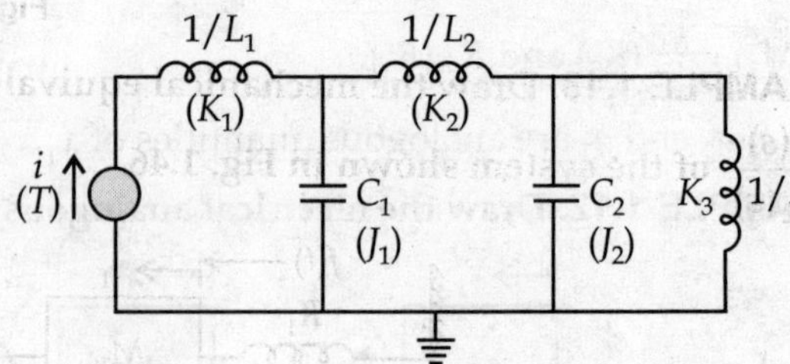

Fig. 1.42.

Solution : The analogous circuit shown in Fig. 1.42.

1.16. MECHANICAL EQUIVALENT NETWORK

The equivalent mechanical network is useful to determine the transfer function of the system. The following procedure is applied to draw the mechanical equivalent network.

Step 1 : Draw a reference line. Consider the fixed points as reference.

Step 2 : Corresponding to the displacement x_1, x_2......, select the nodes.

Step 3 : Connect one end of masses to the reference line.

Step 4 : Connect other elements of the system to the nodes.

Step 5 : Apply nodal analysis, write the system equations.

Consider a simple system shown in Fig. 1.21. The mechanical equivalent network is shown in Fig. 1.43.

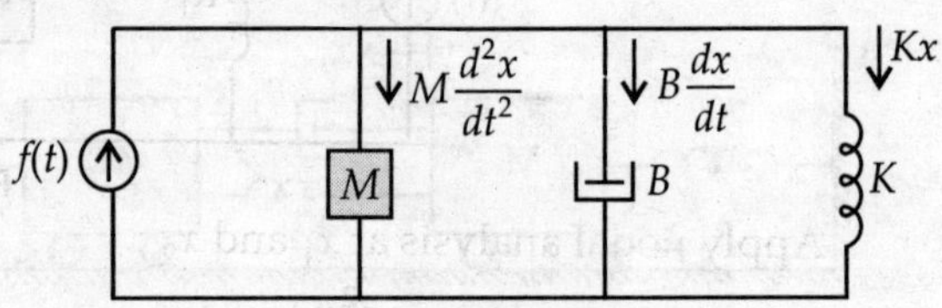

Fig. 1.43.

Apply the nodal analysis the system equation will be

$$f(t) = M\frac{d^2x}{dt^2} + B\frac{dx}{dt} + Kx \qquad ...(1.67)$$

With the help of mechanical equivalent network, we can draw the analogous electrical circuit. (shown in Fig. 1.44).

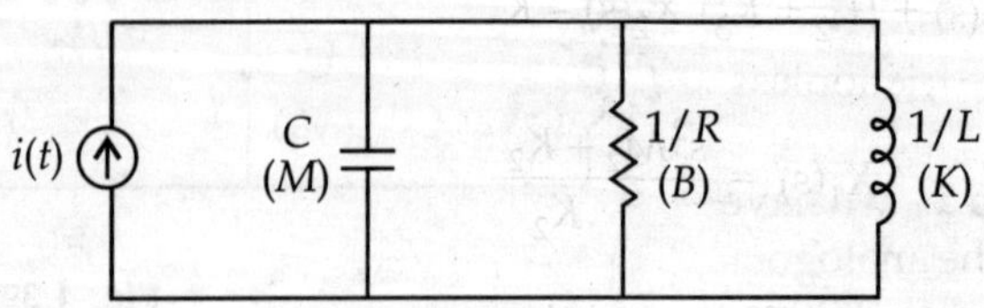

Fig. 1.44.

EXAMPLE 1.17. Draw the mechanical network of the system shown in Fig. 1.37.

Solution : Since, there are two displacements x_1, x_2 we have two nodes.

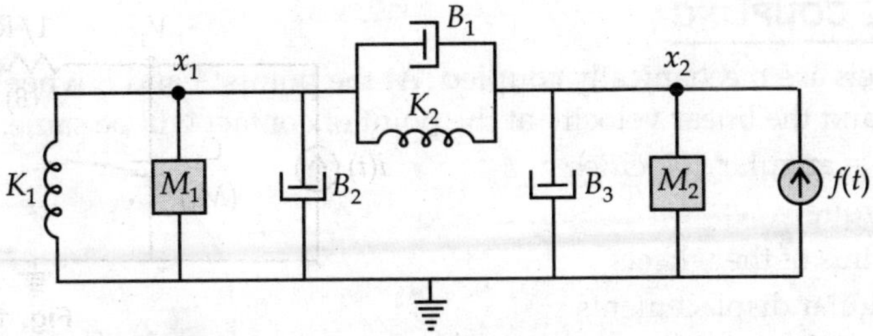

Fig. 1.45.

EXAMPLE 1.18. Draw the mechanical equivalent network, write the system equations and find $\frac{F(s)}{X_2(s)}$ of the system shown in Fig. 1.46.

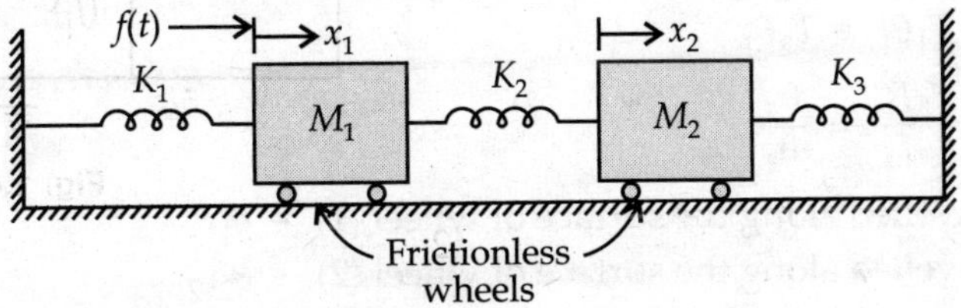

Fig. 1.46.

Solution : Mechanical network is shown in Fig. 1.47.

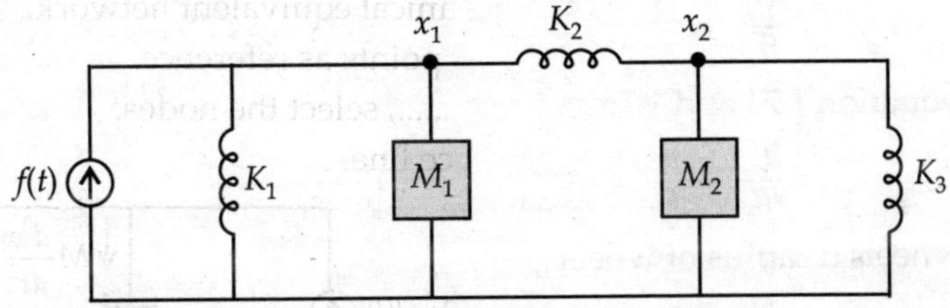

Fig. 1.47.

Apply nodal analysis at x_1 and x_2

$$f(t) = M_1 \frac{d^2 x_1}{dt^2} + K_1 x_1 + K_2 (x_1 - x_2) \quad ...(1.68)$$

$$K_2 (x_2 - x_1) + M_2 \frac{d^2 x_2}{dt^2} + K_3 x_2 = 0 \quad ...(1.69)$$

Laplace transform of equation 1.68 and 1.69

$$F(s) = S^2 M_1 X_1(s) + (K_1 + K_2) X_1(s) - K_2 X_2(s) \quad ...(1.70)$$

$$S^2 M_2 X_2(s) + (K_2 + K_3) X_2(s) - K_2 X_1(s) = 0 \quad ...(1.71)$$

Form equation 1.71

$$X_1(s) = \frac{s^2 M_2 + K_2 + K_3}{K_2} X_2(s) \quad ...(1.72)$$

Put the value of $X_1(s)$ from 1.72 in equation 1.70 and solve for $\frac{F_2(s)}{X_2(s)}$

$$\frac{F_2(s)}{X_2(s)} = \frac{(s^2 M_1 + K_1 + K_2)(s^2 M_2 + K_2 + K_3) - K^2{}_2}{K_2} \quad ...(1.73)$$

Equation. 1.73 is the required answer.

1.17. MECHANICAL COUPLING

In Fig. 1.48 two wheels are mechanically coupled. At the points A and B wheels experiences equal and opposite forces and the linear velocity at the point of contact will be same.

Let ω_1 and ω_2 = angular velocities
T_1 and T_2 = Torque
r_1 and r_2 = radius of the wheels
θ_1 and θ_2 = angular displacements
N_1 and N_2 = No. of teeths on wheels

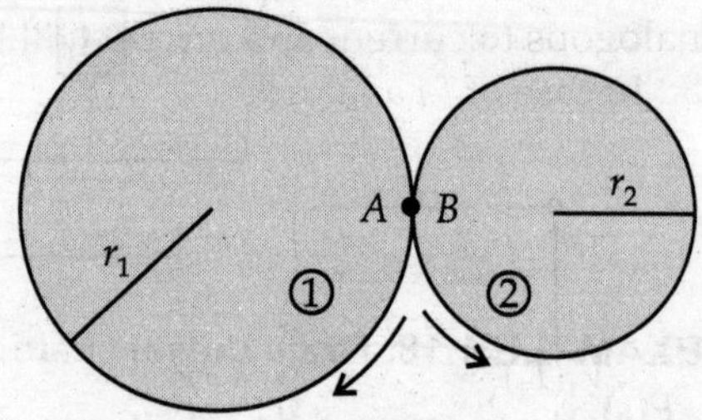

Fig. 1.48.

In Ideal case work done by wheel (1) will be equal to the work done by wheel (2)

Work done by wheel (1) = $T_1\theta_1$
Work done by wheel (2) = $T_2\,\theta_2$

$$\therefore \quad T_1\theta_1 = T_2\theta_2$$

$$\frac{T_1}{T_2} = \frac{\theta_2}{\theta_1} \qquad \text{...(1.74)}$$

Linear distance travelled along the surface of wheel (1) = $\theta_1 r_1$
Linear distance travelled along the surface of wheel (2) = $\theta_2 r_2$
The linear distance will be same

$$\therefore \quad \theta_1 r_1 = \theta_2 r_2$$

$$\frac{\theta_1}{\theta_2} = \frac{r_2}{r_1} \qquad \text{....(1.75)}$$

Combining the equation 1.74 and 1.75

$$\frac{T_1}{T_2} = \frac{\theta_2}{\theta_1} = \frac{r_1}{r_2} \qquad \text{...(1.76)}$$

No. of teeth on wheels α radius of wheels

$$N_1 \;\alpha\; r_1 \qquad \text{...(1.77)}$$

$$N_2 \;\alpha\; r_2 \qquad \text{...(1.78)}$$

combining the equations 1.76, 1.77 and 1.78

$$\frac{T_1}{T_2} = \frac{\theta_2}{\theta_1} = \frac{r_1}{r_2} = \frac{N_1}{N_2} \qquad \text{....(1.79)}$$

Since,

$$\frac{\theta_2}{\theta_1} = \frac{r_1}{r_2}$$

$$\theta_2 = \frac{r_1}{r_2}\theta_1 \qquad \text{...(1.80)}$$

Differentiate equation 1.80

$$\dot{\theta}_2 = \frac{r_1}{r_2}\dot{\theta}_1$$

But

$$\dot{\theta}_2 = \omega_2$$

$$\dot{\theta}_1 = \omega_1$$

$$\therefore \quad \frac{\omega_2}{\omega_1} = \frac{r_1}{r_2} \qquad \text{...(1.81)}$$

Combining the equation 1.79 and 1.81

$$\frac{T_1}{T_2} = \frac{\theta_2}{\theta_1} = \frac{r_1}{r_2} = \frac{N_1}{N_2} = \frac{\omega_2}{\omega_1} \quad ...(1.82)$$

If we consider $r_1 : r_2$ as the turn ratio of an ideal transformer then angular velocity will be analogous to current and torque will be analogous to voltage. This is *f-V* analogy.

In case of *f-i* analogy

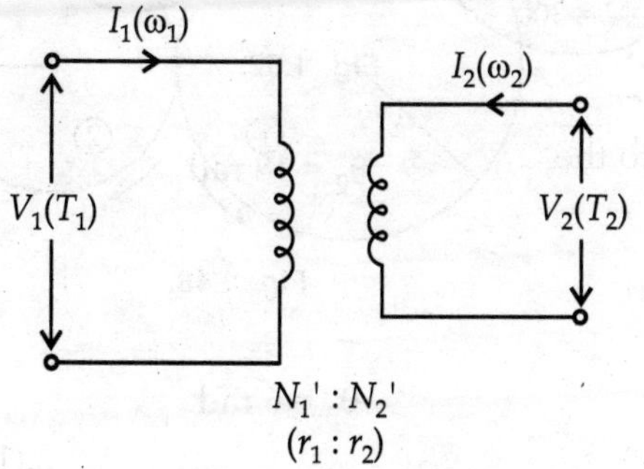

Fig. 1.49.

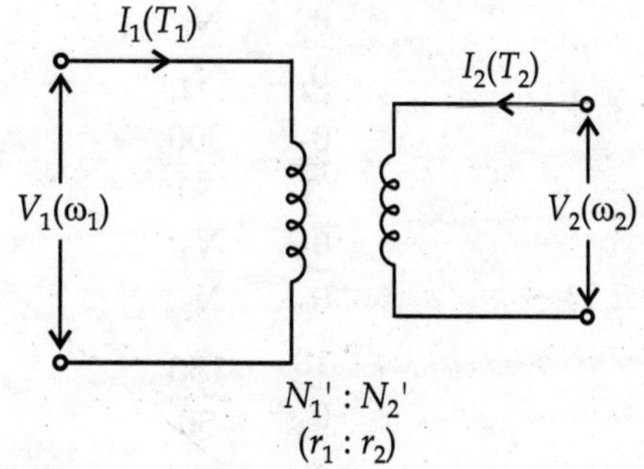

$N_1' : N_2'$ is the turn ratio of the transformer

Fig. 1.50.

EXAMPLE 1.19.

(a) Find the ratio of $r_1 : r_2$

(*b*) If $\omega_1 = 10$ rad/sec. Calculate ω_2

(*c*) If torque for wheel (1) is 5 Nm. What will be the torque for wheel (2)

(*d*) If the displacement of wheel (1) is 20°. What is the displacement of wheel (2).

Solution : We know that

$$\frac{r_1}{r_2} = \frac{N_1}{N_2} = \frac{40}{20} = \frac{2}{1}$$

$\therefore \quad r_1 : r_2 :: 2 : 1$ **Ans.**

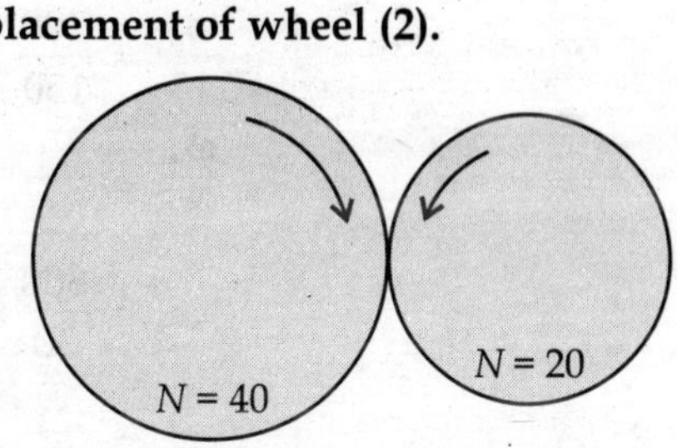

Fig. 1.51.

(*b*) From equation 1.82

$$\frac{N_1}{N_2} = \frac{\omega_2}{\omega_1}$$

$$\frac{40}{20} = \frac{\omega_2}{10} \qquad \therefore \quad \omega_2 = 20 \text{ rad/sec.}$$ **Ans.**

(*c*) from equation 1.82

$$\frac{T_1}{T_2} = \frac{N_1}{N_2}$$

$$\frac{5}{T_2} = \frac{40}{20} \qquad \therefore \quad T_2 = 2.5 \text{ Nm}$$ **Ans.**

(*d*) from equation 1.82

$$\frac{\theta_2}{\theta_1} = \frac{N_1}{N_2}$$

$$\frac{\theta_2}{20} = \frac{40}{20} \qquad \therefore \quad \theta_2 = 40°$$ **Ans.**

EXAMPLE 1.20.

(a) If $\theta_1 = 3$rad. (clockwise) calculate the displacement of wheel 2, 3 and 4.

(*b*) If $\omega_1 = 15$ rad/sec. Calculate ω_3 and ω_4.

(c) If T_1 = 10 Nm. Calculate T_2 and T_3.

(d) If angular acceleration ($\ddot{\theta}_1$) for wheel (1) is 4 rad/sec^2, calculate $\ddot{\theta}_3$.

(e) Find $r_1 : r_2$ and $r_1 : r_3$

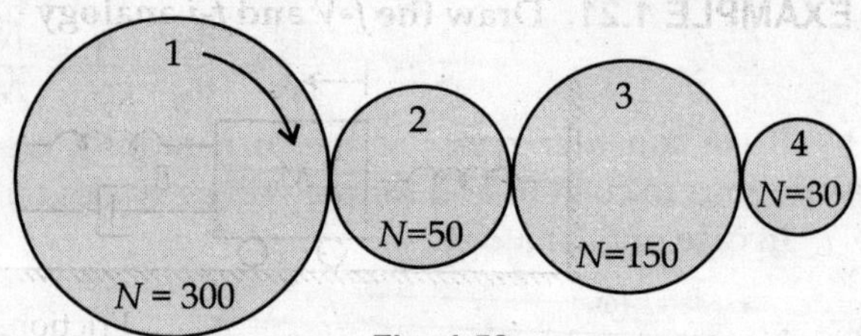

Fig. 1.52.

Solution :

(a) $$\frac{\theta_2}{\theta_1} = \frac{N_1}{N_2}$$

$$\frac{\theta_2}{3} = \frac{300}{50} \qquad \theta_2 = 18 \text{ rad}$$

$$\frac{\theta_2}{\theta_3} = \frac{N_3}{N_2}$$

$$\frac{18}{\theta_3} = \frac{150}{50} \qquad \theta_3 = 6 \text{ rad.}$$

$$\frac{\theta_3}{\theta_4} = \frac{N_4}{N_3}$$

$$\frac{6}{\theta_4} = \frac{30}{150} \qquad \theta_4 = 30 \text{ rad}$$

(b) $$\frac{\omega_3}{\omega_1} = \frac{N_1}{N_3}$$

$$\frac{\omega_3}{15} = \frac{300}{150} \qquad \therefore \omega = 30 \text{ rad/sec.}$$

$$\frac{\omega_4}{\omega_1} = \frac{N_1}{N_4}$$

$$\omega_4 = \frac{300}{30} \times 15 \qquad \therefore \omega_4 = 150 \text{ rad/sec.}$$

(c) $$\frac{T_1}{T_2} = \frac{N_1}{N_2} \text{ or } \frac{10}{T_2} = \frac{300}{50} \qquad \therefore T_2 = 5/3 \text{ Nm}$$

$$\frac{T_1}{T_3} = \frac{N_1}{N_3} \text{ or } \frac{10}{T_3} = \frac{300}{150} \qquad \therefore T_3 = 5 \text{ Nm}$$

(d) $$\frac{\ddot{\theta}_3}{\ddot{\theta}_1} = \frac{N_1}{N_3}$$

$$\frac{\ddot{\theta}_3}{4} = \frac{300}{150} \qquad \ddot{\theta}_3 = 8 \text{ rad/sec}^2 \quad \textbf{Ans.}$$

(e) $$\frac{r_1}{r_2} = \frac{N_1}{N_2} = \frac{300}{50}$$

$$\therefore \quad r_1 : r_2 = 6 : 1 \qquad \textbf{Ans.}$$

$$\frac{r_1}{r_3} = \frac{N_1}{N_3} = \frac{300}{150}$$

$$r_1 : r_3 = 2 : 1 \qquad \textbf{Ans.}$$

EXAMPLE 1.21. Draw the *f*-*V* and *f*-*i* analogy of the mech. system shown in Fig. 1.53.

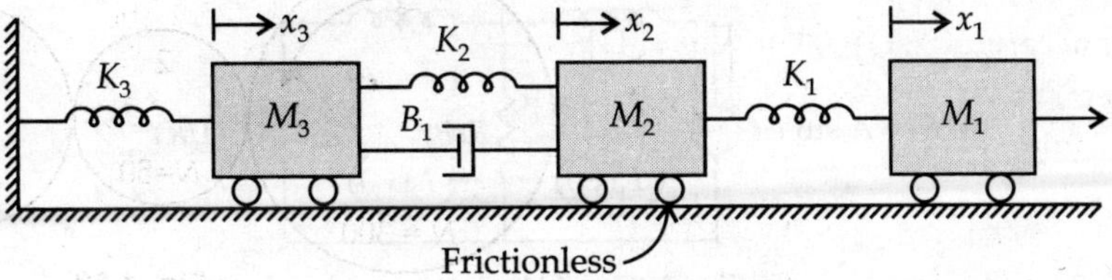

Fig. 1.53.

Solution : Corresponding the displacements x_1, x_2 and x_3 there are three meshes. The first mesh containing a voltage source $V(F)$ and Inductance $L_1(M_1)$, $C_1(1/K_1)$ will be common between two meshes. Similarly $C_2(1/K_2)$ and $R_1(B_1)$ will be common between second and third meshes. The *f*-*V* and *f*-*i* analogous circuit are shown in Fig. 1.54 and 1.55 respectively.

f-*V* analogy :

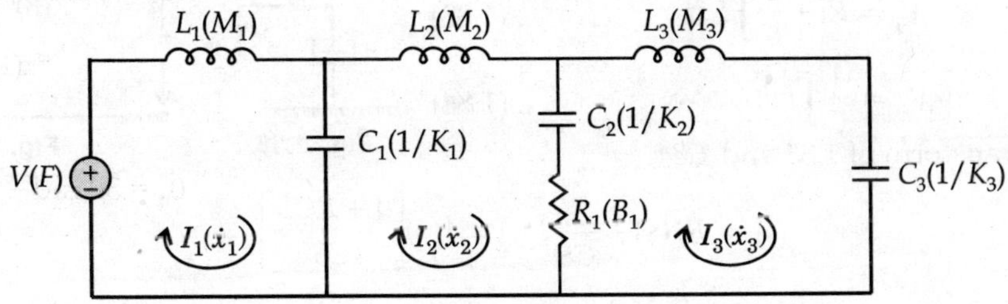

Fig. 1.54.

f-*i* analogy :

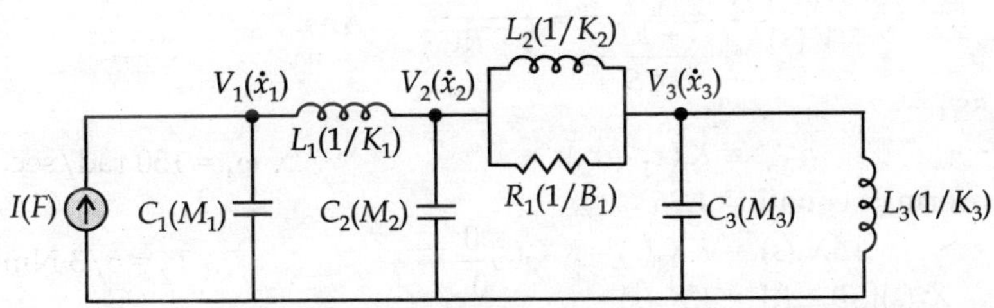

Fig. 1.55.

EXAMPLE 1.22. Draw the electrical analogous circuit of the given mechanical system shown in Fig. 1.56. Use *f*-*V* analogy.

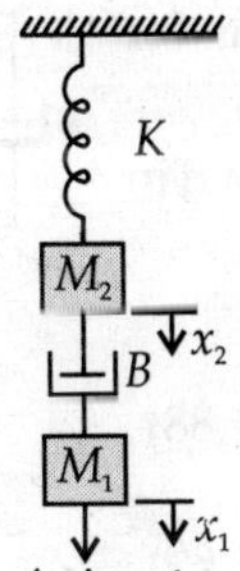

Fig. 1.56.

Solution :

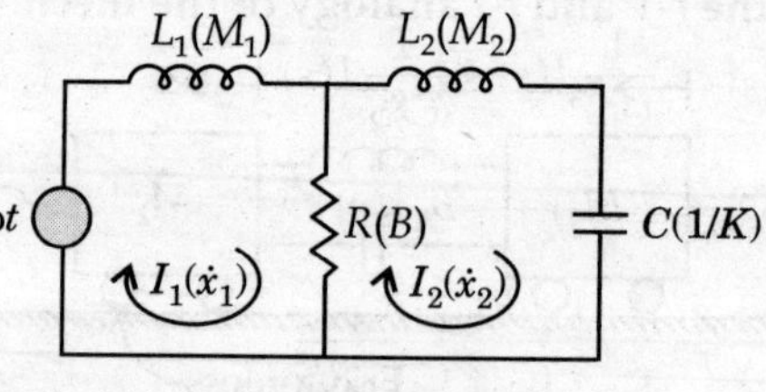

Fig. 1.57. (f-V analogy)

EXAMPLE 1.23. If 'K' is the stiffness of the spring, draw the analogous electrical circuit based on *f-V* analogy and determine the transfer function in each case.

Solution : The electrical analogous circuit is shown in Fig. 1.59.

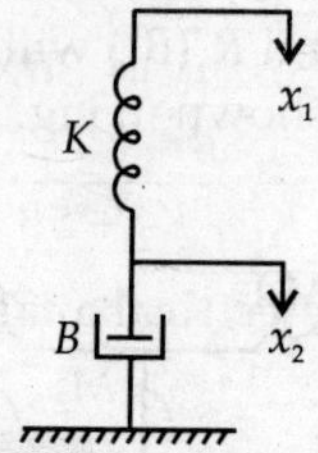

Fig. 1.58.

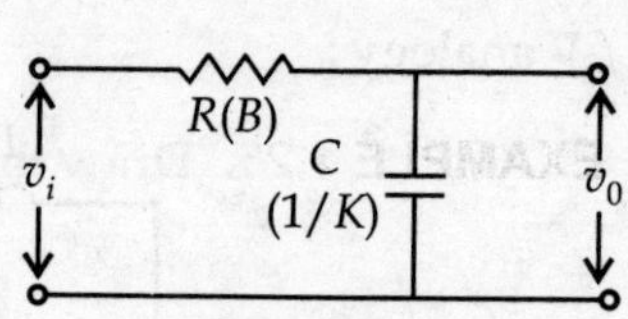

Fig. 1.59.

From Fig. 1.59. :

$$V_i = Ri + \frac{1}{C}\int i\,dt \qquad ...(1.83)$$

$$V_0 = \frac{1}{C}\int i\,dt \qquad ...(1.84)$$

Laplace transform of 1.83 and 1.84

$$V_i(s) = RI(s) + \frac{1}{CS}I(s) = I(s)\left[\frac{1+RCS}{CS}\right] \qquad ...(1.85)$$

$$V_0(s) = \frac{1}{CS}I(s)$$

$$\frac{V_0(s)}{V_i(s)} = \frac{\frac{1}{CS}}{\frac{1+RCS}{CS}} = \frac{1}{1+RCS} \qquad \textbf{Ans.}$$

From Fig. 1.58 :

$$B\dot{x}_2 = K(x_1 - x_2) \qquad ...(1.86)$$

Laplace transform of equation 1.86

$$SBX_2(s) = KX_1(s) - KX_2(s)$$

$$X_2(s)[SB + K] = KX_1(s)$$

$$\therefore \quad \frac{X_2(s)}{X_1(s)} = \frac{K}{K+SB} \qquad \textbf{Ans.} \qquad ...(1.87)$$

EXAMPLE 1.24. Draw the electrical analogous circuit (use *f-V*) analogy and derive their transfer functions.

Solution : Electrical analogous circuit is shown in Fig. 1.61.

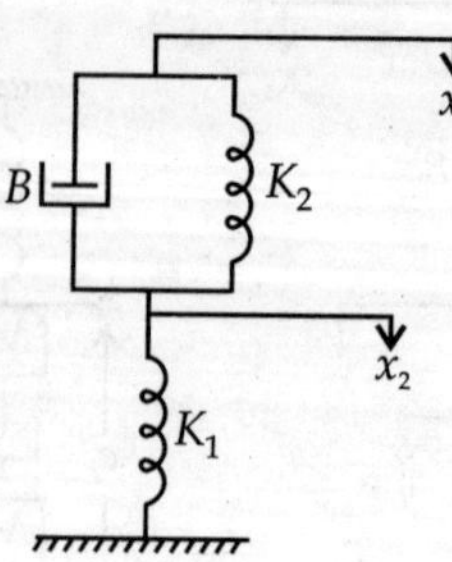

Fig. 1.60.

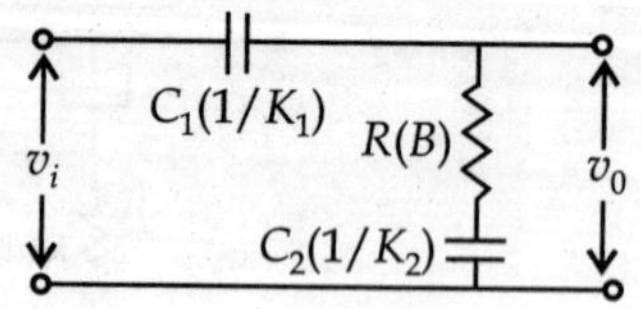

Fig. 1.61.

From Fig. 1.61

$$V_i = \frac{1}{C_1}\int i\,dt + \frac{1}{C_2}\int i\,dt + Ri \qquad ...(1.88)$$

$$V_0 = Ri + \frac{1}{C_2}\int i\,dt \qquad ...(1.89)$$

Laplace transform of 1.88 and 1.89

$$V_i(s) = \frac{1}{C_1S}I(s) + \frac{1}{C_2S}I(s) + RI(s) \qquad ...(1.90)$$

$$V_0(s) = RI(s) + \frac{1}{C_2S}I(s)$$

$$\frac{V_0(s)}{V_i(s)} = \frac{(1+RC_2S)C_1}{C_1+C_2+C_1C_2RS} \qquad ...(1.91)$$

from Fig. 1.60 :

$$K_1x_2 = B\frac{d}{dt}(x_1 - x_2) + K_2(x_1 - x_2) \qquad ...(1.92)$$

Laplace transform of 1.92

$$K_1X_2(s) = SBX_1(s) - SBX_2(s) + K_2X_1(s)\ K_2X_2(s)$$

$$\frac{X_2(s)}{X_1(s)} = \frac{SB+K_2}{K_1+K_2+SB} \qquad ...(1.93)$$

EXAMPLE 1.25. Draw the electrical analogous circuit, use *f-V* and *f-i* analogy.

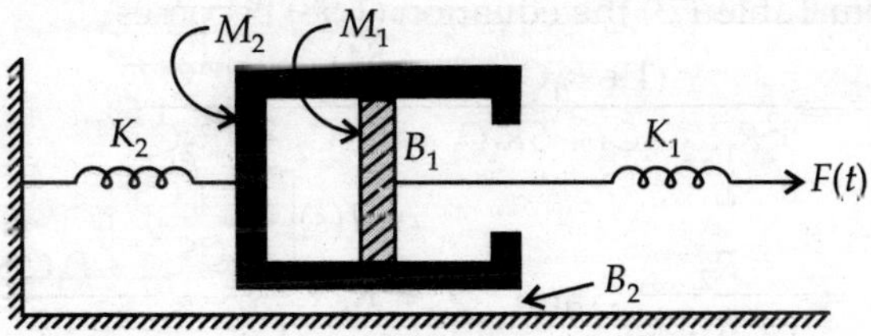

Fig. 1.62.

Solution : Required circuit shown in Fig. 1.63 and 1.64.

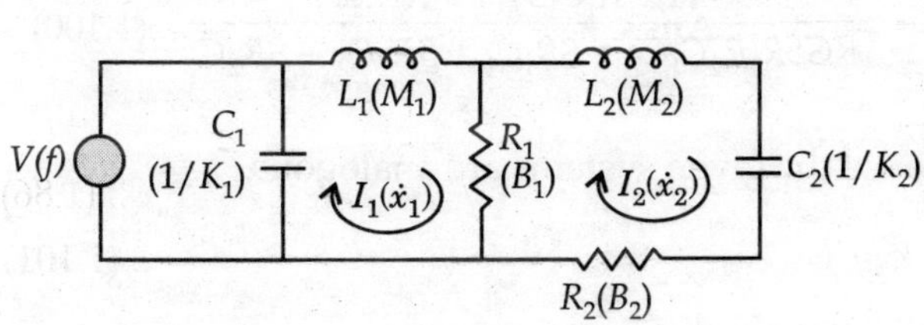

Fig. 1.63. f-V analogous elect. circuit

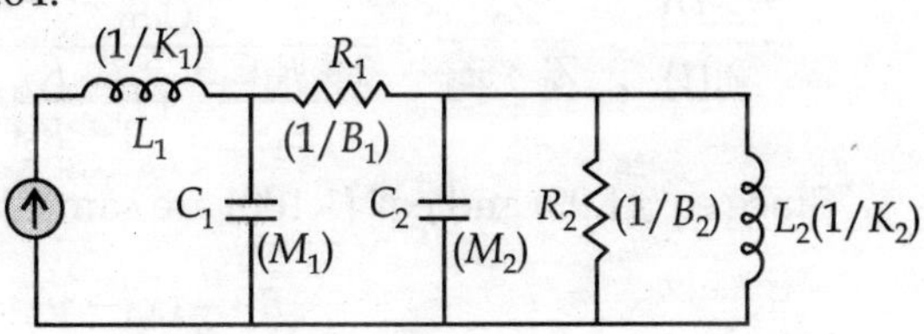

Fig. 1.64. f-i analogous elect. circuit

EXAMPLE 1.26. Show that the system shown in Fig. 1.65 (*a*), 1.65 (*b*) are analogous system.

(RML Univ, Faizabad, 2002)

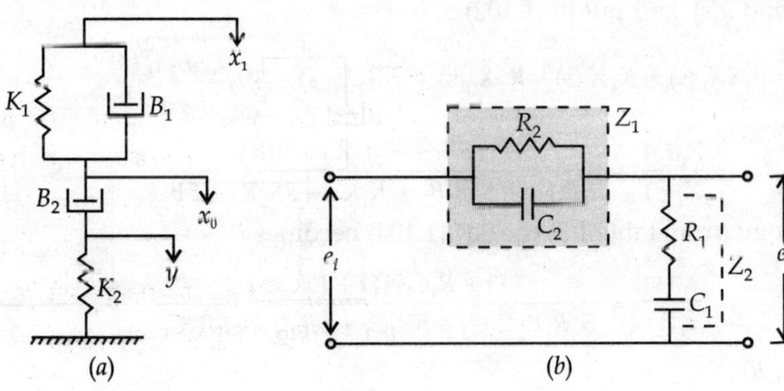

Fig. 1.65.

Solution : From Fig. 1.65 (*a*) :

$$B_1\frac{d}{dt}(x_i - x_0) + K_1(x_i - x_0) = B_2\frac{d}{dt}(x_0 - y) \quad ...(1.94)$$

$$B_2\frac{d}{dt}(x_0 - y) = K_2 y \quad ...(1.95)$$

Laplace transform of 1.94 and 1.95

$$B_1SX_1(s) - B_1SX_i(s) + K_1X_i(s) - K_1X_0(s) = SB_2X_0(s) - SB_2y(s) \quad ...(1.96)$$

$$SB_2X_0(s) - SB_2y(s) = K_2y(s) \quad ...(1.97)$$

from (1.97) find $y(s)$ and put in (1.96)

$$B_1SX_i(s) - B_1SX_0(s) + K_1X_i(s) - K_1X_0(s) = SB_2X_0(s) - SB_2\frac{SB_2x_0(s)}{K_2 + SB_2}$$

$$\therefore \quad \frac{X_0(s)}{X_i(s)} = \frac{(SB_1 + K_1)(K_2 + SB_2)}{SB_1K_2 + S^2B_1B_2 + K_1K_2 + SB_2K_1 + SB_2K_2} \quad ...(1.98)$$

using *f-v* analogy (from Table 1.3) the equation (1.98) becomes

$$\frac{e_0(s)}{e_i(s)} = \frac{(1 + R_1C_1s)(1 + R_2C_2s)}{S^2R_1R_2C_1C_2 + SR_1C_1 + SR_2C_2 + SR_2C_1} \quad ...(1.99)$$

from Fig (1.65*b*) :

$$Z_1 = \frac{R_2}{1 + R_2C_2S} \quad \text{and} \quad Z_2 = R_1 + \frac{1}{C_1S} = \frac{1 + R_1C_1S}{C_1S}$$

$$\frac{e_0(t)}{e_i(t)} = \frac{Z_2}{Z_1 + Z_2} = \frac{\frac{1 + R_1C_1S}{C_1S}}{\frac{1 + R_1C_1S}{C_1S} + \frac{R_2}{1 + R_2C_2S}} = \frac{(1 + R_1C_1S)(1 + R_2C_2S)}{S^2R_1R_2C_1C_2 + SR_1C_1 + SR_2C_2 + SR_2C_1} \quad ...(1.100)$$

Since eq^n (1.99) and eq^n (1.100) are same, hence the given systems are analogous.

$$B_2\frac{d}{dt}(x_0 - y) = K_2 y \quad ...(1.101)$$

Laplace transform of 1.94 and 1.101

$$B_1SX_i(s) - B_1SX_0(s) + K_1X_i(s) - K_1X_0(s) = SB_2X_0(s) - SB_2y(s) \quad ...(1.102)$$

$$SB_2X_0(s) - SB_2y(s) = K_2y(s) \quad ...(1.103)$$

from (1.103) find $y(s)$ and put in (1.102)

$$B_1SX_i(s) - B_1SX_0(s) + K_1X_i(s) - K_1X_0(s) = SB_2X_0(s) - SB_2\cdot\frac{SB_2x_0(s)}{K_2 + SB_2}$$

$$\therefore \quad \frac{X_0(s)}{X_i(s)} = \frac{(SB_1 + K_1)(K_2 + SB_2)}{SB_1K_2 + S^2B_1B_2 + K_1K_2 + SB_2K_1 + SB_2K_2} \quad ...(1.104)$$

using *f-v* analogy (from table 1.3) the eq^n (1.104) becomes

$$\frac{e_0(s)}{e_i(s)} = \frac{(1 + R_1C_1s)(1 + R_2C_2S)}{S^2R_1R_2C_1C_2 + SR_1C_1 + SR_2C_2 + SR_2C_1} \quad ...(1.105)$$

from Fig. (1.65 *b*) :

$$Z_1 = \frac{R_2}{1 + R_2C_2S} \quad \text{and} \quad Z_2 = R_1 + \frac{1}{C_1S} = \frac{1 + R_1C_1S}{C_1S}$$

$$\frac{e_0(t)}{e_i(t)} = \frac{Z_2}{Z_1 + Z_2}$$

$$\frac{e_0(s)}{e_i(s)} = \frac{\dfrac{1+R_1C_1S}{C_1S}}{\dfrac{1+R_1C_1S}{C_1S} + \dfrac{R_2}{1+R_2C_2S}} = \frac{(1+R_1C_1S)(1+R_2C_2S)}{S^2R_1R_2C_1C_2 + SR_1C_1 + SR_2C_2 + SR_2C_1} \quad ...(1.106)$$

Since eq^n (1.105) and eq^n (1.106) are same, hence the given systems are analogous.

BLOCK DIAGRAM

1.18. BLOCK DIAGRAM REPRESENTATION

Any system can be described by a set of differential equations or can be represented by the schematic diagram containing all components and their connections. But for the complicated systems these two methods are not suitable. The block diagram representation is the combination of above two methods. A block may represent a single component or a group of components, but each block is completely characterized by a transfer function. The transfer function is an expression which relates output to input in *s*-domain. Transfer function does not give any information about the internal structure of the system. Once we determine the transfer function, then we can represent the system by the block diagram. Block diagrams are single line diagram, that is the flow of system variables from one bolck to another block is represented by a single line. Fig. 1.66 shows the block diagram representation of a system.

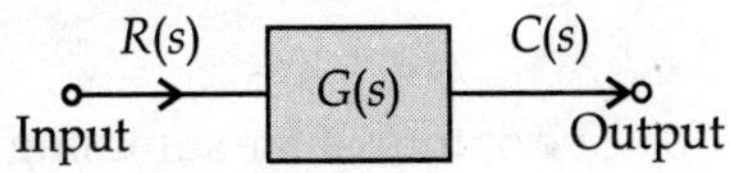

Fig. 1.66.

where, $R(s)$ = Input

$C(s)$ = output

$G(s)$ = transfer function

Then, the system can be represented as

$$C(s) = R(s).\,G(s) \quad ...(1.107)$$

The flow of system variables from one block to another block is represented by the arrow. In addition to this, the sum of the signals or the difference of the signals are represented by a summing point (Fig. 1.67*a*). Application of one input source to two or more blocks is represented by a take off point (Fig 1.67*b*).

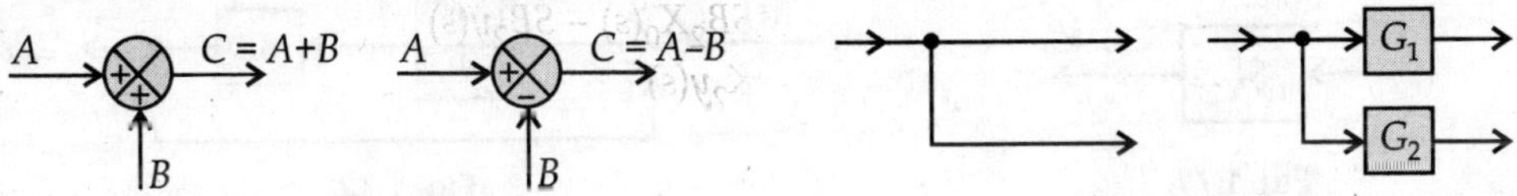

(summing point) (take off point)

Fig. 1.67.

1.19. HOW TO DRAW THE BLOCK DIAGRAM

Consider a simple *R-L* circuit shown in Fig 1.68

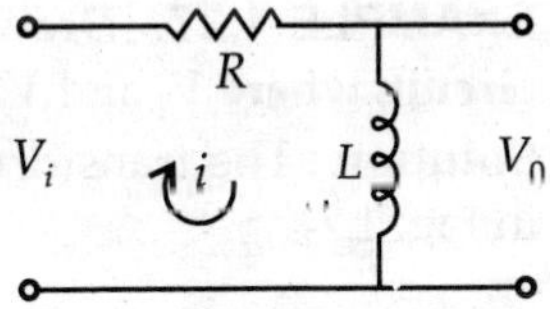

Fig. 1.68.

Apply *KVL*

$$V_i = Ri + L\frac{di}{dt} \quad ...(1.108)$$

$$V_0 = L\frac{di}{dt} \quad ...(1.109)$$

Laplace transform of 1.108 and 1.109 with initial condition zero

$$V_i(s) = I(s)\,R + SL\,I(s)$$

$$V_i(s) = I(s)\,(R + SL) \qquad ...(1.110)$$

$$V_0(s) = SL\,I(s) \qquad ...(1.111)$$

From 1.110 and 1.111

$$\frac{V_0(s)}{V_i(s)} = \frac{sL}{R+sL} \qquad ...(1.112)$$

Form Fig. 1.68

$$i = \frac{V_i - V_0}{R} \qquad ...(1.113)$$

$$V_0 = L\frac{di}{dt} \qquad ...(1.114)$$

Laplace transform of 1.113 and 1.114

$$I(s) = \frac{1}{R}\,[V_i(s) - V_0(s)] \qquad ...(1.115)$$

$$V_0(s) = SL\,I(s) \qquad ...(1.116)$$

For right hand side of equation 1.115, we use a summing point.

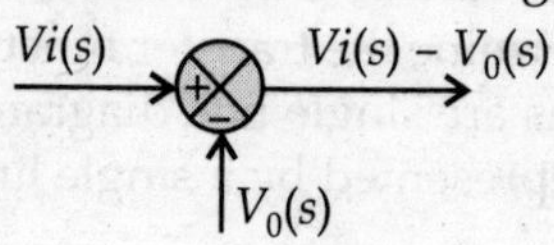

Fig. 1.69

The output of the summing point is given to block and output of the block is $I(s)$ as per eq^n 1.115.

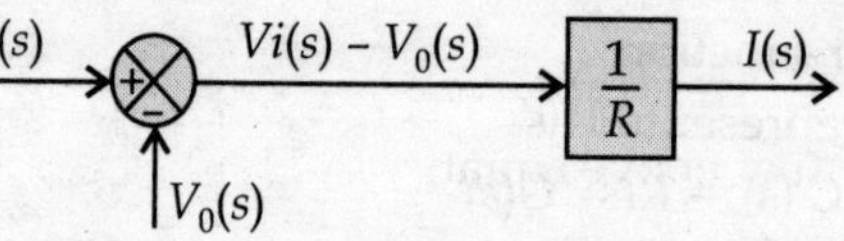

Fig. 1.70.

Form equation 1.116 the output of block $I(s)$ is given to another block containing the element SL and the output of the second block is V_0

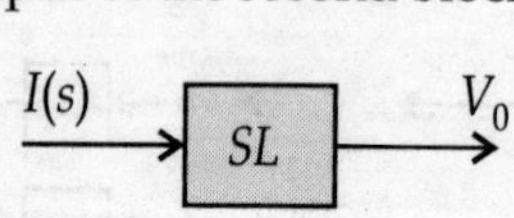

Fig. 1.71.

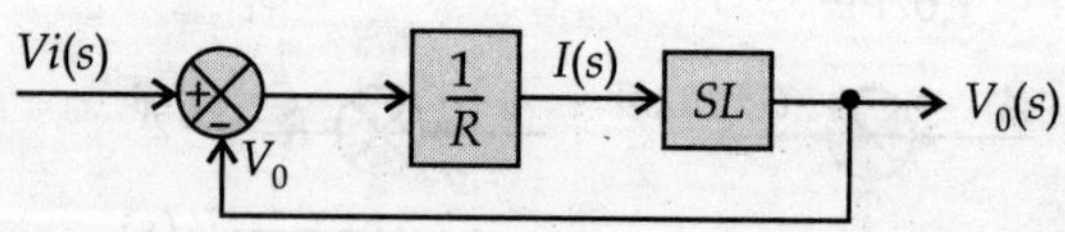

Fig. 1.72.

Combining the Fig 1.70 and 1.71 we get required block diagram.

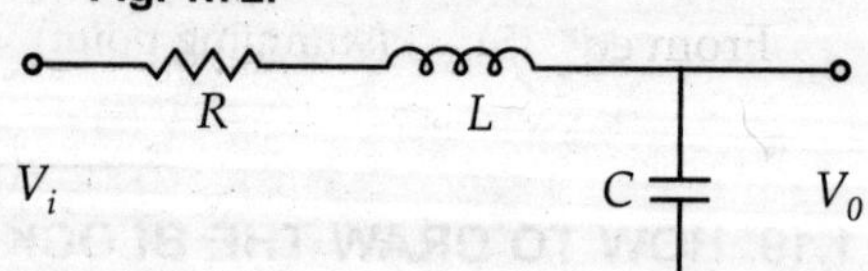

Fig. 1.73

EXAMPLE 1.27. Draw the block diagram of series RLC circuit, where V_i and V_0 are the input and output voltages.

Solution : The transformed network in S-domain is shown in Fig. 1.74.

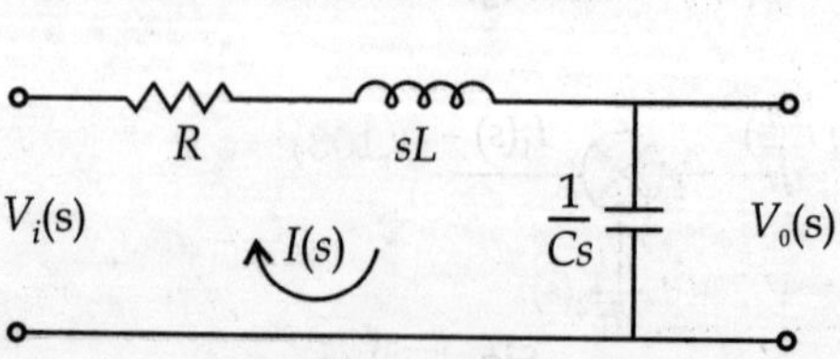

Fig. 1.74.

from 1.74 :
$$I(s) = \frac{V_i(s) - V_0(s)}{R + sL} \qquad ...(1.117)$$

$$V_0(s) = \frac{1}{Cs} I(s) \qquad ...(1.118)$$

For *R.H.S* of equation 1.117, we require a summing point

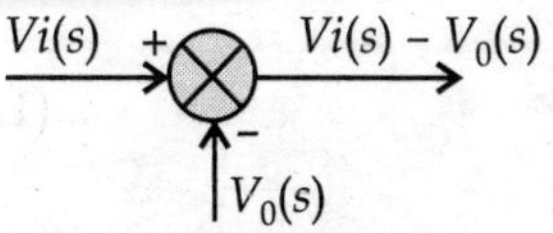

Fig. 1.75.

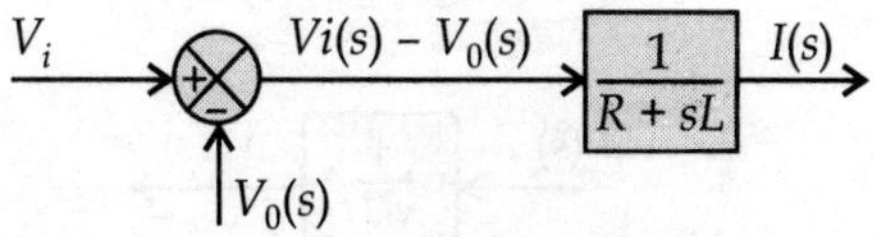

Fig. 1.76.

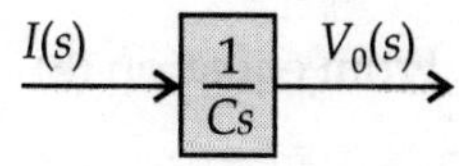

Fig. 1.77.

combining the Fig. 1.76 and 1.77 we get required block diagram.

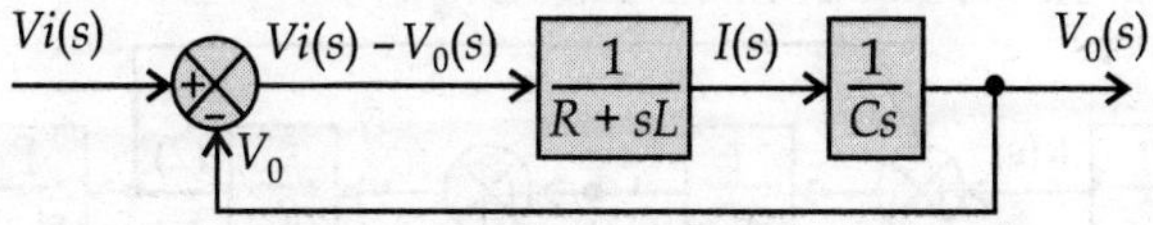

Fig. 1.78. Block diagram of series RLC circuit

EXAMPLE 1.28. Draw the block diagram of the circuit shown in Fig. 1.79.

Solution :

$$i_1(t) = \frac{V_i(t) - V_2(t)}{R_1} \qquad ...(1.119a)$$

$$V_2(t) = \frac{1}{C_1}\int [i_1(t) - i_2(t)]dt \qquad ...(1.119b)$$

$$i_2(t) = \frac{1}{R_2}[V_2(t) - V_0(t)] \qquad ...(1.119c)$$

$$V_0(t) = \frac{1}{C_2}\int i_2\, dt \qquad ...(1.119d)$$

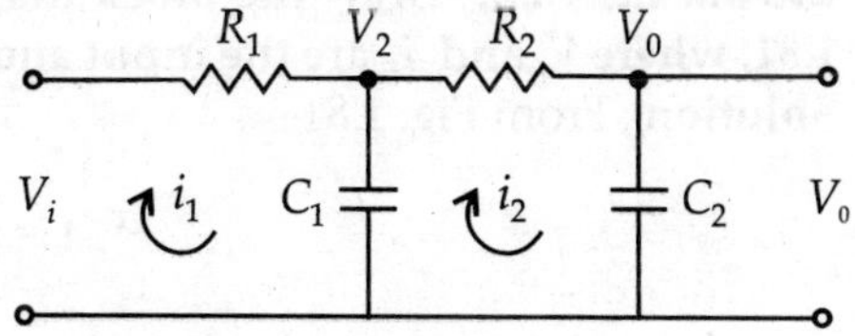

Fig. 1.79.

Take the Laplace transform of above equations

$$I_1(s) = \frac{1}{R_1}[V_i(s) - V_2(s)] \qquad ...(1.119e)$$

$$V_2(s) = \frac{1}{sC_1}[I_1(s) - I_2(s)] \qquad ...(1.119f)$$

$$I_2(s) = \frac{1}{R_2}[V_2(s) - V_0(s)] \qquad ...(1.119g)$$

$$V_0(s) = \frac{1}{sC_2} I_2(s) \qquad ...(1.119h)$$

From eq$^{n.}$ (5)

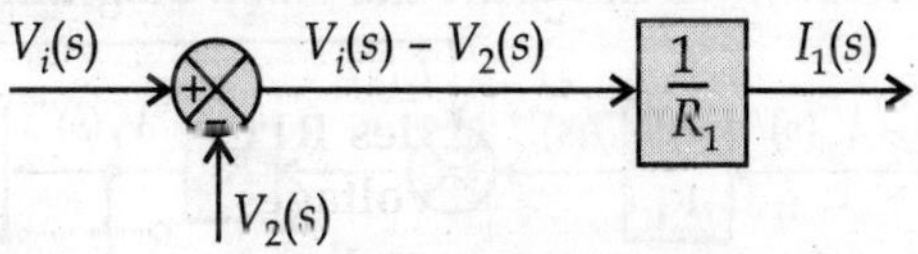

Fig. 1.80 (a)

From eq^{n} (6)

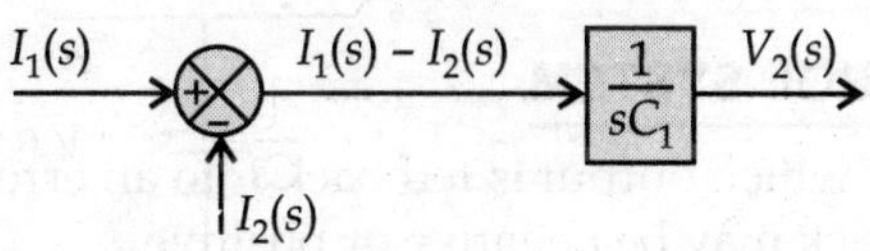

Fig. 1.80 (b)

From equation (7)

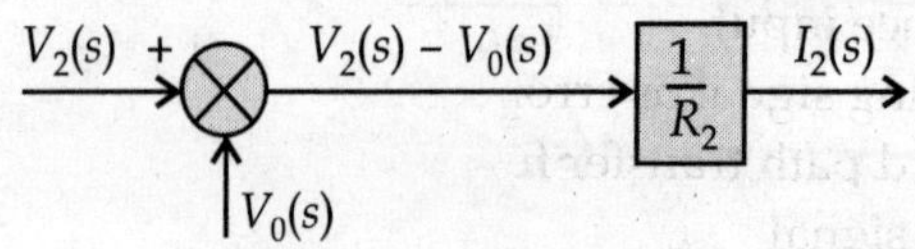

Fig. 1.80 (c)

From equation (8)

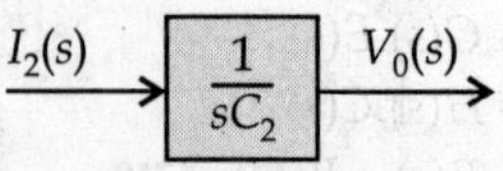

Fig. 1.80 (d)

combining all the Fig. 1.80 *a*, 1.80 *b*, 1.80 *c* and 1.80 *d*

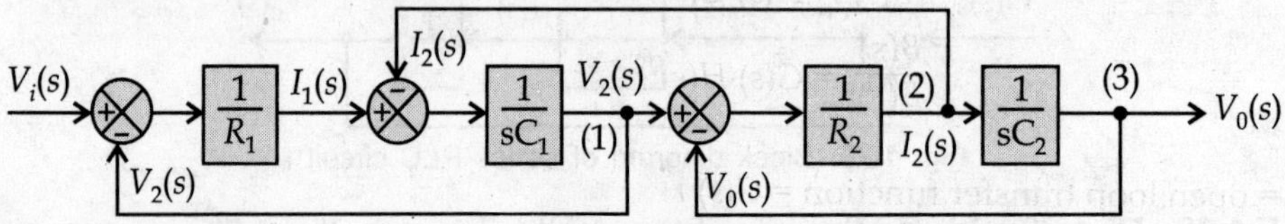

Fig. 1.80 (e)

EXAMPLE 1.29. Draw the block diag. for the circuit shown in Fig. 1.81, where V_i and i_L are the input and output variables respectively.

Solution : From Fig. 1.81.

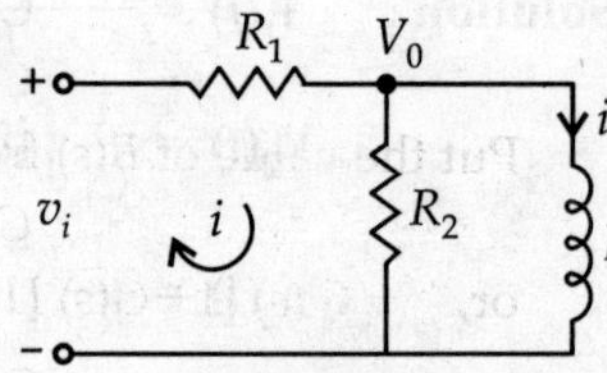

Fig. 1.81.

$$i = \frac{V_i - V_0}{R_1} \quad ...(1.120a)$$

$$i - i_L = \frac{V_0}{R_2} \quad ...(1.120b)$$

$$V_0 = L\frac{di_L}{dt} \quad ...(1.120\ c)$$

laplace transform of above eq^{ns}

$$I(s) = \frac{1}{R_1}\,[V_i(s) - V_0(s)] \quad ...(1.120\ d)$$

$$V_0(s) = R_2\,[I(s) - I_L(s)] \quad ...(1.120\ e)$$

$$V_0(s) = SLI_L(s)$$

$$I_L(s) = \frac{1}{SL}\,V_0(s) \quad ...(1.120\ f)$$

From 1.120 *d*, 1.120 *e* and 1.120 *f*, we can draw the block diagram.

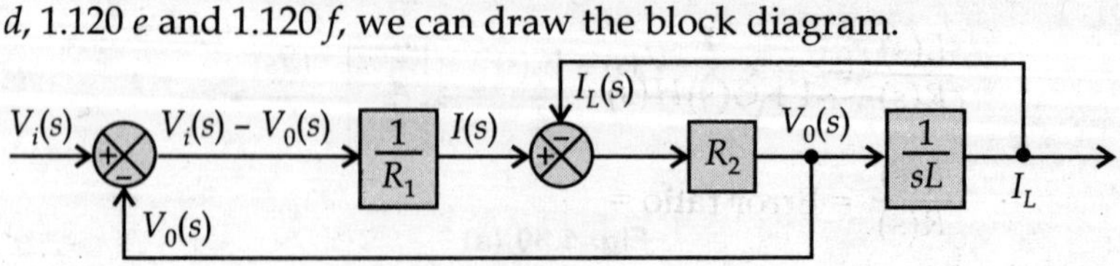

Fig. 1.82.

1.20. CLOSED LOOP CONTROL SYSTEM

A closed loop system is one in which output is fed back into an error detector and compared with the reference input. The feedback may be negative or positive.

Consider a closed loop system shown in Fig. 1.83

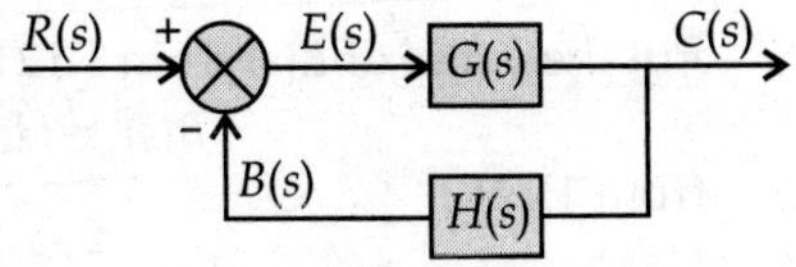

Fig. 1.83

where, $R(s)$ = Reference input
$E(s)$ = Actuating signal or error signal
$G(s)$ = Forward path transfer function
$C(s)$ = output signal
$H(s)$ = Feedback transfer function
$B(s)$ = Feedback signal

From Fig. 1.83

$$C(s) = G(s).E(s) \quad ...(1.121\ a)$$
$$B(s) = H(s).C(s) \quad ...(1.121\ b)$$
$$E(s) = R(s) - B(s) \quad ...(1.121\ c)$$

Put the value of $C(s)$ from 1.121 *a* in 1.121 *b*

$$B(s) = H(s)\cdot G(s)\ E(s)$$
$$\frac{B(s)}{E(s)} = G(s)\cdot H(s)$$
$$\frac{B(s)}{E(s)} = \text{open loop transfer function} = G(s)\ H(s) \quad ...(1.121\ d)$$

Put the value of $E(s)$ from 1.121*c* in 1.121*a*

$$C(s) = G(s)\ [R(s) - B(s)]$$
$$C(s) = R(s).\ G(s) - G(s).\ B(s) \quad ...(1.121\ e)$$

Put the value of $B(s)$ from 1.121 *b* in 1.121 *e*

$$C(s) = R(s)\ G(s) - G(s)\ H(s).\ C(s)$$

or, $$C(s)\ [1 + G(s)\ H(s)] = R(s)\ G(s)$$

or, $$\frac{C(s)}{R(s)} = \frac{G(s)}{1+G(s)\ H(s)} \quad ...(1.121\ f)$$

$$\frac{C(s)}{R(s)} = M(s) = \text{closed loop transfer function} = \frac{G(s)}{1+G(s)\ H(s)}$$

If the feedback is positive, then equation 1.121 *f* becomes

$$\frac{C(s)}{R(s)} = \frac{G(s)}{1-G(s)\ H(s)} \quad ...(1.121\ g)$$

From 1.121*a* put the value of $C(s)$ in equation 1.121 *f*

$$\frac{G(s)\ E(s)}{R(s)} = \frac{G(s)}{1+G(s)H(s)}$$

or, $$\frac{E(s)}{R(s)} = \frac{1}{1+G(s)H(s)} \quad ...(1.121\ h)$$

$$\frac{E(s)}{R(s)} = \text{Error ratio} = \frac{1}{1+G(s)H(s)}.$$

For positive feedback

$$\frac{E(s)}{R(s)} = \frac{1}{1-G(s)H(s)} \quad ...(1.121\ i)$$

Put the value of $C(s)$ from 1.121 *a* in 1.121 *b*

$$B(s) = H(s).\ G(s).\ E(s)$$

Put the value of $E(s)$ from 1.121*c* in above equation

$$B(s) = H(s).\, G(s)\, [R(s) - B(s)] \qquad ...(1.121\, j)$$

from 1.121 *j*

$$\frac{B(s)}{R(s)} = \frac{G(s)\, H(s)}{1 + G(s)\, H(s)} \qquad ...(1.121\, k)$$

$$\frac{B(s)}{R(s)} = \text{Primary feedback ratio} = \frac{G(s)\, H(s)}{1 + G(s)\, H(s)}$$

For positive feedback

$$\frac{B(s)}{R(s)} = \frac{G(s)\, H(s)}{1 - G(s)\, H(s)} \qquad ...(1.121 l)$$

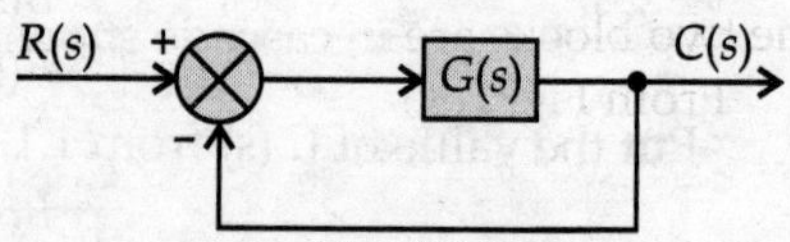

Fig. 1.84.

A unity feedback control system is shown in Fig. 1.84.

For unity feedback control system $H(s) = 1$

$$\therefore \quad \frac{C(s)}{R(s)} = \frac{G(s)}{1 + G(s)} \quad \text{For negative feedback}$$

$$\frac{C(s)}{R(s)} = \frac{G(s)}{1 - G(s)} \quad \text{For positive feedback}$$

1.21. MULTI INPUT-MULTI OUTPUT SYSTEM (MIMO)

When two or more inputs act on a system, each input can be treated independently of the others. Complete output can be obtained by adding the effect of each input.

Consider a two input linear system shown in Fig. 1.85.

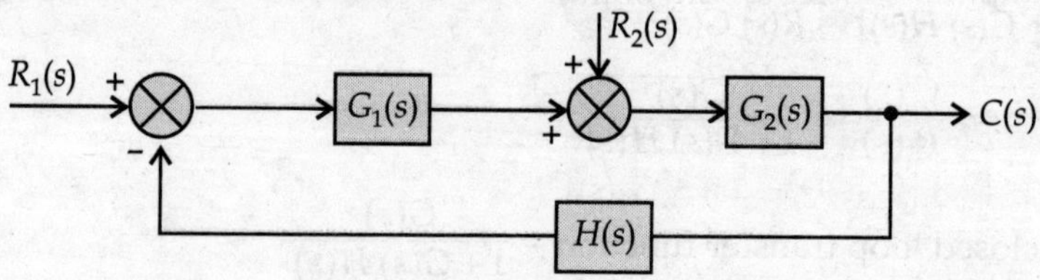

Fig. 1.85.

Step1 : Put $R_1(s) = 0$

$$\frac{C_2(s)}{R_2(s)} = \frac{G_2(s)}{1 + G_1(s) G_2(s) H(s)} \qquad ...(1.122\, a)$$

Step 2: Put $R_2(s) = 0$

$$\frac{C_1(s)}{R_1(s)} = \frac{G_1(s) G_2(s)}{1 + G_1(s) G_2(s) H(s)} \qquad ...(1.122\, b)$$

Hence,

$$C(s) = C_1(s) + C_2(s)$$

$$C(s) = \frac{G_1(s) G_2(s) R_1(s)}{1 + G_1(s) G_2(s) H(s)} + \frac{G_2(s)\, R_2(s)}{1 + G_1(s) G_2(s) H(s)}$$

$$C(s) = \frac{G_2(s)}{1 + G_1(s) G_2(s) H(s)} [R_1(s) G_1(s) + R_2(s)] \qquad ...(1.122\, c)$$

MIMO system can be expressed in matrix form

$$\begin{bmatrix} C_1(s) \\ C_2(s) \\ \vdots \\ C_n(s) \end{bmatrix} = \begin{bmatrix} G_{11}(s) & G_{12}(s) & \cdots \\ \vdots & & \\ \vdots & & \\ G_{nm}(s) & & \end{bmatrix} \begin{bmatrix} R_1(s) \\ R_2(s) \\ \vdots \\ R_r(s) \end{bmatrix}$$

$$C(s) = G(s)\,R(s)$$

where, $G(s)$ = matrix transfer function

1.22. BLOCK DIAGRAM REDUCTION

When a number of blocks are connected, the overall transfer function can be obtained by block diagram reduction technique. The following rules are associated with the block reduction technique.

Rule No. 1. Blocks in cascade

When two or more blocks are in cascade, the resultant block is a product of the individual block transfer function. Consider the two blocks are in cascade shown in Fig. 1.86.

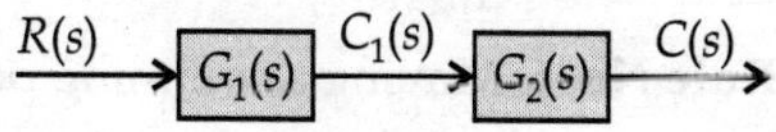

Fig. 1.86

From Fig. 1.86

$$\frac{C_1(s)}{R(s)} = G_1(s) \qquad \text{...(1.123 } a\text{)}$$

$$\frac{C(s)}{C_1(s)} = G_2(s) \qquad \text{...(1.123 } b\text{)}$$

From (1.122 *a*) and (1.123 *b*)

$$\frac{C_1(s)}{R(s)} \cdot \frac{C(s)}{C_1(s)} = \frac{C(s)}{R(s)} = G_1(s)\,.\,G_2(s)$$

The equivalent diagram is shown in Fig 1.87.

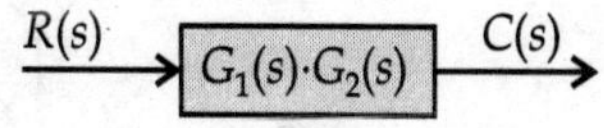

Fig. 1.87

Rule No 2. Blocks in Parallel

When two or more blocks are connected in parallel as shown in Fig. 1.88 *a*, the resultant block is the sum of individual block transfer function.

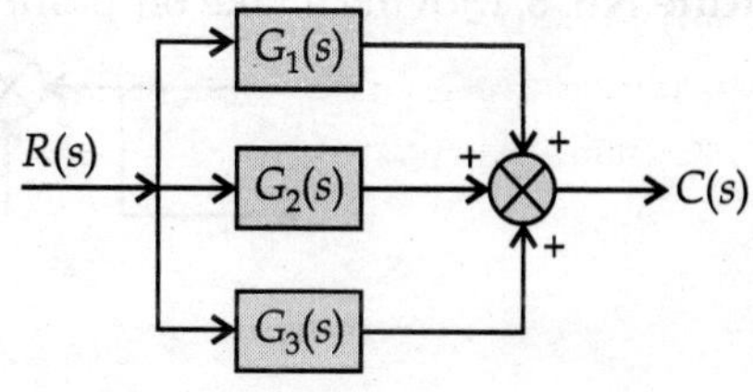

Fig. 1.88 (a)

from Fig. 1.88 *a*

$$C(s) = R(s)\,G_1(s) + R(s)\,G_2(s) + R(s)\,G_3(s)$$
$$= R(s)\,[G_1(s) + G_2(s) + G_3(s)]$$

$$\therefore \quad \frac{C(s)}{R(s)} = G_1(s) + G_2(s) + G_3(s)$$

The equivalent diag. is shown in Fig. 1.88 *b*

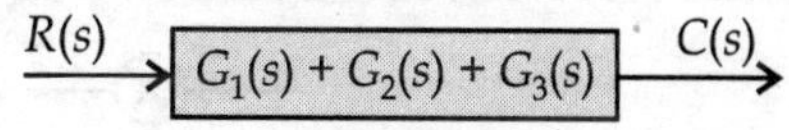

Fig. 1.88 (b)

Rule No 3. Moving a take off point ahead of a block

If a take off point is moved ahead of a block, a block with same transfer function will introduce in the branch of a take off point, as shown in Fig. 1.88 (*c*).

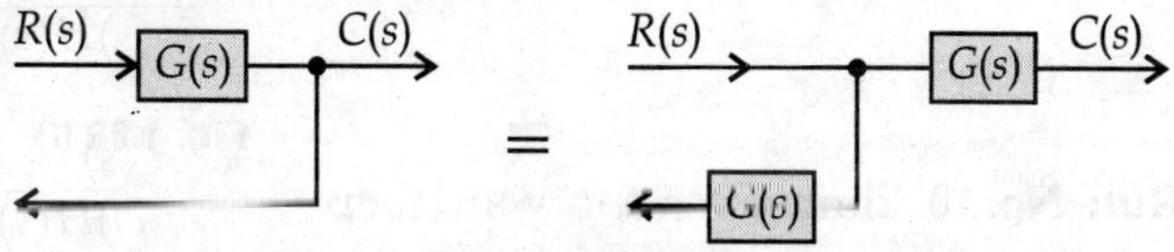

Fig. 1.88 (c)

Rule No. 4. Moving a take off point after the block

If a take off point is moved after the block, a block with the reciprocal of the transfer function is introduced in the branch of a take off point as shown in Fig. 1.88 (*d*).

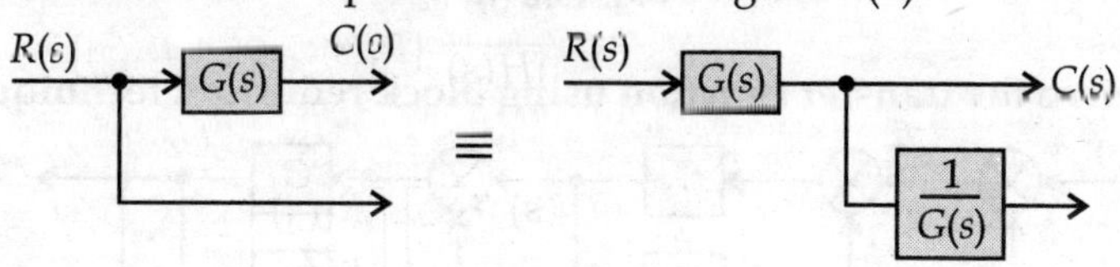

Fig. 1.88 (d)

Rule No. 5. Moving a summing point beyond a block

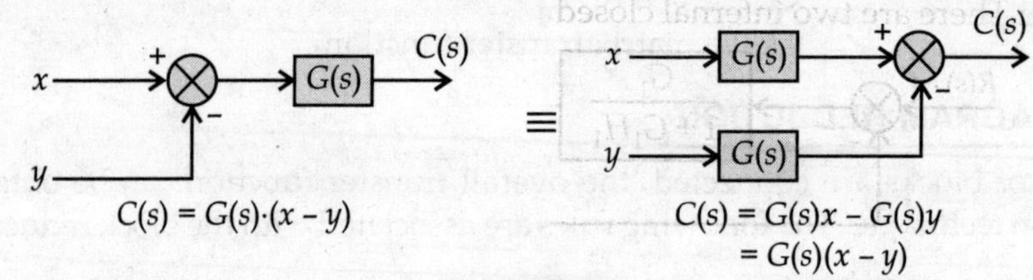

Fig. 1.88 (e)

Rule No. 6. Moving a summing point ahead of a block

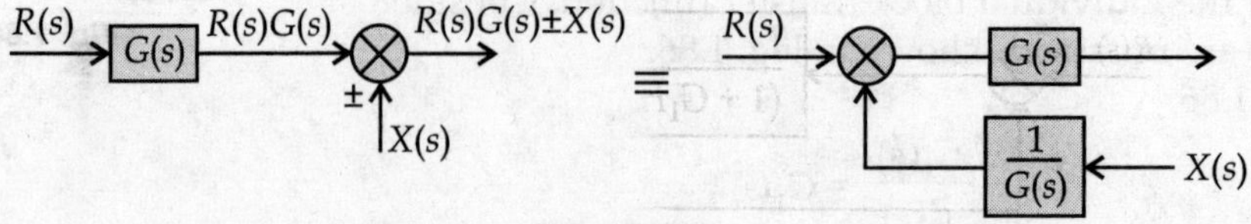

Fig. 1.88 (f)

Rule No. 7. Interchanging two summing points

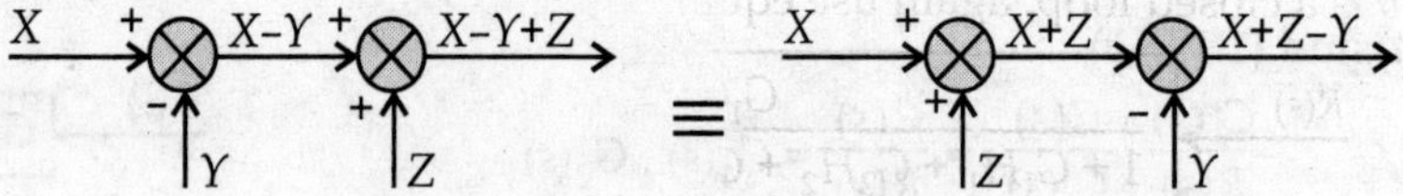

Fig. 1.88 (g)

Rule No. 8. Moving a take off point beyond a summing point

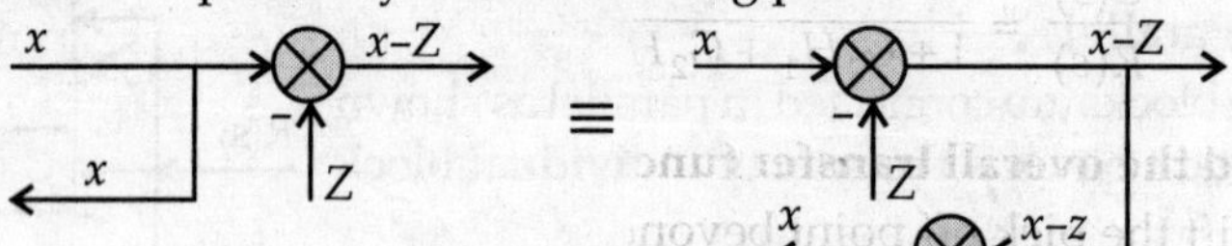

Fig. 1.88 (h)

Rule No. 9. Moving a take off point ahead of a summing point

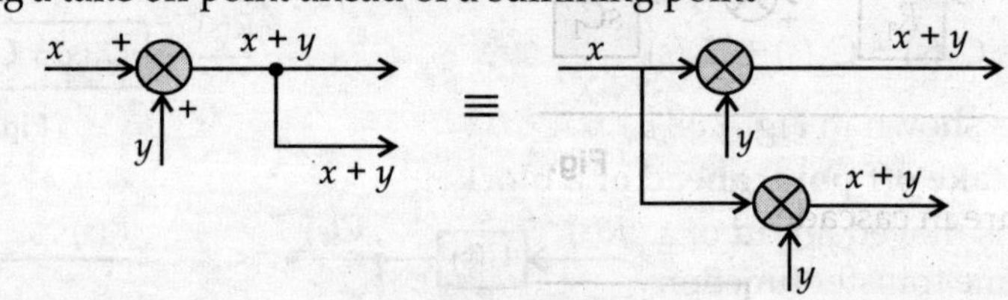

Fig. 1.88 (i)

Rule No. 10. Eliminating a forward loop

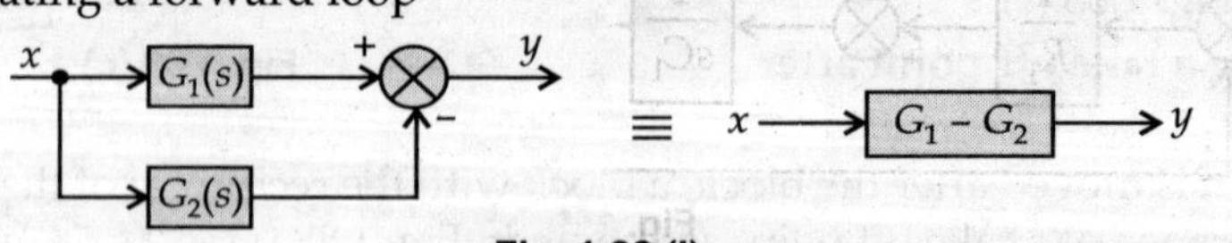

Fig. 1.88 (j)

EXAMPLE 1.30. Derive the transfer function using block reduction technique.

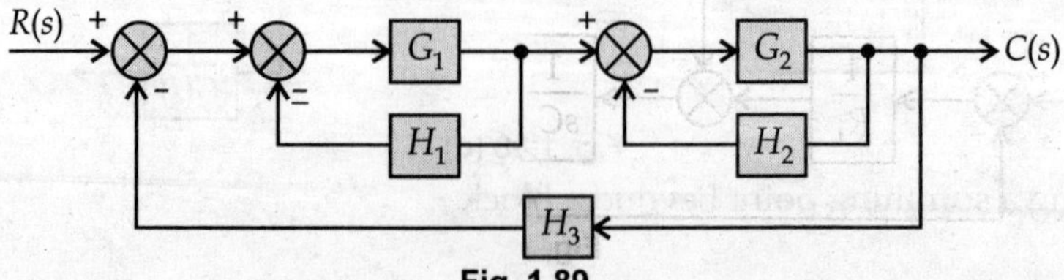

Fig. 1.89.

Solution : Step 1: There are two internal closed loops, remove these loops by using the eq^n 1.121*f*.

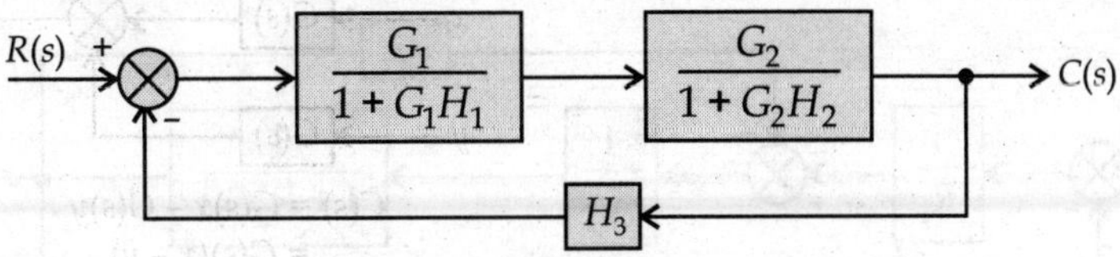

Fig. 1.89 (a)

Step 2: Two blocks are in cascade, use the rule no. 1.

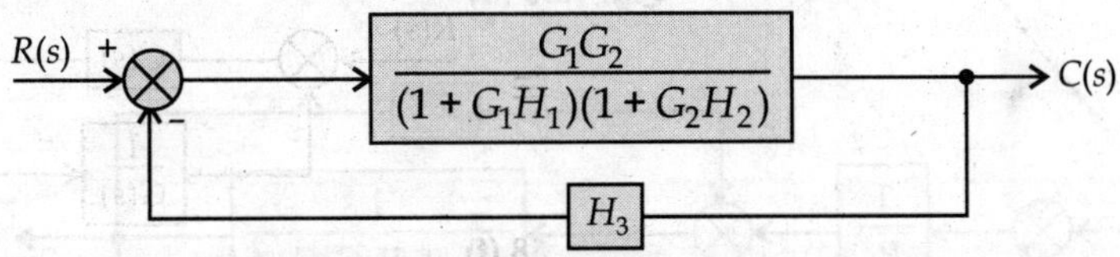

Fig. 1.89(b)

Step 3: Fig. 1.89 *b* is a closed loop, again use equation 1.121 *f*

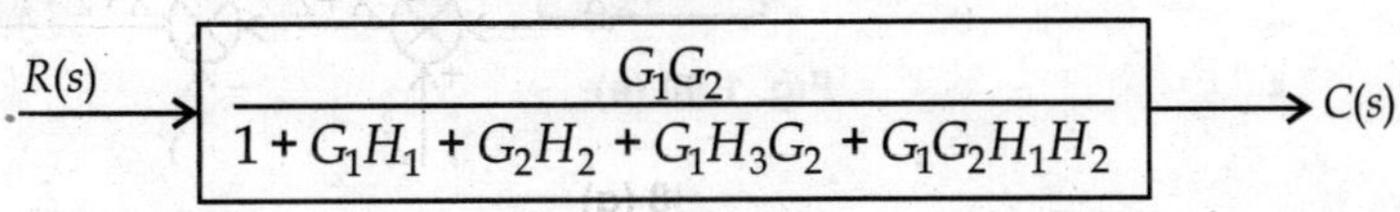

$$\therefore \qquad \frac{C(s)}{R(s)} = \frac{G_1G_2}{1+G_1H_1+G_2H_2+G_1G_2H_3+G_1G_2H_1H_2} \quad \textbf{Ans.}$$

EXAMPLE 1.31. Find the overall transfer function of the system shown in Fig. 1.80 *e*.

Solution : Step 1 : Shift the pick off point beyond the block $1/sC_2$

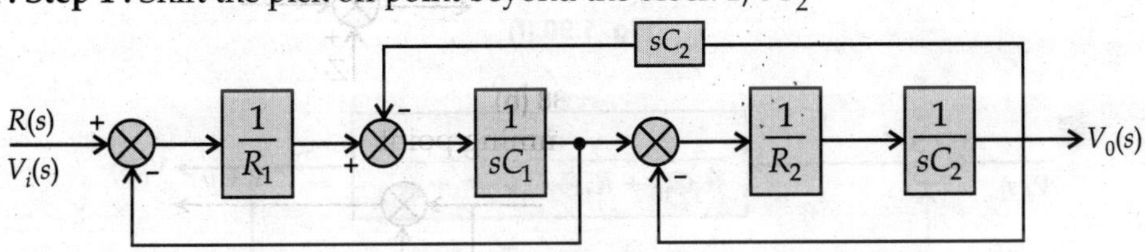

Fig. 1.90 (a)

Step 2 : Two blocks are in cascade

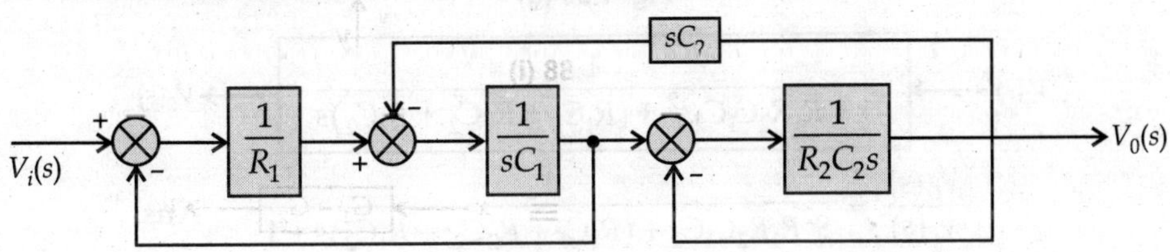

Fig. 1.90 (b)

Step 3 :

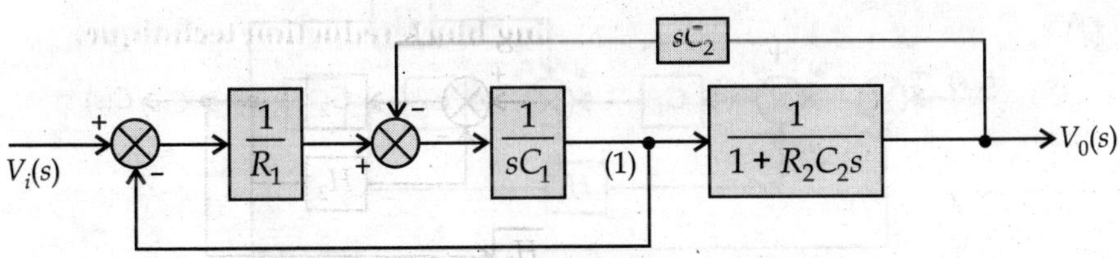

Fig. 1.90 (c)

Step 4 :

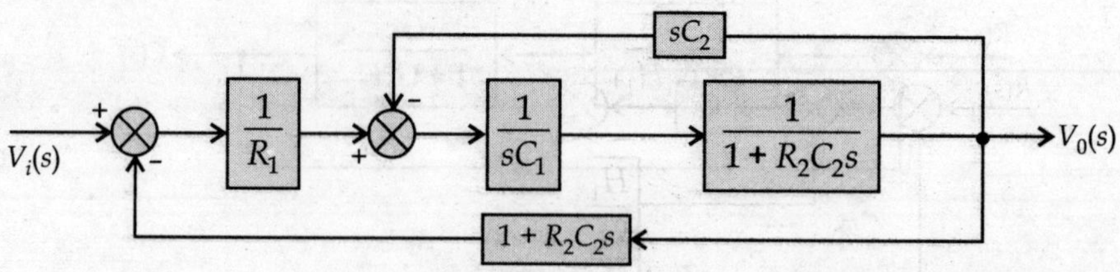

Fig. 1.90 (d)

Step 5 :

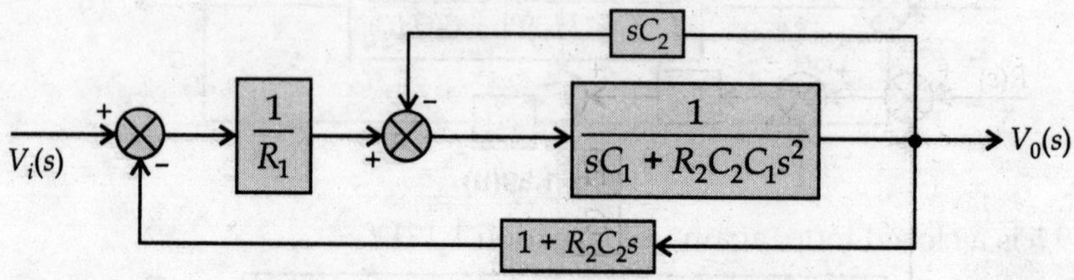

Fig. 1.90 (e)

Step 6 :

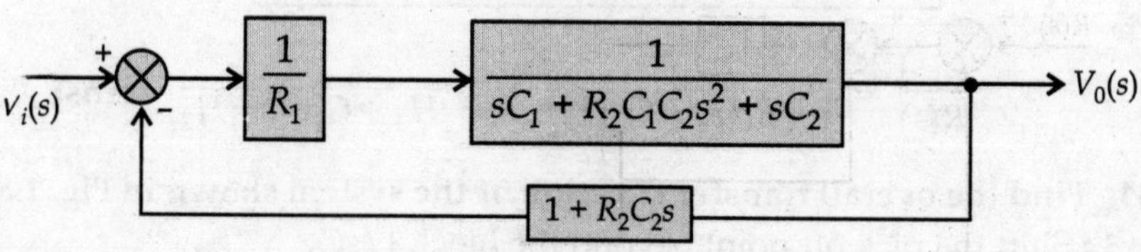

Fig. 1.90 (f)

Step 7 :

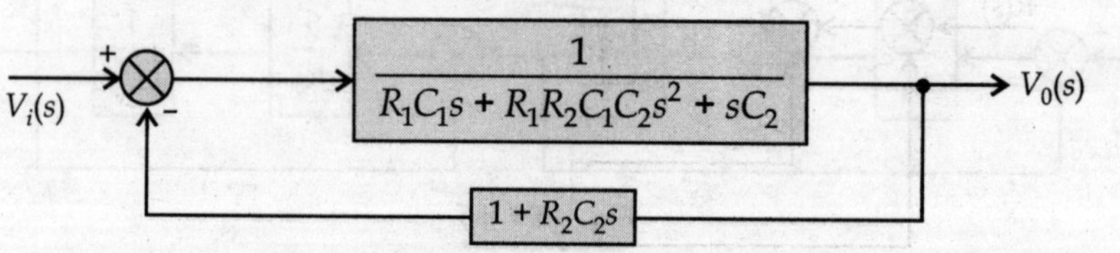

Fig. 1.90 (g)

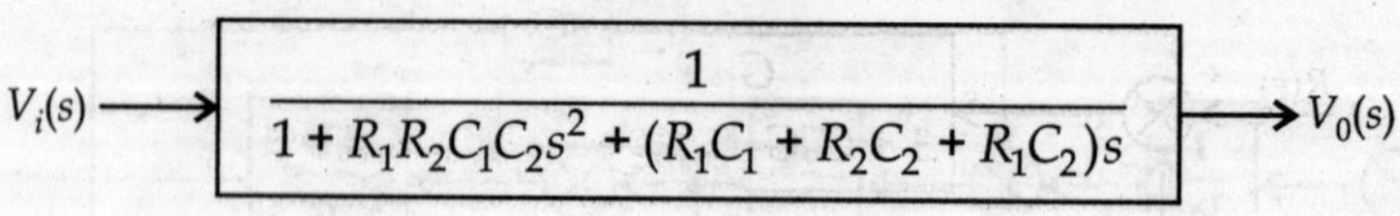

$$\therefore \quad \frac{V_0(s)}{V_i(s)} = \frac{1}{S^2R_1R_2C_1C_2 + (R_1C_1 + R_2C_2 + R_1C_2)s + 1} \quad \textbf{Ans.}$$

EXAMPLE 1.32. Determine the ratio *C*(*s*)/*R*(*s*) for the system shown in Fig. 1.91.

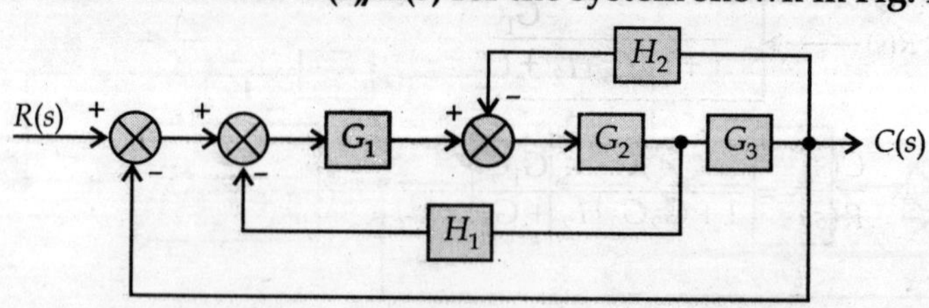

Fig. 1.91.

Solution : Step 1 : Shift the takeoff point beyond the block G_3

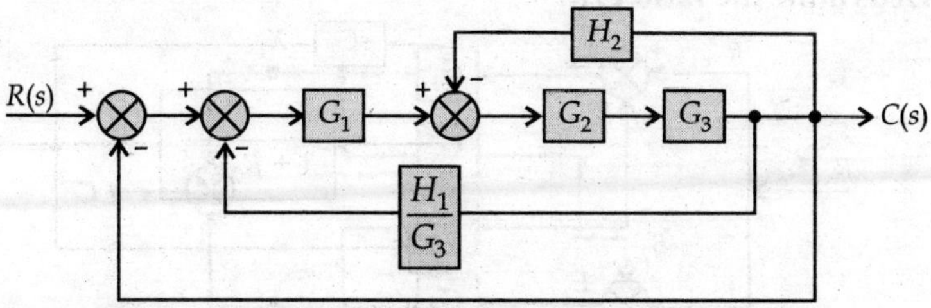

Step 2 : G_2 & G_3 are in cascade

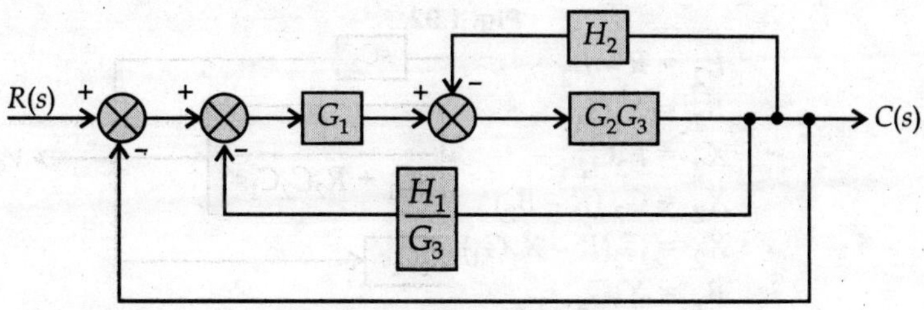

Step 3 :

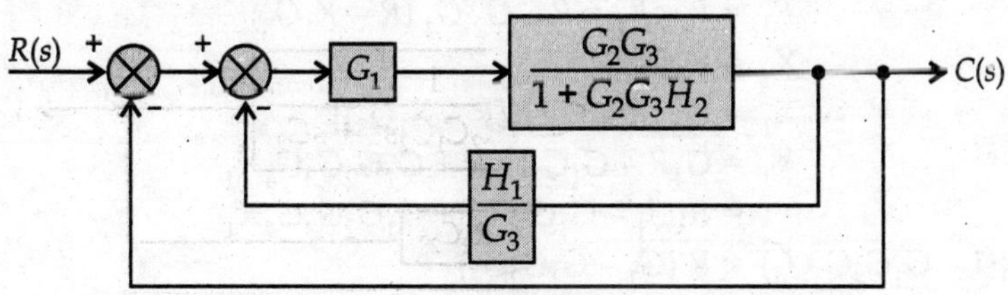

Step 4 :

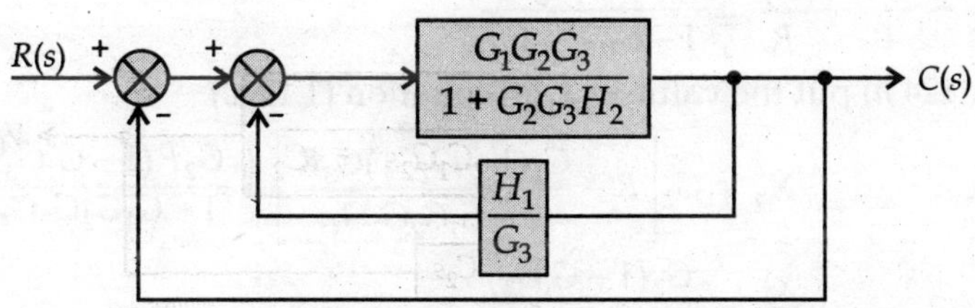

Step 5 :

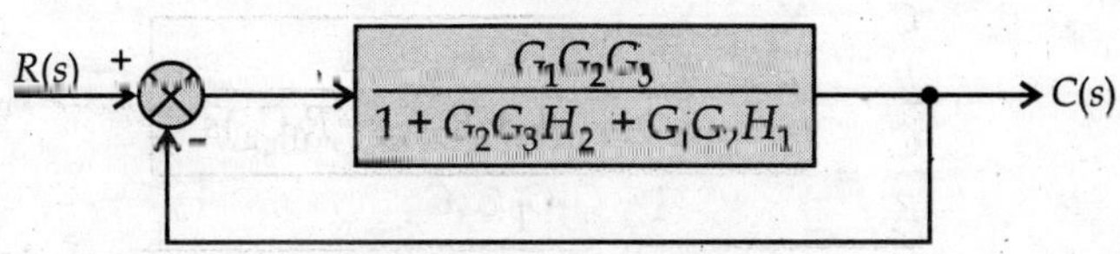

Step 6 :

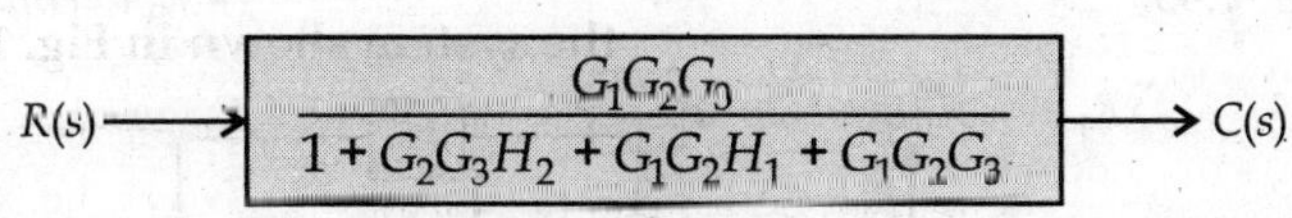

$$\frac{C(s)}{R(s)} = \frac{G_1G_2G_3}{1 + G_2G_3H_2 + G_1G_2H_1 + G_1G_2G_3}$$ **Ans.**

EXAMPLE 1.33. Determine the ratio C/R.

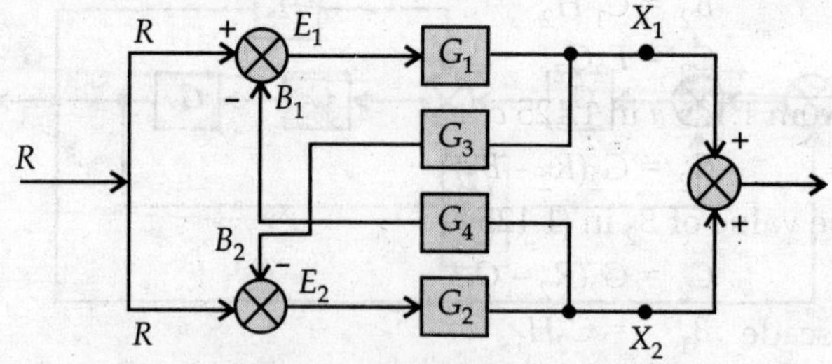

Fig. 1.92.

Solution :

$$E_2 = R - B_2 \quad \text{...(1.124 a)}$$
$$B_2 = X_1G_3 \quad \text{...(1.124 b)}$$
$$X_2 = E_2G_2$$
$$X_2 = G_2\,(R - B_2)$$
$$X_2 = G_2(R - X_1G_3) \quad \text{...(1.124 c)}$$
$$B_1 = X_2G_4$$
$$B_1 = G_4G_2\ (R - X_1G_3) \quad \text{...(1.124 d)}$$
$$E_1 = R - B_1 = R - G_2\,G_4\,(R - X_1G_3) \quad \text{...(1.124 e)}$$
$$X_1 = E_1G_1$$
$$X_1 = G_1\,[R - G_2G_4(R - X_1G_3)] \quad \text{...(1.124 f)}$$
$$\therefore \quad X_1 = G_1R - G_1G_2G_4R + G_1G_2G_3G_4X_1 \quad \text{...(1.124 g)}$$
$$= R(G_1 - G_1G_2G_4) + G_1G_2G_3G_4X_1$$
$$\therefore \quad X_1(1 - G_1G_2G_3G_4) = R\,(G_1 - G_1G_2G_4)$$
$$\frac{X_1}{R} = \frac{G_1 - G_1G_2G_4}{1 - G_1G_2G_3G_4} \quad \text{...(1.124 h)}$$

from equation (1.124 *h*) put the value of x_1 in equation (1.124*c*)

$$X_2 = G_2\left[R - \frac{G_1(1 - G_2G_4)G_3R}{1 - G_1G_2G_3G_4}\right] = \frac{G_2R(1 - G_1G_3)}{1 - G_1G_2G_3G_4} \quad \text{...(1.124 i)}$$
$$\therefore \quad \frac{x_2}{R} = \frac{G_2(1 - G_1G_3)}{1 - G_1G_2G_3G_4}$$
$$\therefore \quad \frac{C}{R} = \frac{X_1}{R} + \frac{X_2}{R}$$
$$\frac{C}{R} = \frac{G_1 + G_2 - G_1G_2G_4 - G_1G_2G_3}{1 - G_1G_2G_3G_4} \quad \textbf{Ans.}$$

EXAMPLE 1.34. Determine the transfer function C_1/R_1, C_2/R_2, C_1/R_2 and C_2/R_1 from the block diagram shown in Fig. 1.93. *(R.M.L. Univ. 2002, Control System)*

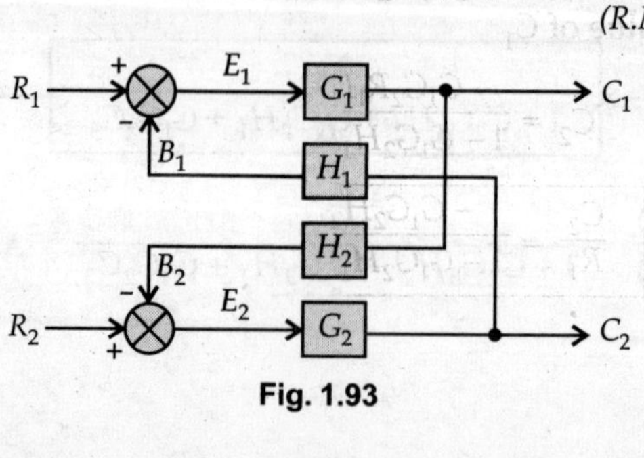

Fig. 1.93

Solution :

$$E_2 = R_2 - B_2 \quad ...(1.125\ a)$$

$$B_2 = C_1 H_2 \quad ...(1.125\ b)$$

$$C_2 = E_2 G_2 \quad ...(1.125\ c)$$

put the value of E_2 from 1.125 *a* in 1.125 *c*

$$C_2 = G_2(R_2 - B_2) \quad ...(1.125\ d)$$

from (1.125 *b*) put the value of B_2 in (1.125 *d*)

$$C_2 = G_2(R_2 - C_1H_2) \quad ...(1.125\ e)$$

$$B_1 = C_2H_1 \quad ...(1.125\ f)$$

put the value of C_2 from (1.125 *e*) in (1.125 *f*)

$$B_1 = G_2H_1 (R_2 - C_1H_2) \quad ...(1.125\ g)$$

$$E_1 = R_1 - B_1$$

$$E_1 = R_1 - G_2H_1(R_2 - C_1H_2) \quad ...(1.125\ h)$$

Also, $C_1 = E_1G_1$ . ..(1.125 *i*)

from (1.125 *h*) put the value of E_1 in (1.125 *i*)

$$C_1 = G_1 [R_1 - G_2H_1 (R_2 - C_1H_2)] \quad ..(1.125\ j)$$

$$C_1(1 - G_1G_2H_1H_2) = G_1R_1 - G_1G_2H_1R_2 \quad ...(1.125\ k)$$

When $R_1 = 0$ from (1.125 *k*)

$$\boxed{\frac{C_1}{R_2} = \frac{-G_1G_2H_1}{1 - G_1G_2H_1H_2}} \quad \textbf{Ans.} \quad ...(1.125\ l)$$

from (1.125 *l*) put the value of C_1 in (1.125 *e*)

$$C_2 = G_2\left[R_2 + \frac{G_1G_2H_1H_2R_2}{1 - G_1G_2H_1H_2}\right]$$

$$\therefore \quad \boxed{\frac{C_2}{R_2} = \frac{G_2}{1 - G_1G_2H_1H_2}} \quad \textbf{Ans.} \quad ...(1.125\ m)$$

When $R_2 = 0$, (1.125 *k*) becomes

$$C_1 = G_1 [R_1 + C_1 G_2H_1H_2]$$

$$C_1 = G_1R_1 + C_1G_1G_2H_1H_2$$

$$\therefore \quad \boxed{\frac{C_1}{R_1} = \frac{G_1}{1 - G_1G_2H_1H_2}} \quad \textbf{Ans.} \quad ...(1.125\ n)$$

When $R_2 = 0$, (1.125 *e*) becomes

$$C_2 = -G_2C_1H_2$$

from (1.125 *n*) put the value of C_1

$$\therefore \quad C_2 = \frac{-G_1G_2R_1H_2}{1 - G_1G_2H_1H_2}$$

$$\boxed{\frac{C_2}{R_1} = \frac{-G_1G_2H_2}{1 - G_1G_2H_1H_2}} \quad \textbf{Ans.} \quad ...(1.125\ n')$$

EXAMPLE 1.35. Determine the ratio $C(s)/R(s)$.

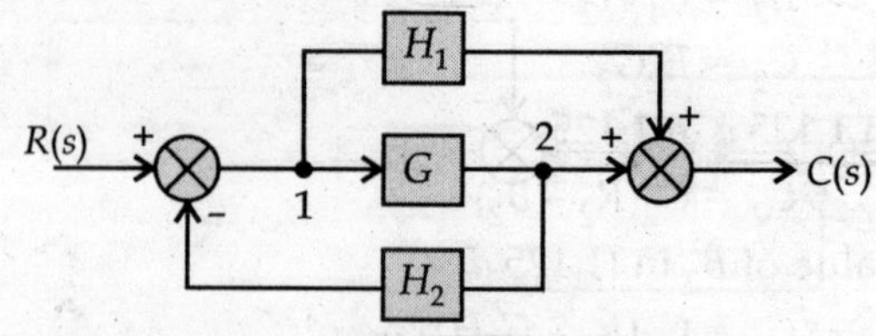

Fig. 1.94.

Solution : Step 1 : Shift the takeoff point beyond block G.

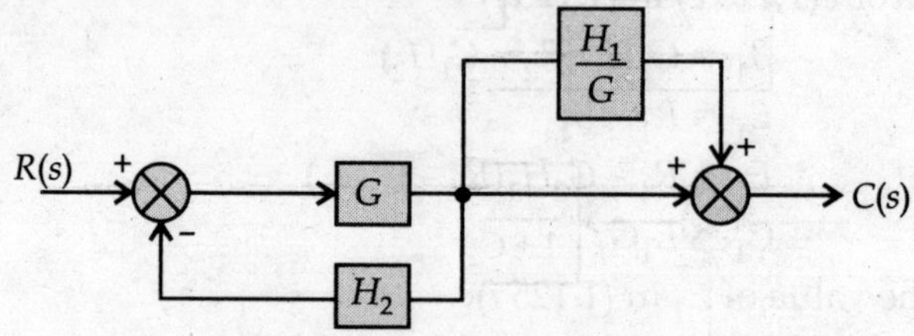

Step 2 :

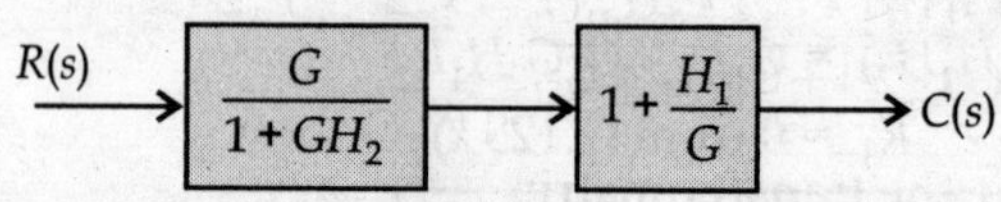

Step 3 :

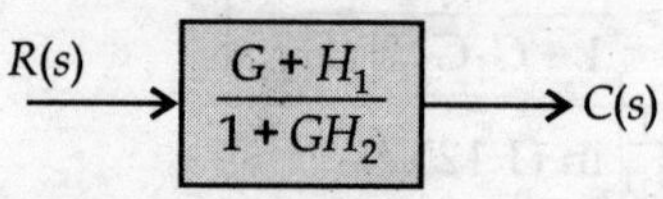

$$\therefore \qquad \frac{C(s)}{R(s)} = \frac{G+H_1}{1+GH_2} \qquad \textbf{Ans.}$$

EXAMPLE 1.36. Find the ratio $C(s)/R(s)$ of the system shown in Fig. 1.95.

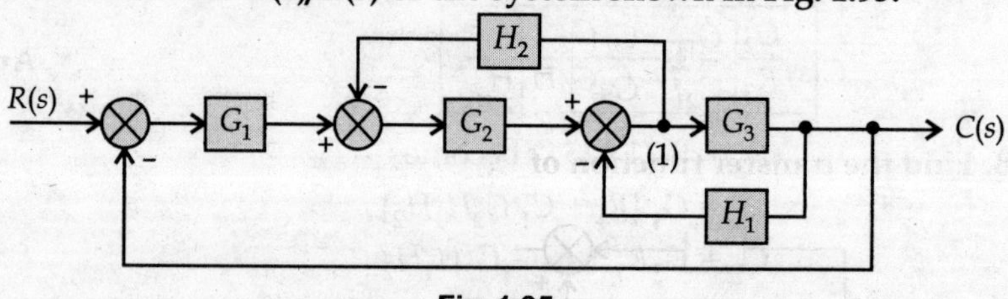

Fig. 1.95.

Solution : Step 1 : Shift the take off point beyond block G_3.

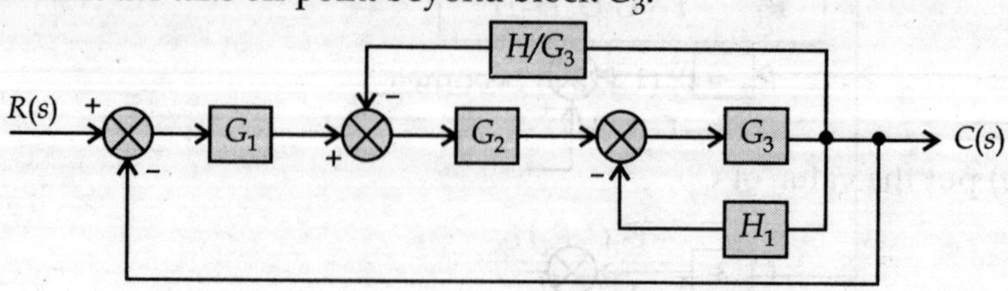

Step 2 :

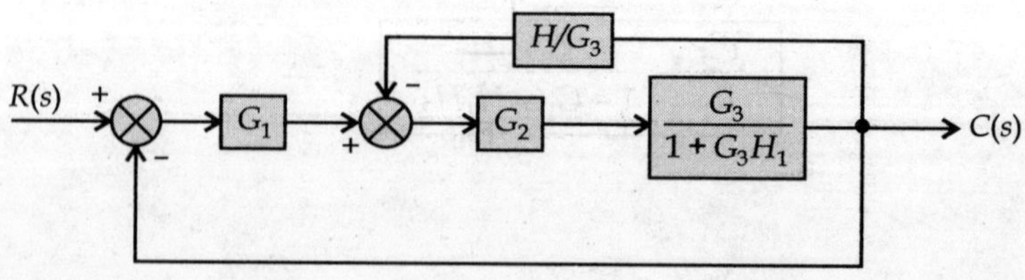

Step 3 : Blocks are in cascade

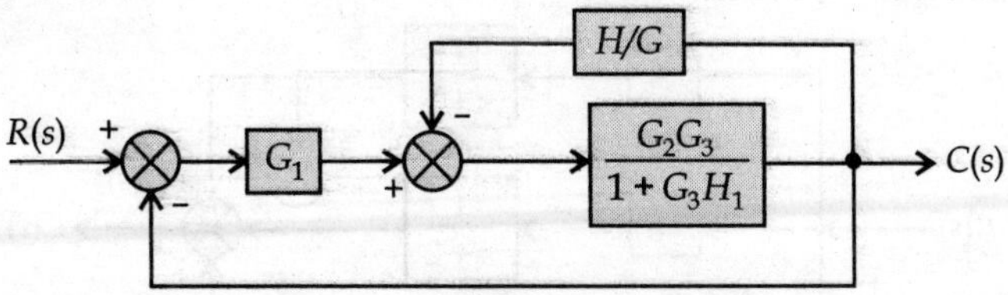

Step 4 :

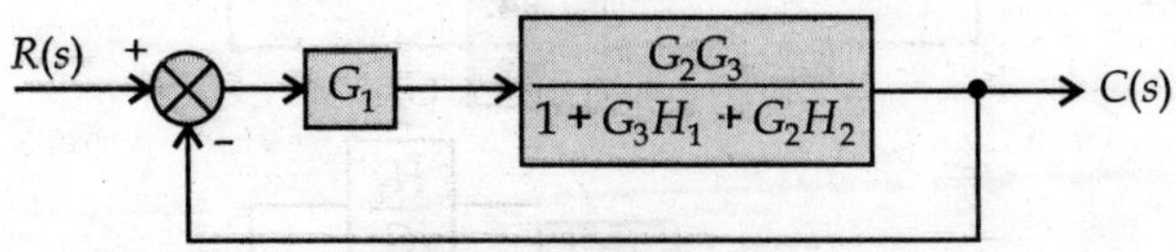

Step 5 :

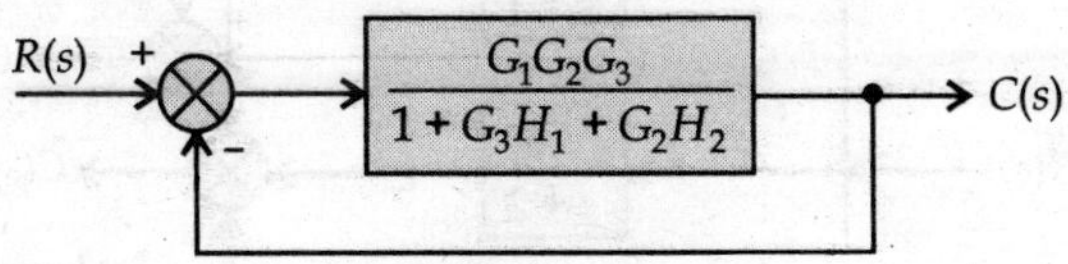

$$\therefore \quad \frac{C(s)}{R(s)} = \frac{G_1G_2G_3}{1+G_3H_1+G_2H_2+G_1G_2G_3} \quad \textbf{Ans.}$$

EXAMPLE 1.37 Determine the transfer matrix MIMO system of example 1.34.

Solution :

$$G_{11} = \left.\frac{C_1}{R_1}\right|_{R_2=0}, \quad G_{21} = \left.\frac{C_2}{R_1}\right|_{R_2=0}$$

$$G_{12} = \left.\frac{C_1}{R_2}\right|_{R_1=0}, \quad G_{22} = \left.\frac{C_2}{R_2}\right|_{R_1=0}$$

$$\therefore \quad G(s) = \begin{bmatrix} G_{11} & G_{12} \\ G_{21} & G_{22} \end{bmatrix} = \begin{bmatrix} G_1 & -G_1G_2H_1 \\ -G_1G_2H_2 & G_2 \end{bmatrix} \frac{1}{1-G_1G_2H_1H_2} \quad \textbf{Ans.}$$

EXAMPLE 1.38. Find the transfer function of the system shown in Fig. 1.96.

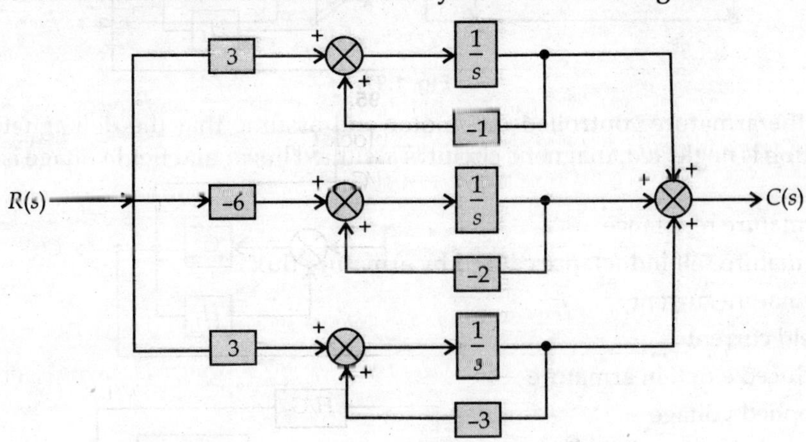

Fig. 1.96.

Solution : Step 1 :

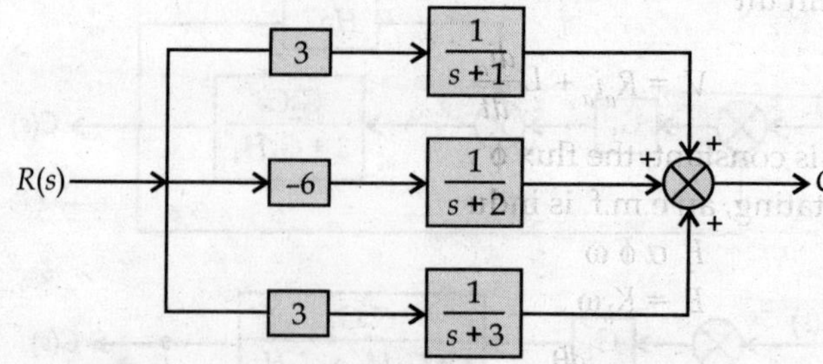

Step 2 :

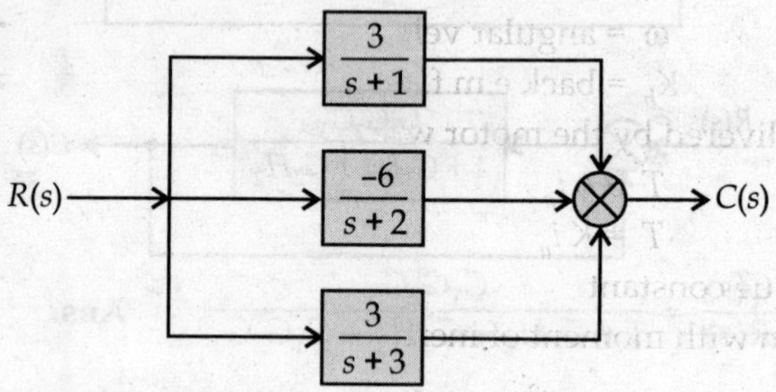

Step 3 :

$$\frac{C(s)}{R(s)} = \frac{3}{s+1} + \frac{-6}{s+2} + \frac{3}{s+3} = \frac{6}{s^3+6s^2+11s+6}$$ **Ans.**

1.23. BLOCK DIAGRAM AND TRANSFER FUNCTION OF D.C MOTOR

1.23.1 Armature Controlled D.C. Motor

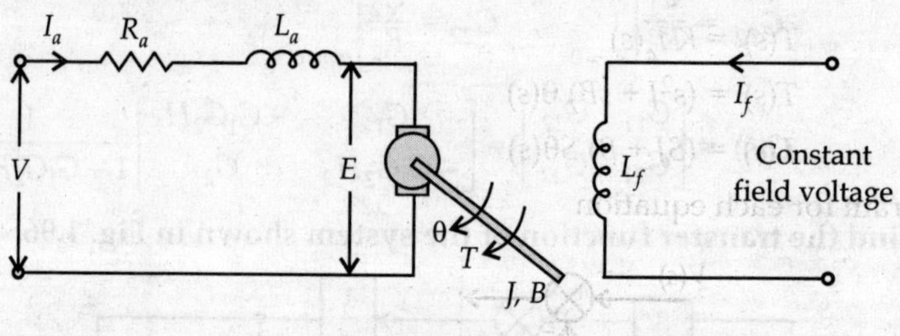

Fig. 1.97.

Consider the armature controlled d.c. motor and assume that the demagnetizing effect of armature reaction is neglected, magnetic circuit is assumed linear and field voltage is constant *i.e.*, I_f = constant.

Let R_a = Armature resistance

L_a = Armature self inductance caused by armature flux

i_a = Armature current

i_f = Field current

E = Induced e.m.f. in armature

V = Applied voltage

T = Torque developed by the motor

θ = Angular displacement of the motor shaft

J = Equivalent moment of inertia of motor shaft and load referred to the motor

B = equivalent coefficient of friction of motor and load referred to the motor

Apply KVL in armature circuit

$$V = R_a i_a + L\frac{di_a}{dt} + E \qquad ...(1.126)$$

Since, field current I_f is constant, the flux ϕ will be constant

When armature is rotating, an e.m.f. is induced

$$E \propto \phi\,\omega$$

$$E = K_b \omega$$

or,

$$E = K_b \frac{d\theta}{dt} \qquad ...(1.127)$$

where, ω = angular velocity

K_b = back e.m.f. constant

Now, the torque T delivered by the motor will be the product of armature current and flux

$$T \propto \phi\, i_a$$

$$T = K\, i_a \qquad ...(1.128)$$

where K = motor torque constant

The dynamic equation with moment of inertia and coeff. of friction will be

$$T = J\frac{d^2\theta}{dt^2} + B\frac{d\theta}{dt} \qquad ...(1.129)$$

Take the laplace transform of equations 1.126, 1.127, 1.128 and 1.129

$$V(s) - E(s) = I_a(s)\,(R_a + SL_a)$$

$$E(s) = K_b\, S\theta(s)$$

$$T(s) = K\, I_a(s)$$

$$T(s) = (s^2 J + sB)\,\theta(s)$$

$$T(s) = (SJ + B)\, S\theta(s)$$

The block diagram for each equation

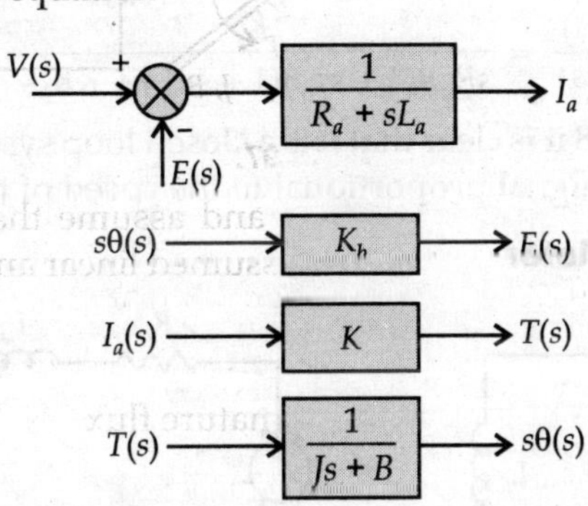

Combine all four block diagrams

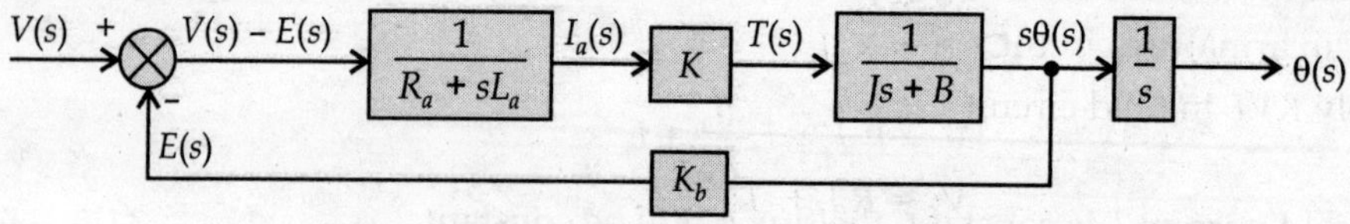

Fig. 1.98. Block diagram of armature controlled dc motor

Now determine the transfer function by block reduction method.

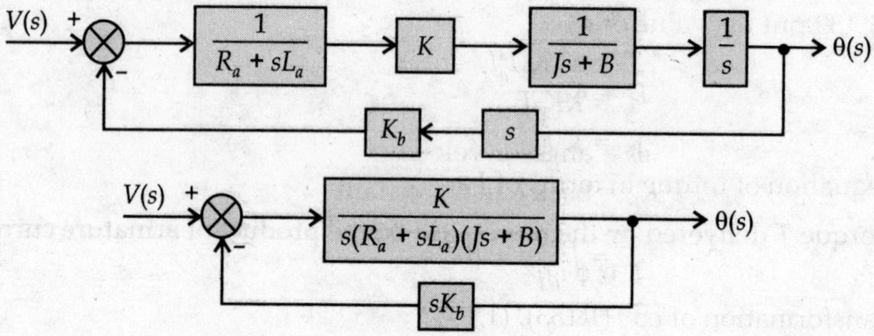

$$\frac{\theta(s)}{V(s)} = \frac{K}{(Ra+sLa)(Js+B)s+KK_bs} \quad ...(1.130)$$

Equation 1.130 can be written as

$$\frac{\theta(s)}{V(s)} = \frac{K}{R_a\left(1+s\frac{L_a}{R_a}\right)sB\left(1+s\cdot\frac{J}{B}\right)+KK_bs}$$

Put $\frac{L_a}{R_a} = \tau_a$ time constant of armature circuit

$\frac{J}{B} = \tau_m$ = mechanical time constant

equation 1.130 becomes

$$\frac{\theta(s)}{V(s)} = \frac{K}{sR_aB(1+s\tau_a)(1+s\tau_m)+KK_bs} \quad ...(1.131)$$

From the block diagram 1.98 it is clear that it is a closed loop system. The effect of the back e.m.f. is represented by the feedback signal proportional to the speed of the motor.

1.23.2. Field Controlled D.C. Motor

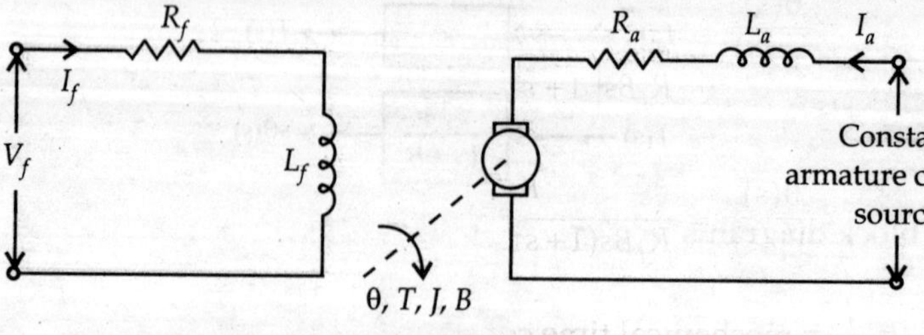

Fig. 1.99.

1. A constant current I_a is fed to the armature.
2. Flux is proprotional to the field current.

$$\phi \propto I_f$$

$$\phi = K_f I_f \qquad ...(1.132)$$

3. Apply *KVL* in field circuit

$$V_f = R_f I_f + L_f \frac{d}{dt} I_f \qquad ...(1.133)$$

4. Torque developed by the motor is proportional to the flux and armature current.

$$T \propto \phi\, I_a$$

From eq^n 1.132 put the value of ϕ

$$T = K' K_f I_a I_f$$

$$T = K K_f I_f \qquad ...(1.134)$$

where $\quad K = K' I_a$

Dynamic equation of torque in terms of J and B

$$T = J\frac{d^2\theta}{dt^2} + B\frac{d\theta}{dt} \qquad ...(1.135)$$

Laplace transformation of eq^n (1.133), (1.134) and (1.135)

$$V_f(s) = R_f I_f(s) + sL_f I_f(s)$$

$$= I_f(s)\,[R_f + sL_f]$$

$$I_f = \frac{V_f(s)}{R_f + sL_f} \qquad ...(1.136)$$

$$T(s) = KK_f I_f(s) \qquad ...(1.137)$$

$$T(s) = \theta(s)\,[S^2 J + sB] \qquad ...(1.138)$$

Put the value of $I_f(s)$ from (1.136) in (1.137)

$$T(s) = K \cdot K_f \cdot \frac{V_f(s)}{R_f + sL_f} \qquad ...(1.139)$$

From equation (1.138) and (1.139)

$$\theta(s)\,[s^2 J + sB] = K \cdot K_f \cdot \frac{V_f(s)}{R_f + sL_f}$$

$$\therefore \quad \frac{\theta(s)}{V_f(s)} = \frac{KK_f}{s(sJ + B)(R_f + sL_f)} \qquad ...(1.140)$$

Equation 1.140 can be written as

$$\frac{\theta(s)}{V_f(s)} = \frac{KK_f}{R_f Bs\left(1 + s\frac{J}{B}\right)\left(1 + s\frac{L_f}{R_f}\right)}$$

$$\frac{\theta(s)}{V_f(s)} = \frac{KK_f}{R_f Bs(1 + s\tau_m)(1 + s\tau_f)}$$

where $\quad \tau_m = \frac{J}{B}$ = mechanical time constant

$\tau_f = \frac{L_f}{R_f}$ = time constant for field circuit

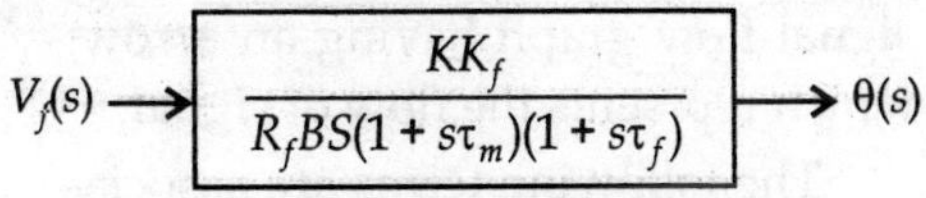

Fig. 1.100. Block diagram of field controlled d.c. motor

Rule	Original diagram	Equivalent diagram
1. Blocks in cascade	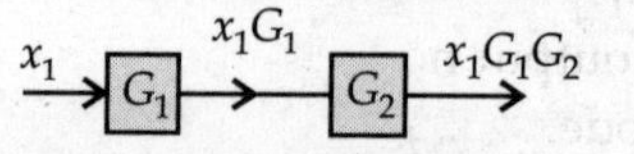	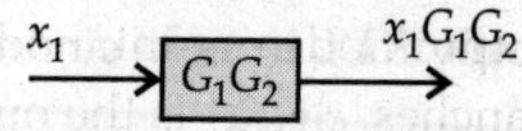
2. Moving a summing point ahead of a block	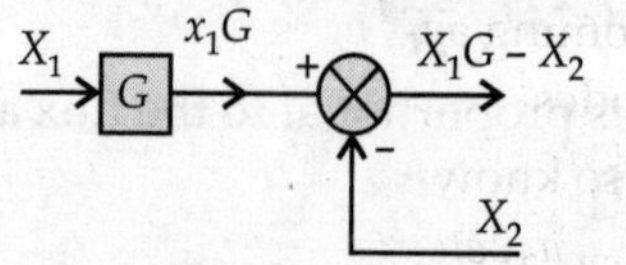	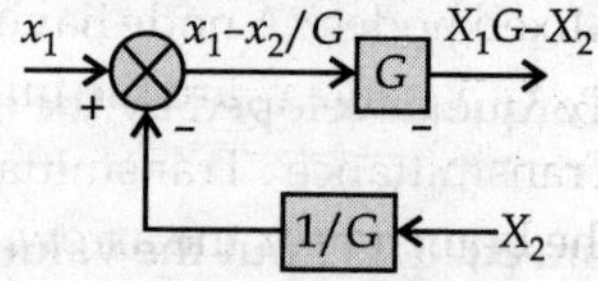
3. Moving a summing point beyond a block	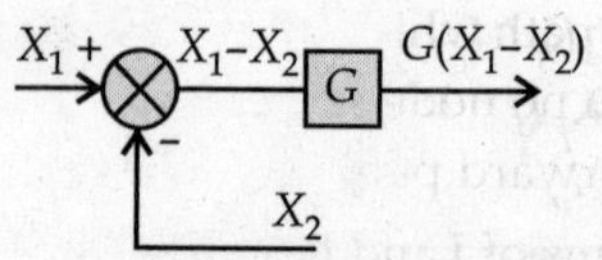	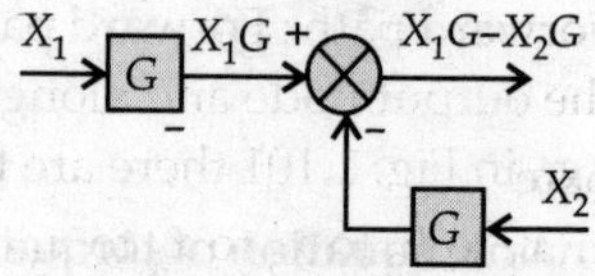
4. Moving a take-off point ahead of a block	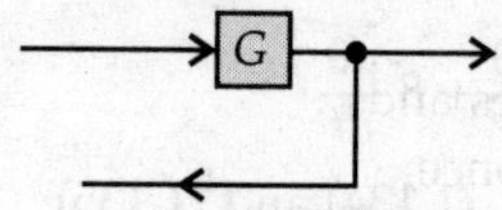	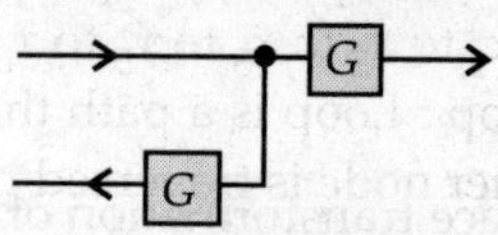
5. Moving a take-off point beyond a block	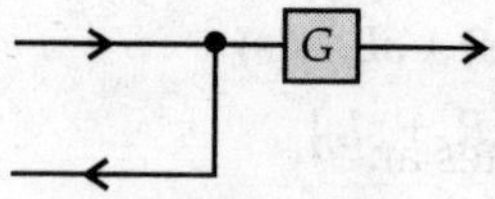	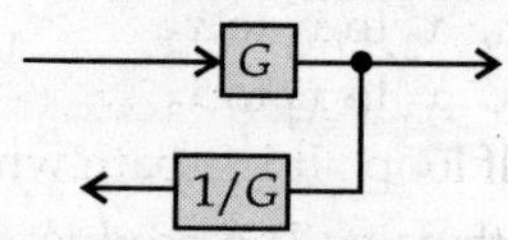
6. Rearrangement of summing point	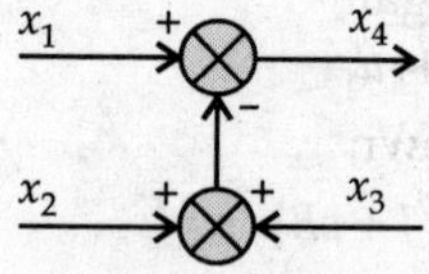	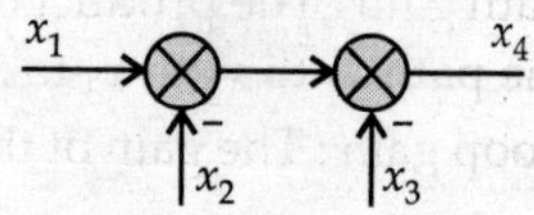
7. Eliminating a forward loop	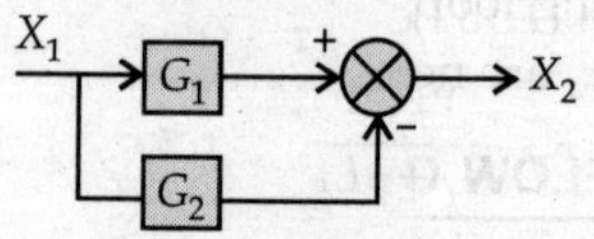	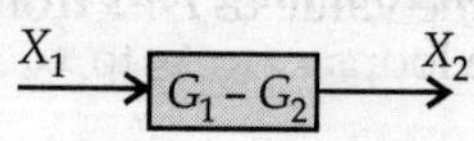
8. Eliminating feedback loop	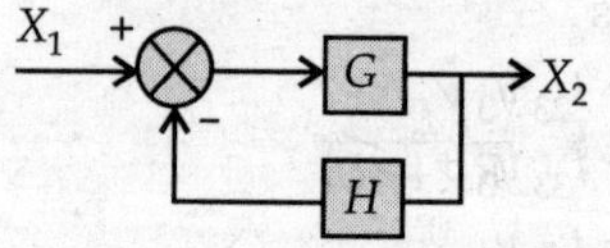	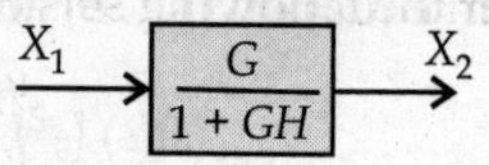

1.24. SIGNAL FLOW GRAPH

The process of block diagram reduction technique is time consuming because at every stage modified block diagram is to be redrawn. A simple method was developed by S.J. Mason which is known as signal flow graph. This method is very simple and does not require any reduction technique. Signal flow graph is applicable to the linear systems.

A signal flow graph is a diagram which represents a set of simultaneous equations. Signal flow graph consists of nodes and these nodes are connected by a directed line called branches. Every branch of signal flow graph having an arrow, which reprsents the flow of signal.

The following terms are associated with the signal flow graph.

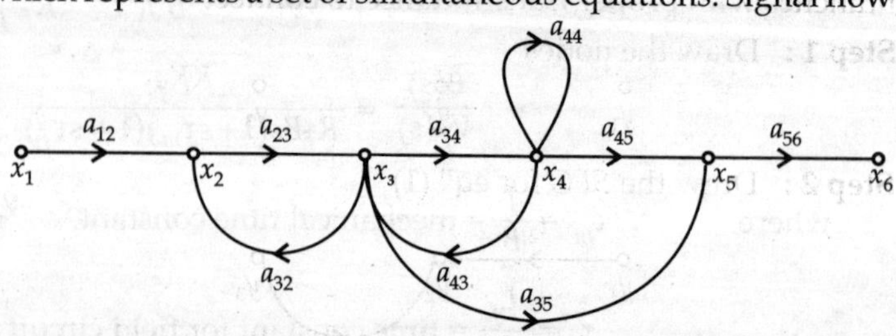

Fig. 1.101

1. Input node or source node : An input node is a node which has only outgoing branches. *e.g.*, x_1 is the input node.
2. Output node or sink node : An output node is a node that has only one or more incoming branches, *e.g.*, x_6 is the output node.
3. Mixed nodes : A node having incoming and outgoing branches is known as mixed nodes. *e.g.*, x_2, x_3, x_4, and x_5 are the mixed nodes.
4. Transmittance : Transmittance also known as transfer function, which is normally written on the branch near the arrow, *e.g.*, a_{12}, a_{23} *etc.*
5. Forward path : Forward path is a path which originates from the input node and terminates at the output node and along which no node is traversed more than once.
   e.g., in Fig. 1.101 there are two forward paths.
   1. x_1 to x_2 to x_3 to x_4 to x_5 to x_6
   2. x_1 to x_2 to x_3 to x_5 to x_6
6. Loop : Loop is a path that originates and terminates on the same node and along which no other node is traversed more than once.
   e.g., x_2 to x_3 to x_2
   x_3 to x_4 to x_3
7. Self loop : It is a path which originates and terminates on the same node. *e.g.*, x_4 to x_4
8. Path gain : The product of the branch gains along the path is called path gain. *e.g.*, the gain of the path x_1 to x_2 to x_3 to x_4 to x_5 to x_6 is $a_{12}\, a_{23}\, a_{34}\, a_{45}\, a_{56}$
9. Loop gain : The gain of the loop is known as loop gain. *e.g.*, the gain of the loop x_2 to x_3 to x_2 is $a_{23}\, a_{32}$.
10. Non-touching loops : Non touching loops having no common nodes branch and paths. *e.g.*, the loops x_2 to x_3 to x_2, and x_4 to x_4 are non touching loops.

1.25. CONSTRUCTION OF SIGNAL FLOW GRAPH FROM EQUATIONS

Consider the following sets of equations

$$\begin{aligned} y_2 &= t_{21}\, y_1 + t_{23}\, y_3 \\ y_3 &= t_{32}\, y_2 + t_{33}\, y_3 + t_{31}\, y_1 \\ y_4 &= t_{43}\, y_3 + t_{42}\, y_2 \\ y_5 &= t_{54}\, y_4 \\ y_6 &= t_{65}\, y_5 + t_{64}\, y_4 \end{aligned}$$

where y_1 is the input and y_6 is the output.

First of all draw the nodes. In the given example there are six nodes. From the first equation it is clear that the y_2 is the sum of two signals. Similarly, y_3 is the sum of three signals & so on. Insert the branches with proper transmittance to connect the nodes.

Step 1 : Draw the nodes

Step 2 : Draw the *SFG* for eq^n (1)

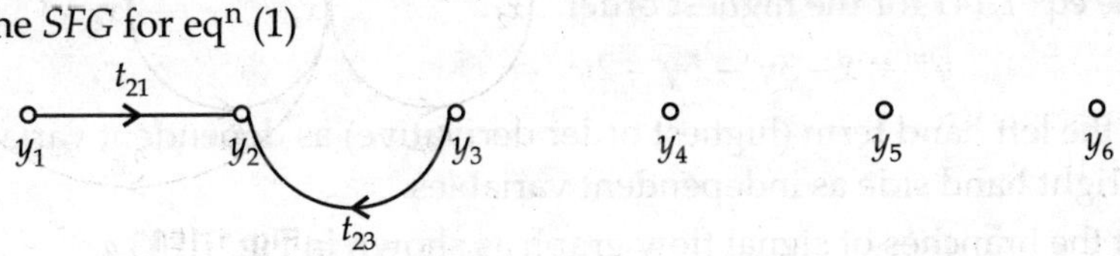

Step 3 : Draw the *SFG* for eq^n (2)

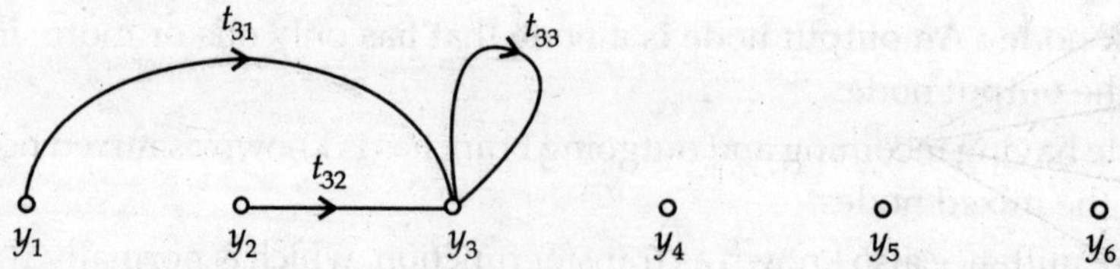

Step 4 : Draw *SFG* for eq^n (3)

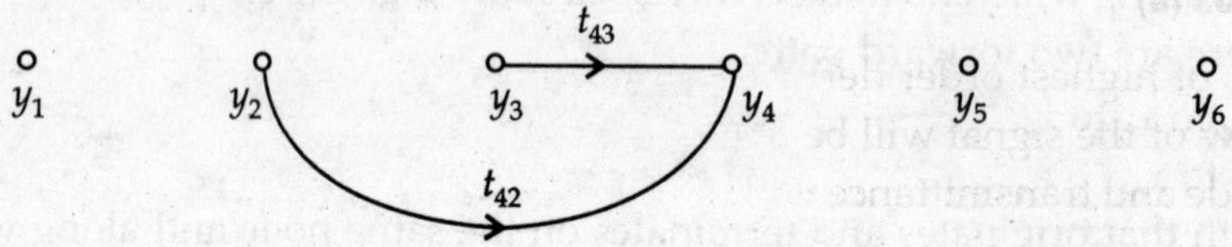

Step 5 : Draw *SFG* for eq^n (4)

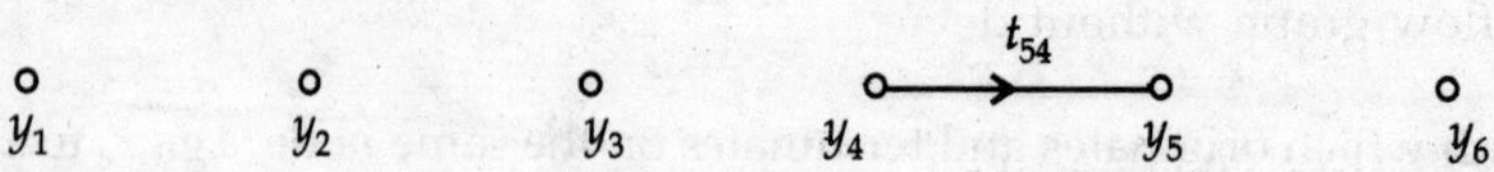

Step 6 : Draw *SFG* for eq^n (5)

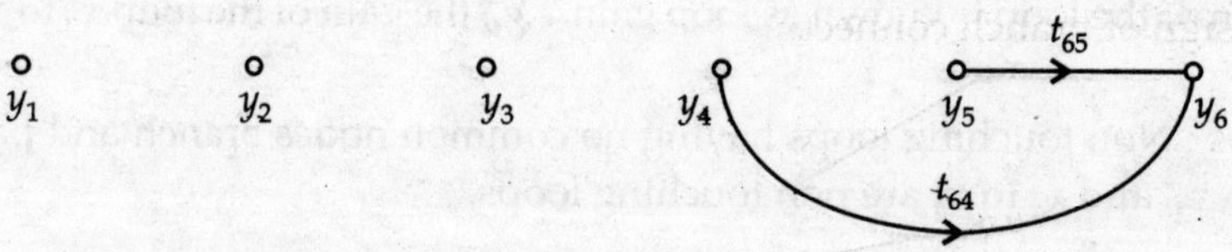

Step 7 : Draw the complete signal flow graph with the help of above graphs.

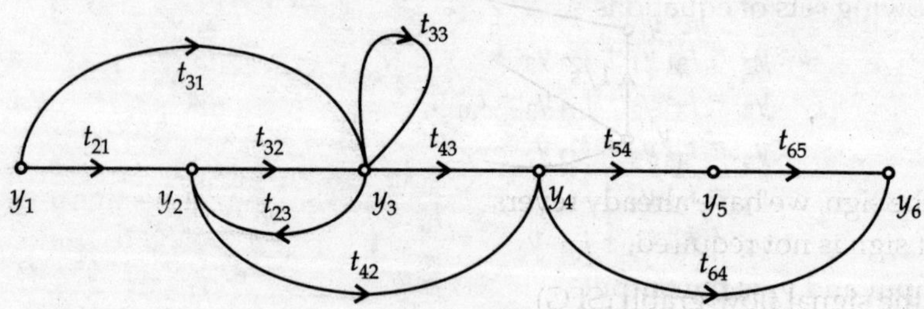

Fig. 1.102.

1.26. SIGNAL FLOW GRAPH FOR DIFFERENTIAL EQUATIONS

Consider the following defferential equation

$$y''' + 3y'' + 5y' + 2y = x \qquad ...(1.141)$$

Step 1 : Solve the eq^n 1.141 for the highest order

$$y''' = x - 3y'' - 5y' - 2y$$

Step 2 : Consider the left hand term (highest order derivative) as dependent variable and all other terms on right hand side as independent variables.

Construct the branches of signal flow graph as shown in Fig. 1.103 *a*.

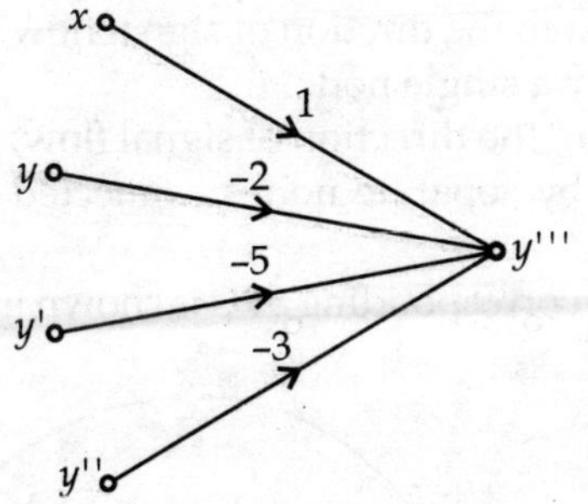

Fig. 1.103 (a)

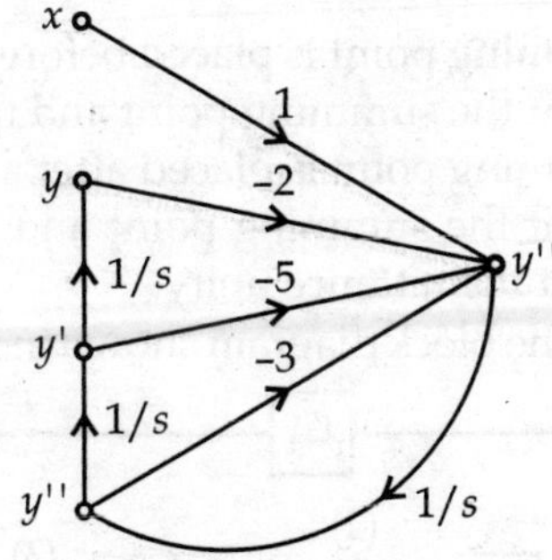

Fig. 1.103 (b)

Step 3 : Connect the nodes of highest order derivative to the node whose order is lower than this and so on. The flow of the signal will be from higher node to the lower order node and transmittance will be $^1/_s$ as shown in Fig. 1.103 *b*.

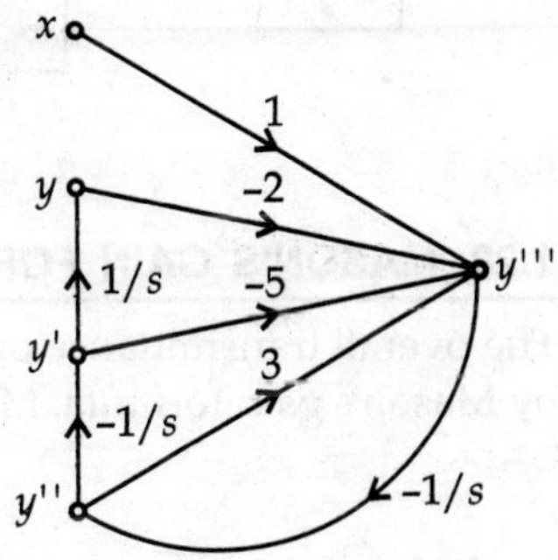

Step 4 : Reverse the sign of a branch connecting the pth node to the qth node of a signal flow graph without disturbing the transfer function.

Consider the Fig. 1.103 *b*, reverse the sign of the branch connecting y''' to y'', it is necessary to reverse the sign of all remaining branches entering as well as leaving the qth node.

Similarly, reverse the sign of branch connecting y'' to y'

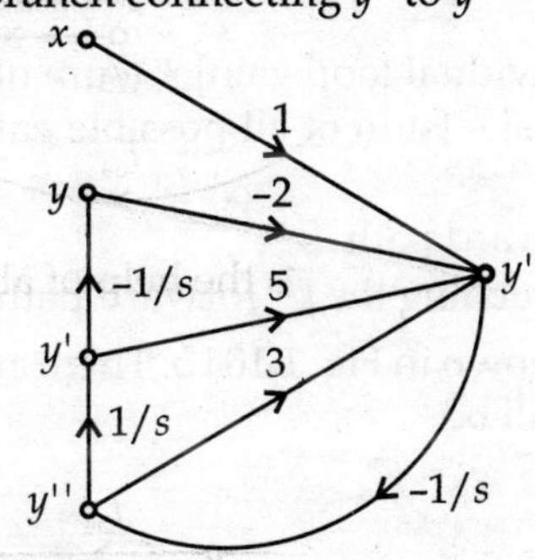

By reversing the sign, we have already reverse the sign of branch connecting y' to y and therefore further reversal of sign is not required

Step 5 : Redraw the signal flow graph (SFG)

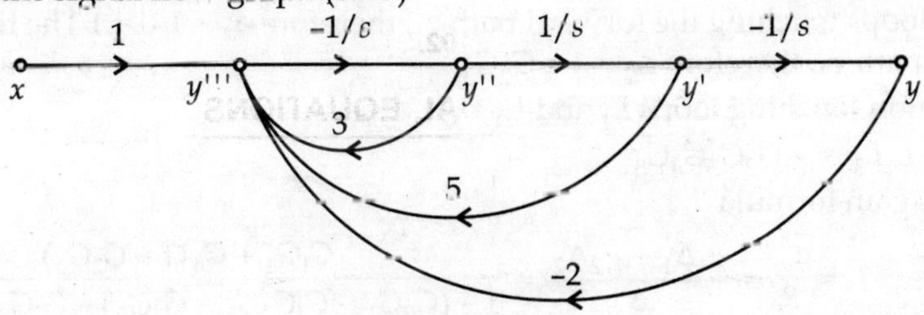

Fig. 1.103 (c)

1.27. CONSTRUCTION OF SIGNAL FLOW GRAPH FROM BLOCK DIAGRAM

Rules: 1. All variables, summing points and take off points are represented by nodes.

2. If a summing point is placed before a take-off point in the dirction of signal flow, in such case represent the summing point and take-off point by a single node.
3. If a summing point is placed after a take-off point in the direction of signal flow, in such case, represent the summing point and take-off point by separate nodes connected by a branch having transmittance unity.

Consider the block diagram shown in Fig. 1.104 *a*, the corresponding *SFG* is shown in Fig. 1.104 *b*.

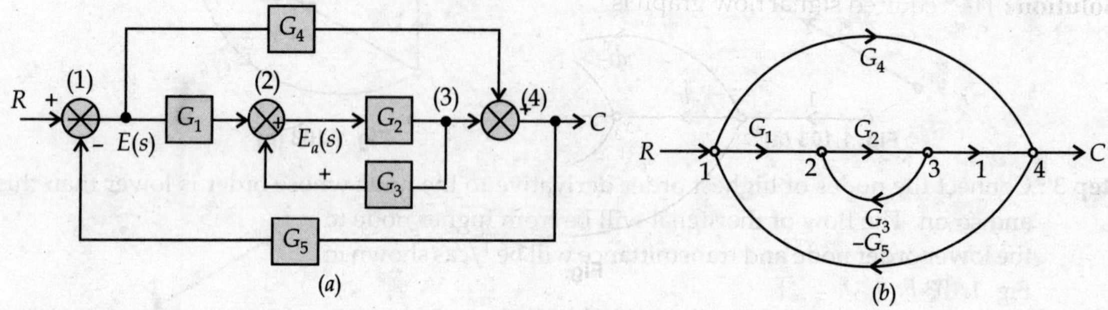

Fig. 1.104.

1.28. MASON'S GAIN FORMULA

The overall transmittance or graph transmittance between the source node and sink node is given by Mason's gain formula. Mason's gain formula for signal flow graph is as follows.

$$T = \frac{\Sigma g_k \Delta_k}{\Delta} \qquad ...(1.142)$$

where, T = transfer function

Δ = 1 – [sum of all individual loop gain] + [sum of all possible gain products of two non-touching loops] – [sum of all possible gain products of three non-touching loops] + ...

g_k = gain of the kth forward path

Δ_k = the part of Δ not touching the kth forward path.

Consider the signal flow graph shown in Fig. 1.104 *b*. There are two forward paths (*i*) 1-2-3-4 (*ii*) 1-4 therefore gain of the two paths will be

$$g_1 = G_1G_2 \text{ and } g_2 = G_4$$

There are three individual loops

$$L_1 = G_2G_3 \qquad (2\text{-}3\text{-}2)$$

$$L_2 = -G_1G_2G_5 \qquad (1\text{-}2\text{-}3\text{-}4\text{-}1)$$

$$L_3 = -G_4G_5 \qquad (1\text{-}4\text{-}1)$$

Since all three loops touching the forward path g_1, therefore Δ_1 = 1-0 = 1 The first loop L_1 do not touch the forward path g_2, therefore $\Delta_2 = 1 - G_2G_3$

There are two non touching loops L_1 and L_3

$$\therefore \quad L_1L_3 = -G_2G_3G_4G_5$$

Apply mason's gain formula

$$T = \frac{C}{R} = \frac{g_1\Delta_1 + g_2\Delta_2}{\Delta} = \frac{G_1G_2 + G_4(1-G_2G_3)}{1-(G_2G_3 - G_1G_2G_5 - G_4G_5) - (-G_2G_3G_4G_5)}$$

$$\frac{C}{R} = \frac{G_1G_2 + G_4 - G_2G_3G_4}{1 - G_2G_3 + G_1G_2G_5 + G_4G_5 - G_2G_3G_4G_5}$$

Note : where $\Delta = 1 - (L_1 + L_2 + L_3) + (L_1L_3)$

EXAMPLE 1.39. Draw the signal flow graph for the following set of equations.

$$x_2 = x_1 + ax_5$$
$$x_3 = bx_2 + cx_4$$
$$x_4 = dx_2 + ex_3$$
$$x_5 = fx_4 + gx_3$$
$$x_6 = x_5$$

Solution : The required signal flow graph is

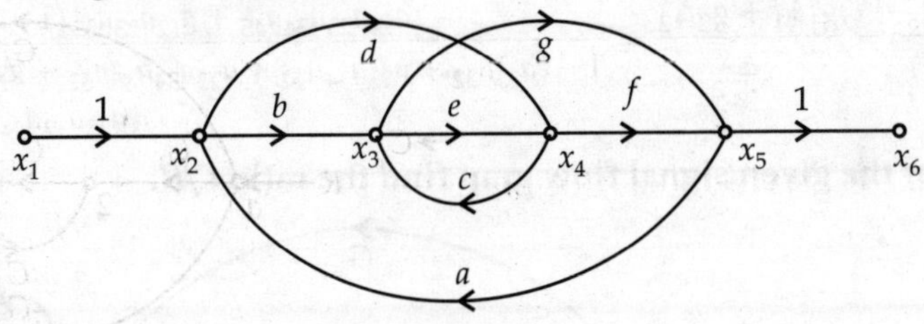

Fig. 1.105.

EXAMPLE 1.40. For the system represented by the given equations find the transfer function x_5/x_1 by the help of signal flow graph technique.

$$x_2 = a_{12}x_1 + a_{32}x_3 + a_{42}x_4 + a_{52}x_5$$
$$x_3 = a_{23}x_2$$
$$x_4 = a_{34}x_3 + a_{44}x_4$$
$$x_5 = a_{35}x_3 + a_{45}x_4$$

where x_1 is the input variable and x_5 is the output variable. *(R.M.L. University Faizabad, 2002)*

Solution : There are two forward paths (1) x_1 to x_2 to x_3 to x_4 + to x_5
(2) x_1 to x_2 to x_3 to x_5

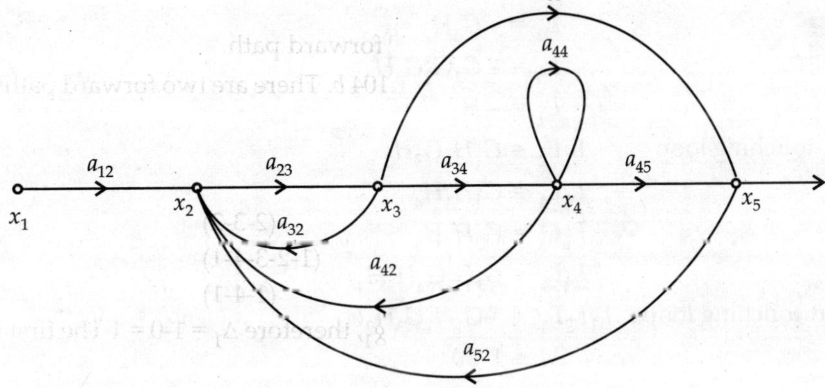

Fig. 1.106.

The gain of the paths $g_1 = a_{12}\, a_{23}\, a_{34}\, a_{45}$ and $g_2 = a_{12}\, a_{23}\, a_{35}$

Gain of individual loops

$$L_1 = a_{23}\, a_{32}$$
$$L_2 = a_{23}\, a_{34}\, a_{42}$$
$$L_3 = a_{23}\, a_{34}\, a_{45}\, a_{52}$$
$$L_4 = a_{23}\, a_{35}\, a_{52}$$
$$L_5 = a_{44}$$

Gain of two non-touching loops

$$L_1L_5 = a_{23}\, a_{32}\, a_{44}$$

$$L_4\, L_5 = a_{23}\, a_{35}\, a_{52}\, a_{44}$$

Since, all the loops touches the forward path (1) $\therefore\ \Delta_1 = 1 - 0 = 1$

loop L_5 do not touch the second forward path $\therefore\ \Delta_2 = 1 - a_{44}$

$$\Delta = 1 - (L_1 + L_2 + L_3 + L_4 + L_5) + (L_1L_5 + L_4L_5)$$

$$\therefore \quad \frac{x_5}{x_1} = \frac{g_1\Delta_1 + g_2\Delta_2}{\Delta} = \frac{a_{12}a_{23}a_{34}a_{45} + a_{12}a_{23}a_{35}(1 - a_{44})}{1 - (a_{23}a_{32} + a_{23}a_{34}a_{42} + a_{23}a_{34}a_{45}a_{52} + a_{23}a_{35}a_{52} + a_{44}) + (a_{23}a_{32}a_{44} + a_{23}a_{35}a_{52}a_{44})}$$

EXAMPLE 1.41. For the given signal flow grap find the ratio *C/R*.

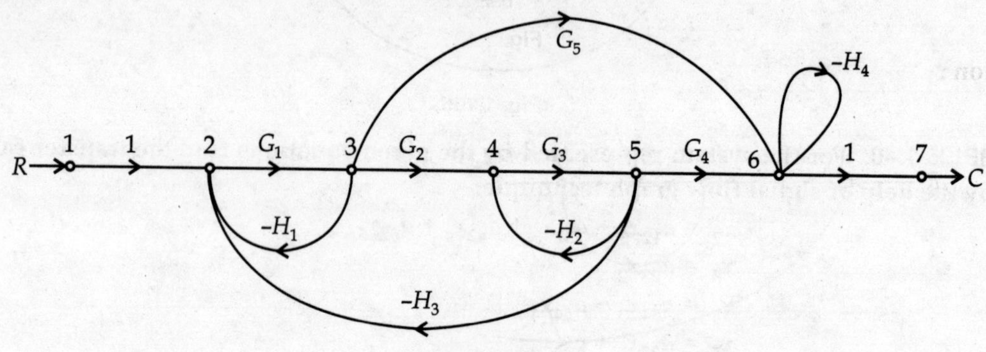

Fig. 1.107.

The gain of the forward paths $g_1 = G_1G_2G_3G_4$

$$g_2 = G_1G_5$$

Individual loops

$$L_1 = -G_1H_1$$

$$L_2 = -G_3H_2$$

$$L_3 = -G_1G_2G_3H_3$$

$$L_4 = -H_4$$

Two non-touching loop

$$L_1L_2 = G_1H_1G_3H_2$$

$$L_1L_4 = G_1H_1H_4$$

$$L_2L_4 = G_3H_2H_4$$

$$L_3L_4 = G_1G_2G_3H_3H_4$$

Three non-touching loops $L_1L_2L_4 = -G_1H_1G_3H_2H_4$

$$\Delta_1 = 1 - 0$$

$$\Delta_2 = 1 + G_3H_2$$

$$\Delta = 1 - (L_1 + L_2 + L_3 + L_4) + (L_1L_2 + L_1L_4 + L_2L_4 + L_3L_4) - (L_1L_2L_4)$$

$$\therefore \quad \frac{c}{R} = \frac{g_1\Delta_1 + g_2\Delta_2}{\Delta}$$

$$\frac{C}{R} = \frac{G_1G_2G_3G_4 + G_1G_5(1 + G_3H_2)}{1 + G_1H_1 + G_3H_2 + G_1G_2G_3H_3 + H_4 + G_1G_3H_1H_2 + G_1H_1H_4 + G_3H_2H_4 + G_1G_2G_3H_3H_4 + G_1G_3H_1H_2H_4}$$

EXAMPLE 1.42. Obtain signal flow graph representation for a system whose block diagram is given below and using Mason's gain formula determine the ratio $\frac{C}{R}$.

(R.M.L. Univ. Faizabad 2001 Linear system theory)

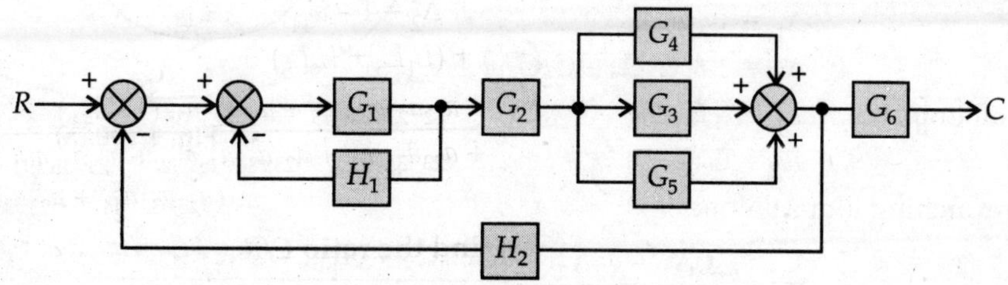

Fig. 1.108.

Solution :

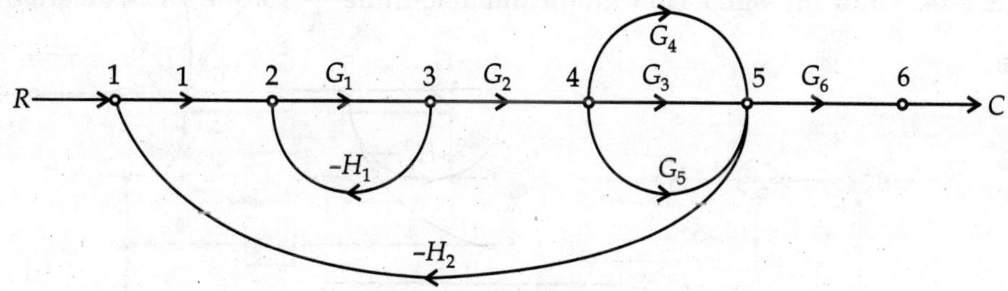

Fig. 1.108 (a)

The signal flow graph is shown in Fig. 1.108 *a*. Consider the SFG

$$g_1 = G_1G_2G_3G_6 \qquad \Delta_1 = 1 - 0 = 1$$
$$g_2 = G_1G_2G_4G_6 \qquad \Delta_2 = 1 - 0 = 1$$
$$g_3 = G_1G_2G_5G_6 \qquad \Delta_3 = 1 - 0 = 1$$

Individual loops

$$L_1 = -G_1H_1 \qquad L_2 = -G_1G_2G_3H_2$$
$$L_3 = -G_1G_2G_4H_2 \qquad L_4 = -G_1G_2G_5H_2$$

Two non-touching loops = None

$$\Delta = 1 - (L_1 + L_2 + L_3 + L_4)$$

$$\therefore \quad \frac{C}{R} = \frac{g_1\Delta_1 + g_2\Delta_2}{\Delta} = \frac{G_1G_2G_3G_6 + G_1G_2G_4G_6 + G_1G_2G_5G_6}{1 + G_1H_1 + G_1G_2G_3H_2 + G_1G_2G_4H_2 + G_1G_2G_5H_2}$$

EXAMPLE 1.43. Obtain the Transfer function for $\frac{C(s)}{R(s)}$ given block diagram using block diagram reduction technique or signal flow graph (Mason's gain formula).

(R.M.L. Univ. Faizabad 2001, Control System)

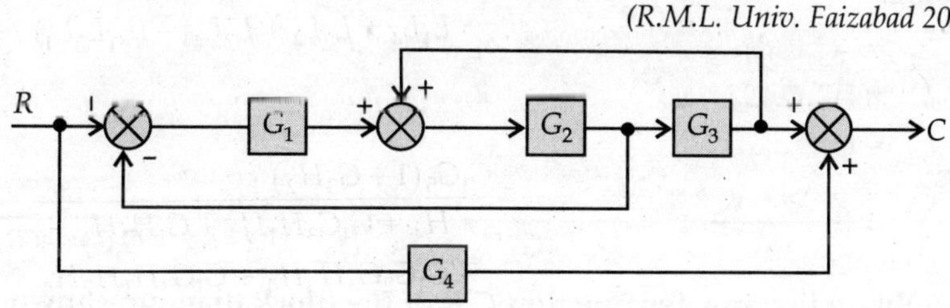

Fig. 1.109.

Solution : By signal flow graph SFG is shown in Fig 1.109 *a*.

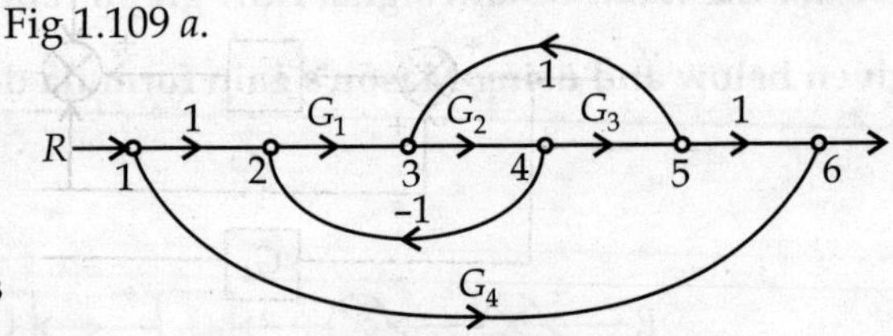

Fig. 1.109 (a)

Gain of forward paths : $g_1 = G_1G_2G_3$

$g_2 = G_4$

$\Delta_1 = 1$

$\Delta_2 = 1 + G_1G_2 - G_2G_3$

Individual loops : $L_1 = - G_1G_2$

$L_2 = G_2G_3$

Two non-touching loops : None

$$\therefore \quad \frac{C}{R} = \frac{g_1\Delta_1 + g_2\Delta_2}{\Delta} = \frac{G_1G_2G_3 + G_4(1 + G_1G_2 - G_2G_3)}{1 + G_1G_2 - G_2G_3}$$

EXAMPLE 1.44. Draw the signal flow graph and determine $\frac{C}{R}$ for the block diagram shown in Fig. 1.110. *(KNIT, Sultanpur, 1999-2000)*

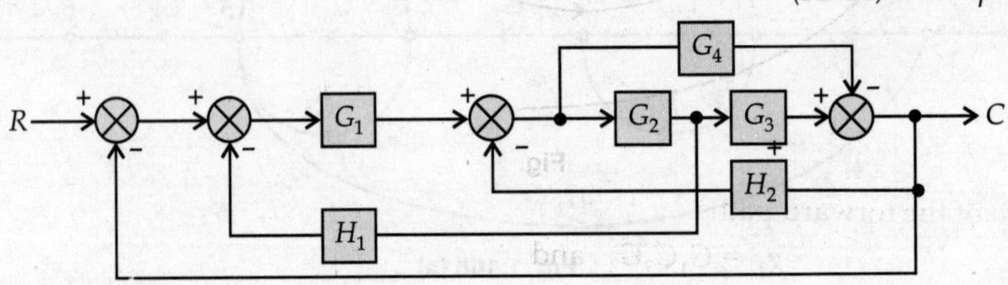

Fig. 1.110.

Solution : The *SFG* is shown in Fig. 1.110 *a*

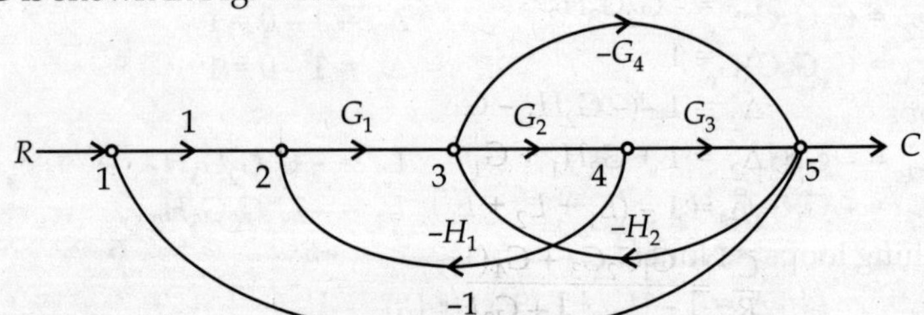

Fig. 1.110 (a)

The gain of the forward paths

$g_1 = G_1G_2G_3$ $\Delta_1 = 1$

$g_2 = - G_1G_4$ $\Delta_2 = 1$

Individual loops

$L_1 = -G_1G_2H_1$ $L_2 = -G_2G_3H_2$

$L_3 = -G_1G_2G_3$ $L_4 = G_1G_4$

$L_5 = G_4H_2$

$$\frac{C}{R} = \frac{g_1\Delta_1 + g_2\Delta_2}{\Delta} = \frac{G_1G_2G_3 - G_1G_4}{1 + G_1G_2H_1 + G_2G_3H_2 + G_1G_2G_3 - G_1G_4 - G_4H_2}$$

EXAMPLE 1.45. Obtain the transfer function *C/R* of the block diagram shown in Fig. 1.111. *(KNIT, Sultanpur)*

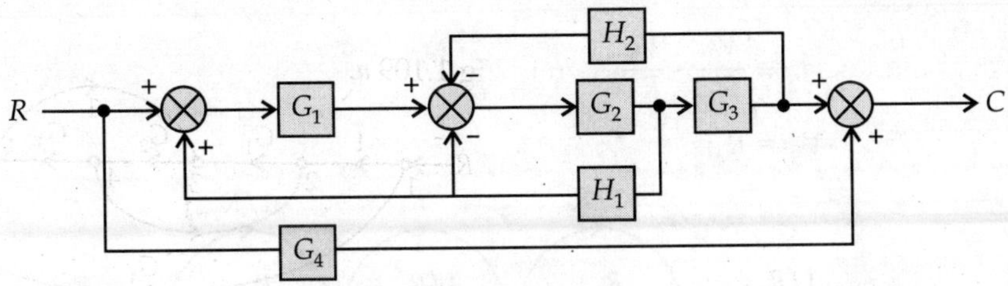

Fig. 1.111.

Solution :

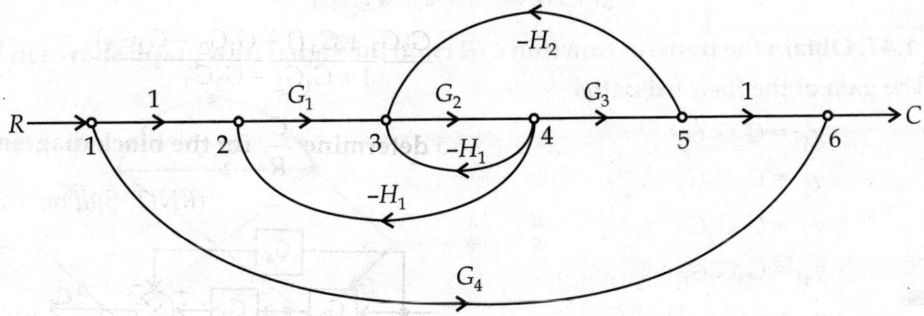

Fig. 1.111 (a)

The gain of the forward paths

$$g_1 = G_1G_2G_3 \quad \text{and} \quad g_2 = G_4$$

Individual loops

$$L_1 = -G_2H_1$$

$$L_2 = -G_1G_2H_1$$

$$L_3 = -\,G_2G_3H_2$$

$$\Delta_1 = 1$$

$$\Delta_2 = 1 -(-\,G_2H_1 - G_1G_2H_1 - G_2G_3H_2)$$

$$\Delta_2 = 1 + G_2H_1 + G_1G_2H_1 + G_2G_3H_2$$

$$\Delta = 1 - (L_1 + L_2 + L_3) = 1 + G_2H_1 + G_1G_2H_1 + G_2G_3H_2$$

$$\frac{C}{R} = \frac{G_1G_2G_3 + G_4(1 + G_2H_1 + G_1G_2H_1 + G_2G_3H_2)}{1 + G_2H_1 + G_1G_2H_1 + G_2H_2G_3}$$

EXAMPLE 1.46 Draw the signal flow graph for the network shown in Fig. 1.112, take V_3 as output node.

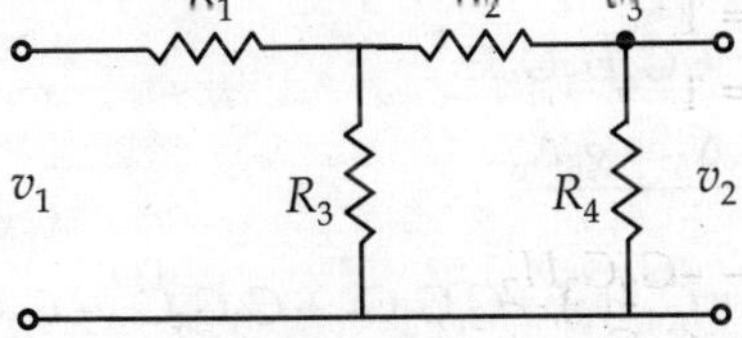

Fig. 1.112

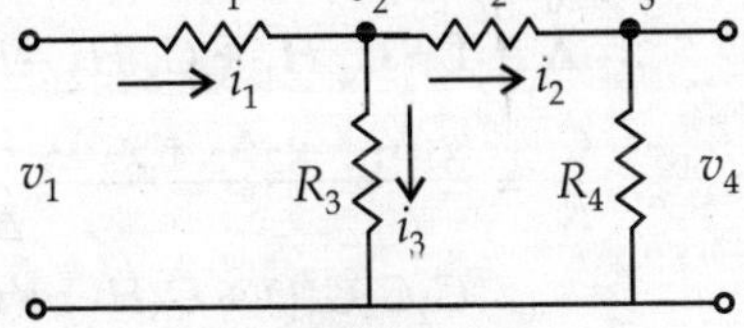

Fig. 1.112 (a)

Solution : From 1.112 *a* : $i_1 = \dfrac{V_1}{R_1} - \dfrac{V_2}{R_1}$

$$V_2 = i_3\,R_3 = R_3\,(i_1 - i_2) = R_3 i_1 - R_3 i_2$$

$$i_2 = \frac{V_2}{R_2} - \frac{V_3}{R_2}$$

$$V_3 = R_4 i_2$$

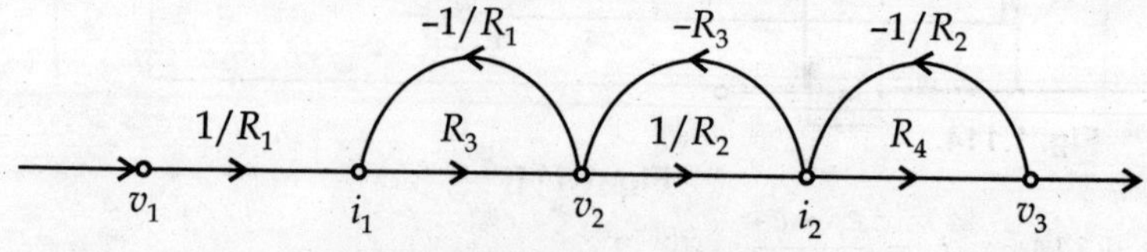

Fig. 1.112 (b) Signal flow graph.

EXAMPLE 1.47. Obtain the transfer function C/R from the signal flow graph shown in Fig. 1.113.

Solution : The gain of the forward paths

$$g_1 = G_2G_4G_6$$
$$g_2 = G_3G_5G_7$$
$$g_3 = G_2G_1G_7$$
$$g_4 = G_3G_8G_6$$
$$g_5 = -G_2G_1H_2G_8G_6$$
$$g_6 = -G_3G_8H_1G_1G_7$$

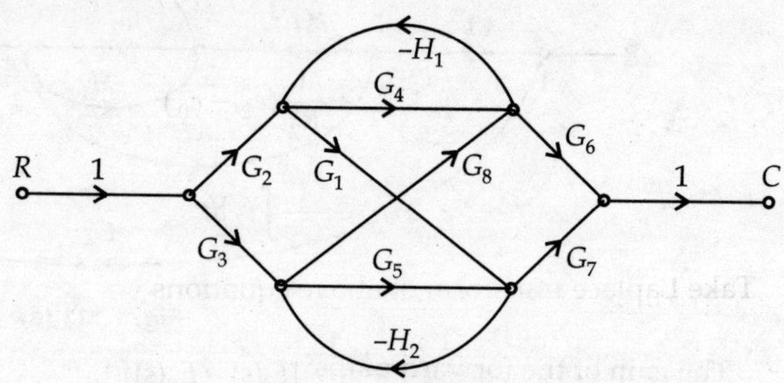

Fig. 1.113.

Individual loops :

$$L_1 = -G_4H_1$$
$$L_2 = -G_5H_2$$
$$L_3 = H_1G_1H_2G_8$$

Two non-touching loops :

$$L_1L_2 = G_4H_1G_5H_2$$
$$\Delta_1 = 1 + G_5H_2$$
$$\Delta_2 = 1 + G_4H_1$$
$$\Delta_3 = 1$$
$$\Delta_4 = 1$$
$$\Delta_5 = 1$$
$$\Delta_6 = 1$$
$$\Delta = 1 + G_4H_1 + G_5H_2 - H_1G_1H_2G_8 + G_4H_1G_5H_2$$

$$\therefore \quad \frac{C}{R} = \frac{g_1\Delta_1 + g_2\Delta_2 + g_3\Delta_3 + g_4\Delta_4 + g_5\Delta_5 + g_6\Delta_6}{\Delta}$$

$$\frac{C}{R} = \frac{G_2G_4G_6(1+G_5H_2) + G_3G_5G_7(1+G_4H_1) + G_1G_2G_7 + G_3G_8G_6 - G_2G_1H_2G_8G_6 - G_3G_8H_1G_1G_7}{1 + G_4H_1 + G_5H_2 - H_1G_1H_2G_8 + G_4H_1G_5H_2}$$

Ans.

EXAMPLE 1.48. Draw the block diagram and signal flow graph and find out the transfer function of the circuit shown in Fig. 1.114.

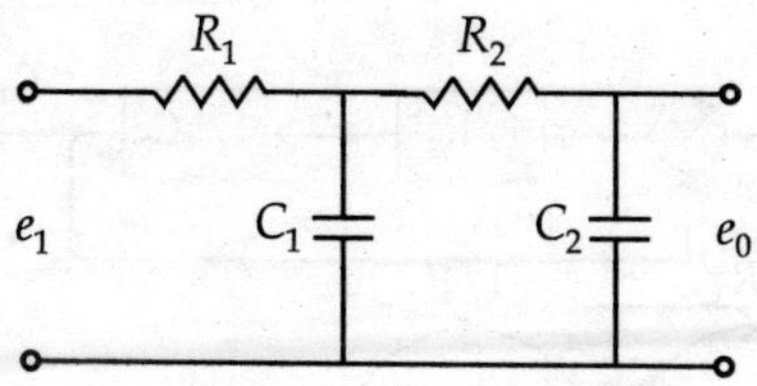

Fig. 1.114.

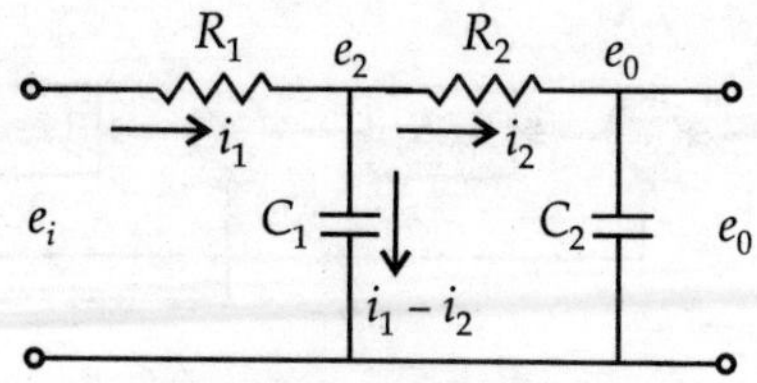

Fig. 1.114 (a)

Solution : From Fig. 1.114 *a* : $i_1 = \dfrac{e_i - e_2}{R_1}$

$$e_2 = \frac{1}{C_1}\int (i_1 - i_2)\,dt$$

$$i_2 = \frac{1}{R_2}(e_2 - e_0)$$

$$e_0 = \frac{1}{C_2}\int i_2\,dt$$

Take Laplace transform of above equations

$$I_1(s) = \frac{1}{R_1}\,[E_i(s) - E_0(s)]$$

$$E_2(s) = \frac{1}{SC_1}\,[I_1(s) - I_2(s)]$$

$$I_2(s) = \frac{1}{R_2}\,[E_2(s) - E_0(s)]$$

$$E_0(s) = \frac{1}{SC_2}\,[I_2(s)]$$

On the basis of above equations draw the block diagram

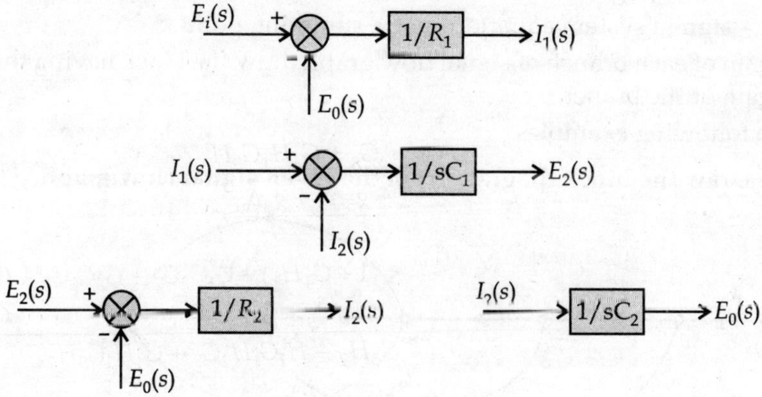

Fig. 1.114 (b) Block diagram

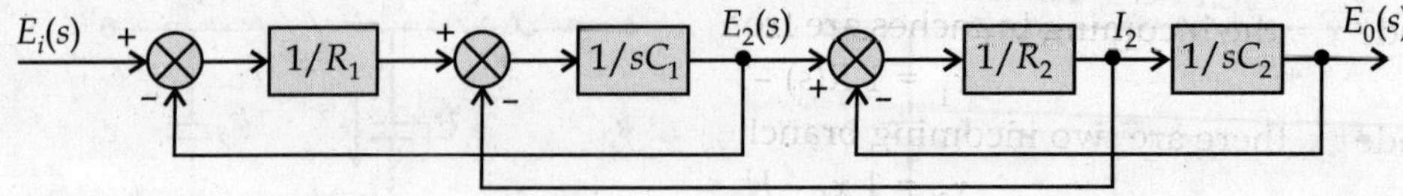

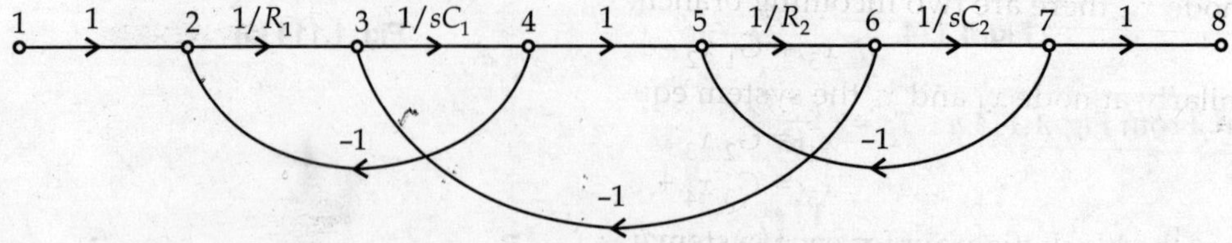

Fig. 1.114 (c). Signal flow graph

$$g_1 = \frac{1}{R_1R_2C_1C_2s^2} \qquad \Delta_1 = 1$$

$$L_1 = -\frac{1}{R_1C_1S} \qquad L_2 = -\frac{1}{R_2C_1s} \qquad L_3 = -\frac{1}{R_2C_2s}$$

Two non-touching loops

$$L_1L_3 = \frac{1}{s^2R_1R_2C_1C_2}$$

$$\frac{C}{R} = \frac{1/R_1R_2C_1C_2s^2}{1+\frac{1}{R_1C_1s}+\frac{1}{R_2C_1s}+\frac{1}{R_2C_2s}+\frac{1}{s^2R_1R_2C_1C_2}}$$

$$\frac{C}{R} = \frac{1}{1+R_1C_1s+R_1C_2s+R_2C_2s+s^2R_1R_2C_1C_2}$$ **Ans.**

1.29. BLOCK DIAGRAM FROM SIGNAL FLOW GRAPH

Step 1 : For given signal flow graph, write the system equations.

Step 2 : At each node consider the incoming branches only.

Step 3 : Add all incoming signals algebraically at a node.

Step 4 : For + or – sign in system equations use a summing point.

Step 5 : For the gain of each branch of signal flow graph draw the block having the same transfer function as the gain of the branch.

Consider the following examples :

EXAMPLE 1.49. Draw the block diagram from the given signal flow graph.

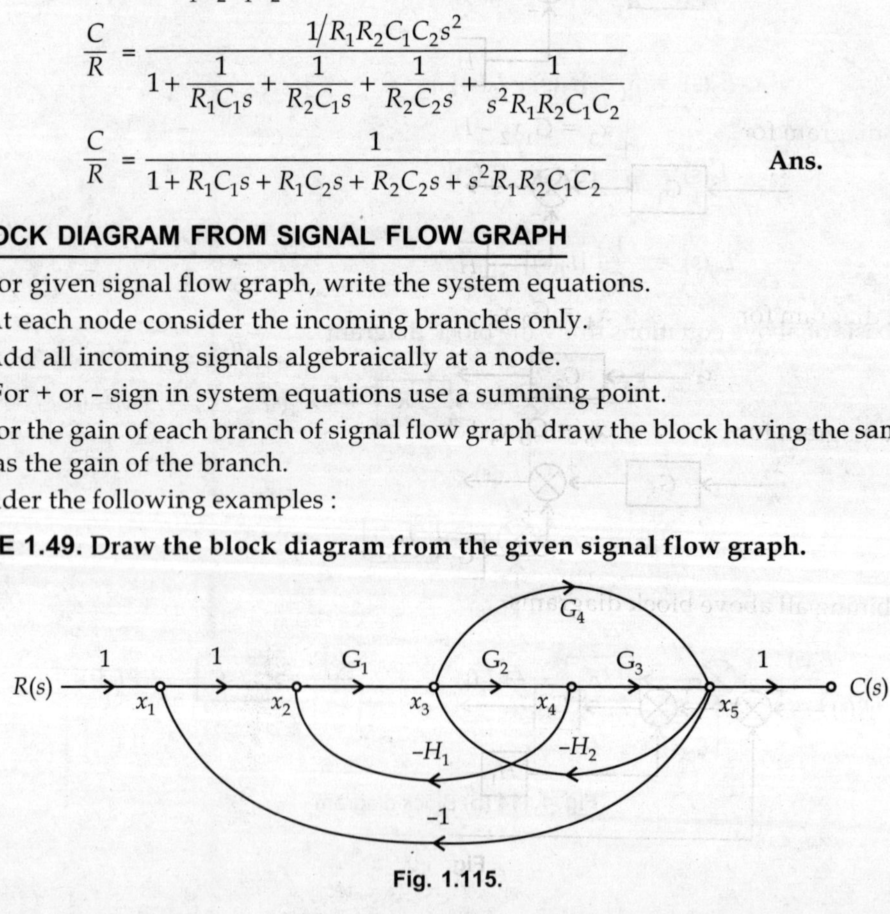

Fig. 1.115.

Solution : The system equations are:

at node x_1, the incoming branches are from $R(s)$ and x_5

$$x_1 = 1\,R(s) - 1 \cdot x_5$$

at node x_2, there are two incoming branches

$$x_2 = 1 \cdot x_1 - H_1 x_4$$

at node x_3, there are two incoming branches

$$x_3 = G_1 x_2 - H_2 x_5$$

Similarly at node x_4 and x_5 the system equations are:

$$x_4 = G_2 x_3$$

$$x_5 = G_3 x_4 + G_4 x_3$$

Draw the block diagram for each system equation

For $\qquad x_1 = R(s) - x_5$

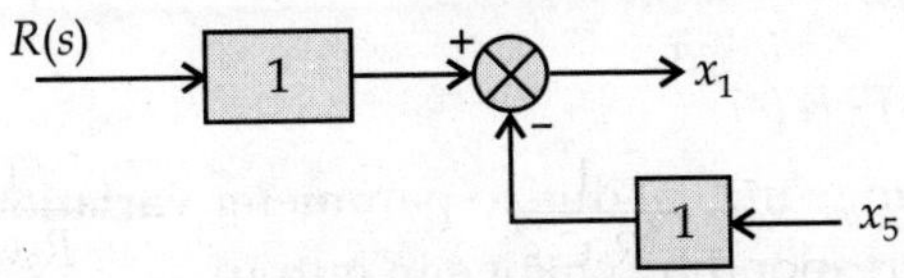

Block diagram for $\qquad x_2 = x_1 - H_1 x_4$

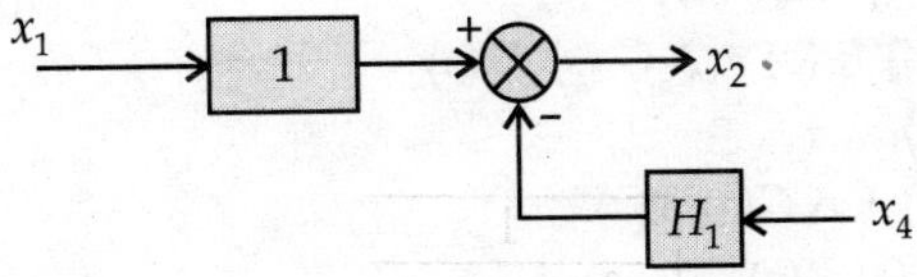

Block diagram for $\qquad x_3 = G_1 x_2 - H_2 x_5$

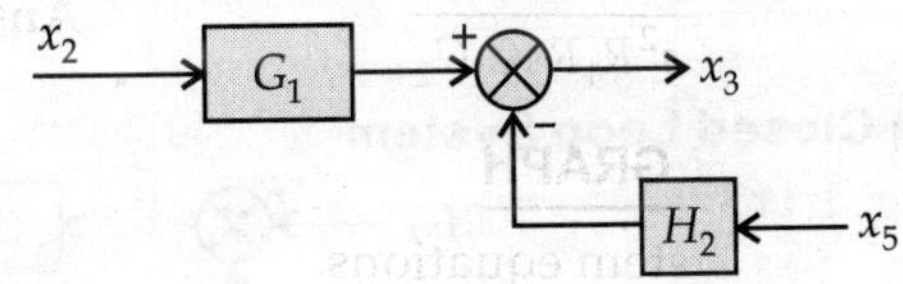

Block diagram for $\qquad x_4 = G_2 x_3$

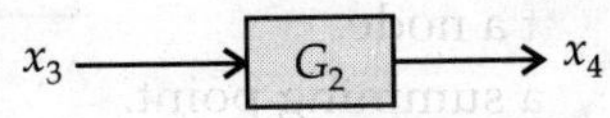

Block diagram for $\qquad x_5 = G_3 x_4 + G_4 x_3$

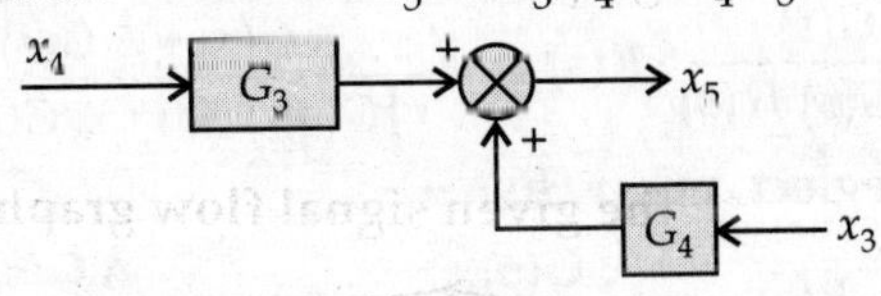

Combining all above block diagrams

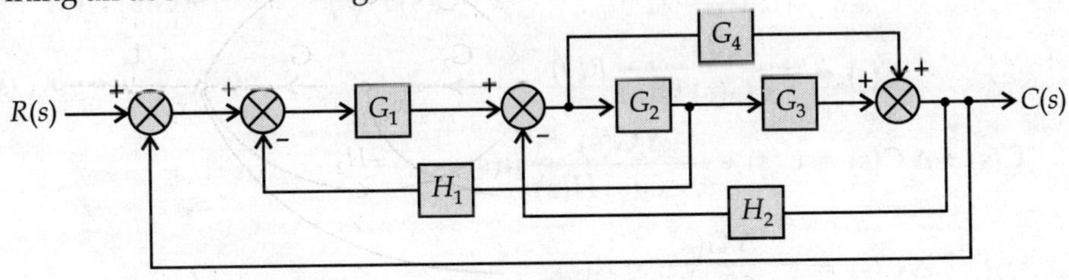

Fig. 115 (a)

1.30. EFFECT OF PARAMETER VARIATIONS

In control systems, the feedback reduces the error, also reduces the sensitivity of the system to parameter variations. The parameter may vary due to some change in conditions. The variation in parameter affects the performance of the system. So, it is necessary to make the system in sensitive to such parameter variations.

1.30.1 Effect of Parameter Variations in Open Loop Control System

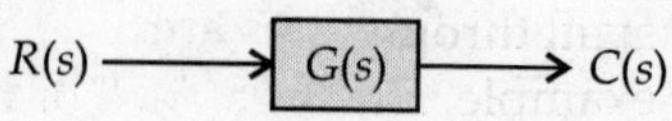

Fig. 1.116.

The open loop control system is shown in Fig 1.116. From Fig. 1.116

$$\frac{C(s)}{R(s)} = G(s)$$

or, $$C(s) = G(s) \cdot R(s) \quad ...(1.143)$$

Let $\Delta G(s)$ = Change in $G(s)$ due to parameter variations

$\Delta C(s)$ = Corresponding change in output

From equation 1.143

$$C(s) + \Delta C(s) = [G(s) + \Delta G(s)]\, R(s)$$

$$C(s) + \Delta C(s) = G(s)\, R(s) + \Delta G(s)\, R(s)$$

Since, $G(s)\, R(s) = C(s)$

$$C(s) + \Delta C(s) = C(s) + \Delta G(s)\, R(s)$$

or, $$\Delta C(s) = \Delta G(s)\, R(s) \quad ...(1.144)$$

Equation 1.144 gives the change in output due to parameter variations in G(s) in open loop system.

1.30.2. Effect of Parameter Variations in Closed Loop System

The closed loop system is shown in Fig. 1.117

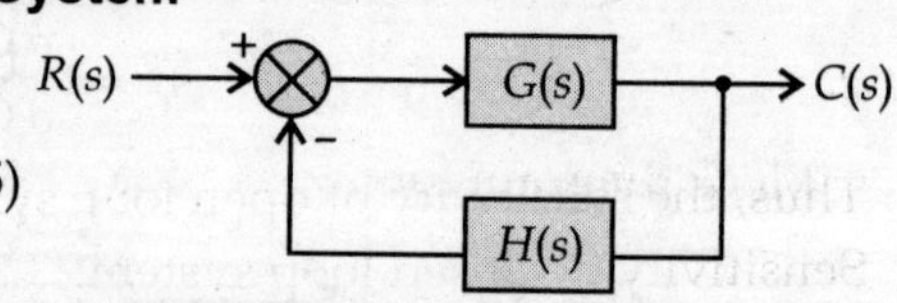

Fig. 1.117

$$\frac{C(s)}{R(s)} = \frac{G(s)}{1 + G(s)\, H(s)} \quad ...(1.145)$$

or, $$C(s) = \frac{G(s)}{1 + G(s)\, H(s)}\, R(s)$$

$$C(s) + \Delta C(s) = \frac{G(s) + \Delta G(s)}{1 + G(s) + \Delta G(s)\, H(s)} \cdot R(s) = \frac{G(s) + \Delta G(s)}{1 + [G(s)\, H(s) + \Delta G(s)]\, H(s)} \cdot R(s)$$

Since, $\Delta G(s)\, H(s) << [1 + G(s)\, H(s)]$, neglect $\Delta G(s)\, H(s)$

$$\therefore \quad C(s) + \Delta C(s) = \frac{G(s) + \Delta G(s)}{1 + G(s)\, H(s)}\, R(s) = \frac{G(s)}{1 + G(s)\, H(s)}\, R(s) + \frac{\Delta G(s)}{1 + G(s)\, H(s)} \cdot R(s)$$

Since, $$C(s) = \frac{G(s)}{1 + G(s)\, H(s)}\, R(s)$$

$$C(s) + \Delta C(s) = C(s) + \frac{\Delta G(s)}{1 + G(s)\, H(s)}\, R(s)$$

or, $$\Delta C(s) = \frac{\Delta G(s)}{1 + G(s)\, H(s)}\, R(s) \quad ...(1.146)$$

Equation 1.146 gives the change in output due to parameter variations in $G(s)$ in a closed loop system.

Generally $[G(s)\,H(s)] >> 1$

from equation 1.140 it is clear that the change in output is reduced due to parameter variations in $G(s)$ by $[1 + G(s)\,H(s)]$. But in open loop system there is no reduction because of no feedback.

1.30.3 Effect of Feedback on Sensitivity

The parameters of any control system changes with the change in environment conditions. Also these parameters cannot be constant throughout the life. These parameter variations affects the performance of the system. For example, the resistance of the winding of a motor changes due to the change in temperature during its operation.

So, a control system should be insensitive to the parameter variations. Let P is a gain parameter that may vary due to the variations in parameters 'R' of the system. The sensitivity of the system parameter P to the parameter R is

$$S = \frac{\%\text{ change in R}}{\%\text{ change in P}}$$

$$S_P^R = \frac{d\,(\ln R)}{d\,(\ln P)} = \frac{\frac{1}{R}\,\partial R}{\frac{1}{F}\,\partial P} = \frac{\partial\, R/R}{\partial\, P/P}$$

In general 'R' may be the output variable and 'P' may be the gain, the feedback factor, etc.

let $T(s)$ = Overall transfer function

$G(s)$ = Forward path transfer function is varying

Then, sensitivity will be

$$S_G^T = \frac{\partial T(s)/T(s)}{\partial G(s)/G(s)} \qquad ...(1.147)$$

For open loop system $T(s) = G(s)$

$$\therefore \quad S_G^T = \frac{\partial\, G(s)/G(s)}{\partial\, G(s)/G(s)} = 1$$

Thus, the sensitivity of open loop system is unity.

Sensitivity of closed loop system:

$$T(s) = \frac{G(s)}{1+G(s)\,H(s)} \qquad ...(1.148)$$

$$\frac{\partial T(s)}{\partial G(s)} = \frac{[1+G(s)\,H(s)]\cdot 1 - G(s)\,H(s)}{[1+G(s)\,H(s)]^2} = \frac{1}{[1+G(s)\,H(s)]^2}$$

Sensitivity is given by from equation 1.147

$$S_G^T = \frac{G(s)}{T(s)}\cdot\frac{\partial\, T(s)}{\partial\, G(s)}$$

put the values of $T(s)$ and $\partial\, T(s)/\partial\, G(s)$

$$= \frac{G(s)}{\frac{G(s)}{1+G(s)\,H(s)}}\cdot\frac{1}{[1+G(s)\,H(s)]^2}$$

$$S_G^T = \frac{1}{1+G(s)\,H(s)} \qquad ...(1.149)$$

From equation (1.149) the sensitivity is reduced due to the feedback by a factor $1/1 + G(s)$ $H(s)$ as compared to open loop system.

Sensitivity due to the variation in $H(s)$:

from equation 1.148

$$\frac{\partial\, T(s)}{\partial\, H(s)} = -\frac{[G(s)]^2}{[1+G(s)\,H(s)]^2}$$

$$\therefore \quad S_H^T = \frac{H(s)}{T(s)} \cdot \frac{\partial\, T(s)}{\partial\, H(s)} = \frac{H(s)}{\dfrac{G(s)}{1+G(s)\,H(s)}} \cdot \frac{-[G(s)]^2}{[1+G(s)\,H(s)]^2} = \frac{-G(s)\,H(s)}{1+G(s)\,H(s)} \qquad ....(1.150)$$

From equation 1.149 and 1.148 it is clear that the closed loop system is more sensitive to variations in feedback path parameters than variations in forward path variations.

1.30.4 Effect of Feedback on Overall Gain

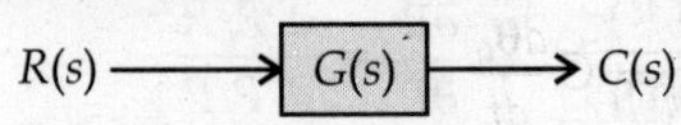

Fig. 1.118.

The overall transfer function of open loop system shown in Fig. 1.118 is

$$\frac{C(s)}{R(s)} = G(s)$$

The overall transfer function of closed loop system shown in Fig. 1.117 is

$$\frac{C(s)}{R(s)} = \frac{G(s)}{1+G(s)\,H(s)}$$

For negative feedback the gain $G(s)$ is reduced by a factor $\dfrac{1}{1+G(s)\,H(s)}$. So due to negative feedback overall gain of the system reduces.

1.30.5. Effect of Feedback on Stability

Consider the open loop system with overall transfer function

$$G(s) = \frac{K}{s+T}$$

The pole is located at $s = -T$

Now, consider closed loop system with unity negative feedback, then overall transfer function is given by

$$\frac{C(s)}{R(s)} = \frac{K}{s+(T+K)}$$

Now, closed loop pole is located at $s = -(T+K)$

Thus, feedback controls the time response by adjusting the location of the poles. The stability depends upon the location of poles. Thus we can say the feedback affects the stability. Feedback can improve the stability or may be harmful to stability if it is not properly design and apply.

1.31. THERMAL SYSTEMS

In thermal systems, there is transfer of heat from one substance to another substance. The thermal system can be analysed in terms of resistance and capacitance. Let us consider a simple thermal system shown in Fig. 1.119.

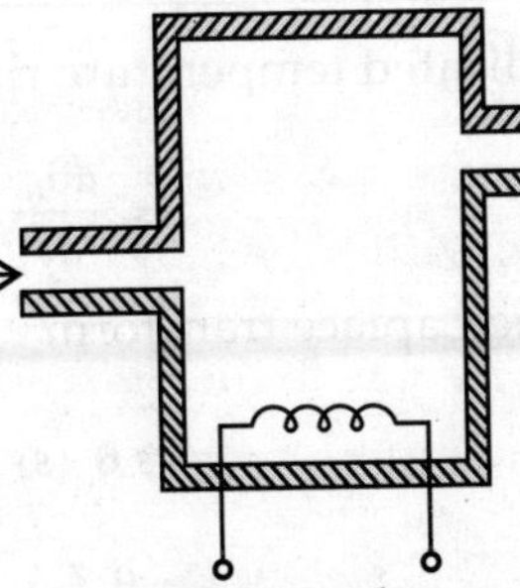

Fig. 1.119.

1.31.1. Heat Transfer System

Suppose, there is no heat store in the insulation and water having uniform temperature.

Let, θ_i = temperature of inlet water (°C)
θ_o = temperature of outlet water (°C)
θ = temperature of the surroundings
q = rate of heat flow from heating elements (J/S)
q_i = rate of heat flow to the water
q_t = rate of heat flow through tank insulation
C = thermal capacitance (J/°C)
R = thermal resistance (°C/J-S^{-1})

Rate of heat flow for the water in tank

$$q_i = C\frac{d\theta_0}{dt} \quad \text{...(1.151)}$$

The rate of heat flow from water to the surrounding through insulation

$$q_t = \frac{\theta_0 - \theta}{R} \quad \text{...(1.152)}$$

$$q = q_i + q_t \quad \text{...(1.153)}$$

$$q = C\frac{d\theta_0}{dt} + \frac{\theta_0 - \theta}{R} \quad \text{...(1.154)}$$

neglect the term $\frac{\theta}{R}$

$$q = C\frac{d\theta_0}{dt} + \frac{\theta_0}{R} \quad \text{...(1.155)}$$

Take Laplace transform of (1.155)

$$Q(s) = sC\,\theta_o(s) + \theta_o(s)$$

$$\text{Transfer function} = \frac{\theta_0(s)}{Q(s)}$$

$$\frac{\theta_0(s)}{Q(s)} = \frac{R}{1+sCR} \quad \text{...(1.156)}$$

∴ Time constant of thermal system is RC.

1.31.2. Thermometer

Consider the Fig. 1.120 A thermometer is immersed in a tub containing water.

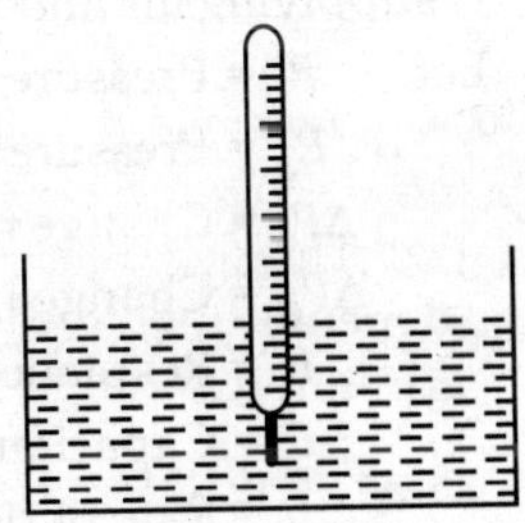
Fig. 1.120.

Let θ_i = temperature of the water in tub.
θ_o = temperature indicated by thermometer

then, rate of heat flow to the thermometer

$$\frac{dq}{dt} = \frac{\theta_i - \theta_0}{R} \quad \text{...(1.157)}$$

The indicated temperature rises at the rate

$$\frac{d\theta_0}{dt} = \frac{1}{C}\frac{dq}{dt} \quad \text{and} \quad \therefore \quad \frac{d\theta_0}{dt} = \frac{1}{C}\left(\frac{\theta_i - \theta_0}{R}\right) \qquad \text{...(1.157}a\text{)}$$

Take the Laplace transform

$$s\,\theta_o(s) = \frac{1}{RC}[\theta_i(s) - \theta_0(s)]$$

$$\therefore \quad \frac{\theta_0(s)}{\theta_i(s)} = \frac{1}{1+sRC} \qquad \text{...(1.158)}$$

Let $\theta_i(s)$ = unit step input

$$\theta_o(s) = \frac{1}{s(1+sRC)}$$

Take inverse Laplace

$$\therefore \quad \theta_o(t) = 1 - e^{-t/RC} \qquad \text{...(1.159)}$$

$\therefore$ Time constant = RC

EXAMPLE 1.50. A thermometer has a time constant of 15.33 sec. It is quickly taken from a temperature 0°C to a water bath having a temp. 100°C. What temperature will be indicated after 60 sec ?

Solution : The thermometer is subjected to a step input of 100°C

$$\therefore \quad \theta_i(s) = \frac{100}{S}$$

$\therefore$ from equation 1.158 $\quad \theta_0(s) = \dfrac{1}{1+sRC}\theta_i(s)$

$$\theta_0(s) = \frac{100}{s(1+sRC)} = \frac{100}{s(1+sT)} \quad \text{where } T \text{ is the time constt}$$

$$\theta_0(t) = 100\,(1 - e^{-t/T})$$

$$\theta_0(t) = 100\,(1 - e^{-60/15.33})$$

$$\theta_0(t) = 98°\text{C}$$

Temperature indicated by thermometer after 60 sec. = 98°C.

1.32. PNEUMATIC SYSTEM

Consider the Fig. 1.121. In Fig. 1.121 a source is supplying air and the air is stored in vessel.

Let, P_i = Pressure of air of the source N/m^2

P_o = Pressure of air in the vessel N/m^2

ΔP_i = Change in air pressure of source

ΔP_0 = Change in air pressure of vessel

R = Resistance to air flow into the vessel

C = Capacitance of the vessel

q = Rate of flow of air due to the differential pressure

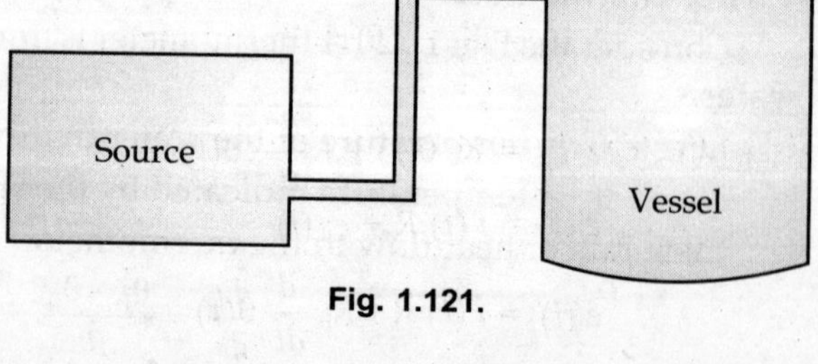

Fig. 1.121.

$$\therefore \quad q = \frac{\Delta P_i - \Delta P_0}{R} \quad ...(1.160)$$

The volume of air stored in vessel increases the pressure inside the vessel.

$$\therefore \quad \text{Volume } V = C\Delta P_0 \quad ...(1.161)$$

$$\therefore \quad \text{Rate of storage in vessel} = \frac{dV}{dt} = C\frac{d\Delta P_0}{dt} \quad ...(1.162)$$

$$\text{Volume of air in flow } V = \int q\,dt = \int \frac{\Delta P_i - \Delta P_0}{R}\,dt \quad ...(1.163)$$

$$\text{Rate of air in flow} = \frac{dv}{dt} = \frac{\Delta P_i - \Delta P_0}{R} \quad ...(1.164)$$

Rate of air in flow = Rate of storage of air in vessel

$$C\frac{d\Delta P_0}{dt} = \frac{\Delta P_i - \Delta P_0}{R} \quad ...(1.165)$$

Take Laplace transform

$$sC\Delta P_0 = \frac{1}{R}[\Delta P_i(s) - \Delta P_0(s)]$$

$$\frac{\Delta P_0}{\Delta P_i} = \frac{1}{1+RCs} \quad ...(1.166)$$

Time constant = RC

1.33. TRANSFER FUNCTION OF MOVING COIL VOLTMETER

Let R = Resistance of coil

$e(t)$ = Voltage to be measured

$i(t)$ = Current flowing through the coil

$\theta(s)$ = Deflection

K_s = Spring constant

J = Moment of Inertia

K_b = Back e.m.f. constant

K_T = Torque constant

T_d = Torque Developed

e_b = Back e.m.f.

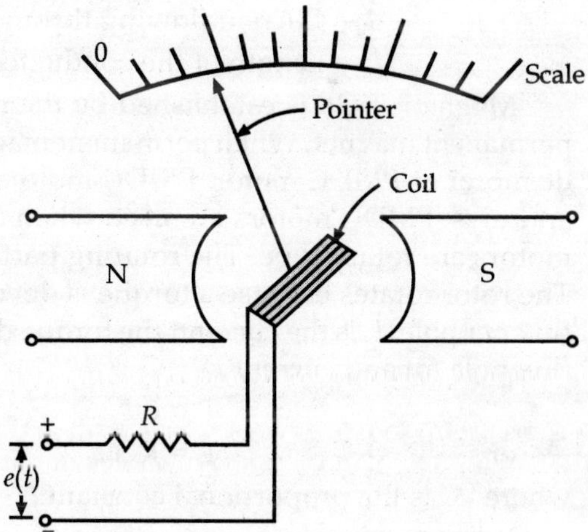

Fig. 1.122. Moving Coil Voltmeter

Deflecting Torque $T_d \propto i(t)$

$$T_d = K_T\, i(t)$$

$$e_b = K_b \frac{d}{dt}\theta(t)$$

$$T_d = K_s\,\theta(t) + J\frac{d^2}{dt^2}\theta(t)$$

$$e(t) = i(t)\,R + e_b(t)$$

$$e(t) = i(t)\,R + K_b\frac{d}{dt}\theta(t)$$

Laplace transform

$$T_d(s) = K_T I(s)$$
$$T_d(s) = K_s \theta(s) + s^2 J \theta(s)$$
$$E(s) = I(s) R + K_b s\theta(s)$$
$$K_T I(s) = K_s \theta(s) + s^2 J \theta(s)$$
$$I(s) = \theta(s)\left[\frac{K_s + s^2 J}{K_T}\right]$$
$$\therefore \quad E(s) = R\,\theta(s)\left[\frac{K_s + s^2 J}{K_T}\right] + K_b s\theta(s)$$
$$\therefore \quad \frac{\theta(s)}{E(s)} = \frac{K_T}{JRs^2 + K_b K_T s + K_s R} \quad \textbf{Ans.}$$

1.34. DC MOTORS IN CONTROL SYSTEMS

A dc motor converts direct current electrical power into mechanical power. DC motors are widely used in industries.

The fundamental principle upon which dc motors are based is that whenever a current carrying conductor is placed inside a magnetic field, a mechanical force is experienced by that conductor. The force is described by the Lornetz force law and direction of this force is governed by Fleming's left hand rule. The mechanical force experienced by each conductor is given by

$$F = BIL \text{ Newton}$$

where, B = Magnetic field strength in Tesla

I = Current flowing through the conductor in Ampere

L = Length of the conductor in metre

Magnetic field is established by the magnet. The magnet may be electromagnet or may be permanent magnet. When permanent magnet is used then motor is known as permanent magnet dc motor or PMDC motor. PMDC motors are commonly used in automobiles, washer, blowers and toys. PMDC motors are used when there is no need of speed control by field. Most of the motors are rotary type. The rotating part is called rotor and stationary part is known as stator. The rotor rotates because a torque is developed about the rotor axis. If 'T' is the torque, ϕ is the flux per pole, I_a is the current, the torque developed by a d.c. motor is directly proportional to the flux pole (ϕ) and current I

$$\therefore \quad T \propto \phi I_a$$

or

$$T = K \phi I_a$$

where 'K' is the proportional constant.

When the conductors moves in the magnetic field and cut the magnetic flux, an e.m.f. is induced in them. This e.m.f. is known as *back e.m.f.* or counter e.m.f. The back emf is given by

$$E_b = K \phi \omega$$

where, E_b is the back emf in volts, ϕ is the flux and ω is the shaft velocity in rad/sec. of the motor.

1.35. EQUIVALENT CIRCUIT OF PMDC MOTOR

In PMDC motors the field is produced by permanent magnet, so there is no need to draw field coils in the equivalent circuit of PMDC motor.

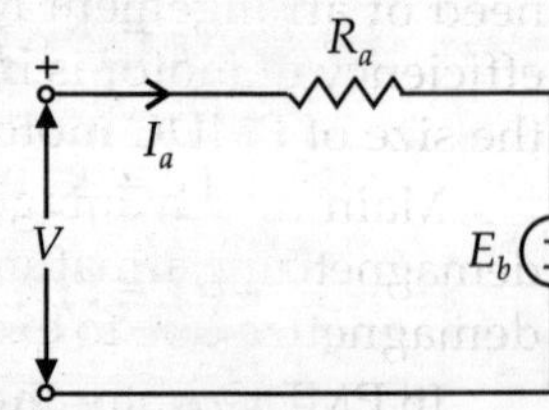

Fig. 1.123.

Apply *KVL*

$$V = I_a R_a + E_b$$

where 'V' is the supply voltage, 'E_b' is the back emf, 'I_a' is armature current and 'R_a' is armature resistance. This is the motor equation. From this equation it is clear that the back emf of the motor is always less than the supply voltage.

1.36. CONSTRUCTION OF PERMANENT MAGNET DC MOTOR (PMDC MOTOR)

In permanent magnet dc motor, the field poles are made of permanent magnet. PMDC motor consists of two parts: (*i*) stator, and (*ii*) armature.

Stator is a stationary part. It is hollow steel cylinder. An even number of magnets are mounted in the inner surface of the cylinder. Stator provides the support to the permanent magnet and low reluctance path to the magnetic flux. Generally, field coils are not provided in the permanent magnet dc motor but sometimes it is found that they are used along with permanent magnet. This is because if permanent magnets lose their strength, then this strength can be compensate by field excitation. Permanent magnets are arranged in such a way that the N-pole and S-pole of each magnet are alternatively faced towards armature.

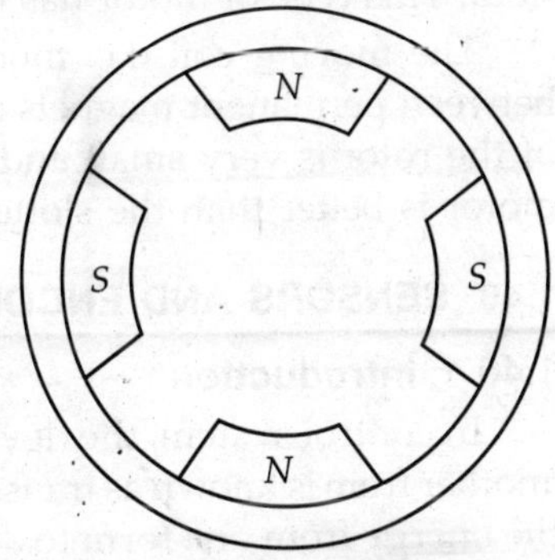

Fig. 1.124.

The rotating part of permanent magnet dc motor is called rotor. This is also known as armature. Armature consists of laminated cylinder known as armature core mounted on the shaft. The armature core has slots on its outer surface. The laminations are insulated from each other to reduced eddy current loss in armature of permanent magnet dc motor. The insulated conductors are put in the slots of armature core. These armature conductors are connected in a suitable manner which give rise to armature winding. The end terminals are connected to the commutator placed on the motor shaft. The commutator rotates with armature is made of segments insulated from each other and from the shaft. Current is collected by the two or more carbon or graphite brushes mounted on the commutator with spring pressure.

1.37. APPLICATIONS OF PERMANENT MAGNET DC MOTOR

Small permanent magnet dc motors are demanded by the manufacturers of toys, automobiles, medical equipments, computer hardware, office equipments, instruments for measurement and control, robots, catering equipment, ticketing machines, kitchen equipment like microwave oven, dishwashers, mixers, etc., vacuum cleaners, pumps, toys.

Permanent magnet dc motors are also used in computer printer, scanners, plotters, photocopy machines, elevators, electric boats, submarines, air compressors, electric wheelchairs, drill machines, hammers, polishers, grinders, saws, etc. These motors are also having applications in renewable energy systems.

1.38. ADVANTAGES AND DISADVANTAGES OF PMDC MOTOR

There are some advantages and disadvantages of PMDC motor. In PMDC motors there is no need of arrangement for field excitation. There is no power required for excitation and hence efficiency of motor is improved. Since in PMDC motor no field coils are required which reduces the size of PMDC motor. These motors are cheaper and economical.

Main disadvantage of PMDC is that magnetic strength of the field may get weak due to demagnetizing armature reaction. Also there is possibilities that poles may be partial or permanently demagnetize due to excessive armature current at starting, reversal or overloading of the motor.

In PMDC motor the field in air gap cannot be controlled externally. So efficient speed control of DC motor is difficult.

1.39. TYPES OF PERMANENT MAGNET DC MOTORS

According to the construction of armature, PMDC motors are classified as Iron-core PMDC motors, surface wound PMDC motors and moving-coil PMDC motors.

In iron-core PMDC motors laminated rotor having slots. The armature conductors are placed in these slots. This type of motors having high inductance, high rotor inertia and low cost.

In surface wound dc motors there is no slots in rotor and armature conductors are on the surface of the rotor. The conductors are placed in air gap between rotor and permanent magnet field. This type of motor has low inductance.

The moving coil d.c. motor has armature winding fixed to an insulating cylinder rotating between permanent magnets and laminated core. In moving coil dc motor the moment of interia of the rotor is very small and also armature inductance is very small. Efficiency of moving coil motor is better than the slotted rotor motor.

1.40. SENSORS AND ENCODERS IN CONTROL SYSTEMS

1.40.1. Introduction

In control system, the devices which are used to convert the process variables in one form to another from is known as transducers. Transducers can also be defined as a device which transforms the energy from one form to another, for example, a thermocouple converts the heat energy into electrical voltage. In control system the following devices are used as a transducer.

1. Potentiometers
2. D.C. servomotors
3. A.C. servomotors
4. Synchros
5. Stepper motors
6. Magnetic amplifier
7. Tachogenerators
8. Gyroscope
9. Differential transformer

1.40.2. Potentiometers

A potentiometer is a simple device which is used for mechanical displacement either linear or angular. Thus, a potentiometer is an electromechanical transducer which converts the mechanical energy into electrical energy. The input to the device is in the form of linear mechanical displacement or rotational mechanical displacement, when the voltage is applied across the fixed terminals, the output voltage is proportional to the displacement.

Consider the Fig. 1.125 (translational potentiometer).

Let E_i = input voltage

E_0 = output voltage

x_i = displacement from zero position

x_t = total length of translational potentiometer

R = Total resistance of potentiometer

Under ideal condition the output voltage E_0 is given by

$$E_0 = \frac{x_i}{x_t} E_i \quad \text{...(1.167)}$$

Equation (1.167) shows a linear relationship shown in Fig. 1.126.

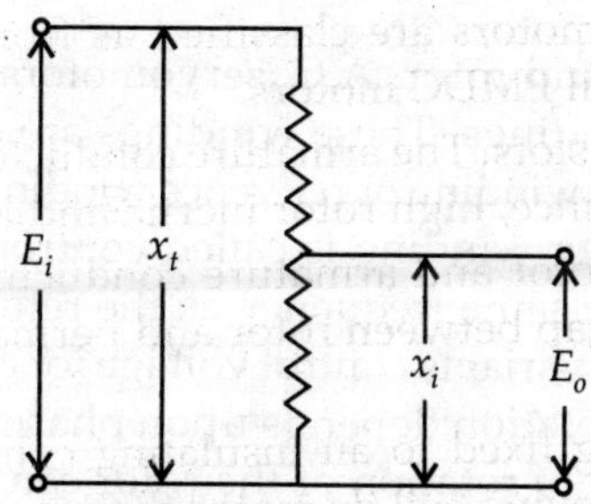

Fig. 1.125.

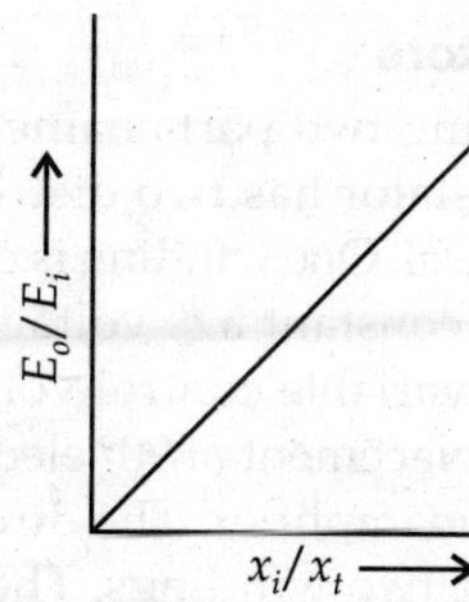

Fig. 1.126.

Similarly for rotational motion, the output voltage E_0 is given by

$$E_0 = E_i \cdot \frac{\theta_i}{\theta_t} \qquad \text{...(1.168)}$$

where, θ_i = input angular displacement (degree or radians)

θ_t = total travel of wiper (degree or radians)

Fig. 1.127(*a*) Shows an arrangement of error sensing transducer. In this arrangement two potentiometers are connected in parallel. The output voltage taken across the variable terminals of the two potentiometer. The output voltage E_0 is given by

$$E_o = K(\theta_1 - \theta_2) \qquad \text{...(1.169)}$$

E_i is the voltage applied, θ_1 and θ_2 are the angular displacement of the wiper, K is constant and is known as sensitivity. The block diagram is shown in Fig. 1.127(*b*)

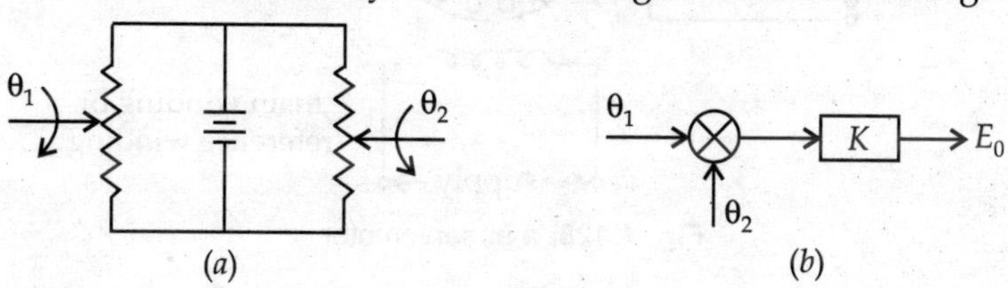

Fig. 1.127.

1.41. SERVOMOTORS

Servomotors are used in feedback control systems. Servomotors have low rotor inertia and high speed of response. The servomotors are also known as *control motors*. The servomotors which are used in feedback control system should have linear relationship between electrical control signal and rotor speed, torque speed characteristic should be linear, the response of the servomotor should be fast and inertia should be low.

1.41.1. Types of Servomotors

The servomotors are classified as :

(*i*) A.C. servomotors

(*ii*) D.C. servomotors

(*iii*) Special servomotors

D.C. servomotors are further classified as armature controlled d.c. servomotors and field control d.c. servomotors.

1.41.2. A.C. Servomotors

These motors having two parts namely, stator and rotor. A.C. servomotors are two phase induction motor. The stator has two distributed windings. These windings are displaced from each other by 90° electrical. One winding is called main winding or reference winding. The reference winding is excited by constant a.c. voltage. The other winding is called control winding. This winding is excited by variable control voltage of the same frequency as the reference winding, but having a phase displacement of 90° electrical. The variable control voltage for control winding is obtained from a servoamplifier. The direction of rotation depends upon phase relationship of voltages applied to the two windings. The direction of rotation of the rotor can be reversed by reversing the phase difference between control voltage and reference voltage.

The rotor of a.c. servomotors are of two types : (*a*) squirrel cage rotor, (*b*) drag cup type rotor. The squirrel cage rotor having large length and small diameter, so its resistance is very high. The air gap of squirrel cage is kept small. In drag cup type there are two air gaps. For the rotor a cup of non-magnetic conducting material is used. A stationary iron core is placed between the conducting cup to complete the magnetic circuit. The resistance of drag cup type is high and therefore having high starting torque. Generally aluminium is used for cup. Fig. 1.128. Shows the schematic diagram of two phase a.c. servomotor and 1.129(*a*), 1.129 (*b*) shows the two types of rotor.

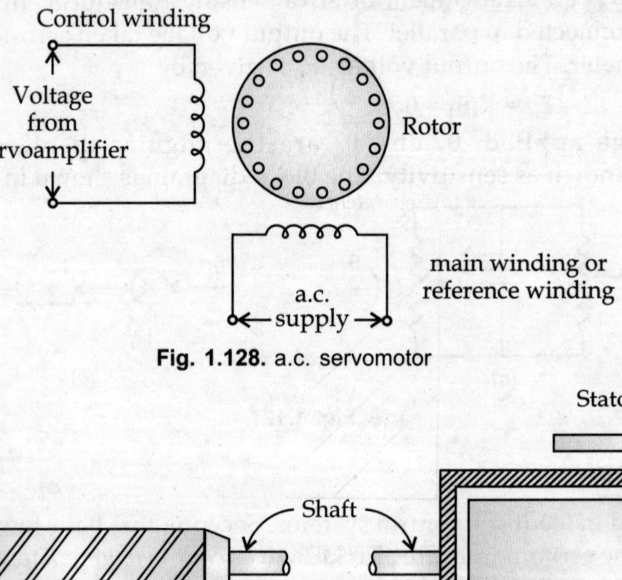

Fig. 1.128. a.c. servomotor

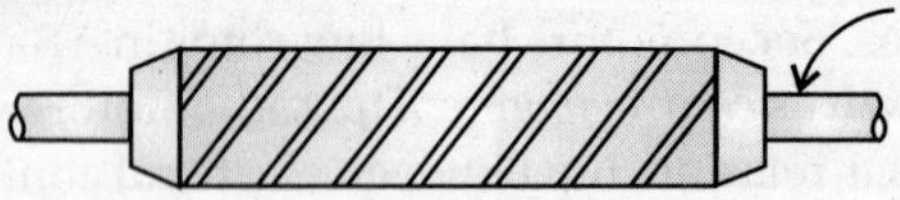

Fig. 1.129(a) Squirrel cage rotor

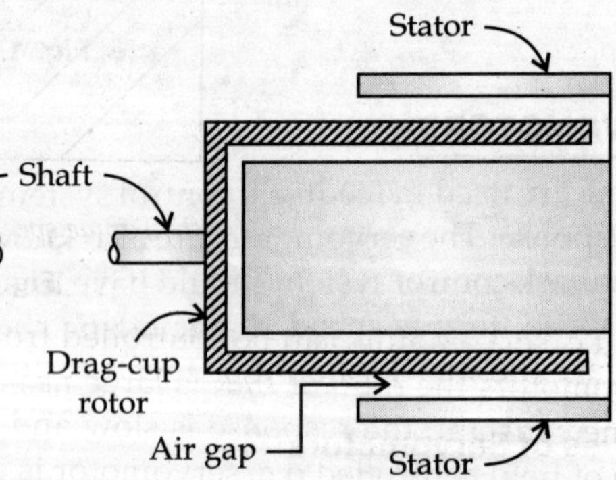

Fig. 1.129(b) Drag cup type rotor

1.41.3. Torque-speed Characteristic

The torque speed characteristic of two phase induction motor depends upon the ratio of reactance to resistance. For high resistance and low reactance, the characteristic is linear and for large ratio of X to R it becomes non-linear as shown in Fig. 1.130(*a*). The torque-speed characteristics for various control voltages are almost linear as shown in Fig. 1.130(*b*).

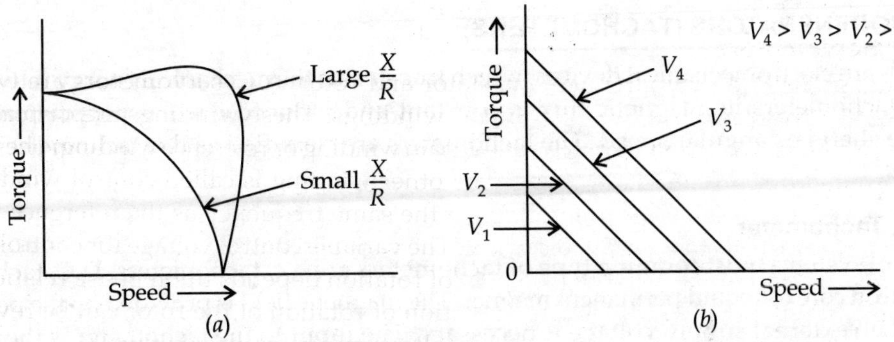

Fig. 1.130.

1.41.4. D.C. Servomotors

D.C. servomotors are separately excited or permanent magnet d.c. servomotors. The armature of d.c. servomotor has a large resistance, therefore torque speed characteristic is linear. The torque speed characteristic shows in Fig. 1.131 (*b*). Fig 1.131 (*a*) shows the schematic diagram of separately excited d.c. servomotor.

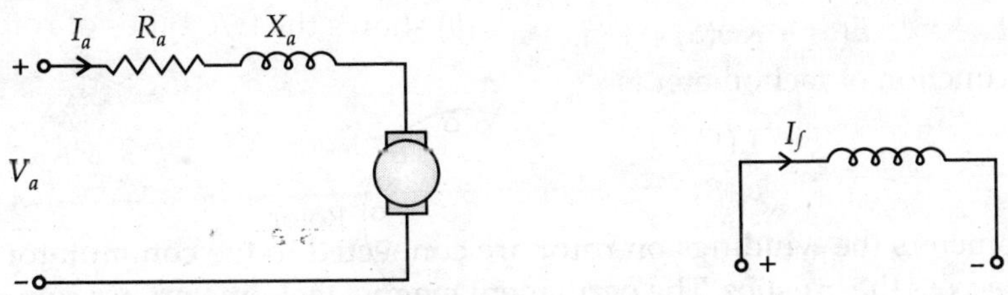

(*a*) *Separately excited d.c. servometer*

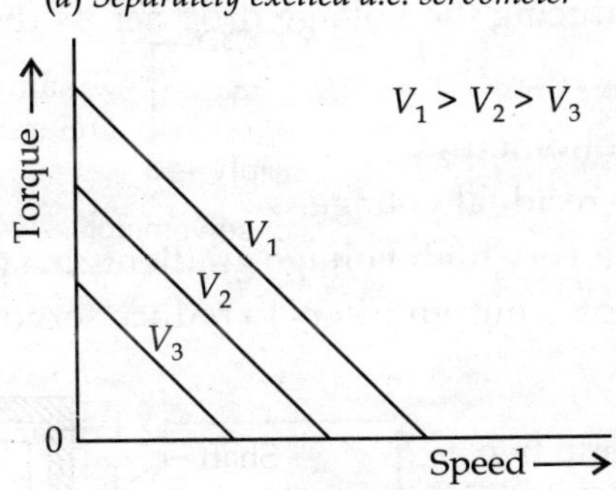

(*b*) *Torque speed characteristic*

Fig. 1.131.

The d.c. servomotors can be controlled from armature side or from field. In field controlled d.c. servomotors the ratio of L/R is large *i.e.*, the time constant for field circuit is large. Due to large time constant, the response is slow and therefore they are not commonly used. Transfer function of field controlled d.c. servomotor is given in Chapter 1. The speed of the motor can be controlled by adjusting the voltage applied to the armature. In armature controlled d.c. servomotor the time constant is small and hence the response is fast. The efficiency is better than the field controlled motor. The transfer function of armature controlled d.c. servomotor is derived in Chapter 1.

1.41.5. Application of Servomotors

Servomotors are widely used in radars, electromechanical actuators, computers, machine tools, tracking and guidance system, process controllers and robots.

1.42. TACHOGENERATORS (TACHOMETERS)

Tachometers are electromechanical devices, which transforms the mechanical energy into electrical energy. In tachometers its magnetic flux is constant and induced e.m.f is proportional to the speed of the shaft i.e., angular speed. The tachometers are classified as d.c. tachometers and a.c. tachometers.

1.42.1. D.C. Tachometer

In control systems most common type of tachometers are d.c. tachometers. D.C. tachometers contains an iron core rotor and permanent magnet. The magnetic field is provided by the permanent magnet and no external supply voltage is necessary. The input to the tachometer is the speed of the shaft and the output is voltage which is proportional to the angular speed of the shaft.

$$\therefore \quad e = K\omega(t) \qquad \text{...(1.170)}$$

where, e = tachometer generator voltage

K = tachometer sensitivity

ω = angular speed of the shaft.

Laplace transform of equation 1.170

$$E(S) = K\omega(S) \qquad \text{...(1.171)}$$

$\therefore$ Transfer function of tachometer is

$$K = \frac{E(S)}{\omega(S)} \qquad \text{...(1.172)}$$

In d.c. tachometers the windings on rotor are connected to the commutator and the output voltage is taken across the brushes. The permanent magnet tachometers are compact and reliable but having high inertia. For reducing the voltage drop across the brushes, metal brushes with silver tips are used.

1.42.2. Advantages of D.C. Tachometers

1. At zero speed there is no residual voltage.
2. It is possible to generate a very high voltage gradients in small size.
3. It can be used with high pass output filters to reduce servo velocity lags.

1.42.3. A.C. Tachometers

A.C. tachometers are similar to two phase induction motor. The schematic diagram of a.c. tachometer is shown in Fig. 1.132. Here two stator windings are placed in quadrature with each other and rotor is short circuited. The rotor of tachometer is a thin aluminium cup. The two stator windings are known as primary winding and secondary winding. Primary winding also known as reference winding.

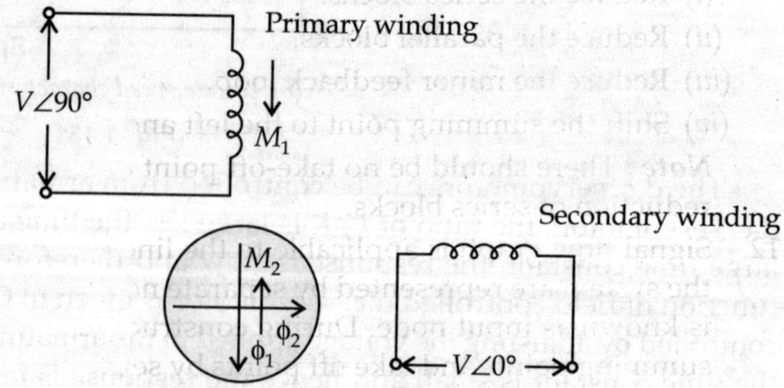

Fig. 1.132. A.C. tachometer

The primary winding carries a voltage $V\angle 90°$. This voltage is fixed in magnitude at fixed frequency known as carrier frequency. The coil axis of primary winding is called direct axis. When the current in primary winding flows a pulsating field M_1 is produced along direct axis. When the rotor is stationary an e.m.f. is induced in it. Since the motor is short circuited eddy current flows in it. Due to this eddy current a pulsating flux M_2 is produced opposite to the M_1. Therefore a pulsating flux ϕ_1 is produced due to the vector sum of M_1 and M_2. Since the secondary winding is in quadrature no e.m. f. will produced in it due to ϕ_1. Now if the rotor rotates it cuts the flux and an e.m.f is induced in it, due to this induced e.m.f a current will flow in the rotor and a pulsating flux will appear along the coil axis of secondary winding. This flux ϕ_2 will induce the voltage in secondary winding. The magnitude of the output voltage will be proportional to the rotor speed.

SUMMARY

1. The open loop control system is without feedback. In open loop system the control action is independent of the desired output.
   Closed loop control systems are known as feedback control systems. In closed loop control systems the control action is dependent on the desired output.
2. The elements of the closed loop systems are command, reference input, error detectors, control element, controlled system and feedback element.
3. Transfer function is defined as the ratio of laplace transform of the output to the laplace transform of input with all initial conditions are zero.
4. Poles are the values of 's' which when substituted in the denominator of a transfer function, make the transfer function value as infinity.
5. Zeros are the values of 's' which when substituted in the numerator of a transfer function, make the transfer function value as zero.
6. The characteristic equation can be obtained by equating the denominator polynomial of the transfer function to zero.
7. The highest power of 's' in the characteristic equation is called the order of the system.
8. The number of polses at the origin defines the type of system.
9. Block diagram is a pictorial representation of the complicated control system.
10. Block diagram can be reduced by reduction rules.
11. The reduction of blocks is possible by the following sequence :
    (i) Reduce the series blocks.
    (ii) Reduce the parallel blocks.
    (iii) Reduce the minor feedback loop
    (iv) Shift the summing point to the left and take-off point to the right, as far as possible.
    Note : There should be no take-off point or summing point between the blocks during the reduction of series blocks.
12. Signal flow graph is applicable to the linear systems. In signal flow graph the variables of the system are represented by separate nodes. The node which has only outgoing branches is known as input node. During construction of signal flow graph represent all variables, summing points and take off points by separate nodes. If a summing point is placed before a take off point represent summing and take off point by a single node. If summing point is placed after take off point, represent summing point and take off point by separate nodes.
13. The resultant transfer function of the system can be obtained from signal flow graph by Mason's gain formula.
    Mason's gain formula is

$$T = \frac{\Sigma g_k \Delta_K}{\Delta}$$

where T = transfer function

Δ = 1 – (sum of all individual loop gain) + (sum of all possible gain products of two non-touching loops) – (sum of all possible gain products of three non-touching loops) + ...

g_K = gain of forward path

Δ_K = the part of Δ not touching the Kth forward path.

14. The motion takes place along a straight line is known as translational motion.
15. Rotational motion of a body is the motion about a fixed axis.
16. The main elements of any mechanical system with translational motions are mass, spring and friction. The mass M cannot store energy, hence cannot cause any change in displacement. Spring stores the potential energy which causes in displacement if it is between two moving surfaces. But cannot cause change in displacement, if its one end is connected to a rigid support.
17. The elements of rotational systems are inertia (J), damping coefficient (B) and torsional stiffness K.
18. For every mechanical system, there is analogous electrical system.
19. D'Alembert's principle states that, the algebraic sum of externally applied forces and the forces resisting motion in any given direction is zero.
20. For mechanical network, analogous electrical network can be obtained by using $f - v$ and $f - i$ analogy.
21. **Force – Voltage analogy :** In this method force is analogous to voltage. Similarly $M \to L, K \to \frac{1}{C}, B \to R$, displacement x charge q

    Force-current analogy : In this method force is analogous to current

    $M \to C, B \to \frac{1}{R}, K \to \frac{1}{L}, X \to \phi.$

    Mechanical Rotational system : (*a*) Force-voltage analogy :

    $T \to E, J \to L, B \to R, K \to \frac{1}{C}, \theta \to q$

    (*b*) Force-current analogy:

    $$T \to I, J \to C, B \to \frac{1}{R}, K \to \frac{1}{L}, \theta \to \phi$$

    Mechanical Coupling: $\frac{T_1}{T_2} = \frac{\theta_2}{\theta_1} = \frac{r_1}{r_2} = \frac{\omega_2}{\omega_1} = \frac{N_1}{N_2}$

22. For feedback control system, servomotor should have following properties
    (*i*) It should be reversible.
    (*ii*) Response should be fast.
    (*iii*) Relationship between electrical control system and rotor speed should be linear.
    (*iv*) It should have linear torque-speed characteristic.

23. A.C. Servomotor has following features
   (*i*) light in weight
   (*ii*) Reliable and stable
   (*iii*) Maintenance free
   (*iv*) large R to X ratio
   (*v*) Smooth & noise free

24. Field controlled D.C. motor has following features.
   (*i*) It has large time constant.
   (*ii*) It is open loop system.
   (*iii*) Control circuit is simple.

Table 1.7. Comparison between A.C. and D.C. servomotors

S.No.	A.C. Servomotor	D.C. Servomotor
1.	Low power output	High power output
2.	Efficiency is less	High efficiency
3.	No brushes & slip rings hence maintenance free	Frequent maintenance required
4.	No radio frequency noise	Brushes produce radio frequency noise
5.	Smooth operation	Noisy operation

25. **Definitions Related to Stepper motor**
   (*i*) **Holding Torque:** It is defined as the maximum static torque that can be applied to the shaft of an excited motor without causing a continuous rotation.
   (*ii*) **Detent Torque:** It is defined as the maximum static torque that can be applied to the shaft of an unexcited motor without causing a continuous rotation.
   (*iii*) **Step Angle:** It is defined as the angular displacement of the rotor in response to each pulse.
   (*iv*) **Critical Torque:** It is defined as the maximum load torque at which rotor does not move when an exciting winding is energized. This is also known as pullout torque.
   (*v*) **Limiting Torque:** It is defined for a given pulsing rate or stepping rate measured in pulses per second, as the maximum load torque at which motor follows the control pulses without missing any step. This is also called as pull in torque.
   (*vi*) **Synchronous Stepping Rate:** It is defined as the maximum rate at which the motor can step without missing steps. The motor can start, stop or reverse at this rate.
   (*vii*) **Slewing Rate :** It is defined as the maximum rate at which the motor can step unidirectionally. The slewing rate is much higher than the synchronous stepping rate. Motor will not be able to stop or reverse without missing steps at this rate.

Stepper motors are used in computer peripheral systems such as printers, tape drives etc., process control systems, robotic control systems, watches etc.

Tachnometers or Tachogenerator is an electro-mechanical unit which generated an electrical output proportional to the speed of the shaft. Tachometers are classified as D.C. Tachometers and A.C. Tachometers.

Potentiometer is an electromechanical transducer that converts mechanical energy into electrical energy. The input to the device is in the form of a mechanical displacement either linear or rotational.

Different type of potentiometers are single turn and multiturn potentiometer, linear motion potentiometers and wire wound potentiometers.

EXERCISE

1.1. Determine the transfer function of the given circuit.

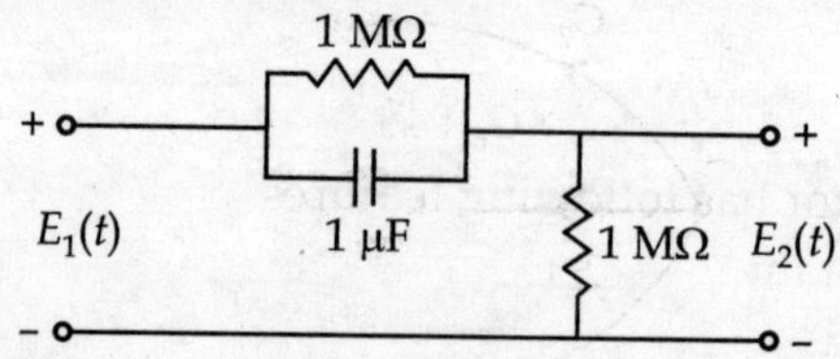

1.2. Determine the transfer function.

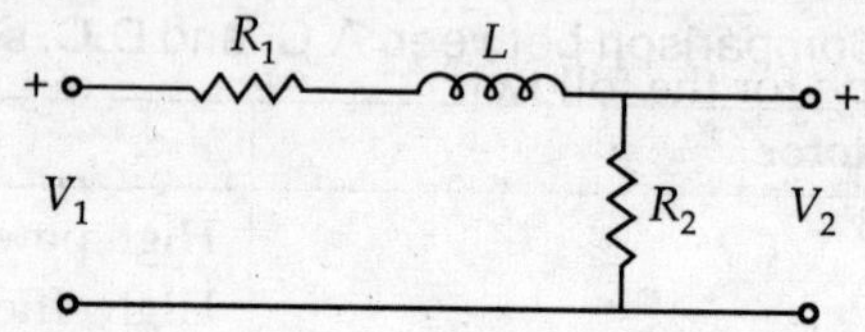

1.3. Derive the transfer function $\frac{V_2(s)}{V_1(s)}$.

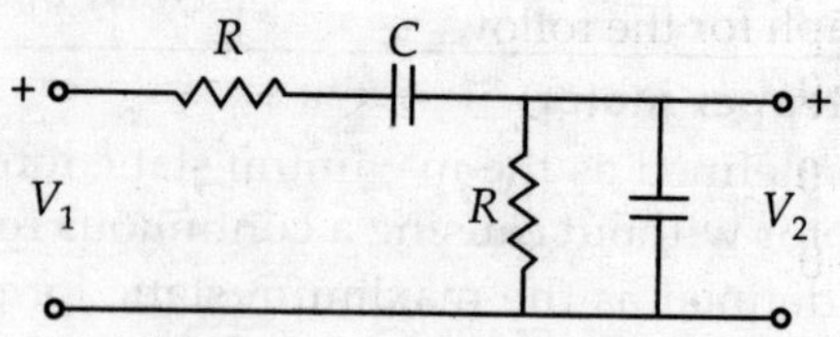

1.4. Determine the transfer function $\frac{V_2(s)}{V_1(s)}$.

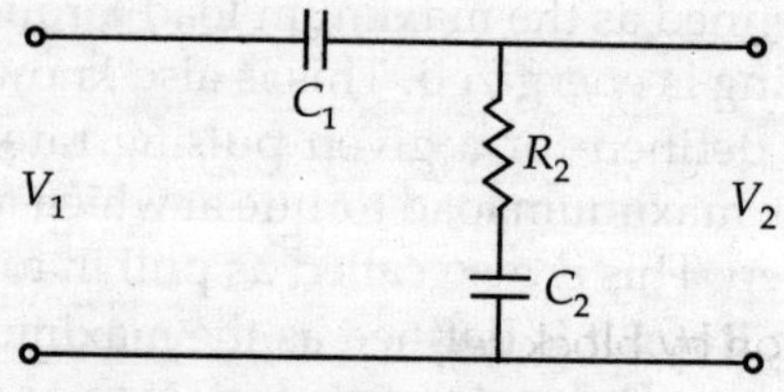

1.5. Find out the transfer function if $L = 1H$ and $C = 1F$.

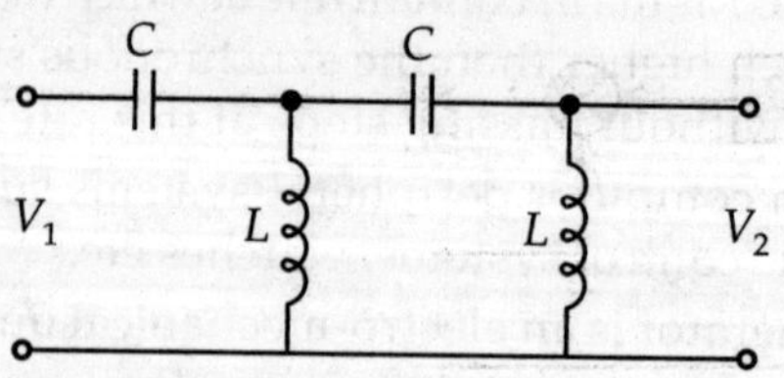

1.6. Draw the signal flow graph of the electrical circuit shown in Fig. 1.79, example 1.28 and find out the transfer function by using the Mason's gain formula.

1.7. Draw the signal flow graph of the block diagram. Shown in Fig. 1.91 example 1.32 and determine the transfer function.

1.8. Using Mason's gain formula, determine the ratio $\frac{C}{R}$.

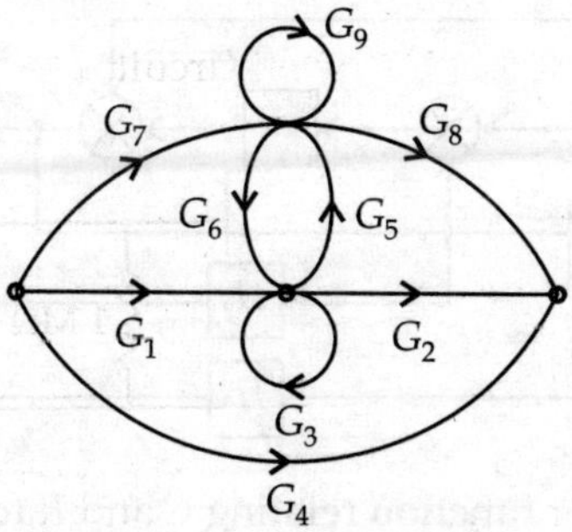

1.9. Draw the signal flow graph for the following equations

(i) $\frac{d^2y}{dt^2}+10\frac{dy}{dt}-5y=3x$ (ii) $y'''' + 6y''' + 5y'' + 3y' + 2y = x$

(iii) $\frac{d^2y}{dt^2}+3y=x$

1.10. Draw the signal flow graph for the following equations

(i) $x_2 + 5x_3 - 2x_1 = 0$

$x_3 + 2x_4 - 4x_2 = 0$

$x_4 - 8x_3 = 0$

(ii) $x_2 = 4x_1 + 2x_3 + 2x_2$

$x_3 = 6x_1 + 5x_2 + 2x_3$

$x_4 = 2x_2 + 2x_3$

(iii) $x_1 - x_2 - 2x_3 - 5x_4 = 0$

$2x_2 - 3x_3 - 5x_4 = 0$

$7x_1 - 3x_3 - 2x_4 = 0$

1.11. Findout the transfer function by block reduction method

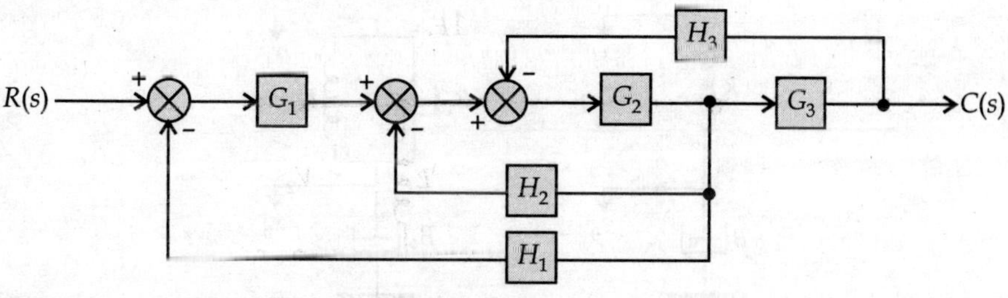

1.12 For the system shown in Fig. obtain $\frac{C(s)}{R(s)}$ by

(i) Block diagram reduction technique

(ii) Signal flow graph

(KNIT, Sultanpur, 1998-99)

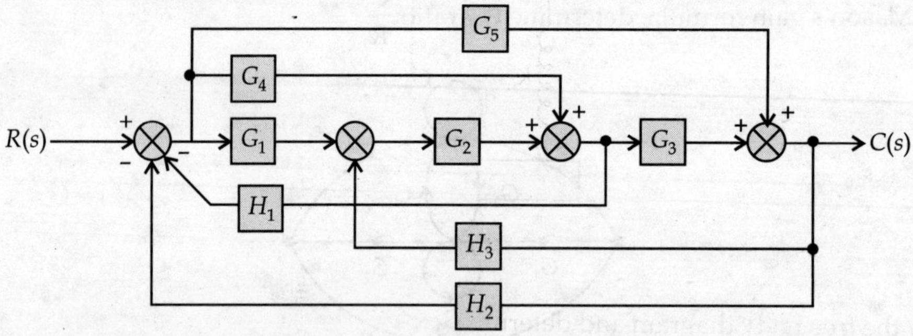

1.13. Determine the overall transfer function relating C and R for the system whose block diagram is given below by block reduction method. *(R.M.L. University Faizabad LST 2001)*

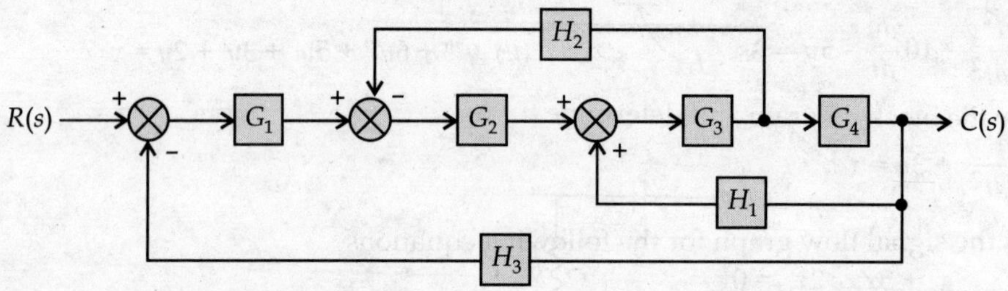

1.14. Determine the transfer function C/R for the system shown in Fig. 1.108 (example 1.42) by block reduction technique.

1.15. Determine the transfer function of the given system

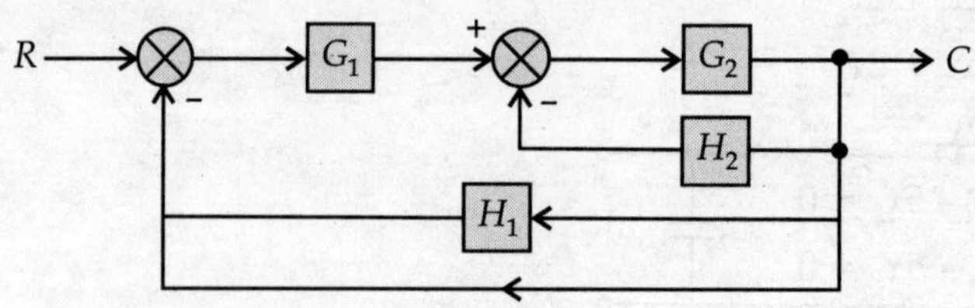

1.16. Draw the electric analog circuit and find $\frac{X_0(s)}{Xi(s)}$

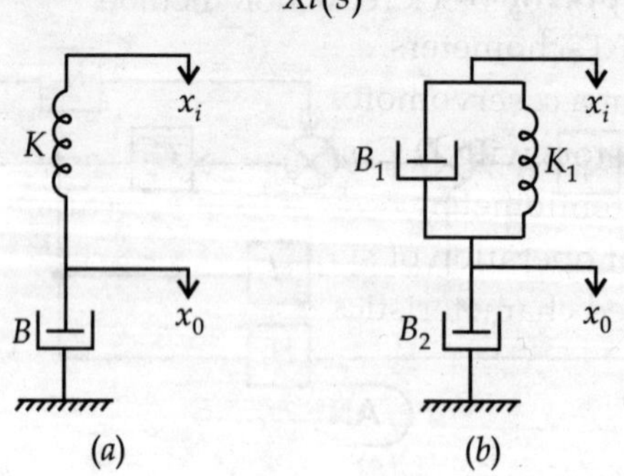

1.17. Determine the system equation $\frac{X(s)}{F(s)}$ where $M = 10$ kg, $B = 30$ N/m/s, $K = 20$ N/m.

(R.M.L. University Faizabad 2001)

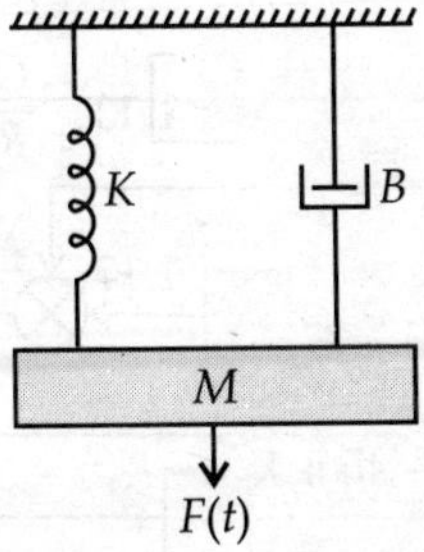

1.18. Draw the free body diagram and determine $\frac{X_2(S)}{F(S)}$ for the system shown.

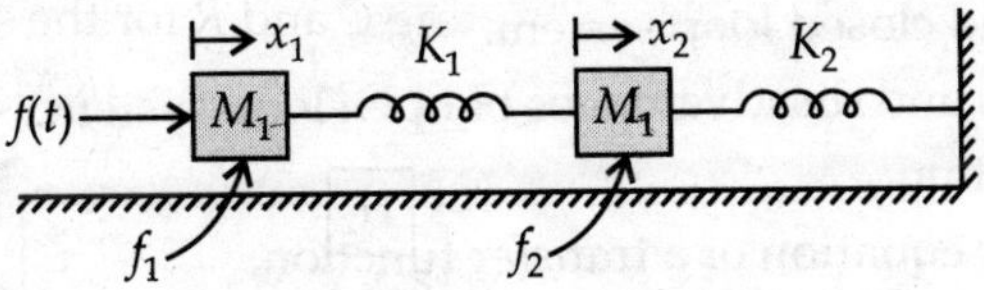

1.19. Draw the block diagram and determine the transfer function.

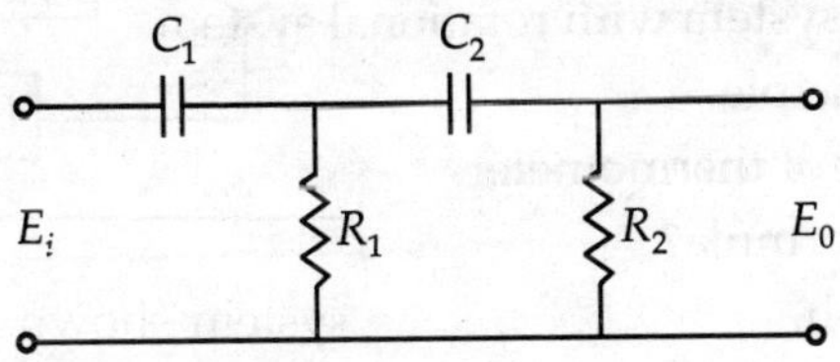

1.20. Determine the transfer function of the given circuits

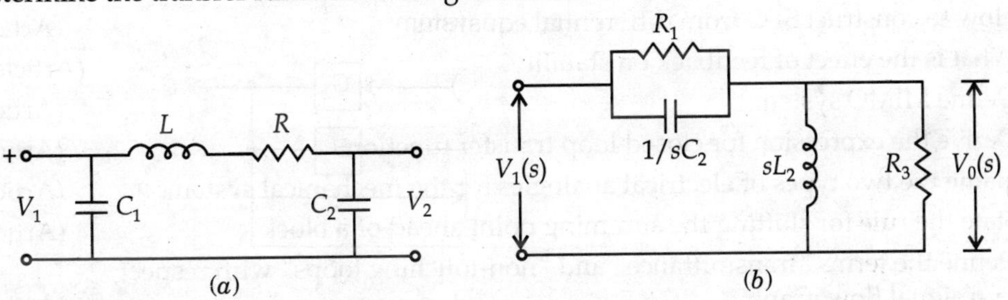

1.21. Write short note on stepper motor.
1.22. Explain the working of Tachometers.
1.23. State the applications of a.c. servomotor.
1.24. Compare A.C. servomotors with D.C. servomotors.
1.25. Write short note on potentiometer.
1.26. Explain the principle of operation of synchro error detector.
1.27. Sketch the Torque-speed characteristics of A.C. servomotor.

ANSWERS

1.1. $\frac{s+1}{s+2}$

1.2. $\frac{V_2(s)}{V_1(s)} = \frac{R_2}{(R_1+R_2)+sL}$

1.3. $\frac{V_2(s)}{V_1(s)} = \frac{SRC}{S^2R^2C^2+3SRC+1}$

1.4. $\frac{V_2(s)}{V_1(s)} = \frac{C_1+RC_1C_2s}{C_1+C_2+RC_1C_2s}$

1.5. $\dfrac{V_2(s)}{V_1(s)} = \dfrac{s^4}{s^4+3s^2+1}$

1.11. $\dfrac{C(s)}{R(s)} = \dfrac{G_1G_2G_3}{1+G_2H_2+G_2G_3H_3+G_1G_2H_1}$

1.15. $\dfrac{C(s)}{R(s)} = \dfrac{G_1G_2}{1+G_1G_2+G_2H_2+G_1G_2H_1}$

1.17. $\dfrac{X_2(s)}{F(s)} = \dfrac{1}{10s^2+30s+20s}$

1.18. $\dfrac{X_2(s)}{F(s)} = \dfrac{K_1}{(s^2M_2+sf_2+K_1+K_2)(s^2M_1+sf_1+K_1)-K_1^2}$

SEMI-OBJECTIVE TYPE QUESTIONS

(i) Define open loop and closed loop system. (Articles 1.2 and 1.3)

(ii) Give the advantages and disadvantages of open loop system. (Article 1.3)

(iii) Define transfer function. (Article 1.6)

(iv) Define characteristic equation of a transfer function. (Article 1.8)

(v) Define poles and zeros of a transfer function. (Article 1.9)

(vi) Compare translational system with rotational system. (Table 1.2)

(vii) Define D'Alembert principle. (Article 1.13)

(viii) Write the time constant of thermometer. (Article 1.31.2)

(ix) What is Mason's gain formula? (Article 1.28)

(x) Define signal flow graph. (Article 1.24)

(xi) Define loop, self loop, path gain, loop gain for SFG. (Article 1.24)

(xii) How to construct SFG from differential equation. (Article 1.26)

(xiii) What is the effect of feedback on stability. (Article 1.30.5)

(xiv) Define MIMO system. (Article 1.21)

(xv) Derive the expression for closed loop transfer function. (Article 1.20)

(xvi) Name the two types of electrical analogies for the mechanical system. (Article 1.15)

(xvii) State the rule for shifting the summing point ahead of a block. (Article 1.22)

(xviii) Define the terms "transmittance" and "non-touching loops" with respect to a signal flow graph. (Article 1.24)

(xix) Name the major parts of a closed loop control system. (Article 1.5)

(xx) What are the basic elements in thermal system? (Article 1.31)

(xxi) Write short note on potentiometers.

(xxii) Distinguish between A.C. servomotor and D.C. servomotor.

(xxiii) State applications of servomotor.

(xxiv) What are the advantages of D.C. tachometers.

(xxv) Write a short note on multistack variable reluctance stepper motor.

(xxvi) What are the major applications of stepper motor.

(xxvii) Write short note on synchro transmitter.

(xxviii) Draw the torque-slip characteristic of D.C. servomotor.

(xxix) Draw the torque-slip characteristic of A.C. servomoter.

(xxx) Write short note on synchro control transformer.

(xxxi) What is a synchro transmitter?

(*xxxii*) Draw the torque-speed characteristics of a two phase servomotor.

(*xxxiii*) What are the two varieties of stepper motors?

(*xxxiv*) Explain the use of synchros as transmission of position over long distances.

(*xxxv*) Distinguish between d.c motor and d.c. servomotors.

(*xxxvi*) What is the function of tachogenerator?

(*xxxvii*) Gives the applications of A.C. tachogenerator.

(*xxxviii*) Explain the use of synchro pair as an error detector.

(*xxxix*) What are the applications of synchro?

(*xl*) Give a trade name of synchro.

□□□

2 TIME DOMAIN ANALYSIS

2.1. INTRODUCTION

Any system containing energy storing element like inductor, capacitor, mass and inertia, etc., posses certain energy. These energy storing elements are the part of the control system and cannot be avoided. If the energy state of the system is disturbed then it takes a certain time to change from one state to another state. This disturbance sometimes occurs at input, sometimes occurs at output and sometimes at both ends. The time required to change from one state to another state is known as transient time and the values of current and voltages during this period is called *transient response*. These transient may have oscillations which may be either sustained or decaying in nature. This will depend upon the parameters of the system. For any system we obtain a linear differential equation. The solution of linear differential equation gives the response of the system.

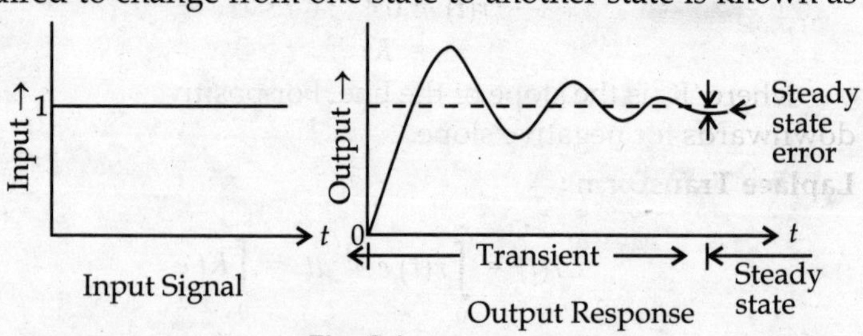

Fig. 2.1.

Thus, the time response of a control system is divided into two parts: (*a*) transient response, (*b*) steady state response.

From the Fig. 2.1 it is clear that the transient response is the part of the response which goes to zero as time increases and steady state response is the part of the total response after transient has died. If the steady state response of the output does not match with the input then the system has steady state error.

2.2. TEST INPUT SIGNALS FOR TRANSIENT ANALYSIS

For the analysis of time response of a control system, the following input signals are used.

1. Step Function

Consider an independent voltage source is in series with a switch 's'. When switch was open the voltage at terminals 1 – 2 is zero. Mathematically

$$V(t) = 0 \qquad -\infty < t < 0$$

When switch is closed at $t = 0$

$$V(t) = K \qquad 0 < t < \infty$$

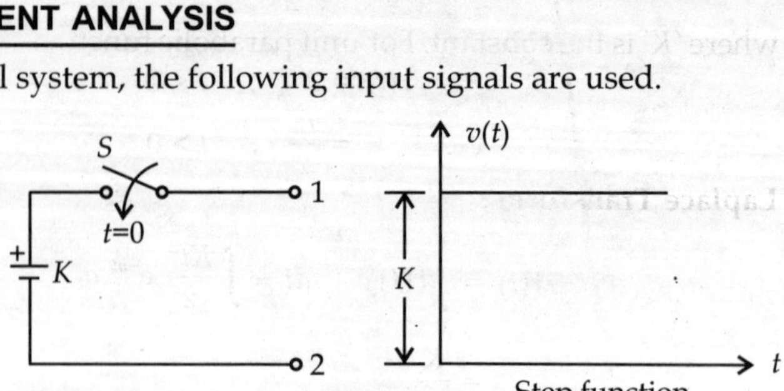

Fig. 2.2.

Combining above two equations

$$V(t) = 0 \quad ; \quad -\infty < t < 0$$
$$= K \quad ; \quad 0 < t < \infty$$

A unit step function is denoted by $u(t)$ and is defined as

$$u(t) = 0 \quad ; \quad t \le 0$$
$$= 1 \quad ; \quad 0 \le t$$

Laplace Transform: let $f(t)$ be defined in the interval $0 \le t \le \infty$ Laplace transform is obtained by multiplying $f(t)$ by e^{-st} and integrate between the limit 0 *to* ∞.

$$£ f(t) = \int_0^\infty f(t)e^{-st} = \int_0^\infty u(t)e^{-st}\,dt = \int_0^\infty 1.\,e^{-st}\,dt = \left[\frac{e^{-st}}{-s}\right]_0^\infty = \frac{1}{s}$$

Step function is also called displacement function. Step function can be described as sudden application of input signal to a system.

If input is $R(s)$, then

$$R(S) = \frac{1}{s}$$

2. Ramp Function

Ramp function starts from origin and increases or decreases linearly with time, as shown in Fig. 2.3. Let $r(t)$ be the ramp function then

$$r(t) = 0 \quad ; \quad t < 0$$
$$= Kt \quad ; \quad t > 0$$

where 'K' is the slope of the line. For positive value of 'K' the slope is upward and the slope is downwards for negative slope.

Laplace Transform:

$$£ r(t) = \int_0^\infty r(t)\,e^{-st}\,dt = \int_0^\infty Kt\,e^{-st}\,dt = \frac{K}{s^2}$$

$$R(S) = \frac{K}{s^2}$$

For, unit ramp K = 1

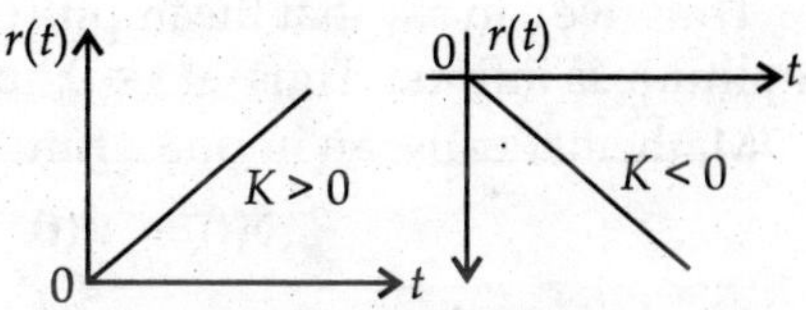

Fig. 2.3. Ramp function

3. Parabolic Function

The value of $r(t)$ is zero for $t < 0$ and is quadratic function of time for $t > 0$. The parabolic function is defined as

$$r(t) = 0 \quad ; \quad t < 0$$
$$= \frac{Kt^2}{2} \quad ; \quad t > 0$$

where 'K' is the constant. For unit parabolic function $K = 1$. The unit parabolic function is defined as

$$r(t) = 0 \quad ; \quad t < 0$$
$$= \frac{t^2}{2} \quad ; \quad t > 0$$

Laplace Transform:

$$£ r(t) = \int_0^\infty r(t)\,e^{-st}\,dt = \int_0^\infty \frac{Kt^2}{2}\,e^{-st}\,dt$$
$$= \frac{K}{s^3}$$

$$R(S) = \frac{K}{s^3}$$

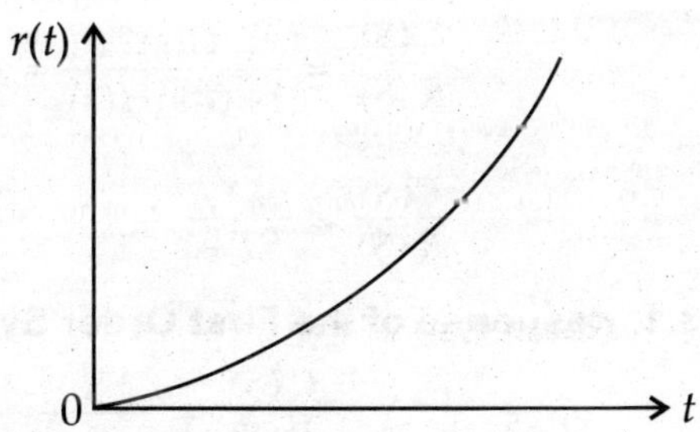

Fig. 2.4. Parabolic function

The parabolic function is shown in Fig. 2.4

4. Impulse Function

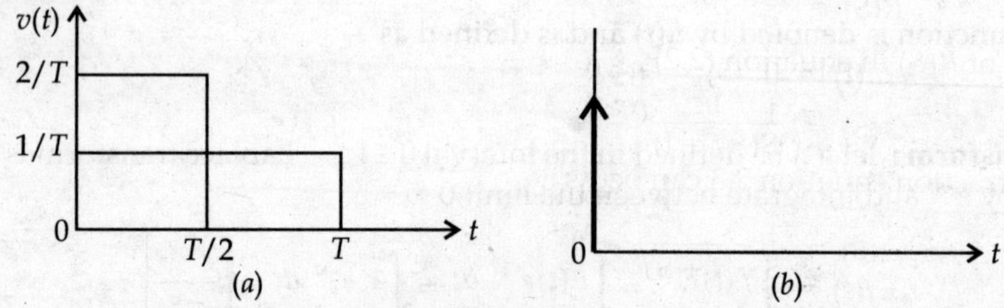

Fig. 2.5.

Consider the Fig. 2.5, the first pulse has a width T and height $\frac{1}{T}$ such that area of the pulse is $T \times \frac{1}{T} = 1$. If we halve the duration and double the amplitude we get second pulse. The area under the second pulse is also unity. We can say the duration of the pulse approaches zero, the amplitude approaches infinity but the area of the pulse is unity. The pulse for which the duration tends to zero and amplitude tends to infinity is called the *impulse function*. Impulse function is also known as delta function,

A unit impulse function is defined as

$$\delta(t) = \begin{bmatrix} 0 & ; & t \neq 0 \\ \infty & ; & t = 0 \end{bmatrix} \qquad \int_{-\infty}^{\infty} \delta(t)\, dt = 1$$

Thus, we can say that the impulse function has zero value everywhere except at $t = 0$, where the amplitude is infinite. Thus, at $t = 0$, unit impulse has infinite amplitude and area equal to unity.

Mathematically, an impulse function is the derivative of a step function *i.e.*

$$\delta(t) = \dot{u}(t)$$

$$£\,\delta(t) = £\frac{d}{dt}[u(t)] = s\,£\text{ (unit step function)}$$

$$= s \cdot \frac{1}{s} = 1$$

2.3. TIME RESPONSE OF A FIRST ORDER SYSTEM

Consider a first order system with unity feedback as shown in Fig. 2.6.

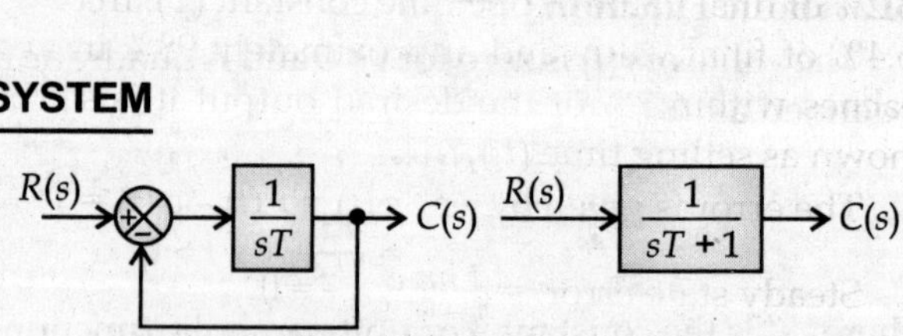

Fig. 2.6. Block diagram of first order system

$$G(S) = \frac{1}{sT}$$

$$H(S) = 1$$

$$\frac{C(S)}{R(S)} = \frac{G(s)}{1 + G(s)H(s)} = \frac{1/ST}{1 + \frac{1}{ST} \cdot 1} = \frac{1/ST}{(ST+1)/ST} = \frac{1}{ST+1}$$

$$\therefore \quad \frac{C(S)}{R(S)} = \frac{1}{ST+1} \qquad \text{...(2.1)}$$

2.3.1. Response of the First Order System with Unit Step Input

$$\frac{C(S)}{R(S)} = \frac{1}{ST+1}$$

$$\therefore \quad C(S) = \frac{1}{ST+1} R(S) \qquad \text{...(2.2)}$$

Since, input is the unit step function

$$\therefore \quad R(S) = \frac{1}{S}$$

Put the value of $R(S)$ in equation (2.2)

$$C(S) = \frac{1}{S(ST+1)} \qquad ...(2.3)$$

After partial fraction equation 2.3 can be written as

$$C(S) = \frac{1}{S} - \frac{T}{1+ST} \qquad ...(2.4)$$

or,

$$C(S) = \frac{1}{S} - \frac{1}{S+\frac{1}{T}} \qquad ...(2.5)$$

Inverse Laplace of equation 2.5

$$£^{-1} C(S) = £^{-1}\frac{1}{S} - £^{-1}\frac{1}{S+1/T}$$

$$C(t) = 1 - e^{-t/T} \qquad ...(2.6)$$

When $t = T$ in equation 2.6

$$C(t) = 1 - e^{-T/T} = 1 - e^{-1} = 0.632 \text{ or } 63.2\ \%$$

where, 'T' is known as time constant and it is defined as the time required for the signal to attain 63.2% of final or steady state value. Time constant indicates how fast the system reaches the final value. Smaller the time constant, faster is the system response. A large time constant corresponds to a sluggish system (slow moving).

Since, the output increases exponentially from zero to final value the slope of the curve at $t = 0$ is

$$\left.\frac{dc}{dt}\right|_{t=0} = \frac{1}{T} e^{-t/T}$$

at $t = 0$

$$\frac{dc}{dt} = \frac{1}{T} e^{\circ} = \frac{1}{T}$$

From the exponential curve it is clear that, the magnitude of the output response equals to 63.2% of final value in one time constant (T). In two time constant the magnitude of output reaches 86.4% of final value and approximately 98% in four time constant ($4T$). When the actual output reaches within 2% of the desired output it is said that steady state has reached. The time $4\ T$ is known as setling time (t_s).

The error is given by $\quad e(t) = r(t) - c(t) = 1 - (1 - e^{-t/T}) = e^{-t/T}$

Steady state error $= \lim_{t \to \infty} e^{-t/T} = 0$

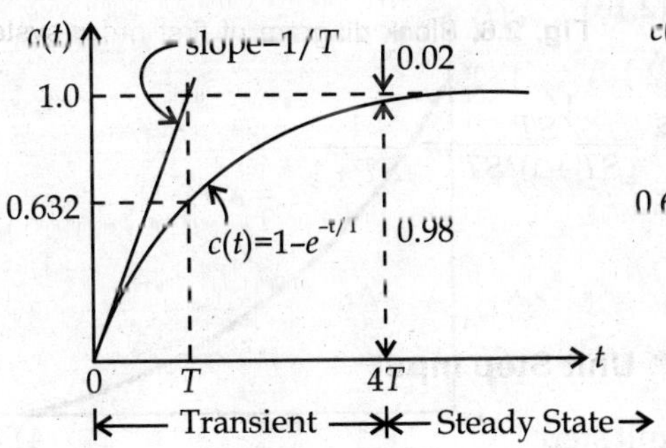

Exponential curve

Fig. 2.7.

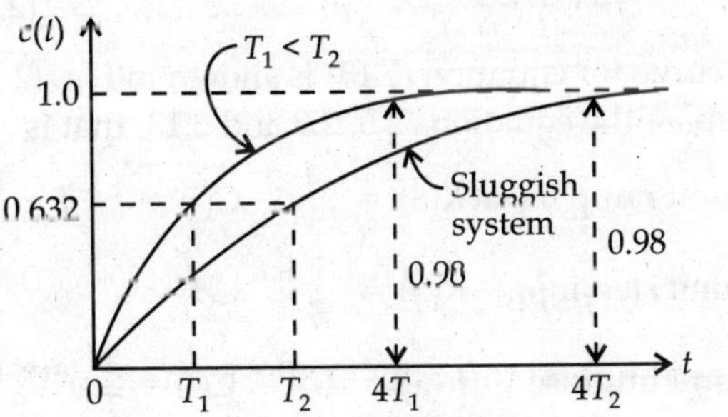

Exponential curves for two systems having different time constants

Fig. 2.8.

2.3.2. Response of the First Order System with Unit Ramp Function

Since, input is the unit ramp $R(S) = \frac{1}{S^2}$

Put the value of $R(S)$ in equation 2.2

$$C(S) = \frac{1}{ST+1} \cdot \frac{1}{S^2} = \frac{1}{S^2(1+ST)} \quad ...(2.7)$$

After partial fraction equation 2.7 can be written as

$$C(S) = \frac{1-ST}{S^2} + \frac{T^2}{1+ST} = \frac{1}{S^2} - \frac{T}{S} + T \cdot \frac{1}{S+\frac{1}{T}} \quad ...(2.8)$$

Inverse laplace of equation 2.8

$$£^{-1} C(S) = £^{-1}\frac{1}{S^2} - £^{-1}\frac{T}{S} + £^{-1}T \cdot \frac{1}{S+1/T}$$

$$C(t) = t - T + T.e^{-t/T} \quad \text{For} \quad t \geq 0 \quad ...(2.9)$$

The error signal $e(t)$ will be

$$e(t) = r(t) - c(t) = t - [t - T + Te^{-t/T}]$$

$$e(t) = T(1 - e^{-t/T}) \quad ...(2.10)$$

$$\text{Steady state error} = \lim_{t \to \infty}(T - T\,e^{-t/T})$$

$$= T \quad ...(2.11)$$

The steady state error is equal to 'T', where 'T' is the time constant of the system. For smaller time constant steady state error will be small and the speed of the response will increase.

2.3.3. Response of the First Order System with Unit Impulse Function

Input is unit impulse function *i.e.*, $R(S) = 1$

Put the value of R(S) in equation 2.2

$$C(S) = \frac{1}{ST+1} \cdot R(S) = \frac{1}{ST+1} \cdot 1$$

or,

$$C(S) = \frac{1}{T} \cdot \frac{1}{S+1/T} \quad ...(2.12)$$

Inverse laplace of 2.12

$$£^{-1} C(S) = £^{-1}\left[\frac{1}{T} \cdot \frac{1}{S+1/T}\right]$$

$$C(t) = \frac{1}{T}e^{-t/T} \quad ...(2.13)$$

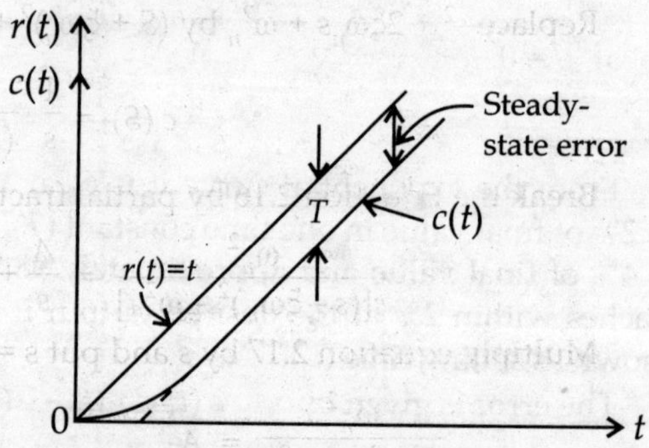

Fig. 2.9. Unit Ramp response of the system

The curve for equation (2.13) is shown in Fig. (2.10)

Compare the equations 2.6, 2.9 and 2.13, that is

For unit ramp input $R(S) = \frac{1}{S^2}$ $\quad C(t) = t - T + T\,e^{-t/T}$

For unit step input $R(S) = \frac{1}{S}$ $\quad C(t) = 1 - e^{-t/T}$

For unit impulse $R(S) = 1$ $\quad C(t) = \frac{1}{T}e^{-t/T}$

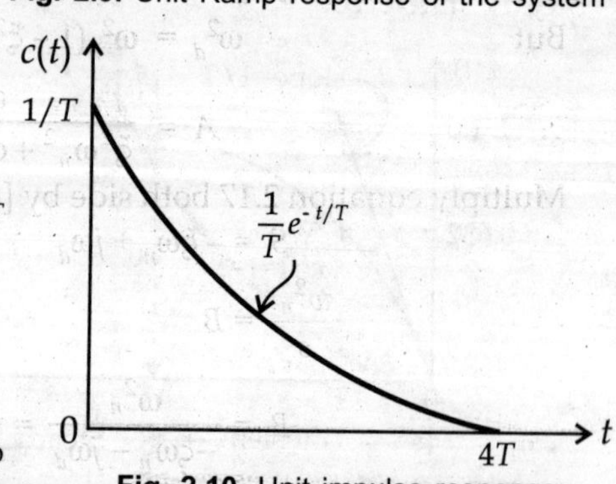

Fig. 2.10. Unit impulse response of the system

From above three equations, it is clear that the unit step input is the derivative of unit ramp input and unit impulse input is the derivative of unit step input. Thus, the response to the derivative of an input signal can be obtained by differentiating the response of the system to the original signal. This is the property of the linear time-invariant systems.

2.4. TIME RESPONSE OF SECOND ORDER SYSTEM

The block diagram of second order control system is shown in Fig. 2.11.

$$\frac{C(S)}{R(S)} = \frac{G(S)}{1+G(S)H(S)}$$

$$G(S) = \frac{\omega_n^2}{s(s+2\xi\omega_n)}, \quad H(S) = 1$$

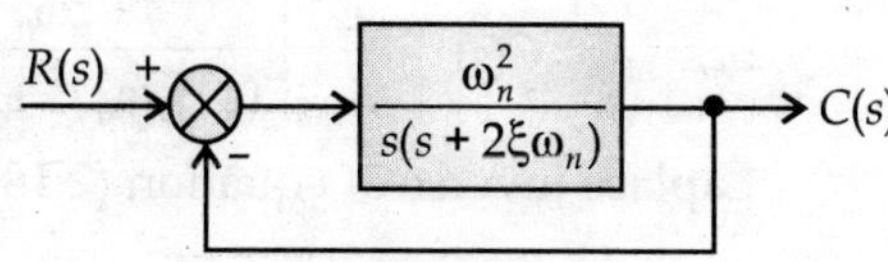

Fig. 2.11.

$$\therefore \quad \frac{C(S)}{R(S)} = \frac{\omega_n^2/s(s+2\xi\omega_n)}{1+\dfrac{\omega_n^2}{s(s+2\xi\omega_n)}} = \frac{\omega_n^2}{s^2+2\xi\omega_n s+\omega_n^2} \qquad \text{...(2.14)}$$

2.4.1. Time Response of Second Order System with Unit Step Input

From equation (2.14)

$$C(S) = \frac{\omega_n^2}{S^2+2\xi\omega_n S+\omega_n^2}.R(S)$$

For unit step input $\quad R(S) = \frac{1}{s}$

$$\therefore \quad c(s) = \frac{1}{s}\cdot\frac{\omega_n^2}{s^2+2\xi\omega_n s+\omega_n^2} \qquad \text{...(2.15)}$$

Replace $s^2 + 2\xi\omega_n s + \omega_n^2$ by $(S + \xi\omega_n)^2 + \omega_n^2(1 - \xi^2)$

$$c(S) = \frac{1}{s}\cdot\frac{\omega_n^2}{(s+\xi\omega_n)^2+\omega_n^2(1-\xi^2)} \qquad \text{...(2.16)}$$

Break the Equation 2.16 by partial fraction, put $\omega_d^2 = \omega_n^2(1 - \xi^2)$

$$\frac{\omega_n^2}{s[(s+\xi\omega_n)^2+\omega_d^2]} = \frac{A}{s}+\frac{B}{(s+\xi\omega_n)^2+\omega_d^2} \qquad \text{...(2.17)}$$

Multiply equation 2.17 by s and put s = 0

$$\frac{\omega_n^2}{\xi^2\omega_n^2+\omega_d^2} = A$$

But $\quad \omega_d^2 = \omega_n^2(1 - \xi^2) = \omega_n^2 - \omega_n^2\xi^2$

$$\therefore \quad A = \frac{\omega_n^2}{\xi^2\omega_n^2+\omega_n^2-\xi^2\omega_n^2} = 1 \qquad \therefore \quad \boxed{A = 1}$$

Multiply equation 2.17 both side by $[(S + \xi\omega_n)^2 + \omega_d^2]$ and put

$$S = -\xi\omega_n + j\omega_d$$

$$\frac{\omega_n^2}{s} = B$$

$$B = \frac{\omega_n^2}{-\xi\omega_n - j\omega_d} = \frac{-\omega_n^2}{\xi\omega_n + j\omega_d} = \frac{-\omega_n^2(\xi\omega_n - j\omega_d)}{(\xi\omega_n + j\omega_d)(\xi\omega_n - j\omega_d)} = -(\xi\omega_n - j\omega_d)$$

put $\quad -j\omega_d = s + \xi\omega_n$

$$B = -(\xi\omega_n + s + \xi\omega_n) = -(S + 2\xi\omega_n)$$

Equation 2.16 can be written as

$$C(S) = \frac{1}{s} - \frac{s+2\xi\omega_n}{(s+\xi\omega_n)^2+\omega^2{}_d} = \frac{1}{s} - \frac{s+\xi\omega_n+\xi\omega_n}{(s+\xi\omega_n)^2+\omega^2{}_d}$$

or,
$$C(s) = \frac{1}{s} - \left[\frac{s+\xi\omega_n}{(s+\xi\omega_n)^2+\omega^2{}_d} + \frac{\xi\omega_n}{\omega_d}\cdot\frac{\omega_d}{(s+\xi\omega_n)^2+\omega^2{}_d}\right] \quad \text{...(2.18)}$$

Laplace inverse of equation (2.18)

$$C(t) = 1 - \left[e^{-\xi\omega_n t}.\cos\omega_d t + \frac{\xi\omega_n}{\omega d}e^{-\xi\omega_n t}.\sin\omega_d t\right] \quad \text{...(2.19)}$$

Put $\omega_d = \omega_n\sqrt{1-\xi^2}$

$$C(t) = 1 - e^{-\xi\omega_n t}\left[\cos\omega_d t + \frac{\xi}{\sqrt{1-\xi^2}}\cdot\sin\omega_d t\right] = 1 - \frac{e^{-\xi w_n t}}{\sqrt{1-\xi^2}}\left[\sqrt{1-\xi^2}.\cos\omega_d t + \xi\sin\omega_d t\right]$$

Put $\sqrt{1-\xi^2} = \sin\phi$

$\therefore$ $\cos\phi = \xi, \tan\phi = \dfrac{\sqrt{1-\xi^2}}{\xi}$

$$\therefore \quad C(t) = 1 - \frac{e^{-\xi\omega_n t}}{\sqrt{1-\xi^2}}[\sin\phi\cdot\cos\omega_d t + \cos\phi\sin\omega_d t] = 1 - \frac{e^{-\xi\omega_n t}}{\sqrt{1-\xi^2}}\sin(\omega_d t + \phi)$$

Put the values of ω_d and ϕ *i.e.*

$$\omega_d = \omega_n\sqrt{1-\xi^2}, \ \phi = \tan^{-1}\frac{\sqrt{1-\xi^2}}{\xi}$$

$$C(t) = 1 - \frac{e^{-\xi\omega_n t}}{\sqrt{1-\xi^2}}\sin\left[\left(\omega_n\sqrt{1-\xi^2}\right)t + \tan^{-1}\frac{\sqrt{1-\xi^2}}{\xi}\right] \quad \text{...(2.20)}$$

The error signal for the system

$$e(t) = r(t) - c(t)$$

$$\therefore \quad e(t) = \frac{e^{-\xi\omega_n t}}{\sqrt{1-\xi^2}}\sin\left[\left(\omega_n\sqrt{1-\xi^2}\right)t + \tan^{-1}\frac{\sqrt{1-\xi^2}}{\xi}\right] \quad \text{...(2.21)}$$

The steady state value of $c(t)$

$$e_{ss} = \lim_{t\to\infty} c(t) = 1$$

Therefore at steady state there is no error between input and output.

ω_n = natural frequency of oscillation or undamped natural frequency

ω_d = damped frequency of oscillation

$\xi\omega_n$ = damping factor or actual damping or damping coefficient.

(*a*) Underdamped case ($0 < \xi < 1$)

From the expression 2.20, it is clear that the time constant is $1/\xi\omega_n$ and the response having damped oscillation with overshoot and undershoot. Such response is known as underdamped response. The response is shown in Fig. 2.12.

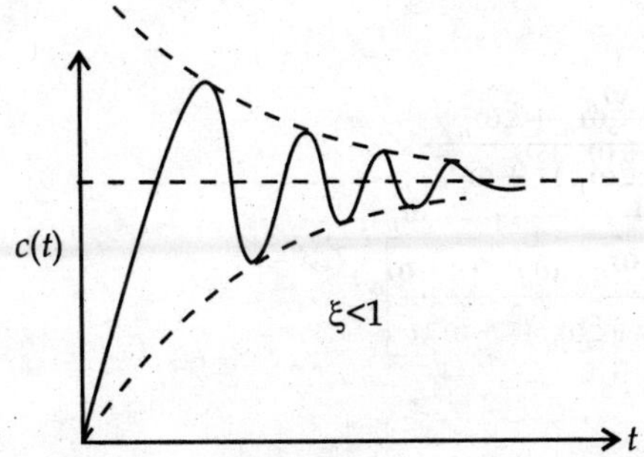

Fig. 2.12.

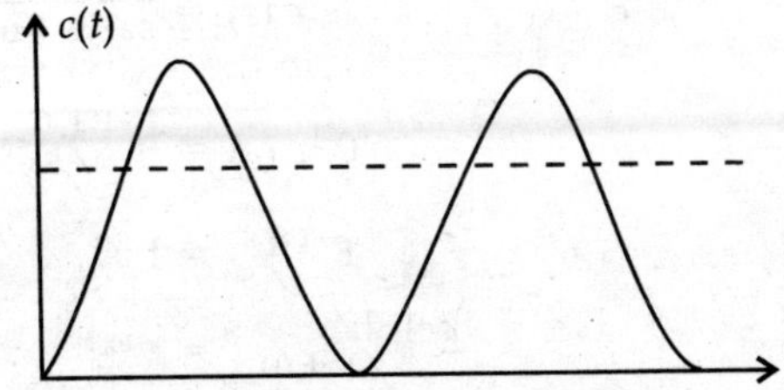

Fig. 2.13. Sustained oscillation

(b) When ξ = 0, undamped case

The expression 2.20 will be

$$C(t) = 1 - \sin(\omega_n t + \pi/2)$$

$$C(t) = 1 - \cos\omega_n t \quad \text{...(2.22)}$$

Thus at ω_n the system will oscillate (with ξ = 0). The damped frequency always less than the undamped frequency (ω_n) because of factor ξ. If the system having certain value of ξ then it is not possible to measure undamped natural frequency experimentally. The observed frequency is the damped frequency (ω_d) which is equal to $\omega_n\sqrt{1-\xi^2}$. The response is shown in Fig. 2.13.

(c) ξ = 1 critically damped case :

Put ξ = 1 in equation 2.15

$$C(S) = \frac{1}{s}\cdot\frac{\omega^2{}_n}{s^2+2\omega_n s+\omega^2{}_n} = \frac{1}{s}\cdot\frac{\omega_n{}^2}{(s+\omega_n)^2} \quad \text{...(2.23)}$$

Break the equation 2.23 by partial fraction

$$\frac{\omega_n{}^2}{s(s+\omega_n)^2} = \frac{A}{s}+\frac{B}{s+\omega_n}+\frac{C}{(s+\omega_n)^2} \quad \text{...(2.24)}$$

Multiply both the sides by $(s + \omega_n)^2$ and put $s = -\omega_n$

$$\frac{\omega^2{}_n}{s} = \frac{A(s+\omega_n)^2}{s}+B(s+\omega_n)+c \quad \text{...(2.25)}$$

$$\boxed{C = -\omega_n}$$

Multiply both the sides by 's' and put s = 0

$$\frac{\omega^2{}_n}{(s+\omega_n)^2} = A+\frac{Bs}{s+\omega_n}+\frac{cs}{(s+\omega_n)^2}$$

$$\boxed{A = 1}$$

Differentiate equation 2.25 and put $s = -\omega_n$

$$\frac{-\omega_n{}^2}{s^2} = A\,\frac{s+\omega_n}{s}+B$$

$$\boxed{B = -1}$$

$$\therefore \quad \frac{\omega^2{}_n}{s(s+\omega_n)^2} = \frac{1}{s}-\frac{1}{s+\omega_n}-\frac{\omega_n}{(s+\omega_n)^2} \quad \text{...(2.25a)}$$

Inverse laplace of equation 2.25

$$c(s) = \frac{1}{s} - \frac{1}{s+\omega_n} - \frac{\omega_n}{(s+\omega_n)^2} \quad ...(2.26)$$

$$£^{-1}c(s) = £^{-1}\,{}^{1}/_{s} - £^{-1}\frac{1}{s+\omega_n} - £^{-1}\frac{\omega_n}{(s+\omega_n)^2} \quad ...(2.27)$$

$$£^{-1}\,{}^{1}/_{s} = 1$$

$$£^{-1}\,{}^{1}/_{s+\omega_n} = e^{-\omega_n t}$$

Inverse laplace of $\frac{\omega_n}{(s+\omega_n)^2}$ can be obtained by the method of residues.

$$£^{-1}\frac{\omega_n}{(s+\omega_n)^2} = \frac{1}{\lfloor 2-1}\frac{d}{ds}\left[(s+\omega_n)\,e^{st}\frac{\omega_n}{(s+\omega_n)^2}\right]_{s=-\omega_n}$$

$$= \omega_n\,[e^{st}.t]_{s=-\omega n}$$

$$= t\omega_n\,e^{-\omega_n t}$$

Put all values in equation (2.27)

$$\therefore \quad £^{-1}\,c(s) = c(t) = 1 - e^{-\omega_n t} - t\,\omega_n e^{-\omega_n t}$$

$$= 1 - e^{-\omega_n t}(1 + \omega_n t) \quad ...(2.28)$$

From the expression 2.20, it is clear that $\xi\omega_n$ is the actual damping for $\xi = 1$, the actual damping $= \omega_n$. This actual damping when $\xi = 1$ is known as CRITICAL DAMPING. At the value of critical damping the oscillations just disappeared. The ratio of actual damping to the critical damping is known as damping ratio.

i.e.,

$$\frac{\text{Actual damping}}{\text{critical damping}} = \frac{\xi\omega_n}{\omega_n} = \xi$$

From equation 2.28, the time constant of the system $= 1/\xi\omega_n$

The response of the equation 2.28 is shown in Fig. 2.14.

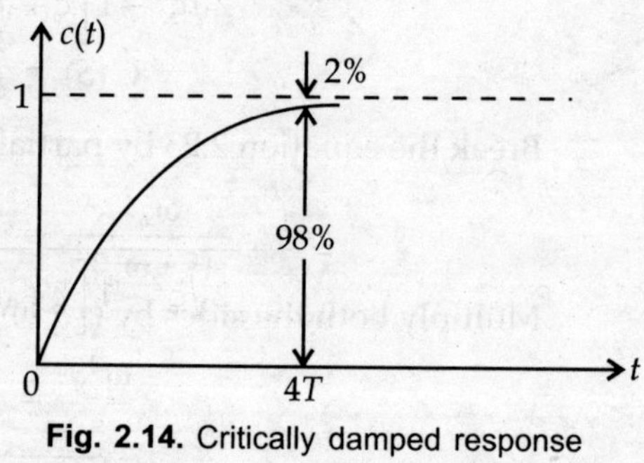

Fig. 2.14. Critically damped response

(*d*) $\xi > 1$ overdamped case

For $\xi > 1$, equation 2.16 can be written as

$$c(s) = \frac{1}{s}\cdot\frac{\omega_n^2}{(s+\xi\omega_n)^2 - \omega_n^2(\xi^2-1)} \quad ...(2.29)$$

put

$$\omega_d^2 = \omega_n^2(\xi^2 - 1)$$

$$c(s) = \frac{1}{s}\cdot\frac{\omega_n^2}{(s+\xi\omega_n)^2 - \omega_d^2} \quad ...(2.30)$$

Break the equation (2.30) by partial fraction

Equation 2.30 can be written as

$$c(s) = \frac{\omega_n^2}{s(s+\xi\omega_n+\omega_d)(s+\xi\omega_n-\omega_d)}$$

$$\frac{\omega_n^2}{s(s+\xi\omega_n+\omega_d)(s+\xi\omega_n-\omega_d)} = \frac{A}{s} + \frac{B}{s+\xi\omega_n+\omega_d} + \frac{C}{s+\xi\omega_n-\omega_d} \quad ...(2.31)$$

Multiply both the sides by 's' and put s = 0

$$\boxed{A = 1}$$

Multiply both the sides of equation 2.31 by $s + \xi\omega_n + \omega_d$ and put $S = -\xi\omega_n - \omega_d$

$$\frac{\omega_n^{\ 2}}{s(s+\xi\omega_n-\omega_d)} = B$$

$$\therefore \quad B = \frac{\omega_n^{\ 2}}{(-\xi\omega_n-\omega_d)(-\xi\omega_n-\omega_d+\xi\omega_n-\omega_d)} = \frac{\omega_n^{\ 2}}{(\xi\omega_n+\omega_d)(2\omega_d)}$$

Put $\omega^2_d = \omega^2_n(\xi^2 - 1)$ and simplify for B

$$B = \frac{1}{2\xi\sqrt{\xi^2-1}+2(\xi^2-1)} = \frac{1}{2\xi\sqrt{\xi^2-1}+2\sqrt{\xi^2-1}\sqrt{\xi^2-1}} = \frac{1}{2\sqrt{\xi^2-1}\left[\xi+\sqrt{\xi^2-1}\right]}$$

Similarly, $C = \dfrac{-1}{2\sqrt{\xi^2-1}\left[\xi-\sqrt{\xi^2-1}\right]}$

Put the values of A, B and C in 2.31

$$c(s) = \frac{1}{s} + \frac{1}{2\sqrt{\xi^2-1}\left[\xi+\sqrt{\xi^2-1}\right]\left[s+\xi\omega_n+\omega_d\right]} - \frac{1}{2\sqrt{\xi^2-1}\left[\xi-\sqrt{\xi^2-1}\right]\left[s+\xi\omega_n-\omega_d\right]}$$

Put the value of ω_d.

$$c(s) = \frac{1}{s} + \frac{1}{2\sqrt{\xi^2-1}\left[\xi+\sqrt{\xi^2-1}\right]\left[s+\xi\omega_n+\omega_n\sqrt{\xi^2-1}\right]} - \frac{1}{2\sqrt{\xi^2-1}\left[\xi-\sqrt{\xi^2-1}\right]\left[s+\xi\omega_n-\omega_n\sqrt{\xi^2-1}\right]}$$

$$= \frac{1}{s} + \frac{1}{2\sqrt{\xi^2-1}\left[\xi+\sqrt{\xi^2-1}\right]\left[s+\omega_n\left(\xi+\sqrt{\xi^2-1}\right)\right]} - \frac{1}{2\sqrt{\xi^2-1}\left[\xi-\sqrt{\xi^2-1}\right]\left[s+\omega_n\left(\xi-\sqrt{\xi^2-1}\right)\right]} \quad ...(2.32)$$

Inverse laplace of (2.32)

$$c(t) = 1 + \frac{e^{-\left(\xi+\sqrt{\xi^2-1}\right)\omega_n t}}{2\sqrt{\xi^2-1}\left(\xi+\sqrt{\xi^2-1}\right)} - \frac{e^{-\left(\xi-\sqrt{\xi^2-1}\right)\omega_n t}}{2\sqrt{\xi^2-1}\left(\xi-\sqrt{\xi^2-1}\right)} \quad ...(2.33)$$

From equation 2.33, we get two time constant

$$T_1 = \frac{1}{\left(\xi+\sqrt{\xi^2-1}\right)\omega_n}, \quad T_2 = \frac{1}{\left(\xi-\sqrt{\xi^2-1}\right)\omega_n}$$

From equation 2.33 we observe that when ξ is greater than one there are two exponential term, the first term has a time constant T_1 which is smaller than the time constant of other exponential term (having time constant T_2), in other words we can say that the first exponential term decaying much faster than other exponential term. So, for time response we can neglect it, then

$$C(t) = 1 - \frac{e^{-\left(\xi - \sqrt{\xi^2-1}\right)\omega_n t}}{2\sqrt{\xi^2-1}\left(\xi - \sqrt{\xi^2-1}\right)} \quad ...(2.34)$$

and, time constant

$$= \frac{1}{(\xi - \sqrt{\xi^2-1})\omega_n} \quad ...(2.35)$$

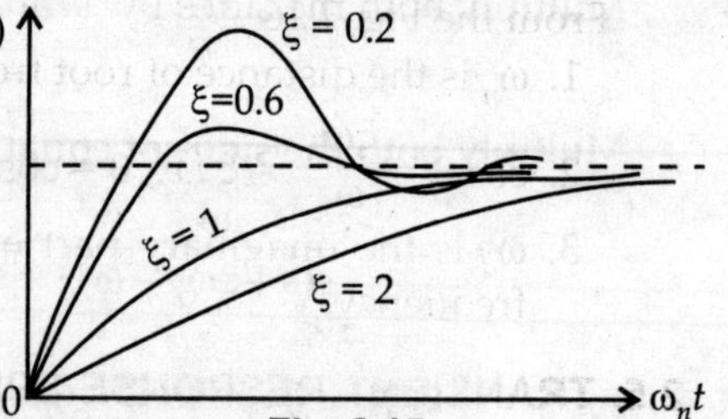

Fig. 2.15.

For different values of ξ the curves of $C(t)$ is shown in Fig. 2.15.
Form the curve it is clear that the overdamped systems are sluggish.

2.4.2. Location of Roots of Characteristic Equation and Time Response

The characteristic equation of

$$\frac{C(S)}{R(S)} = \frac{\omega^2{}_n}{s^2 + 2\xi\omega_n s + \omega^2{}_n}$$

is $\quad s^2 + 2\xi\omega_n s + \omega^2{}_n = 0 \quad ...(2.36)$

The roots of equation 2.36 $S_1 = -\xi\omega_n + j\omega_n\sqrt{1-\xi^2}$ and $S_2 = -\xi\omega_n - j\omega_n\sqrt{1-\xi^2}$

The real part of the roots ($-\xi\omega_n$) represents the damping and imaginary part $\left(\omega_n\sqrt{1-\xi^2}\right)$ represents the damped frequency.

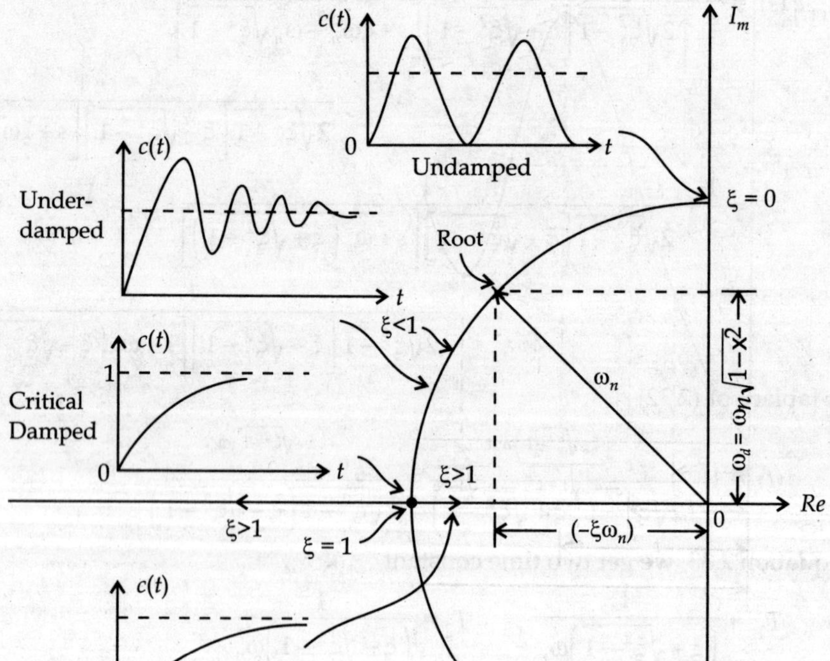

Fig. 2.16.

From the Fig. 2.16

1. ω_n is the distance of root from origin.
2. $\cos\theta = \frac{\xi\omega_n}{\omega_n} = \xi$ *i.e.*, $\theta = \cos^{-1}\xi$ (when the roots are in left-half of s-plane).
3. ω_d is the imaginary part of the root and is known as damped frequency or conditional frequency.

2.5. TRANSIENT RESPONSE SPECIFICATIONS OF SECOND ORDER SYSTEM

The performance of a control system are express in terms of the transient response to a unit step input because it is easy to generate. The transient response of a control system to a unit step input depends upon the initial conditions. Consider a second order system with unit step input and the system initially at rest *i.e.*, all initial conditions are zero. The following are the common transient response characteristics.

1. Delay time (t_d)
2. Rise time (t_r)
3. Peak time (t_p)
4. Maximum overshoot (M_p)
5. Settling time (t_s)
6. Steady state error (e_{ss})

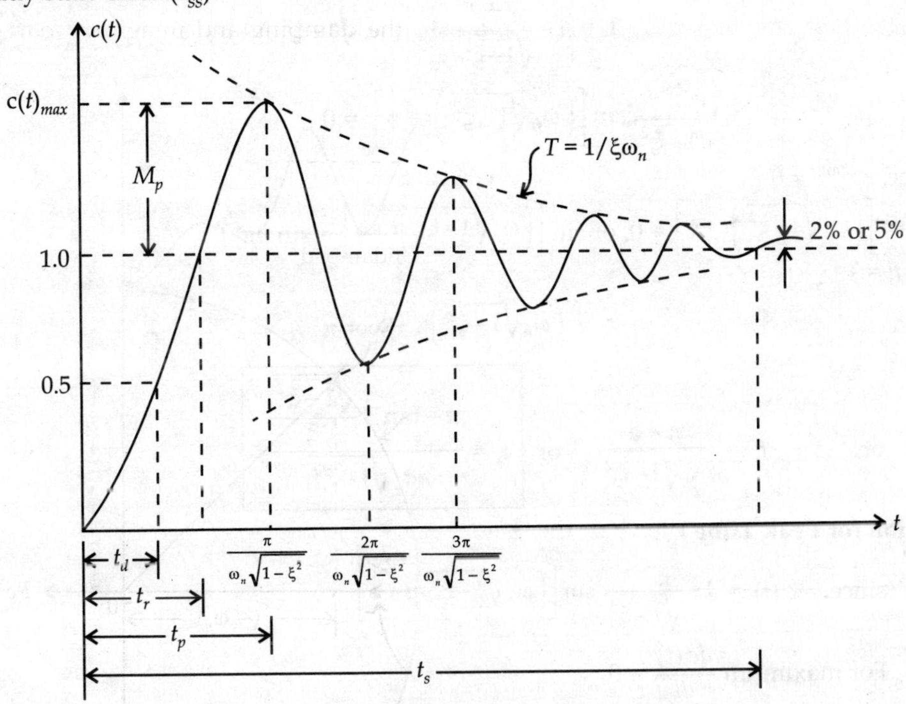

Fig. 2.17.

1. **Delay Time (t_d) :** The delay time is the time required for the response to reach 50% of the final value in first time.
2. **Rise Time (t_r) :** It is the time required for the response to rise from 10% to 90% of its final value for overdamped systems and 0 to 100% for underdamped systems.
3. **Peak Time (t_p) :** The peak time is the time required for the response to reach the first peak of the time response or first peak overshoot.

4. **Maximum Ocershoot (M_p) :** It is the normalized difference between the peak of the time response and steady output. The maximum percent overshoot is defined by

$$\text{Maximum percent overshoot} = \frac{C(t_p) - C(\infty)}{C(\infty)} \times 100$$

5. **Settling Time (t_s) :** The settling time is the time required for the response to reach and stay within the specified range (2% to 5%) of its final value.
6. **Steady State Error (e_{ss}) :** It is the difference between actual output and desired output as time 't' tends to infinity.

$$e_{ss} = \lim_{t\to\infty}[r(t) - C(t)]$$

Expression for Rise Time(t_r) :

From the expression 2.20

$$C(t) = 1 - \frac{e^{-\xi\omega_n t}}{\sqrt{1-\xi^2}} \sin\left[\left(\omega_n\sqrt{1-\xi^2}\right)t + \phi\right]$$

$$\text{where } \phi = \tan^{-1}\frac{\sqrt{1-\xi^2}}{\xi}$$

let response reaches 100% of desired value. Put $c(t) = 1$

$$1 = 1 - \frac{e^{-\xi\omega_n t}}{\sqrt{1-\xi^2}} \sin\left[\left(\omega_n\sqrt{1-\xi^2}\right)t + \phi\right]$$

or,
$$\frac{e^{-\xi\omega_n t}}{\sqrt{1-\xi^2}} \sin\left[\left(\omega_n\sqrt{1-\xi^2}\right)t + \phi\right] = 0$$

Since, $e^{-\xi\omega nt} \neq 0$

$$\therefore \quad \sin\left[\left(\omega_n\sqrt{1-\xi^2}\right)t + \phi\right] = 0, \text{ or } \sin\left[\left(\omega_n\sqrt{1-\xi^2}\right)t + \phi\right] = \sin n\pi$$

Put $n = 1$

$$\therefore \quad \left(\omega_n\sqrt{1-\xi^2}\right)t_r + \phi = \pi$$

$$\text{or,} \quad t_r = \frac{\pi - \phi}{\omega_n\sqrt{1-\xi^2}} \quad \text{or} \quad \boxed{t_r = \frac{\pi - \tan^{-1}\frac{\sqrt{1-\xi^2}}{\xi}}{\omega_n\sqrt{1-\xi^2}}} \qquad \text{...(2.37)}$$

Expression for Peak Time t_p :

$$\text{since,} \quad c(t) = 1 - \frac{e^{-\xi\omega_n t}}{\sqrt{1-\xi^2}} \sin\left[\left(\omega_n\sqrt{1-\xi^2}\right)t + \phi\right]$$

$$\text{For maximum } \frac{dc(t)}{dt} = 0$$

$$\frac{dc(t)}{dt} = \frac{-e^{-\xi\omega_n t}}{\sqrt{1-\xi^2}} \cos\left[\left(\omega_n\sqrt{1-\xi^2}\right)t + \phi\right]\omega_n\sqrt{1-\xi^2} + \sin\left[\left(\omega_n\sqrt{1-\xi^2}\right)t + \phi\right]\frac{\xi\omega_n}{\sqrt{1-\xi^2}}e^{-\xi\omega_n t} \qquad \text{...(2.38)}$$

Since $e^{-\xi\omega nt} \neq 0$

Equation 2.38 can be writen as

$$\cos\left[\left(\omega_n\sqrt{1-\xi^2}\right)t + \phi\right]\sqrt{1-\xi^2} = \sin\left[\left(\omega_n\sqrt{1-\xi^2}\right)t + \phi\right]\xi \qquad \text{...(2.39)}$$

Put $\sqrt{1-\xi^2} = \sin\phi$ & $\xi = \cos\phi$

Equation 2.39 becomes

$$\cos\left[\left(\omega_n\sqrt{1-\xi^2}\right)t+\phi\right]\sin\phi = \sin\left[\left(\omega_n\sqrt{1-\xi^2}\right)t+\phi\right]\cos\phi$$

or,
$$\sin\left[\left(\omega_n\sqrt{1-\xi^2}\right)t+\phi\right]\cos\phi-\cos\left[\left(\omega_n\sqrt{1-\xi^2}\right)t+\phi\right]\sin\phi = 0$$

or,
$$\sin\left[\omega_n\sqrt{1-\xi^2}\right]t = 0$$

the time to various peaks

$$\left(\omega_n\sqrt{1-\xi^2}\right)t_p = n\pi$$

where $n = 0, 1, 2, 3, ...$

Maximum overshoot identified by putting $n = 1$, therefore the peak time to the first overshoot

$$t_p = \frac{\pi}{\omega_n\sqrt{1-\xi^2}} \quad \text{or} \quad \boxed{t_p = \frac{\pi}{\omega_n\sqrt{1-\xi^2}}} \qquad ...(2.40)$$

The first minimum (undershoot) occurs at $n = 2$

$$t_{min} = \frac{2\pi}{\omega_n\sqrt{1-\xi^2}} \qquad ...(2.41)$$

Expression for Maximum Overshoot : M_P

$$C(t) = 1-\frac{e^{-\xi\omega_n t}}{\sqrt{1-\xi^2}}\sin\left[\left(\omega_n\sqrt{1-\xi^2}\right)t+\phi\right] \qquad ...(2.42)$$

Maximum overshoot occurs at peak time *i.e.* $t = t_p$

Put $t = t_p = \dfrac{\pi}{\omega_n\sqrt{1-\xi^2}}$ in equation 2.42

$$C(t) = 1-\frac{e^{-\xi\omega_n\frac{\pi}{\omega_n\sqrt{1-\xi^2}}}}{\sqrt{1-\xi^2}}\sin\left[\omega_n\sqrt{1-\xi^2}\cdot\frac{\pi}{\omega_n\sqrt{1-\xi^2}}+\phi\right] = 1-\frac{e^{-\frac{\xi\pi}{\sqrt{1-\xi^2}}}}{\sqrt{1-\xi^2}}\sin(\pi+\phi) \qquad ...(2.43)$$

Since $\phi = \tan^{-1}\dfrac{\sqrt{1-\xi^2}}{\xi}$ then $\sin\phi = \sqrt{1-\xi^2}$

and $\sin(\pi+\phi) = -\sin\phi$

$$\therefore \quad C(t) = 1+\frac{e^{-\pi\xi/\sqrt{1-\xi^2}}}{\sqrt{1-\xi^2}}(\sin\phi)$$

$$C(t)_{max} = 1+e^{-\frac{\pi\xi}{\sqrt{1-\xi^2}}}$$

$$\therefore \quad M_P = C(t)_{max}-1$$

$$M_P = e^{-\pi\xi/\sqrt{1-\xi^2}}$$

$$\% M_P = e^{-\frac{\pi\xi}{\sqrt{1-\xi^2}}} \times 100 \quad \text{....(2.44)}$$

Settling Time : t_s :

As shown in the Fig. 2.17, the curves for $1 \pm \frac{e^{-\xi\omega_n t}}{\sqrt{1-\xi^2}}$ are the envelope curves of the transient response for unit step input. The time constant of these envelope curves is $\frac{1}{\xi\omega_n}$. The speed of the decay depends upon the time constant. The settling time for a second order system is approximately four times the time constant $(1/\xi\omega_n)$

$$\therefore \qquad t_s = \frac{4}{\xi\omega_n} \quad \text{...(2.45)}$$

For over damped system, the settling time t_s becomes large because of sluggish start. From 2.45, the settling time is inversly proportional to the product of ξ and ω_n. So, for permissible maximum overshoot, the value of ξ is known therefore the settling time can be determined by undamped natural frequency ω_n.

EXAMPLE 2.1. When a second order control system is subjected to a unit step input, the values of $\xi = 0.5$ and $\omega_n = 6$ rad/sec. Determine the rise time, peak time, settling time and peak overshoot.

Solution : Given that $\xi = 0.5$

$\omega_n = 6$ rad/sec.

1. Rise Time :

$$t_r = \frac{\pi - \tan^{-1}\frac{\sqrt{1-\xi^2}}{\xi}}{\omega_n\sqrt{1-\xi^2}} = \frac{\pi - \tan^{-1}\frac{\sqrt{1-(0.5)^2}}{0.5}}{6\sqrt{1-(0.5)^2}} = \frac{3.14-1.047}{5.19} = 0.403 \text{ sec.}$$

2. Peak Time:

$$t_p = \frac{\pi}{\omega_n\sqrt{1-\xi^2}} = \frac{\pi}{6\sqrt{1-(0.5)^2}} = 0.605 \text{ sec.}$$

3. Settling Time :

$$t_s = \frac{4}{\xi\omega_n} = \frac{4}{0.5\times 6} = 1.33 \text{ sec.} \quad \text{(for 2\%)}$$

4. Maximum Overshoot :

$$M_P = e^{\frac{-\pi\xi}{\sqrt{1-\xi^2}}} \times 100 = e^{-\frac{\pi\times 0.5}{\sqrt{1-(0.5)^2}}} \times 100 = 0.163 \times 100 = 16.3\%$$

2.6. TIME RESPONSE OF SECOND ORDER SYSTEM WITH UNIT IMPULSE INPUT

Since, input is unit impulse R(S) = 1

$$\therefore \qquad C(s) = \frac{\omega_n^2}{s^2 + 2\xi\omega_n s + \omega_n^2} \quad \text{...(2.46)}$$

Break the equation 2.46 by partial fraction

$$\frac{\omega^2_n}{s^2 + 2\xi\omega_n s + \omega_n^2} = \frac{K_1}{s - s_1} + \frac{K_1^*}{s - s_2} \quad \text{...(2.47)}$$

characteristic equation $s^2 + 2\xi\omega_n s + \omega_n^2 = 0$

Roots of the characteristic equation $s_1 = -\xi\omega_n + j\omega_n\sqrt{1-\xi^2}$

$$s_2 = -\xi\omega_n - j\omega_n\sqrt{1-\xi^2}$$

Put $\omega_d = \omega_n\sqrt{1-\xi^2}$

$\therefore \quad s_1 = -\xi\omega_n + j\omega_d \quad$ and $\quad s_2 = -\xi\omega_n - j\omega_d$

From equation (2.47)

$$K_1 = \left.\frac{\omega_n^2}{s-s_2}\right|_{s=s_1} = \frac{\omega_n^2}{-\xi\omega_n + j\omega_d + \xi\omega_n + j\omega_d} = \frac{\omega_n^2}{2j\omega_d}$$

$$K_1^* = \frac{\omega_n^2}{-j2\omega_d}$$

Therefore equation (2.47) becomes

$$c(s) = \frac{\omega_n^2}{s^2 + 2\xi\omega_n s + \omega_n^2} = \frac{\omega_n^2}{j2\omega_d}\cdot\frac{1}{s-s_1} - \frac{\omega_n^2}{j2\omega_d}\cdot\frac{1}{s-s_2} \qquad ...(2.48)$$

Inverse Laplace of equation 2.48

$$c(t) = \frac{\omega_n^2}{j2\omega_d}[e^{s_1 t} - e^{s_2 t}] = \frac{\omega_n^2}{j2\omega_d}[e^{(-\xi\omega_n + j\omega_d)t} - e^{(-\xi\omega_n - j\omega_d)t}]$$

$$= \frac{\omega_n^2}{j2\omega_d}[e^{-\xi\omega_n t}\cdot e^{j\omega_d t} - e^{-\xi\omega_n t}\cdot e^{-j\omega_d t}] = \frac{\omega_n^2 e^{-\xi\omega_n t}}{j2\omega_d}[e^{j\omega_d t} - e^{-j\omega_d t}]$$

$$= \frac{\omega_n^2 e^{-\xi\omega_n t}}{j2\omega_d}[\cos\omega_d t + j\sin\omega_d t - \cos\omega_d t + j\sin\omega_d t]$$

$$= \frac{\omega_n^2 e^{-\xi\omega_n t}}{j2\omega_d}[j2\sin\omega_d t] = \frac{\omega_n^2 e^{-\xi\omega_n t}}{\omega_d}\sin\omega_d t \qquad ...(2.49)$$

Put the value of ω_d in equation 2.49

$$C(t) = \frac{\omega_n^2 e^{-\xi\omega_n t}}{\omega_n\sqrt{1-\xi^2}}\sin\left(\omega_n\sqrt{1-\xi^2}\right)t = \frac{\omega_n}{\sqrt{1-\xi^2}}e^{-\xi\omega_n t}\sin\left(\omega_n\sqrt{1-\xi^2}\right)t \qquad ...(2.50)$$

***(a)* $\xi < 1$:**

$$C(t) = \frac{\omega_n}{\sqrt{1-\xi^2}}e^{-\xi\omega_n t}\sin\left(\omega_n\sqrt{1-\xi^2}\right)t \qquad ...(2.51)$$

***(b)* $\xi = 1$:**

$$C(t) = \lim_{\xi\to 1}\frac{\omega_n}{\sqrt{1-\xi^2}}e^{-\xi\omega_n t}\sin\left(\omega_n\sqrt{1-\xi^2}\right)t \quad \text{or} \quad C(t) = t.e^{-\omega_n t}.\omega_n^2 \qquad ...(2.52)$$

***(c)* $\xi > 1$:**

Equation 2.46 can be written

$$C(S) = \frac{\omega^2{}_n}{(s+\xi\omega_n)^2 - \omega^2{}_n(\xi^2-1)} \qquad ...(2.53)$$

Break the equation 2.53 by partial fraction put $\omega^2{}_d = \omega^2{}_n(\xi^2 - 1)$

$$\frac{\omega_n^2}{(s+\xi\omega_n)^2 - \omega^2{}_d} = \frac{\omega^2{}_n}{(s+\xi\omega_n+\omega_d)(s+\xi\omega_n-\omega_d)} = \frac{A}{(s+\zeta\omega_n+\omega_d)} + \frac{B}{(s+\xi\omega_n-\omega_d)} \qquad ...(2.54)$$

From (2.54) $$A = \left.\frac{\omega_n^2}{s+\xi\omega_n-\omega_d}\right|_{s=-\xi\omega_n-\omega_d} = \frac{-\omega_n}{2\sqrt{\xi^2-1}}$$

Similarly, $$B = \frac{\omega_n}{2\sqrt{\xi^2-1}}$$

$$C(S) = \frac{\omega_n^2}{(s+\xi\omega_n)^2-\omega^2_d} = \frac{-\omega_n}{2\sqrt{\xi^2-1}}\cdot\frac{1}{s+\xi\omega_n+\omega_d}+\frac{\omega_n}{2\sqrt{\xi^2-1}}\cdot\frac{1}{s+\xi\omega_n-\omega_d} \quad ...(2.55)$$

Take inverse Laplace

$$C(t) = \frac{\omega_n}{2\sqrt{\xi^2-1}}\left[-e^{-(\xi\omega_n+\omega_d)t}+e^{-(\xi\omega_n-\omega_d)t}\right]$$

$$\therefore \quad C(t) = \frac{\omega_n}{2\sqrt{\xi^2-1}}\left[e^{-\left(\xi\omega_n-\omega_n\sqrt{\xi^2-1}\right)t}-e^{-\left(\xi\omega_n+\omega_n\sqrt{\xi^2-1}\right)t}\right]$$

$$C(t) = \frac{\omega_n}{2\sqrt{\xi^2-1}}\left[e^{-\left(\xi-\sqrt{\xi^2-1}\right)\omega_n t}-e^{-\left(\xi+\sqrt{\xi^2-1}\right)\omega_n t}\right] \quad ...(2.56)$$

Let the maximum overshoot for the unit impulse response of the underdamped system occurs at $t = t_p$.

$$\frac{dC(t)}{dt} = 0$$

Differentiate equation 2.51 put $\omega_d = \omega_n\sqrt{1-\xi^2}$

$$\frac{\omega_n}{\sqrt{1-\xi^2}}\left[e^{-\xi\omega_n t}\cos\omega_d t\,.\omega_d+\sin\omega_d t.\,e^{-\xi\omega_n t}(-\xi\omega_n)\right] = 0$$

$e^{-\xi\omega_n t} \neq 0$

$$\omega_d\cos\omega_d t = \xi\,\omega_n\sin\omega_d t$$

$$\tan\omega_d t = \frac{\omega_d}{\xi\omega_n} = \frac{\omega_n\sqrt{1-\xi^2}}{\xi\omega_n} = \frac{\sqrt{1-\xi^2}}{\xi}$$

$$\omega_d t = \tan^{-1}\frac{\sqrt{1-\xi^2}}{\xi}$$

$$t = \frac{\tan^{-1}\frac{\sqrt{1-\xi^2}}{\xi}}{\omega_d} \quad \text{and} \quad \boxed{t_p = \frac{\tan^{-1}\frac{\sqrt{1-\xi^2}}{\xi}}{\omega_n\sqrt{1-\xi^2}}} \quad ...(2.57)$$

Expression for maximum overshoot.

Since $\tan\omega_d t = \frac{\sqrt{1-\xi^2}}{\xi}$, then $\sin\omega_d t = \sqrt{1-\xi^2}$

$$C(t) = \frac{\omega_n}{\sqrt{1-\xi^2}}e^{-\xi\omega_n t_p}\sin\omega_d t = \frac{\omega_n}{\sqrt{1-\xi^2}}e^{-\xi\omega_n\cdot\frac{\tan^{-1}\frac{\sqrt{1-\xi^2}}{\xi}}{\omega_n\sqrt{1-\xi^2}}}\cdot\sqrt{1-\xi^2}$$

$$\boxed{C(t) = \omega_n\, e^{-\frac{\xi}{\sqrt{1-\xi^2}}\tan^{-1}\frac{\sqrt{1-\xi^2}}{\xi}}} \quad ...(2.58)$$

2.7. TIME RESPONSE OF SECOND ORDER SYSTEM WITH UNIT RAMP INPUT

$$\frac{C(S)}{R(S)} = \frac{\omega_n^2}{s^2 + 2\xi\omega_n s + \omega^2_n} \quad \text{...(2.59)}$$

Input $\quad R(S) = \frac{1}{s^2}$

$$\therefore \quad C(S) = \frac{\omega_n^2}{s^2(s^2 + 2\xi\omega_n s + \omega_n^2)} \quad \text{...(2.60)}$$

Break the equation 2.60 by partial fraction

$$\frac{\omega_n^2}{s^2(s^2 + 2\xi\omega_n s + \omega_n^2)} = \frac{A}{s} + \frac{B}{s^2} + \frac{cs + D}{s^2 + 2\xi\omega_n s + \omega_n^2} \quad \text{...(2.61)}$$

$$\therefore \quad \omega_n^2 = A\,s\,(s^2 + 2\xi\omega_n s + \omega^2_n) + B\,(s^2 + 2\xi\,\omega_n s + \omega^2_n) + (cs + D)\,s^2$$

Equating the coefficient of s^3 :

$$A + C = 0 \quad \text{...(2.62)}$$

Equating the coefficient of s^2 :

$$2\xi\omega_n A + B + D = 0 \quad \text{...(2.63)}$$

Equating the coefficient of s :

$$A\omega^2_n + 2\xi\omega_n B = 0 \quad \text{...(2.64)}$$

Equating the absolute terms :

$$B\omega^2_n = \omega_n^2 \quad \text{...(2.65)}$$

From equations (2.62) , (2.63) , (2.64) , and (2.65)

$$A = -\frac{2\xi}{\omega_n}, C = \frac{2\xi}{\omega_n} \text{ and } B = 1, D = -1 + 4\xi^2$$

Put all these values in equation (2.61)

$$C(S) = \frac{-2\xi}{\omega_n s} + \frac{1}{s^2} + \frac{\frac{2\xi}{\omega_n}s + (4\xi^2 - 1)}{s^2 + 2\xi\omega_n s + \omega_n^2} \quad \text{...(2.66)}$$

$$= \frac{1}{s^2} - \frac{2\xi}{\omega_n s} + \frac{2\xi}{\omega_n}\cdot\frac{(s + \xi\omega_n - \xi\omega_n)}{(s + \xi\omega_n)^2 + \omega_n^2(1 - \xi^2)} + \frac{4\xi^2 - 1}{(s + \xi\omega_n)^2 + \omega_n(1 - \xi^2)}$$

$$= \frac{1}{s^2} - \frac{2\xi}{\omega_n s} + \frac{2\xi}{\omega_n}\cdot\frac{(s + \xi\omega_n)}{(s + \xi\omega_n)^2 + \omega^2{}_n(1 - \xi^2)} - \frac{2\xi}{\omega_n}\,\frac{\xi\omega_n}{(s + \xi\omega_n)^2 + \omega_n^2(1 - \xi^2)} + \frac{4\xi^2 - 1}{(s + \xi\omega_n)^2 + \omega_n(1 - \xi^2)}$$

$$C(S) = \frac{1}{s^2} - \frac{2\xi}{\omega_n s} + \frac{2\xi}{\omega_n}\cdot\frac{(s + \xi\omega_n)}{(s + \xi\omega_n)^2 + \omega_n^2(1 - \xi^2)} + \frac{2\xi^2 - 1}{(s + \xi\omega_n)^2 + \omega_n^2(1 - \xi^2)}$$

Put $\quad \omega^2_d = \omega_n^2\,(1 - \xi^2)$

$$C(S) = \frac{1}{s^2} - \frac{2\xi}{\omega_n s} + \frac{2\xi}{\omega_n}\,\frac{(s + \xi\omega_n)}{(s + \xi\omega_n)^2 + \omega^2_d} + \frac{2\xi^2 - 1}{(s + \xi\omega_n)^2 + \omega^2_d}\cdot\frac{\omega_d}{\omega_d}$$

$$C(S) = \frac{1}{s^2} - \frac{2\xi}{\omega_n s} + \frac{2\xi}{\omega_n}\cdot\frac{s + \xi\omega_n}{(s + \xi\omega_n)^2 + \omega_d^2} + \frac{2\xi^2 - 1}{\omega_d}\cdot\frac{\omega_d}{(s + \xi\omega_n)^2 + \omega^2_d} \quad \text{...(2.67)}$$

Inverse Laplace of equation (2.67)

$$C(t) = t - \frac{2\xi}{\omega_n} + \frac{2\xi}{\omega_n} \cdot e^{-\xi\omega_n t} \cos\omega_d t + \frac{2\xi^2 - 1}{\omega_d} \cdot e^{-\xi\omega_n t} \sin\omega_d t \quad ...(2.68)$$

$$= t - \frac{2\xi}{\omega_n} + \frac{2\xi\sqrt{1-\xi^2}}{\omega_n\sqrt{1-\xi^2}} e^{-\xi\omega_n t} \cos\omega_d t + \frac{2\xi^2 - 1}{\omega_n\sqrt{1-\xi^2}} e^{-\xi\omega_n t} \sin\omega_d t$$

$$= t - \frac{2\xi}{\omega_n} + \frac{e^{-\xi\omega_n t}}{\omega_n\sqrt{1-\xi^2}} \left[2\xi\sqrt{1-\xi^2} \cos\omega_d t + (2\xi^2 - 1) \sin\omega_d t \right]$$

Put

$$2\xi\sqrt{1-\xi^2} = \sin\phi$$

$$2\xi^2 - 1 = \cos\phi$$

$$C(t) = t - \frac{2\xi}{\omega_n} + \frac{e^{-\xi\omega_n t}}{\omega_n\sqrt{1-\xi^2}} [\sin\phi \cos\omega_d t + \cos\phi \sin\omega_d t]$$

$$= t - \frac{2\xi}{\omega_n} + \frac{e^{-\xi\omega_n t}}{\omega_n\sqrt{1-\xi^2}} \sin(\omega_d t + \phi)$$

$$C(t) = t - \frac{2\xi}{\omega_n} + \frac{e^{-\xi\omega_n t}}{\omega_n\sqrt{1-\xi^2}} \sin\left[\left(\omega_n\sqrt{1-\xi^2} \right) t + \phi \right] \quad ...(2.69)$$

Where, $\phi = \cos^{-1}(2\xi^2 - 1)$

or, $\phi = \tan^{-1} \dfrac{2\xi\sqrt{1-\xi^2}}{2\xi^2 - 1}$

Error signal

$$e(t) = r(t) - C(t)$$

$$r(t) = t$$

$$\therefore \quad e(t) = \frac{2\xi}{\omega_n} - \frac{e^{-\xi\omega_n t}}{\omega_n\sqrt{1-\xi^2}} \sin\left[\left(\omega_n\sqrt{1-\xi^2} \right) t + \phi \right] \quad ...(2.70)$$

$$e_{ss} = \lim_{t\to\infty} e(t) = \frac{2\xi}{\omega_n}$$

$$\therefore \text{ steady state error } e_{ss} = \frac{2\xi}{\omega_n} \quad ...(2.71)$$

2.8. TIME RESPONSE OF HIGHER ORDER CONTROL SYSTEM

The input, output and transfer function are given by

$$C(s) = R(s) \cdot T(s)$$

where, $C(S)$ = Output (Laplace transform) of the system
$R(S)$ = Laplace transform of input
$T(S)$ = Transfer function of the system

$$T(S) = \frac{K(s^m + a_1 s^{m-1} + ... + a_m)}{(s^n + b_1 s^{n-1} + ... b_n)}$$

$$\therefore \quad C(S) = R(s)\cdot\frac{K(s^m + a_1 s^{m-1} + \ldots + a_m)}{s^n + b_1 s^{n-1} + \ldots + b_n}$$

Apply Heaviside's partial fraction to obtain the partial fraction.

$$C(S) = \frac{K_1}{(s+p_1)} + \frac{K_2}{(s+p_2)} + \ldots + \frac{K_n}{(s+p_n)}$$

For time response take inverse Laplace of each part of above equation

$$£^{-1}\,C(S) = £^{-1}\frac{K_1}{s+p_1} + £^{-1}\frac{K_2}{s+p_2} + \ldots + £^{-1}\frac{K_n}{(s+p_n)}$$

$$C(t) = K_1 e^{-p_1 t} + K_2 e^{-p_2 t} + \ldots + K_n e^{-p_n t}$$

This is the required result.

EXAMPLE 2.2. The closed loop transfer function is given by

$$G(s) = \frac{s(s^2+9s+19)}{s^3+7s^2+14s+8}$$

Determine the response of the system when a unit step is applied at input.

Solution :

$$C(S) = R(S)\,G(S)$$

$$C(S) = \frac{1}{s}\cdot\frac{s(s^2+9s+19)}{s^3+7s^2+14s+8} = \frac{s^2+9s+19}{s^3+7s^2+14s+8}$$

Break the equation by partial fraction

$$C(S) = \frac{11}{3(s+1)} - \frac{5}{2(s+2)} - \frac{1}{6(s+4)}$$

Take inverse Laplace of each term, we get

$$C(t) = \frac{11}{3}e^{-t} - \frac{5}{2}e^{-2t} - \frac{1}{6}e^{-4t} \quad \textbf{Ans.}$$

EXAMPLE 2.3. A second order system has a transfer function given by

$$G(s) = \frac{25}{s^2+8s+25}$$

If the system initially at rest is subjected to a unit step input at t = 0, the second peak in the response will occur at

(*a*) π sec. (*b*) $\pi/3$ sec. (*c*) $2\pi/3$ sec. (*d*) $\pi/2$ sec. *(GATE 1991)*

Solution :

$$R(S) = \frac{1}{s}$$

$$C(S) = R(s)\cdot G(s)$$

$$C(S) = \frac{25}{s(s^2+8s+25)} \qquad \ldots(2.72)$$

compare the equation 2.72 with

$$s^2 + 2\xi\omega_n s + \omega_n^2 = 0$$

$$2\xi\omega_n = 8$$

$$\omega_n = \sqrt{25} = 5 \text{ rad/sec.}$$

$$\therefore \quad \xi = 0.8$$

$$\text{Required time} = \frac{3\pi}{\omega_n\sqrt{1-\xi^2}} = \frac{3\pi}{5\sqrt{1-(0.8)^2}} = \pi \text{ sec.} \quad \textbf{Ans.}$$

EXAMPLE 2.4. Consider the system as shown in Fig. 2.18 Determine the value of a such that the damping ratio is 0.5. Also obtain the values of rise time and maximum overshoot M_p in its step response. *(GATE 1998)*

Solution :

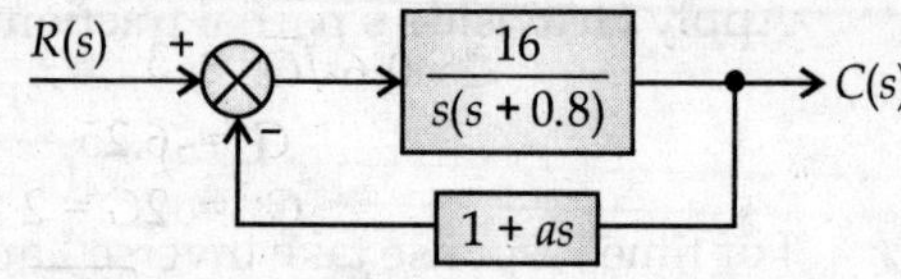

Fig. 2.18.

$$\frac{C(S)}{R(S)} = \frac{\dfrac{16}{s(s+0.8)}}{1+\dfrac{16(1+as)}{s(s+0.8)}}$$

$$= \frac{16}{s^2+(0.8+16a)s+16} \quad \text{...(2.73)}$$

Characteristic equation of 2.73

$$s^2 + (0.8 + 16a)s + 16 = 0 \quad \text{...(2.74)}$$

compare the equation (2.74) with $s^2 + 2\xi\omega_n s + \omega_n^2 = 0$

$$\omega_n^2 = 16 \qquad \therefore \omega_n = 4 \text{ rad/sec.}$$

$$2\xi\omega_n = 0.8 + 16a$$

$$2 \times 0.5 \times 4 = 0.8 + 16a$$

$$\therefore \qquad \boxed{a = 0.2} \qquad \textbf{Ans.}$$

Rise time

$$t_r = \frac{\pi - \tan^{-1}\dfrac{\sqrt{1-\zeta^2}}{\zeta}}{\omega_n\sqrt{1-\zeta^2}} = \frac{\pi - \tan^{-1}\dfrac{\sqrt{1-(0.5)^2}}{0.5}}{4\sqrt{1-(0.5)^2}}$$

$$t_r = 0.605 \text{ sec.} \qquad \textbf{Ans.}$$

Maximum overshoot

$$M_P = e^{\frac{-\pi\xi}{\sqrt{1-\xi^2}}} \times 100 = e^{\frac{-\pi(0.5)}{\sqrt{1-(0.5)^2}}} \times 100 = 16.31\% \qquad \textbf{Ans.}$$

EXAMPLE 2.5. The block diagram of a feedback system is shown in Fig. 2.19 (*a*)

(*a*) Find closed loop transfer function

(*b*) Find the minimum value of *G* for which the step response of the system would exhibit an overshoot as shown in Fig. 2.19 (*b*).

(*c*) For *G* equal to twice this minimum value, find the time period '*T*' indicated in Fig. 2.19 (*b*) *(GATE 2000)*

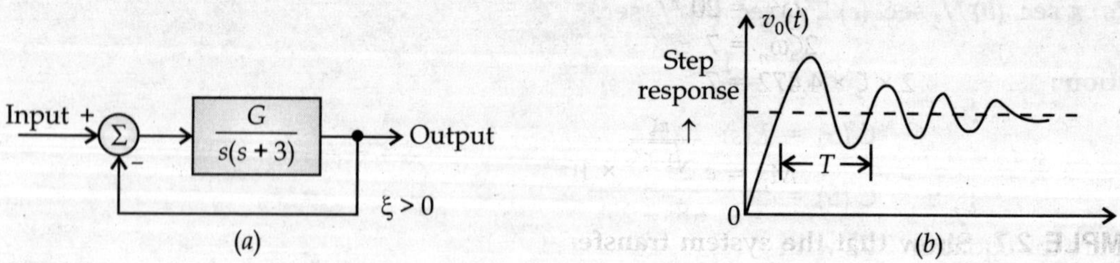

Fig. 2.19.

Solution : Closed Loop Transfer Function

(a) $$\frac{C(S)}{R(S)} = \frac{G/s(s+3)}{1+\dfrac{G}{s(s+3)}} = \frac{G}{s^2+3s+G} \qquad \textbf{Ans.}$$

(b) Characteristic equation $s^2 + 3s + G = 0$

compare with $s^2 + 2\zeta\omega_n s + \omega_n^2 = 0$

$$2\zeta\omega_n = 3$$

$$\omega_n = \sqrt{G}$$

$$\zeta = 0.6 \text{ (assume)}$$

$$2 \times 0.6\sqrt{G} = 3$$

$$\therefore \quad G = 6.25$$

$$G' = 2G = 2 \times 6.25 = 12.5$$

$$\omega_n = \sqrt{12.5} = 3.53 \text{ rad/sec.}$$

$$\zeta = \frac{3}{2\omega_n} = \frac{3}{2 \times 3.53} = 0.424$$

$$\omega_d = \omega_n\sqrt{1-\zeta^2} = 3.53\sqrt{1-(0.424)^2} = 3.197$$

$$\frac{2\pi}{T} = 3.197$$

$$\therefore \quad T = 1.96 \text{ sec.} \qquad \textbf{Ans.}$$

EXAMPLE 2.6. The open loop transfer function of a servo system with unity feedback is given by

$$G(S) = \frac{10}{(s+2)(s+5)}$$

Determine the damping ratio, undamped natural frequency of oscillation. What is the percentage overshoot of the response to a unit step input. *(R.M.L. University, Faizabad 2001)*

Solution : Given that

$$G(S) = \frac{10}{(s+2)(s+5)}$$

$$H(S) = 1$$

characteristic equation = $1 + G(S)\,H(S) = 0$

$$1 + \frac{10}{(s+2)(s+5)} = 0$$

or, $$s^2 + 7s + 20 = 0$$

compare with $s^2 + 2\zeta\omega_n s + \omega_n^2 = 0$

$$\omega_n^2 = 20 \qquad \therefore \boxed{\omega_n = 4.472 \text{ rad/sec.}}$$

$$2\zeta\omega_n = 7$$

$$2 \times \zeta \times 4.472 = 7 \qquad \boxed{\zeta = 0.7826}$$

$$M_p = e^{-\frac{\pi\zeta}{\sqrt{1-\zeta^2}}} \times 100 = e^{-\frac{\pi(0.7826)}{\sqrt{1-(0.7826)^2}}} \times 100 \quad \text{or} \quad \boxed{M_p = 1.92\%}$$

EXAMPLE 2.7. Show that the system transfer function $Y(s)/X(s)$ has a zero in the right half s-plane. Obtain $y(t)$ when $x(t)$ is a unit step for the system shown in Fig. 2.20.

(KNIT Sultanpur, 1999-2000)

Solution : The transfer function is given by

$$\frac{Y(S)}{X(S)} = \frac{6}{s+2} - \frac{4}{s+1} \qquad \text{(Blocks in parallel)}$$

$$\frac{Y(S)}{X(S)} = \frac{2(s-1)}{(s+1)(s+2)}$$

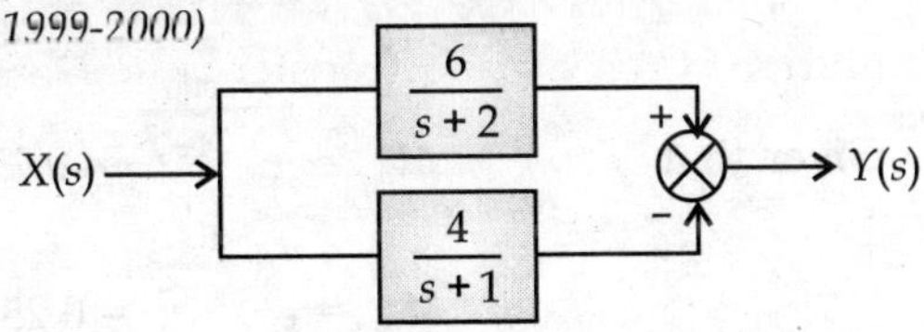

Fig. 2.20.

From above equation it is clear that there is a zero at s = 1 on right half of s-plane.

$$X(t) = 1$$

$$\therefore \quad X(S) = \frac{1}{s}$$

$$\therefore \quad Y(S) = \left[\frac{6}{s+2} - \frac{4}{s+1}\right]\frac{1}{s} = \frac{6}{s(s+2)} - \frac{4}{s(s+1)}$$

Break the equation by partial fraction

$$Y(S) = 6\left[\frac{1}{2s} - \frac{1}{2(s+2)}\right] - 4\left[\frac{1}{s} - \frac{1}{s+1}\right] = \frac{3}{s} - \frac{3}{s+2} - \frac{4}{s} + \frac{4}{s+1}$$

$$Y(S) = -\frac{1}{s} - \frac{3}{s+2} + \frac{4}{s+1}$$

Inverse laplace of above equation

$$\pounds^{-1}Y(S) = \pounds^{-1}(-1/s) + \pounds^{-1}(-3/s+2) + \pounds^{-1}(4/s+1) = -1 - 3e^{-2t} + 4e^{-t}$$

$$y(t) = 4e^{-t} - 3e^{-2t} - 1 \qquad \textbf{Ans.}$$

EXAMPLE 2.8. The open loop transfer function of a unity feedback system is given by

$$G(S) = \frac{K}{s(1+sT)}$$

where 'K' and 'T' are positive constants. By what factor should the amplifier gain be reduced so that the peak overshoot of unit step response of the system is reduced from 75% to 25%.

(KNIT Sultanpur, 1998-99)

Solution :

$$\frac{C(S)}{R(S)} = \frac{G(s)}{1+G(s)H(s)} = \frac{\frac{K}{s(1+sT)}}{1+\frac{K}{s(1+sT)}\cdot 1} = \frac{K}{s^2T+s+K} = \frac{K/T}{s^2+\frac{1}{T}s+\frac{K}{T}}$$

Characteristic equation : $s^2 + \frac{1}{T}s + \frac{K}{T} = 0$

compare with : $s^2 + 2\zeta\omega_n s + \omega^2_n = 0$

$$2\zeta\omega_n = \frac{1}{T} \qquad \therefore \quad \zeta = \frac{1}{2T\omega_n}$$

$$\omega^2_n = \frac{K}{T} \qquad \therefore \quad \omega_n = \sqrt{\frac{K}{T}}$$

$$\therefore \quad \zeta = \frac{1}{2T\sqrt{\frac{K}{T}}} = \frac{1}{2\sqrt{KT}}$$

Given that $\quad M_{p_1} = e^{-\frac{\pi\zeta_1}{\sqrt{1-\zeta^2}}} = 0.75$

$$M_{p_2} = e^{-\frac{\pi\zeta_2}{\sqrt{1-\zeta^2}}} = 0.25$$

$$e^{-\frac{\pi\zeta_1}{\sqrt{1-\zeta^2}}} = 0.75 \quad \therefore \quad \frac{\pi\zeta_1}{\sqrt{1-\zeta_1^2}} = 0.287$$

$$e^{\frac{-\pi\zeta}{\sqrt{1-\zeta^2}}} = 0.25 \qquad \frac{\pi\zeta_2}{\sqrt{1-\zeta_2^2}} = 1.386$$

$$\frac{\pi\zeta_1}{\sqrt{1-\zeta_1^2}} \cdot \frac{\sqrt{1-\zeta_2^2}}{\pi\zeta_2} = \frac{0.287}{1.386} = 0.207$$

$$\frac{\zeta_1}{\zeta_2}\frac{\sqrt{1-\zeta_2^2}}{\sqrt{1-\zeta_1^2}} = 0.207$$

$$\frac{1}{2\sqrt{K_1T}} \cdot 2\sqrt{K_2T}\frac{\sqrt{1-\left(\frac{1}{2\sqrt{K_2T}}\right)^2}}{\sqrt{1-\left(\frac{1}{2\sqrt{K_1T}}\right)^2}} = 0.207$$

Solve the above equation we get

$$\boxed{K_1 = 20\,K_2} \quad \textbf{Ans.}$$

EXAMPLE 2.9. For given series RLC circuit, determine the undamped natural frequency and damping ratio of the circuit. Assume initial conditions are zero. $R = 1K$, $L = 10$ mH, $C = 0.01\mu F$

Solution : Find the transfer function and compare the characteristic equation with $s^2 + 2\zeta\omega_n s + \omega_n^2 = 0$ we get

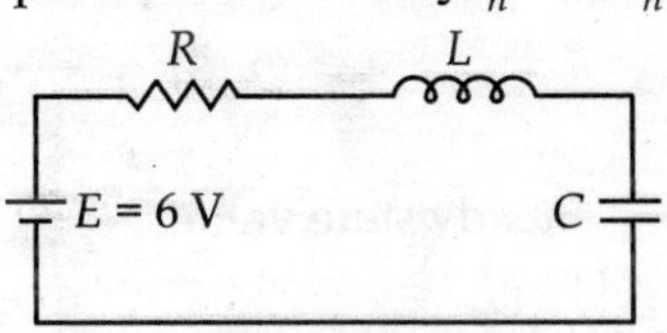

Fig. 2.21.

$$\zeta = \frac{R}{2}\sqrt{\frac{C}{L}} = 0.5$$

$$\omega_n = \frac{1}{\sqrt{LC}} = 10^5 \text{ rad/sec.} \qquad \textbf{Ans.}$$

EXAMPLE 2.10. Determine the value of 'K' and 'H' of the closed loop system so that the maximum overshoot in unit step response is 25% and the peak time is 2 sec. Assume $J = 1$ kg -m^2.

Solution :

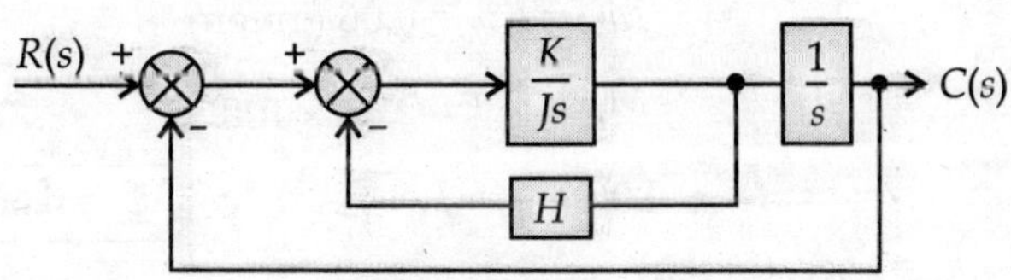

Fig. 2.22.

$$\frac{C(S)}{R(S)} = \frac{K}{Js^2 + KHs + K}$$

Put $J = 1$

$$\frac{C(S)}{R(S)} = \frac{K}{s^2 + KHs + K}$$

characteristic equation $s^2 + KHs + K = 0$, compare with $s^2 + 2\zeta\omega_n s + \omega_n^2 = 0$

$$\omega_n = \sqrt{K}$$

$$2\zeta\omega_n = KH$$

Given that $M_p = 0.25$ *i.e.,* $e^{\frac{-\pi\zeta}{\sqrt{1-\zeta^2}}} = 0.25$

$\therefore$ $\zeta = 0.404$

Peak time $t_p = \dfrac{\pi}{\omega_n\sqrt{1-\zeta^2}}$

$$\omega_n\sqrt{1-\zeta^2} = \frac{\pi}{t_p} = \frac{\pi}{2}$$

$$\omega_n\sqrt{1-(0.404)^2} = \frac{\pi}{2} \qquad \therefore\ \omega_n = 1.72 \text{ rad/sec.}$$

$$\omega_n = \sqrt{K} \qquad \therefore\ K = \omega_n^2 = (1.72)^2 = 2.95\ Nm.$$

$$2\zeta\omega_n = KH \qquad \therefore\ H = \frac{2\zeta\omega_n}{K} = \frac{2\times 0.404\times 1.72}{2.95}$$

$H = 0.471$

$$K = 2.95\ Nm$$

$$H = 0.471 \quad \textbf{Ans.}$$

EXAMPLE 2.11. A second order control system is represented by a transfer function given below.

$$\frac{Q_0(S)}{T(S)} = \frac{1}{Js^2+fs+K}$$

where $Q_0(S)$ is the proportional output and 'T' is the input torque A step input 10 Nm is applied to the system and test results are given below.

(*a*) $M_P = 6\%$ (*b*) $t_p = 1$ sec. (*c*) The steady state error of the output is 0.5 radian. Determine the value of J, K and f. *(R.M.L, University Faizabad 2002)*

Solution : Given $T(S) = \dfrac{10}{s}$

$$Q_0(S) = \frac{10}{s(Js^2+fs+K)}$$

$$\text{Steady state value} = \lim_{s\to 0} s\cdot Q_0(s)$$

$$0.5 = \lim_{s\to 0} s\cdot\frac{10}{s(Js^2+fs+K)}$$

$$0.5 = \frac{10}{k} \quad \therefore \quad \boxed{K = 20}$$

$$M_P = 6\% = 0.06 \text{ (given)}$$

$$M_P = e^{-\pi\zeta/\sqrt{1-\zeta^2}} \times 100$$

$$0.06 = e^{-\pi\zeta/\sqrt{1-\zeta^2}} \quad \therefore \quad \boxed{\zeta = 0.667}$$

$t_p = 1$ sec. (given)

$$t_p = \frac{\pi}{\omega_n\sqrt{1-\zeta^2}}$$

or
$$1 = \frac{\pi}{\omega_n\sqrt{1-(0.667)^2}}$$

$$\omega_n = 4.21 \text{ rad/sec.}$$

characteristic equation

$Js^2 + fs + K = 0$

$s^2 + \frac{f}{J}s + \frac{K}{J} = 0$ compare with $s^2 + 2\zeta\omega_n s + \omega_n^2 = 0$

$\omega_n^2 = \frac{K}{J}$ $\therefore J = \frac{K}{\omega_n^2} = \frac{20}{(4.21)^2} = 1.128$ $\boxed{J = 1.128}$

$2\zeta\omega_n = \frac{f}{J}$ $\therefore f = 2\zeta\omega_n J = 2 \times 0.667 \times 4.21 \times 1.128$

$= 6.34$ $\boxed{f = 6.34}$

EXAMPLE 2.12. A feedback system is described by the following transfer function

$$G(S) = \frac{12}{s^2 + 4s + 16}, H(s) = Ks$$

The damping factor of the system is 0.8. Determine the overshoot of the system, and the value of *K*. *(Pune University)*

Solution :

$$\frac{C(S)}{R(S)} = \frac{G(S)}{1 + G(S)H(S)} = \frac{16}{s^2 + (4 + 16K)s + 16}$$

characteristic equation $s^2 + (4 + 16K)s + 16 = 0$, compare with

$$s^2 + 2\zeta\omega_n s + \omega_n^2 = 0$$

$$\omega_n^2 = 16 \quad \therefore \omega_n = 4 \text{ rad/sec.}$$

$$2\zeta\omega_n = 4 + 16K$$

$$2 \times 0.8 \times 4 = 4 + 16K \quad \therefore K = 0.15$$

$$\therefore \quad \frac{c(s)}{R(s)} = \frac{16}{s^2 + (4 + 16 \times 0.15)s + 16}$$

$$\omega_n^2 = 16$$

$$2\zeta\omega_n = 6.4 \quad \therefore \zeta = 0.8$$

$$M_p = e^{-\frac{\pi\xi}{\sqrt{1-\xi^2}}} \times 100 = e^{-\frac{\pi 0.8}{\sqrt{1-(0.8)^2}}} \times 100$$

$$M_p = 1.5\% \quad \textbf{Ans.}$$

EXAMPLE 2.13. The open loop transfer function of a unity feedback control system is given by

$$G(s) = \frac{K}{s(1 + sT)}$$

By what factor the amplifier gain *K* should be multiplied so that the damping ratio is increased from 0.3 to 0.9.

Solution :

$$\frac{c(s)}{R(s)} = \frac{G(s)}{1 + G(s)H(s)} = \frac{K/s(1+sT)}{1 + \frac{K}{s(1+sT)} \cdot 1} = \frac{K}{s^2T + s + K} = \frac{K/T}{s^2 + \frac{s}{T} + \frac{K}{T}}$$

characteristic equation $s^2 + \frac{1}{T}s + \frac{K}{T} = 0$ compare with $s^2 + 2\zeta\omega_n s + \omega_n^2 = 0$

$$2\zeta\omega_n = \frac{1}{T}$$

$$\omega_n^2 = \frac{K}{T} \qquad \therefore \; \omega_n = \sqrt{\frac{K}{T}}$$

$$2\zeta\sqrt{\frac{K}{T}} = \frac{1}{T} \quad \text{or} \qquad \zeta = \frac{1}{2T}\sqrt{\frac{T}{K}} = \frac{1}{2\sqrt{KT}}$$

Damping ratio $\zeta_1 = 0.3 \;;\quad \zeta_2 = 0.9$

$$\zeta_1 = \frac{1}{2\sqrt{K_1 T}}, \quad \zeta_2 = \frac{1}{2\sqrt{K_2 T}}$$

$$\frac{\zeta_1}{\zeta_2} = \frac{1}{2\sqrt{K_1 T}} \cdot 2\sqrt{K_2 T} = \frac{\sqrt{K_2}}{\sqrt{K_1}} \quad \text{or,} \quad \frac{K_2}{K_1} = \left(\frac{\zeta_1}{\zeta_2}\right)^2$$

$$\frac{K_2}{K_1} = \left(\frac{0.3}{0.9}\right)^2 = \frac{1}{9} \quad \text{or} \quad 9K_2 = K_1 \qquad \boxed{K_1 = 9K_2}$$

Hence, the gain K_1 at which $\zeta = 0.3$ should be multiplied by 1/9 to increase the damping ratio from 0.3 to 0.9.

EXAMPLE 2.14. The system shown in Fig. 2.23 *a* when subjected to a unit step input, the output response is shown in Fig 2.23 *b*. Determine the value of *K* and *T* from the response curve.

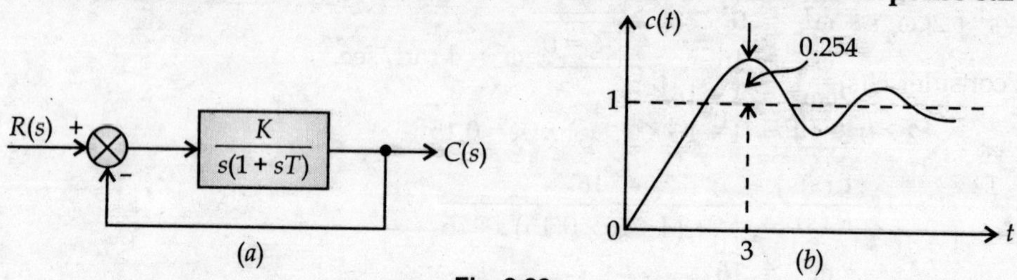

Fig. 2.23.

Solution :

$$\frac{C(S)}{R(S)} = \frac{\dfrac{K}{s(1+sT)}}{1+\dfrac{K}{s(1+sT)}} = \frac{K}{s^2T+s+K} = \frac{K/T}{s^2+\dfrac{1}{T}s+\dfrac{K}{T}}$$

characteristic equation

$$s^2 + \frac{1}{T}s + \frac{K}{T} = 0$$

compare with $s^2 + 2\zeta\omega_n s + \omega_n^2 = 0$

$$\omega_n = \sqrt{\frac{K}{T}} \quad \text{and} \quad 2\zeta\omega_n = \frac{1}{T}$$

$$M_p = 0.254 \text{ (given)}$$

$$M_p = e^{\frac{-\pi\zeta}{\sqrt{1-\zeta^2}}} \times 100$$

$$0.254 = e^{-\frac{\pi\zeta}{\sqrt{1-\zeta^2}}} \qquad \therefore \; \zeta = 0.4$$

$$t_p = 3 \text{ sec. (given)}$$

$$t_p = \frac{\pi}{\omega_n\sqrt{1-\zeta^2}} \text{ or } 3 = \frac{\pi}{\omega_n\sqrt{1-(0.4)^2}} \therefore \omega_n = 1.14$$

$$2\zeta\omega_n = \frac{1}{T} \quad \text{or,} \quad T = \frac{1}{2\zeta\omega_n} = \frac{1}{2\times0.4\times1.14} = 1.09 \qquad \boxed{T = 1.09}$$

$$\omega_n = \sqrt{\frac{K}{T}} \quad \text{or,} \quad \omega^2{}_n = \frac{K}{T} \text{ or } K = \omega^2{}_n T = 1.42 \qquad \boxed{K = 1.42}$$

EXAMPLE 2.15. A system having a forword path transfer function $G(S) = 16/s(s+1)$ and unity feedback. Determine the value of undamped natural frequency, damping ratio. If tachometer feedback is introduced, the feedback path transfer function becomes $(1 + Ks)$. What should be the value of K to obtain damping ratio of 0.6. Also calculate the percentage peak overshoot, first undershoot, t_p and settling time within the 2% of final value.

Solution : The characteristic equation with unity feedback

$$1+G(S) = 0 \qquad \therefore\ H(S) = 1$$

$$1+\frac{16}{s\,(s+1)} = 0$$

or $$s^2 + s + 16 = 0$$

compare with $s^2 + 2\zeta\omega_n s + \omega^2{}_n = 0$

$$\omega^2{}_n = 16 \qquad \boxed{\omega_n - 4 \text{ rad/sec.}}$$

$$2\zeta\omega_n = 1 \qquad \boxed{\zeta = 0.125}$$

Now, consider $H(s) = 1 + Ks$

$$1+G(S)\,H(S) = 0$$

$$1+\frac{16}{s(s+1)}(1+Ks) = 0$$

or, $$s^2 + (1 + 16K)\,s + 16 = 0$$

$$\therefore \quad \omega_n{}^2 = 16 \qquad \omega_n = 4 \text{ rad/sec.}$$

$$2\zeta\omega_n = 1 + 16K$$

$$2\times0.6\times4 = 1+16K \qquad \therefore \quad \boxed{K = 0.2375}$$

$$M_P = e^{\frac{-\pi\zeta}{\sqrt{1-\zeta^2}}} \times 100$$

When $$\zeta = 0.125,\ M_P = e^{\frac{-\pi(0.125)}{\sqrt{1-(0.125)^2}}} \times 100 = 67.3\%.$$

When $$\zeta = 0.6,\ M_P = e^{\frac{-\pi(0.6)}{\sqrt{1-(0.6)^2}}} \times 100 = 9.48\%.$$

First undershoot $$= e^{-2\pi\zeta/\sqrt{1-\zeta^2}} = e^{-2\pi(0.6/\sqrt{1-(0.6)^2}} = 0.009$$

Peak time :

When $$\zeta = 0.125$$

$$t_p = \frac{\pi}{\omega_n\sqrt{1-\zeta^2}} = \frac{\pi}{4\sqrt{1-(0.125)^2}} = 0.79 \text{ sec.}$$

when $$\zeta = 0.6$$

$$t_p = \frac{\pi}{4\sqrt{1-(0.6)^2}} = 0.98 \text{ sec.}$$

Settling time t_s :

$$t_s = \frac{4}{\zeta\omega_n}$$

when $\zeta = 0.125,\ t_s = \frac{4}{0.125 \times 4} = 8 \text{ sec.}$

when $\zeta = 0.6,\ t_s = \frac{4}{4 \times 0.6} = 1.67 \text{ sec.}$

EXAMPLE 2.16. In example 2.13, by what factor the time constant 'T' should be multiplied so that the damping ratio is reduced from 0.8 to 0.2.

Solution :

$$\zeta_1 = 0.8$$
$$\zeta_2 = 0.2$$
$$\zeta_1 = \frac{1}{2\sqrt{T_1 K}}, \quad \zeta_2 = \frac{1}{2\sqrt{T_2 K}}$$

T_1 = time constant for ζ_1
T_2 = time constant for ζ_2

$$\frac{\zeta_1}{\zeta_2} = \frac{0.8}{0.2} = 4 = \sqrt{\frac{T_2}{T_1}}$$
$$\frac{T_2}{T_1} = 16$$
$$\boxed{T_2 = 16\, T_1}$$

Hence, original time constant T_1 should be multiplied by 16, then the damping ratio becomes 0.2.

EXAMPLE 2.17. Consider the system shown in Fig. 2.24. Determine the value of k such that the damping ratio ζ is 0.5. Then obtain the rise time (t_r), peak time (t_p), maximum overshoot M_p, and settling time t_s in the unit step response.

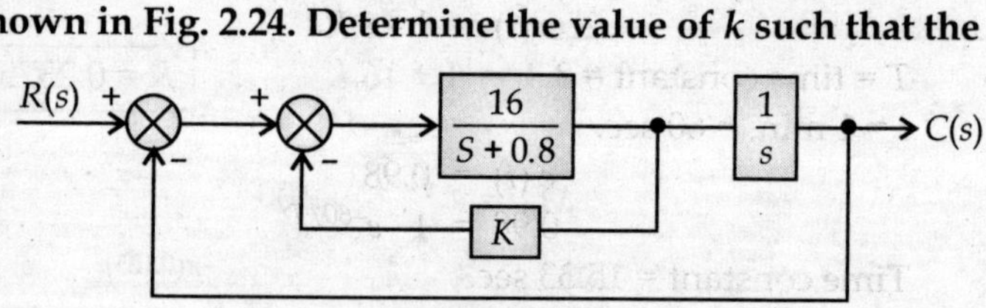

Fig. 2.24.

Solution : Step 1. Find $\frac{c(s)}{R(s)}$ by block reduction method.

$$\frac{c(s)}{R(s)} = \frac{16}{s^2 + (0.8 + 16k)s + 16}$$

Characteristic equation $s^2 + (0.8 + 16k)\,s + 16 = 0$, compare with $s^2 + 2\zeta\omega_n s + \omega_n^2 = 0$

$$\omega_n^2 = 16 \quad \therefore \omega_n = 4 \text{ rad/sec.}$$
$$2\zeta\omega_n = 0.8 + 16k$$

Step 2 : $2 \times 0.5 \times 4 = 0.8 + 16k \quad \boxed{\therefore k = 0.2}$

Rise time $$(t_r) = \frac{\pi - \tan^{-1}\frac{\sqrt{1-\zeta^2}}{\zeta}}{\omega_n\sqrt{1-\zeta^2}} = \frac{\pi - \tan^{-1}\frac{\sqrt{1-0.5^2}}{0.5}}{4\sqrt{1-0.5^2}} = 0.605 \text{ sec.}$$

Peak time $(t_p) = \dfrac{\pi}{\omega_n\sqrt{1-\zeta^2}} = \dfrac{\pi}{4\sqrt{1-0.5^2}} = 0.906$ sec.

Maximum overshoot $(M_P) = e^{\frac{-\pi\zeta}{\sqrt{1-\xi^2}}} \times 100 = e^{\frac{-\pi 0.5}{\sqrt{1-0.5^2}}} \times 100 = 16.3\%$

Settling time $(t_s) = \dfrac{4}{\xi\omega_n} = \dfrac{4}{0.5\times4} = 2$ sec. **Ans.**

EXAMPLE 2.18. A system shown in Fig. 2.25, determine the value of *K* and *k* such that the system has a damping ratio ξ of 0.7 and an undamped natural frequency w_n of 4 rad/sec.

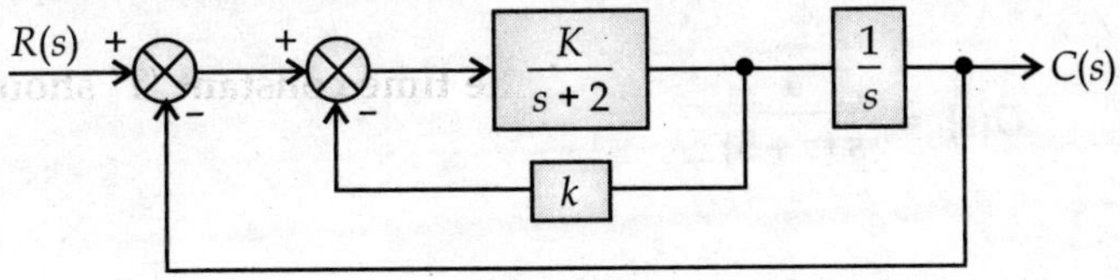

Fig. 2.25.

Solution : Find $\dfrac{c(s)}{R(s)}$ by block reduction method.

$$\frac{c(s)}{R(s)} = \frac{K}{s^2 + (2+Kk)\,s + K}$$

characteristic equation $s^2 + (2 + Kk)\,s + K = 0$ compare with $s^2 + 2\zeta\omega_n s + \omega^2_n = 0$

$$\omega^2_n = K \qquad \therefore K = 16 \qquad \boxed{K = 16}$$

$$2\zeta\omega_n = 2 + Kk$$

$$2 \times 0.7 \times 4 = 2 + 16k \qquad \therefore \qquad \boxed{k = 0.225}$$

EXAMPLE 2.19. A thermometer requires 1 min. to indicate 98% of the response to a step input. Assuming the thermometer to be a first order system, find the time constant.

Solution : For first order system

$$c(t) = 1 - e^{-t/T}$$

T = time constant = ?

t = 1 min. = 60 sec.

$$c(t) = 0.98$$

$$0.98 = 1 - e^{-60/T} \qquad \therefore T = 15.33 \text{ sec.}$$

Time constant = 15.33 sec

EXAMPLE 2.20. Consider the Fig. 2.26. The damping ratio of the system is 0.158 and the undamped natural frequency is 3.16 rad/sec. To improve the relative stability (Fig. 2.26 *b*) we employ tachometer feedback. Determine the value of K_h so that the damping ratio of the system is 0.5.

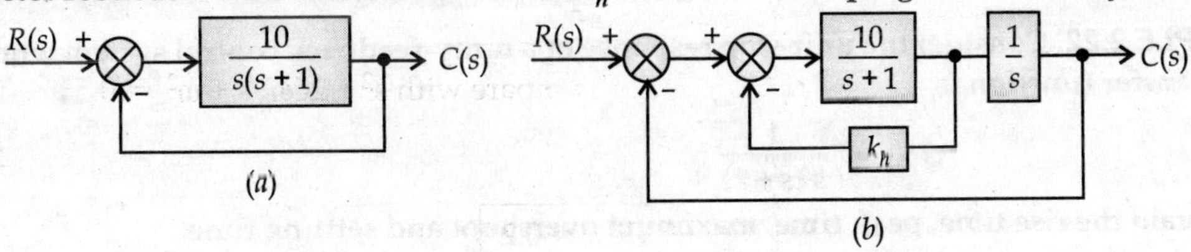

Fig. 2.26.

Solution : From Fig. 2.26 (*b*) we can determine $\dfrac{c(s)}{R(s)}$ by block reduction method.

$$\frac{c(s)}{R(s)} = \frac{10}{s^2+(1+10K_h)s+10}$$

characteristic equation $s^2 + (1 + 10 K_h) s + 10 = 0$ compare with $s^2 + 2\zeta\omega_n s + \omega_n^2 = 0$

$$2\zeta\omega_n = 1 + 10 K_h$$

$$\omega_n^2 = 10 \qquad \therefore \omega_n = 3.16 \text{ rad/sec.}$$

$$2 \times 0.5 \times 3.16 = 1 + 10 K_h$$

$$\therefore \quad K_h = 0.216 \qquad \boxed{K_h = 0.216}$$

EXAMPLE 2.21. Obtain the unit step response of a unity feedback system whose openloop transfer function is

$$G(s) = \frac{4}{s(s+5)}$$

Solution :

$$\frac{c(s)}{R(s)} = \frac{G(s)}{1+G(s)H(s)} = \frac{\frac{4}{s(s+5)}}{1+\frac{4}{s(s+5)}\cdot 1}$$

$$c(s) = \frac{4}{s^2+5s+4}$$

$$\therefore \quad c(s) = \frac{4}{s^2+5s+4} R(s)$$

Given $\quad R(s) = \frac{1}{s}$

$$\therefore \quad c(s) = \frac{4}{s(s^2+5s+4)} = \frac{4}{s(s+1)(s+4)}$$

Break the equation by partial fraction

$$\frac{4}{s(s+1)(s+4)} = \frac{A}{s} + \frac{B}{s+1} + \frac{C}{s+4}$$

$$A = 1, B = -\frac{4}{3}, c = \frac{1}{3}$$

$$c(s) = \frac{1}{s} - \frac{4}{3(s+1)} + \frac{1}{3(s+4)}$$

Inverse laplace of above equation

$$\pounds^{-1} c(s) = \pounds^{-1} 1/s - \frac{4}{3}\pounds^{-1}\frac{1}{s+1} + \frac{1}{3}\pounds^{-1}\frac{1}{s+4}$$

$$c(t) = 1 - \frac{4}{3}e^{-t} + \frac{1}{3}e^{-4t} \quad \textbf{Ans.}$$

EXAMPLE 2.22. Consider the unit-step response of a unity-feedback control system whose open loop transfer function is

$$G(s) = \frac{1}{s(s+1)}$$

Obtain the rise time, peak time, maximum overshoot and settling time.

Solution : Characteristic equation $1 + G(s) H(s) = 0$

$$1 + \frac{1}{s(s+1)} \cdot 1 = 0$$

$$s^2 + s + 1 = 0$$

Compare with $s^2 + 2\xi\omega_n s + \omega^2_n = 0$

$$\omega^2_n = 1 \qquad \therefore \omega_n = 1 \text{ rad/sec.}$$

$$2\zeta\omega_n = 1 \qquad \therefore \zeta = 0.5$$

Rise time
$$t_r = \frac{\pi - \tan^{-1}\frac{\sqrt{1-\zeta^2}}{\zeta}}{\omega_n\sqrt{1-\zeta^2}} = \frac{\pi - \tan^{-1}\frac{\sqrt{1-0.5^2}}{0.5}}{1\sqrt{1-0.5^2}} = 2.41 \text{ sec.}$$

Peak time
$$t_p = \frac{\pi}{\omega_n\sqrt{1-\zeta^2}} = \frac{\pi}{1\sqrt{1-0.5^2}} = 3.625 \text{ sec.}$$

Maximum overshoot
$$M_P = e^{\frac{-\pi\zeta}{\sqrt{1-\zeta^2}}} \times 100 = e^{-\frac{\pi 0.5}{\sqrt{1-0.5^2}}} \times 100 = 16.31\%$$

Settling time
$$t_s = \frac{4}{\zeta\omega_n} = \frac{4}{0.5\times 1} = 8 \text{ sec.}$$

EXAMPLE 2.23. Consider the closed loop system given by

$$\frac{c(s)}{R(s)} = \frac{\omega_n^2}{s^2 + 2\zeta\omega_n s + \omega_n^2}$$

Determine the values of ζ and ω_n so that the system responds to a step input with approximately 5% overshoot and with a settling time of 2 sec.

Solution : Given that

$$t_s = 2 \text{ sec.}$$

$$M_P = 5\% = 0.05$$

$$M_P = e^{\frac{-\pi\zeta}{\sqrt{1-\zeta^2}}}$$

$$0.05 = e^{-\pi\zeta/\sqrt{1-\zeta^2}} \qquad \therefore \boxed{\zeta = 0.689}$$

$$t_s = \frac{4}{\zeta\omega_n} \quad \text{or,} \quad \omega_n = \frac{4}{\zeta t_s}$$

$$\therefore \qquad \omega_n = \frac{4}{0.689\times 2} = 2.9 \text{ rad/sec.} \qquad \boxed{\omega_n = 2.9 \text{ rad/sec.}}$$

EXAMPLE 2.24. Fig. 2.27 is a block diagram of a space-vehicle attitude control system. Assuming the time constant T of the controller to be 3 sec. and the ratio of torque to inertia K/J to be $\frac{2}{9}$ rad²/sec² find the damping ratio of the system.

Solution : The characteristic equation $1 + G(s)\,H(s) = 0$

$$1 + \frac{K(Ts+1)}{Js^2} = 0$$

or,
$$J s^2 + 3\,Ks + K = 0$$

or,
$$s^2 + 3\frac{K}{J}s + \frac{K}{J} = 0$$

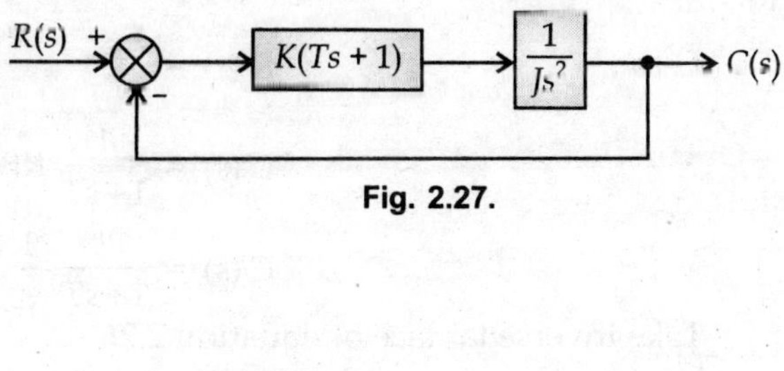

Fig. 2.27.

compare with $s^2 + 2\zeta\omega_n s + \omega_n^2 = 0$

$$\therefore \quad \omega_n^2 = \frac{K}{J}$$

$$2\zeta\omega_n = 3\frac{K}{J}$$

$$\omega_n^2 = \frac{2}{9}$$

$$\boxed{\omega_n = 0.471 \text{ rad/sec}}$$

$$2 \times \zeta \times 0.471 = 3 \cdot \frac{2}{9}$$

$$\therefore \quad \boxed{\zeta = 0.707} \quad \textbf{Ans.}$$

EXAMPLE 2.25. Measurement conducted on a servomechanism show the system response to be

$$c(t) = 1 + 0.2\, e^{-60\,t} - 1.2\, e^{-10\,t}$$

when subjected to a unit step input.

(*a*) Obtain the expression for the closed loop transfer function.

(*b*) Determine the undamped natural frequency and damping ratio of the system.

Solution :

$$c(t) = 1 + 0.2\, e^{-60\,t} - 1.2\, e^{-10\,t}$$

$$c(s) = \frac{1}{s} + \frac{0.2}{s+60} - \frac{1.2}{(s+10)} = \frac{600}{s(s+10)(s+60)}$$

(*a*) Transfer function $= \dfrac{c(s)}{R(s)} = \dfrac{\dfrac{600}{s(s+10)(s+60)}}{1/s} = \dfrac{600}{(s+10)(s+60)}$

(*b*) Characterisc equation = $s^2 + 70\,s + 600$, compare with $s^2 + 2\xi\omega_n\, s + \omega_n^2 = 0$

compare with $\quad 2\zeta\omega_n = 70$

$$\omega_n^2 = 600$$

$$\therefore \quad \omega_n = 24.49 \text{ rad/sec.}$$

$$2 \times \zeta \times 24.49 = 600$$

$$\therefore \quad \zeta = 1.42$$

$$\frac{c(s)}{R(s)} = \frac{600}{(s+10)(s+60)}$$

$$\omega_n = 24.49 \text{ rad/sec.}$$

$$\zeta = 1.42 \quad \textbf{Ans.}$$

2.9. EFFECT OF FEEDBACK ON TIME CONSTANT OF A CONTROL SYSTEM

Consider the open loop transfer function as shown in Fig. 2.28.

$$\frac{C(s)}{R(s)} = \frac{1}{1+sT} \qquad \text{...(2.75)}$$

Let $\quad R(s) = \dfrac{1}{s}$

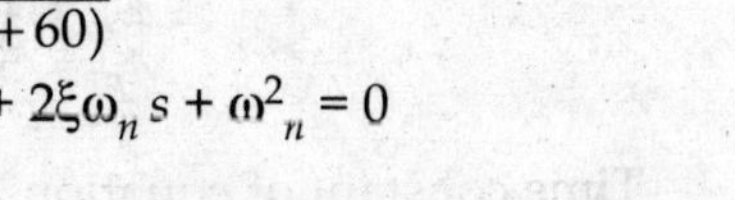

Fig. 2.28.

$$\therefore \quad C(s) = \frac{1}{1+sT} R(C)$$

$$C(s) = \frac{1}{1+sT} \cdot \frac{1}{s} \qquad \text{...(2.76)}$$

Take inverse laplace of equation 2.76

$$£^{-1} C(s) = £^{-1}\left[\frac{1}{s(1+sT)}\right]$$

$$C(t) = 1 - e^{-t/T} \quad ...(2.77)$$

In equation (2.77), 'T' is known as time constant

Now consider the closed loop system as shown in Fig. 2.29.

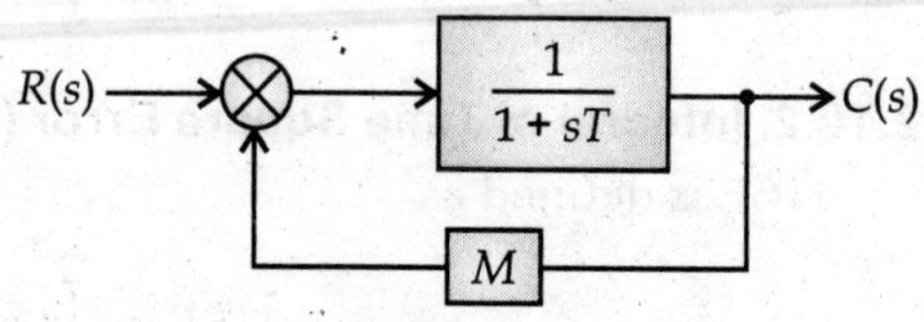

Fig. 2.29.

Suppose $H(s) = M$, then overall transfer function is given by

$$\frac{C(s)}{R(s)} = \frac{\frac{1}{1+sT}}{1+\frac{1}{1+sT}M} = \frac{1}{1+sT+M} = \frac{1/T}{s+\left(\frac{1+M}{T}\right)}$$

Let $R(s) = \frac{1}{s}$

$$\therefore \quad C(s) = \frac{1/T}{s+\left(\frac{1+M}{T}\right)} \quad ...(2.78)$$

Taking inverse Laplace of equation 2.78

$$£^{-1} C(s) = £^{-1}\left[\frac{1/T}{s+\left(\frac{1+M}{T}\right)}\right] = \frac{1}{1+M}\left[T - e^{-\frac{t}{T/1+M}}\right] \quad ...(2.79)$$

Time constant of equation 2.79

$$\text{Time constant} = \frac{T}{1+M}$$

Thus for positive values of M, the time constant $\frac{T}{1+M}$ is less than time constant 'T', It means that the time constant of closed loop system is less than the time constant of open loop system. Less time constant means response is faster. Therefore, feedback improves the time response of the system.

2.10. PERFORMANCE INDICES

Design of control system is to meet the specifications like phase margin, gain margin and steady state etc. These are essential specifications for optimization. Any solution which is optimum for one application may not be optimum for another application. So, a performance index is required. In many cases the performance index is a function of the error $e(t)$ between actual and ideal responses. If $f(e)$ is the function of error then performance indices is given by

$$\text{Performance Index (P.I.)} = \int_0^\infty e^2(t)\, dt$$

This number indicates the "goodness" of the performance. There are many performance indices.

2.10.1. Integral Square Error (ISE)

In closed loop system the error signal is the difference between input signal and the feedback signal. The object is to reduce the error. The error signal is taken as performance index to be optimized. The integral square error is defines as

$$\text{ISE} = \int_0^{\infty} e^2(t)\, dt$$

2.10.2. Integral of Time Square Error (ITSE)

ITSE is defined as

$$\text{ITSE} = \int_0^{\infty} t\, e^2(t)\, dt$$

2.10.3. Integral of Absolute Value of Error (IAE)

$$\text{IAE} = \int_0^{\infty} |\, e(t)\,|\, dt$$

2.10.4. Integral of Time Multiplied by Absolute Value of Error (ITAE)

This is defined as

$$\text{ITAE} = \int_0^{T} t\,|\, e(t)\,|\, dt$$

SUMMARY

Time response consists of two parts (*a*) transient response, (*b*) steady state response. Transient response is the part of the response which goes to zero as time increases and steady state response is the part of the total response after transient has died.

Test Input signals :

1. Step function is also known as displacement function. Laplace transform is $\frac{1}{s}$.
2. Ramp function : Ramp function starts from origin and increases or decreases linearly with time. Laplace transform is $\frac{1}{s^2}$.
3. Parabolic function : Laplace transform is $\frac{1}{s^3}$.
4. Impulse function: Impulse function has zero value everywhere except at $t = 0$ where the amplitude is infinite.

   Step function is the derivative of ramp function.

   $$£\frac{d}{dt}\text{ (unit ramp function)} = s\cdot\frac{1}{s^2} = \frac{1}{s} = \text{step function}$$

   Impulse function is the derivative of step function

   $$£\frac{d}{dt}\text{ (unit step function)} = s\cdot\frac{1}{s} = 1$$

   Ramp function is the derivative of parabolic function

   $$£\frac{d}{dt}\text{ (parabolic function)} = s\cdot\frac{1}{s^3} = \frac{1}{s^2} = \text{ramp function}$$

Transient response depends upon :

(*a*) Location of closed loop poles in s-plane

(*b*) Order of the system.

Response of first order system with unit step input is given by

$$c(t) = 1 - e^{-t/T}$$

Response of first order system with unit ramp input is given by

$$c(t) = t - T + T e^{-t/T}$$

Response of first order system with unit impulse input is given by

$$c(t) = \frac{1}{T} e^{-t/T}$$

The closed loop transfer function of second order system is

$$\frac{C(s)}{R(s)} = \frac{\omega_n^2}{s^2 + 2\zeta\omega_n s + \omega_n^2}$$

where ω_n = natural frequency of oscillations

ζ = damping ratio

Response of second order system with unit step input is given by

$$C(t) = 1 - \frac{e^{-\zeta\omega_n t}}{\sqrt{1-\zeta^2}} \sin\left[\left(\omega_n\sqrt{1-\zeta^2}\right) t + \tan^{-1}\frac{\sqrt{1-\zeta^2}}{\zeta}\right]$$

The system is under damped for $0 < \zeta < 1$

The system is critical damped for $\zeta = 1$

The system is overdamped for $\zeta > 1$

The system is undamped for $\zeta = 0$

Delay time (t_d) : The delay time is the time required for the response to reach 50% of the final value in first time.

Rise time (t_r) : It is time required for the response to rise from 10% to 90% of its final value for overdamped systems and 0 to 100% for underdamped systems.

Peak time : The peak time is time required for the response to reach the first peak of the time response of first peak overshoot.

Maximum overshoot : It is the normalized difference between the peak of the time response and steady output.

Setting time (t_s) : Setting time is time required for the response to reach and stay within get specified range (2% to 5%) of its final value.

Steady state error : Steady state error is the difference between actual output and desired output as time '*t*' tends to infinity.

$$e_{SS} = \lim_{S\to\infty} [r(t) - c(t)]$$

Rise time is given by

$$t_r = \frac{\pi \quad \tan^{-1}\frac{\sqrt{1-\zeta^2}}{\zeta}}{\omega_n\sqrt{1-\zeta^2}}$$

Peak time is given by

$$t_p = \frac{2\pi}{\omega_n\sqrt{1-\zeta^2}}$$

Maximum overshoot is given by

$$\% M_p = e^{-\frac{\pi\zeta}{\sqrt{1-\zeta^2}}} \times 100$$

Settling time is given by

$$t_s = \frac{4}{\zeta\omega_n}$$

Response of second order system with unit impulse is given by

$$c(t) = \frac{\omega_n}{\sqrt{1-\zeta^2}} e^{-\zeta\omega_n t} \sin\left(\omega_n\sqrt{1-\zeta^2}\right)t$$

Response of second order system with unit ramp input

$$c(t) = t - \frac{2\zeta}{\omega_n} + \frac{e^{-\zeta\omega_n t}}{\omega_n\sqrt{1-\zeta^2}} \sin\left[\left(\omega_n\sqrt{1-\zeta^2}\right)t + \phi\right]$$

EXERCISE

1. A unity feedback system is characterised by an open loop transfer function

$$G(s) = \frac{K}{s(s+20)}$$

Determine the gain K so that the system will have a damping ratio of 0.6. For this value of K calculate settling time, peak overshoot and time to peak overshoot for a unit step input.

2. A feedback system is described by the following transfer function

$$G(s) = \frac{20}{s^2+4s+25}, \quad H(s) = Ks$$

The damping factor of the system is 0.8. Determine the overshoot of the system.

3. The forward path transfer function of a system is given by

$$G(s) = \frac{4}{s(s+6)}$$

Obtain an expression for unit step response of the system. Assume unity feedback system.

4. The open loop transfer function of a servosystem with unity feedback system is given by

$$G(s) = \frac{K}{s(s+2)(s+6)}$$

Determine the damping ratio, undamped natural frequency, maximum overshoot for unit step input and $K = 10$.

5. The open loop transfer function of a unity feedback system is $G(s) = \frac{K}{s(1+sT)}$. By what factor the gain K be reduced so that the overshoot is reduced from 60% to 20%.

6. The block diagram of a position control system with velocity feedback is shown in Fig. 2.30. Determine the value of 'a' so that the step response has a maximum overshoot of 20%.

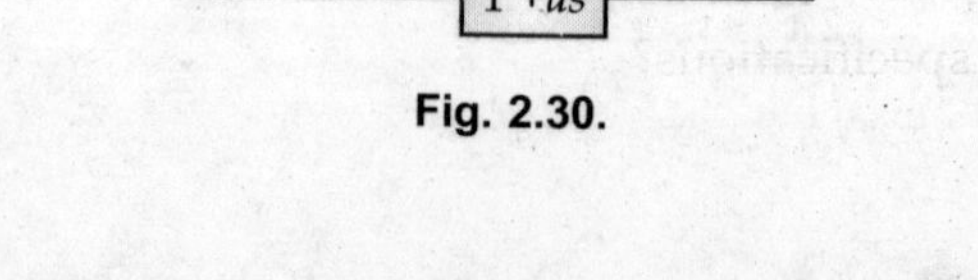

Fig. 2.30.

7. For the closed loop system given by

$$\frac{C(s)}{R(s)} = \frac{w_n^2}{s^2 + 2\xi w_n s + w_n^2}$$

Determine the values of ξ and w_n so that the system responds to a unit step input with 10% overshoot and with settling time of 4 sec.

8. For a unity feedback system with $G(s) = \frac{4}{s(s+2)}$ obtain an expression for unit step response of the system.

9. Determine the time response $y(t)$ for

(a) $Y(s) = \frac{4}{s(s+1)(s+2)}$ (b) $Y(s) = \frac{4}{s(s-1)(s-2)}$

10. What are the time domain quantities which characterise a transient response ? Derive an expression for percentage overshoot of a second order system. *(A.M.U., Aligarh, B.E., 1988-89)*

11. The open loop transfer function of a unity feedback system is

$$G(s) = \frac{K}{s+1}$$

Find the steady state error for an input of a unity feedback system as

$$r(t) = 1 + t + \frac{t^2}{2}$$

(A.M.U., Aligarh, B.E., 1988-89)

SEMI-OBJECTIVE TYPE QUESTIONS

(*i*) Define transient response. (Article 2.1)

(*ii*) Sketch various test input signals for transient analysis. (Article 2.2)

(*iii*) Write mathematical expressions for various test input signals. (Article 2.2)

(*iv*) Derive the expression for response of the first order system with unit step input. (Article 2.3.1)

(*v*) Define delay time, rise time and peak time. (Article 2.5)

(*vi*) Derive the expression for maximum overshoot (M_p) (Article 2.5)

(*vii*) For a system having transfer function

$$\frac{C(s)}{R(s)} = \frac{64}{s^2 + 5s + 64}$$

determine: (*i*) ω_n (*ii*) ξ (*iii*) ω_d (**Ans.** ω_n = 8 rad/sec, ξ = 0.3125, ω_d = 7.6 rad/sec)

(*viii*) Draw a typical transient response output for an underdamped second order system and show the locations of all the transient response specifications. (Article 2.5)

(*ix*) For a first order system find out the output of the system when the input applied to the system is a unit ramp. (Article 2.3.2)

(*x*) What is steady state error? (Article 2.1)

(*xi*) What is the effect of feedback on time constant of a control system? (Article 2.9)

(*xii*) Explain the relationship between damping ratio and % overshoot.

(*xiii*) What are time domain specifications?

□□□

3

ERROR ANALYSIS

3.1. CLASSIFICATION OF CONTROL SYSTEM

Consider the open loop transfer function

$$G(s)H(s) = \frac{K(1+sT_1)(1+sT_2)\ldots}{s^m(1+sT_a)(1+sT_b)+\ldots} \qquad \ldots(3.1)$$

In equation (3.1), the poles are at $s = -\frac{1}{T_a}$, $s = -\frac{1}{T_b}$ ... and zeros are at $s = {}^{-1}/_{T1}$, $s = {}^{-1}/_{T2}$... The equation having a term s^m in denominator, 'm' is the number of poles at the origin.

A system having no pole at origin of the 's' plane, is said to be type '0' (zero) system *i.e.*, $m = 0$

If $m = 1$ *i.e.*, 's', it means the system has a pole at origin of the s-plane and is said to be type '1' (one) system.

A system is called type '2' system if $m = 2$. and so on.

3.2. STEADY STATE ERROR

The steady state error is the difference between the input and output of the system during steady state. For accuracy the steady state error should be minimum.

Consider a closed loop control system shown in Fig. (3.1)

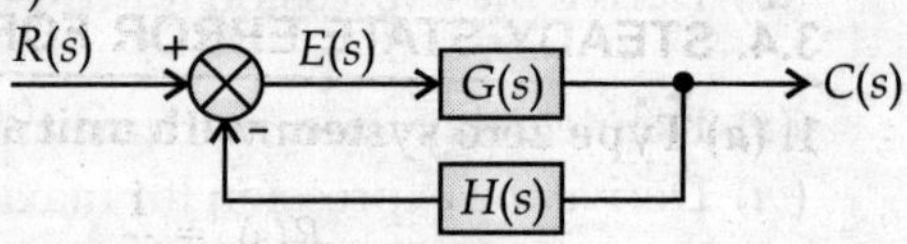

Fig. 3.1.

$$\frac{E(s)}{R(s)} = \frac{1}{1+G(s)H(s)}$$

or,

$$E(s) = \frac{R(s)}{1+G(s)H(s)} \qquad \ldots(3.2)$$

The steady state error of the system is obtained by applying final value theorem.

$$e_{ss} = \lim_{t\to\infty} e(t) = \lim_{s\to 0} s.E(s) \qquad \ldots(3.3)$$

$$e_{ss} = \lim_{s\to 0} s \cdot \frac{R(s)}{1+G(s)H(s)} \qquad \ldots(3.4)$$

For unity feedback system $H(s) = 1$

$$e_{ss} = \lim_{s\to 0} s \cdot \frac{R(s)}{1+G(s)} \qquad \ldots(3.5)$$

From the equation (3.4) or (3.5) it is clear that the steady state error depends on the input and open loop transfer function.

3.3. STATIC ERROR COEFFICIENTS

(*a*) Static-Position Error Constant (or Coefficient) K_p

The steady state error is given by equation 3.4

$$e_{ss} = \lim_{s\to 0} s\cdot\frac{R(s)}{1+G(s)H(s)}$$

For unit step input $R(s) = \frac{1}{s}$, the steady state error is given by

$$e_{ss} = \lim_{s\to 0} s\cdot\frac{1}{s}\frac{1}{1+G(s)H(s)} = \frac{1}{1+\lim_{s\to 0} G(s)H(s)} = \frac{1}{1+K_P} \quad ...(3.6)$$

K_P = static position error constant = $\lim_{s\to 0} G(s)H(s)$

(b) Static Velocity Error Constant (or Coefficient) K_v

$$e_{ss} = \lim_{s\to 0} s.R(s)\cdot\frac{1}{1+G(s)H(s)}$$

Steady state error with a unit ramp input is given by $[R(s) = 1/s^2]$

$$e_{ss} = \lim_{s\to 0} s\cdot\frac{1}{s^2}\frac{1}{1+G(s)H(s)} = \lim_{s\to 0}\frac{1}{s+sG(s)H(s)} = \lim_{s\to 0}\frac{1}{s\,G(s)H(s)}$$

$$e_{ss} = \frac{1}{K_v} \quad ...(3.7)$$

where $Kv = \lim_{s\to 0} s\,G(s)H(s)$ static velocity error coeff.

(c) Static acceleration error constant K_a

The steady state error of the system with unit parabolic input is given by

$$R(s) = \frac{1}{s^3}$$

$$e_{ss} = \lim_{s\to 0} s\cdot\frac{1}{s^3}\cdot\frac{1}{1+G(s)H(s)} = \lim_{s\to 0}\frac{1}{s^2+s^2G(s)H(s)} = \lim_{s\to 0}\frac{1}{s^2G(s)H(s)} = \frac{1}{K_a} \quad ...(3.8)$$

where $K_a = \lim_{s\to 0} s^2\,G(s)\,H(s)$ = static acceleration constant.

3.4. STEADY STATE ERROR FOR DIFFERENT TYPE OF SYSTEMS

1. (a) Type zero system with unit step input

$$R(s) = \frac{1}{s}$$

From equation (3.1)

$$G(s)\,H(s) = \frac{K(1+sT_1)\,(1+sT_2)...}{(1+sT_a)\,(1+sT_b)...} \quad ...(3.9)$$

$$K_P = \lim_{s\to 0} G(s)H(s) = \lim_{s\to 0}\frac{K(1+sT_1)\,(1+sT_2)...}{(1+sT_a)\,(1+sT_b)...} = \text{K}$$

From equation (3.6)

$$e_{ss} = \frac{1}{1+K_P} = \frac{1}{1+K} \qquad \boxed{e_{ss} = \frac{1}{1+K}}$$

Hence, for type zero system the static position error constant K_P is finite.

(b) Type '0' system with unit ramp input

$$K_v = \lim_{s\to 0} s\cdot G(s)H(s) = \lim_{s\to 0} s\cdot\frac{K(1+sT_1)\,(1+sT_2)...}{(1+sT_a)\,(1+sT_b)...} = 0$$

$$\therefore \quad e_{ss} = \frac{1}{Kv} = \infty \qquad \boxed{e_{ss} = \infty}$$

(c) type '0' system with unit parabolic input

$$Ka = \lim_{s\to 0} s^2 G(s)H(s) = \lim_{s\to 0} s^2 \frac{K(1+sT_1)(1+sT_2)...}{(1+sTa)(1+sT_b)...}$$

$$Ka = 0$$

$$\therefore \quad e_{ss} = \frac{1}{Ka} = \infty \qquad \boxed{e_{ss} = \infty}$$

For type '0' system, the steady state error is infinite for ramp and parabolic inputs. Hence, the ramp and parabolic inputs are not acceptable.

2(*a*) Type '1' system with unit step input (*m* = 1)

$$G(s)\,H(s) = \frac{K(1+sT_1)(1+sT_2)...}{s(1+sTa)(1+sT_b)...}$$

$$K_p = \lim_{s\to 0} G(s)H(s) = \lim_{s\to 0} \frac{K(1+sT_1)(1+sT_2)...}{s(1+sT_a)(1+sT_b)...} = \infty$$

$$e_{ss} = \frac{1}{1+K_p} = 0 \qquad \boxed{e_{ss} = 0}$$

(*b*) Type '1' system with unit ramp input

$$K_v = \lim_{s\to 0} s\cdot G(s)H(s) = \lim_{s\to 0} s\cdot\frac{K(1+sT_1)(1+sT_2)...}{s(1+sT_a)(1+sT_b)...}$$

$$K_v = K$$

$$e_{ss} = \frac{1}{K_v} = \frac{1}{K} \qquad \boxed{e_{ss} = \frac{1}{K}}$$

(c) Type '1' system with unit parabolic input

$$K_a = \lim_{s\to 0} s^2 G(s)H(s) = \lim_{s\to 0} s^2 \frac{K(1+sT_1)(1+sT_2)...}{s(1+sT_a)(1+sT_b)...} = 0$$

$$\therefore \quad e_{ss} = \frac{1}{K_a} = \infty \qquad \boxed{e_{ss} = \infty}$$

Hence, from above relations for type '1' system, it is clear that for type '1' system step input and ramp inputs are acceptable and parabolic input is not acceptable.

3 (a) Type '2' system with unit step input

$$G(s)\,H(s) = \frac{K(1+sT_1)(1+sT_2)...}{s^2(1+sT_a)(1+sT_b)...}$$

$$K_p = \lim_{s\to 0} G(s)H(s) = \lim_{s\to 0} \frac{K(1+sT_1)(1+sT_2)...}{s^2(1+sT_a)(1+sT_b)...} = \infty$$

$$e_{ss} = \frac{1}{1+K_p} = 0 \qquad \boxed{e_{ss} = 0}$$

(*b*) Type '2' system with unit ramp input

$$K_v = \lim_{s\to 0} sG(s)H(s)$$

$$= \lim_{s\to 0} s \cdot \frac{K(1+sT_1)(1+sT_2)\ldots}{s^2(1+sT_a)(1+sT_b)\ldots}$$

$$= \infty$$

$$\therefore \quad e_{ss} = \frac{1}{K_v} = 0 \quad \boxed{e_{ss} = 0}$$

(c) Type '2' system with unit parabolic input

$$K_a = \lim_{s\to 0} s^2 G(s)H(s) = \lim_{s\to 0} s^2 \cdot \frac{K(1+sT_1)(1+sT_2)\ldots}{s^2(1+sT_a)(1+sT_b)\ldots} = K$$

$$\therefore \quad e_{ss} = \frac{1}{K_a} = \frac{1}{K} \quad e_{ss} = \boxed{\frac{1}{K}}$$

Hence, for type '2' system all three inputs (step, ramp and parabolic) are acceptable. From Table 3.1, the diagonal elements are the finite values of steady state error.

Table 3.1.

	Type '0' system	Type '1' system	Type '2' system
Unit Step Input	$\frac{1}{1+k}$	0	0
Unit Ramo Input	∞	$\frac{1}{k}$	0
Unit Parabolic Input	∞	∞	$\frac{1}{k}$

EXAMPLE 3.1. The open loop transfer function of unity feedback system is given by

$$G(s) = \frac{50}{(1+0.1s)(s+10)}$$

Determine the static error coefficients K_p, K_v and K_a.

Solution :

$$K_P = \lim_{s\to 0} G(s)H(s) = \lim_{s\to 0} \frac{50}{(1+0.1s)(s+10)} = 5$$

$$K_v = \lim_{s\to 0} s\,G(s)H(s) = \lim_{s\to 0} s \cdot \frac{50}{(1+0.1s)(s+10)} = 0$$

$$K_a = \lim_{s\to 0} s^2 \frac{50}{(1+0.1s)(s+10)} = 0$$

EXAMPLE 3.2. The forward path transfer function of a unity feedback control system is given by

$$G(s) = \frac{5(s^2+2s+100)}{s^2(s+5)(s^2+3s+10)}$$

Determine the step, ramp and parabolic error coefficients. Also determine the type of the system.

Solution :

$$K_P = \lim_{s\to 0} G(s)H(s) = \lim_{s\to 0} \frac{5(s^2+2s+100)}{s^2(s+5)(s^2+3s+10)}$$

$$K_P = \infty \quad \boxed{K_P = \infty}$$

$$K_v = \lim_{s\to 0} sG(s)H(s) = \lim_{s\to 0} s\cdot\frac{5(s^2+2s+100)}{s^2(s+5)(s^2+3s+10)}$$

$$K_v = \infty \qquad \boxed{K_v=\infty}$$

$$K_a = \lim_{s\to 0} s^2G(s)H(s) = \lim_{s\to 0} s^2\cdot\frac{5(s^2+2s+100)}{s^2(s+5)(s^2+3s+10)}$$

$$= 10 \qquad \boxed{K_a = 10}$$

In denominator the value of $m = 2$. Hence, the given system is type '2' system.

EXAMPLE 3.3. The block diagram of an electronic pacemaker is given in Fig. 3.2. Determine the steady state error for unit ramp input when K = 400. Also, determine the value of K for which the steady state error to a unit ramp will be 0.02.

Solution : Given that $K = 400$

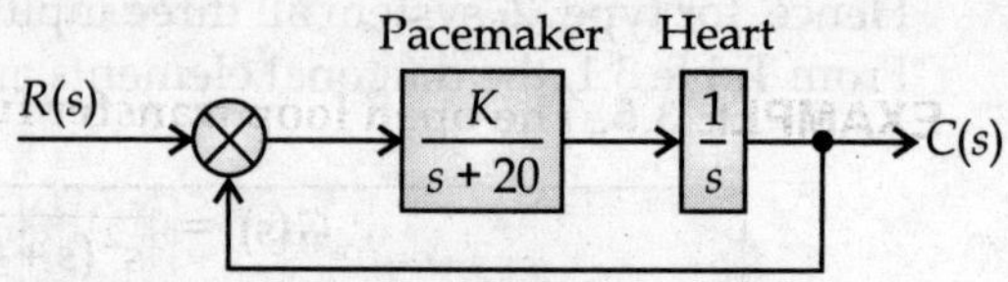

Fig. 3.2.

$$R(s) = \frac{1}{s^2}$$

$$H(s) = 1$$

$$\therefore \quad G(s)\,H(s) = \frac{K}{s(s+20)}$$

Steady state error is given by $e_{ss} = \lim_{s\to 0} s\cdot\frac{R(s)}{1+G(s)H(s)}$

$$e_{ss} = \lim_{s\to 0} s\cdot\frac{1}{s^2}\,\frac{1}{1+\frac{K}{s(s+20)}} = \lim_{s\to 0}\frac{s+20}{s(s+20)+400} = 0.05 \qquad \textbf{Ans.}$$

Now $e_{ss} = 0.02$ (given)

$$e_{ss} = \lim_{s\to 0} s\cdot\frac{1}{s^2}\,\frac{1}{1+\frac{K}{s(s+20)}}$$

$$0.02 = \lim_{s\to 0}\frac{s+20}{s(s+20)+K}$$

$$0.02 = \frac{20}{K} \qquad \therefore \quad \boxed{K = 1000} \quad \textbf{Ans.}$$

EXAMPLE 3.4. For a unity feedback control system the forward path transfer function is given by

$$G(s) = \frac{20}{s(s+2)(s^2+2s+20)}$$

Determine the steady state error of the system. When the inputs are : (*i*) 5, (*ii*) 5*t*, (*iii*) $\frac{3t^2}{2}$.

Solution :

(*i*) $r(t) = 5 \quad \therefore R(s) = \frac{5}{s}$

$$e_{ss} = \lim_{s\to 0} s\cdot E(s) = \lim_{s\to 0} s\cdot\frac{R(s)}{1+G(s)H(s)}$$

$$e_{ss} = \lim_{s\to 0} s\cdot\frac{5}{s}\cdot\frac{1}{1+\frac{20}{s(s+2)(s^2+2s+20)}} = \lim_{s\to 0}\frac{5s(s+2)(s^2+2s+20)}{s(s+2)(s^2+2s+20)+20}$$

$\therefore \quad e_{ss} = 0$

(ii) $\quad R(s) = \dfrac{5}{s^2}$

$$e_{ss} = \lim_{s\to 0} s\cdot\frac{5}{s^2}\cdot\frac{1}{1+\dfrac{20}{s(s+2)(s^2+2s+20)}} = \lim_{s\to 0} s\cdot\frac{5}{s^2}\cdot\frac{s(s+2)(s^2+2s+20)}{s(s+2)(s^2+2s+20)+20}$$

$$e_{ss} = 10$$

(iii) $\quad R(s) = \dfrac{3}{s^3}$

$$e_{ss} = \lim_{s\to 0} s\cdot\frac{3}{s^3}\cdot\frac{s(s+2)(s^2+2s+20)}{s(s+2)(s^2+2s+20)+20}$$

$$e_{ss} = \infty$$

EXAMPLE 3.5. The open loop transfer function of a unity feedback system is given by

$$G(s) = \frac{108}{s^2(s+4)(s^2+3s+12)}$$

Find the static error coefficients and steady state error of the system when subjected to an input given by

$$r(t) = 2 + 5t + 2t^2$$

Solution :

$$K_P = \lim_{s\to 0} G(s) = \lim_{s\to 0}\frac{108}{s^2(s+4)(s^2+3s+12)} = \infty$$

$$K_v = \lim_{s\to 0} s\cdot G(s) = \lim_{s\to 0} s\cdot\frac{108}{s^2(s+4)(s^2+3s+12)} = \infty$$

$$K_a = \lim_{s\to 0} s^2 G(s) = \lim_{s\to 0} s^2\cdot\frac{108}{s^2(s+4)(s^2+3s+12)} = \frac{108}{48}$$

$$r(t) = 2 + 5t + 2t^2$$

$$R(s) = \frac{2}{s} + \frac{5}{s^2} + \frac{4}{s^3}$$

$$\therefore \quad e_{ss} = \frac{R_1}{1+K_P} + \frac{R_2}{K_v} + \frac{R_3}{K_a} = \frac{2}{1+\infty} + \frac{5}{\infty} + \frac{4\times 48}{108} = 1.77$$

$\boxed{e_{ss} = 1.77}$ **Ans.**

EXAMPLE 3.6. Determine the type of the following unity feedback systems for which the forward path transfer function are given

(a) $G(s) = \dfrac{K}{(s+10)(s+5)}$ (b) $G(s) = \dfrac{20}{s(1+0.1s)(1+0.05s)}$

(c) $G(s) = \dfrac{20(s+2)}{s^3(s+2)(s^2+5s+5)}$ (d) $G(s) = \dfrac{K}{s(s+1)(s+2)}$

Solution : (a) Since in denominator the power of s is zero *i.e.,* $m = 0$ hence it is type zero system.

(b) type one system

(c) $m = 3$, type '3' system

(d) $m = 1$, type '1' system

EXAMPLE 3.7. A servomechanism is designed to keep a radar antena pointed at a flying aeroplane. If the aeroplane is flying with a velocity of 600 km/hr, at a range of 2 km. and the maximum tracking error is to be within 0.1°, determine the required velocity error coefficient. *(GATE-1994)*

Solution : The block diag. of servo mechanism is shown in Fig. 3.3.

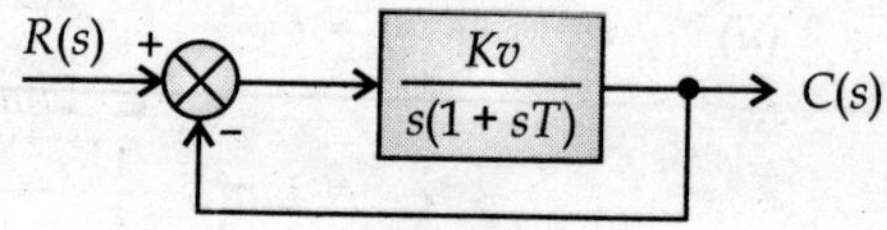

Fig. 3.3.

$$\text{linear velocity} = 600 \text{ km./hr}$$

$$r = 2\text{km.}$$

$$\text{Angular velocity } \omega = \frac{600}{2} = 300 \text{ rad}/hr.$$

$$= \frac{300}{3600} \text{ rad/sec.} = 0.083 \text{ rad/sec.}$$

$$R(s) = \frac{0.083}{s^2}$$

$$e_{ss} = \lim_{s\to 0} s\cdot R(s)\cdot\frac{1}{1+G(s)H(s)} = \lim_{s\to 0} s\cdot\frac{0.083}{s^2}\cdot\frac{1}{1+\dfrac{K_v}{s(1+sT)}}$$

$$e_{ss} = \lim_{s\to 0}\frac{0.083}{s}\cdot\frac{s(1+sT)}{s(1+sT)+K_v} = \frac{0.083}{K_v} = 0.1° \text{ given}$$

$$0.1 = \frac{0.083}{K_v}$$

$$\therefore \quad K_v = 0.83 \text{ per degree}$$

$\therefore$ Required velocity error coefficient $K_v = 0.83$ per degree

EXAMPLE 3.8. For the system shown in Fig. 3.4. Determine K_P and e_{ss} for unit step input.

Solution : $K_P = \lim_{s\to 0} G(s)H(s) = \lim_{s\to 0}\dfrac{1}{(s^2+s+2)(s+1)} = \dfrac{1}{2}$

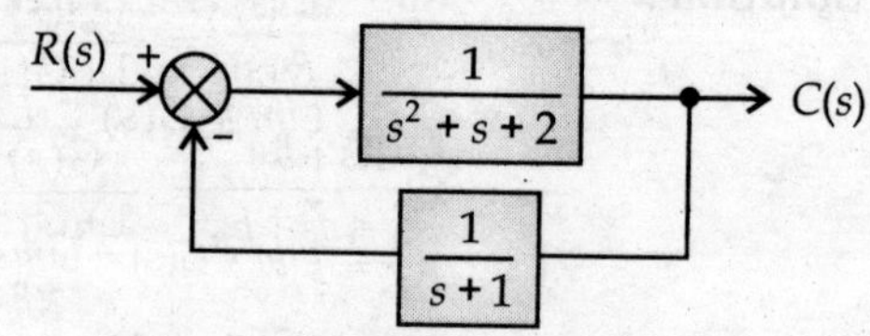

Fig. 3.4.

$$e_{ss} = \frac{1}{1+K_P} = \frac{1}{1+\frac{1}{2}} = \frac{1}{3/2} = \frac{2}{3}$$

$\boxed{K_P = 1/2}$ $\boxed{e_{ss} = 2/3}$ **Ans.**

EXAMPLE 3.9. For a given system shown in Fig. 3.5, determine the actuating signal $E_a(s)$. Also determine the position error constant for unit step input.

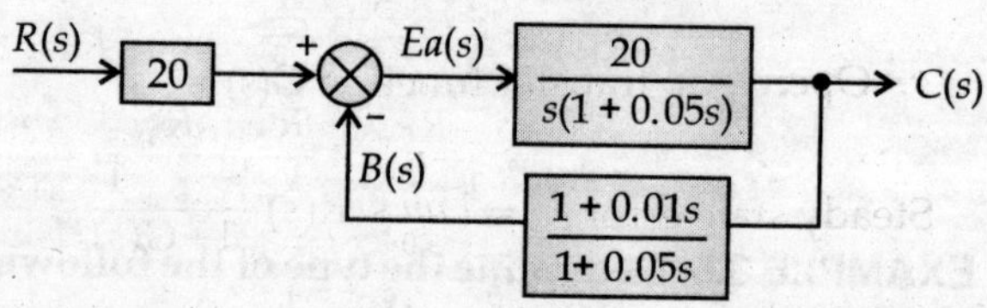

Fig. 3.5.

Solution :

$$E_a(s) = 20\,R(s) - B(s)$$

$$C(s) = \frac{20Ea(s)}{s(1+0.05s)}$$

$$B(s) = C(s)\cdot\frac{1+0.01s}{1+0.05s} = \frac{20(1+0.01s)E_a(s)}{s(1+0.05s)^2}$$

$$\therefore \quad E_a(s) = 20R(s) - \frac{20(1+0.01s)E_a(s)}{s(1+0.05s)^2}$$

$$E_a(s)\left[1+\frac{20(1+0.01s)}{s(1+0.05s)^2}\right] = 20\,R(s)$$

$$\therefore \quad E_a(s) = \frac{20R(s)}{1+\dfrac{20(1+0.01s)}{s(1+0.05s)^2}} = \frac{20s(1+0.05s)^2\cdot R(s)}{s(1+0.05s)^2+20(1+0.01s)} \quad \textbf{Ans.}$$

For step Input $\quad R(s) = \dfrac{1}{s}$

$$E_a(s) = \frac{20(1+0.05s)^2}{s(1+0.05s)^2+20(1+0.01s)}$$

$$\therefore \quad e_{ss} = \lim_{s\to 0} s\cdot E(s) = \lim_{s\to 0} s\cdot\frac{20(1+0.05s)^2}{s(1+0.05s)^2+20(1+0.01s)} = 0$$

But $\quad e_{ss} = \frac{1}{1+K_P} \quad$ or, $\quad 0 = \frac{1}{1+K_P}$

$\therefore \quad \boxed{K_P = \infty}$ **Ans.**

EXAMPLE 3.10. Consider a unity feedback control system with the closed loop transfer function

$$\frac{C(s)}{R(s)} = \frac{Ks+b}{s^2+as+b}$$

Determine the open loop transfer function. Show that the steady state error in the unit ramp input response is given by

$$e_{ss} = \frac{a-k}{b}$$

Solution : For closed loop control system

$$\frac{C(s)}{R(s)} = \frac{G(s)}{1+G(s)H(s)}$$

$$H(s) = 1$$

$$\frac{C(s)}{R(s)} = \frac{G(s)}{1+G(s)}$$

$$\frac{Ks+b}{s^2+as+b} = \frac{G(s)}{1+G(s)}$$

$$\therefore \quad G(s) = \frac{Ks+b}{s^2+s(a-K)}$$

$\therefore$ Open loop transfer function $G(s) = \frac{Ks+b}{s^2+s(a-k)}$ **Ans.**

Steady state error $e_{ss} = \lim_{s\to 0} s\cdot R(s)\cdot\frac{1}{1+G(s)H(s)}$

$$e_{ss} = \lim_{s\to 0} s\cdot\frac{1}{s^2}\cdot\frac{1}{1+\frac{Ks+b}{s^2+s(a-K)}\cdot 1} = \lim_{s\to 0}\frac{1}{s}\cdot\frac{s^2+s(a-K)}{s^2+s(a-K)+Ks+b}$$

$$= \lim_{s\to 0}\frac{1}{s}\,\frac{s[s+a-K]}{s^2+s(a-K)+Ks+b} = \frac{a-K}{b}$$

$$e_{ss} = \frac{a-K}{b} \quad \textbf{Proved.}$$

3.5. DYNAMIC ERROR COEFFICIENTS

For the steady state error, the static error coefficients gives the limited information.

The error function is given by

$$\frac{E(s)}{R(s)} = \frac{1}{1+G(s)H(s)} \qquad ...(3.10)$$

For unity feedback system

$$\frac{E(s)}{R(s)} = \frac{1}{1+G(s)} \qquad ...(3.11)$$

The equation 3.11 can be expressed in polynomial form (ascending power of 's')

$$\frac{E(s)}{R(s)} = \frac{1}{K_1}+\frac{1}{K_2}s+\frac{1}{K_3}s^2+..... \qquad ...(3.12)$$

or,

$$E(s) = \frac{1}{K_1}R(s)+\frac{1}{K_2}sR(s)+\frac{1}{K_3}s^2R(s).... \qquad ...(3.13)$$

Take inverse laplace of equation 3.13, the error given by

$$e(t) = \frac{1}{K_1}r(t)+\frac{1}{K_2}\dot{r}(t)+\frac{1}{K_3}\ddot{r}(t)+.... \qquad ...(3.14)$$

Steady state error is given by

$$e_{ss} = \lim_{s\to 0} sE(s)$$

Now, suppose if the input is unit step, *i.e.*, $R(s) = \frac{1}{s}$ then from 3.13

$$e_{ss} = \lim_{s\to 0} s\cdot\left[\frac{1}{K_1}\cdot\frac{1}{s}+\frac{1}{K_2}s\cdot\frac{1}{s}+\frac{1}{K_3}s^2\cdot\frac{1}{s}+...\right] = \frac{1}{K_1}$$

Similarly, for other test signal we can find steady state error. K_1, K_2, K_3 ... are known as "Dynamic error coefficients".

EXAMPLE 3.11. Find the dynamic error coefficients of the unity feedback system whose forward transfer function is $G(s) = \frac{200}{s(s+5)}$. Find the steady state error of the system for the input $4t^2$.

Solution : Given that

$$G(s) = \frac{200}{s(s+5)}$$

$$H(s) = 1$$

$$\therefore \quad \frac{E(s)}{R(s)} = \frac{1}{1+G(s)H(s)} = \frac{s^2+5s}{s^2+5s+200} = \frac{5s+s^2}{200+5s+s^2}$$

$$\begin{array}{r|l} & 0.025s+0.004375s^2-0.0000218s^4 \\ 200+5s+s^2 & 5s+s^2 \\ & 5s \pm 0.125s^2 \pm 0.025s^3 \\ \hline & 0.875\,s^2 - 0.025\,s^3 \\ & 0.875\,s^2 \mp 0.022\,s^3 \pm 0.004375\,s^4 \\ \hline & -0.004375\,s^4 \\ & \mp 0.004375\,s^4 \mp 0.0001\,s^5 \mp 0.0000218\,s^6 \\ \hline & 0.0001s^5 + 0.0000218\,s^6 \end{array}$$

$$\therefore \quad \frac{E(s)}{R(s)} = 0.025\,s + 0.004375\,s^2 - 0.0000218\,s^4 .....$$

or

$$E(s) = 0.025\,sR(s) + 0.00437\,5s^2\,R(s) - 0.0000218\,s^4\,R(s) .....$$

Take inverse laplace

$$e(t) = 0.025\,\dot{r}(t) + 0.004375\,\ddot{r}(t) - 0.0000218\,\ddddot{r}(t)$$

Given that $r(t) = 4t^2$

$\dot{r}(t) = 8t$

$\ddot{r}(t) = 8$

$\dddot{r}(t) = 0$

$\therefore \quad e(t) = 0.025 \times 8t + 0.004375 \times 8$

$e(t) = 0.2t + 0.035$

The steady state error is given by.

$$\lim_{t \to \infty} e(t) = \lim_{t \to \infty} (0.2t + 0.035)$$

The dynamic error coefficients are given by

$$K_1 = \frac{1}{0.025} = 40 \quad \text{and} \quad K_2 = \frac{1}{0.004375} = 228.57$$

$$K_3 = \infty \quad \text{and} \quad K_4 = 45871.55$$

EXAMPLE 3.12. Find the dynamic error coefficients of the unity feedback system whose forward path transfer function $G(s) = \dfrac{10}{s(s+1)}$.

Find the steady state error to the input $r(t) = P_0 + P_1 t + P_2 t^2$

Solution : $G(s) = \dfrac{10}{s(s+1)}$, $H(s) - 1$

$$\frac{E(s)}{R(s)} = \frac{1}{1+G(s)H(s)} = \frac{s+s^2}{10+s+s^2} = 0.1s + 0.09s^2 - 0.019\,s^3 \ldots$$

$\therefore \quad E(s) = 0.1s\,R(s) + 0.09\,s^2\,R(s) - 0.019s^3\,R(s) \ldots\ldots$

Inverse laplace $\quad e(t) = 0.1\,\dot{r}(t) + 0.09\,\ddot{r}(t) - 0.019\,\dddot{r}(t) \ldots\ldots$

$r(t) = P_0 + P_1 t + P_2 t^2$

$\dot{r}(t) = P_1 + 2P_2 t$

$\ddot{r}(t) = 2P_2$

$\dddot{r}(t) = 0$

$\therefore \quad e(t) = 0.1\,(P_1 + 2P_2 t) + 0.09 \times 2P_2 = 0.1\,(P_1 + 2P_2 t) + 0.18\,P_2$

Steady state error is given by

$$\lim_{t \to \infty} e(t) = \lim_{t \to \infty} 0.1(P_1 + 2P_2 t) + 0.18P_2$$

Dynamic error coefficient are

$K_1 = 10$; $\quad K_2 = 11.1$; $\quad K_3 = -52.63$

SUMMARY

The steady state error is the difference between input and output of the system during steady state. For accuracy steady state error should be minimum.

$$e_{SS} = \lim_{s \to 0} sE(s) = \lim_{s \to 0} s \cdot \frac{R(s)}{1+G(s)\,H(s)}$$

$$G(s)\,H(s) = \frac{K(1+sT_1)\,(1+sT_2 \ldots}{s^m (1+sT_a)\,(1+sT_b) \ldots}$$

m = type of system

if $m = 0$, system is type zero

$m = 1$, system is type one etc.

K_P = static position error constant

$= \lim_{S \to 0} G(s)\, H(s)$ to be used for step input

K_v = Static velocity coefficient = $\lim_{S \to 0} sG(s)\, H(s)$ to be used for ramp input

K_a = static acceleration error constant

$= \lim_{s \to 0} s^2 G(s)\, H(s)$ to be used for parabolic input

	Type zero system	Type one system	Type 2 system
Unit step input	$\frac{1}{1+K}$	0	0
Unit ramp input	∞	$\frac{1}{K}$	0
Unit parabolic input	∞	∞	$\frac{1}{K}$

SEMI-OBJECTIVE TYPE QUESTIONS

(*i*) State how type and order of a control system is determined? (Article 3.1)

(*ii*) Define steady-state error. (Article 3.2)

(*iii*) What are static error coefficients? (Article 3.3)

(*iv*) Derive the values of static error coefficients and steady-state error for a type 1 system with unit step input. (Article 3.4)

(*v*) Define D_p, K_v and K_a. (Articles and 3.4)

(*vi*) What is dynamic error coefficient? (Article 3.5)

(*vii*) The open loop transfer function of unity feedback system is given by

$$G(s) = \frac{100}{(1+0.1s)(s+10)}$$

Determine K_p, K_v & K_a (**Ans.** 10, 0 and 0)

(*viii*) What is velocity error?

(*ix*) What is positional error coefficient?

(*x*) Define static error constant

(*xi*) Write down the error series giving the generalised error coefficients.

□□□

4 FREQUENCY DOMAIN ANALYSIS

4.1. INTRODUCTION

The magnitude and phase relationship between sinusoidal input and steady state output of a system is known as frequency response. It is independent of the amplitude and phase of the input signal. If input signal is

$$r(t) = X \sin \omega t$$

the output can be written as

$$c(t) = Y \sin (\omega t + \theta)$$

The main advantage is that the frequency response of a stable open loop system can be obtained experimentally. Apply a sine wave at the input of the system and measure the ratio of the magnitude of output and input as well as the phase difference. Another advantage of this approach is that it is simple and accurate. We can determine experimentally the transfer function of the complicated system. Frequency response tests are not suitable for the systems having large time constant because for each frequency it takes long time to reach the steady state.

$$G(s) = \frac{C(s)}{R(s)} \quad \text{or,} \quad G(j\omega) = \frac{C(j\omega)}{R(j\omega)}$$

$$\text{or,} \quad C(s) = R(s).\, G(s) \quad \text{or,} \quad C(j\omega) = R(j\omega)\, G(j\omega)$$

The output is the product of input and $|G(s)|$

For stable system, the sinusoidal output is of the same frequency as input but the ampitude and phase of the output will be different from the input.

$$\therefore \quad G(j\omega) = M \angle \phi$$

where M is the ratio of amplitudes of output and input sinusoids and is called the gain or magnitude ratio.

ϕ is angle by which output leads the input. M & ϕ are the function of angular frequency ω.

4.2. POLAR PLOT

The polar plot of a sinusoidal transfer function $G(j\omega)$ is a plot of the magnitude of $C(j\omega)$ versus the phase angle of $G(j\omega)$ on polar coordinates as 'ω' is varied from zero to infinity. The polar plot, therefore is the locus of vectors $|G(j\omega)| \angle G(j\omega)$ as ω is varied from zero to infinity. Thus we can express the frequency response function $G(j\omega)$ in the polar form $Me^{j\phi(\omega)}$ and plot the vector$Me^{j\phi(\omega)}$ in the G-plane as 'ω' varies from zero to infinity. In polar plots the magnitude of $G(j\omega)$ is plotted as the distance from the origin while the phase angle is measured from positive real axis. Positive phase angle is measured counter clockwise while negative phase angle is measured clockwise from the positive real axis. The polar plot is often called the Nyquist Plot.

The advantage in using a polar plot that it depicts the frequency response characteristics of a system over the entire frequency range in a single plot. The disadvantage is that the plot does not indicate the contributions of each individual factor of the open loop transfer function.

4.3. PROCEDURE TO SKETCH THE POLAR PLOT

Step 1 : Determine the transfer function $G(s)$ of the system.

Step 2 : Put $S = j\omega$ in the transfer function to obtain $G(j\omega)$

Step 3 : At $\omega = 0$ and $\omega = \infty$ calculate $|G(j\omega)|$. by $\lim_{\omega\to 0}|G(j\omega)|$ and $\lim_{\omega\to\infty}|G(j\omega)|$.

Step 4 : Calculate the phase angle of $G(j\omega)$ at $\omega = 0$ and $\omega = \infty$ by $\lim_{\omega\to 0}\angle G(j\omega)$ and $\lim_{\omega\to\infty}\angle G(j\omega)$

Step 5 : Rationalize the function $G(j\omega)$ and separate the real and imaginary parts.

Step 6 : Equate the imaginary part $I_m|G(j\omega)|$ to zero and determine the frequencies at which plot intersects the real axis and caluculate the value $G(j\omega)$ at the point of intersection by substituting the determined value of frequency in the expression of $G(j\omega)$.

Step 7 : Equate the real part $\text{Re}|G(j\omega)|$ to zero and determine the frequencies at which plots intersects the imaginary axis and calculate the value of $G(j\omega)$ at the point of intersection by substituting the determined value of frequency in the rationalized expression of $G(j\omega)$.

Step 8 : Sketch the polar plot with the help of above information.

1. Type 'Zero' System

$$G(s) = \frac{K}{(1+ST_1)(1+ST_2)}$$

Step 1 : Put $S = j\omega$

$$G(j\omega) = \frac{K}{(1+j\omega T_1)(1+j\omega T_2)} = \frac{K}{\sqrt{1+(\omega T_1)^2}\sqrt{1+(\omega T_2)^2}} \angle -\tan^{-1}\omega T_1 - \tan^{-1}\omega T_2$$

Step 2 : Taking the limit for the magnitude of $G(j\omega)$.

$$\lim_{\omega\to 0}|G(j\omega)| = \lim_{\omega\to 0}\frac{K}{\sqrt{1+(\omega T_1)^2}\sqrt{1+(\omega T_2)^2}} = K$$

$$\lim_{\omega\to\infty}|G(j\omega)| = \lim_{\omega\to\infty}\frac{K}{\sqrt{1+(\omega T_1)^2}\sqrt{1+(\omega T_2)^2}} = 0$$

Step 3 : Taking the limit for the phase angle of $G(j\omega)$

$$\lim_{\omega\to 0}\angle G(j\omega) = \lim_{\omega\to 0}\angle -\tan^{-1}\omega T_1 - \tan^{-1}\omega T_2 = 0$$

$$\lim_{\omega\to\infty}\angle G(j\omega) = \lim_{\omega\to\infty}\angle -\tan^{-1}\omega T_1 - \tan^{-1}\omega T_2 = -180^\circ \,*$$

* ***Alternative :***

Note : Polar plot will intersect –90° axis so put

$$\angle G(j\omega) = -90\ ; \quad -\tan^{-1}\omega T_1 - \tan^{-1}\omega T_2 = -90^\circ$$

taking tan on both side

$$\frac{\omega T_1 + \omega T_2}{1-\omega T_1\omega T_2} = \infty$$

$$\therefore \quad 1 - \omega_2 T_1 T_2 = 0\ ; \quad \text{or} \quad \omega = 1/\sqrt{T_1T_2}$$

Put $\omega = \frac{1}{\sqrt{T_1T_2}}$ in $\frac{K}{(1+sT_1)(1+sT_2)}$ *i.e.,* $\frac{K}{(1+j\omega T_1)(1+j\omega T_2)}$ and simplify $|G(j\omega)| = \frac{K\sqrt{T_1T_2}}{T_1+T_2}$

Step 4 : Separating the real and imaginary parts of $G(j\omega)$

$$G(j\omega) = \frac{K}{(1+j\omega T_1)(1+j\omega T_2)} \cdot \frac{(1-j\omega T_1)(1-j\omega T_2)}{(1-j\omega T_1)(1-j\omega T_2)}$$

$$G(j\omega) = \frac{K(1-\omega^2 T_1 T_2)}{1+\omega^2 T_1^2+\omega^2 T_2^2+\omega^4 T_1^2 T_2^2} - j\frac{K\omega(T_1+T_2)}{1+\omega^2 T_1^2+\omega^2 T_2^2+\omega^4 T_1^2 T_2^2}$$

Equating the real part to zero.

$$\frac{K(1-\omega^2 T_1 T_2)}{1+\omega^2 T_1^2+\omega^2 T_2^2+\omega^4 T_1 T_2} = 0$$

$$\therefore \quad \omega^2 = \frac{1}{T_1 T_2} \quad \text{or} \quad \omega = \frac{1}{\sqrt{T_1 T_2}} \quad \& \quad \omega = \pm\infty$$

The frequency at which plot intersects the imaginary axis is $\frac{1}{\sqrt{T_1 T_2}}$.

For positive values of frequencies the polar plot intersects the imaginary axis at $\omega = \infty$

$$\omega = \frac{1}{\sqrt{T_1 T_2}} \quad \text{and} \quad \omega = \infty$$

Value of $G(j\omega)$ when

$$\omega = \frac{1}{\sqrt{T_1 T_2}}$$

$$G(j\omega) = 0 - j\frac{K\frac{1}{\sqrt{T_1 T_2}}(T_1+T_2)}{1+\frac{1}{T_1 T_2}T_2^2+\frac{1}{T_1 T_2}T_1^2+\frac{1}{T_1^2 T_2^2}\cdot T_1^2 T_2^2}$$

$$= -j\frac{K\frac{T_1+T_2}{\sqrt{T_1 T_2}}}{2+\frac{T_2}{T_1}+\frac{T_1}{T_2}}$$

$$= -j\frac{K\frac{T_1+T_2}{\sqrt{T_1 T_2}}}{\frac{(T_1+T_2)^2}{T_1 T_2}}$$

$$= -j\frac{K\sqrt{T_1 T_2}}{T_1+T_2}$$

$$|C(j\omega)| = \frac{K\sqrt{T_1 T_2}}{T_1+T_2} \quad \& \quad \angle G(j\omega) = -90°$$

$$\therefore \text{ When} \quad \omega = \frac{1}{\sqrt{T_1 T_2}} \quad G(j\omega) = \frac{K\sqrt{T_1 T_2}}{T_1+T_2}\angle -90°$$

$$\omega = \infty \quad G(j\omega) = 0 \angle -180°$$

Step 5 : Equating the imaginary part to zero

$$\frac{K\omega(T_1+T_2)}{1+\omega^2T_1^2+\omega^2T_2^2+\omega^4T_1T_2}=0$$

$\therefore \quad \omega = 0 \ \& \ \omega = \pm\infty$

When $\quad \omega = 0 \quad |G(j\omega)| = K \quad \angle G(j\omega) = 0°$

$\qquad \omega = \infty \quad |G(j\omega)| = 0 \quad \angle G(j\omega) = 0°$

The coordinates of the intersection points 'A' and '0'

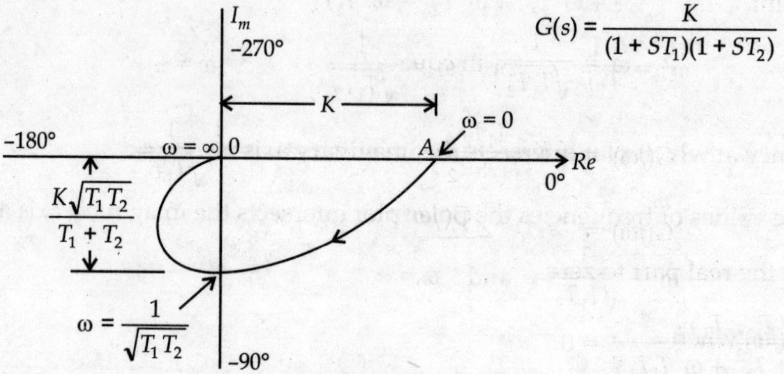

Fig. 4.1.

2. Type 'One' System

$$G(s) = \frac{K}{S(1+ST_1)(1+ST_2)}$$

Step 1 : Put $\quad s = j\omega$

$$G(j\omega) = \frac{K}{j\omega(1+j\omega T_1)(1+(j\omega T_2)}$$

$$G(j\omega) = \frac{K}{\omega\sqrt{1+(\omega T_1)^2}\sqrt{1+(\omega T_2)^2}} \angle -90° - \tan^{-1}\omega T_1 - \tan^{-1}\omega T_2$$

Step 2 : Taking the limit for the magnitude of $G(j\omega)$

$$\lim_{\omega\to 0}|G(j\omega)| = \lim_{\omega\to 0}\frac{K}{\omega\sqrt{1+(\omega T_1)^2}\sqrt{1+(\omega T_2^2)}} = \infty$$

$$\lim_{\omega\to\infty}|G(j\omega)| = \lim_{\omega\to\infty}\frac{K}{\omega\sqrt{1+(wT_1)^2}\sqrt{1+(wT_2)^2}} = 0$$

Step 3 : Taking the limit for the phase angle of $G(j\omega)$

$$\lim_{\omega\to 0}\angle G(j\omega) = \lim_{\omega\to 0}\angle -90° - \tan^{-1}\omega T_1 - \tan^{-1}\omega T_2 = -90°$$

$$\lim_{\omega\to\infty}\angle G(j\omega) = \lim_{\omega\to\infty}\angle -90° - \tan^{-1}\omega T_1 - \tan^{-1}\omega T_2 = -270°$$

Step 4 : Separating the real and imaginary parts

$$G(j\omega) = \frac{K}{j\omega(1+j\omega T_1)(1+j\omega T_2)}$$

$$= \frac{-\omega K(T_1+T_2)}{\omega+\omega^3(T_1^2+T_2^2+\omega^2T_1^2T_2^2)} + \frac{j(K\omega^2T_1T_2-K)}{\omega+\omega^3(T_1^2+T_2^2+\omega^2T_1^2T_2^2)} \qquad \text{...(A)}$$

Equate the imaginary part equal to zero

$$\frac{K\omega^2 T_1 T_2 - K}{\omega + \omega^3 (T_1^2 + T_2^2 + \omega^2 T_1^2 T_2^2)} = 0$$

$$\therefore \qquad \omega = \frac{1}{\sqrt{T_1 T_2}} = \frac{1}{\sqrt{T_1 T_2}} \ \& \ \omega = \pm\infty$$

The frequency at the point of intersection on real axis is $\frac{1}{\sqrt{T_1 T_2}}$. Now calculate the value of $G(j\omega)$ at this point.

Put $\omega = \frac{1}{\sqrt{T_1 T_2}}$ in equation (A)

$$G(j\omega) = -K\frac{T_1 T_2}{T_1 + T_2} \qquad \angle G(j\omega) = 0°$$

$$G(j\omega) = \infty \qquad \angle G(j\omega) = 0°$$

Step 5 : Equate the real part to zero

$$\frac{-\omega K(T_1 + T_2)}{\omega + \omega^3 (T_1^2 + T_2^2 + \omega^2 T_1 T_2)} = 0$$

$$\therefore \qquad \omega = \infty$$

For positive values of frequencies the polar plot intersects the imaginary axis at $\omega = \infty$

$$\therefore \qquad G(j\omega) = 0 \angle -270°$$

Polar plot is shown in Fig. 4.2.

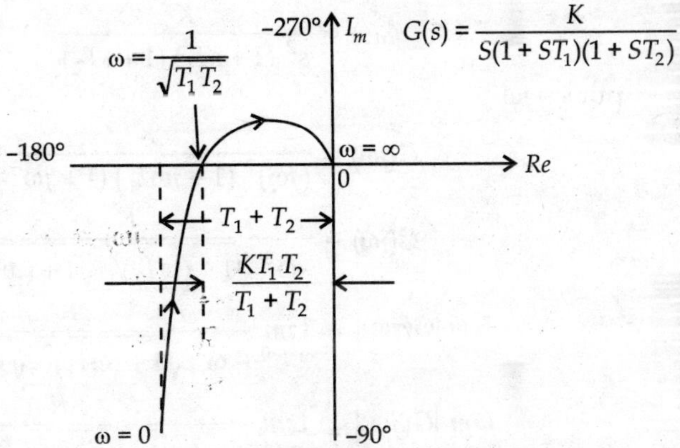

Fig 4.2.

From the polar plot it is clear that in type one system the $j\omega$ term in denominator contributes – 90° to the total phase angle. At $\omega = 0$, the magnitude is infinity and phase angle – 90°. At $\omega = \infty$, the magnitude becomes zero and curve converges to origin. At low frequency, the polar plot is asymptotic to a line parallel to negative imaginary axis.

3. Type 'Two' System

$$G(s) = \frac{K}{s^2(1+sT_1)}$$

Put $s = j\omega$

$$G(j\omega) = \frac{K}{(j\omega)^2 (1 + j\omega T_1)} = \frac{K}{-\omega^2 \sqrt{1 + (\omega T_1)^2}} \angle -180° - \tan^{-1} \omega T_1$$

$$\lim_{\omega \to 0} |G(j\omega)| = \lim_{\omega \to 0} \frac{K}{-\omega^2 \sqrt{1+(\omega T_1)^2}} = \infty$$

$$\lim_{\omega \to \infty} |G(j\omega)| = \lim_{\omega \to \infty} \frac{K}{-\omega^2 \sqrt{1+(\omega T_1)^2}} = 0$$

$$\lim_{\omega \to 0} \angle -180° - \tan^{-1} \omega T_1 = -180°$$

$$\lim_{\omega \to \infty} \angle G(j\omega) = \lim_{\omega \to \infty} \angle -180° - \tan^{-1} \omega T_1 = -270°$$

The presence of S^2 in the denominator contributes a constant – 180° to the angle of $G(j\omega)$ for all frequencies.

The polar plot is a smooth curve whose angle decreases continuously from – 180° to – 270°. The plot is shown in Fig. 4.3.

From the polar plot it is clear that at $\omega = 0$, magnitude is infinity and phase angle – 180°, at $\omega = \infty$ magnitude is zero and at low frequencies the polar is asymptotic to a line parallel to negative real axis.

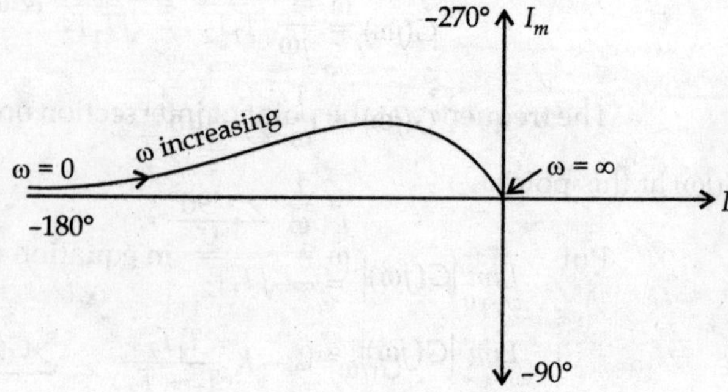

Fig. 4.3.

Introduction to Additional Pole

$$G(s) = \frac{K}{s^2 (1 + sT_1)(1 + sT_2)}$$

put $s = j\omega$

$$G(j\omega) = \frac{K}{(j\omega)^2 (1 + j\omega T_1)(1 + j\omega T_2)}$$

$$G(j\omega) = \frac{K}{-\omega^2 \sqrt{1 + (wT_1)^2} \sqrt{1 + (wT_2)^2}} \angle -180° - \tan^{-1} \omega T_1 - \tan^{-1} \omega T_2$$

$$\lim_{\omega \to 0} |G(j\omega)| = \lim_{\omega \to 0} \frac{K}{-\omega^2 \sqrt{1 + (\omega T_1)^2} \sqrt{1 + (\omega T_2)^2}} = \infty$$

$$\lim_{\omega \to \infty} |G(j\omega)| = \lim_{\omega \to \infty} \frac{K}{-\omega^2 \sqrt{1 + (\omega T_1)^2} \sqrt{1 + (wT_2)^2}} = 0$$

$$\lim_{\omega \to 0} \angle G(j\omega) = \lim_{\omega \to 0} \angle -180° - \tan^{-1} \omega T_1 - \tan^{-1} \omega T_2 = -180°$$

$$\lim_{\omega \to \infty} \angle G(j\omega) = \lim_{\omega \to \infty} \angle -180° - \tan^{-1} \omega T_1 - \tan^{-1} \omega T_2 = -360°$$

Separate the real and imaginary parts of $G(j\omega)$.

The curve intersects the imaginary axis and 'ω' at this point is $1/\sqrt{T_1 T_2}$.

The angle of the curve decreases-continuously from –180° to – 360°.

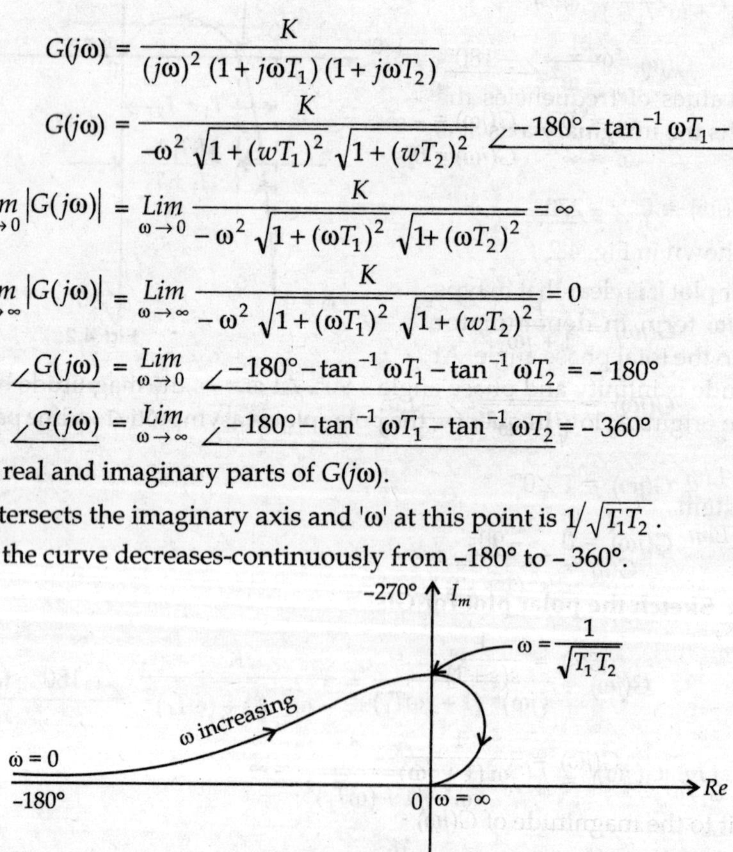

Fig. 4.4.

4.4. POLAR PLOTS OF SOME STANDARD TYPE FUNCTION

1. $$G(s) = \frac{1}{s}$$

Put $s = j\omega$

$$G(j\omega) = \frac{1}{j\omega}$$

$$G(j\omega) = \frac{1}{\omega} \angle -\tan^{-1} \infty$$

$$= \frac{1}{\omega} \angle -90°$$

$$\lim_{w \to 0} |G(j\omega)| = \infty$$

$$\lim_{\omega \to \infty} |G(j\omega)| = 0$$

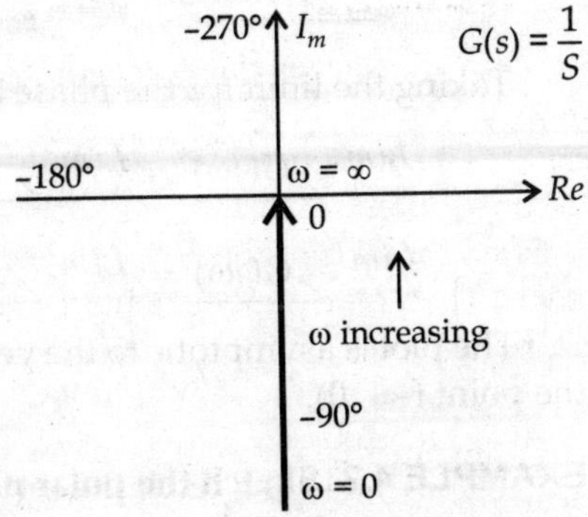

Fig. 4.5.

2. $$G(s) = \frac{1}{s^2}$$

Put $S = j\omega$

$$G(j\omega) = \frac{1}{(j\omega)^2}$$

or $$G(j\omega = \frac{1}{-\omega^2} \angle -180°$$

$$\omega = 0 \quad G(j\omega) = -\infty$$

$$\omega = \infty \quad G(j\omega) = 0$$

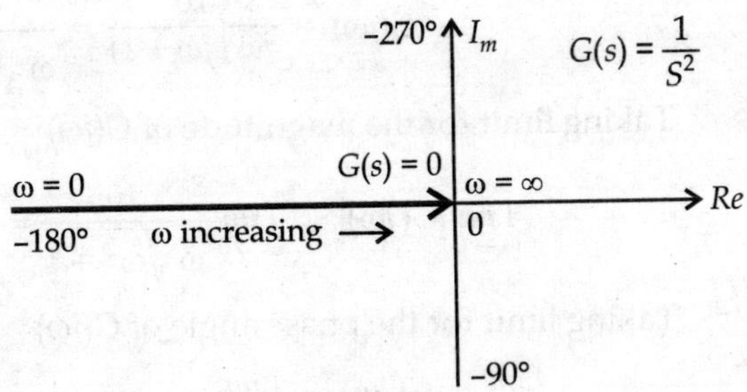

Fig. 4.6.

3. $$G(s) = \frac{1}{S+1}$$

Put $S = j\omega$

$$G(j\omega) = \frac{1}{1 + j\omega}$$

$$G(j\omega) = \frac{1}{\sqrt{1+\omega^2}} \angle -\tan^{-1} \omega$$

$$\lim_{\omega \to 0} G(j\omega) = 1 \angle 0°$$

$$\lim_{\omega \to \infty} G(j\omega) = 0 \angle -90°$$

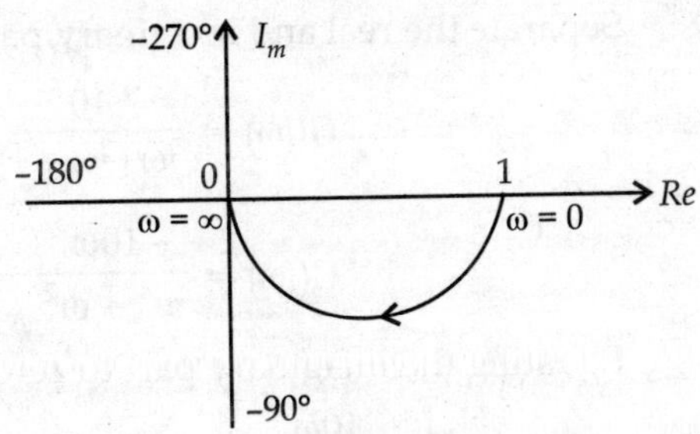

Fig. 4.7.

EXAMPLE 4.1. Sketch the polar plot for $G(s) = 1/s(s + 1)$

Solution : $$G(s) = \frac{1}{s(s+1)}$$

Put $S = j\omega$

$$G(j\omega) = \frac{1}{j\omega\,(1 + j\omega)} = \frac{1}{\omega\sqrt{1+\omega^2}} \angle -90° - \tan^{-1}\omega$$

Apply limit to the magnitude of $G(j\omega)$

$$\lim_{\omega \to 0} |G(j\omega)| = \lim_{\omega \to 0} \frac{1}{\omega\sqrt{1+\omega^2}} = \infty$$

$$\lim_{\omega \to \infty} |G(j\omega)| = \lim_{\omega \to \infty} \frac{1}{\omega\sqrt{1+\omega^2}} = 0$$

Taking the limit for the phase angle of $G(j\omega)$

$$\lim_{\omega \to 0} \angle G(j\omega) = \lim_{\omega \to 0} \angle -90^\circ - \tan^{-1}\omega = -90^\circ$$

$$\lim_{\omega \to \infty} \angle G(j\omega) = \lim_{\omega \to \infty} \angle -90^\circ - \tan^{-1}\omega = -180^\circ$$

The plot is asymptotic to the vertical line passing through the point (– 1, 0).

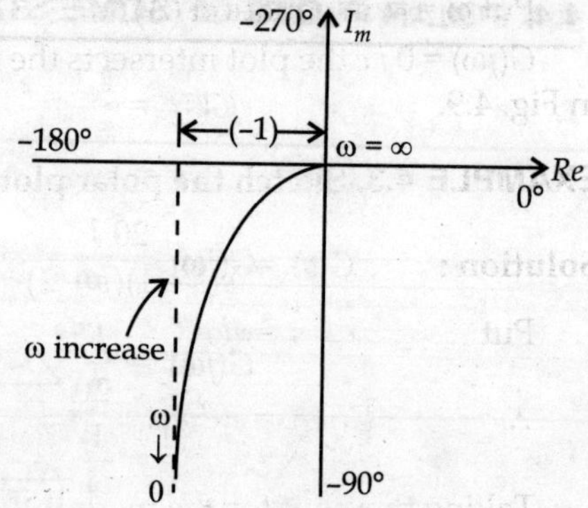

Fig. 4.8.

EXAMPLE 4.2. Sketch the polar plot of G(s) = 10/s(s + 1).

Solution : $G(s) = \dfrac{10}{s(s+1)}$

Put $s = j\omega$

$$G(j\omega) = \frac{10}{j\omega\,(j\omega+1)} = \frac{10}{\omega\sqrt{\omega^2+1}} \angle -90^\circ - \tan^{-1}\omega$$

Taking limit for the magnitude of $G(j\omega)$

$$\lim_{\omega \to 0} |G(j\omega)| = \lim_{\omega \to 0} \frac{10}{\omega\sqrt{\omega^2+1}} = \infty = \lim_{\omega \to \infty} \frac{10}{\omega\sqrt{\omega^2+1}} = 0$$

Taking limit for the phase angle of $G(j\omega)$

$$\lim_{\omega \to 0} \angle G(j\omega) = \lim_{\omega \to 0} \angle -90^\circ - \tan^{1}\omega = -90^\circ = \lim_{\omega \to \infty} \angle -90^\circ - \tan^{1}\omega$$

$$= -90^\circ - 90^\circ = -180^\circ$$

Separate the real and imaginary part of $G(j\omega)$

$$G(j\omega) = \frac{10}{j\omega\,(j\omega+1)} = \frac{10}{-\omega^2 + j\omega} = \frac{10}{-\omega^2 + j\omega} \cdot \frac{-\omega^2 - j\omega}{-\omega^2 - j\omega} = \frac{10\,(-\omega^2 - j\omega)}{\omega^4 + \omega^2}$$

$$G(j\omega) = \frac{-10\omega^2}{w^4 + \omega^2} - j\,\frac{10\omega}{\omega^4 + \omega^2} \quad \text{...(A)}$$

Equating the imaginary part to zero

$$\frac{10\omega}{\omega^4 + \omega^2} = 0 \qquad \therefore \omega = \infty$$

Put $\omega = \infty$ in equation (A)

$G(j\omega) = 0$, *i.e.*, plot intersect the real axis at the origin

Equating the real part to zero

$$\frac{-10\omega^2}{\omega^4 + \omega^2} = 0$$

or, $$\frac{-10}{\omega^2 + 1} = 0$$

$$\therefore \quad \omega = \infty$$

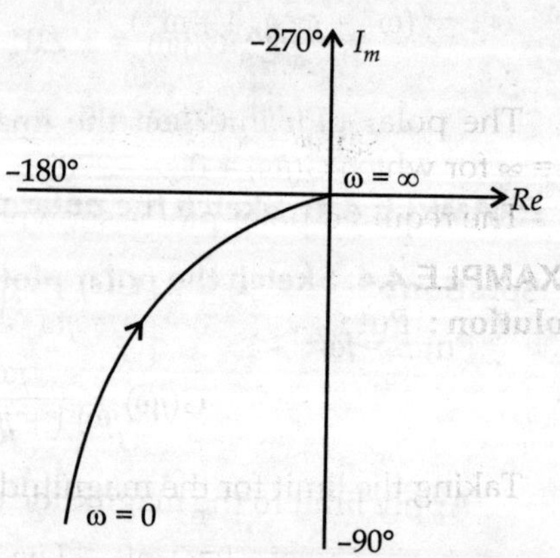

Fig. 4.9.

Put $\omega = \infty$ in equation (*A*)

$G(j\omega) = 0$ *i.e* the plot intersects the imaginary axis at the origin. The required polar plot is shown in Fig. 4.9.

EXAMPLE 4.3. Sketch the polar plot for $G(s) = 20/s(s + 1)(s + 2)$.

Solution : $$G(s) = \frac{20}{s(s+1)(s+2)}$$

Put $s = j\omega$

$$\therefore \quad G(j\omega) = \frac{20}{j\omega\,(1+j\omega)\,(j\omega+2)} = \frac{20}{\omega\sqrt{1+\omega^2}\sqrt{4+\omega^2}} \angle -90° - \tan^{-1}\omega - \tan^{-1}\omega/2$$

Taking the limit for the magnitude of $G(j\omega)$

$$\lim_{\omega\to 0}|G(j\omega)| = \lim_{\omega\to 0}\frac{20}{\omega\sqrt{1+\omega^2}\sqrt{4+\omega^2}} = \infty = \lim_{\omega\to\infty}\frac{20}{\omega\sqrt{1+\omega^2}\sqrt{4+\omega^2}} = 0$$

Taking the limit for the phase angle of $G(j\omega)$

$$\lim_{\omega\to 0}\angle G(j\omega) = \lim_{\omega\to 0}\angle -90° - \tan^{-1}\omega - \tan^{-1}\omega/2 = -90°$$

$$\lim_{\omega\to \infty}\angle G(j\omega) = \lim_{\omega\to \infty}\angle -90° - \tan^{-1}\omega - \tan^{-1}\omega/2 = -270°$$

Separating the real and imaginary part of $G(j\omega)$

$$G(j\omega) = \frac{20}{j\omega\,(1+j\omega)(2+j\omega)} = \frac{-60\omega^2}{(\omega^4+\omega^2)(4+\omega^2)} + j\frac{20(\omega^3-2\omega)}{(\omega^4+\omega^2)(4+\omega^2)}$$

Equating the imaginary part to zero

$$\frac{20(\omega^3-2\omega)}{(\omega^4+\omega^2)(4+\omega^2)} = 0 \qquad \therefore\ \omega = \pm\sqrt{2} \text{ and } \omega = \pm\infty$$

For the positive values of frequencies polar plot intersects the real axis at $\omega = \sqrt{2}$ and $\omega = \infty$ for which

$$G(j\omega) = \frac{-10}{3}\angle 0° \text{ and } 0\angle 0°$$

Equating the real part to zero

$$\frac{-60\,\omega^2}{(\omega^4+\omega^2)(4+\omega^2)} = 0$$

$$\therefore \qquad \omega - \infty$$

The polar plot intersect the imaginary axis at $\omega = \infty$ for which $G(j\omega) = 0\angle -270°$.

The required polar plot is shown in Fig. 4.10.

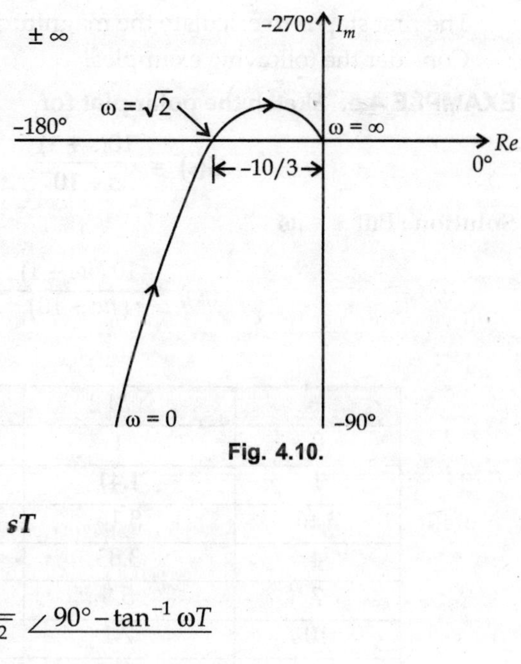

Fig. 4.10.

EXAMPLE 4.4. Sketch the polar plot for $G(s) = s/1 + sT$

Solution : Put $s = j\omega$

$$G(j\omega) = \frac{j\omega}{1+j\omega T} = \frac{\omega}{\sqrt{1+\omega^2T^2}} \angle 90° - \tan^{-1}\omega T$$

Taking the limit for the magnitude of $G(j\omega)$

$$\underset{\omega\to 0}{Lim}|G(j\omega)| = \underset{\omega\to 0}{Lim}\frac{\omega}{\sqrt{1+\omega^2T^2}} = 0 = \underset{\omega\to\infty}{Lim}\frac{\omega}{\sqrt{1+\omega^2T^2}} = \frac{1}{T}$$

Taking the limit for the phase angle of $G(j\omega)$

$$\lim_{\omega\to 0} \angle G(j\omega) = \lim_{\omega\to 0} \angle 90° - \tan^{-1}\omega T = 90° = \lim_{\omega\to\infty} \angle 90° - \tan^{-1}\omega T = 0°$$

Separating the real and imaginary terms of $G(j\omega)$

$$G(j\omega) = \frac{j\omega}{1+j\omega T}\,\frac{1-j\omega T}{1-j\omega T} = \frac{\omega^2 T}{1+\omega^2T^2} + j\frac{\omega}{1+\omega^2T^2}$$

equate the imaginary part to zero

$$\frac{\omega}{1+\omega^2T^2} = 0 \qquad \therefore \omega = \infty$$

Polar plot intersects the real axis at $\omega = \infty$ at which

$$G(j\omega) = \frac{1}{T}\angle 0°$$

Equate the real part to zero

$$\frac{\omega^2 T}{1+w^2T^2} = 0 \qquad \therefore \omega = 0$$

Polar plot intersects the imaginary axis at $\omega = 0$ at which $G(j\omega) = 0 \angle 90°$

The required polar plot is given below.

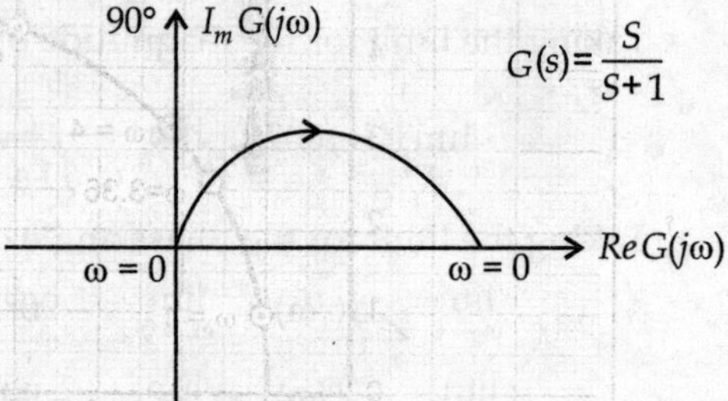

Fig. 4.11.

4.5. ALTERNATIVE METHOD

The other mtehod to draw the polar plot is the tip of the vector $M(\omega)\,e^{j\Phi(\omega)}$ in the G-plane as 'ω' varies from zero to infinity.

The first step is to calculate the magnitude ratio M and phase shift ϕ for different values of 'ω'. Consider the following examples.

EXAMPLE 4.5. Sketch the polar plot for

$$G(s) = \frac{10(s+1)}{s+10}$$

(R.M.L. University Faizabad 2001)

Solution : Put $s = j\omega$

$$\therefore \quad (j\omega) = \frac{10\,(j\omega+1)}{(jw+10)} = \frac{10\sqrt{1+\omega^2}}{\sqrt{\omega^2+100}}, \angle G(j\omega) = \tan^{-1}\omega - \tan^{-1}\frac{\omega}{10}$$

Table 4.1.

ω	*M*	φ	*M*cos φ + *jM*sin° φ.
0	1	0°	1 + *j*0
1	1.41	39.29°	1 . 09 + *j* 0.89
3.16	3.16	54.9°	1.81 + *j* 2.58
4	3.83	54.2°	2.24 + *j* 3.106
7	5.8	46.86°	3.96 + *j* 4.2
10	7.1	39.3°	5.5 + *j*4.5
20	8.96	23.7°	8.2 + *j* 3.6
50	9.81	10.16°	9.67 + *j* 1.73
100	9.95	5.14°	9.9 + *j* 0.89
200	9.99	2.58°	9.97 + *j* 0.45

The corresponding polar plot is shown on the graph paper in Fig. 4.12.

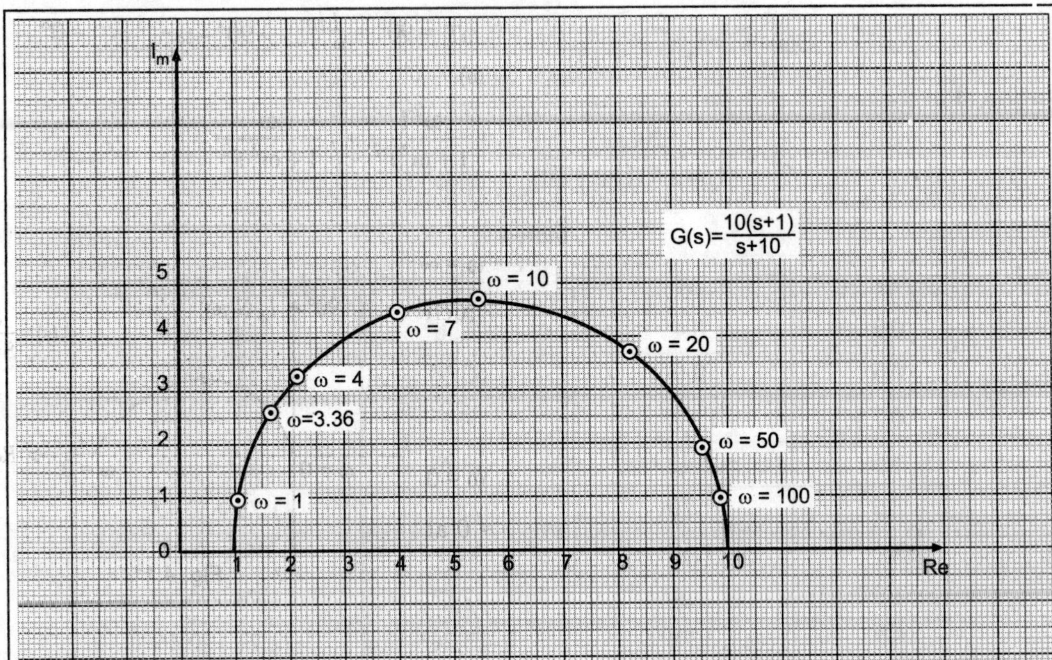

Fig. 4.12.

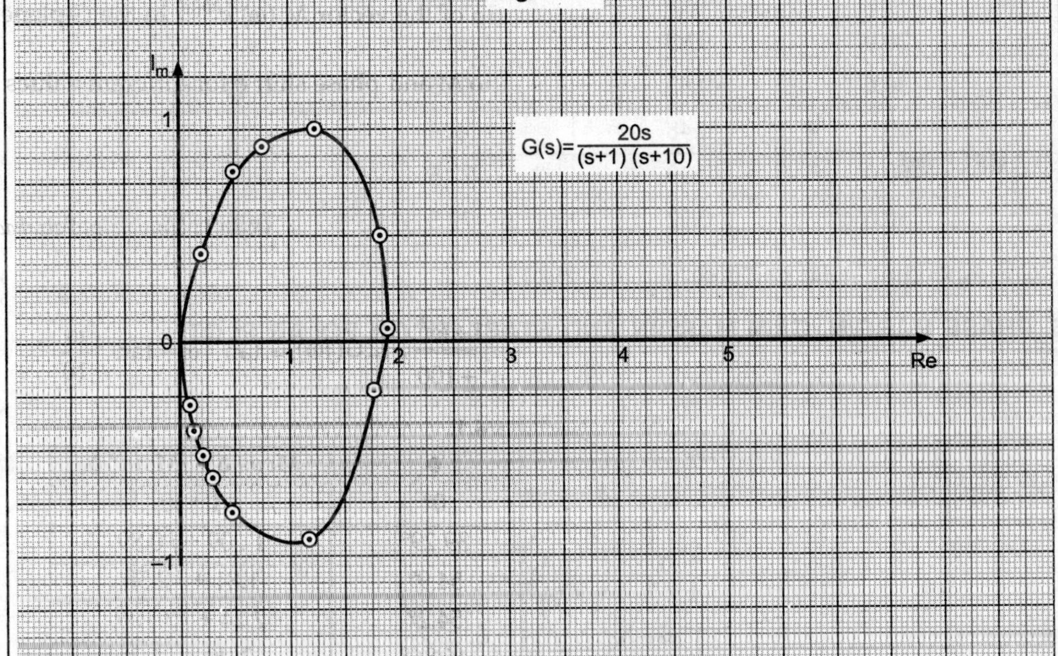

Fig. 4.13.

EXAMPLE 4.6. Sketch the polar plot of

$$G(s) = \frac{20s}{(s+1)(s+10)}$$

Solution : Put $s = j\omega$

$$G(j\omega) = \frac{20(j\omega)}{(j\omega+1)(j\omega+10)}$$

Table 4.2.

ω	M	ϕ	Mcos ϕ + jMsin° ϕ.
0.2	0.4	77.56°	0.08 + j0.4
0.5	0.9	60.58	0.44 + j0.78
0.7	1.147	51.0	0.72 + j0.89
1.0	1.41	39.28	1.09 + j0.9
2.0	1.79	15.3	1.72 + j0.47
3.0	1.82	1.74	1.8 + j0.05
4.0	1.8	–7.8°	1.78 – j0.24
5.0	1.78	–15.26°	1.71 – j0.46
10.0	1.41	–39.3	1.09 – j0.9
20.0	0.89	–60.5	0.43 – j0.77
30.0	0.63	–69.5	0.22 – j0.6
40.0	0.48	–74.56	0.12 – j0.5
50.0	0.39	–77.5	0.08 – j0.4
60.0	0.32	–79.5	0.05 – j0.3

Corresponding polar plot is shown in Fig. 4.13.

4.6. PHASE MARGIN, GAIN MARGIN AND STABILITY ON POLAR PLOT

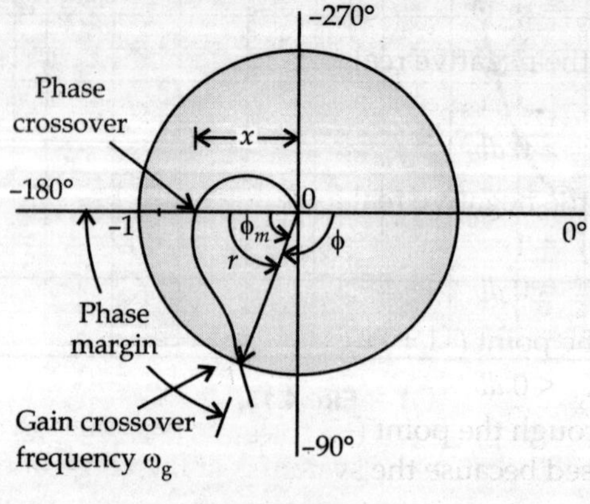

Fig. 4.14.

The stability of control system is measured from the measurement of quantities like 'phase margin' and 'gain margin'. These quantities are useful to design the compensating network for unstable systems, *i.e.*, for the improvement of stability of an existing control system.

In above Fig. 4.14 polar plot intersects the negative real axis at a distance of 'x' from the origin. The frequency where it crosses the negative real axis is called "phase cross-over frequency." The point where the polar plot intersects the negative real axis is called "phase crossover point."

Phase Margin

The phase margin is that amount of additional phase lag at the gain crossover frequency required to bring the system to the verge of instability.

A circle with radius equal to unity is drawn and it intersects the polar plot at point 'A' where the frequency is ω_g. This frequency is called "gain cross over frequency." In other words the gain cross over frequency is the frequency at which $|G(j\omega)|$, the magnitude of the open loop transfer function is unity.

The phase margin is equal to 180° plus the angle of $G(j\omega)$ at the gain crossover point, *i.e.*,

$$\phi_m = 180° + \phi$$

$$\{r - \phi + 180 = 360, \quad r = 360 + \phi - 180 = 180 + \phi\}$$

The angle from the negative real axis to the line OA is the phase margin. The phase margin is positive for $\phi_m > 0$ and negative $\phi_m < 0$. For minimum phase system* to be stable, the phase margin must be positive. For unstable system the gain crossover point would not be found in the third quadrant of $G(j\omega)$ plane. When phase crossover point is to the left of $(-1 + j\,0)$ the phase margin is negative and the closed loop system is unstable.

Gain Margin

The gain margin is the reciprocal of the magnitude $|G(j\omega)|$ at the frequency at which the phase angle is – 180°.

$$\therefore \text{ Gain margin } (GM)\ K_g = \frac{1}{|G(jw_c)|}$$

In terms of decibles

$$K_g\ dB = 20 \log K_g = -20 \log |G(j\omega_c)|$$

(*i*) If plot does not intersects the negative real axis $|G(jw_c)| = 0$

$$\therefore \quad K_g = \infty$$

(*ii*) If the plot intersects the negative real axis between the origin and $(-1 + j\,0)$

$0 < |G(j\omega_c)| < 1$

$$\therefore \quad K_g > 0\ db$$

(*iii*) If the plot intersects the negative real axis at $(-1 + j0)$

$$|G(j\omega_C)| = 1$$

$$\therefore K_g = 0\ db$$

(*iv*) If the plot encloses the point $(-1 + j\,0)$ $|G(jwc)| > 1$

$$\therefore \quad K_g < 0\ db$$

When the plot passes through the point $(-1 + j0)$, the gain margin is zero it means that the loop gain can no longer be increased because the system is at the verge of instability.

---

* minimum phase system is describe in Art. 4.9.

If the gain margin is positive it means that the system is stable and negative gain margin means the system is unstable.

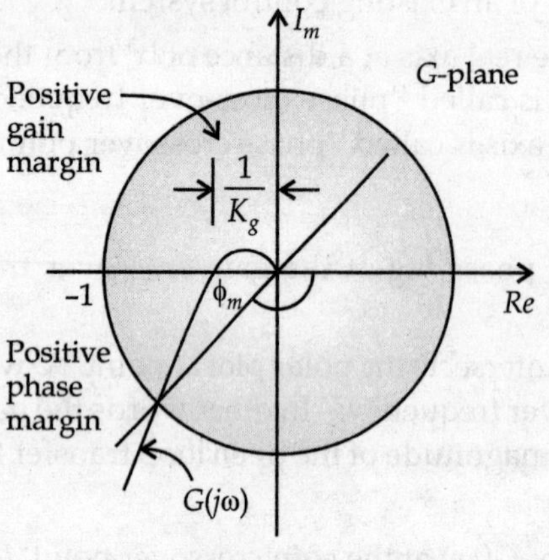

Stable system

Fig. 4.15.

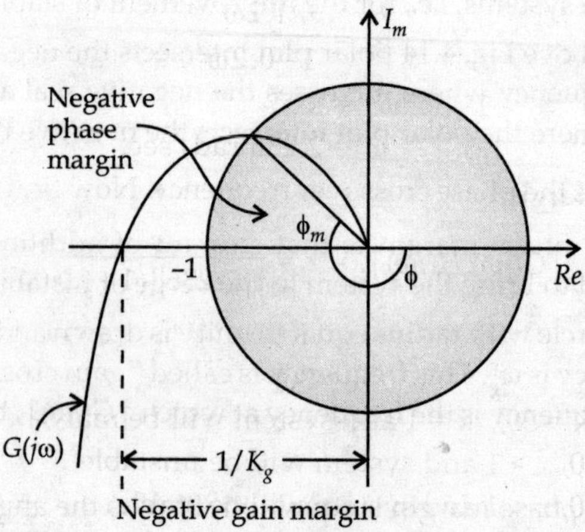

Unstable stable

Fig. 4.16.

The polar plots can also be used to determine the stability of the system. To determine the stability of the system first draw the polar plot for given transfer function. If the critical point ($-1 + j0$) is within the plot, it means the system is unstable. If the polar plot passes through the pint ($-1 + j0$) the system is marginally stable and if the critical point is outside the polar plot, the system is stable. For example, for $G(s)\,H(s) = \dfrac{K}{s(s+1)\,(s+5)}$ sketch the polar plot and determine the range of K for stability.

$$G(j\omega)\,H(j\omega) = \frac{K}{j\omega\,(1+j\omega)\,(5+j\omega)}$$

$\omega = 0$, $\quad |G(j\omega)\,H(j\omega)| = \infty$ and $\angle G(j\omega)\,H(j\omega) = -90°$

$\omega = \infty$, $\quad |G(j\omega)\,H(j\omega)| = 0$ and $\angle G(j\omega)\,H(j\omega) = -270°$

Sketch the polar plot

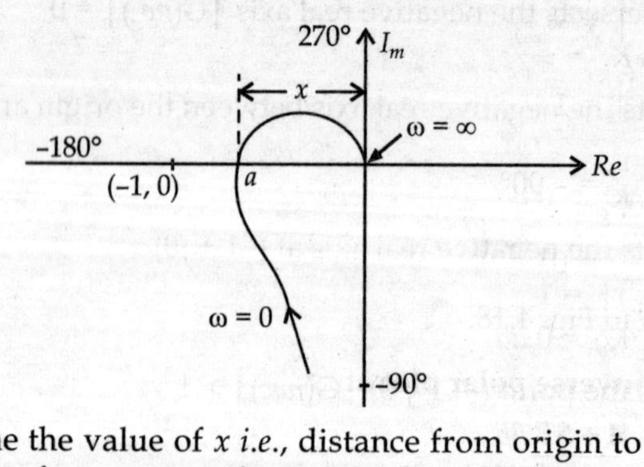

For stability determine the value of x *i.e.*, distance from origin to 'a'. The point 'a' is called *phase crossover point* and the frequency at this point 'a' is called phase crossover frequency.

At this point $\quad \phi = -180°$

$\therefore \quad -180° = -90° - \tan^{-1}\omega - \tan^{-1} 0.2\,\omega$

$$90° = \tan^{-1} \omega + \tan^{-1} 0.2\omega$$

$$\tan 90° = \tan (\tan^{-1} \omega + \tan^{-1} 0.2 \omega)$$

$$\infty = \frac{1.2\omega}{1-0.2\omega^2}$$

or, $1 - 0.2\omega^2 = 0$

or, $\omega = 2.236$ rad/sec.

This is the phase crossover frequency. Now find the magnitude of $G(j\omega)\, H(j\omega)$ at this frequency.

$$|G(j\omega)\, H(j\omega)| = x = \frac{K}{|j2.236||(1+j2\cdot 236)||5+j2.236|}$$

$$x = \frac{K}{30}$$

$\therefore$ if $K = 30$, $x = 1$ and system will be marginally stable

$K > 30$, $x > 1$ and system will be unstable.

$K < 30$, $x < 1$ and system will be stable.

4.7. INVERSE POLAR PLOT

The inverse polar plot of $G(j\omega)$ is a graph of $\frac{1}{G(j\omega)}$ as a function of ω.

For example, if $G(j\omega) = 1/j\omega$ then $G(j\omega)^{-1} = j\omega$

$$G(j\omega)^{-1} = \omega \angle 90°$$

$$\lim_{\omega \to 0} \left|G(j\omega)^{-1}\right| = 0$$

$$\lim_{\omega \to \infty} \left|G(j\omega)^{-1}\right| = \infty$$

The inverse polar plot is shown in the Fig. 4.17.

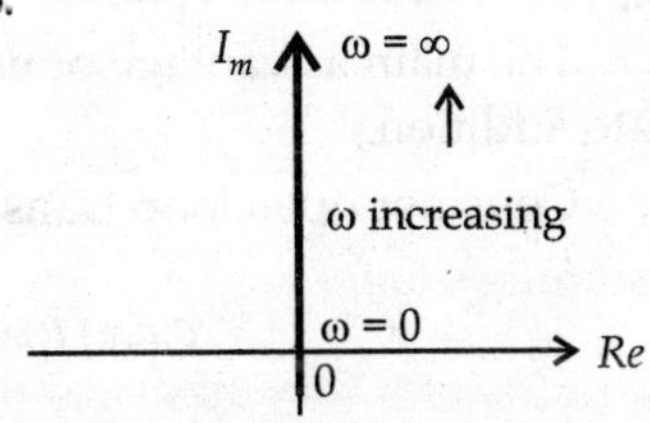

Fig. 4.17. Inverse polar plot of 1/jω

EXAMPLE 4.7. Sketch the inverse polar plot of $G(j\omega) = j\omega T/1 + j\omega T$

Solution :

$$G(j\omega)^{-1} = \frac{1}{G(j\omega)} = \frac{1+j\omega T}{j\omega T} = \frac{1}{j\omega} + 1$$

$$\lim_{\omega \to 0} \left|G(j\omega)^{-1}\right| = \infty$$

$$\lim_{\omega \to \infty} \left|G(j\omega)^{-1}\right| = 1$$

$$\lim_{\omega \to 0} \angle G(j\omega)^{-1} = -90°$$

$$\lim_{\omega \to \infty} \angle G(j\omega)^{-1} = 0°$$

The polar plot is shown in Fig. 4.18.

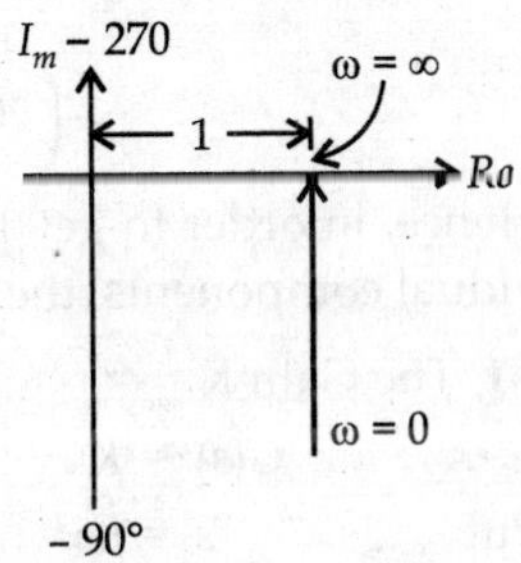

Fig. 4.18. Inverse plot of jωT/1+jωT

EXAMPLE 4.8. Sketch the inverse polar plot of

$$G(s) = \frac{1+ST}{ST}$$

Solution :

$$G(s)^{-1} = \frac{1}{G(s)} = \frac{sT}{1+sT}$$

Put $s = j\omega$

$$G(j\omega)^{-1} = \frac{1}{G(j\omega)} = \frac{j\omega T}{1+j\omega T}$$

$$\lim_{\omega \to 0} \left|G(j\omega)^{-1}\right| = 0 \qquad \lim_{\omega \to 0} \angle G(j\omega)^{-1} = 90°$$

$$\lim_{\omega \to \infty} \left|G(j\omega)^{-1}\right| = 1 \qquad \lim_{\omega \to \infty} \angle G(j\omega)^{-1} = 0°$$

Inverse polar plot is shown in Fig. 4.19.

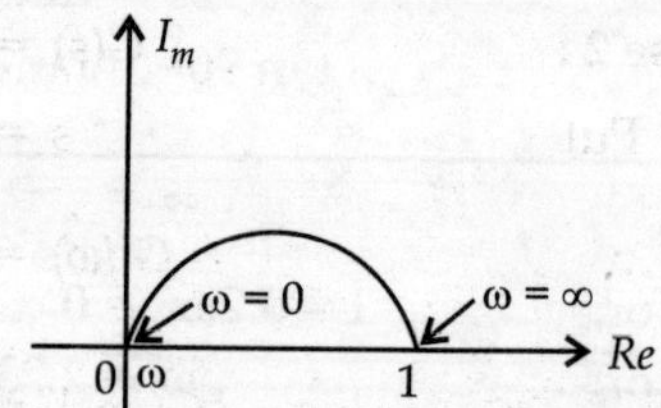

Fig. 4.19. Inverse polar plot of $\frac{1+ST}{ST}$

4.8. BODE PLOT

Bode plot is a graphical representation of the transfer function for determining the stability of the control system. Bode plot consists of two separate plots. One is a plot of the logarithm of the magnitude of a sinusoidal transfer function, the other is a plot of the phase angle, both plots are plotted against the frequency. The curves are drawn on semilog graph paper, using the log scale for frequency and linear scale for magnitude (in decibles) or phase angle (in degrees). The magnitude is represented in decibles. Thus, Bode plot consists of

(*i*) $20 \log_{10} |G(j\omega)|$ V_s $\log\omega$.

(*ii*) Phase shift V_s $\log\omega$

The main advantage of using Bode plot is that multiplication of magnitudes can be converted into addition.

Consider open loop transfer function of a closed loop control system

$$G(s)\,H(s) = \frac{K(1+sT_a)(1+sT_b)....}{s^N(1+sT_1)(1+sT_2)....}$$

Put $S = j\omega$

$$G(j\omega)\,H(j\omega) = \frac{K(1+j\omega T_a)(1+j\omega T_b).......}{(j\omega)^N(1+j\omega T_1)(1+j\omega T_2)......}$$

$$20\log_{10}|G(j\omega)\;H(j\omega)| = \left(20\log K + 20\log\sqrt{1+\omega^2T^2{}_a} + 20\log\sqrt{1+\omega^2T^2{}_b}\right)....)$$

$$-\left(20N\log\omega + 20\log\sqrt{1+\omega^2T_1^2} + 20\log\sqrt{1+\omega^2T_2^2}\right)....)$$

Hence, in order to get $|G(j\omega)\,H(j\omega)|$ we will have to obtain the individual plots and adding individual components, the resultant can be obtained. Suppose, $H(s) = 1$

Case 1. The Gain K

$$G(s) = K$$

Put $s = j\omega$

$$G(j\omega) = K$$

$$20\log_{10}|G(j\omega)| = 20\log_{10}K \qquad ...(4.1)$$

$$\text{Phase angle } \phi = \angle G(j\omega) = 0° \qquad ...(4.2)$$

From equations 4.1 and 4.2 it is clear that the magnitude is independent of $\log_{10}\omega$ and phase angle always zero. The plots are shown in Fig. 4.20.

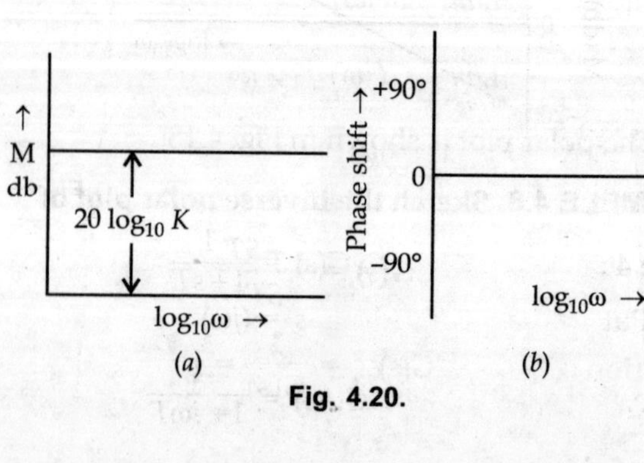

Fig. 4.20.

Case 2 : $G(s) = \frac{1}{S^N}$

Put $s = (j\omega)^N$

$\therefore \quad G(j\omega) = \frac{1}{(j\omega)^N}$

$$20 \log_{10} |G(j\omega)| = 20 \log_{10} \frac{1}{(j\omega)^N} = 20 \log_{10} (j\omega)^{-N} = -20 N \log_{10} (\omega)$$

$$\angle G(j\omega)^N = -90 N°$$

where $N = 1,2,3, \ldots\ldots\ldots$

The plot M Vs $\log_{10} \omega$ is a straight line. For $N = 1$ the line has a slope of –20 *db*/ decade and angle – 90°. For $N = 2$, the slope of the line will be –40 *db*/ decade and angle will be – 180° and so on.

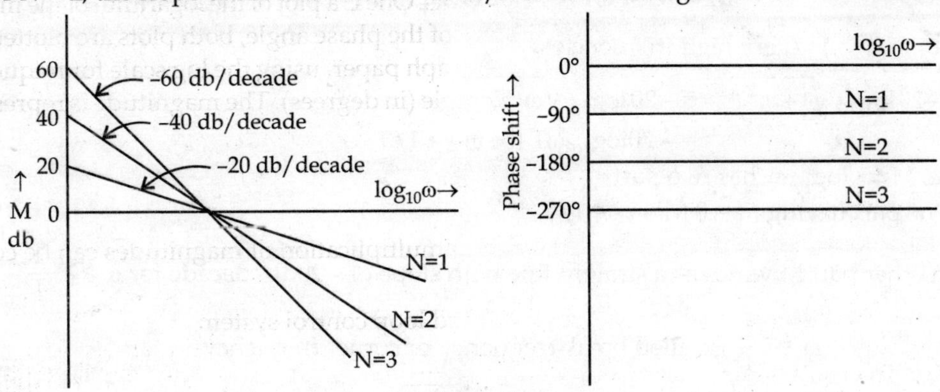

Fig. 4.21.

Case 3 : $G(s) = S$

Put $S = j\omega$

$G(j\omega) = j\omega$

$M = 20 \log_{10} |G(j\omega)| = 20 \log_{10} \omega$

$\angle G(j\omega) = +90°$

The plot M V_s log_{10} ω is a straight line having a slope of + 20 *db*/dec. and angular phase shift of + 90°.

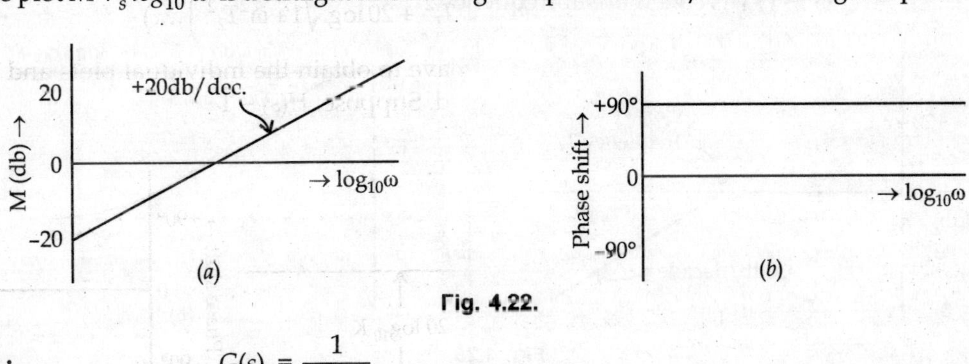

Fig. 4.22.

Case 4 : $G(s) = \frac{1}{1+sT}$

Put $s = (j\omega)$

$\therefore \quad G(j\omega) = \frac{1}{1+j\omega T}$

$$|G(j\omega)| = \frac{1}{\sqrt{1+\omega^2T^2}}$$

$$20\log_{10}|G(j\omega)| = 20\log_{10}\left[\frac{1}{\sqrt{1+\omega^2T^2}}\right] = 20\log_{10}1 - 20\log_{10}\sqrt{1+\omega^2T^2}$$

$$= -20\log_{10}\sqrt{1+\omega^2T^2} \qquad \because \log_{10}1 = 0$$

Put the different values of ω, we will get $|G(j\omega)|$ consider following two cases.

(*a*) For $\omega T << 1$ (very low frequencies)

$$-20\log_{10}\sqrt{1+\omega^2T^2} = -20\log_{10}\sqrt{1} = 0$$

$$\therefore \qquad M = 0 \text{ for } \omega T << 1 \text{ or } \omega \le \frac{1}{T}$$

(*b*) For $\omega T >> 1$ (very high frequencies)

$$-20\log_{10}\sqrt{1+\omega^2T^2} = -20\log_{10}\sqrt{\omega^2T^2}$$

$$= -20\log_{10}\omega T \text{ for } \omega >> 1/T$$

Hence, M *Vs* $\log_{10}\omega$ has two parts

(*i*) One part having $M = 0$ for $\omega << 1/T$

(*ii*) In other part M varies as a straight line with slope of – 20db/decade for $\omega >> \frac{1}{T}$

$\omega = \frac{1}{T}$ is called break frequency or corner frequency

$$M = -20\log_{10}\omega T = -20(\log_{10}\omega + \log_{10}T)$$

$$M = -20\log_{10}\omega - 20\log_{10}T$$

$$= -20\log_{10}\omega + 20\log_{10}1/T \qquad \text{...(4.3)}$$

The above two parts of the graph intersect 0 *db* axis is determined by equating the eq^n (4.3) to zero

$$0 = -20\log_{10}\omega + 20\log_{10}1/T$$

$$\therefore \qquad \omega = 1/T \text{ is called break frequency.}$$

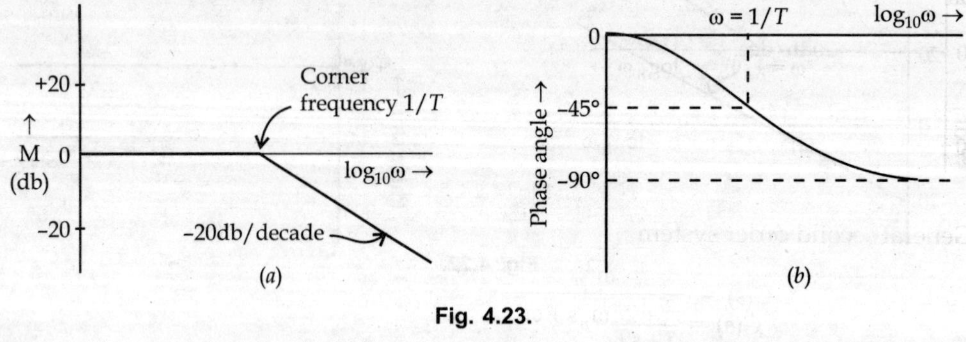

Fig. 4.23.

Case 5 : $G(s) = (1+sT)$

Put $s = j\omega$

$$G(j\omega) = (1+j\omega T)$$

$$|G(j\omega)| = \sqrt{1+\omega^2T^2}$$

$$20\log_{10}|G(j\omega)| = 20\log_{10}\sqrt{1+\omega^2T^2}$$

(i) When $\omega T << 1$

$$M = 20\log_{10}\sqrt{1} = 0\ db$$

(ii) When $\omega T >> 1$

$$M = 20\log_{10}\omega T$$

$$M = 20\log_{10}\omega T = 20\log_{10}\frac{\omega}{1/T} = 20\log_{10}\omega - 20\log_{10}\frac{1}{T}$$

equate the above equation to zero

$$0 = 20\log_{10}\omega - 20\log_{10}\frac{1}{T}$$

$$\therefore \qquad \omega = \frac{1}{T}\ \text{corner frequency.}$$

Thus, the two parts of the graph intersects the '0' db axis at $\omega = \frac{1}{T}$. The second part is a straight line having the slope of + 20 db/decade.

Phase Angle Plot

$$\phi = \angle G(j\omega) = \tan^{-1}\omega T$$

(i) At very low frequencies ωT is very very small

$$\phi = \tan^{-1}(0) = 0°$$

(ii) at $\omega T = 1$

$$\phi = \tan^{-1}1 = 45°$$

(iii) at very high frequencies

$$\phi = \tan^{-1}(\infty) = 90°$$

Thus, the value of ϕ gradually changes from 0° to 90° as ω increases from 0 to very high values.

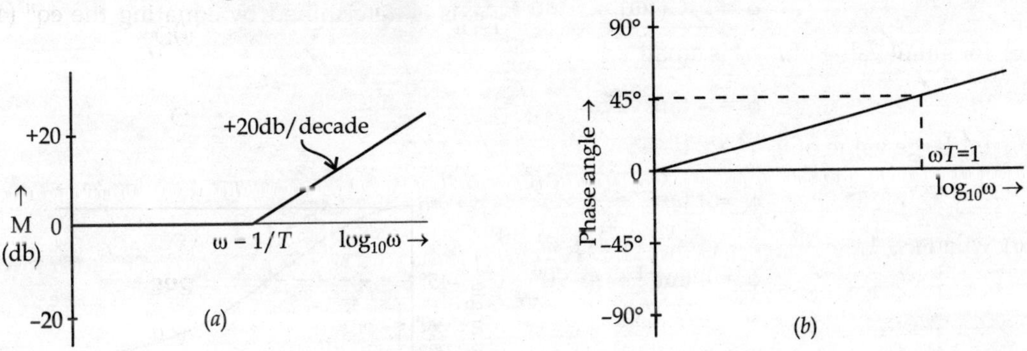

Fig. 4.24.

Case 6 : General second order system

$$G(s) = \frac{\omega^2{}_n}{s^2+2\xi\omega_n s+\omega^2{}_n}$$

Put $s = j\omega$

$$G(s) = \frac{\omega^2{}_n}{(jw)^2+2\zeta\omega_n(j\omega)+\omega^2{}_n} = \frac{\omega_n{}^2}{-\omega^2+j2\zeta\omega_n\omega+\omega_n{}^2}$$

$$G(s) = \frac{w_n^2}{\omega_n^2 - \omega^2 + j2\zeta\omega_n\omega} = \frac{1}{1-\left(\frac{\omega}{\omega_n}\right)^2 + j2\zeta\frac{\omega}{\omega_n}}$$

$$20\log_{10}|G(j\omega)| = 20\log_{10}\left|\frac{1}{1-\left(\frac{\omega}{\omega_n}\right)^2 + j2\zeta\frac{\omega}{\omega_n}}\right| = -20\log_{10}\sqrt{\left\{1-\left(\frac{\omega}{\omega_n}\right)^2\right\}^2 + 4\zeta^2\left(\frac{\omega}{\omega_n}\right)^2}$$

Suppose $\frac{\omega}{\omega_n} = u$

$\therefore$ $$20\log_{10}|G(j\omega)| = M = -20\log_{10}\sqrt{(1-u^2)^2 + 4\zeta^2u^2}$$

Consider the two cases

1. $u << 1$ *i.e.,* $\frac{\omega}{\omega_n} << 1$

$$M = -20\log_{10}\sqrt{1} = 0 \text{ db.}$$

2. $u >> 1$ *i.e.,* $\frac{\omega}{\omega_n} >> 1$

$$M = -20\log_{10}\sqrt{(u^2)^2} = -20\log_{10}u^2 = -40\log_{10}u$$

So, it is a straight line having slope of – 40 db/dec. and passing through the point u

Therefore, the asymptotic plot consists of

(*i*) $M = 0$ $\quad u << 1$

(*ii*) $M = -40\log_{10}u$ $\quad u >> 1$

Phase Angle Plot

$$\phi = \angle G(j\omega) = -\tan^{-1}\frac{2\xi u}{1-u^2}$$

(*i*) for small value of u, u^2 is small

$\therefore$ $$\phi = -\tan^{-1}2\xi u$$

(*ii*) for large value of u, $u^2 >> 1$

$\therefore$ $$\phi = +\tan^{-1}\frac{2\zeta}{u}$$

(*iii*) when $u = 1$

$$\phi = -\tan^{-1}\infty = -90°$$

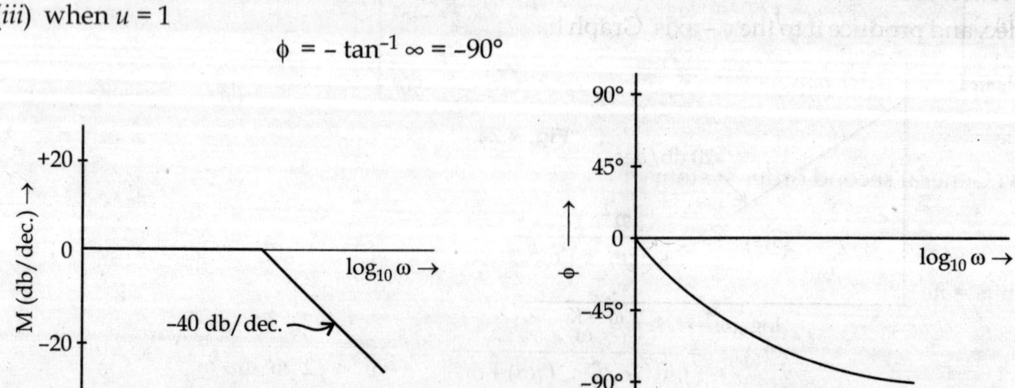

Fig. 4.25.

Initial Slope of Bode Plot

Let $$G(s)\,H(s) = \frac{K}{S^N}$$

Put $$s = j\omega$$

$$G(j\omega)\,H(j\omega) = \frac{K}{(j\omega)^N}$$

$$20\log_{10}|G(j\omega)\,H(j\omega)| = 20\log_{10}\left|\frac{K}{(jw)^N}\right| = 20\log_{10}K - 20\,N\log_{10}\omega \quad ...(4.4)$$

1. For N = 0 (Type zero system)

   $20\log_0 |G(j\omega)\,H(j\omega)| = 20\log_{10}K.$

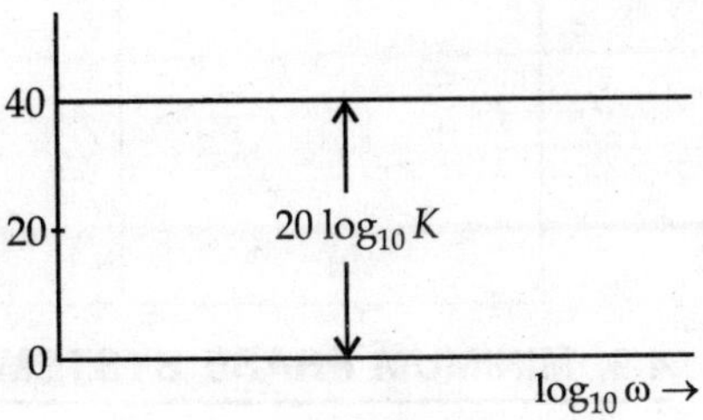

Fig. 4.26.

This is a straight line. The graph is shown in Fig. 4.26.

2. For $N = 1$ (type one system)

   Put $N = 1$ in equation (4.4)

   $$20\log_{10}|G(j\omega)\,H(j\omega)| = 20\log_{10}K - 20\log_{10}\omega$$

   Intersection with 0 db axis

   $$0 = 20\log_{10}K - 20\log_{10}\omega$$

   $$\therefore \quad K = \omega$$

   locate $\omega = K$ on 0 db axis and at this point draw *a* line of – 20 db/decade produce it till it intersect the y-axis that will be the starting point on Bode plot.

3. For $N = 2$ (type two system)

   Put $N = 2$ in equation (4.4)

   $$20\log_{10}|G(j\omega)\,H(j\omega)| = 20\log_{10}K - 20.\,2\log_{10}\omega = 20\log_{10}K - 40\log_{10}\omega$$

   Intersection with 0 db axis

   $$0 = 20\log_{10}K - 40\log_{10}\omega$$

   $$20\log_{10}K = 40\log_{10}\omega$$

   $$20\log_{10}K = 20\log_{10}\omega^2$$

   $$\omega^2 = K$$

   $$\omega = \sqrt{K}$$

Hence, graph intersect the 0 db axis at $\omega = \sqrt{K}$. Locate $\omega = \sqrt{K}$ on 0 db axis and draw a line – 40 db/dec. and produce it to the y – axis. Graph having the slope of – 40 *db*/decade is shown in Fig. 4.27.

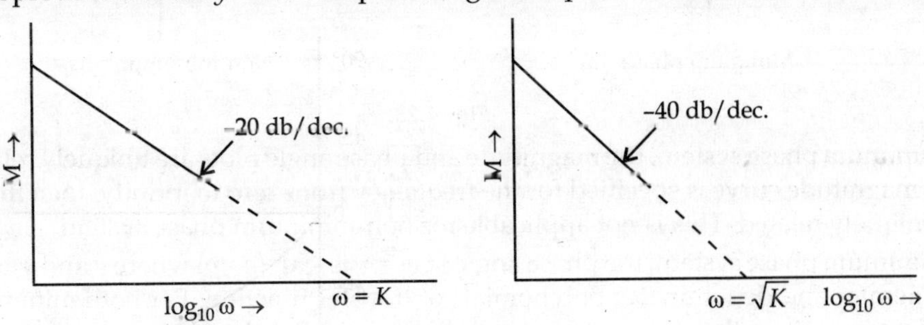

Fig. 4.27.

Table 4.3.

Type of the system N	Initial slope 0 db Axis	Intersection with
0	0 db/decade	Parallel to 0 axis
1	– 20 db/dec.	$= K$
2	– 40 db/dec.	$= \sqrt{K}$
3	– 60 db/dec.	$= K^{1/3}$
⋮	⋮	⋮
N	–20N db/dec.	$K^{1/N}$

4.9. MINIMUM PHASE SYSTEMS AND NON-MINIMUM PHASE SYSTEMS

The transfer functions having no poles and zeros in the right half *s*-plane are called minimum phase transfer functions. Systems with minimum phase transfer functions are called minimum phase systems.

The transfer functions having poles and/or zeros in the right half *s*-plane are called non-minimum phase transfer functions. Systems with non-minimum phase transfer functions are called non-minimum-phase systems.

let

$$G_1(j\omega) = \frac{1+j\omega T_A}{1+j\omega T_B} \quad ...(4.5)$$

$$G_2(j\omega) = \frac{1-j\omega T_A}{1+j\omega T_B} \quad ...(4.6)$$

The transfer function given by equation 4.5 is a minimum phase transfer function and transfer function given by *eq*^{n} 4.6 is non-minimum phase type transfer function.

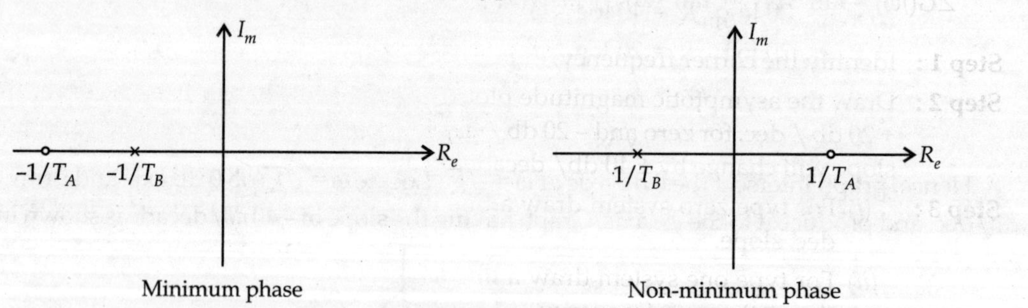

Fig. 4.28.

For minimum phase system, the magnitude and phase angle plots are uniquely related. It means that if the magnitude curve is specified for the frequency from zero to infinity, then the phase angle curve is uniquely related. This is not applicable for non-minimum phase system.

For minimum phase system, the phase angle at $\omega = \infty$ is $-90°\ (q - p)$ where p and q are the degrees of the numerator and denominator polynomials of transfer function. For both minimum and non-minimum phase system the slope is $-20\ (q - p)$ db/decade. But the phase angle for non-minimum system is differ from $-\ 90°\ (q - p)$. Thus it is possible to determine whether the system is minimum phase or non-minimum phase system. If the slope of the log-magnitude curve as ω approaches infinity is –20 (q - p) db/dec. and the phase angle is equal to –90° (q - p) at $\omega = \infty$ then the system is minimum phase otherwise not.

Non-minimum phase systems are slow in response. In control system excessive phase lag should be avoided.

4.10. PROCEDURE FOR DRAWING THE BODE PLOTS

Consider the transfer function

$$G(s) = \frac{K(1+sT_a)(1+sT_b)\ldots\ldots}{S^N(1+sT_1)(1+sT_2)\ldots\ldots\left[1+2\zeta\left(\frac{\omega}{\omega_n}\right)+\left(j\frac{\omega}{\omega_n}\right)^2\right]} \quad \ldots(4.7)$$

where N is the number of poles at the origin, *i.e.*, N defines the type of system.

For type zero system $K = K_p$

For type one system $K = K_v$

For type two system $K = K_a$

In above transfer function put $s = j\omega$

$$G(j\omega) = \frac{K(1+j\omega T_a)(1+j\omega T_b)\ldots..}{(j\omega)^N(1+j\omega T_1)(1+j\omega T_2)\ldots..\left[1+2\zeta(\omega/\omega_n)+(j\omega/\omega_n)^2\right]} \quad \ldots(4.8)$$

$$20\log_{10}|G(j\omega)| = 20\log K + 20\log\sqrt{1+\omega^2T_a^{\,2}} + 20\log\sqrt{1+\omega^2T_b^{\,2}} + \ldots.. - 20N\log\omega$$

$$-20\log\sqrt{1+\omega^2T_1^2} - 20\log\sqrt{1+\omega^2T_2^2} - 20\log\sqrt{\left[1-\left(\frac{\omega}{\omega_n}\right)^2\right]^2 + 4\xi^2\left(\frac{\omega}{\omega_n}\right)^2} \quad \ldots(4.9)$$

Phase angle

$$\angle G(j\omega) = \tan^{-1}\omega T_a + \tan^{-1}\omega T_b + \ldots.\ N(90°) - \tan^{-1}\omega T_1 - \tan^{-1}\omega T_2 \ldots.\ \tan^{-1}\left[\frac{2\zeta\omega\omega_n}{\omega^2{}_n - \omega^2}\right] \quad \ldots(4.10)$$

Step 1 : Identify the corner frequency.

Step 2 : Draw the asymptotic magnitude plot. The slope will change at each corner frequency by + 20 db / dec. for zero and – 20 db / dec for pole. For complex conjugate pole and zero the slope will change by ± 40 db/ decade.

Step 3 : (*i*) For type zero system draw a line upto first (lowest) corner frequency having 0 db dec. slope.

(*ii*) For type one system draw a line having slope – 20 db/ dec. upto $\omega = K$. Mark first (lowest) corner frequency.

(*iii*) For type two system draw the line having slope – 40 db/ dec. upto $\omega = \sqrt{K}$ and so on. Mark first corner frequency.

Step 4 : Draw a line upto second corner frequency by adding the slope of next pole or zero to the previous slope and so on.

Step 5 : Calculate phase angle for different values of ω from the equation 4.10 and join all points.

4.11. PHASE MARGIN AND GAIN MARGIN

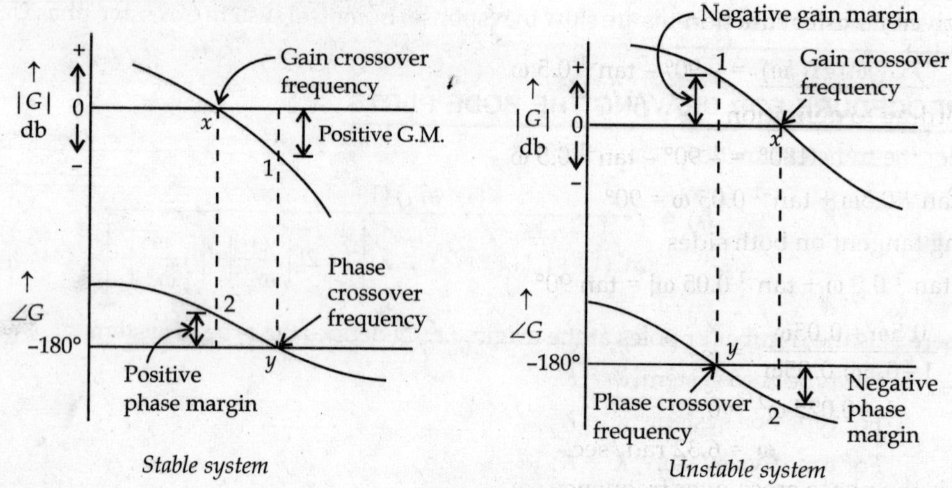

Fig. 4.29.

Positive gain margin means the system is stable and negative gain margin means the system is unstable. For minimum phase system both phase margin and gain margin must be positive for the system to be stable.

The point at which the magnitude curve crosses the 0 db line is the gain crossover frequency. The phase crossover frequency is the point where the phase curve crosses the 180° line.

Gain Margin : Gain margin is defined as the margin in gain allowable by which gain can be increased till system reaches on the verge of instability. Mathematically gain margin is defined as the reciprocal of the magnitude of the $G(j\omega)\ H(j\omega)$ at phase cross-over frequency.

$$\therefore \qquad \text{G.M} = \frac{1}{|G(j\omega)\ H(j\omega)|_{\omega = \omega_{c_2}}}$$

where ω_{c_2} = phase cross-over frequency.

Generally, G.M is expressed in decibels

$$\therefore \text{ In decibels} \qquad \text{G.M} = 20 \log \frac{1}{|G(j\omega)\ H(j\omega)|_{\omega = \omega_{c_2}}}$$

$$\text{or,} \qquad \text{G.M.} = -20 \log_{10} |G(j\omega)\ H(j\omega)|_{\omega = \omega_{c2}}$$

Phase Margin : For gain the additional phase lag can be introduced without affecting the magnitude plot. Therefore, phase margin can be defined as the amount of additional phase lag which can be introduced in the system till system reaches on the verge of instability is called as phase margin (P.M.). Mathematically phase margin can be defined as

$$\text{P.M.} = \left[\angle G(j\omega)\ H(j\omega)\big|_{\omega = \omega_{c_1}} \right] - (-180^\circ) = 180^\circ + \angle G(j\omega)\ H(j\omega)\big|_{\omega = \omega_{c_1}}$$

where ω_{c_1} = Gain cross – over frequency.

The above expressions for gain margin and phase margin can be directly used for mathematical determination of G.M. and P.M. For example, for the given transfer function $G(s)\ H(s) = \dfrac{2}{s\,(1+0.5s)\,(1+0.05s)}$ determine phase cross-over frequency, gain margin, gain cross-over frequency and phase margin.

According to definition, the phase cross-over frequency is the frequency at which phase angle of $G(j\omega)\,H(j\omega)$ is $-180°$.

For given transfer function

$$\angle G(j\omega)\,H(j\omega) = -90° - \tan^{-1} 0.5\,\omega - \tan^{-1} 0.05\,\omega$$

According to definition

$$-180° = -90° - \tan^{-1} 0.5\,\omega - \tan^{-1} 0.05\,\omega$$

or, $\tan^{-1} 0.5\omega + \tan^{-1} 0.05\,\omega = 90°$

taking tangent on both sides

$$\tan\,[\tan^{-1} 0.5\,\omega + \tan^{-1} 0.05\,\omega] = \tan 90°$$

$$\frac{0.5\omega + 0.05\omega}{1 - 0.5\omega . 0.05\omega} = \infty$$

or, $\quad 1 - 0.025\,\omega^2 = 0$

$$\omega = 6.32 \text{ rad/sec.}$$

This is the phase cross-over frequency.

$$\therefore \quad \omega_{c_2} = 6.32 \text{ rad/sec.}$$

Now calculate magnitude at this frequency

$$G\,(j\omega)\,H\,(j\omega) = \frac{2}{j6.32\,(1 + j0.5 \times 6.32)\,(1 + j0.05 \times 6.32)}$$

$$\left|G(j\omega)\,H(j\omega)\right|_{\omega = \omega_{c_2}} = 0.0918$$

$$\therefore \quad \text{G.M.} = -20 \log_{10} \left|G(j\omega)\,H(j\omega)\right|_{\omega = \omega_{c_2}}$$

$$\text{G.M.} = -20 \log_{10} 0.0918 = 20.74 \text{ db.}$$

At gain cross-over frequency the magnitude of $G(j\omega)\,H(j\omega)$ is unity.

$$|\,G(j\omega)\,H(j\omega)\,| = \frac{2}{\omega\sqrt{1 + 0.25\omega}\,\sqrt{1 + 0.0025\omega}}$$

or $$1 = \frac{2}{\omega\sqrt{1 + 0.25\omega}\,\sqrt{1 + 0.0025\omega}}$$

$$\therefore \quad \omega = 2 \text{ rad/sec.}$$

This is the gain cross-over frequency

$$\therefore \quad \omega_{c_1} = 2 \text{ rad/sec.}$$

$$G(j\omega)\,H(j\omega)\Big|_{\omega = \omega_{c_1}} = \frac{2}{j2\,(1 + j\,1)\,(1 + j\,0.1)} = 0.707\,\angle{-140.71°}$$

$$\angle G(j\omega)\,H(j\omega)\Big|_{\omega = \omega_{c_1}} = -140.71°$$

According to definition

$$\text{P.M.} = 180° + \angle G(j\omega)\,H(j\omega)\Big|_{\omega = \omega_{c_1}} = 180° - 140.71°$$

$$\text{P.M.} = 39.28°$$

From Bode's plot, phase margin and gain margin can be determined as follows.

Draw a line downwards from gain crossover point 'x' till it intersects phase angle plot. This point of intersection is represented by '2' the distance between point '2' and – 180° line is the phase margin. If point '2' on phase angle plot is above – 180° line, phase margin is positive and if point '2' is below – 180° line, phase margin is negative (as shown in Fig. 4.29).

From phase cross-over point 'y' draw a line upwards till it intersects magnitude plot at point '1'. The difference between magnitude corresponding to point '1' and '0' db is gain margin. If point '1' is below 0 db G.M. is positive and if point '1' is above 0 db line G.M. is negative as shown in Fig. 4.29.

For positive G.M. and P.M. the system is stable and the gain cross-over frequency < phase cross-over frequency. If gain cross-over frequency > phase cross-over frequency both P.M. and G.M. will be negative and the system is unstable. If gain cross-over frequency = phase cross-over frequency, the system is marginally stable.

The transfer function changes with the change of temperature and pressure. The gain of the system is also affected by supply voltage, supply frequency, air pressures. If the gain is high, the gain margin is low and for step response the settling time is large. If the gain is low G.M. & P.M. will be high and settling time will be more, *i.e.*, the system response will be sluggish and also the rise time and steady state error will be high.

4.12. DETERMINATION OF K_p, K_v AND K_a FROM BODE PLOT

(*a*) Static Position Error Constant K_P :

Static position error constant is defined for type zero system and is given by

$$K_P = \lim_{s \to 0} G(s)\, H(s)$$

Put $s = j\omega$

$$K_P = \lim_{j\omega \to 0} G(j\omega)\, H(j\omega)$$

From equation (4.4)

$$20 \log_{10} |G(j\omega)\, H(j\omega)| = 20 \log_{10} K \quad \because N = 0$$

Consider the Bode plot shown in Fig. 4.26, the magnitude of $G(j\omega)\, H(j\omega)$ is $20 \log K$ which is K_P

$$\therefore \quad K_P = 20 \log K$$

(*b*) Static Velocity Error Constant K_v :

Static velocity error constant is defined for type one system and is given by

$$K_v = \lim_{s \to 0} s\, G(s)\, H(s)$$

From equation (4.4) put $N = 1$

$$20 \log_{10} |G(j\omega)\, H(j\omega)| = 20 \log_{10} K - 20 \log_{10} \omega$$

$$0 \text{ db} = 20 \log_{10} K - 20 \log_{10} \omega$$

$$K = \omega$$

This is static velocity error constant

$$K_v = \omega$$

For type one system, the initial slope of Bode plot (magnitude) is – 20 db/decade. So, draw a line having slope – 20 db/dec. which intersect 0 db line. The frequency at which initial slope of a line is – 20 db/dec. intersects 0 db line is the value of static velocity error constant (K_v) as shown in Fig. 4.27.

(c) Acceleration Error Constant K_a :

Acceleration error constant is defined for type two system and is given by

$$K_a = \lim_{s \to 0} s^2 G(s) H(s)$$

For type two system the initial slope of Bode plot is – 40 db/dec.

From equation 4.4.

$20 \log_{10} |G(j\omega) H(j\omega)| = 20\log_{10} K - 20.2 \log_{10} \omega$

Extend the line having slope – 40 db/dec. to intersect 0 db line

$\therefore \quad 0 = 20 \log_{10} K - 40 \log_{10} \omega$

or, $\quad \omega = \sqrt{K} \quad$ or, $\quad K = \omega^2$

This is the acceleration error constant.

So, the frequency at which initial line of slope – 40 db/decade intersects 0 db line, gives the value of square root of acceleration constant. as shown in Fig. 4.27.

$\therefore \quad \omega = \sqrt{K_a} \quad$ or, $\quad K_a = \omega^2$.

4.13. BODE PLOT WITH TRANSPORTATION LAG

Transportation lag is also known as time delay. Due to the various reasons sometimes it is necessary to stop the action for some time. Such time delay is called transportation lag. In many systems such as hydraulic, pneumatic and thermal pure delay occurs. In laplace domain the transportation lag is given by e^{-Ts}, where 'T' is the delay in seconds.

$\therefore \quad G(s) = e^{-Ts}$

put $s = j\omega$

$$G(j\omega) = e^{-j\omega T}$$

we know that

$$e^{-j\theta} = \cos\theta - j \sin \theta$$

$\therefore \quad G(j\omega) = \cos \omega T - j \sin \omega T$

$$|G(j\omega)| = \sqrt{\cos^2 \omega T + \sin^2 \omega T} = 1$$

Therefore, the magnitude is always equal to unity for all values of ω.

In db = 20 log | G(jω) | = 20 log 1 = 0

In other words we can say that the introduction of time delay has no effect on magnitude plot.

Phase Angle of Transportation Lag

$$\angle G(j\omega) = \tan^{-1}\left[-\frac{\sin \omega T}{\cos \omega T}\right] = -\tan^{-1}(\tan \omega T) = -\omega T \text{ (radians)}$$

ω is in radians/sec. and t is in seconds, therefore phase lag is in radians.

$$\angle G(j\omega) = -57.3\ \omega T \text{ degree} \qquad (\because \pi \text{ radians} = 180°)$$

The phase angle linearly varies with the frequency ω.

For $T = 1$

ω	– 57.3 ωT
0.1	– 5.73°
0.5	– 28.65°
1.0	– 114.6°
5.0	– 286.5°
10	– 573°

EXAMPLE 4.9. Sketch the Bodes plot for the transfer function

$$G(s) = \frac{1000}{(1+0.1s)(1+0.001s)}$$

Determine the : ***(a)*** **P.M**

(b) **Gain margin**

(c) **Stability of the system.**

Solution : Step 1 : Put $s = j\omega$

$$G(j\omega) = \frac{1000}{(1+j0.1\omega)(1+j0.001\omega)}$$

The given transfer function is of type 'o' system. Therefore the initial slope of the Bode plot is 0 db/decade. The starting point is given by.

$$20 \log_{10} K = 20 \log_{10} 1000 = 60 \text{ db}$$

corner frequencies $\omega_1 = \frac{1}{0.1} = 10$ rad/sec.

$$\omega_2 = \frac{1}{0.001} = 1000 \text{ rad/sec.}$$

Step 2 : Mark the starting point 60 db on *y*-axis and draw a line of slop 0 db/decade upto first corner frequency.

Step 3 : From first corner frequency to second corner frequency draw a line with slope (0 – 20 = – 20 db/decade).

Step 4 : From second corner frequency to next corner frequency (if given) draw a line having the slope –20 + (–20) = –40 db/decade.

Step 5 : The magnitude plot is complete and now draw the phase plot by calculating the phase at different frequencies (as given in Table).

Step 6 : From the bode plot

From the point of intersection of magnitude curve with 0 db axis draw a line on phase curve. This line cuts the phase curve at – 154°

$$\therefore \quad P.M. = -154 - (-180)$$

$$= +26°$$

Step 7 : Gain margin $G.M. = \infty$

Since, *P.M.* = + 26° and gain margin = ∞, the system is inherently stable.

Table 4.4

ω	– Arg (1 + j0.1ω) – $\tan^{-1}$ 0.1ω	–Arg (1 + j0.001ω) –$\tan^{-1}$ 0.001 ω	Resultant
50	– 78.6°	– 2.86°	– 81.46°
100	– 84.2°	– 5.7°	– 90°
150	– 86.2°	– 8.5°	– 94°
200	– 87.13°	– 11.3°	– 98°
500	– 88.85°	– 26.56°	– 115.4°
800	– 89.28°	– 38.65°	– 127.93°
1000	– 89.48°	– 45°	– 134.42°
2000	– 89.72°	– 63.43°	– 153.15°
3000	– 89.8°	– 71.56°	– 161.36°
5000	– 89.88°	– 78.69°	– 168.57°
8000	– 89.92°	– 82.87°	– 172.79°

EXAMPLE 4.10. Sketch the Bode plot for the transfer function.

$$G(s) = \frac{1000}{s(1+0.1s)(1+0.001s)}$$

Determine : (*i*) Gain cross over frequency

(*ii*) Phase crossover frequency

(*iii*) G.M and P.M

(*iv*) Stability of the given system

Solution : Step 1 :

Put $s = j\omega$

$$G(j\omega) = 1000/ (j\omega)\ (1 + j\ 01\omega)\ (1+ j\ 0.001\omega)$$

Step 2 : Draw the magnitude curve

corner frequencies $\omega_1 = 1/0.1 = 10$ rad/sec.

$\omega_2 = 1/0.001 = 1000$ rad/sec.

Initial slope of the curve will be – 20 db/decade due to $^1/_s$ term. On ω – axis mark $\omega = K = 1000$. From this point draw a line having the slope – 20 db/decade to meet y-axis. This will be the starting point. From the starting point to the first corner frequency the slope will be – 20 db/decade, then from corner frequency 10 the slope will be – 20 + (–20) = – 40 db/dec. upto 1000.

After 1000, the slope will be – 40 + (–20) = – 60 db/decade.

Step 3 : Draw the phase curve

$$\phi = -90° - \tan^{-1} 0.1\omega - \tan^{-1} 0.001\omega$$

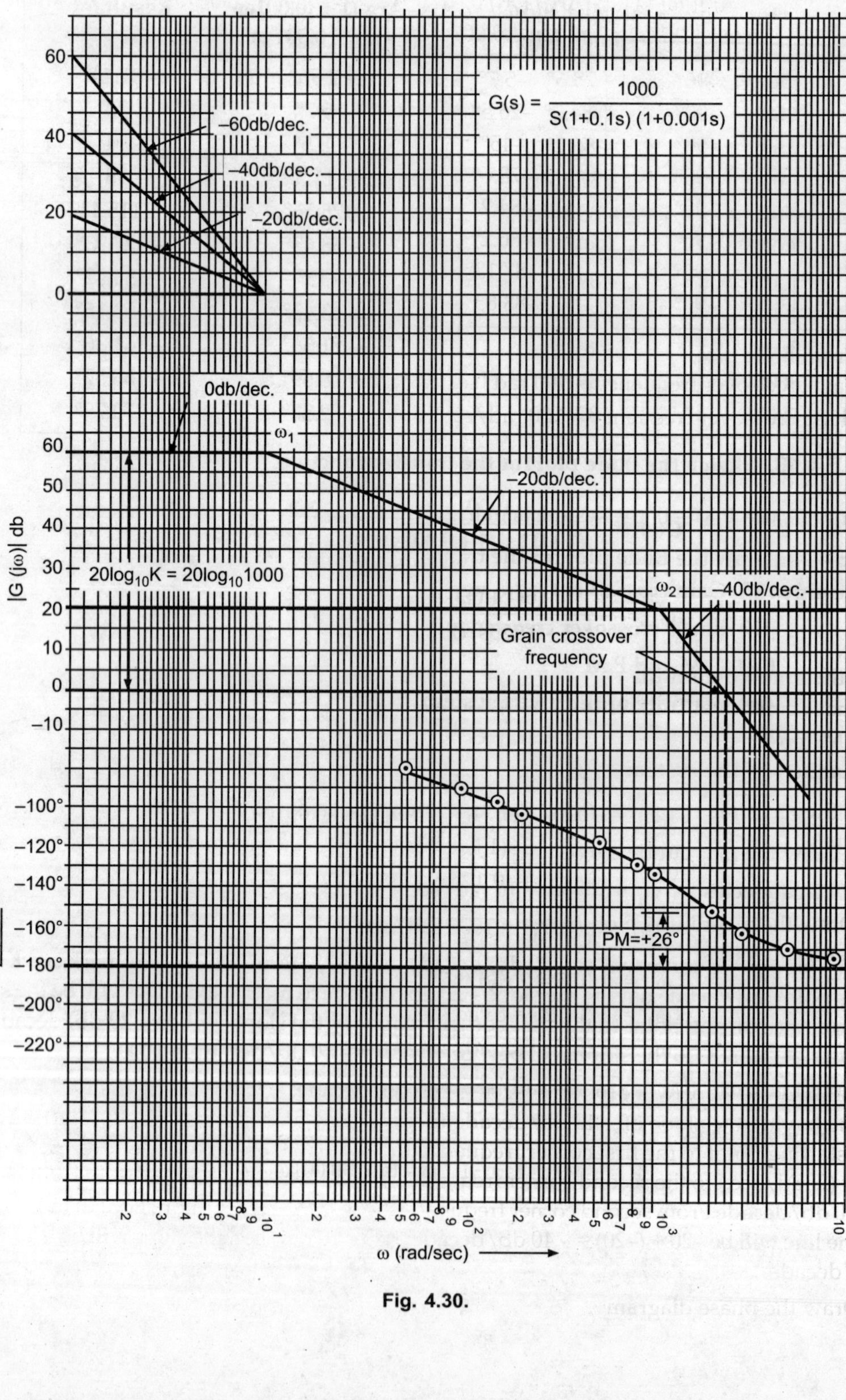

60
40
20
0
–60db/dec.
–40db/dec.
–20db/dec.
G(s) = 1000 / S(1+0.1s) (1+0.001s)
0db/dec.
ω1
–20db/dec.
ω2
–40db/dec.
20log10K = 20log10 1000
Grain crossover frequency
|G (jω)| db
60
50
40
30
20
10
0
–10
–100°
–120°
–140°
–160°
–180°
–200°
–220°
/G (jω) deg.
PM=+26°
ω (rad/sec)

Fig. 4.30.

Table 4.5.

ω	$Arg(j\omega)$ $-tan^{-1}\ 0.1\omega$	$Arg\ (1 + j\ 0.\ 1\omega)$ $-tan^{-1}\ 0.001\omega$	$-\ Arg\ (1 + j\ 0.001\omega)$ ϕ	Resultant
1	−90°	−5.7°	−0.06°	−95.7°
5	−90°	−26.5°	−0.28°	−116.5°
10	−90°	−45°	−0.57°	−135.63°
50	−90°	−78.6°	−2.86°	−171.46°
100	−90°	−84.2°	−5.7°	−179.9°
150	−90°	−86.2°	−8.5°	−184°
200	−90°	−87.13°	−11.3°	−188°
500	−90°	−88.85°	−26.56°	−205.41°

Step 4 : From Bode plot

(*i*) Gain cross over frequency = 100 rad/sec.

(*ii*) Phase crossover frequency = 100 rad/sec.

(*iii*) G.M = 0

P.M = 0

(*iv*) Since G.M = P.M = 0, the system is marignally stable.

EXAMPLE 4.11. Draw the Bode plot for the transfer function

$$G(s) = \frac{16(1+0.5s)}{S^2(1+0.125s)\,(1+0.1s)}$$

From the graph determine :

(*i*) Phase cross over frequency **(*ii*) Gain cross over frequency**

(*iii*) P.M **(*iv*) G.M**

(*v*) Stability of the system

Solution : Step 1 : Put $s = j\omega$

$$G(\omega) = \frac{16(1+j0.5\omega)}{(j\omega)^2(1+j0.125\omega)(1+jo.1\omega)}$$

Step 2 : Draw the magnitude plot

Corner frequencies

$$\omega_1 = 1/0.5 = 2 \text{ rad/sec.}$$

$$\omega_2 = 1/0.125 = 8 \text{ rad/sec.}$$

$$\omega_3 = 1/0.1 = 10 \text{ rad/sec.}$$

At ω-axis mark $\omega = \sqrt{K} = \sqrt{16} = 4$ rad/sec. because this is a type two system. From 4 rad/sec. draw a line having the slope of – 40 db/decade to meet the y-axis. This will be the starting point From the starting point to the first corner frequency the slope will be – 40 db/dec. from first corner frequency (2 rad/sec.) to the second corner frequency (8 rad/sec.) the slope of the line will be −40 + (+20) = −20 db/decade From second corner frequency to the third corner frequency (10 rad/sec.) the slope of the line will be −20 + (−20) = – 40 db/decade. After 10 rad/sec. the slope will be – 40 + (−20) = −60 db/decade.

Step 3 : Draw the phase diagram

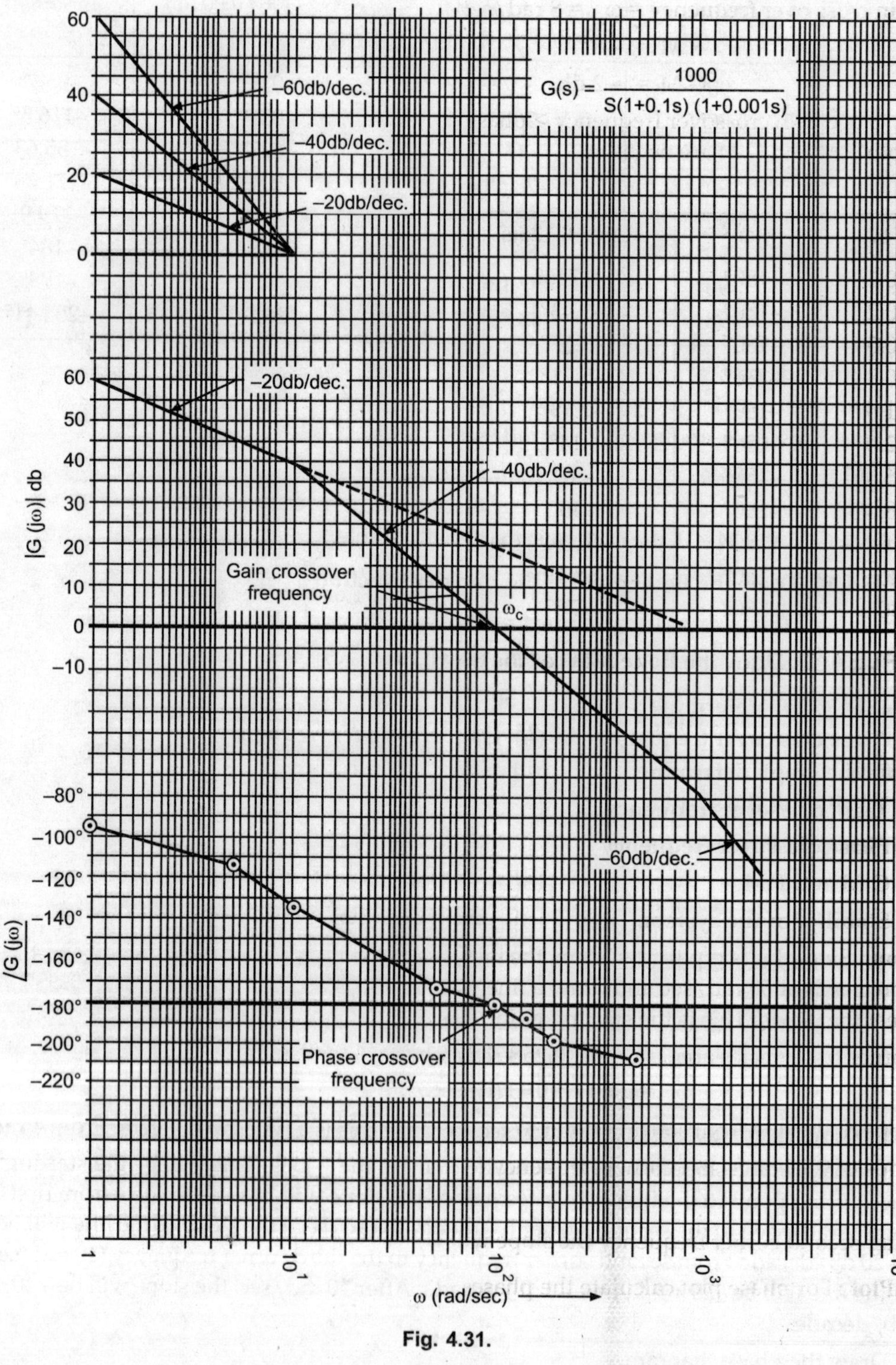
60
40
20
0
–60db/dec.
–40db/dec.
–20db/dec.
G(s) = 1000 / S(1+0.1s) (1+0.001s)
60
50
40
30
20
10
0
–10
|G (jω)| db
–20db/dec.
–40db/dec.
Gain crossover frequency
ωc
–60db/dec.
–80°
–100°
–120°
–140°
–160°
–180°
–200°
–220°
/G (jω)
Phase crossover frequency
1
10^1
10^2
10^3
10^4
ω (rad/sec)

Fig. 4.31.

Step 4 : Phase cross over frequency = 6.5 rad/sec (ω_{c2})

Gain cross over frequency = ω_{c1} = 8 rad/sec.

$$P.M = -8°$$

$$G.M = -2 \text{ db}$$

(*v*) Since gain crossover frequency > phase cross over frequency hence the system is unstable. Also both *P.M* and *G.M* are negative.

Table 4.6.

ω	$-(j\omega)^2$	$-\tan^{-1} 0.125\omega$	$-\tan^{-1} 0.1\omega$	$+\tan^{-1} 0.5\omega$	Resultant
0.1	–180°	–0.716°	–0.57°	+2.86°	–178.42°
0.2	–180°	–1.43°	–1.15°	+5.71°	–176.86°
0.5	–180°	–3.57°	–2.86°	+14°	–172.43°
0.8	–180°	–5.71°	–4.57°	+21.8°	–168.48°
1.0	–180°	–7.13°	–5.71°	+26.56°	–166.28°
2.0	–180°	–14°	–11.3°	+ 45°	–160.3°
5.0	–180°	–32°	–26.56°	+ 68.19°	–170.37°
8.0	–180°	–45°	–38.66°	+76°	–187°
10.0	–180°	–51.34°	–45°	+78.69°	–197.65°
20.0	–180°	–68.19°	–63.43°	+ 84.28°	–227.34°
30.0	–180°	– 75°	–71.56°	+86.18°	–240.38°

EXAMPLE 4.12. Draw the Bode plot for the transfer function

$$G(s) = \frac{50}{s(1+0.25s)(1+0.1s)}$$

From the graph determine

1. **Gain crossover frequency**
2. **Phase crossover frequency**
3. **G.M and P.M.**
4. **Stability of the system.**

Solution : Since this is type one system, the initial slope of the line will be – 20 db/decade. Mark the point on ω-axis at 50 rad/sec and draw a line with slope –20 db/dec. from the point at 50, this line will meet the y-axis. This is the starting point.

Corner frequencies are $\omega_1 = 1/0.25 = 4$ rad/sec.

$\omega_2 = 1/0.1 = 10$ rad/sec.

Slope of the line from starting point to I corner freqency = – 20 db/dec.

Slope of the line from I corner frequency to II = –20 + (–20)

= – 40 db/dec.

After second corner frequency the slope will be – 40 + (–20) = – 60 db/dec.

Phase Plot : For phase plot calculate the phases at different frequencies.

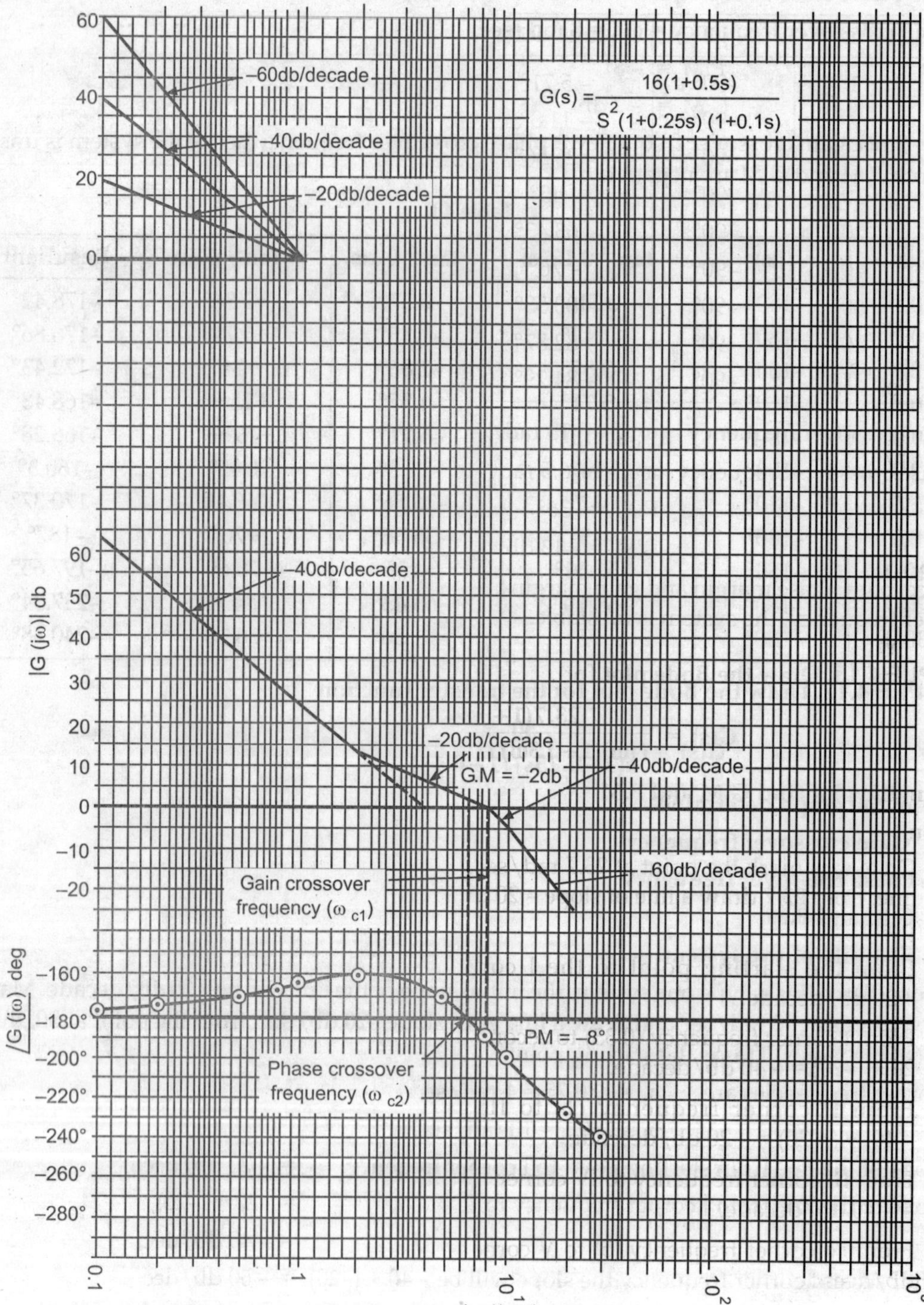

G(s) = 16(1+0.5s) / S^2(1+0.25s) (1+0.1s)
–60db/decade
–40db/decade
–20db/decade
–40db/decade
–20db/decade
G.M = –2db
–40db/decade
–60db/decade
|G (jω)| db
Gain crossover frequency (ω c1)
/G (jω) deg
PM = –8°
Phase crossover frequency (ω c2)
ω (rad/sec)

Fig. 4.32.

Table 4.7.

ω	$j\omega$	$-\tan^{-1} 0.1\omega$	$-\tan^{-1} 0.25\omega$	Resultant
1	-90°	- 5.71°	- 14.03°	- 109.74°
2	-90°	- 11.3°	- 26.56°	- 127.86°
5	-90°	- 26.56°	- 51.34°	- 168°
10	-90°	- 45°	- 68.19°	- 203.19°
20	-90°	- 63.43°	- 78.69°	- 232.12°
40	-90°	- 76°	- 84.28°	- 250.28°
60	-90°	- 80.53°	- 86.18°	- 256.71°
80	-90°	- 82.87°	- 87.13°	- 260°
100	-90°	- 84.29°	- 87.7°	- 262°

Gain crossover frequency (ω_{c1}) = 13 rad/sec.

Phase crossover frequency (ω_{c2}) = 6.5 rad/sec.

P.M = -36°

G.M = - 13 db

Since, both phase margin and gain margin are negative and gain crossover frequency > phase crossover frequency, the system is un stable.

EXAMPLE 4.13. Draw the Bode plot for

$$G(s) = \frac{23.7(1+j\omega)(1+j0.2\omega)}{(j\omega)(1+j3\omega)(1+j0.5\omega)(1+j0.1\omega)}$$

From the plot find G.M and P.M

Solution :

Step 1 : On ω-axis mark the point at 23.7 rad/sec. Since in denominator ($j\omega$) term is having power one, from 23.7 draw a line of slope – 20 db/decade to meet y-axis. This will be the starting point.

From the starting point to the I corner frequency (0.33) the slope of the line is – 20 db/decade.

From I corner frequency (0.33) to second corner frequency (1) the slope of the line will be 20 + (–20) = – 40 db/decade.

From II corner frequency (1) to III corner frequency (2) the slope of the line – 40 + (+ 20) = – 20 db/deceade.

From III corner frequency to IV corner frequency, the slope of the line will be – 20 + (–20) = – 40 db/decade

From IV corner frequency (5) to V corner frequency the slope will be – 40 + (+20) = – 20 db/decade.

After V corner frequency, the slope will be – 20 + (–20) = – 40 db/dec.

Step 2 : Draw the phase plot.

Values of ϕ at different frequencies are tabulated.

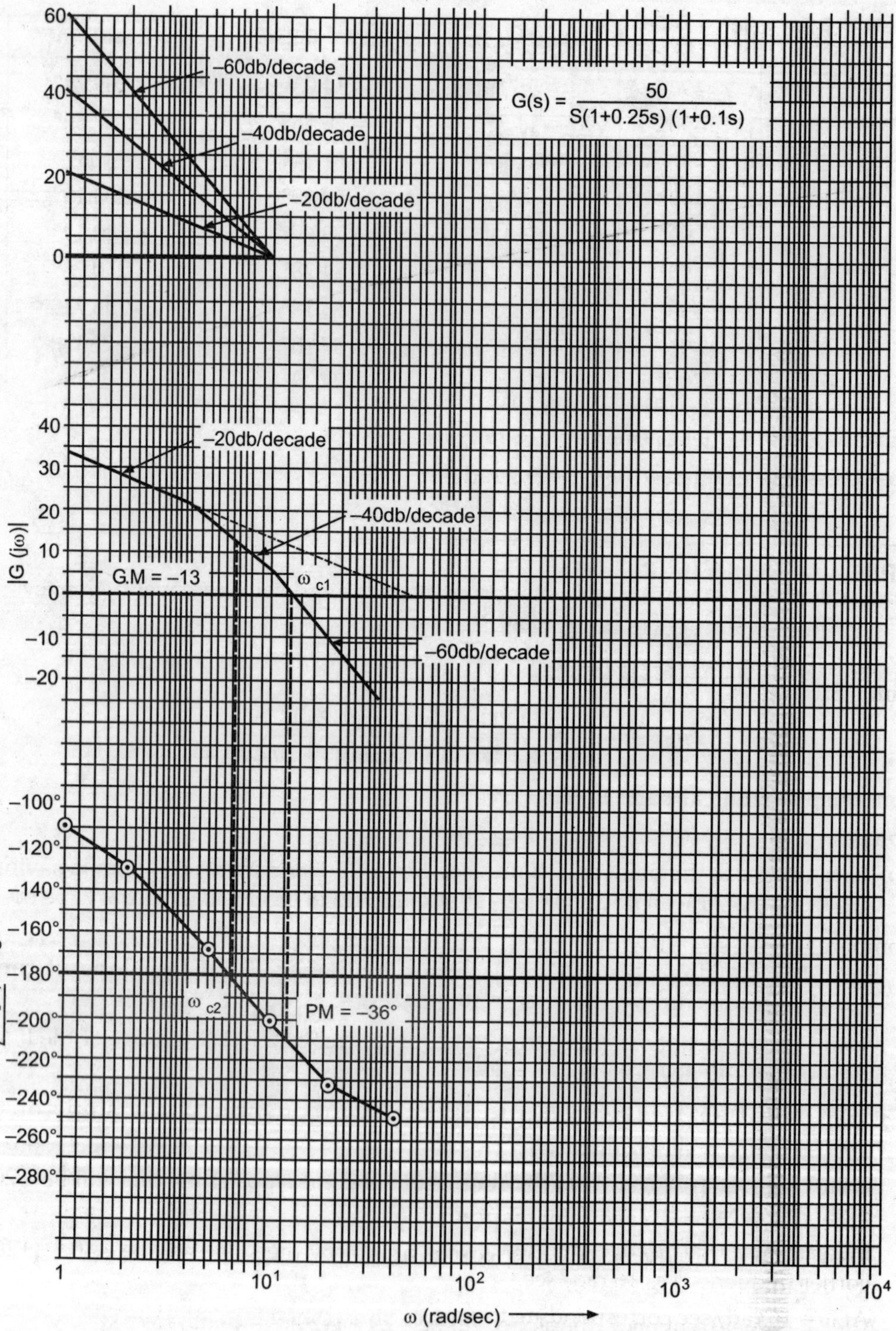
60
40
20
0
–60db/decade
–40db/decade
–20db/decade
G(s) = 50 / S(1+0.25s) (1+0.1s)
40
30
20
10
0
–10
–20
|G (jω)|
–20db/decade
–40db/decade
G.M = –13
ω c1
–60db/decade
–100°
–120°
–140°
–160°
–180°
–200°
–220°
–240°
–260°
–280°
/G (jω) deg
ω c2
PM = –36°
1
10^1
10^2
10^3
10^4
ω (rad/sec)

Fig. 4.33.

Step 3 : From graph

$P.M = +34°$

$G.M = \infty$

Table 4.8.

ω	- tan⁻¹jω	-tan⁻¹3ω	-tan⁻¹ 0.5ω	- tan⁻¹ 0.1ω	tan⁻¹ω	tan⁻¹0.2ω	Resultant
0.1	-90°	-16.7°	-2.86°	-0.57°	+ 5.71°	+1.14°	-103°
0.2	-90°	-31°	-5.71°	-1.14°	+ 11.3°	+2.3°	-114.25°
0.5	-90°	-56.3°	-14.03°	-2.86°	+ 26.56°	+5.71°	-130.92°
0.8	-90°	-67.4°	-21.8°	-4.57°	+ 38.65°	+9.09°	-136.03°
1.0	-90°	-71.56°	-26.56°	-5.71°	+ 45°	+11.3°	-137.5°
2.0	-90°	-80.54°	-45°	-11.3°	+ 63.43°	+ 21.8°	- 141.61°
5.0	-90°	-86.18°	- 68.19°	-26.56°	+ 78.7°	+ 45°	- 147.23°
8.0	-90°	-87.61°	- 76°	- 38.65°	+ 82.87°	+ 58°	- 151.39°
10.0	-90°	- 88°	- 78.7°	- 45°	+ 84.3°	+ 63.4°	- 154.0°
20.0	-90°	- 89°	- 84.3°	- 63.43°	+ 87.13°	+ 76°	- 163.6°

EXAMPLE 4.14. Sketch the Bode plot

$$G(s) = \frac{512(s+3)}{s(s^2+16s+256)}$$

Solution : Step 1

$$G(s) = \frac{512\left(\frac{s}{3}+1\right)3}{s\left[\frac{s^2}{256}+\frac{16s}{256}+1\right]256} = \frac{6(1+s/3)}{s\left[\frac{s^2}{256}+\frac{16}{256}s+1\right]}$$

Compare $\frac{s^2}{256}+\frac{16}{256}s+1$ with $\frac{s^2}{\omega_n{}^2}+\frac{2\zeta}{\omega_n}s+1$

$$\omega^2_n = 256 \therefore \omega_n = 16 \text{ rad/sec.}$$

$$\frac{2\zeta}{\omega_n} = \frac{16}{256} \therefore \xi = 0.5$$

Step 2 : On ω-axis mark the point at 6. From this point draw a line having slope – 20 db/decade, meets the *y*-axis. This is the starting point. From the starting point to the I corner frequency (3 rad/sec.) the slope will be – 20 db/decade.

Since ω = 3 rad/sec. corresponding to a zero, at this point the slope is increased by + 20 db/decade. Hence the slope will be – 20 + (+20) = 0 db/decade upto next corner frequency *i.e.*, 16 rad/sec.

At ω = 16 rad/sec corresponding to the pair of complex poles the slope must be decreased by – 40 db/dec. (0 – 40 = – 40 db/dec.)

Step 3 : Draw the phase plot

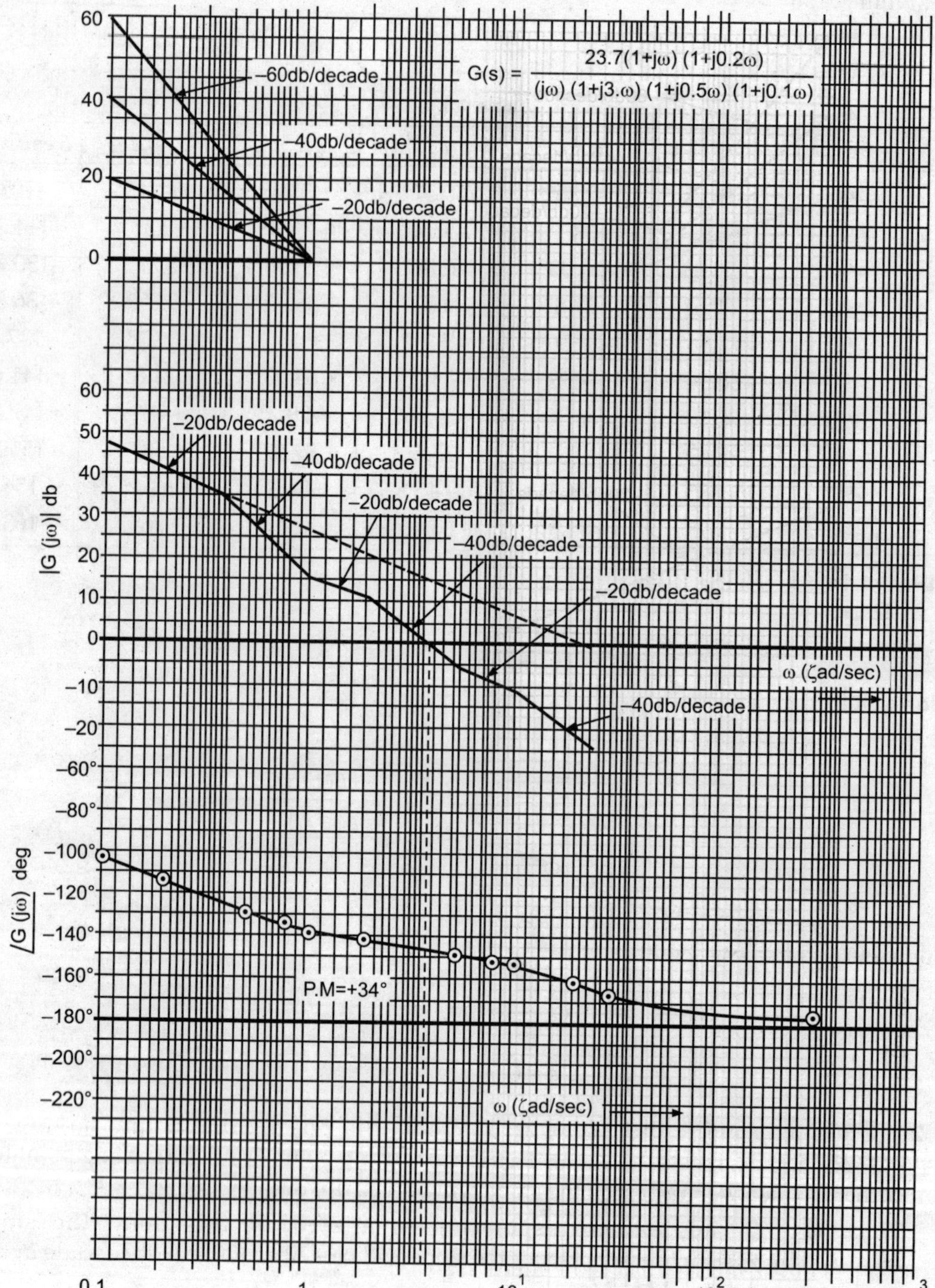
G(s) = 23.7(1+jω) (1+j0.2ω) / (jω) (1+j3.ω) (1+j0.5ω) (1+j0.1ω)
–60db/decade
–40db/decade
–20db/decade
–20db/decade
–40db/decade
–20db/decade
–40db/decade
–20db/decade
–40db/decade
|G (jω)| db
/G (jω) deg
P.M=+34°
ω (ζad/sec)
ω (ζad/sec)
ω (ζad/sec)

Fig. 4.34.

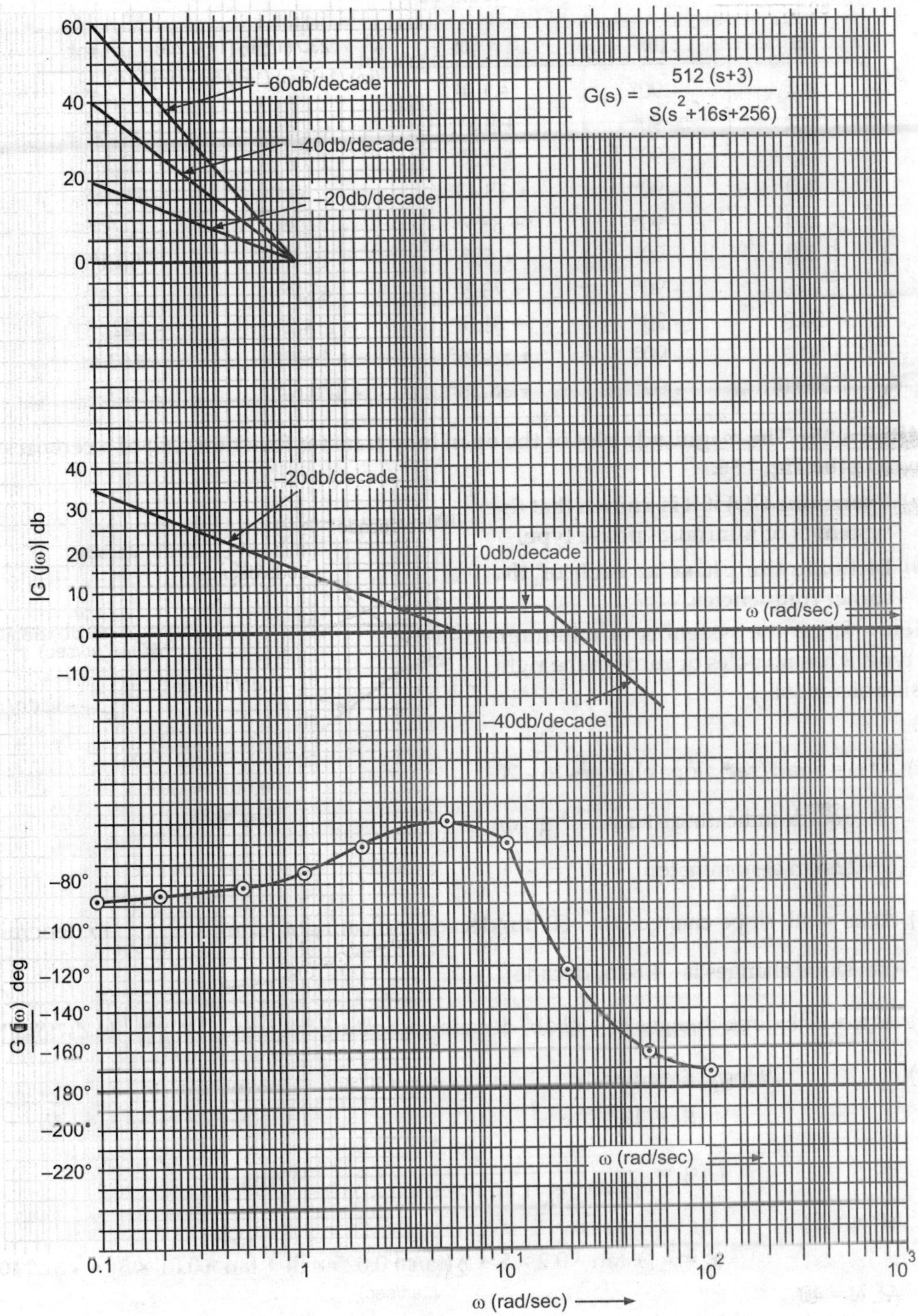

$G(s) = \frac{512 (s+3)}{S(s^2+16s+256)}$
–60db/decade
–40db/decade
–20db/decade
60
40
20
0
–20db/decade
0db/decade
ω (rad/sec)
–40db/decade
40
30
20
10
0
–10
|G (jω)| db
–80°
–100°
–120°
–140°
–160°
180°
–200°
–220°
G (jω) deg
ω (rad/sec)
0.1
1
10¹
10²
10³
ω (rad/sec)

Fig. 4.35.

Table 4.9

ω	jω	3 + jω	$s^2 + 16s + 256$	Resultant
0.1	– 90°	+ 1.9°	– 0.36°	– 88.46°
0.2	– 90°	+ 3.8°	– 0.72°	– 86.92°
0.5	– 90°	+ 9.5°	– 1.79°	– 82.29°
1.0	– 90°	+ 18.43°	– 3.59°	– 75.16°
2.0	– 90°	+ 33.7°	– 7.23°	– 63.53°
5.0	– 90°	+ 59°	– 19.1°	– 50.1°
10.0	– 90°	+ 73.3°	– 45.7°	– 62.4°
20.0	– 90°	+ 81.5°	– 114.2°	– 122.7°
50.0	– 90°	+ 86.56°	– 160°	– 163.4
100.0	– 90°	+ 88.28°	– 170.67°	– 172.4

EXAMPLE 4.15. The magnitude plot of the open loop transfer function $G(s)$ of a certain system is shown in the Fig. 4.36.

(a) Determine $G(s)$ if it is known that the system is of minimum phase type.

(b) Estimate the phase at each of the corner frequencies.

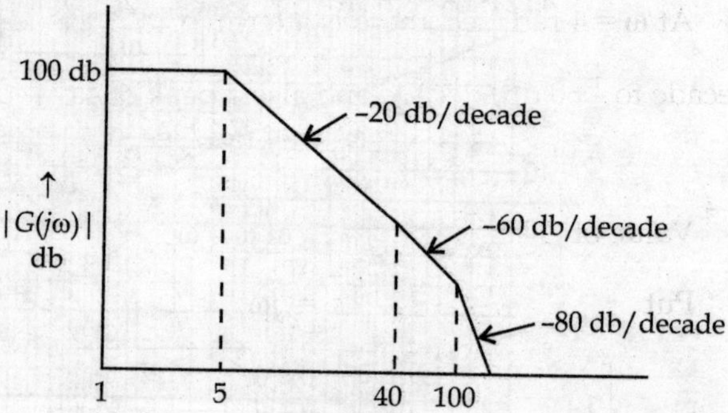

Fig. 4.36.

Solution : Since, the system is of minimum phase type, it has no poles or zeros in the right hand side of s-plane.

(a)

(i) At ω = 5 rad/sec. slope changes to – 20 db/decade indicating a term $\left(1+\frac{s}{5}\right)$ or (1+0.2s) in denominator.

(ii) At ω = 40, slope changes to – 60 db/decade indicating a term $\left(1+\frac{s}{40}\right)^2$ in denominator (net slope changes is - 40 db/decade)

(iii) At ω = 100, slope changes to – 80 db/decade, indicating the term $\left(1+\frac{s}{100}\right)$ in denominator.

(iv) $$20 \log_{10} K = 100$$

$$\therefore \quad K = 10^5$$

$$\therefore \quad G(s) = \frac{10^5}{(1+0.2s)(1+0.025s)^2(1+0.01s)} \quad \textbf{Ans.}$$

(b) At ω = 5

$$\phi_5 = -[+\tan^{-1} 0.2 \times 5 + 2(\tan^{-1} 0.025 \times 5) + \tan^{-1} 0.01 \times 5] = -62.11°$$

At ω = 40

$$\phi_{40} = -[\tan^{-1} 40 \times 0.2 + 2(\tan^{-1} 40 \times 0.025) + \tan^{-1} 40 \times 0.01]$$
$$= -[82.87 + 90 + 21.8]$$
$$\phi_{40} = -194.67°$$

At $\omega = 100$

$$\phi_{100} = -[\tan^{-1} 100 \times 0.2 + (\tan^{-1} 100 \times 0.025)^2 + \tan^{-1} 100 \times 0.01]$$
$$= -[87.13 + 136.39 + 45]$$
$$= -268.53°$$

EXAMPLE 4.16. Find the transfer function of the system from the data given on Bode diagram.

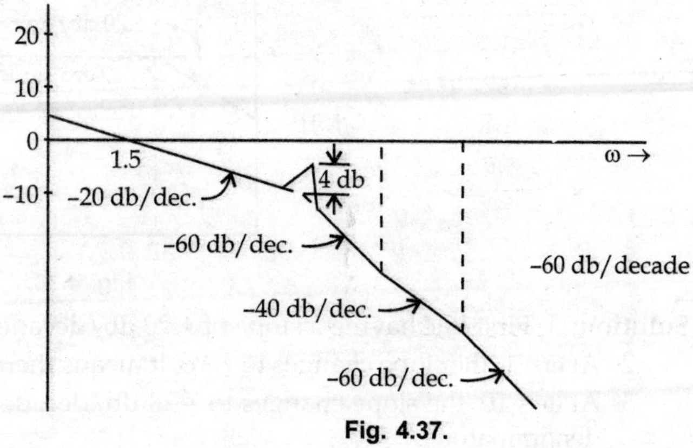

Fig. 4.37.

Solution : First line having a slope of –20 db/decade and it is not passing through $\omega = 1$ rad/sec. it means there is term K/s

At $\omega = 1$ rad/sec. of initial part = 20 log 1.5 = 3.521

$$\therefore \quad 20 \log K = 3.521$$
$$K = 1.5$$

At $\omega = 4$ rad/sec. there is a term $\left[1+\frac{2\zeta}{\omega_n}s+\frac{s^2}{\omega^2{}_n}\right]^{-1}$ because the slope changes from – 20 db/decade to – 60 db/decade and also a peak of 4 db is given.

$$\omega_n = 4 \text{ rad/sec.}$$

Value of $\left[1+\frac{2\zeta}{\omega_n}s+\frac{s^2}{\omega^2{}_n}\right]^{-1}$ at $\omega = \omega_n$

Put $s = j\omega$

$$\left[1+j\frac{2\zeta}{\omega_n}\omega-\frac{\omega^2}{\omega^2{}_n}\right]^{-1} = \left[\sqrt{\left(1-\frac{\omega^2}{\omega^2{}_n}\right)^2+\left(\frac{2\zeta}{\omega_n}\right)^2}\right]^{-1} = \left[\sqrt{\left(1-\frac{4}{4}\right)^2+\left(\frac{2\zeta}{4}4\right)^2}\right]^{-1} = \frac{1}{2\zeta}$$

$$20\log\frac{1}{2\zeta} = 4 \qquad \therefore \xi = 0.316$$

$$\therefore \left[1+\frac{2\zeta}{\omega_n}s+\frac{s^2}{\omega_n{}^2}\right]^{-1} = [1+0.158s+0.0625s^2]^{-1}$$

At $\omega = 10$ rad/sec. the slope changes from – 60 db/dec. to – 40 db/dec indicating the term $\left(1+\frac{s}{10}\right)$ in numerator.

At $\omega = 20$ rad/sec, the slope changes from – 40 db/decade to – 60 db/dec. indicating the term $\left(1+\frac{s}{20}\right)$ in denominator

$$\therefore \quad G(s) = \frac{1.5(1+s/10)}{s(1+0.158s+0.0625s^2)(1+s/20)}$$

$$G(s) = \frac{48(s+10)}{s(s^2+2.52s+16)\,(s+20)}$$

this is the required transfer frnction.

EXAMPLE 4.17. Find the transfer function of the given Bode diagram.

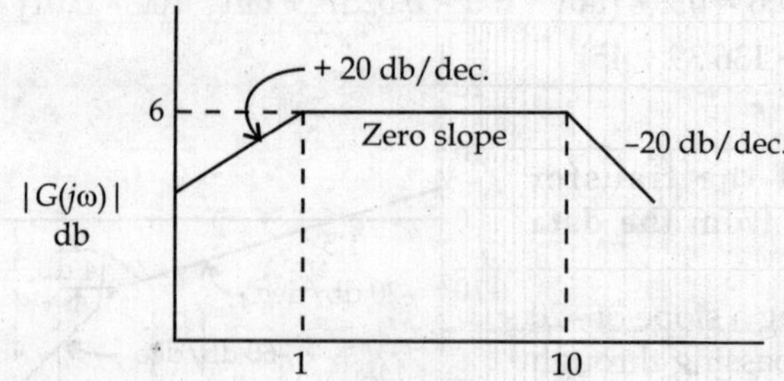

Fig. 4.38.

Solution: 1. First line having a slope of + 20 db/decade. Therefore there is a term 's' in numerator.

2. At $\omega = 1$, the slope changes to zero, it means there is a term $(1 + {}^{s}/_{1}) = (s + 1)$ in denominator.
3. At $\omega = 10$, the slope changes to – 20 db/decade, this indicates there is a term $(1 + {}^{s}/_{10})$ in denominator.
4. $20 \log K = 6$

$$\therefore \quad K = 1.99 \approx 2.0$$

$$\therefore \quad G(s) = \frac{2s}{(s+1)(1+s/10)} = \frac{20s}{(s+1)(s+10)}$$ is the required transfer function.

4.14. FREQUENCY DOMAIN SPECIFICATIONS

(*a*) **Resonant Peak: M_r :** The maximum value of magnitude is known as resonant peak. The magnitude of resonant peak gives the information about the relative stability of the system. A large value of resonant peak implies undesirable transient response.

(*b*) **Resonant Frequency (ω_r) :** The frequency at which magnitude has maxm. value is known as resonant frequency. If ω_r is large, the time response is fast.

(*c*) **Bandwidth :** Bandwidth is defined as the range of frequencies in which the magnitude of close loop does not drop – 3 db. In other words we can say that the bandwidth is defined as the band of frequencies lying between – 3 db points.

(*d*) **Cut-off Frequency :** The frequency at which the magnitude is 3 db below its zero frequency value is called cut-off frequency (ω_b).

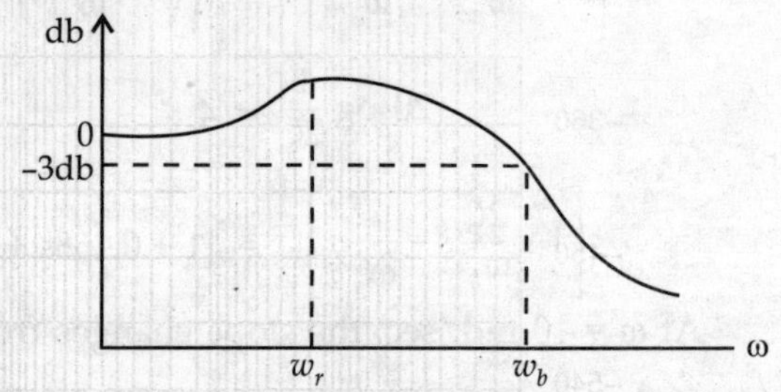

Fig. 4.40. Bode plot

(*e*) **Cut-off rate :** The cut off rate is the slope of the log magnitude curve near the cut off frequency.

4.15. CORRELATION BETWEEN TIME AND FREQUENCY RESPONSE

Consider the second order system

$$\frac{C(s)}{R(s)} = \frac{\omega_n^2}{s^2 + 2\zeta\omega_n s + \omega_n^2} \quad \text{...(4.11)}$$

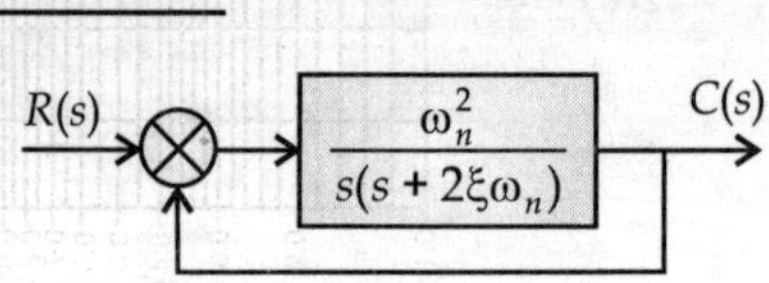

Fig. 4.41.

Put $s = j\omega$ in equation (4.11)

$$\frac{C(j\omega)}{R(j\omega)} = \frac{\omega_n^2}{(j\omega)^2 + 2\zeta\omega_n(j\omega) + \omega_n^2)} = \frac{\omega_n^2}{(\omega_n^2 - \omega^2) + j\,2\zeta\omega_n\omega}$$

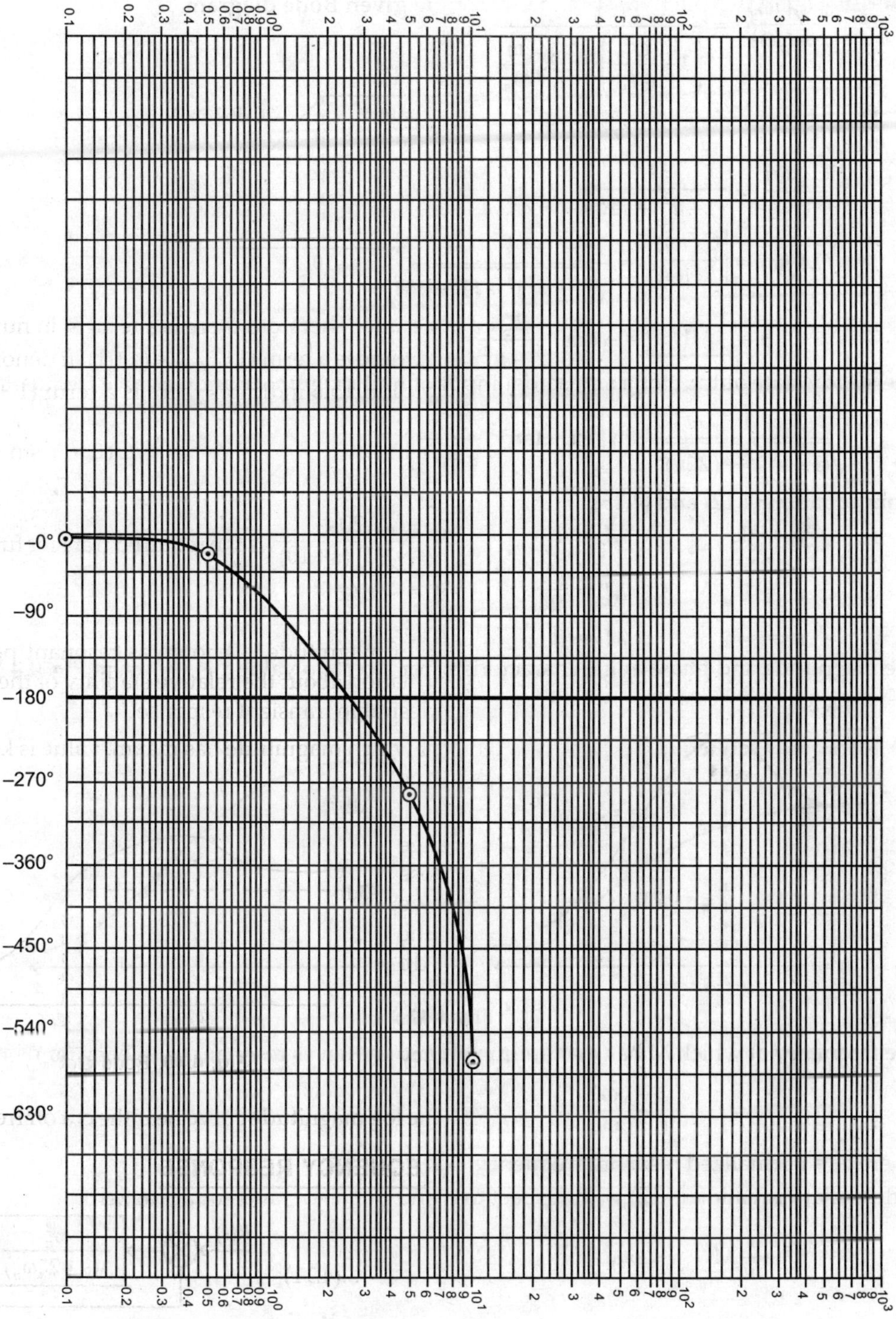

0.1
0.2
0.3
0.4
0.5
0.6
0.7
0.8
0.9
10^0
2
3
4
5
6
7
8
9
10^1
10^2
10^3
–0°
–90°
–180°
–270°
–360°
–450°
–540°
–630°

Fig. 4.39.

or $\quad \dfrac{C(j\omega)}{R(j\omega)} = \dfrac{1}{\left(1-\dfrac{\omega^2}{\omega^2_n}\right)+j\,2\zeta\dfrac{\omega}{\omega_n}}$

let $\dfrac{\omega}{\omega_n} = u$

$$\therefore \quad \frac{C(j\omega)}{R(j\omega)} = \frac{1}{(1-u^2)+j2\zeta u}$$

$$M(j\omega) = \left|\frac{C(j\omega)}{R(j\omega)}\right| = \frac{1}{\sqrt{(1-u^2)^2+(2\zeta u)^2}} \quad \text{...(4.12)}$$

$$\theta = \angle M(j\omega) = -\tan^{-1}\frac{2\zeta u}{1-u^2} \quad \text{...(4.13)}$$

The steady state output for a sinusoidal input of unit magnitude and variable frequency.

$$C(t) = \frac{1}{\sqrt{(1-u^2)^2+(2\zeta u)^2}}\sin\left[\omega t - \tan^{-1}\frac{2\zeta u}{1-u^2}\right] \qquad \because C(t) = Y\sin(\omega t+\theta)$$

From equation (4.12) and (4.13)

if	$u = 0$	$M = 1$	$\theta = 0°$
	$u = 1$	$M = \dfrac{1}{2\zeta}$	$\theta = -\pi/2$
	$u = \infty$	$M = 0$	$\theta = -\pi$

The magnitude and phase angle characteristic for a certain value ξ is shown in Fig. 4.42.

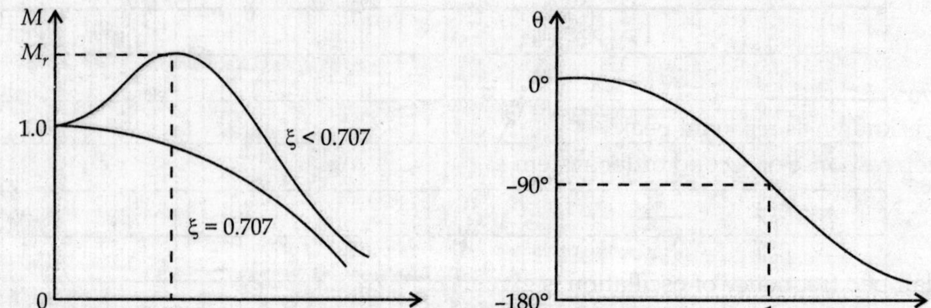

Fig. 4.42.

The frequency at which M has maximum value is known as resonant frequency (ω_r)

$$u_r = \frac{\omega_r}{\omega_n}$$

where u_r = normalized resonant frequency

Differentiate equation (4.12) with respect to u and put $u = u_r$ and equate to zero.

$$\left.\frac{d}{du}M\right|_{u=u_r} = 0$$

$$\left.\frac{d}{du}M\right|_{u=u_r} = -\frac{1}{2}\,[\,(1-u_r^2)^2 + (2\xi\,u_r)^2]^{-3/2}\,[4u^3_r - 4u_r + 8u_r\xi^2] = 0$$

or, $\quad 4u_r^3 - 4u_r + 8u_r\xi^2 = 0$

or, $u_r^3 - u_r + 2u_r \xi^2 = 0$

or, $$u_r = \sqrt{1-2\zeta^2}$$

But $$u_r = \frac{\omega_r}{\omega_n}$$

$$\therefore \quad \frac{\omega_r}{\omega_n} = \sqrt{1-2\zeta^2}$$

$$\boxed{\omega_r = \omega_n \sqrt{1-2\zeta^2}} \qquad \text{....(4.14)}$$

Form equation 4.13.

$$\theta = -\tan^{-1}\frac{2\xi u_r}{1-u^2} = -\tan^{-1}\frac{2\xi\sqrt{1-2\xi^2}}{1-1+2\xi^2}$$

$$\boxed{\theta = -\tan^{-1}\frac{\sqrt{1-2\xi^2}}{\xi}} \qquad \text{...(4.15)}$$

For maximum value of magnitude put $u_r = \sqrt{1-2\xi^2}$ in equation 4.12.

$$M_r = \frac{1}{\sqrt{[1-(1-2\xi^2)]^2 + 4\xi^2(1-2\xi^2)}}$$

$$\boxed{M_r = \frac{1}{2\xi\sqrt{1-\zeta^2}}} \qquad \text{...(4.16)}$$

where M_r is known as resonant peak.

For step response of second order system (for $0 \le \zeta \le 1$) maximum overshoot M_p is given by

$$M_P = e^{-\frac{\pi\xi}{\sqrt{1-\xi^2}}} \qquad \text{...(4.17)}$$

and damped frequency of oscillation is given by

$$\omega_d = \omega_n \sqrt{1-\xi^2} \qquad \text{(4.18)}$$

From equation (4.14) and (4.18)

$$\frac{\omega_r}{\omega_d} = \frac{\omega_n\sqrt{1-2\xi^2}}{\omega_n\sqrt{1-\xi^2}}$$

$$\omega_r = \omega_d\left[\frac{\sqrt{1-2\xi^2}}{\sqrt{1-\xi^2}}\right] \qquad \text{...(4.19)}$$

From above expression if the value of ξ is small, the damped frequency ω_d and resonant frequency are nearly same. Hence for large value of ω_r, the time response is faster.

From equation (4.16) and (4.17) it is clear that both M_r and M_P are the function of ζ. As ζ increases both M_r and M_P decreases, but the M_P is limited to 1 and M_r having very large values for $\zeta < 0.4$ as shown in Fig. (4.42).

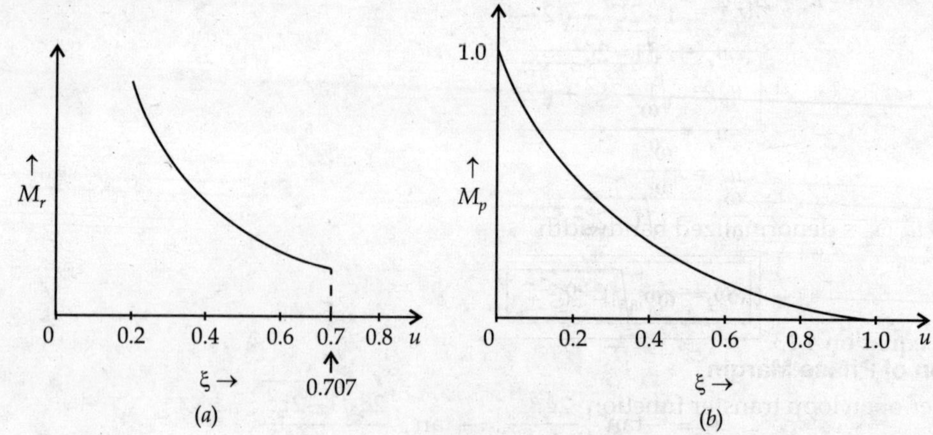

Fig. 4.43.

The bandwidth is also defined as the range of frequencies over which M is equal or greater than $1/\sqrt{2}$.

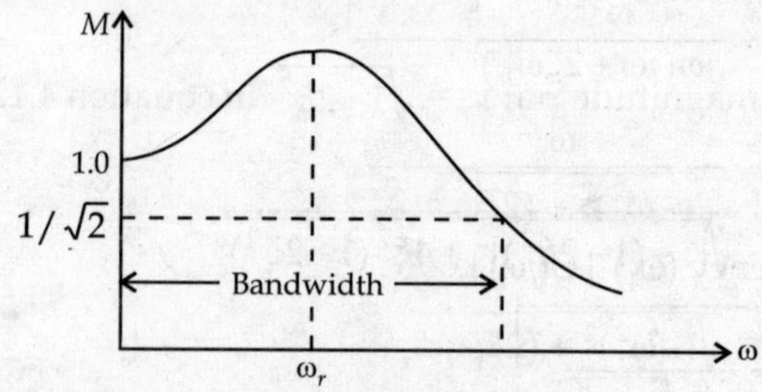

Fig. 4.44.

From equation number (4.12).

$$M = \frac{1}{\sqrt{(1-u^2)^2+(2\zeta u)^2}}$$

let u_b = normalized bandwidth = $\frac{\omega_b}{\omega_n}$ then

$$M = \frac{1}{\sqrt{(1-u^2{}_b)^2+(2\zeta u_b)^2}} = \frac{1}{\sqrt{2}}$$

or, $(1 - u^2{}_b)^2 + (2\,\xi\, u_b)^2 = 2$

$1 + u_b{}^4 - 2u_b{}^2 + 4\,\xi^2\, u_b{}^2 = 2$

or, $u_b{}^4 - 2u_b{}^2 + 4\,\xi^2\, u_b{}^2 - 1 = 0$

or, $u_b{}^4 - 2u_b{}^2\,(1 - 2\,\xi^2) - 1 = 0$

Put $u_b{}^2 = x$

$x^2 - 2x\,(1 - 2\xi^2) - 1 = 0$

$$\therefore \quad x = \frac{+2(1-2\zeta^2) \pm \sqrt{4(1-2\zeta^2)^2+4}}{2} = 1-2\zeta^2 \pm \sqrt{1-4\zeta^2+4\zeta^4+1}$$

$$= 1-2\zeta^2 \pm \sqrt{2-4\zeta^2+4\zeta^4}$$

consider the positive part

$$u_b^2 = 1-2\zeta^2+\sqrt{2-4\zeta^2+4\zeta^4}$$

$\therefore$ $$u_b = \sqrt{1-2\zeta^2+\sqrt{2-4\zeta^2+4\zeta^4}} \quad \text{...(4.20)}$$

But $$u_b = \frac{\omega_b}{\omega_n}$$

$\therefore \omega_b = u_b\,\omega_n$ = denormalized bandwidth

$\therefore$ $$\omega_b = \omega_n\sqrt{1-2\zeta^2+\sqrt{2-4\zeta^2+4\zeta^4}} \quad \text{...(4.21)}$$

Calculation of Phase Margin

consider open loop transfer function

$$G(s) = \frac{\omega_n{}^2}{s(s+2\zeta\omega_n)} \quad \text{...(4.22)}$$

Put $s = j\omega$

$$G(j\omega) = \frac{\omega_n{}^2}{j\omega(j\omega+2\zeta\omega_n)} = \frac{\omega_n{}^2}{-\omega^2+j2\zeta\omega\omega_n}$$

$$|G(j\omega)| = \frac{\omega_n{}^2}{\sqrt{(-\omega^2)^2+(2\zeta\omega\omega_n)^2}}$$

At gain cross over frequency (ω_1) $|G(j\omega)| = 1$

$\therefore$ $$\omega_n{}^2 = \sqrt{(-\omega_1^2)^2+(2\zeta\omega_1\omega_n)^2}$$

$$\omega_1{}^4 + 4\xi^2\,\omega_1{}^2\omega_n{}^2 - \omega_n{}^4 = 0$$

Put $x = \omega_1{}^2$

$$x^2 + (4\xi^2\,\omega_n{}^2)\,x - \omega_n{}^4 = 0$$

$$x = \frac{-4\zeta^2\omega^2{}_n \pm \sqrt{(4\zeta^2\omega^2{}_n)^2+4\omega_n{}^4}}{2}$$

$$x = -2\zeta^2\omega^2{}_n \pm \omega^2{}_n\sqrt{4\zeta^2+1}$$

or, $$\omega_1{}^2 = -2\zeta^2\omega^2{}_n \pm \omega^2{}_n\sqrt{4\zeta^2+1}$$

or, $$\omega_1 = \omega_n\sqrt{\sqrt{4\zeta^4+1}-2\zeta^2}$$

$$\angle G(j\omega_1) = \tan^{-1}\frac{2\zeta\omega_1\omega_n}{-\omega_1{}^2} = +\tan^{-1}\frac{2\zeta\omega_n}{\omega_1}$$

Put the value of ω_1

$$\angle G(jw_1) = +\tan^{-1}\frac{2\zeta\omega_n}{\omega_n\sqrt{\sqrt{4\zeta^4+1}-2\zeta^2}} = +\tan^{-1}\frac{2\zeta}{\sqrt{\sqrt{4\zeta^4+1}-2\zeta^2}}$$

$\therefore$ Phase margin

$$\phi = -180+\tan^{-1}\frac{2\zeta}{\sqrt{\sqrt{4x^4+1}-2\zeta^2}} \quad \text{or,} \quad \phi = \tan^{-1}\frac{2\zeta}{\sqrt{\sqrt{4\zeta^4+1}-2\zeta^2}}$$

$\because$ $$\tan(180+\phi) = \tan\phi \quad \text{...(4.23)}$$

4.16. RELATIVE AND ABSOLUTE STABILITY

There are two types of stability namely, absolute stability and relative stability. Absolute stability means whether the system is stable or not, *i.e.*, the system is stable or unstable. If the system is stable then we determine how stable it is *i.e.*, we measure the degree of stability, it is known as relative stability.

In time domain the relative stability is measured by maximum overshoot and damping ratio. In frequency domain relative stability is measured by resonant peak (M_r).

EXAMPLE 4.18. The forward path transfer function of a unity feedback control system is

$$G(s) = \frac{100}{s(s+6.54)}$$

Find the resonance peak M_r, resonant frequency ω_r and bandwidth of the closed loop system.

Solution : Given that

$$G(s) = \frac{100}{s(s+6.54)}$$

$$H(s) = 1$$

$$\therefore \quad \frac{C(s)}{R(s)} = \frac{G(s)}{1+G(s)\,H(s)} = \frac{100/s(s+6.54)}{1+\dfrac{100}{s(s+6.54)}}$$

$$= \frac{100}{s(s+6.54)+100} = \frac{100}{s^2+6.54s+100}$$

compare with $\dfrac{\omega_n{}^2}{s^2+2\zeta\omega_n s+w_n{}^2}$

$$\omega_n{}^2 = 100 \qquad \boxed{\omega_n = 10 \text{ rad/sec.}}$$

$$2\xi\,\omega_n = 6.54$$

$$2\xi\,10 = 6.54 \qquad \therefore \boxed{\xi = 0.327}$$

$$\omega_r = \omega_n\sqrt{1-2\zeta^2}$$

$$\omega_r = 10\sqrt{1-2(0.327)^2} = 8.86 \text{ rad/sec.}$$

$$M_r = \frac{1}{2\xi\sqrt{1-\xi^2}} = \frac{1}{2(0.327)\sqrt{1-(0.327)^2}}$$

$$\therefore \quad \boxed{M_r = 1.618}$$

Bandwidth $= \omega_n\sqrt{1-2\xi^2+(2-4\xi^2+4\xi^4)^{1/2}}$

$$= 10\sqrt{1-2(0.327)^2+(2-4\times 0.327^2+4\times 0.327^4)^{1/2}}$$

$= 14.34$ rad/sec.

$$\boxed{\text{Bandwidth} = 14.34 \text{ rad/sec.}}$$

EXAMPLE 4.19. The specification on a second order unity feedback control system with closed loop transfer function.

$$\frac{C(s)}{R(s)} = \frac{\omega_n^2}{s^2 + 2\xi\omega_n s + \omega_n^2}$$

are that the maximum overshoot must not exceed 30%. Find the corresponding limiting values of M_r and BW.

Solution : we know that

$$Mp = e^{-\pi\xi/\sqrt{1-\xi^2}} \times 100$$

$$\therefore \quad e^{-\pi\xi/\sqrt{1-\xi^2}} \times 100 = 30$$

$$\text{or,} \quad e^{+\pi\xi/\sqrt{1-\xi^2}} = \frac{100}{30}$$

$$\frac{\pi\xi}{\sqrt{1-\xi^2}} = 1.2 \quad \text{(take the natural log)}$$

$$\text{or,} \quad \pi^2\xi^2 = 1.2^2\,(1-\xi^2) \quad \text{or,} \quad \xi = 0.356$$

$$M_r = \frac{1}{2\xi\sqrt{1-\xi^2}} = \frac{1}{2\times 0.356\sqrt{1-0.356^2}}$$

$$\therefore \quad \boxed{M_r = 1.5}$$

$$\text{Normalized B.W} = \sqrt{1-2\xi^2+(2-4\xi^2+4\xi^4)^{1/2}}$$

$$= \sqrt{1-2\times 0.356^2+(2-4\times 0.356^2+4\times 0.356^4)^{1/2}} = 1.41 \text{ rad/sec.}$$

EXAMPLE 4.20. The closed loop frequency response magnitude versus frequency of a second order system is shown in Fig. 4.45. Find %M_p (maximum overshoot) and peak time.

Solution : From Fig. 4.45 $M_r = 1.4$ & $\omega_r = 3$ rad/sec.

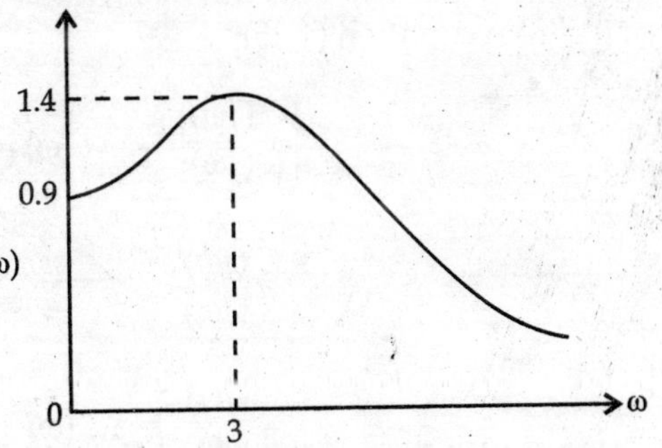

Fig. 4.45.

$$\text{since,} \quad M_r = \frac{1}{2\xi\sqrt{1-\xi^2}}$$

$$\therefore \quad 1.4 = \frac{1}{2\xi\sqrt{1-\xi^2}}$$

$$\xi^2(1-\xi^2) = 0.1275$$

$$\xi^2 - \xi^4 - 0.1275 = 0$$

$$\text{or} \quad \xi^4 - \xi^2 + 0.1275 = 0$$

Put $x^2 = \xi$

$$x^2 - x + 0.1275 = 0$$

$$x = 0.8495,\ 0.1505$$

consider $x = 0.1505$ or $\xi^2 = 0.1505$ $\therefore$ $\boxed{\xi = 0.387}$

$$M_p = e^{-\pi\xi/\sqrt{1-\xi^2}} \times 100 = e^{-\pi 0.387/\sqrt{1-0.387^2}} \times 100 = 26\% \quad \textbf{Ans.}$$

$$\text{Peak time} \quad t_p = \frac{\pi}{\omega_n\sqrt{1-\xi^2}}$$

$$\omega_r = \omega_n\sqrt{1-2\xi^2}$$

or $$3 = \omega_n \sqrt{1-2(0.387)^2}$$

$\therefore$ $$\omega_n = 3.58 \quad \text{rad/sec.}$$

$\therefore$ $$t_p = \frac{\pi}{3.58\sqrt{1-0.387^2}} = 0.95 \text{ sec.}$$

$\boxed{t_p = 0.95 \text{ sec.}}$ **Ans.**

4.17. FREQUENCY RESPONSE OF A CLOSED LOOP SYSTEM

Consider the transfer function

$$\frac{C(s)}{R(s)} = \frac{G(s)}{1+G(s)H(s)}$$

For unity feedback $H(s) = 1$

$\therefore$ $$\frac{C(s)}{R(s)} = \frac{G(s)}{1+G(s)} \qquad ...(4.24)$$

Put $s = j\omega$

$$\frac{C(j\omega)}{R(j\omega)} = \frac{G(j\omega)}{1+G(j\omega)} \qquad ...(4.25)$$

The polar plot of equation 4.25 is shown in Fig. 4.46.

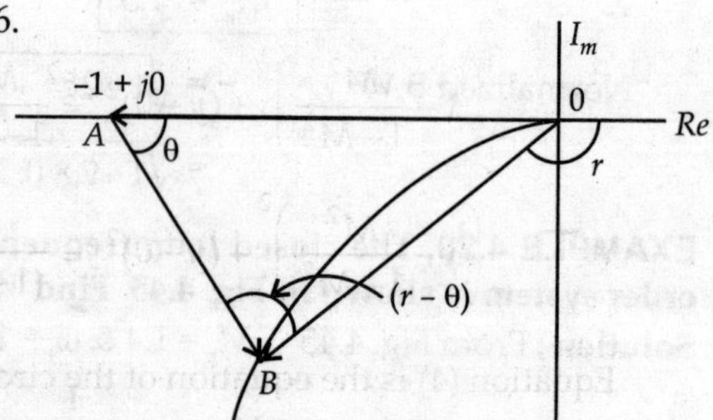

Fig. 4.46.

From Fig. 4.46 $$\overrightarrow{OB} = G(j\omega)$$

$$\overrightarrow{OA} = -1$$

$$\overrightarrow{AB} = \overrightarrow{OB} - \overrightarrow{OA} = G(j\omega) - (-1)$$

$$\overrightarrow{AB} = 1 + G(j\omega)$$

From equation (4.25)

$$\left|\frac{C(j\omega)}{R(j\omega)}\right| = M(\omega) = \frac{\overrightarrow{OB}}{\overrightarrow{AB}}$$

$$\angle\frac{C(j\omega)}{R(j\omega)} = \frac{\angle \overrightarrow{OB}}{\angle \overrightarrow{AB}} = \frac{r}{\theta} = r - \theta$$

$\therefore$ $$\frac{C(j\omega)}{R(j\omega)} = M(\omega)\, e^{j\,\phi(\omega)}$$

where $M(j\omega)$ is the magnitude and $\phi(\omega) = r - \theta$

Frequency response consists of two parts (*i*) magnitude (*ii*) phase angle. Both can be plotted against defferent values of ω.

The constant magnitude loci and constant phase angle can be plotted in complex plane $G(j\omega)$, from these locii we can obtain the frequency response of closed loop system.

4.18. CONSTANT MAGNITUDE CIRCLE (M-CIRCLE)

Let $$G(j\omega) = x + jy$$

Then from equation (4.25)

$$M = M(\omega) = \frac{|C(j\omega)|}{|R(j\omega)|} = \left|\frac{G(j\omega)}{1+G(j\omega)}\right| = \frac{|x+jy|}{|1+x+jy|} = \frac{\sqrt{x^2+y^2}}{\sqrt{(1+x)^2+y^2}} \qquad ...(4.26)$$

Squaring both sides

$$M^2 = \frac{x^2+y^2}{(1+x)^2+y^2}$$

or

$$M^2 [(1+x)^2 + y^2] = x^2 + y^2$$

$$M^2 [1 + 2x + x^2 + y^2] = x^2 + y^2$$

$$M^2 + 2x M^2 + x^2 M^2 + M^2y^2 - x^2 - y^2 = 0$$

$$x^2(M^2 - 1) + y^2 (M^2 - 1) + 2x M^2 + M^2 = 0$$

or,

$$x^2 (1 - M^2) + y^2 (1 - M^2) - 2x M^2 = M^2$$

Divide both the sides by $(1 - M^2)$

$$x^2 - y^2 - 2x\frac{M^2}{1-M^2} = \frac{M^2}{1-M^2}$$

$\left(\frac{M^2}{1-M^2}\right)^2$ add in both sides

$$x^2 + y^2 - 2x\frac{M^2}{1-M^2} + \left(\frac{M^2}{1-M^2}\right)^2 = \frac{M^2}{1-M^2} + \left(\frac{M^2}{1-M^2}\right)^2$$

$$\left(x - \frac{M^2}{1-M^2}\right)^2 + (y-0)^2 = \frac{M^2}{1-M^2} + \frac{M^4}{(1-M^2)^2}$$

$$\left(x - \frac{M^2}{1-M^2}\right)^2 + (y-0)^2 = \frac{M^2}{(1-M^2)^2} \quad ...(4.27)$$

Equation (4) is the equation of the circle with centre $\left(\frac{M^2}{1-M^2}, 0\right)$ and radius $\left(\frac{M}{1-M^2}\right)$

If $M = 1$ Then equation (4.26) becomes

$$(1+x)^2 + y^2 = x^2 + y^2$$

or,

$$x = -\frac{1}{2}$$

This is the equation of the straight line parallel to the y-axis and passing through $-(^1/_2, 0)$ in the $G(j\omega)$ plane.

The constant M locii for different value of M shown in Fig. 4.48. From Fig. 4.48 it is clear that

(a) The locii are symmetrical with respect to $M = 1$

(b) The M-circles for $M > 1$ are on the left side of the line $M = 1$ and for $M < 1$ the constant M-circles are on right side of the line $M = 1$.

The intersection of $G(j\omega)$ plot (Nyquist plot) & constant M locii gives the value of magnitude M. The M-circle which is tangent to the $G(j\omega)$ plot will give the value of resonance peak (M_r) and resonance frequency ω_p (as shown in Fig. 4.47).

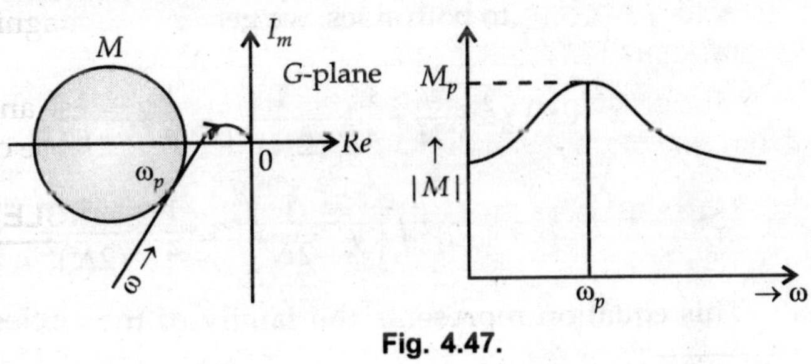

Fig. 4.47.

Table 4.10. For construction of M-circle

S.No.	M	Centre $\frac{M^2}{1-M^2},0$	Radius $\frac{M}{1-M^2}$
1.	0.3	(0.098,0)	0.329
2.	0.5	(0.33,0)	0.666
3.	0.67	(0.814, 0)	1.215
4.	0.766	(1.42, 0)	1.854
5.	0.833	(2.27, 0)	2.72
6.	1.2	(–3.27, 0)	–2.72
7.	1.3	(–2.45, 0)	– 1.88
8.	1.5	(– 1.8, 0)	– 1.2
9.	2.0	(–1.33, 0)	– 0.66
10.	3.0	(–1.125, 0)	– 0.375

4.19. CONSTANT N-CIRCLES (PHASE ANGLE LOCII)

From equation (4.26), the phase shift can be written as

$$\phi = \arg \text{ of } (x + jy) - \arg (1 + x + jy)$$

$$\phi = \tan^{-1}\frac{y}{x} - \tan^{-1}\frac{y}{1+x} \qquad ...(4.28)$$

or,

$$\tan\phi = \frac{\frac{y}{x} - \frac{y}{1+x}}{1 + \left(\frac{y}{x}\right)\left(\frac{y}{1+x}\right)} = \frac{y}{x^2 + x + y^2} \qquad ...(4.29)$$

let

$$\tan\phi = N$$

$\therefore$

$$N = \frac{y}{x^2 + x + y}$$

or,

$$N(x^2 + x + y^2) = y$$

$$x^2 + x + y^2 - \frac{y}{N} = 0$$

Add $\frac{1}{4} + \frac{1}{(2N)^2}$ to both sides, we get

$$x^2 + x + y^2 - \frac{y}{N} + \frac{1}{4} + \frac{1}{(2N)^2} = \frac{1}{4} + \frac{1}{(2N)^2}$$

or,

$$(x + 1/2)^2 + \left(y - \frac{1}{2N}\right)^2 = \frac{1}{4} + \frac{1}{(2N)^2}$$

This equation represents the family of the circles, with centres at $\left(-\frac{1}{2}, \frac{1}{2N}\right)$ and radius $\sqrt{\frac{1}{4} + \frac{1}{4N^2}}$

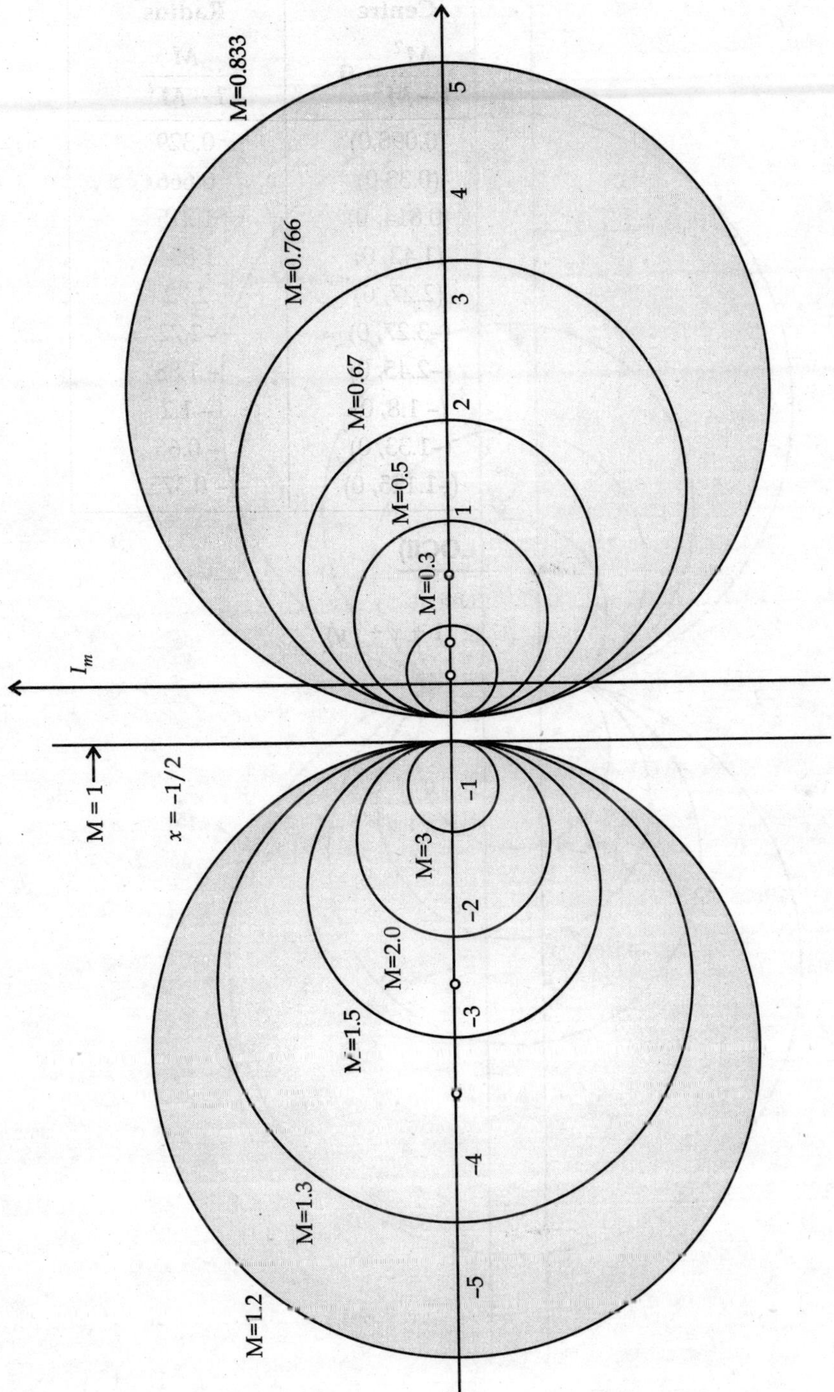

Fig. 4.48. Constt. M-Circles

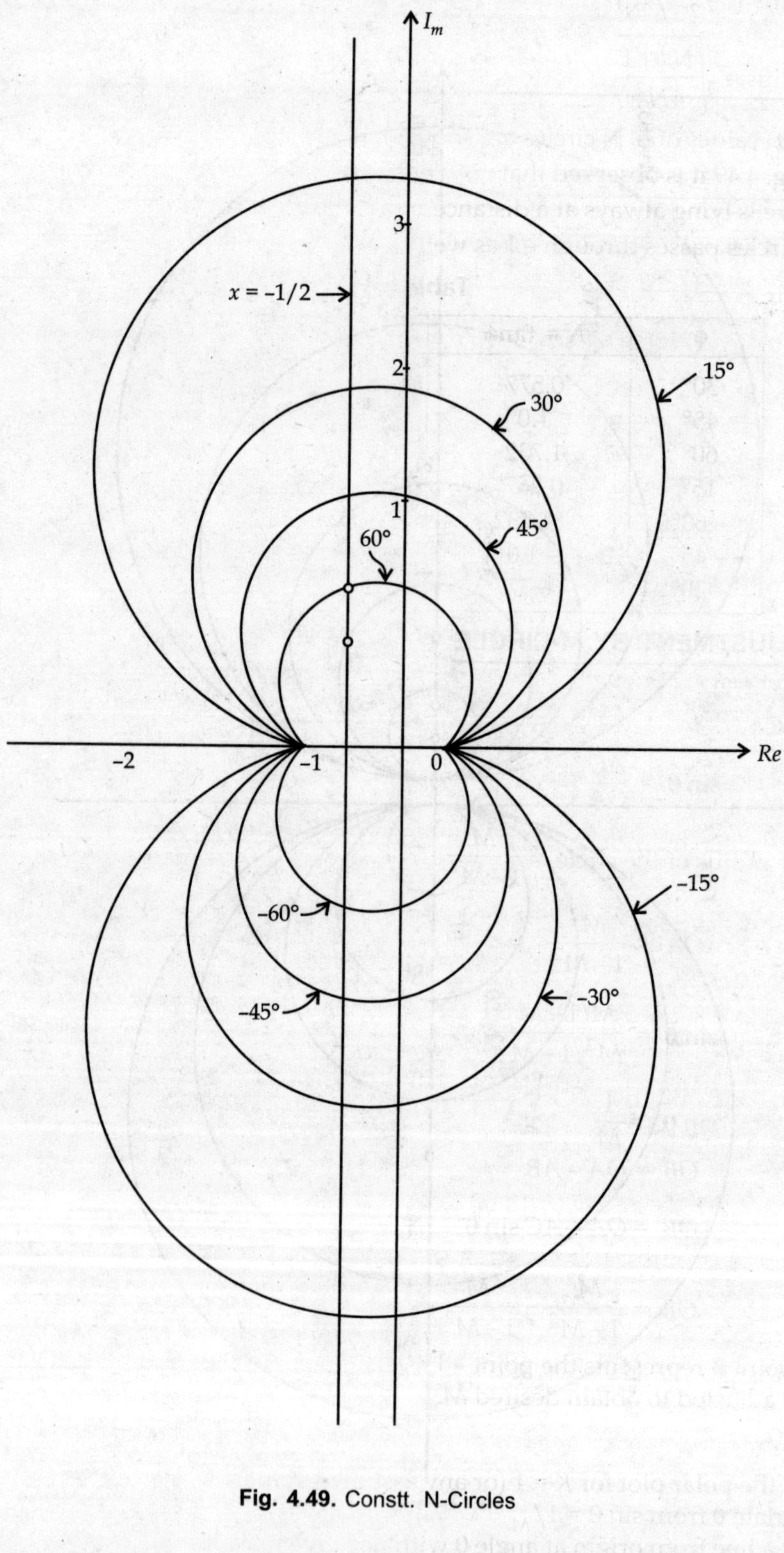

Fig. 4.49. Constt. N-Circles

centre $(^{-1}/_{2}, {}^{1}/_{2N})$

$$\text{Radius} = \sqrt{\frac{N^2+1}{4N^2}}$$

For different values of ϕ, N circles are shown in Fig. 4.49

From the Fig. 4.49 it is observed that

(a) The centre is lying always at a distance $x = -1/2$ and y depends upon the phase shift.

(b) All the circles passes through – 1 as well as 0.

Table 4.11

ϕ	$N = \tan\phi$	centre	Radius
30°	0.577	(–0.5, 0.866)	1.0
45°	1.0	(–0.5, 0.5)	0.7
60°	1.732	–0.5, 0.288	0.577
15°	0.267	0.5, 1.87	1.9
– 60°	– 1.732	– o.5, – 0.288	0.577
– 45°	– 1.0	–0.5, – 0.5	0.7
– 30°	– 0.577	–0.5, – 0.866	1.0

4.20. GAIN ADJUSTMENT BY M-CIRCLE

Consider the Fig. 4.50

From Δ AOC :

$$\sin\theta = \frac{AC}{OA}$$

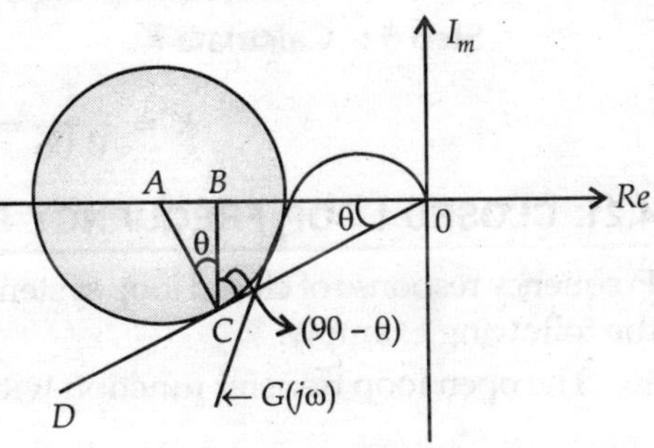

Fig. 4.50.

where AC = radius of the circle $= \dfrac{M}{1-M^2}$

$$OA = \frac{M^2}{1-M^2}$$

$$\therefore \quad \sin\theta = \frac{M/1-M^2}{M^2/1-M^2}$$

$$\sin\theta = \frac{1}{M}$$

$$OB = OA - AB$$

$$OB = OA - AC\sin\theta \qquad \because \frac{AB}{AC} = Sin\,\theta$$

$$\therefore \quad OB = \frac{M^2}{1-M^2} - \frac{M}{1\ \ M^2}\cdot\frac{1}{M} = -1$$

Thus, the point B represents the point $-1 + j\,0$ in the *GH*-plane and gives the factor by which the gain can be adjusted to obtain desired *M*.

PROCEDURE :

Step 1 : Draw the polar plot for $K = 1$ (or any assumed value)

Step 2 : Calculate θ from $\sin\theta = 1/_{M}$

Step 3 : Draw a line from origin at angle θ with negative real axis

Step 4 : Draw a circle with centre on negative real axis, which is tangent to the line *OD* and to the polar plot.

Step 5 : From the point *C* draw a line *BC* perpendicular to the negative real axis.

Step 6 : Required gain can be obtained as

$$K = \frac{\text{assumed value of K}}{\text{length OB}}$$

EXAMPLE 4.21. The open loop transfer function with unity feedback is given by

$$\mathbf{G(s)} = \frac{\mathbf{K}}{\mathbf{s(s+2)(s+4)}}$$

Determine the value of *K* so that *M* = 2.

Solution : Step 1 : Assume K = 1 and draw the polar plot (Fig. 4.51)

Step 2 : $\sin\theta = 1/2 \quad \therefore \theta = 30°$

Draw a line at 30° with negative real axis

Step 3 : Draw the circle that is tangent to both polar plot and line *OD*.

Step 4 : Draw the line from point c perpendicular to the negative real axis intersects at (– 0.06, 0)

Step 5 : Calculate *K*

$$K = \frac{1}{0.06} = 16.66 \approx 17$$

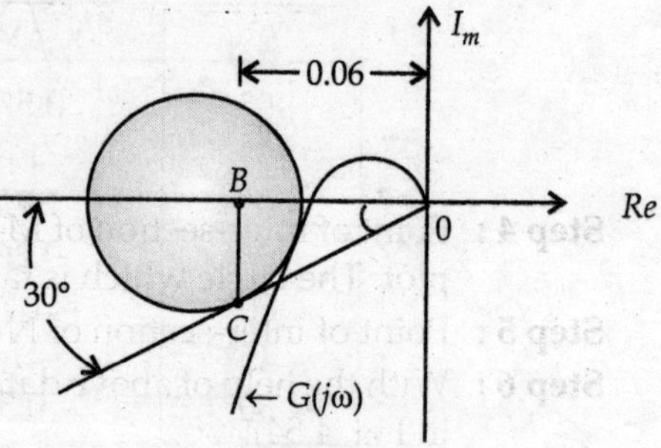

Fig. 4.51.

4.21. CLOSED LOOP FREQUENCY RESPONSE FROM M & N-CIRCLES

Frequency response of closed loop system can be obtained with the help of M and N circles. Consider the following example.

The open loop transfer function with unity feedback system is

$$G(j\omega) = \frac{10}{j\omega(1+j0.2\omega)\,(1+j0.05\omega)}$$

Step 1 : Draw the polar plot of the given transfer function.

Step 2 : Draw *M*-circles

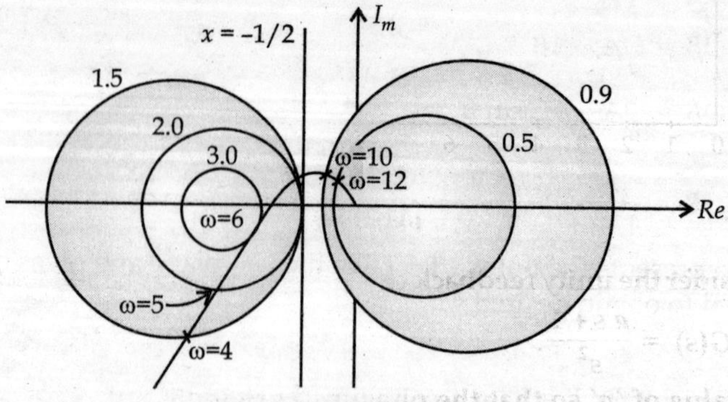

Fig. 4.52.

Step 3 : Draw N-circles

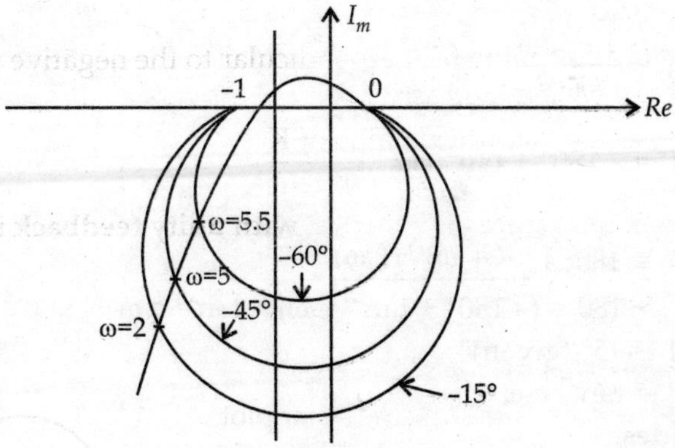

Fig. 4.53.

Step 4 : Point of inter-section of M-circle gives the magnitude and read the frequency from polar plot. The circle which is tangent to the polar plot gives the value of M_P.

Step 5 : Point of inter-section of N-circle gives the phase angle.

Step 6 : With the help of above data plot the frequency response of closed loop system. (as shown in Fig. 4.54).

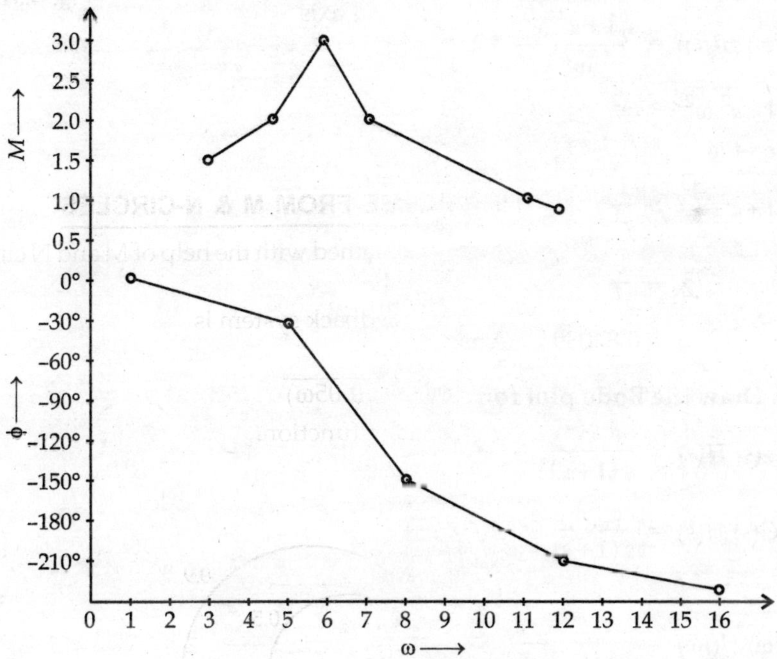

Fig. 4.64.

EXAMPLE 4.22. Consider the unity feedback control system whose open loop transfer function is

$$G(s) = \frac{as+1}{s^2}.$$

Determine the value of '*a*' so that the phase margin is 45°.

Solution : Given that $G(s)\,H(s) = \dfrac{1+as}{s^2}$

Put $s = j\omega$

$$G(j\omega)\,H(j\omega) = \frac{1 + j\,a\omega}{(j\omega)^2}$$

$$\therefore \quad \angle G(j\omega)\,H(j\omega) = -180° + \tan^{-1} a\omega$$

We know that

$$\text{P.M} = 180° + \angle G(j\omega)\,H(j\omega)$$

$$= 180 + (-180° + \tan^{-1} a\,\omega) = \tan^{-1} a\,\omega$$

$$\text{P.M} = 45° \text{ (given)}$$

$$45° = \tan^{-1} a\,\omega$$

taking tan on both sides

$$\tan 45° = \tan(\tan^{-1} a\,\omega)$$

$$a\,\omega = 1$$

$$\omega = 1/a$$

This is the gain crossover frequency ω_{c1}.

At gain crossover frequency

$$|G(j\omega)\,H(j\omega)| = 1$$

$$|G(j\omega)\,H(j\omega)| = \frac{\sqrt{1 + a^2\omega^2}}{\omega^2} = 1$$

$$\sqrt{1 + a^2\omega^2} = \omega^2$$

Put the value of ω

$$\sqrt{1 + a^2 \cdot \frac{1}{a^2}} = \frac{1}{a^2}$$

or,

$$\sqrt{2} = \frac{1}{a^2}$$

$$\therefore \quad a = 0.84089 \qquad \textbf{Ans.}$$

EXAMPLE 4.23. Draw the Bode plot for

$$G(s)\,H(s) = \frac{e^{-0.3s}}{s\,(1+s)}.$$

Solution :

$$G(s)\,H(s) = \frac{e^{-0.3s}}{s\,(1+s)}$$

Put $s = j\omega$

$$G(j\omega)\,H(j\omega) = \frac{e^{-j0.3\omega}}{j\omega\,(1 + j\omega)}$$

Since, there is a pole at origin, the initial slope of magnitude plot will -20db/dec. $e^{-j0.3\omega}$ has no effect on magnitude pole.

The corner frequency $\omega = 1$ rad/sec.

From $\omega = 1$, the slope of next line will be – 20 + (– 20) = – 40db/dec. as there is a simple pole.

PHASE PLOT : $\angle e^{-j0.3\omega} = -57.3 \times 3\omega = -17.19\omega$

$$\angle G(j\omega)\,H(j\omega) = -90° - \tan^{-1}\omega - 17.19\omega$$

Table 4.12.

ω	Arg (jω)	Arg (1 + jω)	– 17.19 ω	Resultant
0.1	– 90°	– 5.7°	– 1.719°	– 97.419°
0.5	– 90°	– 26.56°	– 8.595°	– 125.15°
1.0	– 90°	– 45°	– 17.19°	– 152.19°
2.0	– 90°	– 63.43°	– 34.38°	– 187.81°
4.0	– 90°	– 75.96°	– 68.76°	– 234.72°
5.0	– 90°	– 78.69°	– 85.95°	– 254.64°
8.0	– 90°	– 84.28°	– 171.9	– 346.18°

The magnitude plot and phase plot are shown in Fig. 4.55 From Bode plot:

Gain crossover frequency ω_{c1} = 1 rad/sec.
Phase crossover frequency ω_{c2} = 1.8 rad/sec.
Gain margin = + 10 db
Phase margin = + 28°

4.22. EFFECT OF ADDING A ZERO TO THE FORWARD PATH

Consider the system shown in Fig. 4.41. This system is unity feedback system and having forward path transfer function as

$$G(s) = \frac{\omega_n^2}{s(s+2\zeta\omega_n)} \quad \text{...(4.30)}$$

Now, let a zero (1 + *sT*) is added to the forward path transfer function then the equation (4.30) can be written as

$$G(s) = \frac{\omega_n^2(1+sT)}{s(s^2+2\zeta\omega_n)} \quad \text{...(4.31)}$$

The closed loop transfer function for unity feedback will be

$$\frac{c(s)}{R(s)} = \frac{\omega_n^2(1+Ts)}{s^2+(2\zeta\omega_n+T\omega_n^2)s+\omega_n^2} \quad \text{...(4.32)}$$

From equation (4.32) the bandwidth can be calculated

$$\text{Bandwidth} = \left(-x+\frac{1}{2}\sqrt{x^2+4\omega_n^2}\right)^2 \quad \text{...(4.33)}$$

where, $$x = 4\zeta\omega_n^2+4\zeta\omega_n^3\,T-2\omega_n^2-\omega_n^4T^2 \quad \text{...(4.34)}$$

So, when a zero is added to the forward path transfer function, the bandwidth of the closed loop system will increase.

4.23. EFFECT OF ADDING A POLE TO THE FORWARD PATH

Let a pole (1 + *sT*) is added to the forward path transfer function then equation (4.30) can be written as

$$G(s) = \frac{\omega_n^2}{s(s+2\zeta\omega_n)(1+sT)} \quad \text{...(4.35)}$$

Calculation of Bandwidth is very complicated. Then according to Kuo*, for small values of *T*, the bandwidth of the system is increased but resonant peak Mr also increased. But when *T* becomes

* B.C. Kuo, Automatic Control System, 24th Indian Reprint, p. 556

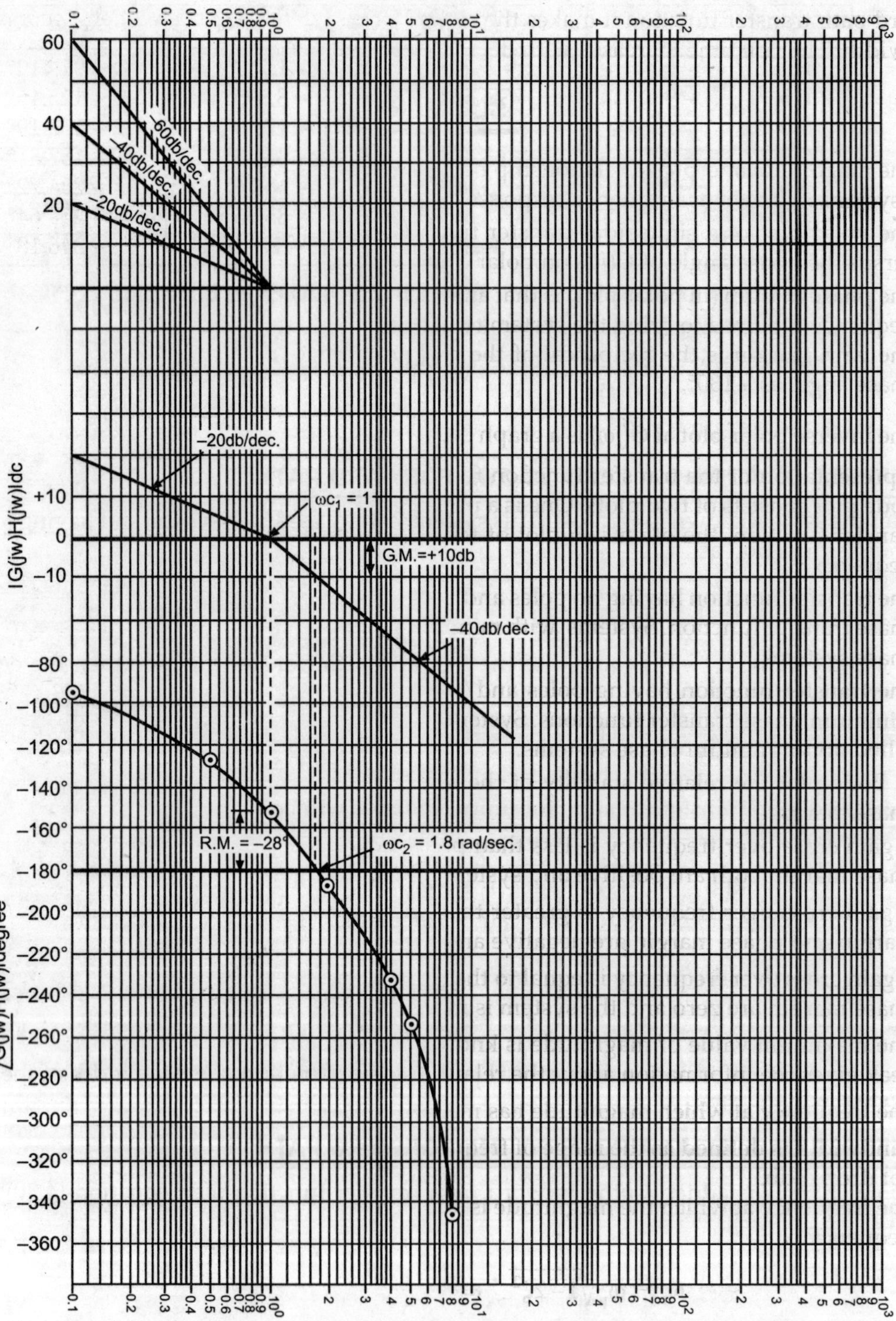
–60db/dec.
–40db/dec.
–20db/dec.
–20db/dec.
ωc1 = 1
G.M.=+10db
–40db/dec.
R.M. = –28°
ωc2 = 1.8 rad/sec.
IG(jw)H(jw)Idc
/G(jw)H(jw)Idegree
60
40
20
+10
0
–10
–80°
–100°
–120°
–140°
–160°
–180°
–200°
–220°
–240°
–260°
–280°
–300°
–320°
–340°
–360°

Fig. 4.55.

large the bandwidth decreased but resonant peak Mr increased. So, when a pole is added to the forward path transfer function it makes the closed loop system less stable when decreasing the bandwidth. The rise time increases with decrease of bandwidth.

SUMMARY

1. The magnitude and phase relationship between sinusoidal input and steady state output of a system is known as frequency response.
2. The polar plot of a sinusoidal transfer function $G(j\omega)$ is a plot of the magnitude of $G(j\omega)$ versus the phase angle of $G(j\omega)$ on polar coordinates as 'ω' varied from zero to infinity.
3. The phase margin (in polar plot) is that amount of additional phase lag at the gain crossover frequency required to bring the system to the verge of instability.
4. The gain margin is the reciprocal of the magnitude $[G(j\omega)]$ at the frequency at which the phase angle as – 180°.
5. The inverse polar plot at $G(j\omega)$ is a graph of $\dfrac{1}{G(j\omega)}$ as a function of ω. Bode plot is a graphical representation of the transfer function for determining the stability of the control system. Bode plot consists of two plots. One is a plot of the logarithm of the magnitude of sinusoidal transfer function, the other is a plot of the phase angle, both plots are plotted against the frequency.
6. The transfer function having no poles and zeros in the right half s-plane are called minimum phase transfer function. Systems with minimum phase transfer function are called minimum phase systems.
7. The transfer function having poles and/or zeros in the right half s-plane are called non-minimum phase transfer functions. Systems with non-minimum phase transfer function are called non-minimum phase systems.
8. In Bode plot the relative stability of the system is determined from the gain margin and phase margin.
9. If gain cross-over frequency is less than phase cross over frequency then gain margin and phase margin both are positive and system is stable.
10. If gain cross over frequency is greater than the phase crossover frequency than both gain margin and phase margin are negative and the system is unstable.
11. If gain cross over frequency is equal to the phase cross-over frequency the gain margin and phase margin are zero and the system is marginally stable.
12. The maximum value of magnitude is known as resonant peak. The magnitude of resonant peak gives the information about the relative stability of the system.
13. The frequency at which magnitude has maximum value is known as resonant frequency.
14. Bandwidth is defined as the range of frequencies in which the magnitude of close loop does not drop -3db.
15. The frequency at which the magnitude is 3db below its zero frequency value is called cut off frequency.

$$\omega_r = \omega_n\sqrt{1-2\zeta^2} \quad \text{and} \quad M_r = \frac{1}{2\zeta\sqrt{1-\zeta^2}}$$

16. There are two types of stability namely absolute stability and relative stability. Absolute stability means whether the system is stable or not. If the system is stable then we measure the degree of stability. It is known as relative stability.

EXERCISE

4.1. Sketch the polar plots for the transfer functions

(*a*) $G(s) = \dfrac{1}{s(1+s)(1+2s)}$ (*b*) $G(s) = \dfrac{10}{s(s+1)(s+4)}$

4.2. Sketch the Bode plots for the following transfer function

(*a*) $\dfrac{20s}{(s+1)(s+10)}$ (*b*) $\dfrac{1000}{s(1+0.1s)(1+0.005s)}$ (*c*) $\dfrac{1500}{s(s+2)(s+30)}$

4.3. Determine the resonant frequency, resonance peak and bandwidth for the system whose transfer function is

$$\frac{C(s)}{R(s)} = \frac{5}{s^2+2s+5}$$

4.4. The forward path transfer function of a unity feedback control system is

$$G(s) = \frac{21.39}{s(s+6.54)}$$

Determine the resonance peak, resonant frequency and normalized bandwidth.

4.5. Repeat the problem 4.19 for maximum overshoot 20%.

SEMI-OBJECTIVE TYPE QUESTIONS

(*i*) Define frequency response. (Article 4.1)
(*ii*) What is a polar plot? (Article 4.2)
(*iii*) Write the procedure to sketch the polar plot. (Article 4.3)
(*iv*) Define minimum phase systems and non-minimum phase systems. (Article 4.9)
(*v*) What is a bode plot? (Article 4.8)
(*vi*) Define resonant peak, resonant frequency and bandwidth. (Article 4.14)
(*vii*) Derive the expression for resonant frequency. (Article 4.15)
(*viii*) Define relative and absolute stability. (Article 4.16)
(*ix*) Construct the constant N-circle. (Article 4.19)
(*x*) Define Phase margin and gain margin. (Article 4.11)
(*xi*) Write a short note on correlation between time domain and frequency domain specification. (Article 4.15)
(*xii*) Write a short note on frequency domain specifications. (Article 4.14)
(*xiii*) With a suitable diagram define: (*a*) Phase crossover frequency, (*b*) gain crossover frequency, (*c*) P.M., (*iv*) gain margin. (Article 4.11)
(*xiv*) What information you get from Bode's plot? (Article 4.8)
(*xv*) Why logarithmic scale is used while plotting the Bode plots? (Article 4.8)
(*xvi*) State the advantages of Bode plots.
(*xvii*) State the centre and radius of constant M-circle.
(*xviii*) What is bandwidth?
(*xix*) What is corner frequency?
(*xx*) What are the frequency domain specifications?

5 STABILITY THEORY

5.1 BOUNDED INPUT-BOUNDED OUTPUT STABILITY CONTINUOUS DATA SYSTEM

5.1.1 Concept of Stability

The concept of stability is very important to analyse and design the system. A system is said to be stable if its response cannot be made to increase indefinitely by the application of a bounded input excitation. If the output approaches towards infinite value for sufficiently large time, the system is said to be unstable.

A linear time invariant (LTI) system is stable if :

1. the system is excited by a bounded input, the output is bounded. (BIBO stability criteria)
2. in the absence of the input, the output tends towards zero (the equilibrium state of the system). This is known as asymptotic stable.

Consider the transfer function

$$\frac{C(s)}{R(s)} = \frac{a_m s^m + a_{m-1} s^{m-1} + \ldots . a_1 s + a_0}{b_n s^n + b_{n-1} s^{n-1} + \ldots . + b_1 s + b_0} \qquad \ldots(5.1)$$

the output is given by

$$C(t) = \int_0^{\infty} g(\tau) r(t-\tau)\, d\tau \qquad \ldots(5.2)$$

where $g(\tau) = £^{-1} G(s)$ = impulse response of the system. So, a system is said to be stable if the impulse response approaches zero for sufficiently large time. If the impulse response approaches infinity for sufficiently large time, the system is said to be unstable. If the impulse response approaches a constant value for sufficiently large time, the system is said to be marginally stable.

The absolute value of equation (5.2) given by

$$|c(t)| = \left| \int_0^{\infty} g(\tau)\, r(t-\tau)\, d\tau \right|$$

As the absolute value of an integral is lesser than the integral of the absolute value of integrand, we have

$$|c(t)| \leq \int_0^{\infty} |g(\tau)| |r(t-\tau)|\, d\tau$$

Bounded input means $|r(t)| \leq M_1 \leq \infty$ and output is bounded implies $|c(t)| \leq M_2 \leq \infty$

For bounded input, bounded output condition

$$|c(t)| \geq M_1 \int_0^{\infty} |g(\tau)|\, d\tau \leq M_2$$

The impulse response $g(t)$ has a nature dependent on location of the poles of the transfer function $G(s)$ and the roots of the characteristic equation.

Thus for bounded input bounded output stability we can say :

1. If all the roots of characteristic equation have negative real part *i.e.*, they lie in the left half of the *s*-plane, then the system is bounded input, bounded output stable and the impulse response to be bounded which finally becomes zero *i.e.*,

$$\int_0^\infty |g(\tau)|\, d\tau \text{ is "finite".}$$

2. If any root of the given characteristic equation has a positive real part *i.e.*, they lies in the right half of *s*-plane, the impulse response $g(t)$ is bounded and therefore $\int_0^\infty |g(\tau)|\, d\tau$ is "infinite", this makes the system unstable.
3. If there repeated roots on the imaginary axis, the impulse response $g(t)$ is unbounded and $\int_0^\infty |g(\tau)|\, d\tau$ is infinite and the system is unstable.
4. If one or more non-repeated roots of the characteristic equation on the imaginary axis, then the impulse response $g(t)$ is bounded but $\int_0^\infty |g(\tau)|\, d\tau$ is infinite, which makes the system unstable.

Thus, for Bounded input Bounded output stability we can say that A linear time invariant system is stable if :

(*i*) The system is excited by a bounded input, output is also bounded

(*ii*) In the absence of input, output must tends to zero irrespective of the initial conditions.

A linear time invariant system is unstable if :

(*i*) For a bounded input it produces unbounded output.

(*ii*) In absence of the input, output may not return to zero. It shows certain output without input.

Consider a system whose closed loop transfer function is

$$\frac{C(s)}{R(s)} = \frac{1000}{(s^2 + 40s + 300)} \quad \textit{i.e.,} \quad \frac{C(s)}{R(s)} = \frac{1000}{(s+10)\,(s+30)}$$

Here, closed loop poles are located in left half of s-plane. Now find out the output response for unit step input *i.e.*, $R(s) = \dfrac{1}{s}$.

By partial fraction $\dfrac{C(s)}{R(s)} = \dfrac{1000}{s(s+10)\,(s+30)}$

$$C(s) = \frac{3.33}{s} - \frac{5}{s+10} + \frac{1.67}{s+30}$$

Take the inverse Laplace

$$C(t) = 3.33 - 5\,e^{-10t} + 1.67\,e^{-30}$$

As $t \to \infty$, the exponential terms will approaches to zero and the output will be steady state.

Such systems are known as absolutely stable system.

Now, consider another system whose transfer function is given by

$$\frac{C(s)}{R(s)} = \frac{3}{(s-1)(s+2)}$$

Here one closed loop pole is located in right half of *s*-plane. For $R(s) = 1/s$ *i.e.*, unit step function the output will be

$$C(s) = \frac{3}{s(s-1)(s+2)}$$

By partial fraction

$$C(s) = \frac{-15}{s} + \frac{1}{s-1} + \frac{0.5}{s+2}$$

Take the inverse laplace

$$C(t) = -1.5 + e^{t} + 0.5\, e^{-2t}$$

As $t \to \infty$, the transients goes to increase in amplitude because one exponential term with positive index is present.

Such systems are known as unstable system.

Hence, it is clear that the stability depends on locations of closed loop poles. For stable system the closed loop poles should lie in the left half of *s*-plane. If the closed loop poles are lies in right half of *s*-plane the system will be unstable.

Relative Stability : If the system is stable we can determine the relative stability by the settling time. If the settling time is less than that of the other system then the system is said to be relatively more stable. The settling time is inversly proportional to the real part of the dominant roots. If the roots are located near imaginary axis, settling time will be large. As roots moves away from the imaginary axis settling time becomes lesser and the system becomes more and more stable.

Asymtotic Stability: Consider an RC series circuit with capacitor initially charged to some voltage. This initial voltage is sufficient to operate the system without any external input. This initial voltage drives the current till capacitor gets fully discharge. The stability of such system under zero input but under initial condition is called asymtotic stability. The asymtotic stability depends on the closed loop poles of the system, *i.e.*, roots of the characteristic equation.

5.2. EFFECT OF LOCATION OF POLES ON STABILITY

(*a*) Poles on negative real axis

Consider a simple pole at $s = -a$ as shown in Fig. 5.1*a*., the corresponding impulse response is given by

$$g(t) = £^{-1}G(s)$$

$$= £^{-1}\frac{K}{s+a}$$

$$= K.e^{-at} \qquad ...(5.3)$$

As the time '*t*' increases, the response approaches zero and the system is stable. The response is shown in Fig. 5.1(*b*).

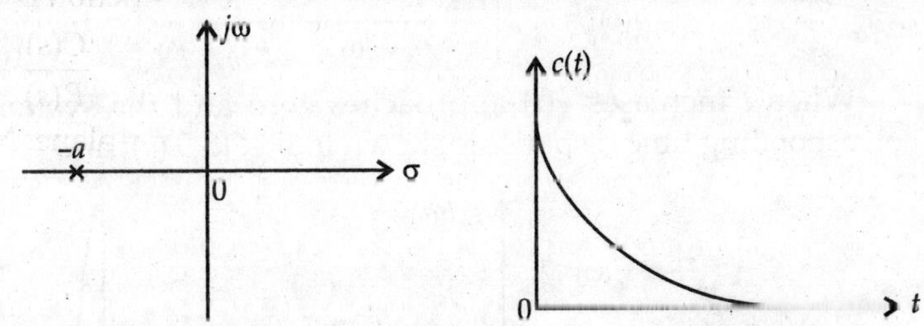

(*a*) Simple plle on negative real axis (*b*) Response

Fig. 5.1.

(*b*) Pole on positive real axis

Consider a system having simple pole on positive real axis at $s = a$, the corresponding impulse response is given by

$$g(t) = £^{-1}\frac{K}{s-a} = K.e^{at} \qquad ...(5.4)$$

The response increases exponentially with time, hence the system is unstable. The simple pole and response are shown in Fig. 5.2 (*a*) and 5.2 (*b*).

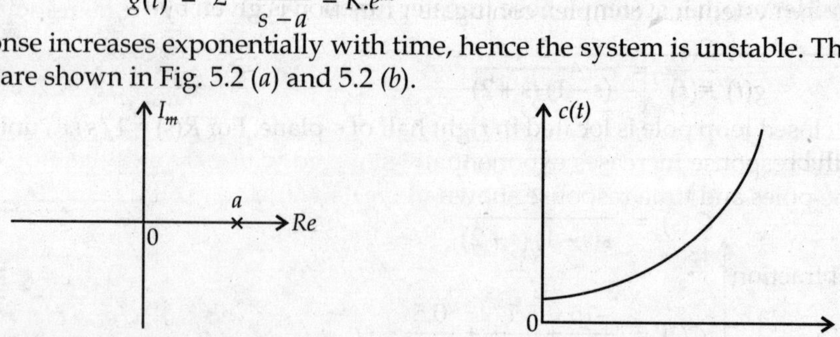

(*a*) Simple pole on positive real axis (*b*) Response

Fig. 5.2.

(*c*) Pole at the origin : Consider a pole at origin

$$\therefore \qquad g(t) = £^{-1}\frac{K}{s} = K \qquad ...(5.5)$$

This is constant value, hence the system is marginally stable. If there are two poles at the origin, the time response would be

$$g(t) = £^{-1}\frac{K}{s^2} = Kt \qquad ...(5.6)$$

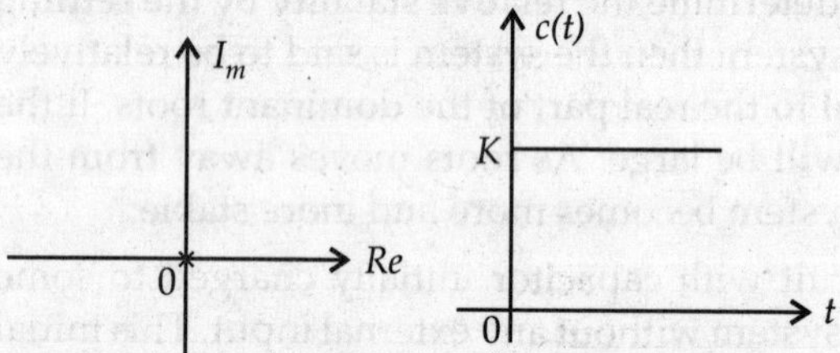

(*a*) Single pole at origin (*b*) Response due to single pole

Fig. 5.3.

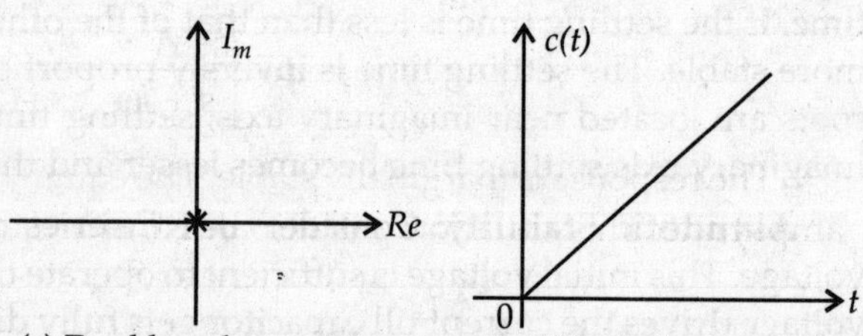

(*a*) Double pole at origin (*b*) Impulse response

Fig. 5.4.

(*d*) Complex pole in the left half of *s*-plane

Let the transfer function has a complex conjugate poles at $s = -\alpha \pm j\omega$. The time response due to the complex conjugate poles is given by

$$g(t) = £^{-1}\left[\frac{K}{s+\alpha-j\omega} + \frac{K}{s+\alpha+j\omega}\right] = £^{-1}\left[\frac{2K\,(s+\alpha)}{(s+\alpha)^2+\omega^2}\right] = 2K\,e^{-\alpha t}\cos\omega\,t \qquad ...(5.7)$$

When t increases $g(t)$ approaches zero and the system is stable. The complex poles and corresponding time response is shown in Fig. [5.5 (*a*)] and [5.5 (*b*)] respectively.

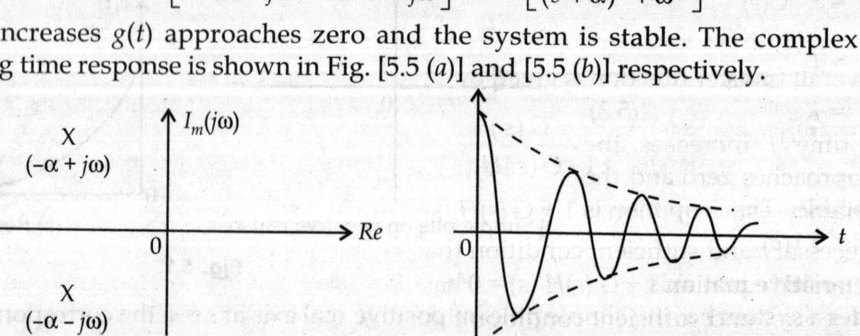

(*a*) (*b*)

Fig. 5.5.

(e) Complex poles in the right half of s-plane

Suppose the system has complex conjugate poles at $s = \alpha \pm j\omega$. The time response is given by

$$g(t) = £^{-1}\left[\frac{A}{s-\alpha-j\omega} + \frac{A}{s-\alpha+j\omega}\right] = £^{-1}\left[\frac{2A(s-\alpha)}{(s-\alpha)^2+\omega^2}\right] = 2\,A\,e^{\alpha t}\cos\omega t \quad ...(5.8)$$

Hence, the response increases exponentially sinusoid with time and therefore the response is unstable. The poles and time response shown in Fig. [5.6(*a*)] and [5.6(*b*)] respectively.

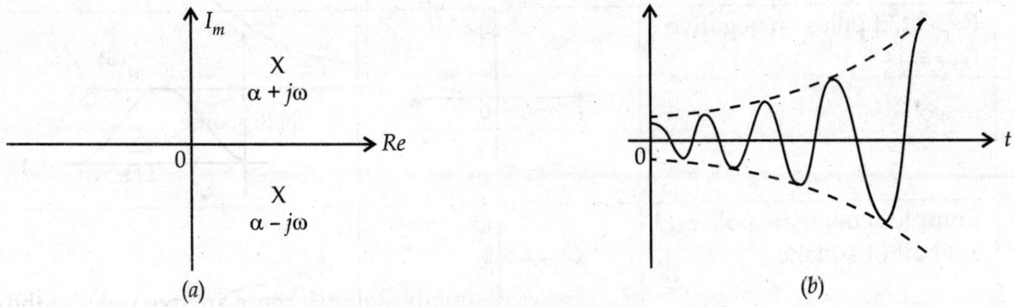

Fig. 5.6.

(f) Poles on jω-axis

If the system having the complex poles on $j\omega$-axis the corresponding time response would be

$$g(t) = £^{-1}\left[\frac{A}{s+j\omega} + \frac{A}{s-j\omega}\right] = £^{-1}\left[\frac{2As}{s^2+\omega^2}\right] = 2A\cos\omega t \quad ...(5.9)$$

The response is marginally stable. The equation (5.9) shows the sustained oscillations of constant amplitude. This situation will also be considered unstable.

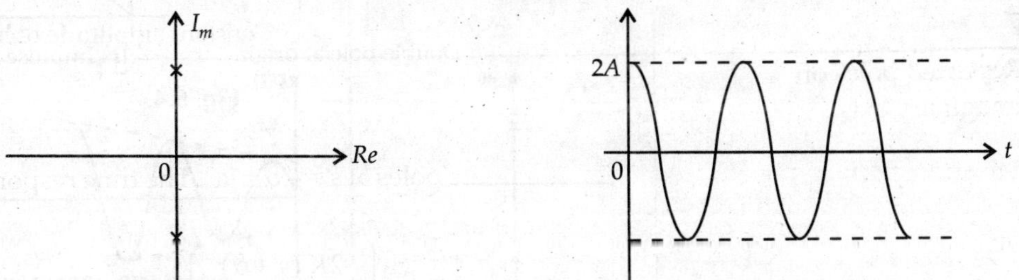

(*a*) Poles on imaginary axis (*b*) Time response

Fig. 5.7.

The overall transfer function is given by

$$\frac{C(s)}{R(s)} = \frac{G(s)}{1+G(s)H(s)} \quad ...(5.10)$$

The characteristic equation is $1 + G(s)H(s) = 0$ ...(5.11)

The necessary and sufficient condition that a feedback system be stable is that all the zeros of the characteristic equation $1 + G(s)H(s) = 0$ have negative real part. or, in terms of poles we can say that the necessary and sufficient condition that a feedback system be stable is that all the poles of overall transfer function have negative real part.

S.No.	Location of poles	Roots in the s-plane	Corresponding impulse response
1.	Single pole on negative real axis	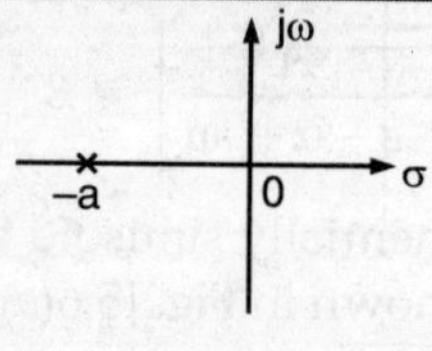	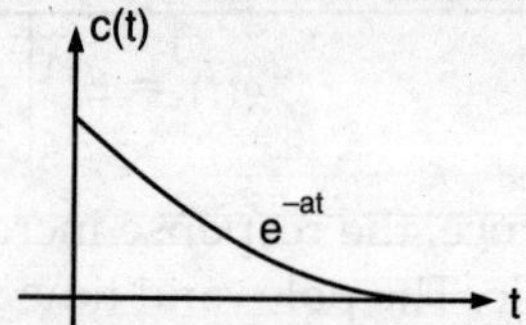
2.	Repeated poles on negative real axis	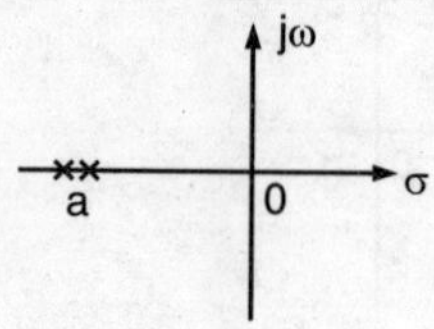	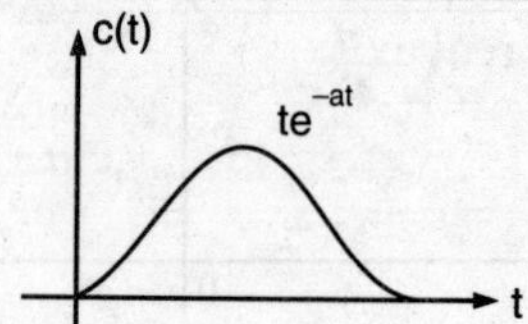
3.	Complex conjugate pole on left half of s-plane	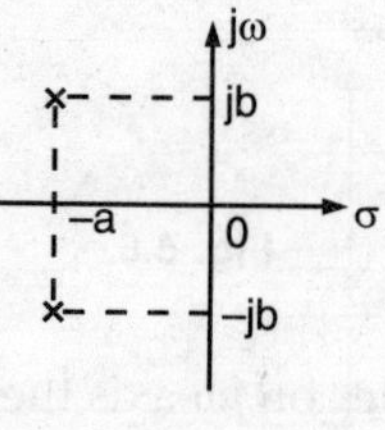	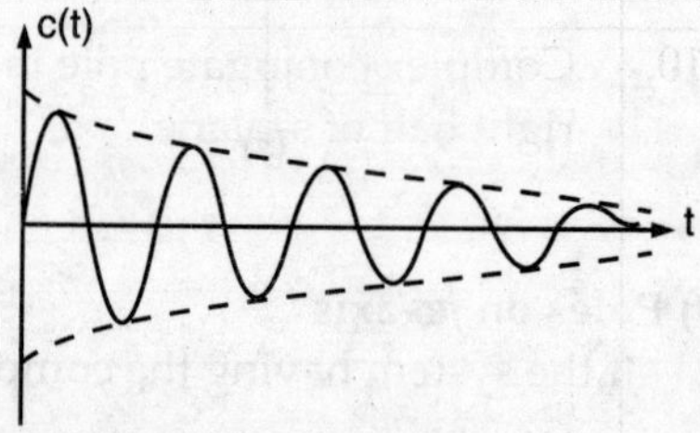
4.	Poles on imaginary axis	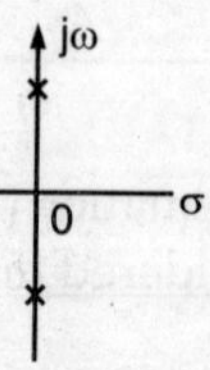	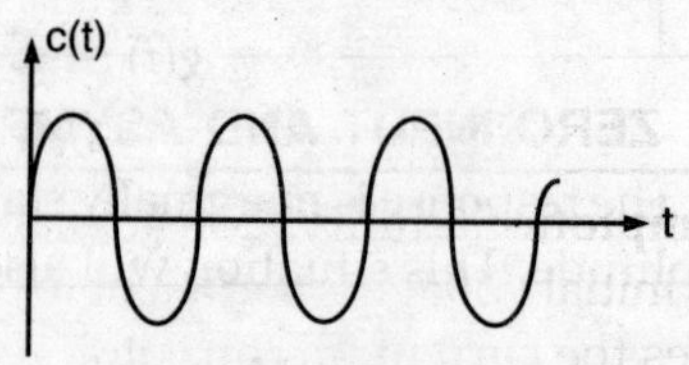 Constant amplitude oscillation
5.	Repeated poles on imaginary axis	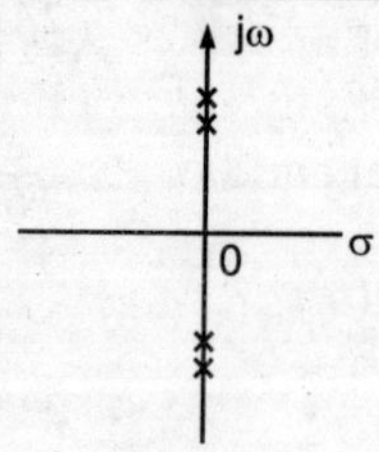	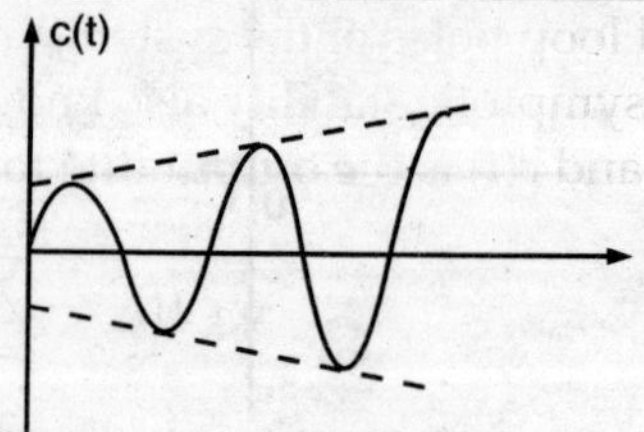 Increasing amplitude
6.	Single pole at origin	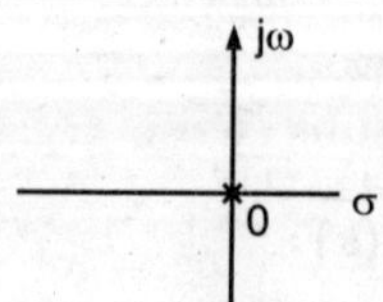	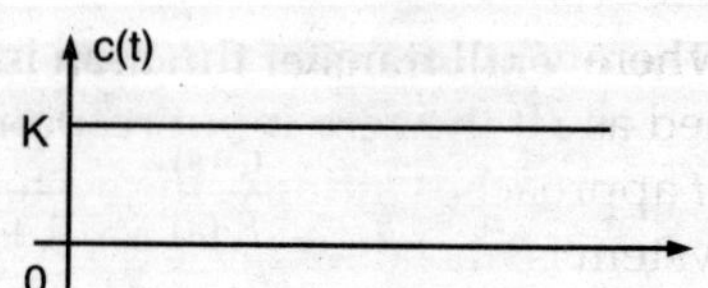
7.	Repeated pole at origin	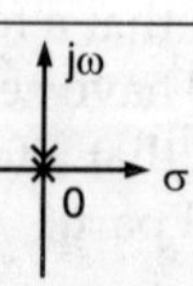	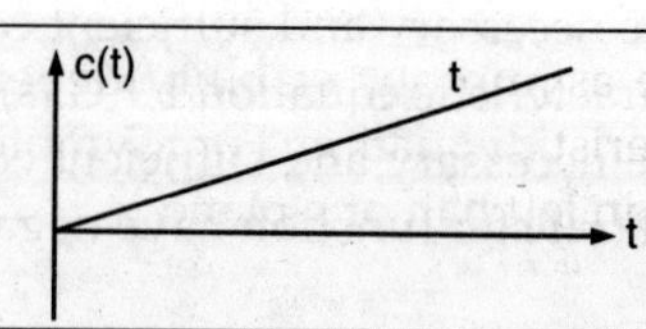

8.	Single pole on positive real axis	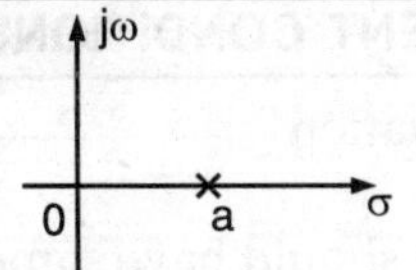	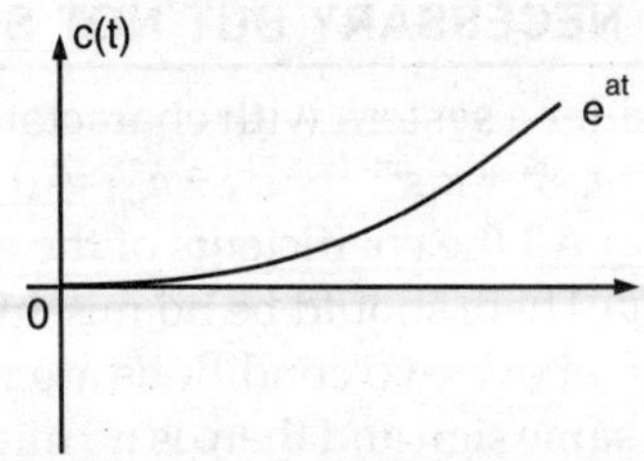
9.	Repeated poles on positive real axis	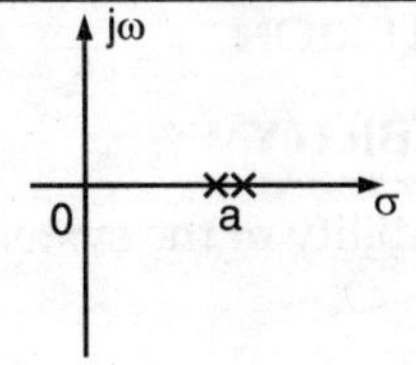	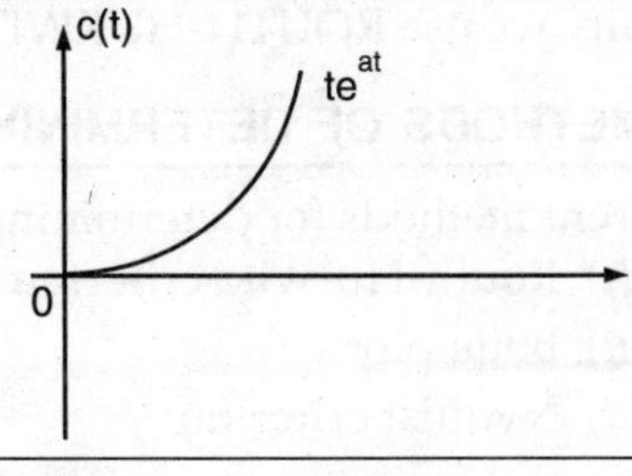
10.	Complex conjugate pole in right half of s-plane	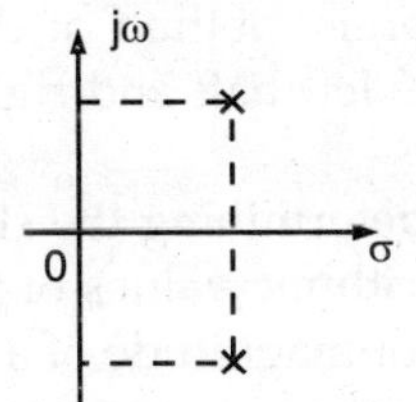	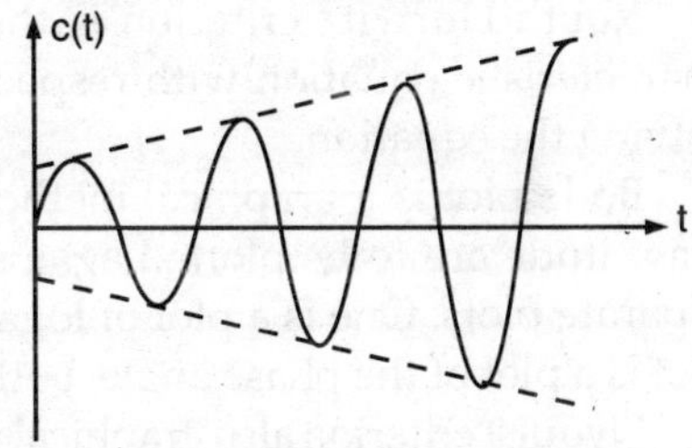

5.3. ZERO INPUT AND ASYMPTOTIC STABILITY OF CONTINUOUS DATA SYSTEMS

Asymptotic Stability: Consider an RC series circuit with capacitor initially charged to some voltage. This initial voltage is sufficient to operate the system without any external input. This initial voltage drives the current till capacitor gets fully discharge. The stability of such system under zero input but under initial condition is called asymptotic stability. The asymtotic stability depends on the closed loop poles of the system, *i.e.*, roots of the characteristic equation.

Asymptotic stability also known as "zero input stability". Consider nth order system with zero input and $c(t)$ is the output due to the initial conditions, mathematically

$$c(t) = \sum_{i=0}^{n-1} g_i(t)\, C^i(t_0)$$

where,

$$C^i(t_0) = \left.\frac{d^i c(t)}{dt^i}\right|_{t=t_0}$$

where $g_i(t)$ denotes the zero input response due to $C^i(t_0)$ so, the zero input stability can be defined as : If the zero input response $c(t)$ subjected to the finite initial conditions, reaches zero as time t approaches infinity, the system is said to be zero-input stable or asymptotic stable otherwise the system is unstable.

$$\lim_{t\to\infty} |c(t)| = 0$$

The asymptotic stability depends on the closed loop poles of the system, *i.e.*, roots of the characteristic equation. For asymptotic stability, all the roots of the characteristic equation must be located in left half of s-plane.

5.4. NECESSARY BUT NOT SUFFICIENT CONDITIONS FOR STABILITY

Consider a system with characteristic equation

$$a_0 s^m + a_1 s^{m-1} + \ldots + a_m = 0 \quad \ldots(5.12)$$

(*a*) All the coefficients of the equation should have same sign,

(*b*) There should be no missing term.

If above two conditions are not satisfied the system will be unstable. But if all the coefficients have same sign and there is no missing term we have no guarantee that the system will be stable. For stability we use ROUTH-HURWITZ CRITERION.

5.5 METHODS OF DETERMINING STABILITY

Different methods for determining the stability of the system are:

(*i*) Routh-Hurwitz criterion

(*ii*) Bode plot

(*iii*) Nyquist criterion

Routh-Hurwitz criterion is the algebraic method of determining the location of poles of a characteristic equation with respect to the left half and right half of the s-plane without actually solving the equation.

Bode plot is a graphical method for determining the stability in which logarithmic values of magnitude are to be plotted against logarithmic values of frequencies. Bode plot consists of two separate plots. One is a plot of logarithm of magnitude of a sinusoidal transfer function, the other plot is a plot of the phase angle, both plots are plotted against the logarithmic values of frequencies.

Nyquist criterion also graphical method for determining the stability of the system. In this method the encirclements of critical point $(-1 + j_o)$ by Nyquist plot is used to determine the stability of the system.

5.6. THE ROUTH-HURWITZ CRITERION

Consider the following characteristic polynomial

$$a_0 s^n + a_1 s^{n-1} + \ldots + a_n = 0 \quad \ldots(5.13)$$

where the coefficients $a_0, a_1 \ldots a_n$ are all of the same sign and none is zero.

Step 1 : Arrange all the coefficients of equation 5.13 in two rows

Row 1	a_0	a_2	a_4	....
Row 2	a_1	a_3	a_5	....

Step 2 : From these two rows form a third row

Row 1	a_0	a_2	a_4	....
Row 2	a_1	a_3	a_5	....
Row 3	b_1	b_3	b_5	....

where, $b_1 = -\dfrac{1}{a_1}\begin{vmatrix} a_0 & a_2 \\ a_1 & a_3 \end{vmatrix}$ and $b_3 = -\dfrac{1}{a_1}\begin{vmatrix} a_0 & a_4 \\ a_1 & a_5 \end{vmatrix}$

Step 3 : From second and third row, form a fourth row

Row 1	a_0	a_2	a_4	....
Row 2	a_1	a_3	a_5	.....
Row 3	b_1	b_3	b_5	....
Row 4	c_1	c_3	c_5	....

where,

$$c_1 = -\frac{1}{b_1}\begin{vmatrix} a_1 & a_3 \\ b_1 & b_3 \end{vmatrix}$$

$$c_3 = -\frac{1}{b_1}\begin{vmatrix} a_1 & a_5 \\ b_1 & b_5 \end{vmatrix}$$

$$\vdots$$

Step 4 : Continue this procedure of forming a new rows.

5.6.1. Statement of Routh-Hurwitz Criterion

Routh-Hurwitz Criterion states that the system is stable if and only if all the elements in the first column have the same algebaric sign. If all elements are not of the same sign then the number of sign changes of the elements in first column equals the number of roots of the characteristic equation in the right half of the s-plane (or equals to the number of roots with positive real parts).

EXAMPLE 5.1. Check the stability of the system whose characteristic equation is given by

$$s^4 + 2s^3 + 6s^2 + 4s + 1 = 0.$$

Solution : Obtain the array of coefficients as follows

s^4	1	6	1
s^3	2	4	
s^2	4	1	
s^1	3.5		
s^0	1		

$$b_1 = -\frac{1}{2}\begin{vmatrix} 1 & 6 \\ 2 & 4 \end{vmatrix} = 4, \quad c_1 = -\frac{1}{4}\begin{vmatrix} 2 & 4 \\ 4 & 1 \end{vmatrix} = 3.5$$

$$b_2 = -\frac{1}{2}\begin{vmatrix} 1 & 1 \\ 2 & 0 \end{vmatrix} = 1, \quad d_1 = -\frac{1}{3.5}\begin{vmatrix} 4 & 1 \\ 3.5 & 0 \end{vmatrix} = 1$$

Since, all the coefficients in the first column are of the same sign (positive), the given equation has no roots with positive real parts. Hence, the system is stable.

EXAMPLE 5.2. Determine the stability of the system whose characteristic equation is given by

$$2s^4 + 2s^3 + s^2 + 3s + 2 = 0.$$

Solution :

s^4	2	1	2
s^3	2	3	
s^2	-2	2	
s^1	5		
s^0	2		

$$b_1 = -\frac{1}{2}\begin{vmatrix} 2 & 1 \\ 2 & 3 \end{vmatrix} = -2, \quad c_1 = -\frac{1}{(-2)}\begin{vmatrix} 2 & 3 \\ -2 & 2 \end{vmatrix} = 5$$

$$b_2 = -\frac{1}{2}\begin{vmatrix} 2 & 2 \\ 2 & 0 \end{vmatrix} = 2, \quad d_1 = -\frac{1}{5}\begin{vmatrix} -2 & 2 \\ 5 & 0 \end{vmatrix} = 2$$

There are two changes of sign in the first column (from 2 to –2 and from –2 to 5), hence there are two roots in the right half of s-plane. The system is unstable.

EXAMPLE 5.3. Determine the stability of the system having following characteristic equation

$$2s^4 + 5s^3 + 5s^2 + 2s + 1 = 0.$$

Solution :

s^4	2	5	1
s^3	5	2	
s^2	4.2	1	
s^1	0.809		
s^0	1		

From the above Routh table :
No. of sign changes in first column = 0
No. of poles on the right hand side of s-plane = 0
Hence, the system is stable.

EXAMPLE 5.4. Check the stability of the system having following characteristic equation.
$s^4 + 2s^3 + 3s^2 + 4s + 5 = 0$.

Solution :

s^4	1	3	5
s^3	2	4	
s^2	1	5	
s^1	–6		
s^0	5		

$$b_1 = -\frac{1}{2}\begin{vmatrix}1 & 3\\ 2 & 4\end{vmatrix} = 1, \qquad b_2 = -\frac{1}{2}\begin{vmatrix}1 & 5\\ 2 & 0\end{vmatrix} = 5$$

$$c_1 = -\frac{1}{1}\begin{vmatrix}2 & 4\\ 1 & 5\end{vmatrix} = -6, \qquad d_1 = -\frac{1}{(-6)}\begin{vmatrix}1 & 5\\ -6 & 0\end{vmatrix} = 5$$

From above table :
No. of sign changes in first column = 2
No. of roots in right half of s-plane = 2
Hence, the system is unstable.

EXAMPLE 5.5. A closed loop control system has the characteristic equation given by
$s^3 + 4.5s^2 + 3.5s + 1.5 = 0$.
Investigate the stability using Routh Hurwitz criterion. *(R.M.L. University Faizabad, 2001)*

Solution :

s^3	1	3.5
s^2	4.5	1.5
s^1	3.17	
s^0	1.5	

No. of sign changes in first column = 0
No. of roots in right half of s-plane = 0
Hence, system is stable.

EXAMPLE 5.6. Check the stability of the system, having following characteristic equation.

Solution: $s^5 + 6s^4 + 3s^3 + 2s^2 + s + 1 = 0$

s^5	1	3	1
s^4	6	2	1
s^3	2.67	0.83	
s^2	0.135	1	
s^1	–18.95		
s^0	1		

No. of sign change in first column = 2
No. of poles on right half of s-plane = 2
Hence, system is unstable.

SPECIAL CASES

Case 1 : If a first column term in any row is zero, but the remaining terms are not zero or there is no remaining term, then multiply the original equation by a factor $(s + a)$ where 'a' is any positive real number. The simplest value of 'a' is 1 (take $a = 1$). Consider the following example.

EXAMPLE 5.7. Investigate the stability
$s^5 + s^4 + 2s^3 + 2s^2 + 3s + 5 = 0$

Solution :

s^5	1	2	3
s^4	1	2	5
s^3	0		
s^2			
s^1			
s^0			

Now, multiply the characteristic equation by $(s + 1)$

$(s + 1)(s^5 + s^4 + 2s^3 + 2s^2 + 3s + 5) = 0$

or, $s^6 + 2s^5 + 3s^4 + 4s^3 + 5s^2 + 8s + 5 = 0$

s^6	1	3	5	5
s^5	2	4	8	
s^4	1	1	5	
s^3	2	−2		
s^2	2	5		
s^1	−7			
s^0	1			

From the above table :

No. of sign change in the first column = 2

No. of roots in the right half of s-plane = 2

Hence, system is unstable.

EXAMPLE 5.8. Apply Routh Hurwitz criterion to the following equation and investigate the stability.

$$s^5 + 2s^4 + 2s^3 + 4s^2 + 11s + 10 = 0$$

Solution :

s^5	1	2	11
s^4	2	4	10
s^3	0		
s^2			
s^1			
s^0			

Since, the third element in first column is zero, so multiply the equation by $(s + 1)$

$$(s + 1)(s^5 + 2s^4 + 2s^3 + 4s^2 + 11s + 10) = 0$$

or, $s^6 + 3s^5 + 4s^4 + 6s^3 + 15s^2 + 21s + 10 = 0$

s^6	1	4	15	10
s^5	3	6	21	
s^4	2	8	10	
s^3	−6	6		
s^2	10	10		
s^1	12			
s^0	1			

From the above table it is clear that there are two changes of sign in first column, therefore there are two roots in the right half of s-plane. The system is unstable.

ALTERNATIVE METHOD

Replace the zero by a small positive quantity ε and continue the procedure. Consider the following example.

EXAMPLE 5.9. Consider the following equation

$s^3 + s + 2 = 0$

Solution :

s^3	1	1
s^2	0	2
s^1		
s^0		

Replace 0 by ε and continue the procedure.

s^3	1	1
s^2	ε	2
s^1	$-\frac{1}{\varepsilon}(2-\varepsilon)$	–
s^0	2	

No. of sign changes in first column = 2

No. of roots in right of *s*-plane = 2

Case 2 : When any one row of Routh table is zero.

When any one row of Routh table is zero, it indicates that the equation has at least one pair of roots which lie radially opposite each other and equidistant from origin. The array can be completed by forming the auxiliary polynomial. The polynomial whose coefficients are the element of the row just above the row of zeros in Routh array is called an auxiliary polynomial. Consider the following example

EXAMPLE 5.10. Consider the equation

$s^5 + 2s^4 + 24s^3 + 48s^2 - 25s - 50 = 0$

Solution :

s^5	1	24	– 25
s^4	2	48	– 50
s^3	0	0	
s^2			
s^1			
s^0			

From above table it is clear that the third row is zero. Hence form the Auxiliary polynomial. $A(s)$

$$A(s) = 2s^4 + 48s^2 - 50$$

$$\frac{dA(s)}{ds} = 8s^3 + 96s$$

Now the Routh array can be written as

s^5	1	24	– 25
s^4	2	48	–50
s^3	8	96	
s^2	24	– 50	
s^1	112.6		
s^0	– 50		

No. of sign changes in first column = 1

No. of roots in right half of *s*-plane = 1

The roots of the equation formed by the auxiliary polynomial

$$2s^4 + 48s^2 - 50 = 0$$

are also the roots of the original equation.

The given equation can be written in factored form as

$$(s+1)(s-1)(s+j5)(s-j5)(s+2) = 0$$

It is clear that the original equation has one root with positive real part.

The roots of the auxiliary equation are also the roots of the characteristic equation because auxiliary equation is the part of characteristic equation. Suppose, original characteristic equation is of order six, *i.e.*, having six roots and auxiliary equation is of order four *i.e.*, having four roots. These four roots are also the roots of the original characteristic equation and remaining (6 – 4 = 2) two roots will always lie to the left half of s-plane and these two roots do not have significant role for stability because they lie far away from imaginary axis. The roots of the auxiliary equations are dominant roots*, the stability can also be determined from the roots of the auxiliary equation.

Consider the following characteristic equation :

$$S^6 + 2S^5 + 8S^4 + 12S^3 + 20S^2 + 16S + 16 = 0$$

S^6	1	8	20	16
S^5	2	12	16	0
S^4	2	12	16	0
S^3	0	0	0	0
S^2				
S^1				
S^0				

Auxiliary equation $A(s) = 2S^4 + 12S^2 + 16 = 0$, $\dfrac{dA(s)}{ds} = 8S^3 + 24S$. Since, the order of the characteristic equation is six and order of auxiliary equation is four, the two roots (6 – 4 = 2) lie to the left half of s-plane and four roots of auxiliary equations are dominant roots.

S^6	1	8	20	16
S^5	2	12	16	0
S^4	2	12	16	0
S^3	8	24	0	0
S^2	6	16	0	
S^1	2.67	0		
S^0	16			

No sign change in first column, the system may be stable, since there is a row of zeros it means the roots are lying on imaginary axis. So solve the auxiliary equation $A(s) = 0$.

$$2S^4 + 12S^2 + 16 = 0$$

put $S^2 = x$

$$\therefore \quad 2x^2 + 12x + 16 = 0$$

or,

$$x^2 + 6x + 8 = 0$$

* dominant roots means the roots are close to the imaginary axis or on the imaginary axis or they are in right half of s-plane

$$(x+2)(x+4) = 0$$

$$x+2 = 0, \qquad x+4 = 0$$

$$x = -2, \qquad x = -4$$

$$S^2 = -2, \qquad S^2 = -4$$

$$S = \pm j\sqrt{2}, \qquad S = \pm j\,2$$

Since, the roots are non-repeated on imaginary axis, hence the system is marginally stable.

No. of roots with positive real part = 0

No. of roots with negative real part = 2

No. of roots with zero real part = 4

Important Note : If sign changes in first column, system is unstable. The number of roots with positive real part of characteristics equation will be equal to the number of sign changes *i.e.*, the roots are located in right half of s-plane. But if no sign changes in first column the system may be stable. In such case stability can be determined by solving auxiliary equation for roots and from the location of roots, stability can be determined.

5.7. APPLICATION OF ROUTH'S STABILITY CRITERION TO CONTROL SYSTEM ANALYSIS

Routh stability criterion is also used for the determination of stability of the linear feedback systems. Consider the following example.

EXAMPLE 5.11. The open loop transfer function of unity feedback system is $\dfrac{K}{s(1+0.4s)(1+0.25s)}$. Find the restriction of K so that the closed loop system is absolutely stable.

(R.M.L. University Faizabad 2001)

Solution : Given that

$$G(s) = \frac{K}{s(1+0.4s)(1+0.25s)}$$

$$H(s) = 1$$

The characteristic equation $1 + G(s)H(s) = 0$

$$1 + \frac{K}{s(1+0.4s)(1+0.25s)} = 0$$

or, $\quad s(1+0.4s)(1+0.25s) + K = 0$

or, $\quad s^3 + 6.5s^2 + 10s + 10K = 0$

s^3	1	10
s^2	6.5	10K
s^1	$\dfrac{65-10K}{6.5}$	
s^0	10K	

For absolute stability, there should be no sign change in the first column, *i.e.*, no root of the characteristic equation should lie in right half of s-plane. This is possible only when

$K > 0$ and $65 - 10K > 0$

Hence, for closed loop stability

$0 < K < 6.5$

5.8. RELATIVE STABILITY ANALYSIS

Routh stability criterion gives the information about the absolute stability. But we are more interested for the relative stability of the system. Relative stability can be examined by shifting the s-plane and then apply Routh stability criterion. The characteristic equation is modified by shifting the origin of s-plane to $s_1 = -\sigma$ by the substitution

$$s = z - s_1$$

Now apply the Routh stability criterion, the number of sign change in the first column is equal to the number of roots to the right of the vertical line $s_1 = -\sigma$ or in other words the roots of the original characteristic equation are more negative than $-\sigma_1$.

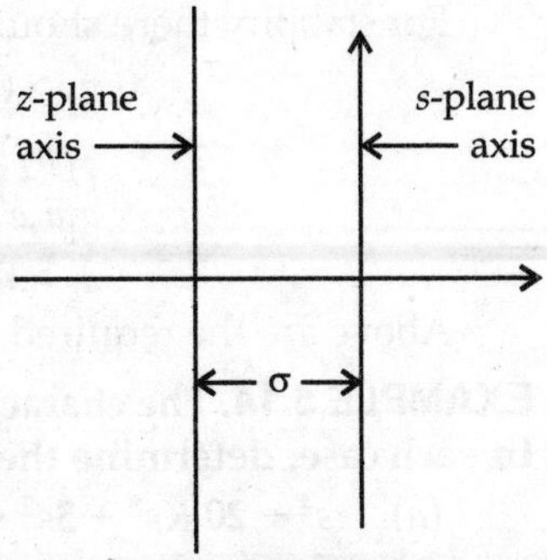

Fig. 5.8.

EXAMPLE 5.12. Check whether all the roots of the equation $s^3 + 7s^2 + 25s + 39 = 0$ have real parts more negative than – 1.

Solution:

s^3	1	25
s^2	7	39
s^1	19.42	
s^0	39	

Since, there is no sign change in the first column, it means all the roots are in the left half of s plane.

Put $s = z - 1$ in the characteristic equation

$$(z-1)^3 + 7(z-1)^2 + 25(z-1) + 39 = 0$$

or,
$$z^3 + 4z^2 + 14z + 20 = 0$$

z^3	1	14
z^2	4	20
z^1	9	
z^0	20	

Since, no sign change in the first column, the roots of the characteristic equation lie in the left of z- plane, it means all the roots of the original equation in s-domain lie to the left of $s = -1$.

ILLUSTRATIVE EXAMPLES

EXAMPLE 5.13. The characteristic equation of a servo system is given by

$$a_0 s^4 + a_1 s^3 + a_2 s^2 + a_3 s + a_4 = 0$$

Determine the conditions which must be satisfied by the coefficient of the characteristic equation for the system to be stable.

Solution:

s^4	a_0	a_2	a_4
s^3	a_1	a_3	
s^2	$\dfrac{a_1a_2 - a_0a_3}{a_1}$	$\dfrac{a_4a_1}{a_1}$	
s^1	$\dfrac{a_3\left[\dfrac{a_1a_2 - a_0a_3}{a_1}\right] - \dfrac{a_1^2a_4}{a_1}}{\dfrac{a_1a_2 - a_0a_3}{a_1}}$		
s^0	a_4		

For stability there should not be any sign change in first column.

$$\therefore \quad a_1 > 0$$
$$a_1 a_2 - a_0 a_3 > 0$$
$$a_1 a_2 a_3 - a_0 a_3^2 - a_1^2 a_4 > 0$$
$$a_4 > 0$$

Above are the required conditions.

EXAMPLE 5.14. The characteristic equations for certain feedback control system are given below. In each case, determine the range of values K for the system to be stable.

(a) $s^4 + 20Ks^3 + 5s^2 + 10s + 15 = 0$

(b) $s^3 + 2Ks^2 + (K+2)s + 4 = 0$

Solution : a. Routh array is

s^4	1	5	15
s^3	$20K$	10	
s^2	$\dfrac{100K-10}{20K}$	15	
s^1	$10-\dfrac{600K^2}{10K-1}$		
s^0	15		

For stability no root should lie in right half of s-plane.

(a) $20K > 0$ *i.e* $K > 0$

(b) $\dfrac{100K-10}{20K} > 0$ or, $5 - \dfrac{1}{2K} > 0$ *i.e* $5 > \dfrac{1}{2K}$ or $K > \dfrac{1}{10}$

(c) $10 - \dfrac{600K^2}{10K-1} > 0$

i.e., $-600K^2 + 100K - 10 > 0$

This gives the complex roots, hence the system is unstable.

(b) $s^3 + 2Ks^2 + (K+2)s + 4 = 0$

s^3	1	$K+2$
s^2	$2K$	4
s^1	$\dfrac{K^2+2K-2}{K}$	
s^0	4	

For stability

$2K > 0$ *i.e.,* $K > 0$

$\dfrac{K^2+2K-2}{K} > 0$ *i.e.,* $K^2 + 2K - 2 > 0$

$K > 0.73$

EXAMPLE 5.15. A system oscillates with frequency ω, if it has poles at s = ± *j*ω and no poles in the right half s-plane. Determine the values of *K* and a so that the system shown in Fig. 5.9 oscillates at a frequency 2 rad/sec.

Solution : The characteristic equation

$$1 + G(s)H(s) = 0$$

$$1 + \frac{K(s+1)}{s^3 + as^2 + 2s + 1} \cdot 1 = 0$$

Fig. 5.9.

or, $s^3 + as^2 + 2s + 1 + Ks + K = 0$

or, $s^3 + as^2 + (K+2)s + (K+1) = 0$

$$\begin{array}{lll} s^3 & 1 & K+2 \\ s^2 & a & K+1 \\ s^1 & K+2-\dfrac{(K+1)}{a} & \\ s^0 & K+1 & \end{array}$$

The system will have sustained oscillation when

$$K+2-\frac{(K+1)}{a} = 0$$

or $a(K+2) - (K+1) = 0$ or $a = \dfrac{K+1}{K+2}$

The auxiliary equation of row 2

$$as^2 + (K+1) = 0$$

$$s^2 = -\frac{K+1}{a}$$

Put $s = j\omega$

$$\therefore \quad (j\omega)^2 = -\frac{K+1}{a}$$

$\omega = 2$ rad/sec. (given)

$$(j2)^2 = -\frac{K+1}{a}$$

$$\therefore \quad a = \frac{K+1}{4}$$

Since,

$$a = \frac{K+1}{K+2}$$

$$\therefore \quad \frac{K+1}{4} = \frac{K+1}{K+2}$$

$$\therefore \quad \boxed{K = 2}$$

$$\therefore \quad a = \frac{K+1}{4} = \frac{2+1}{4} = \frac{3}{4} = 0.75$$

$$\therefore \quad \boxed{a = 0.75} \quad \textbf{Ans.}$$

EXAMPLE 5.16. The open loop transfer function of a unity feedback control system is given by

$$G(s) = \frac{K}{(s+2)(s+4)(s^2+6s+25)}$$

By applying Routh criterion, discuss the stability of the closed loop system as a function of *K*. Determine the value of *K* which will cause sustained oscillations in the closed loop system. What are the corresponding oscillation frequencies ?

Solution : The characteristic equation is given by

$$1 + G(s)H(s) = 0$$

$$1 + \frac{K}{(s+2)(s+4)(s^2+6s+25)} \cdot 1 = 0$$

or, $s^4 + 12s^3 + 69s^2 + 198s + (200+K) = 0$

$$\begin{array}{llll} s^4 & 1 & 69 & 200+K \\ s^3 & 12 & 198 & \\ s^2 & 52.5 & 200+K & \\ s^1 & 198-\dfrac{12(200+K)}{52.5} & & \\ s^0 & 200+K & & \end{array}$$

system will be stable when

$$200 + K > 0 \quad \text{or} \quad K > -200$$

$$198-\frac{12(200+K)}{52.5} > 0 \quad \text{or} \quad K < 666.25$$

oscillations will occur when $K = 666.25$

Auxiliary equation : $52.5\, s^2 + (200 + K) = 0$

$$52.5s^2 + (200 + 666.25) = 0$$

$$52.5s^2 = -866.25$$

$$s^2 = -16.5$$

$$s = \pm j\, 4.06 \text{ or } j\omega = \pm j4.06$$

$\therefore$ Frequency of sustained oscillation = 4.06 rad/sec. **Ans.**

EXAMPLE 5.17. A unity feedback control system is characterized by open loop transfer function

$$G(s) = \frac{K(s+13)}{s(s+3)(s+7)}$$

(a) Using Routh criterion, calculate the range of values of K for the system to be stable.

(b) Check if for K = 1, all these roots of the characteristic equation of the above system have damping factor greater than 0.5.

Solution : The characteristic equation is given by

$$1 + G(s)H(s) = 0$$

$$1+\frac{K(s+13)}{s(s+3)(s+7)} = 0$$

or, $s(s + 3)(s + 7) + K(s + 13) = 0$

or, $s^3 + 10s^2 + (21 + K)s + 13K = 0$

$$\begin{array}{lll} s^3 & 1 & 21+K \\ s^2 & 10 & 13K \\ s^1 & 21-\dfrac{3K}{10} & \\ s^0 & 13K & \end{array}$$

(a) For stability

$$13K > 0 \quad \text{or} \quad K > 0$$

$$21-\frac{3K}{10} > 0 \quad \text{or} \quad K < 70$$

$\therefore$ Range is $\boxed{0 < K < 70}$

$\therefore$ For stability of the system the range of K is $0 < K < 70$

(b) When $K = 1$, the characteristic equation becomes

$$s^3 + 10s^2 + 22s + 13 = 0$$

or, $(s + 7.2)(s^2 + 2.8s + 1.84) = 0$

∴ The roots are $s = -7.2, -1.745, -1.0535$

Since, all the roots are real and negative. Therefore all the roots lie on negative real axis.

$$\text{damping factor} = \cos\theta$$

Since all the roots are lie on negative real axis ∴ $\theta = 180°$

$$\therefore \quad \text{damping ratio} = |\cos 180°| = 1$$

Hence, for all roots damping ratio is greater than 0.5.

or
$$s^3 + 10s^2 + 22s + 13 = 0 ;$$
$$(s+1)(s^2 + 9s + 13) = 0 ; \quad \xi = 1.248$$

Hence, for all roots ξ is greater than 0.5

EXAMPLE 5.18. A feedback system has an open loop transfer function

$$G(s)H(s) = \frac{Ke^{-s}}{s(s^2+5s+9)}$$

Determine by use of Routh criterion, the maximum value of K for the closed loop system to be stable.

Solution : For low frequencies $e^{-s} = (1-s)$

$$\therefore \quad G(s)H(s) = \frac{K(1-s)}{s(s^2+5s+9)}$$

Characteristic equation is given by $1 + G(s).H(s) = 0$

$$1 + \frac{K(1-s)}{s(s^2+5s+9)} = 0$$

$$s(s^2 + 5s + 9) + K - Ks = 0$$
$$s^3 + 5s^2 + 9s + K - Ks = 0$$
$$s^3 + 5s^2 + (9-K)s + K = 0$$

s^3	1	$9-K$
s^2	5	K
s^1	$9-\frac{6K}{5}$	
s^0	K	

For stability $K > 0$

$$9 - \frac{6K}{5} > 0 \quad \text{or} \quad K < 7.5$$

Hence, the range of K is $\boxed{0 < K < 7.5}$ **Ans.**

EXAMPLE 5.19. By means of Routh criterion, determine the stability of the systems represented by the following characteristic equations. For systems found to be unstable, determine the unmber of roots of the characteristic equation in the right half s-plane.

(a) $s^4 + 2s^3 + 8s^2 + 4s + 3 = 0$

(b) $s^4 + 2s^3 + s^2 + 4s + 2 = 0$

(c) $s^6 + 3s^5 + 5s^4 + 9s^3 + 8s^2 + 6s + 4 = 0$

Solution : (a) $s^4 + 2s^3 + 8s^2 + 4s + 3 = 0$

s^4	1	8	3
s^3	2	4	
s^2	6	3	
s^1	3		
s^0	3		

No. of sign change in first column = 0
No. of roots in right half of s-plane = 0
Hence system is stable.

(b) $s^4 + 2s^3 + s^2 + 4s + 2 = 0$

s^4	1	1	2
s^3	2	4	
s^2	–1	2	
s^1	8		
s^o	2		

No. of sign change in first column = 2
No. of roots in right half of s-plane = 2
Hence system is unstable.

(c) $s^6 + 3s^5 + 5s^4 + 9s^3 + 8s^2 + 6s + 4 = 0$

s^6	1	5	8	4
s^5	3	9	6	
s^4	2	6	4	
s^3	0	0		
s^2				
s^1				
s^0				

The elements of fourth row are zero, hence form the auxiliary equation

$$A(s) = 2s^4 + 6s^2 + 4$$

$$\frac{dA(s)}{ds} = 8s^3 + 12s$$

s^6	1	5	8	4
s^5	3	9	6	
s^4	2	6	4	
s^3	8	12		
s^2	3	4		
s^1	4/3			
s^0	4			

Since, no sign change in the first column, hence system is stable
The roots of auxiliary equation are

$$s = \pm j1$$

$$s = \pm j\sqrt{2}$$

∴ Four roots lie on imaginary axis.

EXAMPLE 5.20. Determine the range of values of K (K > 0) such that the characteristic equation

$$s^3 + 3(K + 1)s^2 + (7K + 5)s + (4K + 7) = 0$$

has roots more negative than $s = -1$

Solution : Put $s = z - 1$ in given characteristic equation

$$(z - 1)^3 + 3(K + 1)(z - 1)^2 + (7K + 5)(z - 1) + (4K + 7) = 0$$

or, $z^3 + 3Kz^2 + (K + 2)z + 4 = 0$

z^3	1	$K + 2$
z^2	$3K$	4
z^1	$\dfrac{3K(K+2)-4}{3K}$	
z^0	4	

For stability

$$\frac{3K(K+2)-4}{3K} > 0 \text{ and } 3K > 0 \text{ i.e } K > 0$$

$$3K^2 + 6K - 4 > 0$$

$\therefore$ $K = -2.527, 0.527$

Hence, the range is

$$\boxed{\infty > K > 0.53} \qquad (\because K > 0)$$

EXAMPLE 5.21. The characteristic equation of feedback control system is

$$s^4 + 20s^3 + 15s^2 + 2s + K = 0$$

(a) Determine the range of K for the system to be stable.

(b) Can the system be marginally stable? If so, find the required value of K and the frequency of sustained oscillation. *(GATE 1998)*

Solution :

s^4	1	15	K
s^3	20	2	
s^2	14.9	K	
s^1	$\frac{29.8-20K}{14.9}$		
s^0	K		

(a) For stability $K > 0$

$$29.8 - 20K > 0$$

or, $K < 29.8/20$

$$K < 1.49$$

Hence, range $0 < K < 1.49$

(b) For marginally stable $K = 1.49$

Auxiliary equation $A(s) = 14.9s^2 + 1.49$

$$14.9s^2 = -1.49$$

$$s^2 = -0.1$$

$$s = \pm j\,0.316$$

$\therefore$ $\omega = 0.316$ rad/sec.

$\therefore$ Frequency of sustained oscillation = 0.316 rad/sec.

EXAMPLE 5.22. The characteristic equation for certain feedback control system is given below. Determine the range of value of K for the system to be stable.

$$s^4 + 4s^3 + 13s^2 + 36s + K = 0$$

Solution :

s^4	1	13	K
s^3	4	36	
s^2	4	K	
s^1	36–K		
s^0	K		

For stability $K > 0$

$$36 - K > 0 \text{ or } K < 36$$

Hence, range is $36 > K > 0$ **Ans.**

5.9. ROOT LOCUS

Root locus is a graphical method in which roots of the characteristic equation are plotted in s-plane for the different values of prameter. The locus of the roots of the characteristic equation when gain is varied from zero to infinity is called root locus.

Consider a unity feedback system as shown in Fig. 5.10.

The characteristic equation $1 + G(s)\,H(s) = 0$

$$G(s) = \frac{K}{s\,(s+2)}, \; H(s) = 1$$

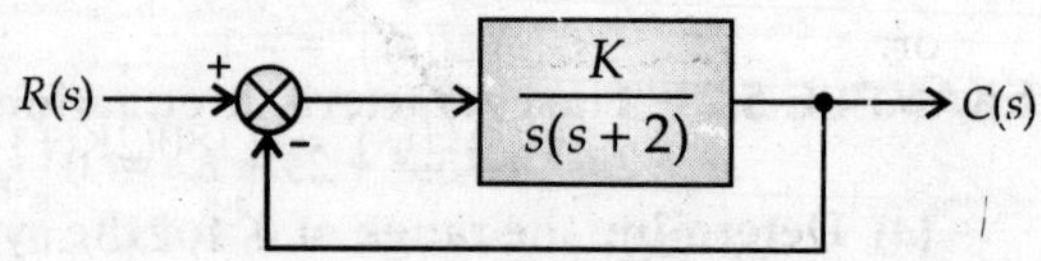

Fig. 5.10.

$$1 + \frac{K}{s\,(s+2)} \cdot 1 = 0 \text{ or,}$$

$$s^2 + 2\,s + K = 0 \qquad \text{...(5.14)}$$

The roots of eq^n (5.14) are

$$s_1 = -1 + \sqrt{1-K}$$

$$s_2 = -1 - \sqrt{1-K}$$

As 'K' is varied, the two roots give the locii in s-plane. For various values of K, the location of the roots are :

1. when $0 < K < 1$, the roots are real and distinct
2. when $K = 0$, the two roots are $s_1 = 0$ and $s_2 = -2$. These are also the open loop poles.
3. when $K = 1$, both roots are real and equal.
4. when $K > 1$, the roots are complex conjugate with real part $= -1$

   when 'K' is varying the root locus is shown in the Fig. 5.11.

   (*a*) when $K = 0$, two branches of root locus starts from $s = 0$ & $s = -2$.

   (*b*) when $K = 1$, both roots meet at $s = -1$

   (*c*) when $K > 1$, the roots breakaway from the real axis and become complex conjugate having negative real part equal to -1.

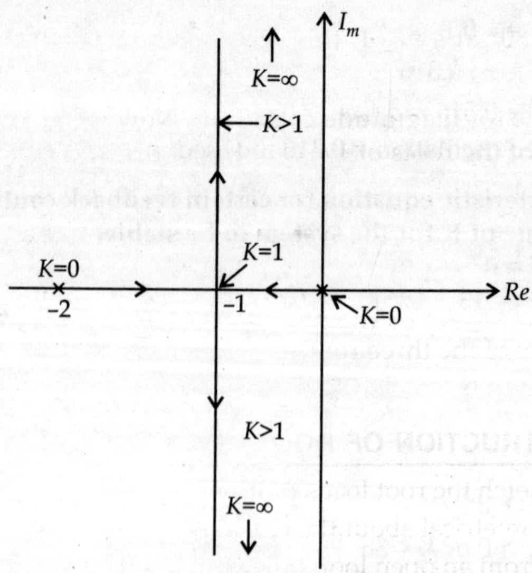

Fig. 5.11.

Consider a closed loop system shown in Fig. 5.11(*a*). The overall transfer function is

$$\frac{C(s)}{R(s)} = \frac{G(s)}{1+G(s)H(s)}$$

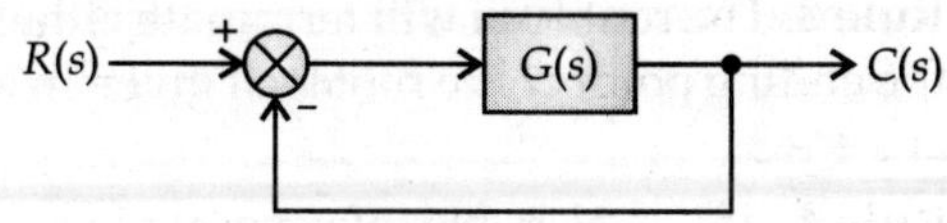

Fig. 5.11 (a)

The characteristic equation is $1 + G(s)\,H(s) = 0$

or, $\quad G(s)\,H(s) = -1$

or, $\quad |G(s)\,H(s)| = -1 \qquad$ ...(5.15)

$$\angle G(s)H(s) = \pm 180(2K+1)$$

$$K = 0, 1, 2, 3, \ldots \qquad \text{...(5.16)}$$

Equations (5.15) and (5.16) are the magnitude and angle conditions. The roots of the characteristic equations must satisfied the above two conditions or in other words the values of '*s*' which satisfies the above said conditions are the roots of the characteristic equation or poles of the closed loop systems. These two conditions are not independent. A plot of the points in 's' plane satisfying the second condition (*eq.* 5.16 known as angle criterion) is the root locus. The value of gain corresponding to a root can be determined from the first condition (*eqn* 5.15 known as magnitude criterion).

Consider a system with $G(s)\,H(s) = \dfrac{K}{s\,(s+4)\,(s+5)}$. Find whether $s = -1$ is on root locus or not using angle condition.

The angle condition is

$$\angle G(s)\,H(s) = \pm 180°\,(2\,K+1).$$

$$\angle G(s)\,H(s)\Big|_{s=-1} = \frac{\angle K + jo}{(-1+jo)\,(3+jo)\,(4+jo)}$$

$$= \frac{0°}{180°.0°.0°} = -180°$$

Since $\angle G(s)\,H(s) = -180°$ at $s = -1$, this satisfies the angle condition, hence the point at $s = -1$ is on the root locus.

Consider above example for magnitude condition. Now we are interested to find the value of *K* at which $s = -1$ is one of the roots of $1 + G(s)\,H(s) = 0$.

$$|G(s)\,H(s)|_{s=-1} = 1$$

$$\frac{|K|}{|-1|\,|-1+4|\,|-1+5|} = 1$$

$$\therefore \quad K = 12.$$

Therefore for $K = 12$, one of the three roots is located at $s = -1$. Remaining two roots can also be easily calculated.

5.10. RULES FOR CONSTRUCTION OF ROOT LOCII

Following are the rules to sketch the root locus plot.

Rule 1. The root locus is symmetrical about the real axis.

Rule 2. The root locii starts from an open loop pole with $K = 0$ *e.g.*, For the system having

$$G(s)H(s) = \frac{K\,(s+3)}{(s+2)} \qquad \text{...(5.17)}$$

Find the starting point of the root locii.

Solution : According to the rule the root locii starts from $s = -2$

Rule 3. The root locii will terminate either on an open loop zeros or on infinity with $K = \infty$ *e.g.*, Find the ending point of the root locii given in eqn. 5.17. According to the rule the root locii will terminate at $s = -3$

Rule 4 . If N = No. of separate locii
P = No. of finite poles
Z = No. of finite zeros then

number of root locii will be equal to the no. of poles if number of poles are more than number of zeros *i.e.*, $P > Z$

$$N = P \text{ if } P > Z$$

If $Z > P$, then number of root locii will be equal to the number of zeros.

If $P = Z$, then No. of root locii = Poles = Zeros.

e.g., Find the number of separate root locii for the system given by the eq^n 5.17

Solution : $P = 1$

$Z = 1$

$\therefore$ $N = 1$

Rule 5. Root Locii on the Real Axis

Any point on the real axis is a part of the root locus if and only if the number of poles and zeros to its right is odd.

Rule 6. Asymptotes

The branches of root locus tend to infinity along a set of straight line called asymptotes. These asymptotes making an angle with real axis and is given by

$$\phi = \frac{(2K+1)180°}{P-Z} \quad \text{where } K = 0, 1, 2, ...$$

The total number of asymptotes = $P–Z$

e.g., If $G(s)H(s) = \dfrac{K}{s(s^2+6s+10)}$ ...(5.18)

$P = 3$

$z = 0$

No. of asymptotes = $P - Z = 3 - 0 = 3$

$K = 0$ $\quad \phi_1 = \dfrac{(2\times0+1)180°}{3} = 60°$

$K = 1$ $\quad \phi_2 = \dfrac{(2\times1+1)180°}{3} = 180°$

$K = 2$ $\quad \phi_2 = \dfrac{(2\times2+1)180°}{3} = 300°$

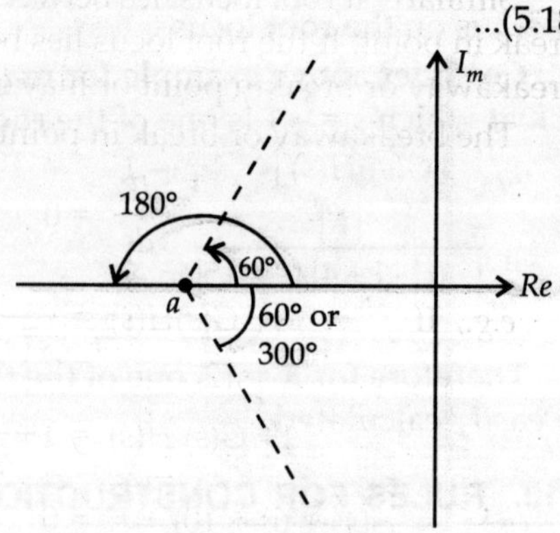

Fig. 5.12.

Rule 7. Centroid of Asymptotes

The point of intersection of asymptotes with real axis is called centroid of asymptotes (σ_A) and is given by

$$\sigma_A = \frac{\text{sum of poles} - \text{sum of zeros}}{P-Z}$$

e.g., Find the centroid of asymptotes of the system given by equation 5.18

Solution : There are three poles at $s_1 = 0$, $s_2 = -3 + j1$, $s_3 = -3 - j1$

No. of zeros = 0

$$\therefore \text{centroid} \qquad \sigma_A = \frac{0-3+j1-3-j1-0}{3} = -2$$

The centroid is shown in Fig. 5.12 by the point '*a*'.

Rule 8. Angle of Departure and Angle of Arrival of the Root Locii

The angle of departure of the root locus from a complex pole is given by

ϕ_d = 180° – sum of angles of vectors drawn to this pole from other poles + sum of angles of vectors drawn to this pole from the zeros.

The angle of arrival at a complex zero is given by

ϕ_a = 180° – sum of angles of vectors drawn to this zero from other zeros + sum of angles of vectors drawn to this zero from poles.

e.g., For

$$G(s)\,H(s) = \frac{K}{s\,(s+6)\,(s^2+4s+13)}$$

Determine the angle of departure from complex poles.

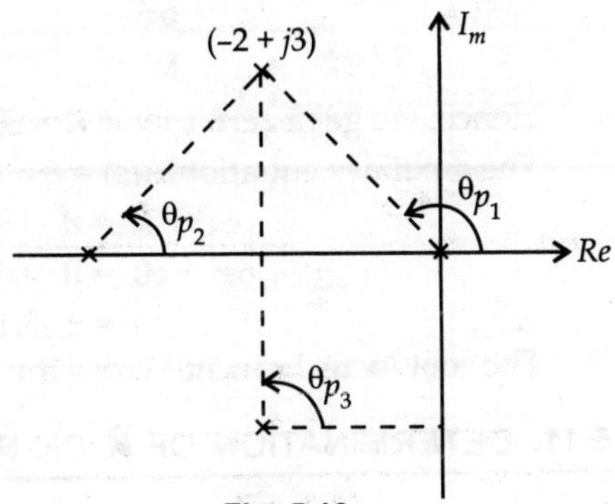

Fig. 5.13.

$$\phi_d = 180° - (\theta_{P_1} + \theta_{P_2} + \theta_{P_3})$$
$$= 180° - (123° + 37° + 90°) = -70°$$

$\therefore$ angle of departure at $(-2 + j3) = -70°$

$\therefore$ angle of departure at $(-2 - j3) = 70°$

Rule No. 9. Breakaway Point of Real Axis

If the root locus lies between two adjacent open loop poles on the real axis then there will be at least one breakaway point, because the roots move towards each other as K is increased and meet at a point. At this point K is maximum. If we increase the value of K between two poles the root locus breaks in two parts.

Similarly if root locus lies between two adjacent zeros on real axis, then there will be at least one break in point. If the root locus lies between an open loop pole and zero, then then there will be no breakaway or breakin point or may be both occur.

The breakaway or break in points can be determined from the roots of

$$\frac{dK}{ds} = 0 \qquad \text{...(5.19)}$$

e.g., If $\quad G(s)H(s) = \dfrac{K}{s\,(s^2+6s+10)}$ Determine the breakaway point

$$1 + G(s)\,H(s) = 1 + \frac{K}{s\,(s^2+6s+10)}$$

$$s(s^2 + 6s + 10) + K = 0$$

or, $\quad K = -s^3 - 6s^2 - 10s$

$$\frac{dK}{ds} = -3s^2 - 12s - 10 = 0$$

or, $\quad 3s^2 + 12s + 10 = 0$

$s_1 = -1.1835$ and $s_2 = -2.815$ are the breakaway points.

Rule No. 10. **The intersection of root locus branches with jω-axis can be determined through Routh Hurwitz criterion.**

e.g., If $G(s)\,H(s) = \dfrac{K}{s\,(s^2+6s+10)}$. **Find the intersection of the root locii with the imaginary axis.**

Solution : The characteristic eq^n $s^3 + 6s^2 + 10s + K = 0$

$$\begin{array}{lll} s^3 & 1 & 10 \\ s^2 & 6 & K \\ s^1 & \dfrac{60-K}{6} & \\ s^o & K & \end{array}$$

Hence, we get a zero row if $K = 60$

The auxiliary equation $A(s) = 6s^2 + K$

$$6s^2 + K = 0$$

$$6s^2 + 60 = 0$$

$$\therefore \quad s = \pm j3.16$$

The root locus branches cross the imaginary axis at $s = \pm j3.16$ for $K = 60$.

5.11. DETERMINATION OF K ON ROOT LOCII

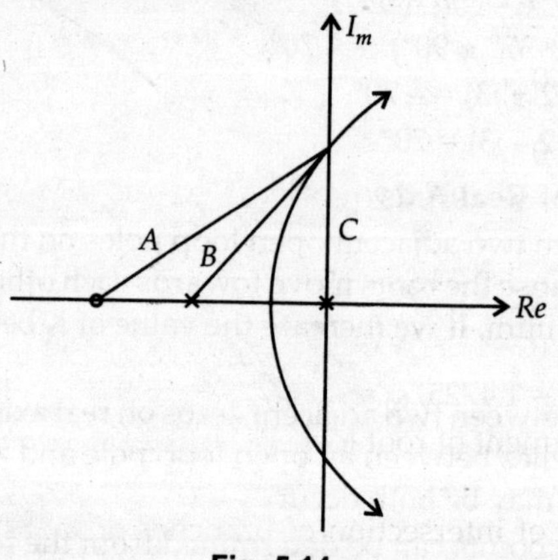

Fig. 5.14.

The value of K can be determined by

$$K = \frac{\text{Product of all vector lengths drawn from the poles of } G(s)\,H(s) \text{ to the point}}{\text{Product of all vector lengths drawn from the zeros of } G(s)\,H(s) \text{ to the point}}$$

e.g., in Fig. 5.14 determine the value of K at the point of intersection of root locus branch with imaginary axis.

$$K = \frac{B.C}{A}$$

EXAMPLE 5.23. **The forward path transfer function of a unity feedback system is given by** $G(s) = \dfrac{K}{s\,(s+4)\,(s+5)}$ **Sketch the root locus as K varies from zero to infinity.**

Solution : Step 1 : Plot the poles and zeros

Poles are at $s = 0, -4$ and -5

No. of poles $P = 3$
No. of zeros $z = 0$

Step 2 : The root locus exists between $s = 0$ and $s = -4$ and to the left of -5. Mark the root locus on real axis.

Step 3 : Number of root locii

$$P = 3$$
$$Z = 0$$

$\therefore$ Number of root locii $(N) = 3$

Step 4 : Centroid of asymptotes

$$\sigma A = \frac{\text{sum of poles} - \text{sum of zeros}}{P - Z} = \frac{(0-4-5)-(0)}{3-0} = -3$$

Step 5 : Angle of asymptotes

$$\phi = \left(\frac{2K+1}{P-Z}\right)180^\circ$$

$$K = 0, \quad \phi_1 = 60^\circ$$
$$K = 1, \quad \phi_2 = 180^\circ$$
$$K = 2, \quad \phi_3 = 300^\circ$$

Step 6 : Caluculation of breakaway point

The characteristic equation $1 + G(s)\,H(s) = 0$

$$1+\frac{K}{s\,(s+4)\,(s+5)} = 0$$

$$s^3 + 9s^2 + 20\,s + K = 0$$

$$\therefore \quad K = -s^3 - 9s^2 - 20s$$

$$\frac{dk}{ds} = -3s^2 - 18\,s - 20 = 0$$

$$3s^2 + 18s + 20 = 0$$

$$s_1 = -1.4725, \; s_2 = -4.5275$$

Since, – 4 to – 5 is not the segment of root locus. Therefore we consider – 1.4725 as a breakaway point.

Step 7 : Determination of point of intersection of branches of root locus with imaginary axis. by Routh Hurwitz

The characteristic eq^n is

$$s^3 + 9s^2 + 20s + K = 0$$

s^3	1	20
s^2	9	K
s^1	$\frac{180-K}{9}$	
s^0	K	

For $K = 180$, the auxiliary eq^n $A(s) = 9s^2 + K$

$$9s^2 + K = 0$$
$$9s^2 + 180 = 0$$
$$s = \pm j\,4.47$$

The complete root locus is shown in Fig. 5.15.

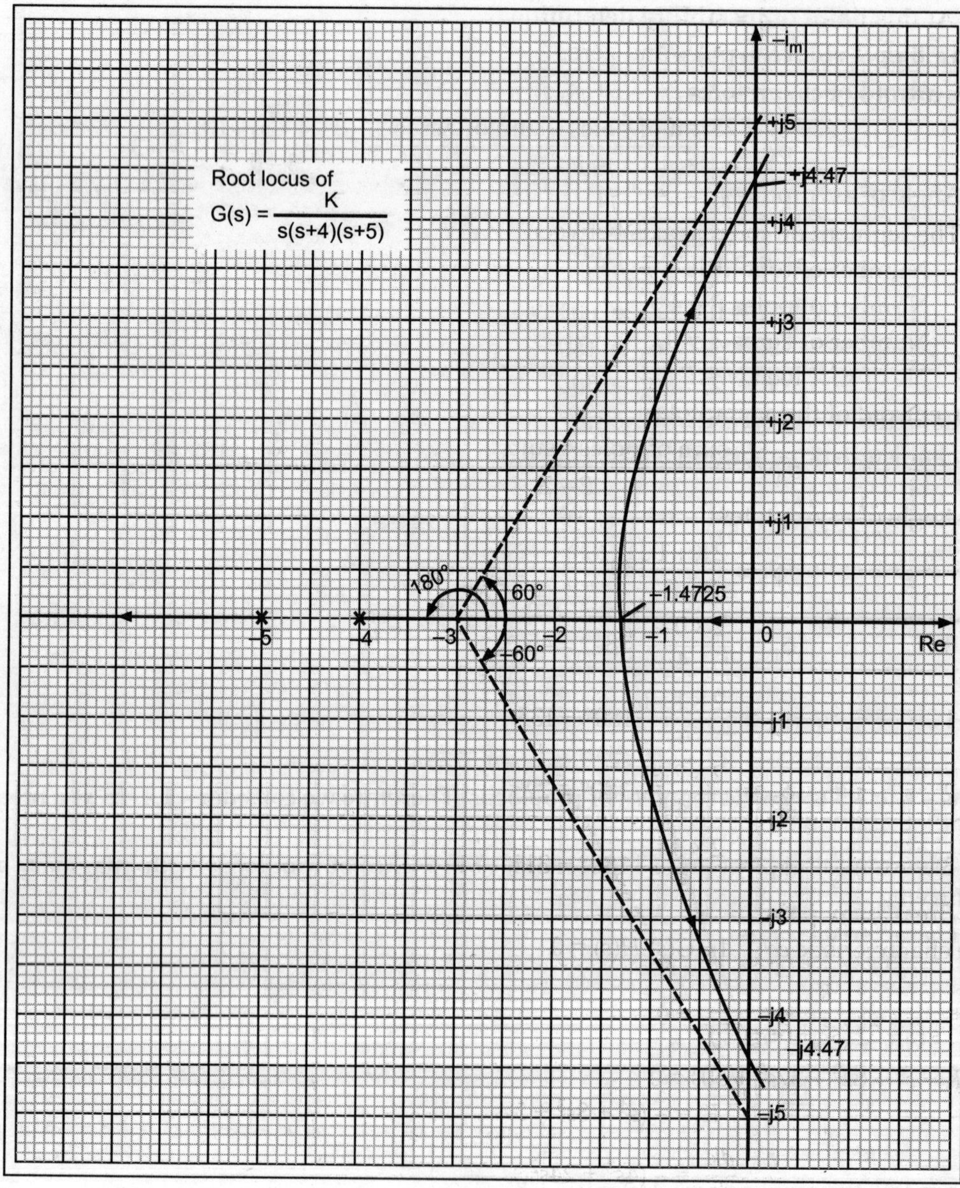
Root locus of
G(s) = K / s(s+4)(s+5)
–jm
+j5
+j4.47
+j4
+j3
+j2
+j1
180°
60°
–60°
–1.4725
–5
–4
–3
–2
–1
0
Re
–j1
–j2
–j3
–j4
–j4.47
–j5

Fig. 5.15.

EXAMPLE 5.24. For a unity feedback system the open loop transfer function is given by

$$G(s) = \frac{K}{s(s+2)(s^2+6s+25)}$$

(a) Sketch the root locus for $0 \le K \le \infty$

(b) At what value of 'K' the system becomes unstable.

(c) At this point of instability determine the frequency of oscillation of the system.

Solution : Step 1 : Plot the poles and zeros

Poles are at $s_1 = 0, s_2 = -2, s_3 = -3 + j4, s_4 = -3 - j4$

$$s^2 + 6s + 25 = 0$$

$$s_3 = -3 + j4$$

$$s_4 = -3 - j4$$

Step 2 : The segment on the real axis between $s = 0$ and $s = -2$ is the part of the root locus.

Step 3 : Number of root locii

Number of poles $P = 4$

Number of zeros $Z = 0$

$\therefore$ No of root locii $N = P = 4$

Step 4 : Centroid of the asymptotes

$$\sigma_A = \frac{\text{sum of poles} - \text{sum of zeros}}{P - Z} = \frac{(0-2-3+j4-3-j4)-(0)}{4-0} = -2$$

Step 5 : Angle of asymptotes

$$\phi = \frac{2K+1}{P-Z}180°$$

$K = 0$ $\qquad \phi_1 = \frac{2\times0+1}{4}180° = 45°$

$K = 1$ $\qquad \phi_2 = \frac{2\times1+1}{4}180° = 135°$

$K = 2$ $\qquad \phi_3 = \frac{2\times2+1}{4}180° = 225°$

$K = 3$ $\qquad \phi_4 = \frac{2\times3+1}{4}180° = 315°$

Step 6 : Breakaway point

The characteristic eq^n $1 + G(s)H(s) = 0$

$$1 + \frac{K}{s(s+2)(s^2+6s+25)} = 0$$

or, $K + s^4 + 8s^3 + 37s^2 + 50s = 0$

$$K = -(s^4 + 8s^3 + 37s^2 + 50s)$$

$$\therefore \quad \frac{dk}{ds} = -(4s^3 + 24s^2 + 74s + 50) = 0$$

or, $4s^3 + 24s^2 + 74s + 50 = 0$

By trial and error method $s = -0.8981$*

---

* Put the values of 's' between 0 and – 2 because the portion between 0 and – 2 is the part of root locus, if there is any breakaway point then it will lie between these points.

Step 7 : Determination of jω crossover (by Routh Hurwitz)

ch. eq^n $s^4 + 8s^3 + 37s^2 + 50s + K = 0$

s^4	1	37	K
s^3	8	50	
s^2	30.75	K	
s^1	$\frac{1537.5-8K}{30.75}$		
s^o	K		

$$\frac{1537.5-8K}{30.75} = 0 \qquad \therefore K = 192.18$$

For $K = 192.18$ Auxiliary $eq^{n.}$

$$30.75\, s^2 + K = 0$$
$$30.75\, s^2 = -192.18$$
$$s = \pm j2.5$$

Step 8 : (*a*) Angle of departure from upper compex pole

$$\phi_d = 180° - (104° + 90° + 127°)$$
$$\phi_d = -141°$$

(*b*) The range of values for stability is $0 < K < 192.18$

The closed loop system becomes unstable for $K < 0$ and $K > 192.18$

(*c*) At this point of instability the gain is $K = 192.18$

$$30.75\, s^2 + 192.18 = 0$$

Put $s = j\omega$

$$-30.75\, \omega^2 + 192.18 = 0$$

$\therefore$ $\omega = 2.5$ rad/sec.

$\therefore$ The frequency of oscillation at the point of instability = 2.5 rad/sec.

The root locus plot is shown in Fig. 5.16.

EXAMPLE 5.25. Consider a unity feedback control system with the following feedforward transfer function.

$$G(s) = \frac{K}{s\,(s^2+4s+8)}$$

Plot the root locii for the system.

Solution : Step 1 : Plot the poles and zeros

$$s^2 + 4s + 8 = 0$$

$$s_2, s_3 = \frac{-4 \pm \sqrt{16-32}}{2} = -2 \pm j2$$

Three poles are at $s_1 = 0$, $s_2 = -2 + j2$ and $s_3 = -2 - j2$

Step 2 : Since there is only one pole at $s = 0$, the entire left half of the real axis is the part of the root locus.

Step 3 : Number of root locii

No. of poles $P = 3$

No. of zeros $Z = 0$

$\therefore$ No. of root locii $N = P = 3$

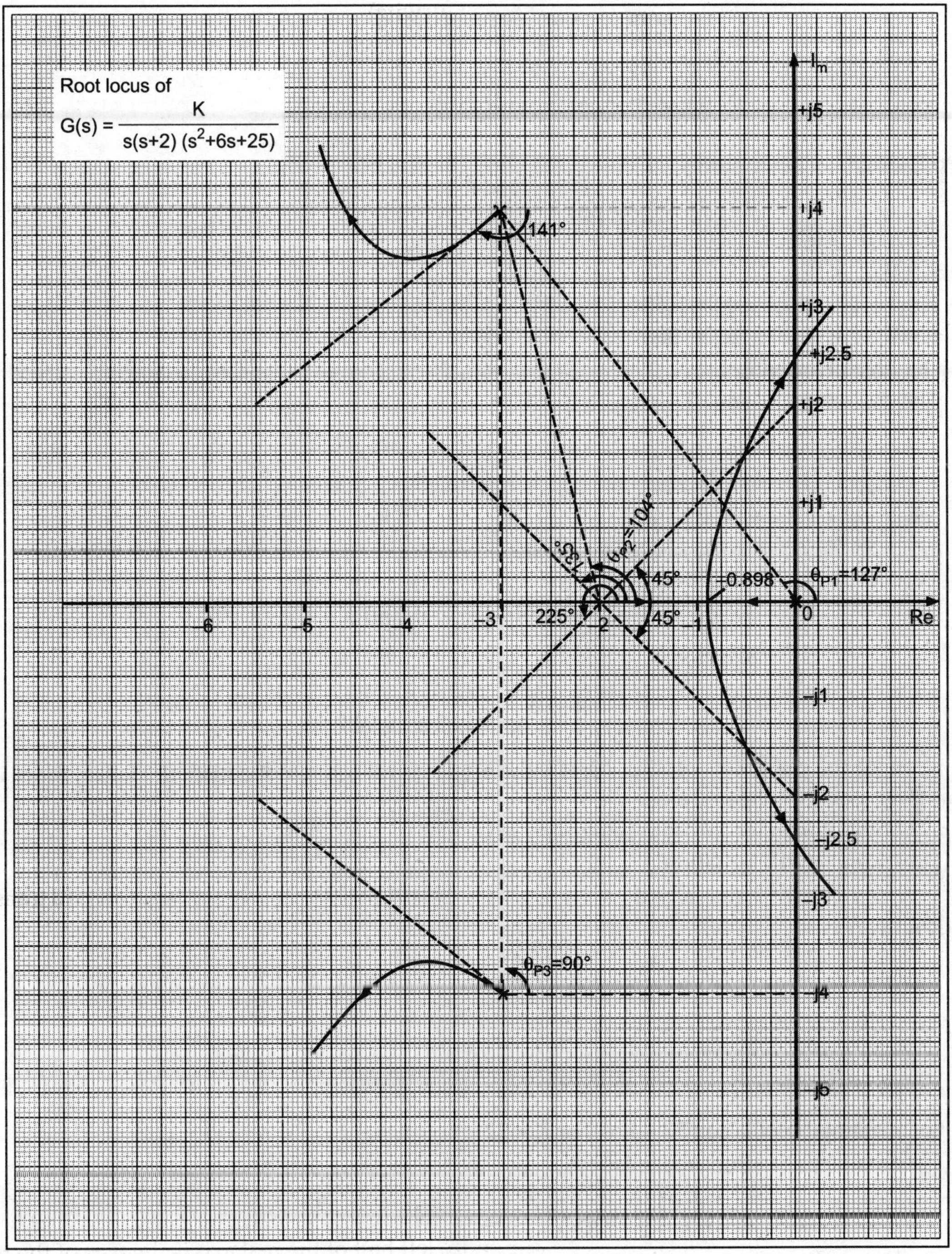

Root locus of
G(s) = K / s(s+2) (s^2+6s+25)
Im
Re
+j5
+j4
+j3
+j2.5
+j2
+j1
–j1
–j2
–j2.5
–j3
–j4
–j5
–6
–5
–4
–3
2
–1
0
141°
θP2=104°
135°
45°
225°
45°
–0.898
θP1=127°
θP3=90°

Fig. 5.16.

Step 4 : Centroid of the asymptotes

$$\sigma_A = \frac{\text{sum of poles} - \text{sum of zeros}}{P - Z}$$

$$= \frac{(0-2+j2-2-j2)-(0)}{3} = -4/3 = -1.33$$

Step 5 : Angle of asymptotes

$$\phi = \frac{2K+1}{P-Z}180°$$

$K = 0$ $\quad\phi_1 = 60°$

$K = 1$ $\quad\phi_2 = 180°$

$K = 2$ $\quad\phi_3 = 300°$

Step 6 : Breakaway point

The characteristic eqn $1 + G(s)\,H(s) = 0$

$$1+\frac{K}{s\,(s^2+4s+8)} = 0$$

$$s^3 + 4s^2 + 8s + K = 0$$

or,

$$K = -(s^3 + 4s^2 + 8s)$$

$$\frac{dK}{ds} = -(3s^2 + 8s + 8)$$

$$3s^2 + 8s + 8 = 0$$

$$\therefore \quad s = \frac{-8 \pm \sqrt{64-96}}{6} = -1.33 \pm j\,0.943$$

Since, at the point s = – 1.33 ± j 0.943, the angle condition is not satisfied. Hence, there is no breakaway point on the real axis.

Step 7 : Point of intersection with imaginary axis.

The characteristic equation

$$s^3 + 4s^2 + 8s + K = 0$$

Routh array is

s^3	1	8
s^2	4	K
s^1	$\frac{32-K}{4}$	
s^0	K	

For sustained oscillation $32 - K = 0$ or $K = 32$

The auxiliary equation $A(s) = 4s^2 + K$

$$4s^2 + 32 = 0$$

$$\therefore \quad s = \pm j\,2.83$$

Step 8 : The angle of departure from the upper complex pole is

$$\phi_d = 180° - (135° + 90°) = -45°$$

The root locus is shown in Fig. 5.17.

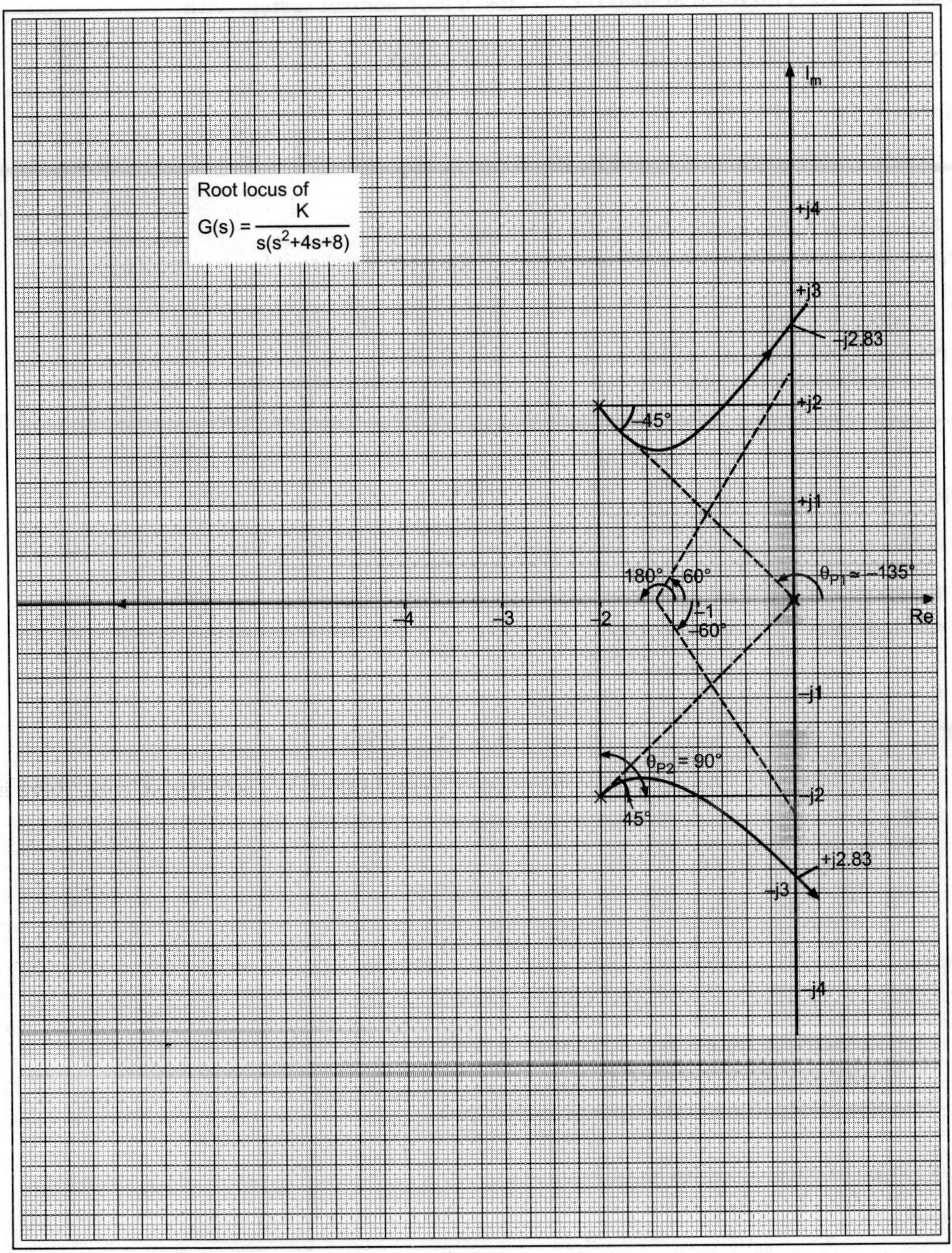
Root locus of
G(s) = K / s(s²+4s+8)
Im
Re
+j4
+j3
–j2.83
+j2
+j1
–j1
–j2
+j2.83
–j3
–j4
–4
–3
–2
–1
45°
180°
60°
–60°
θP1 = –135°
θP2 = 90°
45°

Fig. 5.17.

EXAMPLE 5.26. Plot the root locii for the closed loop control system with

$$G(s) = \frac{K}{s\,(s+1)\,(s^2+4s+5)},\; H(s) = 1.$$

Solution : Step 1 : Plot the poles and zeros $s^2 + 4s + 5$

$$s = \frac{-4 \pm \sqrt{16-20}}{2} = -2 \pm j1$$

Poles are at $s_1 = 0$, $s_2 = -1$, $s_3 = -2 + j1$ and $s_4 = -2 - j1$

Step 2 : The segment between $s = 0$ and $s = -1$ is the part of the root locus on real axis.

Step 3 : Number of root locii

Number of root locii $N = P = 4$

Step 4 : Centroid of the asymptotes

$$\sigma_A = \frac{\text{sum of poles} - \text{sum of zeros}}{P-Z} = \frac{0-1-2+j1-2-j1-0}{4}$$

$$= -5/4 = -1.25$$

Step 5 : Angle of asymptotes

$$\phi = \frac{2K+1}{P-Z}180°$$

$K = 0$ $\phi_1 = 45°$

$K = 1$ $\phi_2 = 135°$

$K = 2$ $\phi_3 = 225°$

$K = 3$ $\phi_4 = 315°$

Step 6 : Breakaway point

The characteristic equation

$$1 + G(s)\,H(s) = 0$$

$$1 + \frac{K}{s\,(s+1)\,(s^2+4s+5)} = 0$$

$$s^4 + 5s^3 + 9s^2 + 5s + K = 0$$

$$K = -(s^4 + 5s^3 + 9s^2 + 5s)$$

$$\frac{dK}{ds} = -[4s^3 + 15s^2 + 18s + 5]$$

$$4s^3 + 15s^2 + 18s + 5 = 0$$

$\therefore$ Breakaway point is $s = -0.4$

Step 7 : Angle of departure at the upper complex pole

$$\phi_d = 180° - (154° + 136° + 90°) = -200°$$

Step 8 : Point of intersection on $j\omega$ axis

$$s^4 + 5s^3 + 9s^2 + 5s + K = 0$$

s^4	1	9	K
s^3	5	5	
s^2	8	K	
s^1	$\frac{40-5K}{8}$		
s^0	K		

At $40 - 5K = 0$ $K = 8$ (for sustained oscillation)

the auxiliary equation $A(s) = 8s^2 + K$

$$8s^2 + 8 = 0$$

$$s = \pm j1$$

The root locus plot is shown in Fig. 5.18.

EXAMPLE 5.27. Sketch the root locii for

$$G(s) = \frac{K(s+1)}{s^2(s+3.6)}$$

$$H(s) = 1$$

Solution : Step 1 : Plot the poles and zeros

Poles are at $s = 0, 0, -3.6$

zeros at $s = -1$

Step 2 : The segment between $s = -1$ and -3.6 is the part of the root locus

Step 3 : Centroid of the asymptotes

$$\sigma_A = \frac{\text{sum of poles} - \text{sum of zeros}}{P-Z}$$

$$= \frac{0+0-3.6+1}{3-1} = -1.3$$

Step 4 : Angle of asymptotes

$$\phi = \frac{2K+1}{P-Z}180^\circ$$

$K = 0$ $\phi_1 = 90^\circ$

$K = 1$ $\phi_2 = 270^\circ$

Step 5 : Breakaway point

characteristic equation $1 + G(s)H(s) = 0$

$$1 + \frac{K(s+1)}{s^2(s+3.6)} = 0$$

$$s^2(s+3.6) + K(s+1) = 0$$

$$K = \frac{s^3 + 3.6s^2}{s+1}$$

$$\frac{dK}{ds} = \frac{(s+1)(3s^2+7.2s) - (s^3+3.6s^2)}{(s+1)^2} = 0$$

or, $s^3 + 3.3s^2 + 3.6s = 0$

$$s(s^2 + 3.3s + 3.6) = 0$$

$$s = 0 \text{ and } s = \frac{-3.3 \pm \sqrt{3.3^2 - 4\times 1\times 3.6}}{2}$$

$$s = 0,\ s = -1.65 \pm j\,0.936$$

Point $s = 0$, is the actual breakaway point

$s = -1.65 \pm j\,0.936$ is a complex quantity, this will not be breakaway point

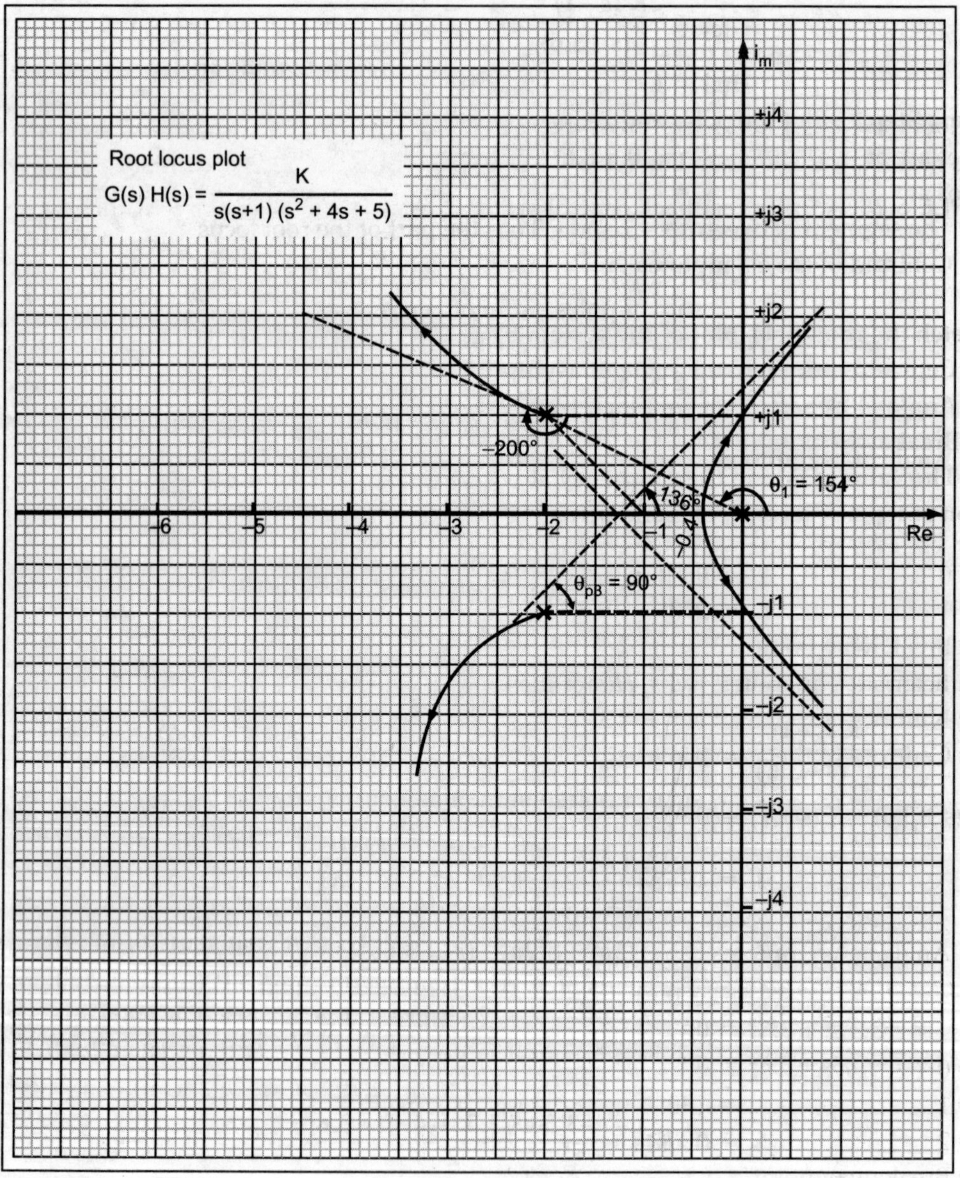
Root locus plot
G(s) H(s) = K / s(s+1) (s² + 4s + 5)
im
+j4
+j3
+j2
+j1
–j1
–j2
–j3
–j4
Re
–6
–5
–4
–3
–2
–1
–0.4
–200°
136°
θ1 = 154°
θpB = 90°

Fig. 5.18.

Step 6 : Point of intersection of root locii on $j\omega$ axis.

The ch. $eq^{n.}$ $s^3 + 3.6s^2 + Ks + K = 0$

s^3	1	K
s^2	3.6	K
s^1	$0.72K$	
s^o	K	

For sustained oscillation $K = 0$

$$A(s) = 3.6\, s^2 + K$$

$$3.6\, s^2 + 0 = 0$$

$$s^2 = 0$$

$\therefore$ Root locus branches do not cross the $j\omega$ axis.

The root locus is shown in Fig. 5.19.

EXAMPLE 5.28. A unity feedback control system has an open loop transfer function :

$$G(s) = \frac{K}{s\,(s^2+4s+13)}$$

Sketch the root locus plot of the system by determining the following

(*a*) Centroid, number and angle of asymptotes.

(*b*) Angle of departure of root locii from the poles

(*c*) Breakaway point if any

(*d*) The value of *K* and the frequency at which the root locii cross $j\omega$ axis.

Solution : Step 1 : Plot the poles and zeros

$$s = 0$$

$$s = \frac{-4 \pm \sqrt{16-52}}{2} = -2 \pm j3$$

Step 2 : From $s = 0$ to the left is the part of the root locus

Step 3 : Number of root locii $N = P = 3$

Step 4 : Centroid of the asymptotes

$$\sigma_A = \frac{\text{sum of poles} - \text{sum of zeros}}{P-Z}$$

$$= \frac{0-2+j3-2\;\; j3\;\; 0}{3-0} = -4/3 = -1.33$$

Step 5 : Angle of asymptotes

$$\phi = \frac{2K+1}{P-Z}180°$$

$K = 0$ $\quad \phi_1 = 60°$

$K = 1$ $\quad \phi_2 = 180°$

$K = 2$ $\quad \phi_3 = 300°$

Step 6 : Breakaway point

The characteristic equation $1 + G(s)\,H(s) = 0$

$$1+\frac{K}{s\,(s^2+4s+13)} = 0$$

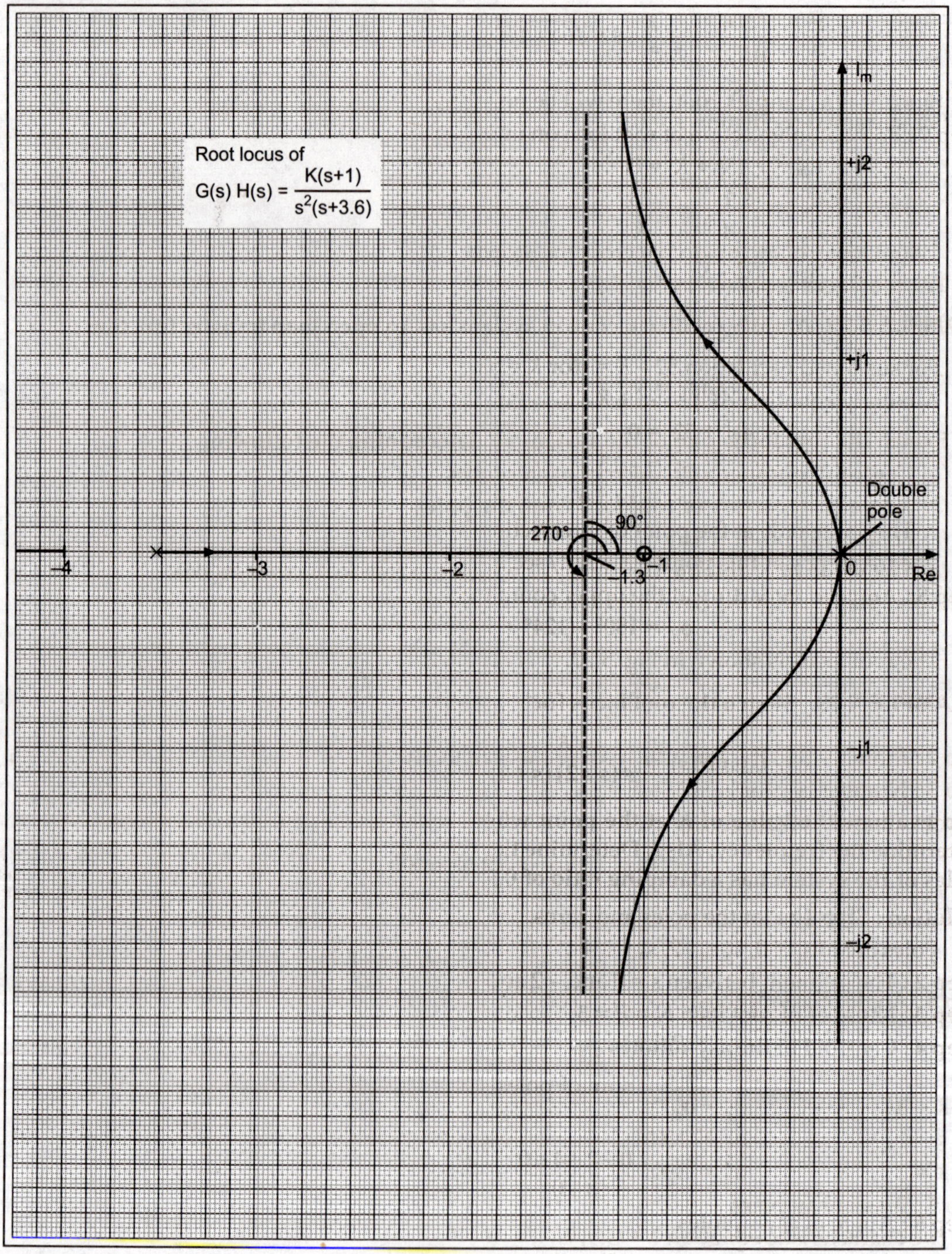
Root locus of
G(s) H(s) = K(s+1) / s^2(s+3.6)
Im
+j2
+j1
–j1
–j2
Double pole
270°
90°
–1.3
–1
–4
–3
–2
0
Re

Fig. 5.19.

$$K = -[s^3 + 4s^2 + 13s]$$

$$\frac{dK}{ds} = -[3s^2 + 8s + 13] = 0$$

$$\therefore \qquad s = \frac{-8 \pm \sqrt{64-156}}{6} = \frac{-8 \pm j\,9.59}{6} = -1.33 \pm j\,1.59$$

Since, it is a complex root, there will be no breakaway point on real axis.

Step 7 : Point of intersection of root locii on imaginary axis characteristic eqn

$$s^3 + 4s^2 + 13s + K = 0$$

$$\begin{array}{lcc} s^3 & 1 & 13 \\ s^2 & 4 & K \\ s^1 & \frac{52-K}{4} & \\ s^o & K & \end{array}$$

For sustained oscillation $52 - K = 0$ or $K = 52$

Auxiliary equation $A(s) = 4s^2 + K$

$$4s^2 + 52 = 0$$

$$s^2 = -13$$

$$s = \pm j\,3.6$$

$\therefore$ The value of K at the point of intersection on imaginary axis = 52

The frequency at this point = 3.6 rad/sec.

Step 8 : Angle of departure at the upper complex pole

$$\phi_d = 180° - (124° + 90°) = -34°$$

The root locus is shown in Fig. 5.20.

EXAMPLE 5.29. Sketch the root locus for

$$G(s)\,H(s) = \frac{K}{s\,(s+2)\,(s+4)}$$

and evaluate the value of K at the point where the root locii crosses the imaginary axis. Also determine the frequency. Also, Determine the value of 'K' so that the dominant pair of complex poles of the system has a damping ratio of 0.5.

Solution : Step 1 : Plot the poles and zeros.

Poles are at $s_1 = 0, s_2 = -2, s_3 = -4$

Step 2 : The segment between $s = 0$ and -2 and from $s = -4$ to the left are the parts of root locus.

Step 3 : Number of root locii $N = P = 3$

Step 4 : Centroid of the asymptotes

$$\sigma_A = \frac{\text{sum of poles} - \text{sum of zeros}}{P - Z}$$

$$= \frac{0 \;\; 2 \;\; 4 \;\; 0}{3-0} = -2$$

Step 5 : Angle of asymptotes

$$\phi = \frac{2K+1}{P-Z}180°$$

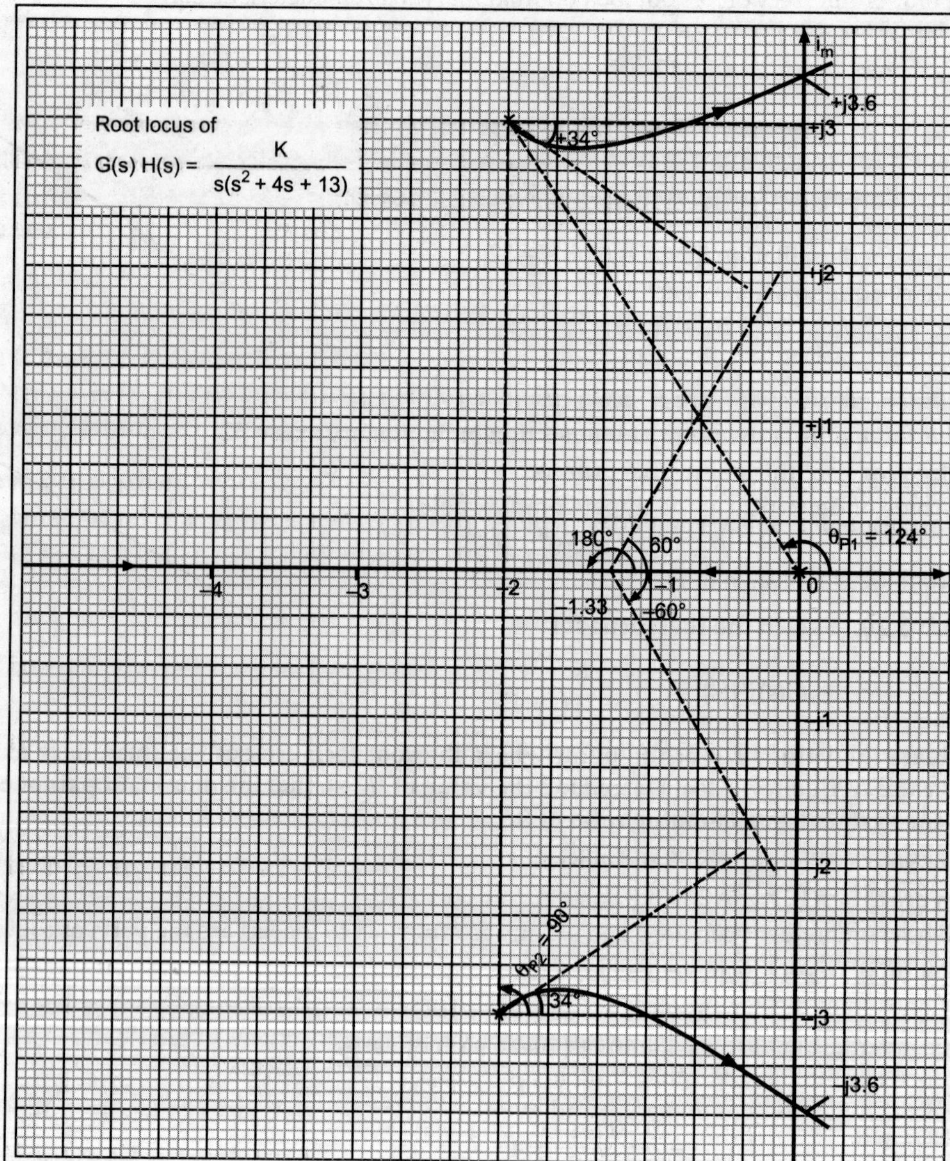

Root locus of
G(s) H(s) = $\frac{K}{s(s^2 + 4s + 13)}$
i_m
+j3.6
+j3
+34°
+j2
+j1
180°
60°
θ_{P1} = 124°
–4
–3
–2
–1
0
–1.33
–60°
–j1
–j2
θ_{P2} = –90°
34°
–j3
j3.6

Fig. 5.20.

$K = 0$ $\phi_1 = 60°$

$K = 1$ $\phi_2 = 180°$

$K = 2$ $\phi_3 = 300°$

Step 6 : Breakaway point

The characteristic equation

$$s(s + 2)(s + 4) + K = 0$$

$$K = -(s^3 + 6s^2 + 8s)$$

$$\frac{dK}{ds} = -(3s^2 + 12s + 8) = 0$$

$$3s^2 + 12s + 8 = 0$$

$$s = -0.85 \text{ and } -3.15$$

Since, – 3.15 is not the part of root locus therefore breakaway point is – 0.85.

Step 7 : Point of intersection of root locii on imaginary axis

characteristic equation

$$s^3 + 6s^2 + 8s + K = 0$$

s^3	1	8
s^2	6	K
s^1	$\frac{48-K}{6}$	
s^o	K	

For sustained oscillation $K = 48$

Auxiliary eq^n $A(s) = 6s^2 + K$

$$6s^2 + 48 = 0$$

$$s = \pm j\,2.8$$

Put $s = j\omega$

$\therefore$ $\omega = 2.8$

$\therefore$ frequency of oscillation = 2.8 rad/sec.

The value of K at the point of intersection of root locii with the imaginary axis = 48

This value of K can also be determined from the graph. Draw the lines from all poles to the point of intersection on imaginary axis (c).

$$K = AC \times BC \times OC$$

from graph AC = 9.8 cm. = 4.9

BC – 6.9 cm. = 3.45

OC = 5.8 cm. = 2.9

$\therefore$ $K = 4.9 \times 3.45 \times 2.9 = 49$

In second quadrant draw a ζ-line at $\theta = \cos^{-1} 0.5 = 60°$ with negative real axis. This line intersect the root locii at point D. The value of 'K' at this point can be obtained as :

Connect all the poles to this point and measure all the distance of all poles to this point

$\therefore$ OD = 2.6 cm. = 1.3

BD = 3.6 cm. = 1.8

AD = 7.2 cm. = 3.6

$\therefore$ $K = 1.3 \times 1.8 \times 3.6 = 8.424$

The root locus is shcwn in the Fig. 5.21.

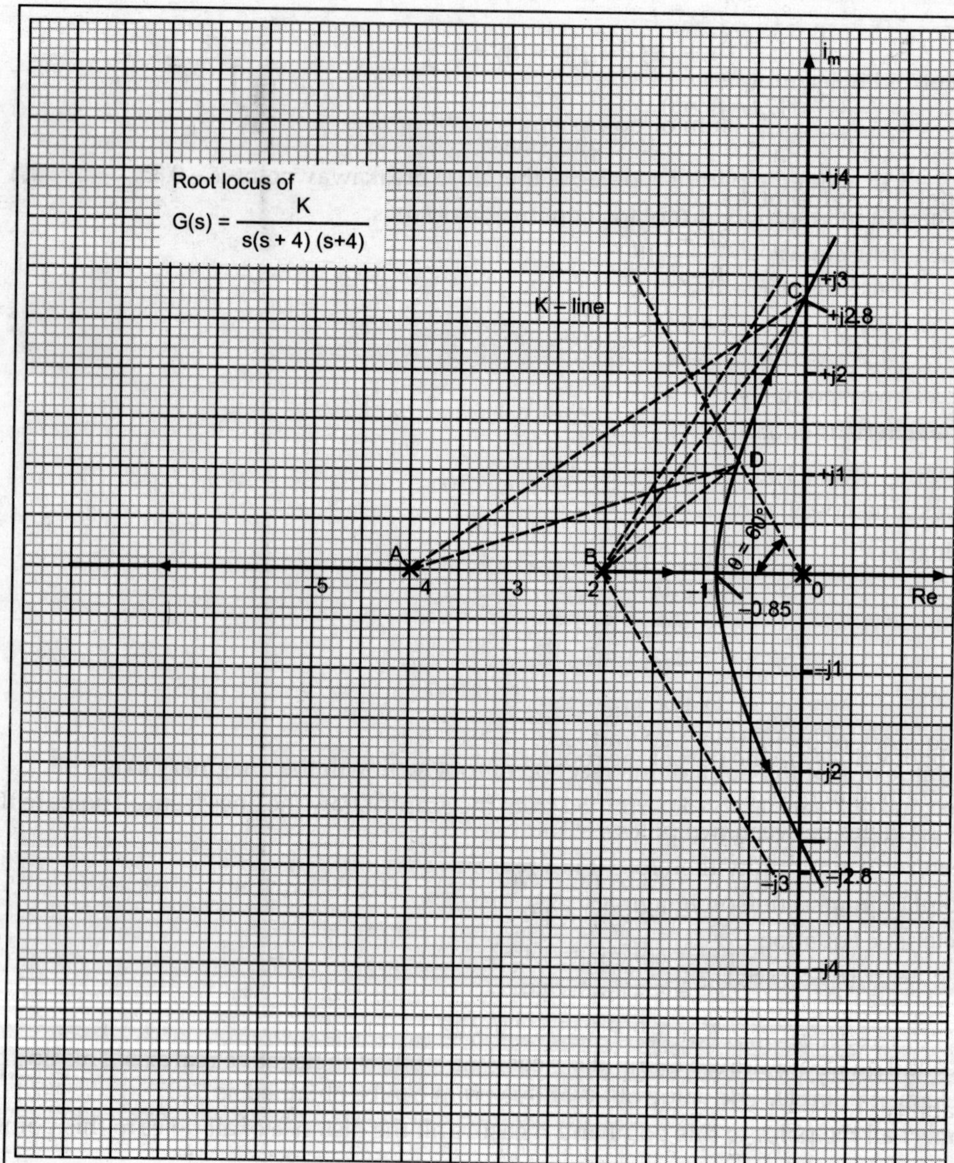
Root locus of
G(s) = K / s(s + 4) (s+4)
jm
+j4
+j3
+j2.8
+j2
+j1
K – line
C
D
A
B
θ = 60°
–5
–4
–3
–2
–1
0
–0.85
Re
–j1
–j2
–j3
–j2.8
–j4

Fig. 5.21.

EXAMPLE 5.30. A unity feedback system has an open loop transfer function

$$G(s) = \frac{K(s+1)}{s\,(s-1)}$$

Sketch the root locus plot with 'K' as variable parameter and show that the loci of complex roots are part of a circle with (–1, 0) as centre and radius = $\sqrt{2}$.

Solution : Step 1 : Plot the poles and zeros

Poles . $s_1 = 0, s_2 = 1$

zeros. $s_3 = -1$

Step 2 : Centroid of asymptotes

$$\sigma_A = \frac{\text{sum of poles} - \text{sum of zeros}}{P-Z} = \frac{0+1-(-1)}{2-1} = 2$$

Step 3 : Angle of asymptotes

$$\phi = \frac{2K+1}{P-Z}180°$$

$$K = 0, \qquad \phi = \frac{2\times 0+1}{2-1}\times 180° = 180°$$

Step 4 : Breakaway point

The characteristic equation

$$1+\frac{K(s+1)}{s\,(s-1)} = 0$$

$$K = -\frac{s\,(s-1)}{s+1}$$

$$\frac{dK}{ds} = s^2 + 2s - 1 = 0$$

$$s = -2.414 \text{ and } 0.414$$

Since, both values are the part of the root locus, hence there will be a breakaway point and other is re-entry or breaking point.

As $s = 0.414$ lies between two poles $s = 0$ and $s = 1$, therefore it is breakaway point and $s = -2.414$ lies between zero and infinity therefore it is a breakin or re-entry point.

Step 5 : Point of intersection on imaginary axis.

characteristic equation

$$s^2 + (K-1)\,s + 1 = 0$$

s^2	1	1
s	$K-1$	0
s^o	1	

For stability $K - 1 > 0$

$K > 1$

Auxiliary eq^n $A(s) = s^2 + 1$

$s = \pm j1$

The root locus is shown in Fig. 5.22.

From the root locus it is clear that centre at (– 1, 0) and radius = $\sqrt{2}$

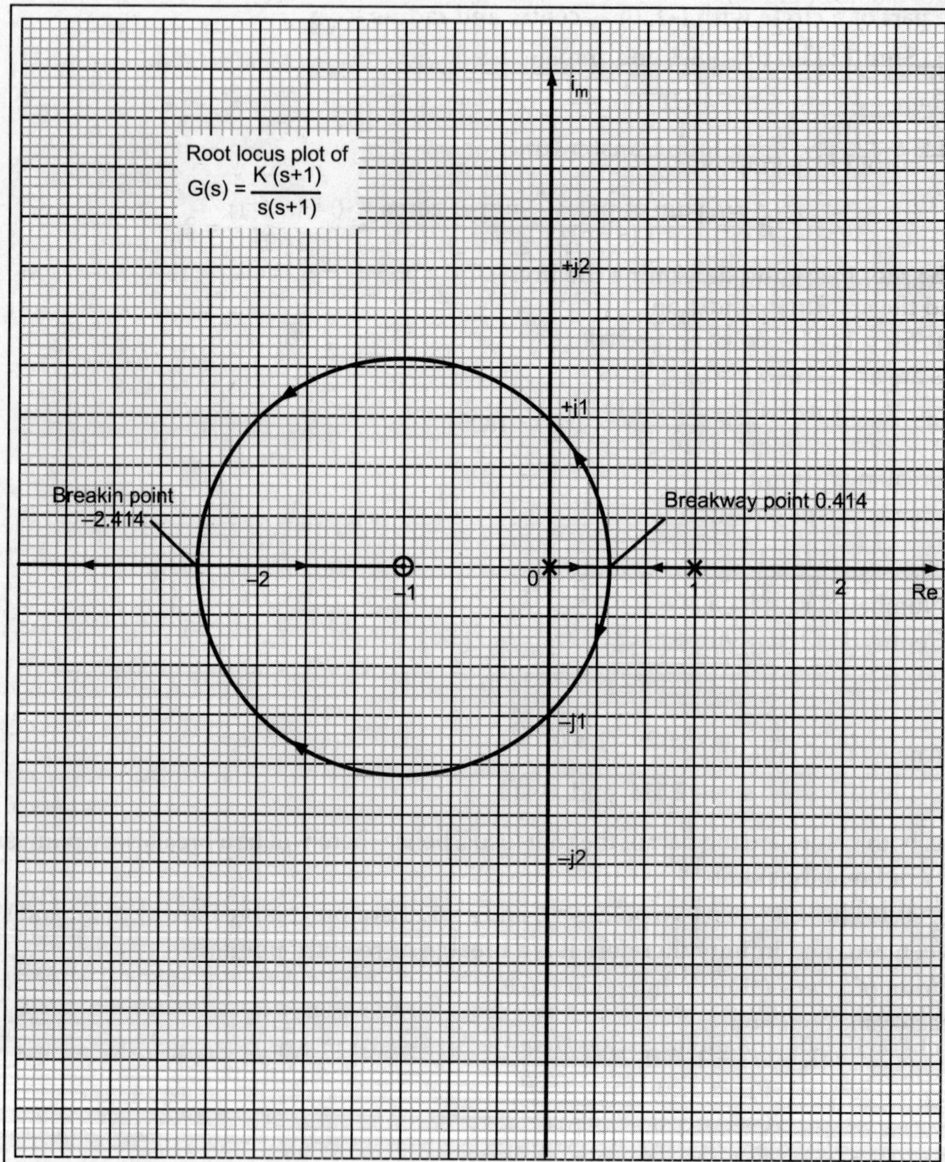
Root locus plot of
G(s) = K (s+1) / s(s+1)
i_m
+j2
+j1
–j1
–j2
Breakin point
–2.414
Breakway point 0.414
–2
–1
0
1
2
Re

Fig. 5.22.

EXAMPLE 5.31. In example 5.30. Is the system stable for all values of *K* ? If not determine the range of *K* for stable system operation. Find also the marginal value of K which causes sustained oscillations and the frequency of these oscillations.

Solution : In example 5.30 step 5 it is clear that the system is not stable for all values of *K*. The range of the *K* for stable operation is

$$1 < K < \infty$$

Marginal value of '*K*' for sustained oscillations = 1.

Frequency of sustained oscillations = 1 rad/sec.

EXAMPLE 5.32. The open loop transfer function of a control system is given by

$$G(s)\,H(s) = \frac{K}{s\,(s+6)\,(s^2+4s+13)}$$

sketch the root locus and determine the breakaway point, the angle of departure from complex poles and the stability condition. *(R.M.L. University Faizabad 2002, LST)*

Solution : Step 1 : Plot poles and zeros.

Poles are at $s_1 = 0,\ s_2 = -6$

$$s^2 + 4s + 13 = 0$$

$$s = \frac{-4 \pm \sqrt{16-52}}{2} = -2 \pm j3$$

$$s_3 = -2 + j3,\ s_4 = -2 - j3$$

Step 2 : The segment between $s = 0$ and $s = -6$ is the part of the root locus.

Step 3 : Number of root locii $N = P = 4$

Step 4 : Centroid of the asymptotes

$$\sigma_A = \frac{\text{sum of poles} - \text{sum of zeros}}{P - Z} = \frac{0-6-2+j3-2+j3-0}{4} = -2.5$$

Step 5 : Angle of asymptotes

$$\phi = \frac{2K+1}{P-Z}180°$$

eK = 0 $\quad \phi_1 = 45°$

K = 1 $\quad \phi_2 = 135°$

K = 2 $\quad \phi_3 = 225°$

K = 3 $\quad \phi_4 = 315°$

Step 6 : Break away point

The characteristic equation $1 + G(s)\,H(s) = 0$

$$1 + \frac{K}{s\,(s+6)\,(s^2+4s+13)}$$

or, $\quad s\,(s+6)\,(s^2+4s+13) + K = 0$

or, $\quad K = -(s^4 + 10s^3 + 37s^2 + 7s) = 0$

$$\frac{dK}{ds} = -(4s^3 + 30s^2 + 74s + 78) = 0$$

$$4s^3 + 30s^2 + 74s + 78 = 0$$

By trial and error method $s = -4.2$

Breakaway point = – 4.2

Step 7 : Point of intersection of root locii on imaginary axis.

characteristic equation is

$$s^4 + 10s^3 + 37s^2 + 78s + K = 0$$

Routh array

s^4	1	37	K
s^3	10	78	
s^2	29.2	K	
s^1	$78 - 0.342K$		
s^o	K		

For sustained oscillations $78 - 0.34\,K = 0 \therefore K = 229.41$

The auxiliary $eq^{n.}$ $\quad A(s) = 29.2\,s^2 + K$

$$29.2\,s^2 + 229.41 = 0$$

$$s = \pm j\,2.8$$

Step 8 : From Routh table for stability

$$K > 0$$

$$78 - 0.342\,K > 0 \text{ or } K < 229.41$$

Step 9 : The angle of departure from upper complex pole is

$$\phi_d = 180° - (124° + 90° + 37°)$$

$$\phi_d = -71°$$

similarly, the angle of departure from lower complex pole is

$$\phi_d = +71°$$

The root locus plot is shown in the Fig. 5.23.

EXAMPLE 5.33. Draw the root locus for a system whose open loop transfer function is given by

$$G(s)\,H(s) = \frac{K}{s\,(s+4)\,(s^2+4s+20)}$$

show all the salient points on the locus. *(R.M.L. University Faizabad 2002)*

Solution : Step 1. Plot poles and zeros

Poles are at $\quad s = 0, s = -4$

$$s^2 + 4s + 20 = 0$$

$$s = -2 \pm j4$$

Step 2 : The segment between s = 0 and s = – 4 is the part of the root locus.

Step 3 : Number of root locii $N = P = 4$

Step 4 : Centroid of asymptotes

$$\sigma_A = \frac{\text{sum of poles} - \text{sum of zeros}}{P - Z} = \frac{0-4-2+j4-2-j4-0}{4} = -2$$

Step 5 : Angle of asymptotes

$$\phi = \frac{2K+1}{P-Z}180°$$

$K = 0$	$\phi_1 = 45°$
$K = 1$	$\phi_2 = 135°$
$K = 2$	$\phi_3 = 225°$
$K = 3$	$\phi_4 = 315°$

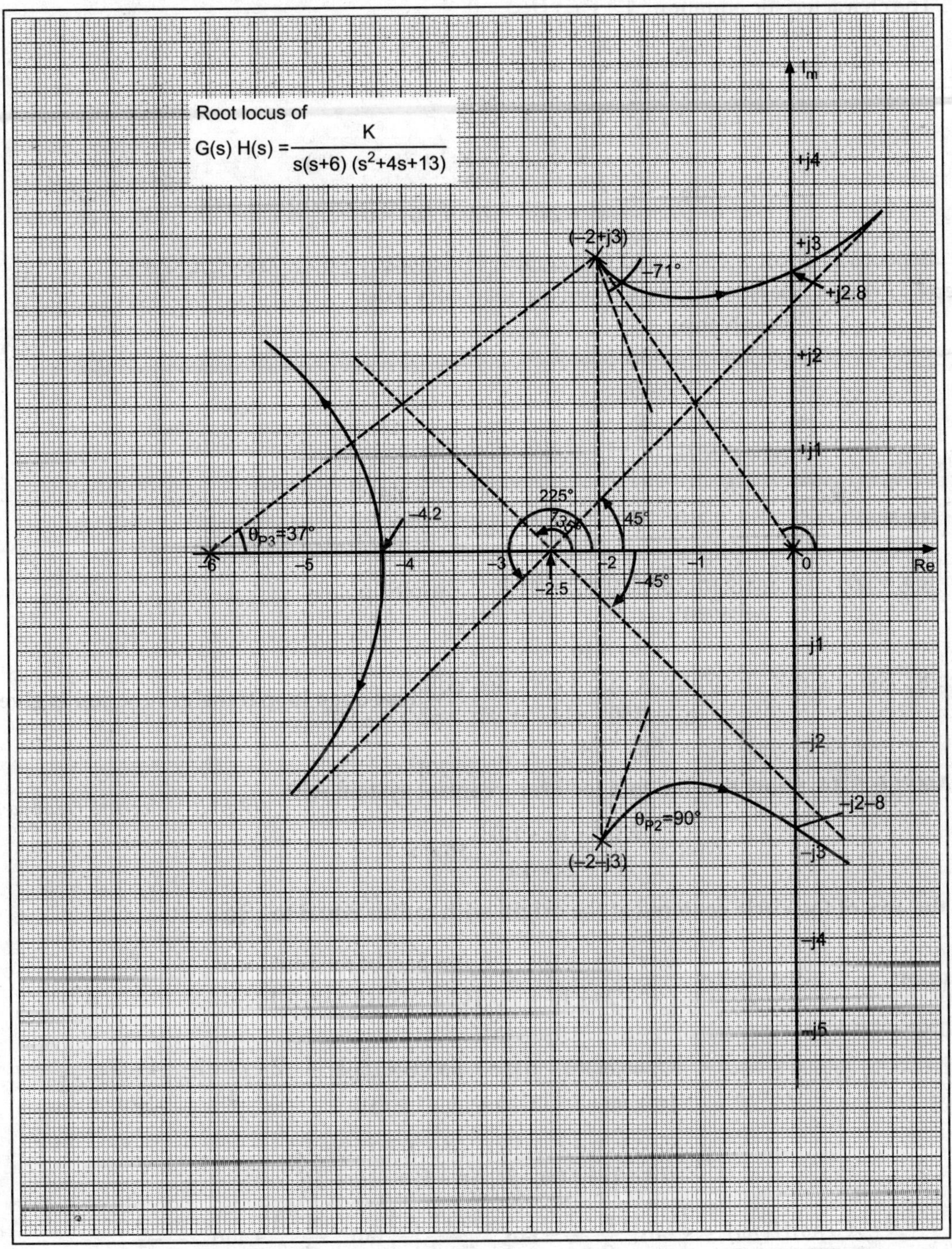
Root locus of
G(s) H(s) = K / s(s+6) (s²+4s+13)
Im
Re
+j4
+j3
+j2.8
+j2
+j1
–j1
–j2
–j2–8
–j3
–j4
–j5
(–2+j3)
(–2–j3)
–71°
225°
45°
–45°
–4.2
–2.5
θP3=37°
θP2=90°
–6
–5
–4
–3
–2
–1
0

Fig. 5.23.

Step 6 : Breakaway point

The characteristic equation . $1 + G(s)\, H(s) = 0$

$$1+\frac{K}{s\,(s+4)\,(s^2+4s+20)} = 0$$

$$s^4 + 8s^3 + 36\,s^2 + 80\,s + K = 0$$

$$K = -(s^4 + 8s^3 + 36s^2 + 80s)$$

$$\frac{dK}{ds} = -(4s^3 + 24s^2 + 72s + 80) = 0$$

Breakaway point is $s = -2$ and two complex breakaway points are $-2 \pm j\,2.45$

Step 7 : Point of intersection of root locii on imaginary axis

Routh array

s^4	1	36	K
s^3	8	80	
s^2	26	K	
s^1	$80 - 0.307K$		
s^0	K		

For stability $K > 0$

$80 - 0.307K > 0$ or $K < 260$

at $K = 260$, the auxiliary eq^n $A(s) = 26s^2 + K$

$$26s^2 + 260 = 0$$

$$s = \pm j\,3.16$$

Step 8 : The angle of departure at upper complex pole

$$\phi_d = 180° - (117° + 90° + 63°) = -90°$$

The root locus is shown in Fig. 5.24.

EXAMPLE 5.34. Show that the root locii for a control system with

$$G(s) = \frac{K(s^2+6s+10)}{s^2+2s+10},\ H(s) = 1$$

are the arcs of the circle centered at the origin with radius equal to $\sqrt{10}$.

Solution :
$$G(s)\,H(s) = \frac{K\,(s^2+6s+10)}{s^2+2s+10} = \frac{K\,(s+3-j1)\,(s+3+j1)}{(s+1-j3)\,(s+1+j3)}$$

Put $s = \sigma + j\omega$

$$G(s)H(s)\big|_{s=\sigma+j\omega} = \frac{K[s+j\omega+3-j1]\,[\sigma+j\omega+3+j1]}{[\sigma+j\omega+1-j3]\,[\sigma+j\omega+1+j3]}$$

$$= \frac{K[(\sigma+3)+j\,(\omega-1)]\,[(\sigma+3)+j\,(\omega+1)]}{[(\sigma+1)+j\,(\omega-3)]\,[(\sigma+1)+j\,(w+3)]}$$

$$\angle G(j\omega)H(j\omega) = \left[\tan^{-1}\frac{\omega-1}{\sigma+3}+\tan^{-1}\frac{\omega+1}{\sigma+3}\right]-\left[\tan^{-1}\frac{\omega-3}{\sigma+1}+\tan^{-1}\frac{\omega+3}{\sigma+1}\right]$$

To satisfy the angle criterion equate the above eq^n to 180°

$$\left[\tan^{-1}\frac{\omega-1}{\sigma+3}+\tan^{-1}\frac{\omega+1}{\sigma+3}\right]-\left[\tan^{-1}\frac{\omega-3}{\sigma+1}+\tan^{-1}\frac{\omega+3}{\sigma+1}\right] = 180$$

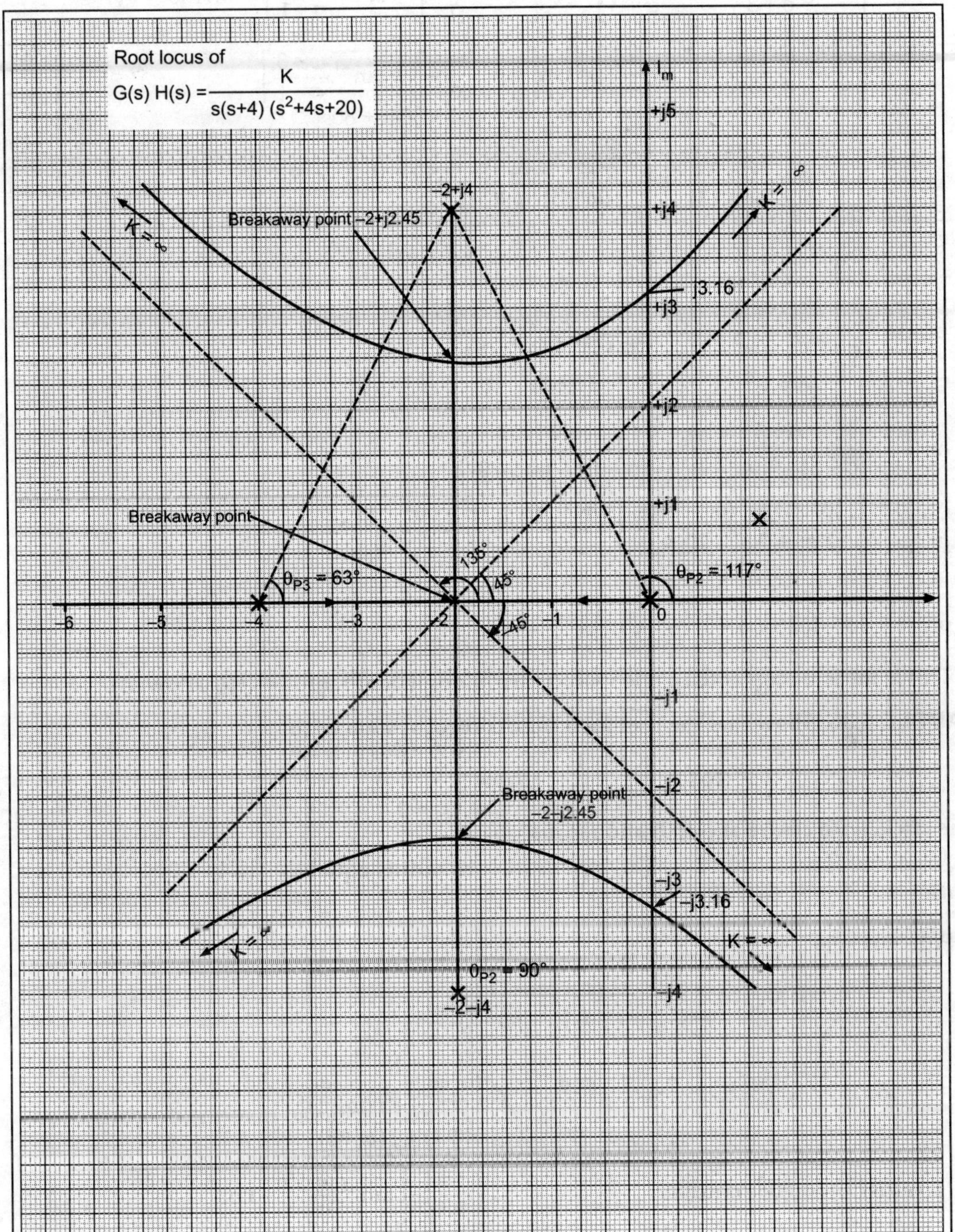
Root locus of
G(s) H(s) = K / s(s+4) (s²+4s+20)
Im
+j5
+j4
+j3
+j2
+j1
−j1
−j2
−j3
−j4
−2+j4
Breakaway point −2+j2.45
K = ∞
j3.16
Breakaway point
135°
45°
45°
θP3 = 63°
θP2 = 117°
−6
−5
−4
−3
−2
−1
0
Breakaway point
−2−j2.45
−j3.16
K = ∞
θP2 = 90°
−2−j4

Fig. 5.24.

Taking tangent of both sides

$$\tan\left[\tan^{-1}\frac{\omega-1}{\sigma+3}+\tan^{-1}\frac{\omega+1}{\sigma+3}\right]-\tan\left[\tan^{-1}\frac{\omega-3}{\sigma+1}+\tan^{-1}\frac{\omega+3}{\sigma+1}\right]=\tan 180°$$

$$\tan\left[\tan^{-1}\frac{\omega-1}{\sigma+3}+\tan^{-1}\frac{\omega+1}{\sigma+3}\right]=\tan\left[\tan^{-1}\frac{\omega-3}{\sigma+1}+\tan^{-1}\frac{\omega+3}{\sigma+1}\right]$$

or, $$\frac{\frac{\omega-1}{\sigma+3}+\frac{\omega+1}{\sigma+3}}{1-\frac{(\omega-1)(\omega+1)}{(\sigma+3)^2}}=\frac{\frac{\omega-3}{\sigma+1}+\frac{\omega+3}{\sigma+1}}{1-\frac{(\omega-3)(\omega+3)}{(\sigma+1)^2}}$$

or, $$\frac{\frac{2\omega}{\sigma+3}}{\frac{(\sigma+3)^2-(\omega^2-1)}{(\sigma+3)^2}}=\frac{\frac{2\omega}{\sigma+1}}{\frac{(\sigma+1)^2-(\omega^2-9)}{(\sigma+1)^2}}$$

or, $$\frac{\sigma+3}{(\sigma+3)^2-(\omega^2-1)}=\frac{\sigma+1}{(\sigma+1)^2-(\omega^2-9)}$$

or, $(\sigma + 1)\,[(\sigma + 3)^2 - (\omega^2 - 1)] = (\sigma + 3)\,[(\sigma + 1)^2 - (\omega^2 - 9)]$

or, $2\sigma^2 + 2\omega^2 - 20 = 0$

or, $\sigma^2 + \omega^2 = 10$

or, $(\sigma - 0)2 + (\omega - 0)^2 = \left(\sqrt{10}\right)^2$

This is the equation of the circle with centre at origin and radius = $\sqrt{10}$ **(Proved).**

EXAMPLE 5.35. Show that the root locii for a control system with

$$G(s)=\frac{K(s+1)}{s(s-1)},\ H(s)=1$$

is the circle with centre at (–1, 0) and radius $\sqrt{2}$

Solution : $$G(s)H(s)=\frac{K(s+1)}{s(s-1)}$$

Put $s = \sigma + j\omega$

$$G(s)H(s)\big|_{s=\sigma+j\omega}=\frac{K(\sigma+j\omega+1)}{(\sigma+j\omega)(\sigma+j\omega-1)}$$

$$\angle G(s)H(s)\big|_{s=\sigma+j\omega}=\angle\sigma+j\omega+1-\angle\sigma+j\omega-\angle\sigma+j\omega-1$$

$$=\tan^{-1}\frac{\omega}{\sigma+1}-\tan^{-1}\frac{\omega}{\sigma}-\tan^{-1}\frac{\omega}{\sigma-1}$$

To satisfy the angle criterion

$$\tan^{-1}\frac{\omega}{\sigma+1}-\tan^{-1}\frac{\omega}{\sigma}-\tan^{-1}\frac{\omega}{\sigma-1}=180°$$

Taking tangent on both sides

$$\tan\left[\tan^{-1}\frac{\omega}{\sigma+1} - \tan^{-1}\frac{\omega}{\sigma-1}\right] - \tan\left[\tan^{-1}\frac{\omega}{\sigma}\right] = \tan 180°$$

or, $$\frac{\dfrac{\omega}{\sigma+1} - \dfrac{\omega}{\sigma-1}}{1+\dfrac{\omega}{\sigma+1}\cdot\dfrac{\omega}{\sigma-1}} = \frac{\omega}{\sigma}$$

or, $$\frac{-2}{\sigma^2 - 1 + \omega^2} = \frac{1}{\sigma}$$

or, $$-2\sigma = \sigma^2 - 1 + \omega^2$$

or, $$\omega^2 + \sigma^2 - 1 + 2\sigma = 0$$

or, $$\omega^2 + \sigma^2 + 2\sigma - 1 + 2 - 2 = 0$$

or, $$\omega^2 + (\sigma + 1)^2 = 2$$

or, $$(\sigma + 1)^2 + (\omega)^2 = \left(\sqrt{2}\right)^2$$

$$(\sigma + 1)^2 + (\omega - 0)^2 = \left(\sqrt{2}\right)^2$$

This is the equation of the circle with centre at (–1, 0) and radius $\sqrt{2}$ **(Proved).**

EXAMPLE 5.36. The open loop transfer function of a system is given by

$$G(s)\,H(s) = \frac{K(s+12)}{s^2(s+20)}$$

sketch the root locus for the system. *(R.M.L, University Faizabad 2001, Control System)*

Solution : Step 1 : Plot the poles and zero

Poles are at $s_1 = 0$, $s_2 = 0$, $s_3 = -20$

zero is at $s_4 = -12$

Step 2 : The segment between $s = -20$ and $s = -12$ is the part of the root locus.

Step 3 : Centroid of asymptotes

$$\sigma_A = \frac{\text{sum of poles} - \text{sum of zeros}}{P - Z} = \frac{0+0-20+12}{3-1} = -4$$

Step 4 : Angle of asymptotes

$$\phi = \frac{2K+1}{P-Z}180°$$

$K = 0$ $\qquad \phi_1 = 90°$

$K = 1$ $\qquad \phi_2 = 270°$

Step 5 : Breakaway point the characteristic eqn

$$1 + \frac{K(s+12)}{s^2(s+20)} = 0$$

or, $$K = -\frac{(s^3 + 20s^2)}{s+12}$$

$$\frac{dK}{ds} = -\left[\frac{(s+12)(3s^2+40s) - (s^3+20s^2)}{(s+12)^2}\right] = 0$$

or, $s^3 + 28s^2 + 240\,s = 0$

$s\,(s^2 + 28s + 240) = 0$

we get $s = 0, -14 \pm j\,6.63$

Breakaway point $s = 0$, points $-14 \pm j\,6.63$ are neither breakaway point nor breakin point, because the corresponding gain values K becomes complex quantities.

Step 6 : Point of intersection of rod locii with imaginary axis.

The characteristic $eq^{n.}$

$$s^3 + 20\,s^2 + Ks + 12\,K = 0$$

Put $s = j\omega$

$$(j\omega)^3 + 20\,(j\omega)^2 + K\,(j\omega) + 12\,K = 0$$

$$(12\,K - 20\omega^2) + j\omega\,(K - \omega^2) = 0$$

if $\omega = 0,\ K = 0$

because of double pole at the origin, the root locus is tangent to the imaginary axis at $\omega = 0$.

The root locus is shown in Fig. 5.25.

EXAMPLE 5.37. Sketch the root locus for

$$G(s) = \frac{K}{s\,(s^2 + 6s + 12)},\ H(s) = 1$$

Solution : Step 1 : Draw the pole zero plot

poles at $s = 0$

$$s^2 + 6s + 12 = 0$$

$$s = -3 \pm j\,1.73$$

Step 2 : The left portion of the negative real axis from origin is the part of the root locus.

Step 3 : Centroid of asymptotes

$$\sigma_A = \frac{\text{sum of poles} - \text{sum of zeros}}{P - Z} = \frac{0 - 3 + j1.73 - 3 - j1.73}{3} = -2$$

Step 4 : Angle of asymptotes

$$\phi = \frac{2K+1}{P-Z}180°$$

$K = 0$ $\phi_1 = 60°$

$K = 1$ $\phi_2 = 180°$

$K = 2$ $\phi_3 = 300°$

Step 5 : Breakaway point

The characteristic equation

$$1 + \frac{K}{s\,(s^2 + 6s + 12)} = 0$$

$$K = -(s^3 + 6s^2 + 12)$$

$$\frac{dK}{ds} = -(3s^2 + 12\,s + 12) = 0$$

$$3s^2 + 12\,s + 12 = 0$$

$s = -2$ is the breakaway point

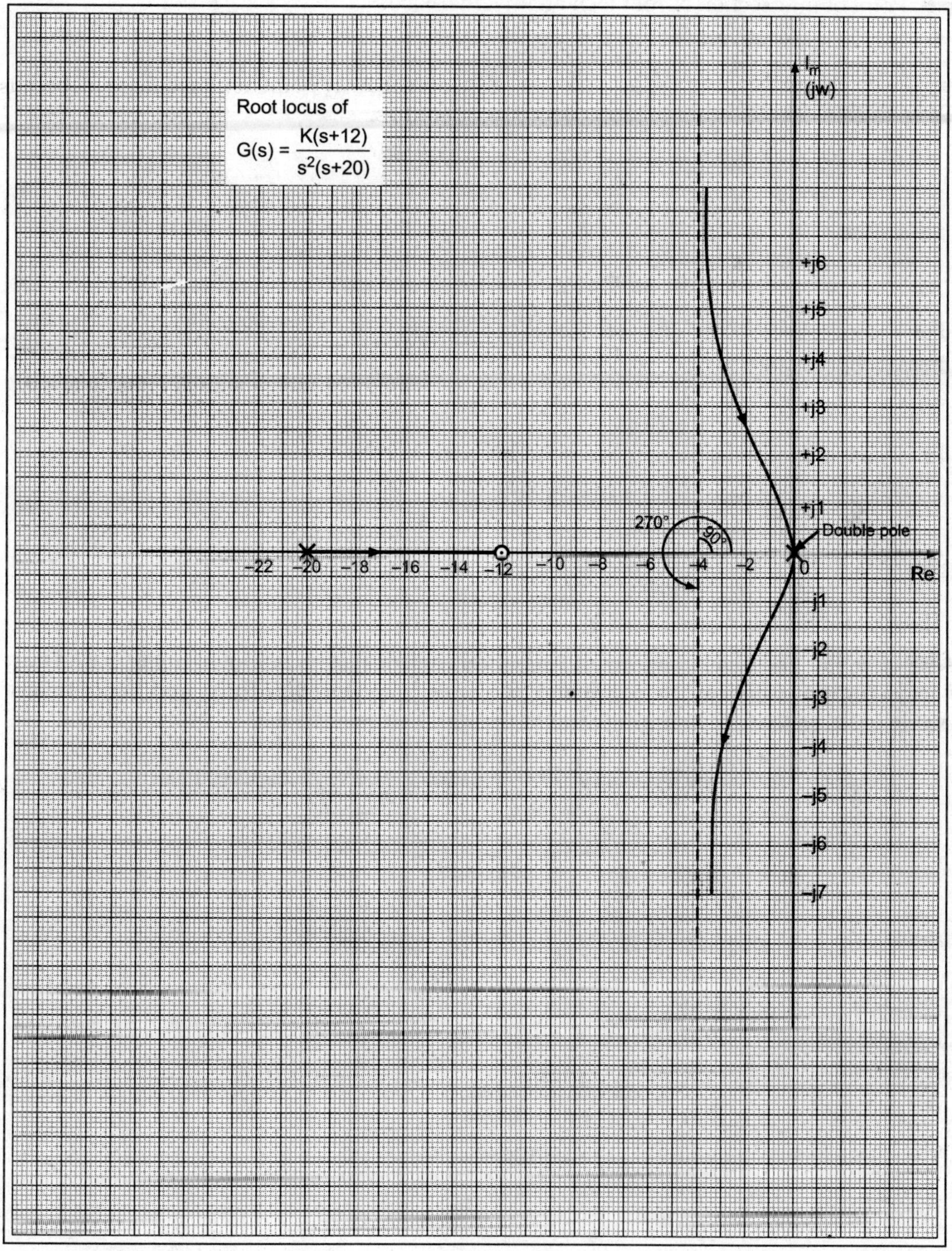
Root locus of
$G(s) = \frac{K(s+12)}{s^2(s+20)}$
Im (jw)
+j6
+j5
+j4
+j3
+j2
+j1
270°
90°
Double pole
−22
−20
−18
−16
−14
−12
−10
−8
−6
−4
−2
0
Re
−j1
−j2
−j3
−j4
−j5
−j6
−j7

Fig. 5.25.

Step 6 : Point of intersection of root locii on imaginary axis

Routh array

$$\begin{array}{lcc} s^3 & 1 & 12 \\ s^2 & 6 & K \\ s^1 & \dfrac{72-K}{6} & \\ s^o & K & \end{array}$$

for stability $K > 0$

$$\frac{72-K}{6} > 0 \quad \text{or } K < 72$$

For $K = 72$

the auxiliary equation $A(s) = 6s^2 + K$

$$6s^2 + 72 = 0$$

$$s = \pm j\,3.46$$

Step 7 : Angle of departure from upper complex pole

$$\phi_d = 180° - (150° + 90°)$$

$$\phi_d = -60°$$

similarly from lower complex pole

$$\phi_d = +60°$$

The root locus is shown in Fig. 5.26.

EXAMPLE 5.38. The block diagram of a position control system is shown in Fig. 5.27. Draw the root locus for the system as the parameter 'α' varies from zero to infinity and hence determine the value of 'α' so that the damping ratio of the dominat poles is 0.6.

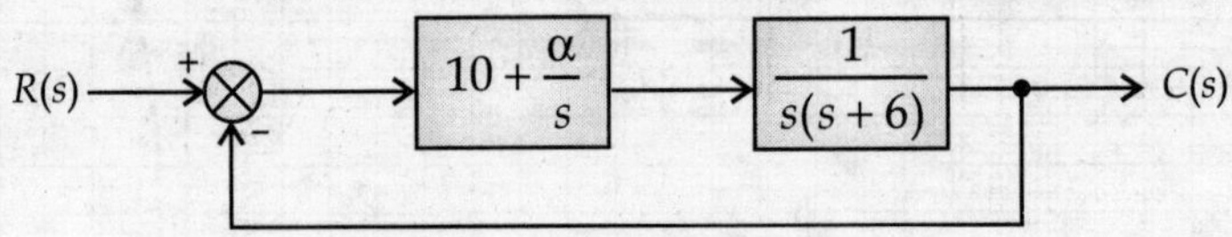

Fig. 5.27.

Solution :
$$G(s) = \frac{10s+\alpha}{s^2\,(s+6)}, \; H(s) = 1.$$

$$\therefore \quad \frac{C(s)}{R(s)} = \frac{G(s)}{1+G(s)\,H(s)}$$

$$\frac{C(s)}{R(s)} = \frac{10s+\alpha}{s^2\,(s+6)+10s+\alpha}$$

The characteristic equation. $s^2\,(s + 6) + 10s + \alpha = 0$

$$\text{or,} \quad 1+\frac{\alpha}{s^2\,(s+6)+10s} = 0$$

let
$$G'(s) = \frac{\alpha}{s^2\,(s+6)+10s} = \frac{\alpha}{s\,(s^2+6s+10)}$$

Step 1 : Poles are at $s_1 = 0, s_2 = -3 + j1, s_3 = -3 - j1$

Step 2 : The entire left half of real axis is the part of root locus.

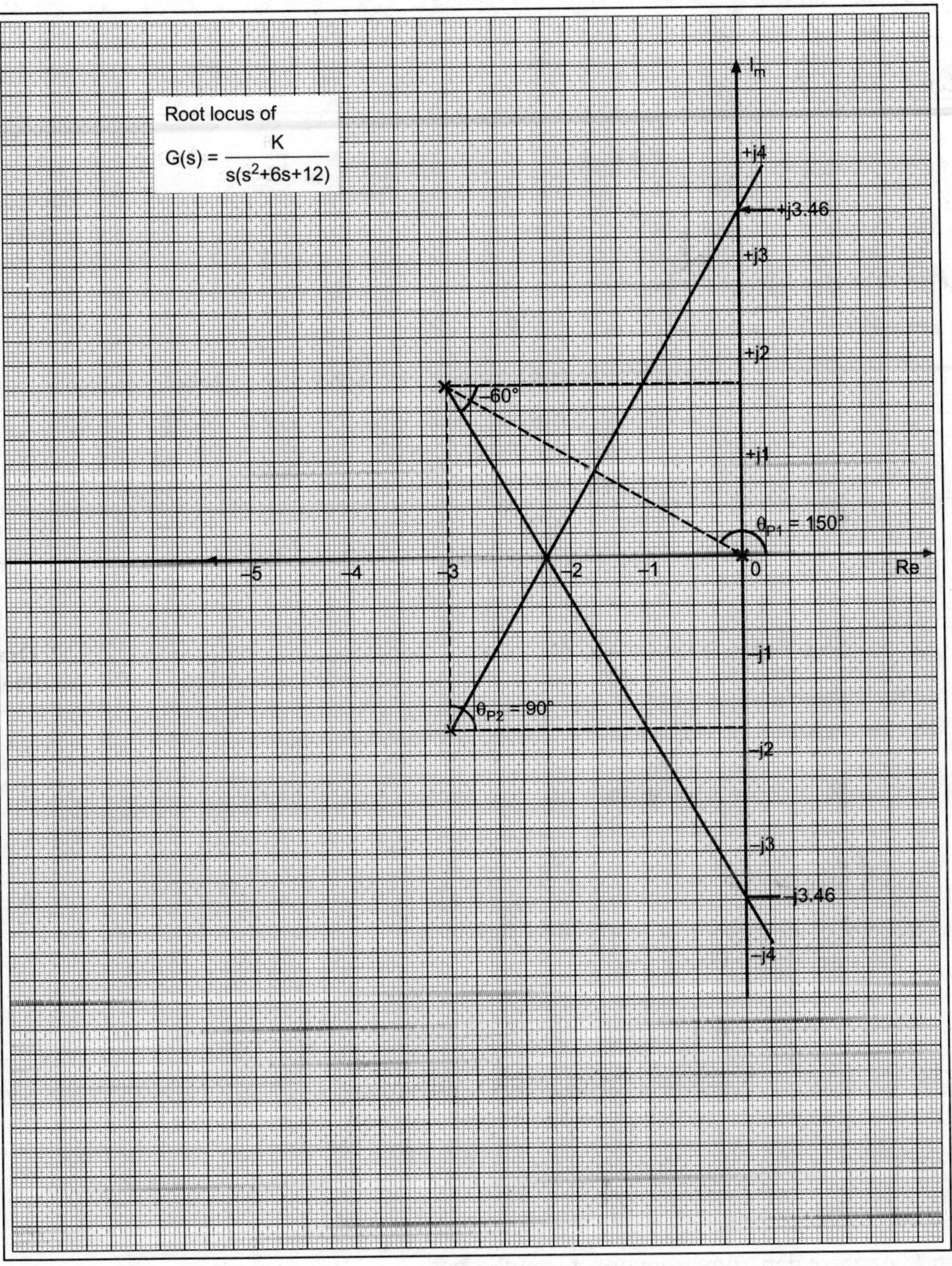
Root locus of
G(s) = K / s(s²+6s+12)
Im
+j4
+j3.46
+j3
+j2
+j1
–j1
–j2
–j3
–j3.46
–j4
–60°
θP1 = 150°
θP2 = 90°
–5
–4
–3
–2
–1
0
Re

Fig. 5.26.

Step 3 : The centroid of the asymptotes

$$\sigma_A = \frac{\text{sum of poles} - \text{sum of zeros}}{P-Z} = \frac{0-3+j1-3-j1}{3} = -2$$

Step 4 : Angle of asymptotes

$$\phi = \frac{2K+1}{P-Z}180^\circ$$

$K = 0 \qquad \phi_1 = 60^\circ$

$K = 1 \qquad \phi_2 = 180^\circ$

$K = 2 \qquad \phi_3 = 300^\circ$

Step 5 : The characteristic equation

$$\alpha = -(s^3 + 6s^2 + 10s)$$

$$\frac{d\alpha}{ds} = -3s^2 - 12s - 10$$

$$s = -1.18, -2.816$$

$$\frac{d^2\alpha}{ds^2} = -6s - 12$$

Put $s = -1.18$ $\qquad \dfrac{d^2\alpha}{ds^2} = +7.08 - 12 = -4.92$

Since, $\dfrac{d^2\alpha}{ds^2}$ is negative for s = – 1.18

∴ Breakaway point = – 1.18

Put $s = -2.816$

$$\frac{d^2\alpha}{ds^2} = +4.896$$

Since $\dfrac{d^2\alpha}{ds^2}$ is positive for $s = -2.816$, therefore $s = -2.816$ is the re-entry point.

Step 6 : Point of intersection of root locii on $j\omega$ axis

$$s^3 + 6s^2 + 10s + \alpha = 0$$

s^3	1	10
s^2	6	α
s^1	$\frac{60-K}{6}$	
s^o	α	

At $K = 60$, auxiliary equation $A(s) = 6s^2 + \alpha$

$$6s^2 + 60 = 0$$

∴ $\qquad s = \pm j\,3.16$

Step 7 : Angle of departure from upper complex pole.

$$\theta_{P1} = 162^\circ$$

$$\theta_{P2} = 90^\circ$$

$$\phi_d = 180^\circ - (\theta_{P1} + \theta_{P2}) = 180^\circ - (162^\circ + 90^\circ) = -72^\circ$$

The root locus plot is shown in the Fig. 5.28.

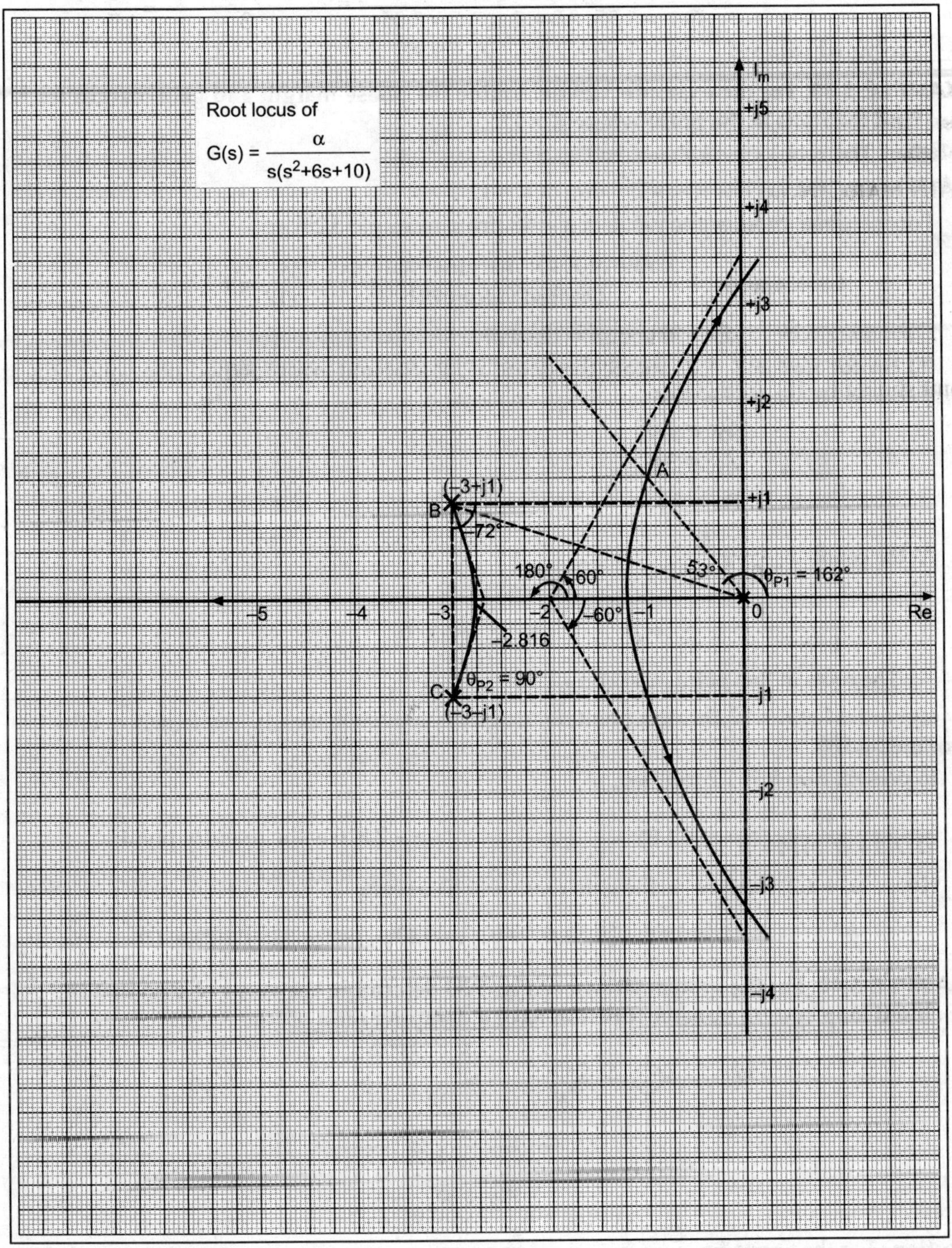

Root locus of
G(s) = $\frac{\alpha}{s(s^2+6s+10)}$
Im
+j5
+j4
+j3
+j2
+j1
−j1
−j2
−j3
−j4
A
B
C
(−3+j1)
(−3−j1)
72°
180°
60°
−60°
53°
θP1 = 162°
θP2 = 90°
−2.816
−5
−4
−3
−2
−1
0
Re

Fig. 5.28.

Step 8 : Draw ζ-line at $\cos^{-1} 0.6 = 53.1°$ from negative real axis. This line intersects the root locus at point *A*. Measure the distance of point *B* and point C from point A.

$$OA = 1.55, AB = 2.1, AC = 3.05$$

$$\therefore \text{ required value of } \quad \alpha = 1.55 \times 2.1 \times 3.05 = 9.927$$

EXAMPLE 5.39. For the example 5.29 determine the phase margin and gain margin for *K* = 12. Also find the roots at the point where *K* = 12.

Solution. The root locus is shown in Fig. 5.29.

GAIN MARGIN. (*a*) The desired value of $K = 12$

For marginal stability the value of $K = 48$ (from example 5.29)

$$\text{Gain margin} = \frac{\text{Value of K for marginal stability}}{\text{Desired value of K}} = \frac{48}{12} = 4$$

$$\text{Gain margin in} \quad db = 20 \log_{10} 4 = 12.04 \; db.$$

(*b*) For the root at $K = 12$, locate the point '*x*' by trial and error method (the value of *K* at this point should be 12). In this case

$$Ox = 1.55$$
$$Ax = 3.8$$
$$Bx = 2.05$$

$$\text{At the point } x: \frac{K}{|0x||Ax||Bx|} = 1$$

$$\frac{K}{1.55 \times 3.8 \times 2.05} = 1 \quad \therefore \; K = 12.07$$

$$\text{The roots at this point} = -0.5 \pm j1.5$$

(*c*) For *P.M* at $K = 12$, locate a point on imaginary axis such that $|G(j\omega) H(j\omega)| = 1$ (by trial and error method)

In present case it is 1.2 rad/sec.

$$\frac{12}{j1.2(2+j1.2)(4+j1.2)} = 1.02$$

Hence $\omega_1 = 1.2$ (rad/sec) is the gain crossover frequency

$$P.M. = 180° - \left(\theta_{p_1} + \theta_{p_2} + \theta_{p_3}\right) = 180° - (90° + 32° + 17°) = 41°$$

EXAMPLE 5.40. For the system shown in Fig. 5.30 determine the value of α so that the poles of the closed loop system transfer function have a damping ratio of 0.6.

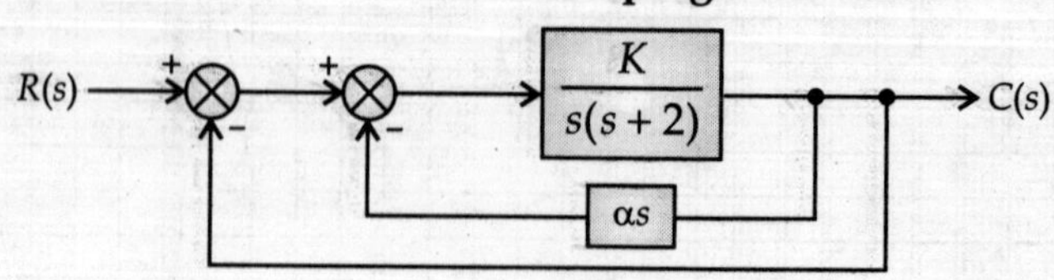

Fig. 5.30.

Solution : For the above fig. $\dfrac{C(s)}{R(s)} = \dfrac{K}{s^2 + 2s + 25\alpha s + 25}$

Characteristic equation.

$$s^2 + 2s + 25\alpha s + 25 = 0$$

$$\text{or,} \quad s^2 + 2s + 25 + 25\,\alpha s = 0$$

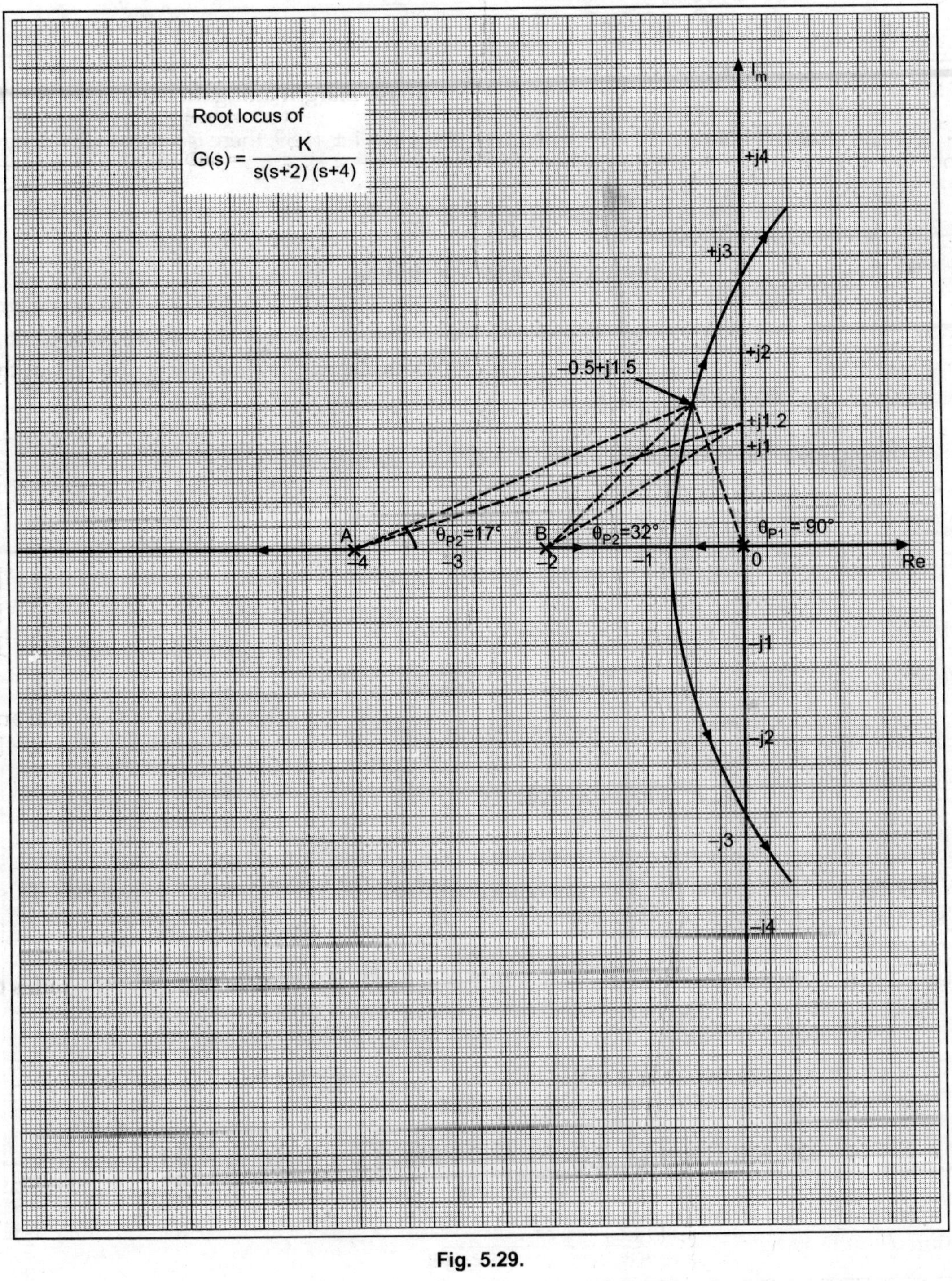
Root locus of
$G(s) = \frac{K}{s(s+2)(s+4)}$
Im
+j4
+j3
+j2
−0.5+j1.5
+j1.2
+j1
A
B
θ_{P2}=17°
θ_{P2}=32°
θ_{P1} = 90°
−4
−3
−2
−1
0
Re
−j1
−j2
−j3
−j4

Fig. 5.29.

or, $$1+\frac{25\alpha s}{s^2+2s+25}=0$$

Put $$K = 25\,\alpha$$

$$1+\frac{Ks}{s^2+2s+25}=0$$

The given system has a zero at s= 0 and two poles at – 1 ± *j*4.89, there is a possibility of circular root locus. Since, the angle condition is

$$\angle\frac{Ks}{s^2+2s+25}=\pm 180°\,(2K+1).$$

We have $\angle s-\angle s+1+j4.89-\angle s+1-j4.89=\pm 180°(2K+1)$

Put $s = \sigma + j\omega$

$$\angle \sigma+1+j(4.89+\omega)+\angle \sigma+1+j(\omega-4.89)=\angle \sigma+j\omega+180°$$

$$\tan^{-1}\frac{\omega+4.89}{\sigma+1}+\tan^{-1}\frac{\omega-4.89}{\sigma+1}=\tan^{-1}\frac{\omega}{\sigma}$$

Taking tangent on both sides we get

$$\frac{\dfrac{\omega+4.89}{\sigma+1}+\dfrac{\omega-4.89}{6+1}}{1-\left(\dfrac{\omega+4.89}{\sigma+1}\right)\left(\dfrac{\omega-4.89}{\sigma+1}\right)}=\frac{\omega}{\sigma}$$

or $$\frac{\dfrac{2\omega}{\sigma+1}}{1-\dfrac{(\omega^2-23.9121)}{(\sigma+1)^2}}=\frac{\omega}{\sigma}$$

or, $$2\sigma(\sigma+1)=(\sigma+1)^2-\omega^2+23.9121$$

or, $$\sigma^2+\omega^2=24.9121$$

or, $$\sigma^2+\omega^2=25$$

This is the equation of the circle* with center at origin and radius is 5.

We require $\zeta = 0.6$, draw. a line at an angle $\cos^{-1}(0.6) = 53.13°$ with negative real axis. This line intersects the root locus at '*x*'. The value of *K* at '*x*' is

$$K=\frac{D_1D_3}{D_2}=\frac{2.2\times 9.1}{5}=4.004$$

Since, $$25\alpha = K$$

$$25\alpha = 4.00 \quad \alpha = 0.16016$$

The root locus is shown in Fig. 5.31.

The root at the point of intersection is $-3 + j\,3.95$.

5.12. EFFECT OF ADDITION OF POLES AND ZEROS ON ROOT LOCUS

Consider $$G(s)\,H(s)=\frac{K}{s(s+4)}$$

---

* For $K > 0$, the portion of the circle lies to the left of the complex poles. For $K < 0$ the portion of the circle lies to the right of the complex poles. Hence, for present case, it is not the root locus, where $K > 0$.

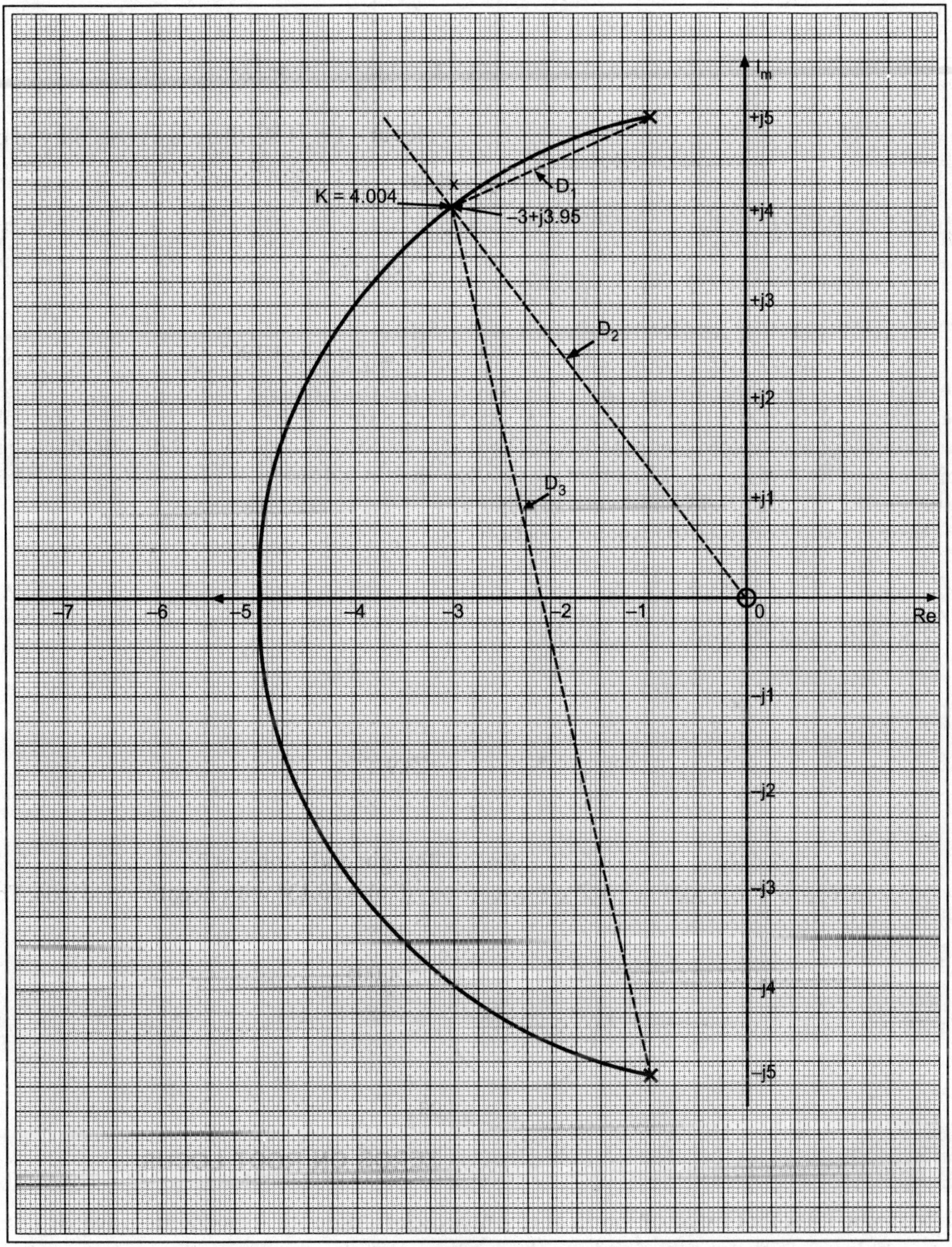

Fig. 5.31.

The root locus is shown in Fig. 5.32(a)

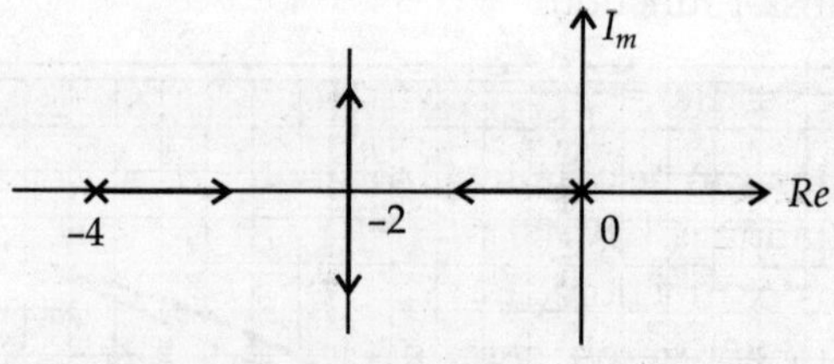

Fig. 5.32.(a)

Now add one pole at $s = -5$, then

$$G(s)\,H(s) = \frac{K}{s(s+4)(s+5)}$$

the corresponding root locus is shown in Fig. 5.15.

So, before addition of pole for any value of K the system is stable but after addition of pole to the left half, the two branches of root locus moves to the right half for some value of K. Therefore, the system will be stable for this value of K, after this value of K the system becomes unstable. So the stability of the system gets restricted. For further addition of poles to left half, the breaking point moves towards right. Therefore we can say that by addition of poles to left half, the root locus shifted towards right hand side and stability of the system decreases.

Now consider open loop transfer function

$$G(s)\,H(s) = \frac{K}{s(s+2)}$$

The root locus is shown in Fig. 5.32(b)

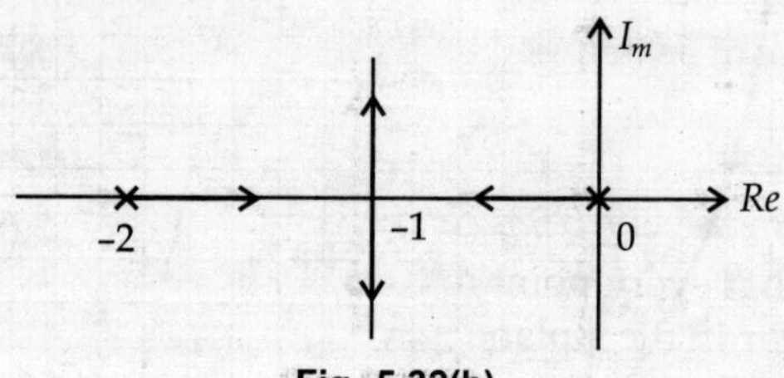

Fig. 5.32(b)

Now add zero to left half at $s = -3$, the root locus is shown in Fig. 5.32(c)

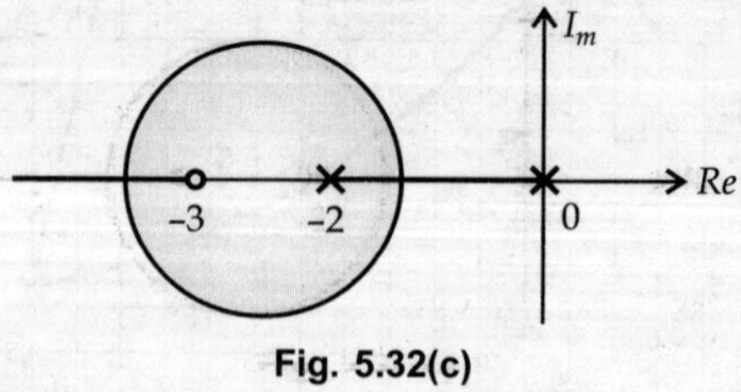

Fig. 5.32(c)

$$G(s)\,H(s) = \frac{K(s+3)}{s(s+2)}$$

It can be seen that by addition of zero towards left, the root locus shifts towards left half. Since root locus shifts towards left half, the relative stability increases.

In conclusion we can say that by addition of poles, the root locus shifts towards imaginary axis and system stability decreases, while by addition of zeros towards left half, the root locus moves away from the imaginary axis and system stability increases.

5.13. INTRODUCTION

Consider the closed loop transfer function

$$\frac{C(s)}{R(s)} = T(s) = \frac{G(s)}{1+G(s)H(s)}$$

The characteristic equation can be obtained by equating the denominator polynomial to zero.

∴ The characteristic equation is $1 + G(s)\,H(s) = 0$

$G(s)\,H(s)$ is the loop transfer function, $L(s)$

where, Loop transfer function zeros : zeros of $L(s)$

Loop transfer function poles : Poles of $L(s)$

Closed loop transfer function poles : zeros of $1 + G(s)\,H(s)$

Zeros of $1 + G(s)\,H(s)$: Roots of the characteristic equation.

For the stability of the system we have already studied the Routh-Hurwitz and root locus technique. For stability of the system, the roots of the characteristic equation should not lie in the right half of the s-plane or on the imaginary axis. In other words, we can say that a system is said to be closed loop stable if the poles of the close loop transfer function or zeros of $1 + G(s)\,H(s)$ are all in the left half of s-plane. Similarly, for open loop stability, a system is said to be open loop stable if the poles of the loop transfer function are all in the left half of s-plane.

Now we are going to study the Nyquist stability criterion. The Nyquist stability criterion is based on the principle of argument. The principle of argument is related with the theory of mapping. So first we will study the mapping technique and principle of argument then Nyquist stability criterion.

5.14. MAPPING

Consider a function $\quad D(s) = s^2 + 1$

Any point in the s-plane can be mapped by locating the values of u and v for the given value of s.

let $s_1 = 2 + j\,4$

$$D(s_1) = D\,(2 + j4) = (2 + j4)^2 + 1 = -11 + j16$$

The mapping is shown in the Fig. 5.33.

Therefore, we can say that every point in the s-plane maps into one and only one point in the $D(s)$ plane. Any close contour in the s-plane maps into the closed contour in the $D(s)$ plane. These are the properties of the mapping.

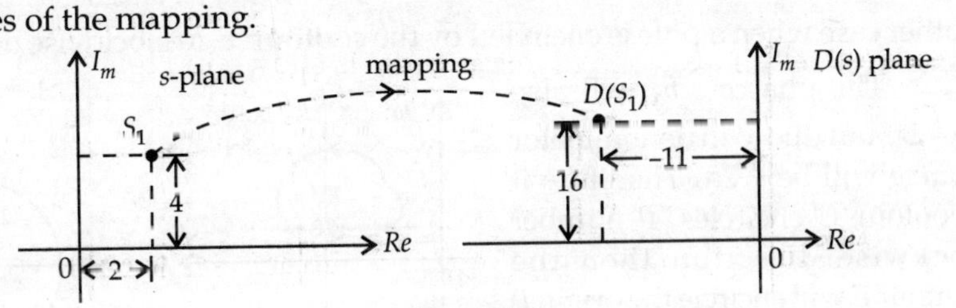

Fig. 5.33.

5.15. MAPPING OF CLOSE CONTOUR AND PRINCIPLE OF ARGUMENT

Consider the characteristic equation

$$D(s) = 1 + G(s)\,H(s) = \frac{K(s-a_1)\,(s-a_2)....}{(s-b_1)\,(s-b_2)....} \quad ...(5.20)$$

where a_1, a_2 ..... are zeros and b_1, b_2, ...are poles. let 's' be a complex variable $s = 6 + j\omega$, then, $D(s)$ will also be complex

$$D(s) = U(s) + jV(s) \qquad \text{...(5.21)}$$

Now choose a closed path 'C' arbitrarily. The contour 'C' in s-plane is mapped onto the $D(s)$ plane as contour Γ.

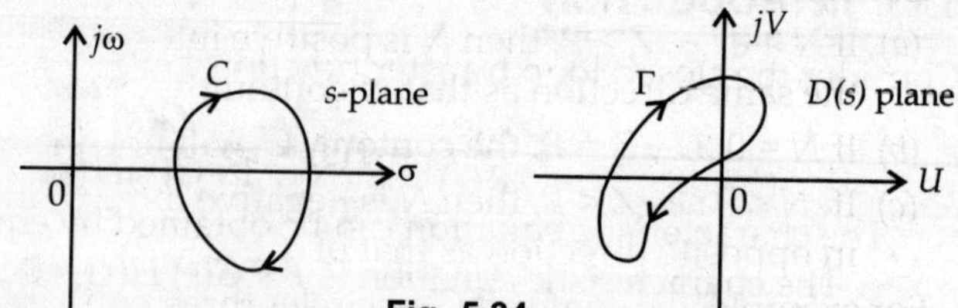

Fig. 5.34.

Now, consider a contour 'C' in s-plane. This contour encircles neither zeros nor poles, then the contour 'Γ' in $D(s)$ plane cannot encircle the origin or the point at infinity. The inside region of the contour 'C' does not have any pole or zero. If we trace the contour 'C' in clockwise direction then the inside region will be on right of the contour. The corresponding region in $D(s)$ plane which contains neither origin nor the point at infinity should also be to the right of the contour Γ.

If the contour 'C' encircles a zero but not poles then the point $D_1(s)$ is given by

$$|D_1(s)| = \frac{|s-a_1||s-a_2|...}{|s-b_1||s-b_2|...}$$

$$\angle D_1(s) = \angle s-a_1 + \angle s-a_1 .... - \angle s-b_1 - \angle s-b_2 \quad i.e. \quad D_1(s) = r\angle\phi$$

It means that the tip of $D_1(s)$ forms a close contour about the origin in clockwise direction*. Similarly, if the contour 'C' encircles 'z' zeros in clockwise direction then the contour in $D(s)$ plane encircles the origin of $D(s)$ plane 'z' times in clockwise direction. In Fig. 5.36 (*a*) the contour 'C' encircles two zeros, the corresponding contour in $D(s)$ plane is shown in Fig. 5.36 (*b*).

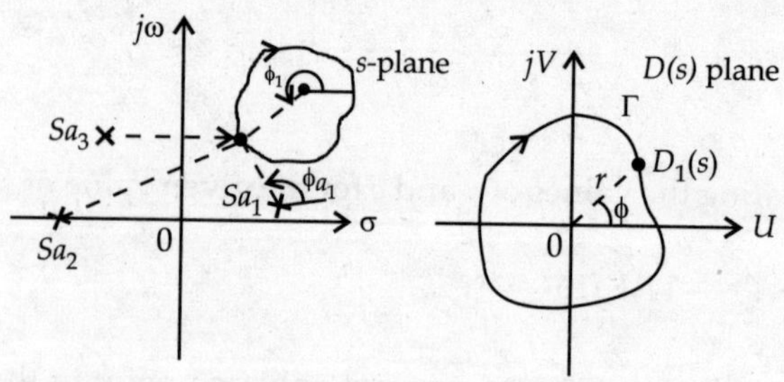

Fig. 5.35.

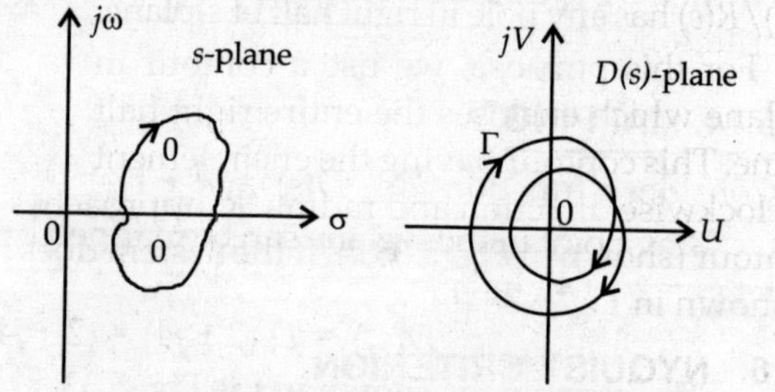

Two zeros are encircled by contour 'C'

Origin encircles two times by contout Γ

Fig. 5.36.

Consider another case when a pole is encircled by the contour 'c' in clockwise direction then the angle will be $\angle{-\phi}$. The phasor s-b_1 will also generate an angle -2π* but due to in denominator the net phase change will be $+ 2\pi$. Therefore it means that if the contour 'C' encircles. 'P' number of poles in clockwise direction then the corresponding contour Γ will encircle the origin P times in counter clockwise direction.

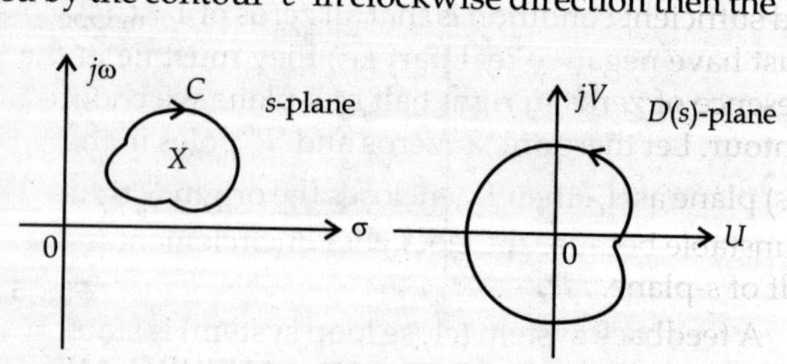

Fig. 5.37.

Now, if contour 'C' encircles both zeros and poles in clockwise direction then the corresponding contour 'Γ' encircles the origin of $D(s)$ plane Z–P = N times in clockwise direction.

This relation between enclosuer of poles and zeros by the contour in s-plane and encirclement of origin by the contour 'Γ' in $D(s)$ plane is known as principle of argument.

If any point traces the curve 'C' in clockwise direction and return to the starting point then that phasor will generate an angle -2π.

Conclusion : $N = Z - P$

(*a*) If $N > 0$ *i.e.*, $Z > P$, then N is positive integer. In this case Γ will encircle the origin N times in the same direction as that of contour 'C'.

(*b*) If $N = 0$, *i.e.*, $Z = P$, the contour 'C' will not encircle the origin.

(*c*) If $N < 0$ *i.e.*, $Z < P$, then N is negative integer, in this case Γ will encircle the origin N times in opposite direction as that of 'C'.

For example, the contour 'C' encircles three poles and one zero in clockwise then the number of encirclement $N = 1-3 = -2$. Then the contour 'Γ' will encircle the origin two times in counterclockwise direction.

5.16. NYQUIST PATH OR NYQUIST CONTOUR

The overall transfer of a system is given by

$$\frac{C(s)}{R(s)} = \frac{G(s)}{1+G(s)H(s)} \quad ...(5.22)$$

The characteristic equation is

$$1 + G(s)\,H(s) = 0 \quad ...(5.23)$$

The main purpose in study the stability of the closed loop system is to determine whether the characteristic equation has any root in the right half of *s*-plane *i.e.*, whether $C(s)/R(s)$ has any pole in right half of *s*-plane.

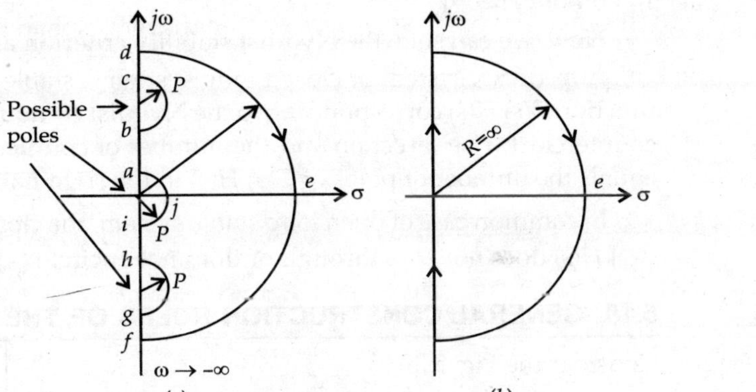

(*a*) Fig. 5.38. (*b*)

For this purpose, we use a contour in *s*-plane which encloses the entire right half plane. This contour having the encirclement in clockwise direction and radius 'R' approaches infinity. This path or contour is known as Nyquist contour (shown in Fig. 5.38 *a*). If the system does not have any pole or zero at origin then the contour is shown in Fig. 5.38 (*b*).

5.16. NYQUIST CRITERION

The characteristic equation is given by

$$D(s) = 1 + G(s)\,H(s) \quad ...(5.24)$$

The zeros of $D(s)$ are the roots of the characteristic equation. For a feedback system the necessary and sufficient condition is that all zeros of $1 + G(s)\,H(s)$ that is the roots of the characteristic equation must have negative real part *i.e.*, they must lie in the left half of *s*-plane. In order to determine the presence of zeros in right half of *s*-plane we choose a contour as shown in Fig. 5.15 called Nyquist contour. Let there are 'Z' zeros and 'P' poles in the right half of *s*-plane. If this contour is mapped in $D(s)$ plane as Γ_D then Γ_D encloses the origin N times (where $N = Z-P$) in clockwise. Hence the system is unstable because the clockwise encirclement is possible only when there are zeros of $D(s)$ in right half of *s*-plane.

A feedback system (close loop system) is stable if and only if there is no zeros of $D(s)$ in the right half of *s*-plane. *i.e.*, $Z = 0$

$$\therefore \quad N = -P$$

Therefore, for a closed loop system to be stable, the number of counter-clockwise encirclement of the origin of $D(s)$ plane by Γ_D should equal the number of right half *s*-plane poles of $D(s)$ which are the poles of open loop transfer function $G(s)\,H(s)$.

Since $\quad D(s) = 1 + G(s)\,H(s)$

or, $\quad G(s)\,H(s) = D(s) - 1$

The contour Γ_D in $D(s)$ plane can be mapped in $G(s)$ $H(s)$ plane. Γ_{GH} by shifting horizontally to the left by one unit. Thus the encirclement of the origin by the contour Γ_D is equivalent to the encirclement of the point (–1+j0) by the contour Γ_{GH} as shown in Fig. 5.39.

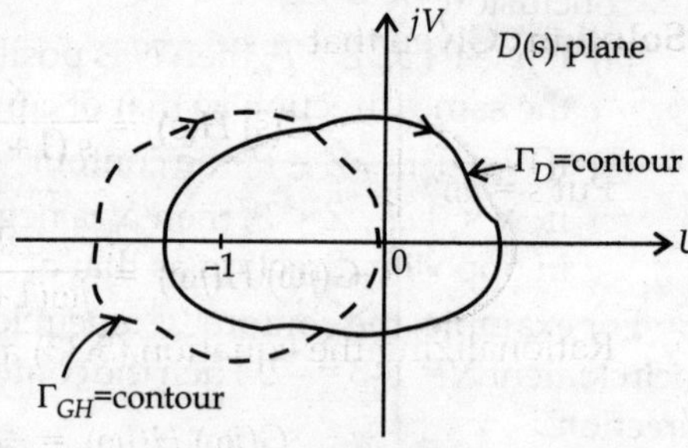

Fig. 5.39.

In most single loop feedback system $G(s)$ $H(s)$ has no poles in the right half plane *i.e.*, $P = 0$ then closed loop system is stable if $N = P = 0$

So, we can say that A closed loop system with $P = 0$ is stable if the net encirclement of the origin of $D(s)$ plane by Γ_D contour is zero.

Now, we can state the Nyquist stability criterion as follows:

A feedback system or closed loop system is stable if the contour Γ_{GH} of the open loop transfer function $G(s)$ $H(s)$ corresponding to the Nyquist contour in the s-plane encircles the point (–1 + jo) in counterclockwise direction and the number of counterclockwise encirclements about the (–1 + jo) equals the number of poles of $G(s)$ $H(s)$ in the right half of s-plane *i.e* with positive real parts.

In common case of open loop stable system, the closed loop system is stable if the contour Γ_{GH} of $G(s)$ $H(s)$ does not pass through or does not encircle (–1 + j0) point, *i.e* net encirclement is zero.

5.18. GENERAL CONSTRUCTION RULES OF THE NYQUIST PATH

Consider the Fig. 5.39

Table 5.1.

Path *ab*	$s = j\omega$	$0 < \omega < \omega_\circ$	...(5.25)
Path *bc*	$s = \lim_{P\to 0} (j\omega_\circ + Pe^{j\theta})$	$-90° \le \theta \le 90°$	...(5.26)
Path *cd*	$s = j\omega$	$\omega_\circ \le \omega \le \infty$	...(5.27)
Path *def*	$s = \lim_{R\to\infty} Re^{j\theta}$	$-90° \le \theta \le 90°$	...(5.28)
Path *fg*	$s = j\omega$	$\infty < \omega < -\omega_\circ$	...(5.29)
Path *gh*	$s = \lim_{P\to 0} (j\omega_\circ + Pe^{j\theta})$	$-90° \le \theta \le 90°$	...(5.30)
Path *hi*	$s = j\omega$	$-\omega_\circ \le \omega \le 0$	...(5.31)
Path *ija*	$s = \lim_{P\to 0} Pe^{j\theta}$	$-90° \le \theta \le 90°$	...(5.32)

Step 1 : Check $G(s)$ for poles on $j\omega$ axis and at the origin.

Step 2 : Using equation 5.25 to equation 5.27 sketch the image of the path $a - d$ in the $G(s)$-plane. If there are no poles on $j\omega$ axis equation (5.26) need not be employed.

Step 3 : Draw the mirror image about the real axis of the sketch resulting from step 2.

Step 4 : Use equation (5.28) plot the image of path def. This path at infinity usually plot into a point in the $G(s)$ -plane.

Step 5 : Use equation (5.32) plot the image of path *ija* (pole at origin).

Step 6 : Connect all curves drawn into the previous steps.

EXAMPLE 5.41. Determine the closed loop stability of a control system whose open loop transfer function is

$$G(s)\,H(s) = \frac{K}{s\,(1+sT)} \qquad \text{(type '1' system)}$$

Solution : Given that

$$G(s)\,H(s) = \frac{K}{s\,(1+sT)}$$

Put $s = j\omega$

$$G(j\omega)\,H(j\omega) = \frac{K}{j\omega(1+j\omega T)} \qquad ...(5.33)$$

Rationalizing the equation (5.33) and separating into real and imaginary parts.

$$G(j\omega)\,H(j\omega) = -\frac{KT}{1+\omega^2T^2} - j\,\frac{K}{\omega(1+\omega^2T^2)} \qquad ...(5.34)$$

$$\lim_{\omega\to 0} |G(j\omega)H(j\omega)| = \infty$$

$$\lim_{\omega\to 0} \angle G(j\omega)H(j\omega) = -90°$$

$$\lim_{\omega\to \infty} |G(jw)H(jw)| = 0$$

$$\lim_{\omega\to \infty} \angle G(j\omega)H(j\omega) = -180°$$

The polar plot will lie in third quadrant.

The Nyquist plot is shown in Fig. 5.40. The part for $0 < \omega < +\infty$ is drawn (1) (2) and for $-\infty < \omega < 0$ is shown by the point (2) (3) which is the mirror image of (1) (2). The semicircular detour around the origin in s-plane is mapped into a semicircular path of infinite radius representing a change of phase from $+\pi/2$. to $-\pi/2$.

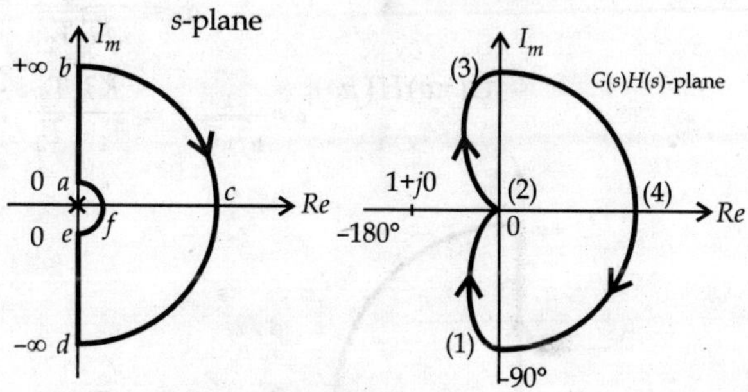

Fig. 5.40. Nyquist plot

As the point $(-1 + jo)$ is not encircled by the plot, $N = 0$

$$N = 0 \qquad P = 0$$

$$\because \quad N = Z - P \qquad \therefore \quad Z = 0$$

The number of zeros or roots of the characteristic equation with positive real part is nil and hence the closed loop system is stable.

EXAMPLE 5.42. Sketch the Nyquist plot and determine the stability of a unity feedback control system.

$$G(s) = \frac{K}{(1+sT_1)\,(1+sT_2)} \qquad \text{(Type O system)}$$

Solution : Given that :

$$G(s)\,H(s) = \frac{K}{(1+sT_1)\,(1+sT_2)}$$

Put $s = j\omega$

$$G(j\omega)\,H(j\omega) = \frac{K}{(1+j\omega T_1)\,(1+j\omega T_2)} \qquad ...(5.35)$$

$$|G(j\omega)\,H(j\omega)| = \frac{K}{\sqrt{1+\omega^2T^2_1}\,\sqrt{1+\omega^2{T_2}^2}} \qquad ...(5.36)$$

$$\angle G(j\omega)H(j\omega) = -\tan^{-1}\omega T_1 - \tan^{-1}\omega T_2 \qquad ...(5.37)$$

$$\lim_{\omega\to 0} |G(j(\omega)H(j\omega)| = \text{K}$$

$$\lim_{\omega\to 0} \angle G(j\omega)H(j\omega) = 0$$

$$\lim_{\omega \to \infty} |G(j\omega)H(j\omega)| = 0$$

$$\lim_{\omega \to \infty} \angle G(j\omega)H(j\omega) = -180°$$

Rationalize the equation (5.35) and separate the real and imaginary parts.

$$\frac{K}{(1+j\omega T_1)(1+j\omega T_2)} = \frac{K(1-\omega^2 T_1 T_2)}{(1+\omega^2 T_1^2)(1+\omega^2 T_2^2)} - j\frac{\omega(T_1+T_2)K}{(1+\omega^2 T_1^2)(1+\omega^2 T_2^2)} \quad ...(5.38)$$

Equate the real part to zero, we get

$$\omega = \frac{1}{\sqrt{T_1 T_2}}$$

$$|G(jw)H(jw)|_{w=\frac{1}{\sqrt{T_1T_2}}} = \frac{KT_1T_2}{T_1+T_2}$$

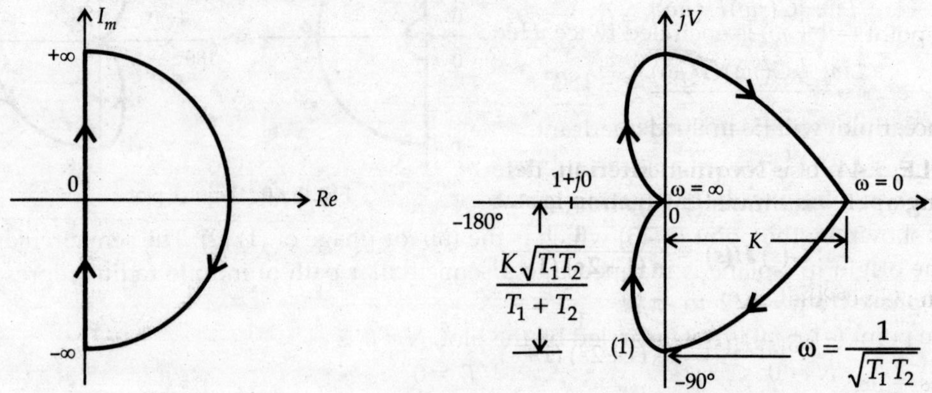

Fig. 5.41.

The plot of $G(j\omega)$ $H(j\omega)$ is shown in Fig. 5.41. The infinite semicircular arc of the Nyquist contour maps into origin. As the point $(-1 + jo)$ is not encircled by the plot

$\therefore$ $N = 0$

$P = 0$

$\therefore$ $Z = 0$

Hence, the system is stable.

EXAMPLE 5.43. Using Nyquist criterion, determine the stability of the feedback system which has the following open loop transfer function.

$$G(s)\,H(s) = \frac{K}{s^2(1+sT)} \quad \text{(Type '2' system)}$$

Solution : Given that

$$G(s)\,H(s) = \frac{K}{s^2(1+sT)}$$

Put $s = j\omega$

$$G(j\omega)\,H(j\omega) = \frac{K}{(j\omega)^2(1+j\omega T)} \quad ...(5.39)$$

Rationalizing the equation 5.39 and separating the real and imaginary part

$$G(j\omega)\,H(j\omega) = \frac{K}{-\omega^2(1+\omega^2T^2)} + j\frac{K}{\omega(1+\omega^2T^2)} \quad ...(5.40)$$

The Nyquist diagram is shown in the Fig. 5.42. Because of the double pole at s = 0, a small semicircular detour at the origin should be made.

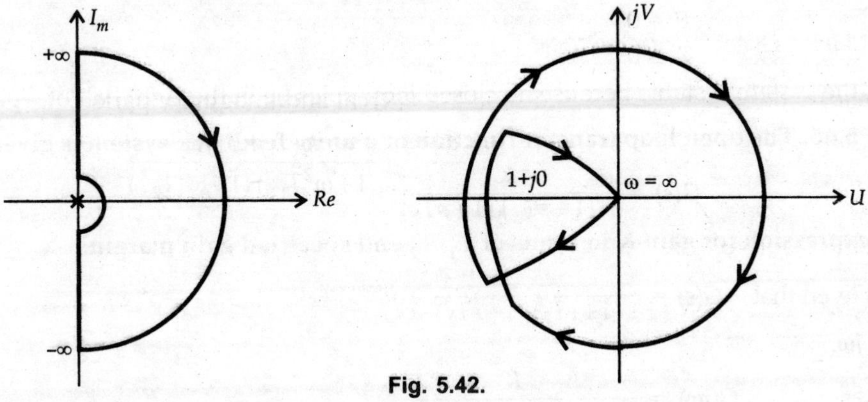

Fig. 5.42.

The point (– 1 + jo) is encircled twice. Hence N = 2

$$P = 0$$

$$\therefore \quad Z = 2$$

Hence, the system is unstable.

EXAMPLE 5.44. Use Nyquist criterion, determine whether the closed loop system having the following open loop transfer function is stable or not.

$$G(s)\,H(s) = \frac{1}{s\,(1+2s)\,(1+s)}$$

Solution : Given that

$$G(s)\,H(s) = \frac{1}{s\,(1+2s)\,(1+s)}$$

Put $s = j\omega$

$$G(j\omega)\,H(j\omega) = \frac{1}{j\omega(1+j2\omega)\,(1+j\omega)} \qquad ...(5.41)$$

Rationalizing the equation 5.41 and separate the real and imaginary part.

$$G(j\omega)\,H(j\omega) = \frac{-3}{(1+4\omega^2)\,(1+\omega^2)} - j\,\frac{1-2\omega^2}{\omega(1+4\omega^2)\,(1+\omega^2)} \qquad ...(5.42)$$

Equate the imaginary part to zero, we get the point of intersection on real axis

$$1 - 2\omega^2 = 0$$

$$\therefore \quad \omega = 0.707$$

$$|G(j\omega)H(j\omega)|_{j\omega = 0.707} = 0.66$$

The Nyquist plot is shown in Fig. 5.43.

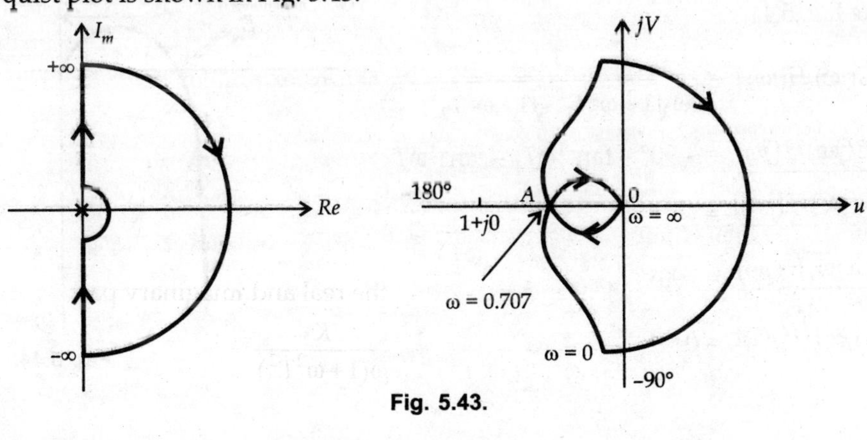

Fig. 5.43.

Since, OA = 0.66, the point (-1 + *jo*) is not encircled

$\therefore \quad N = 0$

$P = 0$

$\therefore \quad Z = 0$

Hence, the system is stable because no closed loop poles lie in the right half of *s*-plane.

EXAMPLE 5.45. The open loop transfer function of a unity feedback system is given by

$$G(s) = \frac{K}{s(1+sT_1)(1+sT_2)}$$

Derive an expression for gain *K* in terms of T_1, T_2 and specified gain margin.

Solution : Given that $G(s) = \dfrac{K}{s(1+sT_1)(1+sT_2)}$

Put $s = j\omega$

$$G(j\omega) = \frac{K}{j\omega(1+j\omega T_1)(1+j\omega T_2)} \quad ...(5.43)$$

Rationalizing the equation 5.43 and separate the real and imaginary parts

$$G(j\omega)\,H(j\omega) = \frac{-K(T_1+T_2)}{(1+\omega^2 T_1^2)(1+\omega^2 T_2^2)} - j\frac{K(1-\omega^2 T_1 T_2)}{\omega(1+\omega^2 T_1^2)(1+w^2 T_2^2)} \quad ..(5.44)$$

Equating the imaginary part to zero of equation 5.44

$$K(1-\omega^2 T_1 T_2) = 0$$

$$\therefore \quad \omega = \frac{1}{\sqrt{T_1 T_2}}$$

$$G(j\omega)H(j\omega)\Big|_{\omega=\frac{1}{\sqrt{T_1T_2}}} = -\frac{KT_1T_2}{T_1+T_2} \quad ...(5.45)$$

Gain margin $\quad Gm = \dfrac{1}{a}$

Here, $\quad a = \dfrac{KT_1T_2}{T_1+T_2}$

$$\therefore \quad Gm = \frac{T_1+T_2}{K(T_1T_2)}$$

$$K = \frac{1}{Gm}\left[\frac{1}{T_1}+\frac{1}{T_2}\right]$$

This is the required expression. The polar plot is shown in the Fig. 5.44.

$$|G(j\omega)\,H(j\omega)| = \frac{1}{\omega\sqrt{1+\omega^2T_1^2}\sqrt{1+\omega^2T_2^2}}$$

$$\angle G(j\omega)H(j\omega) = -90° - \tan^{-1}\omega T_1 - \tan^{-1}\omega T_2$$

$$\lim_{\omega\to 0}|G(j\omega)H(j\omega)| = \infty$$

$$\lim_{\omega\to 0}\angle G(j\omega)H(j\omega) = -90°$$

$$\lim_{\omega\to\infty}|G(j\omega)H(j\omega)| = 0$$

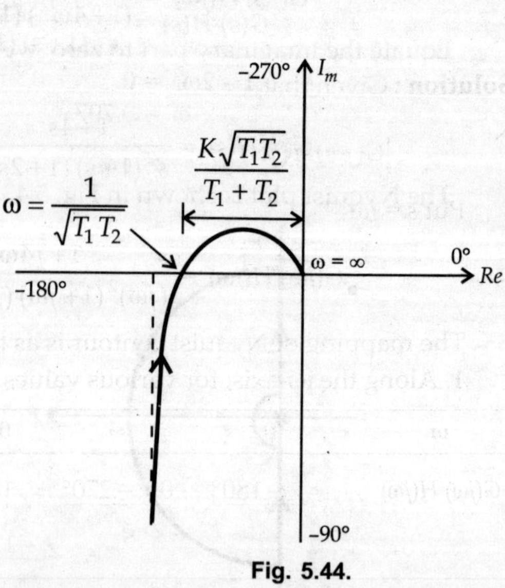

Fig. 5.44.

$$\lim_{\omega \to \infty} \angle G(j\omega)H(j\omega) = -270°$$

EXAMPLE 5.46. Consider a feedback system having the characteristic equation

$$1 + \frac{K}{(s+1)(s+1.5)(s+2)} = 0$$

It is desired that all the roots of the characteristic equation have real parts less than –1. Extend the Nyquist stability criterion to find the largest value of *K*, satisfying this condition.

Solution : Given that

$$G(s)\,H(s) = \frac{K}{(s+1)(s+1.5)(s+2)}$$

Put s = –1 + *j*ω

$$G(-1+j\omega)\,H(-1+j\omega) = \frac{K}{j\omega(0.5+j\omega)(1+j\omega)} \qquad ...(5.46)$$

Rationalizing the equation 5.46 and separate the real and imaginary parts.

$$G(-1+j\omega)\,H(-1+j\omega) = \frac{-1.5K}{(0.25\omega^2)(1+\omega^2)} - j\frac{K(0.5-\omega^2)}{\omega(0.25+\omega^2)(1+\omega^2)}$$

Equate the imaginary term to zero

$$0.5 - \omega^2 = 0$$

$$\therefore \qquad \omega = 0.707 \text{ rad/sec.}$$

$$\left| G(-1+j\omega)H(-1+j\omega) \right|_{\omega=0.707} = 1.33K$$

As per condition given

$$1.33K < 1 \quad \text{or,} \quad K < 0.7518$$

∴ Required largest value of *K* is 0.7518 **Ans.**

EXAMPLE 5.47. Use Nyquist criterion, determine whether the closed loop system having the following open loop transfer function is stable or not. If not how many closed loop poles lie in the right half *s*-plane?

$$G(s)\,H(s) = \frac{1+4s}{s^2(1+s)(1+2s)}$$

(R.M.L. University Faizabad 2002, LST)

Solution : Given that

$$G(s)\,H(s) = \frac{1+4s}{s^2(1+s)(1+2s)}$$

Put s = *j*ω

$$G(j\omega)\,H(j\omega) = \frac{1+j4\omega}{(j\omega)^2(1+j\omega)(1+j2\omega)}$$

The mapping of Nyquist contour is as follows:

1. Along the *j*ω-axis, for various values of 'ω', *G*(*j*ω) *H*(*j*ω)is plotted.

ω	0	∞	0.1	1.0	2.0	4.0	10.0	20.0
G(jω) H(jω)	$\infty\angle -180°$	$0\angle -270°$	105	1.3	0.21	0.03	0.0019	0.00024
			$\angle -175.22°$	-212.47°	-236.5°	-252.4°	-262.8°	-266.4°

$$M = |G(j\omega)\,H(j\omega)| = \frac{\sqrt{1+16\omega^2}}{\omega^2\sqrt{1+\omega^2}\,\sqrt{1+4\omega^2}}$$

$$\angle G(j\omega)H(j\omega) = \tan^{-1}4\omega - 180° - \tan^{-1}\omega - \tan^{-1}2\omega$$

The point of intersection of $G(j\omega)\,H(j\omega)$ with real axis is calculated as

$$\angle G(j\omega)H(j\omega) = -180°$$

$$+\tan^{-1}4\omega - 180° - \tan^{-1}\omega - \tan^{-1}2\omega = -180°$$

or, $\tan^{-1}4\omega = \tan^{-1}\omega + \tan^{-1}2\omega$

taking tangent on both sides

$$4\omega = \frac{\omega+2\omega}{1-2\omega^2}$$

$\therefore$ $\omega = 0.354$ rad/sec.

$$\left|G(j\omega)H(j\omega)\right|_{\omega=0.354} = \frac{1+j4\times 0.354}{(j0.354)^2\,(1+j0.354)\,(1+j2\times 0.354)} = -10.64$$

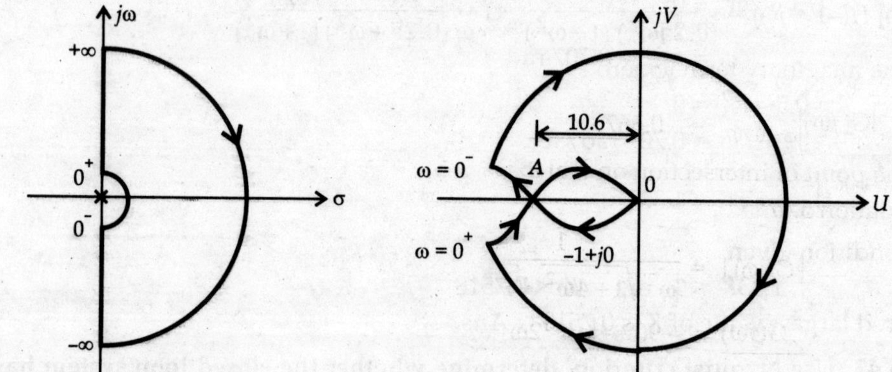

Fig. 5.45.

2. Mapping of infinite semicircle

$$\lim_{R\to\infty}\frac{1+\text{R}e^{j\theta}\cdot 4}{(R^2e^{j2\theta})\,(1+\text{R}e^{j\theta})\,(1+2\text{R}e^{j\theta})} = 0\angle -270°$$

3. Mapping of $j\omega$-axis from $-\infty$ to $\bar{0}$, which is the mirror image of $j\omega$-axis from $\overset{+}{0}$ to $+\infty$

4. Mapping small semicircle at origin

$$\lim_{P\to 0}\left[\frac{1+4\,\text{R}e^{j\theta}}{(P^2e^{j2\theta})\,(1+Pe^{j\theta})\,(1+2Pe^{j\theta}}\right] = \infty(\angle 180° \to \angle 0° \to \angle -180°)$$

The complete Nyquist plot is shown in the Fig. 5.45.

Since, $OA = -10\,.\,64$, hence $(-1+j0)$ encircles two times in clockwise direction.

$\therefore$ $N = 2$

$P = 0$

$N = Z - P$

$2 = Z - 0$

$\therefore$ $Z = 2$

Hence, two roots of the characteristic equation lies in right half of s-plane. The closed loop system is unstable.

EXAMPLE 5.48. A unity feedback system has open loop transfer function.

$$G(s) = \frac{1}{s(1+2s)(1+s)}$$

Sketch Nyquist plot for the system and therefore obtain the gain margin and the phase margin.

(GATE 1992)

Solution : Given that $G(s) = \frac{1}{s(1+2s)(1+s)}$

Put $s = j\omega$

$$G(j\omega) = \frac{1}{j\omega(1+j2\omega)(1+j\omega)} \quad ...(5.47)$$

Rationalizing the equation (5.47) and separate the real and imaginary parts.

$$G(j\omega) = \frac{-3\omega^2}{\omega^2(1+4\omega^2)(1+\omega^2)} - j\frac{\omega(1-2\omega^2)}{\omega^2(1+4\omega^2)(1+\omega^2)}$$

Equate the imaginary part to zero

$$1 - 2\omega^2 = 0$$

$$\therefore \quad \omega = \frac{1}{\sqrt{2}} = 0.707 \text{ rad/sec.}$$

$$|G(j\omega)|_{\omega=0.707} = -0.667$$

This is the point of intersection on real axis.

From equation 5.47

$$|G(j\omega)| = \frac{1}{\omega^2\sqrt{1+4\omega^2}\sqrt{1+\omega^2}}$$

$$\angle G(j\omega) = -90° - \tan^{-1}2\omega - \tan^{-1}\omega$$

$$\lim_{\omega \to 0} |G(j\omega)| = \infty, \qquad \lim_{\omega \to 0} \angle G(j\omega) = -90°$$

$$\lim_{\omega \to \infty} |G(j\omega)| = 0, \qquad \lim_{\omega \to \infty} \angle G(j\omega) = -270°$$

The Nyquist contour and Nyquist plot are shown in Fig. 5.46.

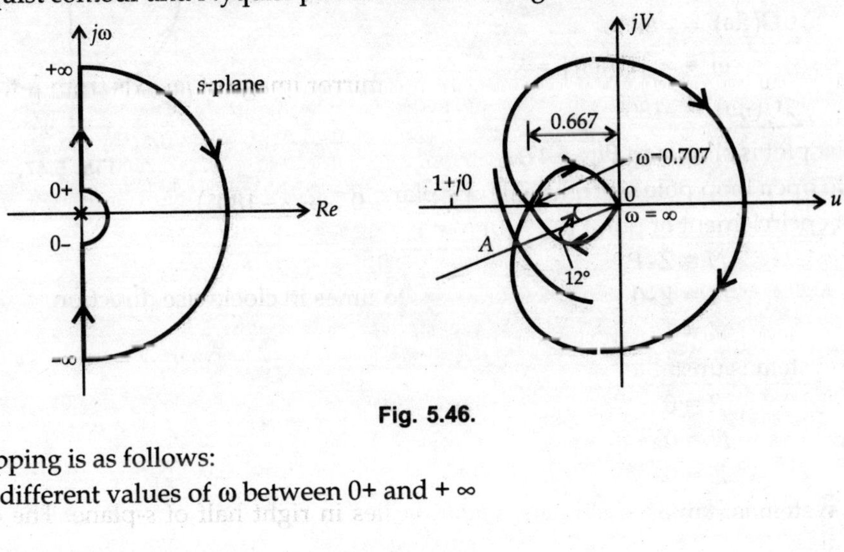

Fig. 5.46.

The mapping is as follows:

1. For different values of ω between 0+ and + ∞

2. Mapping of infinite semicircle

$$\lim_{R \to \infty} \frac{1}{Re^{j\theta}(1+2Re^{j\theta})(1+Re^{j\theta})} = 0 \angle -270°$$

3. Mapping of $j\omega$ axis from 0 to $-\infty$, which is the mirror image of $j\omega$ axis from 0+ to $+\infty$.
4. Mapping of semicircle at origin

$$\lim_{P \to \infty} \left[\frac{1}{Pe^{j\theta}(1+2Pe^{j\theta})(1+Pe^{j\theta})} \right] = \infty \angle -90°$$

$$\text{Gain margin} = 20 \log \frac{1}{0.667} = 3.51$$

For phase margin draw the unit circle which intersect the plot at A. Join the point 0 to A and measure the angle.

$\therefore$ P.M = 12

EXAMPLE 5.49. Draw the Nyquist plot for unity feedback control system, with open loop transfer function

$$G(s) = \frac{K(1-s)}{s+1}$$

Using Nyquist stability criterion, determine the stability of the closed loop system.

Solution : Given that

$$G(s) = \frac{K(1-s)}{(1+s)}$$

Put $s = j\omega$

$$G(j\omega) = \frac{K(1-j\omega)}{(1+j\omega)} \quad |G(j\omega)| = \frac{K\sqrt{1+\omega^2}}{\sqrt{1+\omega^2}}$$

$$\angle G(j\omega) = -2\tan^{-1}\omega$$

$$\omega = 0 \quad |G(j\omega)| = K$$

$$\angle G(j\omega) = 0°$$

$$\omega = 1 \quad |G(j\omega)| = K$$

$$\angle G(j\omega) = -90°$$

$$\omega = \infty \quad |G(j\omega)| = K$$

$$\angle G(j\omega) = -180°$$

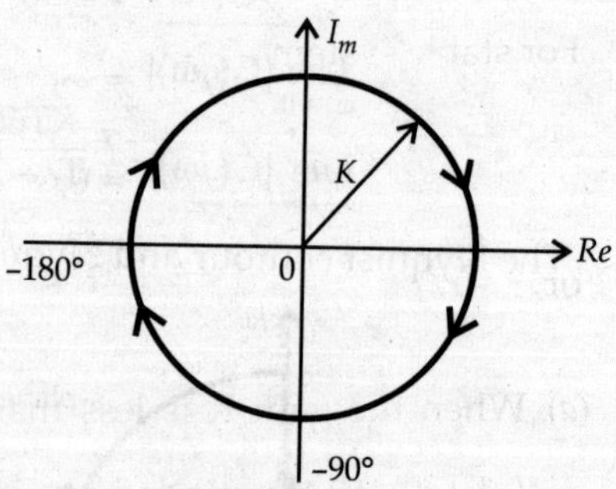

Fig. 5.47.

The Nyquist plot is shown in Fig. 5.47.

If $K > 1$: No open loop pole on right half of s-plane $P = 0$.

Number of encirclement of point $(-1 + j0)$ $N = 1$

$\therefore$ $N = Z - P$

$1 = Z - 0$

$\therefore$ $Z = 1$

Hence, the system is unstable

If $K < 1$: $P = 0$

$N = 0$

$\therefore$ $Z = 0$

Hence, the system is stable.

EXAMPLE 5.50. Using Nyquist criterion investigate the stability of a closed loop control system whose open loop transfer function is given below :

$$G(s)\,H(s) = \frac{K}{s\,(1+sT_1)\,(1+sT_2)}$$

(R.M.L. University Faizabad, 2002, Control Systems)

Solution : Given that

$$G(s)\,H(s) = \frac{K}{s\,(1+sT_1)\,(1+sT_2)}$$

Put $s = j\omega$

$$G(j\omega)\,H(j\omega) = \frac{K}{j\omega(1+j\omega T_1)\,(1+j\omega T_2)} \quad ...(5.48)$$

Rationalizing the equation 5.48 and separate the real and imaginary part

$$G(j\omega)\,H(j\omega) = \frac{-K(T_1+T_2)}{(1+\omega^2T_1^{\,2})\,(1+\omega^2T_2^{\,2})} - j\,\frac{K(1-\omega^2T_1T_2)}{\omega(1+\omega^2T_1^{\,2})\,(1+\omega^2T_2^{\,2})} \quad ...(5.49)$$

The value of 'ω' at the point of intersection of plot with real axis can be obtained by equating the imaginary part to zero.

$$\frac{K(1-\omega^2T_1T_2)}{\omega(1+\omega^2T_1^{\,2})\,(1+\omega^2T_2^{\,2})} = 0$$

or,
$$1 - \omega^2T_1T_2 = 0$$

or,
$$\omega = \frac{1}{\sqrt{T_1T_2}}$$

$$\left|G(j\omega)H(j\omega)\right|_{\omega=\frac{1}{\sqrt{T_1T_2}}} = -\frac{KT_1T_2}{T_1+T_2}$$

For stable system

$$\frac{KT_1T_2}{T_1+T_2} < 1$$

or,
$$K < \frac{T_1+T_2}{T_1T_2}$$

(*a*) When the gain K is less than $\frac{T_1+T_2}{T_1T_2}$, the point $(-1 + j0)$ is not encircled by $G(j\omega)\,H(j\omega)$ plot and the system is stable

(*b*) When $K = \frac{T_1+T_2}{T_1T_2}$, the plot passes through $(-1 + j0)$ which indicates the system has roots on $j\omega$-axis. Hence, the system is marginally stable.

(*c*) When $K > \frac{T_1+T_2}{T_1T_2}$, the plot encircles the point $(-1 + j0)$. Hence the system is unstable.

The plot is shown in Fig. 5.48.

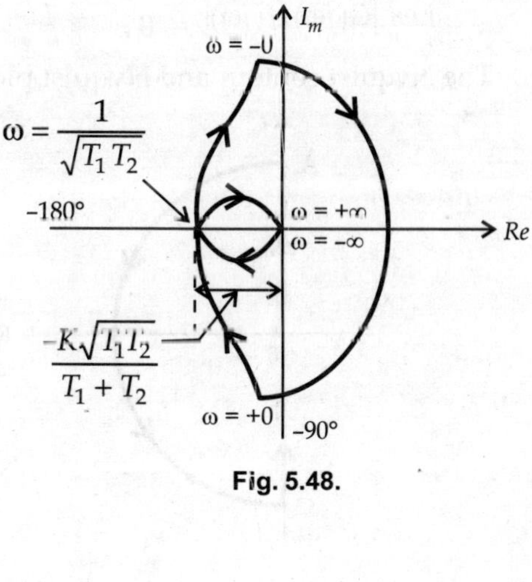

Fig. 5.48.

EXAMPLE 5.51. Consider the transfer function

$$G(s)\,H(s) = \frac{60}{(s+1)\,(s+2)\,(s+5)}$$

Using Nyquist stability criterion determine whether the closed loop system is stable or not.

(KNIT, Sultanpur 1998-99)

Solution : Given that

$$G(s)\,H(s) = \frac{60}{(s+1)\,(s+2)\,(s+5)}$$

Put $s = j\omega$

$$G(j\omega)\,H(j\omega) = \frac{60}{(1+j\omega)\,(2+j\omega)\,(5+j\omega)} \qquad \text{...(5.50)}$$

Rationalizing the equation 5.50 and separate the real and imaginary parts.

$$G(j\omega)\,H(j\omega) = \frac{60\,(10-8\omega^2)}{(1+\omega^2)\,(4+\omega^2)\,(25+\omega^2)} - j\,\frac{60(17\omega-\omega^3)}{(1+\omega^2)\,(4+\omega^2)\,(25+\omega^2)}$$

Equate the real part to zero

$$\frac{60(10-8\omega^2)}{(1+\omega^2)\,(4+\omega^2)\,(25+\omega^2)} = 0$$

$$\omega = 1.25 \text{ rad/sec.}$$

Point of intersection on imaginary axis

$$|G(j\omega)H(j\omega)|_{\omega=1.25} = -3.057$$

Equate the imaginary part to zero

$$\frac{60(17\omega-\omega^3)}{(1+\omega^2)\,(4+\omega^2)\,(25+\omega^2)} = 0$$

$$\therefore \qquad \omega = 4.123 \text{ rad/sec.}$$

Point of intersection of plot on real axis

$$|G(j\omega)H(j\omega)|_{\omega=4.123} = -0.47$$

From equation 5.50

$$\lim_{\omega\to 0} |G(j\omega)H(j\omega)| = 6, \qquad \lim_{\omega\to 0} \angle G(j\omega)H(j\omega) = 0°$$

$$\lim_{\omega\to \infty} |G(j\omega)H(j\omega)| = 0, \qquad \lim_{\omega\to \infty} \angle G(j\omega)H(j\omega) = -270°$$

The Nyquist contour and Nyquist plot are shown in Fig. 5.49.

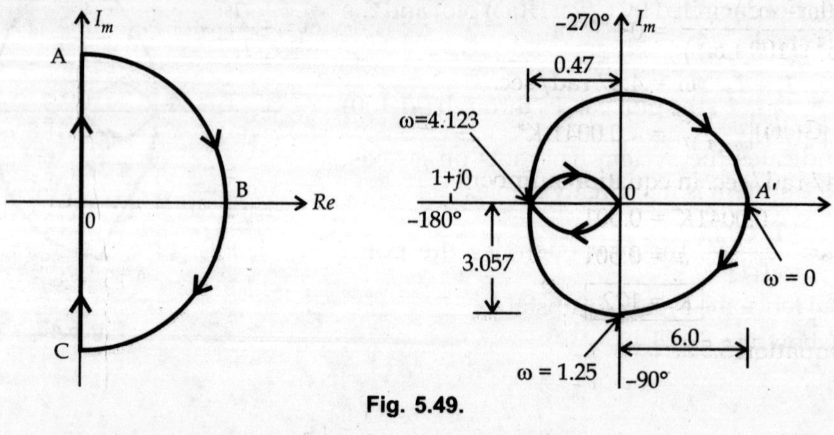

Fig. 5.49.

The mapping is as follows:

1. Part OA in the s-plane is the $j\omega$-axis from $\omega = 0$ to $\omega = \infty$ and maps into the polar plot $A'O$ in GH-plane.
2. The infinite semicircle ABC maps into origin in GH-plane.
3. Part CO which is the negative part of $j\omega$ axis, maps into OA' From Nyquist plot, the point $(-1 + j0)$ is not encircled by the plot

$$\therefore \quad N = 0$$

No pole in the right half

$$\therefore \quad P = 0$$

$$N = Z - P$$

$$\therefore \quad Z = 0$$

Therefore, no zero in right half and hence the system is stable.

EXAMPLE 5.52. The open loop transfer function of a control system is given by

$$G(s) = \frac{K}{s(s+2)(s+10)}$$

Determine the value of K so that the system may be stable with

(a) gain margin equal to 6 db

(b) P.M equal to 45°

Solution :

$$G.M = 20\log_{10}\left(\frac{1}{a}\right)$$

where 'a' is the point of intersection of Nyquist plot on negative real axis.

$$\therefore \quad 6 = 20\log_{10} {}^{1}/_{a}$$

$$\therefore \quad a = 0.501$$

Since,

$$G(s) = \frac{K}{s(s+2)(s+10)}$$

Put $s = j\omega$

$$G(j\omega) = \frac{K}{j\omega(2+j\omega)(10+jw)} \qquad ...(5.51)$$

Rationalizing the equation 5.51 and separate the real and imaginary parts.

$$G(j\omega) = \frac{-12K\omega^2}{\omega^2(4+\omega^2)(100+\omega^2)} - j\frac{K(20\omega-\omega^3)}{\omega^2(4+\omega^2)} \qquad ...(5.52)$$

To get the point of intersection on real axis, equate the imaginary part to zero.

$$\frac{K(20\omega - \omega^3)}{\omega^2(4+\omega^2)(100+\omega^2)} = 0$$

$$\therefore \quad \omega = 4.47 \text{ rad/sec.}$$

$$|G(j\omega)|_{\omega=4.47} = -0.0041K^*$$

Put $\omega = 4.47$ rad/sec. in equation number 5.52

$$0.0041K = 0.501$$

$$\therefore \quad a = 0.501$$

$$\therefore \quad \boxed{K = 122}$$

(*ii*) From equation 5.51

$$P.M = \angle G(j\omega)H(j\omega) + 180°$$

$$45° = -90° - \tan^{-1}\frac{\omega}{2} - \tan^{-1}\frac{\omega}{10} + 180°$$

$$\tan^{-1}\frac{\omega}{2} + \tan^{-1}\frac{\omega}{10} = 45°$$

Taking tangent on both sides

$$\frac{\frac{\omega}{2} + \frac{\omega}{10}}{1 - \frac{\omega}{2} \cdot \frac{\omega}{10}} = 1$$

$\therefore$ $$\omega^2 + 12\omega - 20 = 0$$

$\therefore$ $$\omega = -13.48, 1.48$$

Consider the positive value of ω *i.e.*, $\omega = 1.48$ rad/sec.

$$|G(j\omega| = 1$$

$$\frac{K}{j\omega(2 + j\omega)(10 + j\omega)} = 1$$

$$\frac{K}{j1.48(2 + j1.48)(10 + j1.48)} = 1$$

$$\boxed{K = 37.22}$$

5.19. HOW TO OBTAIN OPEN LOOP TRANSFER FUNCTION G(s) H(s) FROM CHARACTERISTIC EQUATION

The following steps are required to obtain the open loop transfer function $G(s)\ H(s)$.

Step 1 : Write all the terms of 's' together without K.

Step 2 : Write all the terms of K together

Step 3 : Divide the equation by polynomial containing the terms of s only without K.

Step 4 : Compare with $1 + G(s)\ H(s) = 0$

For example consider the characteristic equation

$$S^3 + 10\ S^2 + 21\ S + KS + 13\ K = 0$$

Step 1 : $(S^3 + 10\ S^2 + 21\ S) + K(S + 13) = 0$

Step 2 : Divide the equation by $S^3 + 10\ S^2 + 21\ S$, we get

$$1 + \frac{K(s+13)}{S^3 + 10s^2 + 21s} = 0$$

Step 3 : Compare with $1 + G(s)\ H(s) = 0$

$\therefore$ $$G(s)\ H(s) = \frac{K(s+13)}{S^3 + 10\ s^2 + 21\ s} = \frac{K(s+13)}{s(s+3)(s+7)}$$

SUMMARY

A linear time invariant system is stable if : (*a*) the system is excited by a bounded input, the output is bounded, (*b*) in the absence of the input the output tends towards zero.

A linear time invariant system is unstable if : (*a*) the system is excited by a bounded input, it produces unbounded output, (*b*) in the absence of the input, output may not tends towards zero.

If one or more pairs of simple non-repeated roots are located on the imaginary axis of *s*-plane but there are no roots in right half of *s*-plane the response will be undamped sinusoidal oscillations of constant amplitude and frequency as shown in Fig. 5.7.

A linear time invariant system is said to be conditionally stable if for certain condition of a particular parameter of the system, its output is bounded otherwise for other conditions the output becomes unbounded and system becomes unstable. Such system is called conditionally stable.

Routh-Hurwitz method is used to determine the location of poles of a characteristic equation in left and right half of *s*-plane without actually solving the equation.

Routh-Hurwitz criterion states that the system is stable is stable if and only if all the elements in the first column have the same algebraic sign. If all elements are not of the same sign then the number of sign changes of the elements in first column equals the number of roots of the characteristic equation in the right half of *s*-plane.

Advantages of Routh-Hurwitz criterion:

1. Stability of the system can be determine without solving the equation.
2. Relative stability can be easily determined.
3. By this method we can determine the range of *K* for stability.
4. By this method we can determine the point of intersection of root locus with imaginary axis.
5. For unstable system it gives the number of roots of characteristic equation having positive real part.

Limitations of Routh-Hurwitz criterion:

1. This is applicable only for linear systems.
2. It does not provide the exact location of closed loop poles in left or right half of *s*-plane.
3. It is valid only for real coefficients of the characteristic equation.

Nyquist stability criterion states that a closed loop system is stable if the contour *GH* of the open loop transfer function *G*(*s*) *H*(*s*) corresponding to the Nyquist contour in *s*-plane encircles the point ($-1 + j0$) in counterclockwise direction and the number of counterclockwise encirclement about ($-1 + j0$) equals the number of poles of *G*(*s*) *H*(*s*) in the right half of *s*-plane.

The closed loop system is stable if the contour *GH* of *G*(*s*) *H*(*s*) does not pass through or does not encircles ($-1 + j0$) point *i.e.*, net encirclement is zero.

EXERCISE

5.1. Consider the closed loop feedback control system shown in Fig. 5.50. Using routh Hurwitz criterion, determine the range of *K* for which the system is stable. Find also the number of roots of the characteristic equation that are in the right half of *s*-plane for K = 0.5.

(GATE 1996)

5.2. The characterisic equation of a system is given by

$$s^4 + 20s^3 + 15s^2 + 2s + K = 0$$

(*i*) Determine range of *K*, for system to stable.

(*ii*) Can system be marginally stable ? If so, find the required value of *K* and the frequency of sustained oscillation.

5.3. Apply Routh's criterion, determine the number of roots that lie in the right half-plane of the following characteristic equation

(i) $s^6 + 6s^5 + 10s^4 + 12s^3 + 13s^2 - 18s - 24 = 0$

(ii) $s^3 + 2s^2 + 4s + 8 = 0$

5.4. The characteristic equation for a control system is

$$s(s^2 + 5s + 20) + K(s + 2) = 0$$

Use Routh's criterion to determine the range of value of K such that the system is stable.

5.5. Sketch the Nyquist plot for $0 < \omega < \infty$ and find the phase crossover frequency for

$$G(s)\,H(s) = \frac{200K}{s(s+2)(s+10)}$$

5.6. For open transfer function $G(s)\,H(s) = \dfrac{K(s+2)}{(s+1)(s-3)}$, sketch the Nyquist plot for determine whether the system is stable or not.

5.7. Point out the limitations of Routh-Hurwitz stability criterion. The open loop transfer function of a unity feedback control system is given by $G(s) = \dfrac{K}{s(s^2+8s+T)}$. Using Routh-Hurwitz criterion determine the values of K and T which corresponds to a stable system.

Also determine the new values of K and T if it is required that all the roots of the system characteristic equation lie in the region to the left of the line $s = -1$. *(UPSC 1974)*

(***Hint*** : For second part modified the characterisitc equation by putting $s = s + 1$ in the characteristic equation and then apply Routh-Hurwitz criterion).

ANSWERS

5.1. Stable for $0 < K < 4.67$,

For $K = 0.5$, two roots in right half

5.2. $0 < K < 1.49$, $\omega = 0.316$ rad/sec.

5.5. Phase cross over frequency $= 4.472$ rad/sec.

5.6. Stable if $K > 2$

5.7. $K > 0$ and $T > K/B$

For second part$200 + 10T - K > 0$

$9 + T + K > 0.$

SEMI-OBJECTIVE TYPE QUESTIONS

(i) Define stability (Article 5.1)

(ii) State Routh-Hurwitz criterion (Article 5.4)

(iii) State necessary but not sufficient conditions for stability. (Article 5.3)

(iv) Define marginally stable system. (Article 5.1)

(v) What are the two special cases, which occur in Routh's table? How are those cases handled? (Article 5.4)

(vi) Define relative stability (Article 5.6)

(vii) What is root locus? (Article 5.7)

(viii) What it breakaway point? (Article 5.7)

(ix) Write short note on Nyquist criterion. (Article 5.15)

(x) Write short note on mapping. (Article 5.12)

(xi) State Sylvester's Theorem. (Article 5.18)

(xii) State advantages of Routh-Hurwitz criterion. (Summary)

(xiii) State limitations of Routh-Hurwitz criterion. (Summary)

(xiv) State Lyapunov stability theorem. (Article 5.17.3)

(xv) Write a note on angle and magnitude condition of root locus. (Article 5.7)

(xvi) Explain the term : *(a)* asymptotes, *(b)* centroid. (Article 5.8)

(xvii) Write a short note on effects of adding pole and zero on the root locus. (Article 5.10)

(xviii) State the rule for obtaining the breakaway point in the root locii.

(xix) How to obtain the open gain at any point on the root locii?

(xx) How location of roots of the characteristic equation are related to stability?

(xxi) How will you obtain the angle of departure and angle of arrival of root locii?

(xxii) How will you determine the absolute value of K an any point S_1 on the root locii?

□□□

6 STATE SPACE ANALYSIS OF CONTROL SYSTEM

6.1. ANALYSIS OF SYSTEMS

The procedure for determining the state of a system is called *state variable analysis.* The state of a dynamic system is the smallest set of variables such that the knowledge of these variables at $t = t_0$ with the knowledge of the input for $t \geq t_0$ completely determines the behaviour of the system for any time $t \geq t_0$. This set of variables is called state variables.

To understand the concept of state and state variables consider the R-C circuit as shown in Fig. 6.1 (*a*).

The capacitor get charged by the current $i(t)$ and the voltage at the capacitor is $v(t)$

$$i = C\frac{dv(t)}{dt}$$

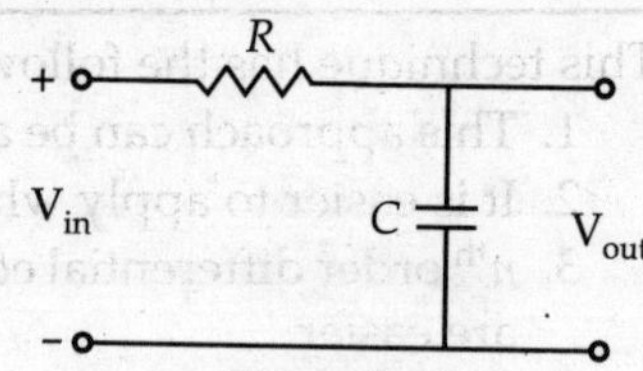

Fig. 6.1 (a).

$$\therefore \quad dv(t) = \frac{1}{C} i\, dt$$

Integrate both sides

$$\int dv(t) = \int \frac{1}{C} i\, dt$$

$$\boxed{v(t) = \frac{1}{C}\int_{-\infty}^{t} i\, dt}$$

$$v(t) = \frac{1}{C}\int_{-\infty}^{t_0} i\, dt + \frac{1}{C}\int_{t_0}^{t} i\, dt$$

where t_0 = initial time

$$\frac{1}{M}\int_{-\infty}^{t_0} i\, dt = v(t_0)$$

$$\therefore \quad v(t) = v_0(t) + \frac{1}{C}\int_{t_0}^{t} i\, dt$$

---

1. If the characteristic of a system does not change with time, then the system is said to be time invariant.

2. A system is said to be a single variable system if and only if it has only one input terminal and only one output terminal. A system is said to be multi variable system if and only if it has more than one input terminal or more than one output terminal.

Also, $$V_{out} = i R_1 + \frac{1}{C}\int_0^t i\,dt + V_0(t)$$

From above equations it is clear that the knowledge of capacitor voltage at $t = 0$ is required to determine the new value V. Thus, for the same input we get different values of the output voltage depending upon the parameter t_0 and value $v_0(t)$.

Therefore, to find the output voltage, the knowledge of the initial capacitor voltage must be known. Only the value of the input voltage is not sufficient to obtain the value of output voltage at any time $t \geq 0$. Thus, initial conditions describe the status or state of the system at $t = t_0$.

The state of the system is the combined effect of the initial conditions of different elements at any time 't'.

In earlier chapters we studied the linear system by transfer function, block diagram etc. The transfer function has some drawbacks, *e.g.*, transfer function is only defined under zero initial conditions and also it is applicable to linear time invariant systems.[1] Therefore due to these limitation state variable approach is developed. This technique can be used for analysis and design of linear and non linear, time invariant or time variant and multi input multi output systems.[2] The state space analysis involves the description of the system in terms of I order differential equations by selecting suitable state variables, the first order derivatives are arranged on left hand side and on right hand side the terms are free from derivatives. The state space techniques have many advantages (Given in next article *i.e.*, 6.2).

6.2. ADVANTAGES OF STATE SPACE TECHNIQUES

This technique has the following advantages.

1. This approach can be applied to linear or non-linear, time variant or time invariant systems.
2. It is easier to apply where the laplace transform cannot be applied
3. n^{th} order differential equations can be expressed as 'n' equation of first order whose solutions are easier.
4. It is a time domain approach.
5. This method is suitable for digital computer computation because this is a time domain approach.
6. The system can be designed for optimal conditions with respect to given performance indices.

6.3. SOME IMPORTANT DEFINITIONS

State : The state of a system at any time 't_0' is the minimum set of numbers $x_1, x_2, \ldots, x_n$ which along with the input to the system for time $t \geq t_0$ is sufficient to determine the behaviour of the system for all $t \geq t_0$. In other words, the state of a system represents the minimum amount of information that we need to know about a system at "t_0" such that its future behaviour can be determined without reference to the input before 't_0'. The state can also be defined as the state of a system at time t_0 is the amount of information at t_0, that, together with input $u(t_0, \infty)$ determines the unique behaviour of the system for all $t \geq t_0$. By the behaviour of the system, we mean all responses, including the state of the system. If the system is a network we mean the voltage and current of every branch of the network.

Consider the network shown in Fig. 6.1 if the initial current through the inductor and initial voltage across the capacitor are known, then for any driving voltage the behaviour of the netwrok can be determined. Hence, the inductor current and capacitor voltage can be considered as the state of the network.

State Variables : The definition is given in Article 6.1

State Vector : If we need n variables to completely describe the behaviour of a given system, then these n state variables may be considered as n component of a vector x. Such a vector is called state vector, A state vector is thus a vector which determines uniquely the system state $x(t)$ for any time $t \geq t_0$, once the state at $t = t_0$ is given and the input $u(t)$ for $t \geq t_0$ is specified.

State space : The n-dimensional space whose coordinate axes consists of the x_1 axis, x_2 axis ......, x_n axis is called state space. Any state can be represented by a point in the state space.

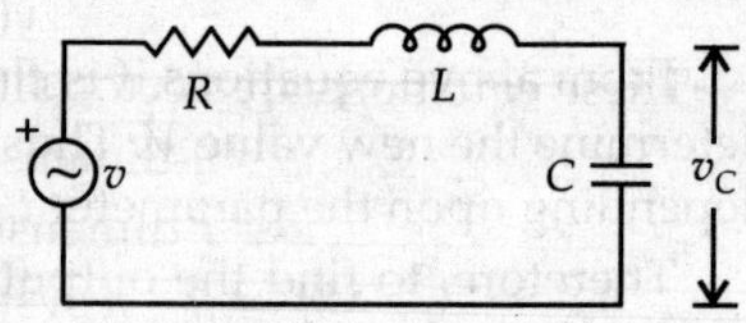

Fig. 6.1. (b)

6.4. STATE SPACE REPRESENTATION

6.4.1. State Space Representation for Electrical Network (Physical Variable Form)

Consider an RLC network shown in Fig. 6.2. Let, the current at time $t = 0$ be $i_L(0)$ and capacitor voltage at time $t = 0$ be $V_c(0)$. Thus, the state of the network at time $t = 0$ is specified by the inductor current and capacitor voltage. Hence, the pair $i_L(0)$, $V_c(0)$ is called the initial state of the network. Similarly at time 't', the pair $i_L(t)$, $V_c(t)$ is called the state of the network at 't'. The variable i_L and V_c are called state variables of the network.

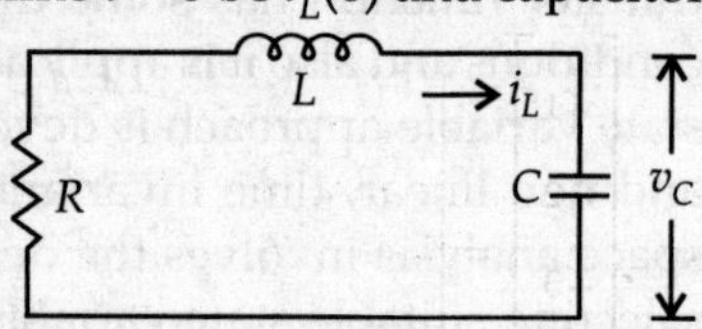

Fig. 6.2.

Apply KVL

$$Ri_L + L\frac{di_L}{dt} + V_c = 0 \qquad ...(6.1)$$

Also,

$$i_c = i_L = C\frac{dV_c}{dt} \qquad ...(6.2)$$

from (6.1)

$$\frac{di_L}{dt} = -\frac{R}{L}i_L - \frac{1}{L}V_c \qquad ...(6.3)$$

$$\frac{dv_c}{dt} = \frac{1}{C}i_L \qquad ...(6.4)$$

Equations of this form are called state equations. In such equations all the variables present are state variables.

Equations 6.3 and 6.4 can be written in matrix form as

$$\begin{bmatrix} \frac{di_L}{dt} \\ \frac{dv_c}{dt} \end{bmatrix} = \begin{bmatrix} -\frac{R}{L} & -\frac{1}{L} \\ \frac{1}{C} & 0 \end{bmatrix} \begin{bmatrix} i_L \\ V_C \end{bmatrix} \qquad ...(6.5)$$

let $\quad x(t) \,\Delta \begin{bmatrix} i_L \\ V_C \end{bmatrix}$ and $A = \begin{bmatrix} -\frac{R}{L} & -\frac{1}{L} \\ \frac{1}{C} & 0 \end{bmatrix}$

then equation 6.5 can be written as

$$\frac{d}{dt}x(t) = A\,x(t)$$

or,

$$\dot{x}(t) = A\,x(t)$$

In the linear time-invariant systems, the general form of state equations are

$$\dot{x}(t) = A\,x(t) + B\,u(t) \quad ...(6.6)$$

$$y(t) = C\,x(t) + D\,u(t) \quad ...(6.7)$$

These equations are vector differential equations where 'x' is the n-dimensional state vector

y = n-dimensional output vector

u = r-dimensional control vector or input vector

A = $n \times n$ system matrix

B = $n \times r$ control matrix

C = $n \times n$ output matrix

In some cases there is no direct connection between input and output so $D\,u(t)$ will not be there.

$$y(t) = C\,x(t) \quad ...(6.8)$$

Equation (6.6) and (6.8) can be expressed as

$$\begin{bmatrix} \dot{x}_1 \\ \dot{x}_2 \\ \dot{x}_3 \\ \vdots \\ \dot{x}_n \end{bmatrix} = \begin{bmatrix} a_{11} & a_{12} & \cdots\cdots & a_{1n} \\ a_{21} & a_{22} & \cdots\cdots & a_{2n} \\ \vdots & & & \\ \vdots & & & \\ a_{n_{n1}} & & & a_{nn} \end{bmatrix} \begin{bmatrix} x_1 \\ x_2 \\ \vdots \\ \vdots \\ x_n \end{bmatrix} + \begin{bmatrix} b_{11} & b_{12} & \cdots\cdots & b_{1r} \\ \vdots & & & \\ \vdots & & & \\ b_{n1} & & & b_{nr} \end{bmatrix} \begin{bmatrix} u_1 \\ u_2 \\ \vdots \\ \vdots \\ u_n \end{bmatrix} \quad ...(6.9)$$

$$\begin{bmatrix} y_1 \\ y_2 \\ \vdots \\ y_n \end{bmatrix} = \begin{bmatrix} C_{11} & C_{12} & \cdots\cdots & C_{1n} \\ \vdots & & & \\ \vdots & & & \\ C_{m1} & \cdots\cdots & \cdots\cdots & C_{mn} \end{bmatrix} \begin{bmatrix} x_1 \\ x_2 \\ \vdots \\ x_n \end{bmatrix} \quad ...(6.10)$$

6.4.2. State Space Representation of *n*th Order Differential Equations

Consider the following examples :

6.4.2 (*a*) For *n*th order differential equation

EXAMPLE 6.1. A system is described by the differential equation

$$\frac{d^3y}{dt^3} + 6\frac{d^2y}{dt^2} + 11\frac{dy}{dt} + 10y = 8\,u(t)$$

where y is the output and u is the input to the system. Obtain state space representation of the system. *(R.M.L., University, Faizabad, 2003)*

Solution : Select the state variables as

$$x_1 = y,\ x_2 = \dot{y} \text{ and } x_3 = \ddot{y} \text{ then}$$

$$\dot{x}_1 = x_2$$

$$\dot{x}_2 = x_3$$

$$\dot{x}_3 = 8\,u(t) - 10\,x_1 - 11x_2 - 6x_3$$

The last equation is obtained from the given equation.

$$\begin{bmatrix} \dot{x}_1 \\ \dot{x}_2 \\ \dot{x}_3 \end{bmatrix} = \begin{bmatrix} 0 & 1 & 0 \\ 0 & 0 & 1 \\ -10 & -11 & -6 \end{bmatrix} \begin{bmatrix} x_1 \\ x_2 \\ x_3 \end{bmatrix} + \begin{bmatrix} 0 \\ 0 \\ 8 \end{bmatrix} u(t) \quad ...(6.11)$$

Compare equation (6.11) with equation (6.6) we get

$$A = \begin{bmatrix} 0 & 1 & 0 \\ 0 & 0 & 1 \\ -10 & -11 & -6 \end{bmatrix}, \quad B = \begin{bmatrix} 0 \\ 0 \\ 8 \end{bmatrix}, \quad x(t) = \begin{bmatrix} x_1 \\ x_2 \\ x_3 \end{bmatrix}$$

6.4.2 (b) : State space representation of nth order Linear system with r Forcing function

Consider the following example :

EXAMPLE 6.2. A system is described by the following differential equation. Represent the system in state space.

$$\frac{d^3x}{dt^3} + 3\frac{d^2x}{dt^2} + 4\frac{dx}{dt} + 4x = u_1(t) + 3u_2(t) + 4u_3(t)$$

and outputs are

$$y_1 = 4\frac{dx}{dt} + 3u_1$$

$$y_2 = \frac{d^2x}{dt^2} + 4u_2 + u_3$$

Solution : Select the state variables as

$$x_1 = x$$
$$\dot{x}_1 = \dot{x} = x_2$$
$$\dot{x}_2 = x_3$$
$$\dot{x}_3 = u_1(t) + 3u_2(t) + 4u_3(t) - 3x_3 - 4x_2 - 4x_1$$

$$\begin{bmatrix} \dot{x}_1 \\ \dot{x}_2 \\ \dot{x}_3 \end{bmatrix} = \begin{bmatrix} 0 & 1 & 0 \\ 0 & 0 & 1 \\ -4 & -4 & -3 \end{bmatrix} \begin{bmatrix} x_1 \\ x_2 \\ x_3 \end{bmatrix} + \begin{bmatrix} 0 & 0 & 0 \\ 0 & 0 & 0 \\ 1 & 3 & 4 \end{bmatrix} \begin{bmatrix} u_1 \\ u_2 \\ u_3 \end{bmatrix}$$

outputs :

$$y_1 = 4x_2 + 3u_1$$
$$y_2 = x_3 + 4u_2 + u_3$$

$$\begin{bmatrix} y_1 \\ y_2 \end{bmatrix} = \begin{bmatrix} 0 & 4 & 0 \\ 0 & 0 & 1 \end{bmatrix} \begin{bmatrix} x_1 \\ x_2 \\ x_3 \end{bmatrix} + \begin{bmatrix} 3 & 0 & 0 \\ 0 & 4 & 1 \end{bmatrix} \begin{bmatrix} u_1 \\ u_2 \\ u_3 \end{bmatrix}$$

6.4.3. State Space Representation for Transfer Function

Consider the following example :

EXAMPLE 6.3. For the given transfer function, obtain the state model.

$$G(s) = \frac{y(s)}{u(s)} = \frac{K}{s^3 + a_3 s^2 + a_2 s + a_1}$$

Solution : This transfer function has no zeros.

$$(s^3 + a_3 s^2 + a_2 s + a_1)\, y(s) = Ku(s)$$

or, $$s^3 y(s) + a_3 s^2 y(s) + a_2 s\, y(s) + a_1 y(s) = Ku(s)$$

Taking inverse laplace

$$\dddot{y}(t) + a_3 \ddot{y}(t) + a_2 \dot{y}(t) + a_1 y(t) = K u(t)$$

or $$\dddot{y}(t) = Ku(t) - a_3 \ddot{y}(t) - a_2 \dot{y}(t) - a_1 y(t)$$

Select the state variables as, first state variable as output

$$y(t) = x_1$$
$$\dot{y}(t) = \dot{x}_1 = x_2$$
$$\ddot{y}(t) = \dot{x}_2 = x_3$$
$$\dddot{y}(t) = \dot{x}_3$$
$$\therefore \quad \dddot{y}(t) = -a_3 x_3 - a_2 x_2 - a_1 x_1 + Ku(t)$$

Rewriting the equations

$$\dot{x}_1 = x_2$$
$$\dot{x}_2 = x_3$$
$$\dot{x}_3 = -a_3 x_3 - a_2 x_2 - a_1 x_1 + Ku(t)$$

$$\begin{bmatrix} \dot{x}_1 \\ \dot{x}_2 \\ \dot{x}_3 \end{bmatrix} = \begin{bmatrix} 0 & 1 & 0 \\ 0 & 0 & 1 \\ -a_1 & -a_2 & -a_3 \end{bmatrix} \begin{bmatrix} x_1 \\ x_2 \\ x_3 \end{bmatrix} + \begin{bmatrix} 0 \\ 0 \\ K \end{bmatrix} u(t)$$

$$y(t) = [1\ 0\ 0]\, x_1(t)$$

BLOCK DIAGRAM :

The block diagram of the given transfer function is shown in Fig. 6.3.

Now consider another case when the transfer function has zeros.

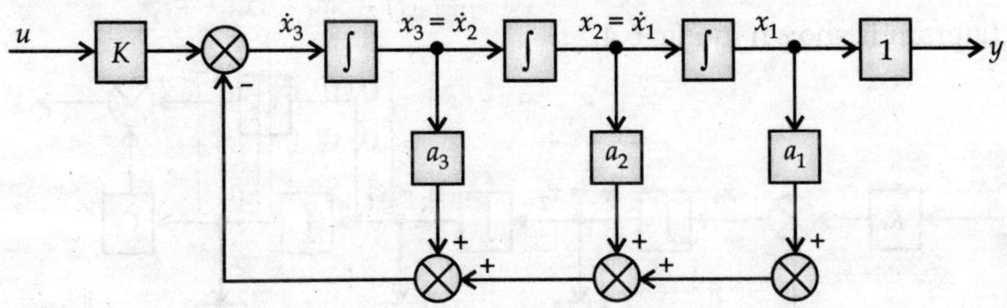

Fig. 6.3.

EXAMPLE 6.4. Obtain the state model for the given transfer function.

$$G(s) = \frac{y(s)}{u(s)} = \frac{K(C_2 s + C_1)}{s^3 + a_3 s^2 + a_2 s + a_1}$$

Solution : Break the transfer function in two parts

$$\frac{y(s)}{u(s)} = \frac{x_1(s)}{u(s)} \cdot \frac{y(s)}{x_1(s)} = \frac{K}{s^3 + a_3 s^2 + a_2 s + a_1} (C_2 s + C_1)$$

Now consider $\dfrac{x_1(s)}{u(s)} = \dfrac{K}{s^3 + a_3 s^2 + a_2 s + a_1}$

$[s^3 + a_3 s^2 + a_2 s + a_1]\, x_1(s) = Ku(s)$

Taking inverse laplace

$$\dddot{x}_1(t) + a_3 \ddot{x}_1(t) + a_2 \dot{x}_1(t) + a_1 x_1(t) = Ku(t)$$
$$\dddot{x}_1(t) = -a_3 \ddot{x}_1(t) - a_2 \dot{x}_1(t) - a_1 x_1(t) + Ku(t)$$

Select the state variables as

$$\dot{x}_1 = x_2$$
$$\ddot{x}_1 = \dot{x}_2 = x_3$$

$$\dddot{x}_1 = \dot{x}_3$$

Rewrite
$$\dot{x}_1 = x_2$$
$$\dot{x}_2 = x_3$$
$$\dot{x}_3 = -a_3 x_3 - a_2 x_2 - a_1 x_1 + Ku$$

$$\begin{bmatrix} \dot{x}_1 \\ \dot{x}_2 \\ \dot{x}_3 \end{bmatrix} = \begin{bmatrix} 0 & 1 & 0 \\ 0 & 0 & 1 \\ -a_1 & -a_2 & -a_3 \end{bmatrix} \begin{bmatrix} x_1 \\ x_2 \\ x_3 \end{bmatrix} + \begin{bmatrix} 0 \\ 0 \\ K \end{bmatrix} u(t)$$

Consider,
$$\frac{y(s)}{x_1(s)} = C_2 s + C_1$$
$$y(s) = x_1(s)\,[C_2\, s + C_1]$$

Take inverse Laplace
$$y(t) = C_1 x_1 + C_2 x_2$$
$$y(t) = [C_1 \;\; C_2 \;\; 0]\, x(t)$$
$$= [C_1 \;\; C_2 \;\; 0] \begin{bmatrix} x_1 \\ x_2 \\ x_3 \end{bmatrix}$$

The block diagram is shown in Fig. 6.4.

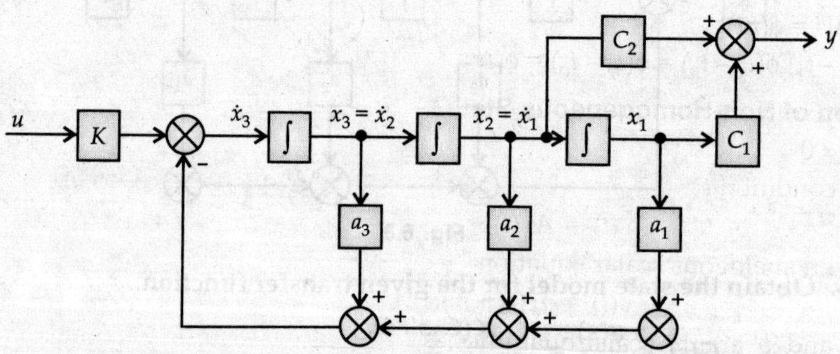

Fig. 6.4.

6.5. SOLUTION OF THE TIME-INVARIANT STATE EQUATION

6.5.1. Solution of Homogeneous State Equation : Laplace Transform Method

We know that
$$\dot{x}(t) = Ax(t) + Bu(t) \qquad \text{...(6.12)}$$
$u(t) = 0$ for unforced response

then
$$\dot{x}(t) = Ax(t) \qquad \text{...(6.13)}$$
let us consider the analogous scalar equation
$$\dot{x}(t) = ax(t) \qquad \text{...(6.14)}$$
Take the Laplace transform of equation 6.14
$$sX(s) - x(0) = aX(s)$$

$$(s - a)\, X(s) = x(0)$$

or, $$X(s) = (s - a)^{-1}\, x(0) \quad \text{...(6.15)}$$

Take the inverse laplace of equation 6.15

$$x(t) = e^{at}\, x(0) \quad \text{...(6.16)}$$

If equation (6.16) is the solution of equation (6.14) then the solution of equation (6.13)

$$x(t) = e^{At}\, x(0)$$

$e^{At} = \phi(t)$ = State Transition Matrix (STM)

$$= I + At + \frac{A^2t^2}{2!} + \frac{A^3t^3}{3!} + ... = \sum_{i=0}^{n} \frac{A^i t^i}{i!} \quad \text{...(6.17)}$$

$$\phi(t) = £^{-1}\phi(s) = £^{-1}[sI - A]^{-1}$$

where $\phi(s)$ = Resolvant matrix

6.5.2. Properties of State Transition Matrices

For time invariant system

$$\dot{x} = Ax \text{ and}$$

$$\phi(t) = e^{At} = I + At + \frac{A^2t^2}{2!} + \frac{A^3t^3}{3!} + ...$$

(*i*) $\phi(0) = e^{A(0)} = I$

(*ii*) $\phi(t) = e^{At} = (e^{-At})^{-1} = [\phi(-t)]^{-1}$

(*iii*) $\phi^{-1}(t) = \phi(-t)$

(*iv*) $\phi(t_1 + t_2) = e^{A(t_1+t_2)} = e^{At_1}\, e^{At_2} = \phi(t_1)\, \phi(t_2) = \phi(t_2)\, \phi(t_1)$

(*v*) $[\phi(t)]^n = \phi(nt)$

(*vi*) $\phi(t_2 - t_1)\, \phi(t_1 - t_0) = \phi(t_2 - t_0) = \phi(t_1 - t_0)\, \phi(t_2 - t_1)$

6.5.3. Solution of Non-Homogeneous State Equations

Here $u(t) \neq 0$

(*a*) Initial condition $t = 0$

$$\dot{x}(t) = Ax(t) + Bu(t) \quad \text{...(6.18)}$$

Consider an analogous scalar equation

$$\dot{x}(t) = ax(t) + bu(t) \quad \text{...(6.19)}$$

where '*a*' and '*b*' are the scalar quantities. Take laplace transform of equation (6.19).

$$sX(s) - x(0) = aX(s) + bu(s)$$

or, $$(s - a)\, X(s) = x(0) + bu(s)$$

Multiplying both the sides by $(s - a)^{-1}$ we get

$$X(s) = (s - a)^{-1}\, x(0) + (s - a)^{-1}\, bu(s) \quad \text{...(6.20)}$$

Take inverse laplace of equation (6.20)

$$x(t) = e^{at}\, x(0) + \int_0^t e^{a(t-\tau)}\, bu(\tau)\, d\tau \quad \text{...(6.21)}$$

Equation (6.21) is the solution of equation (6.19), then the solution of equation (6.18) will be

$$x(t) = e^{At}\, x(0) + \int_0^t e^{A(t-\tau)}\, bu(\tau)\, d\tau \quad \text{...(6.22)}$$

or, $$x(t) = \phi(t)\,x(0) + \int_0^t \phi(t-\tau)\,bu(\tau)\,d\tau \quad ...(6.23)$$

(b) When initial time is given by t_0 i.e $t = t_0$

$$x(t) = \phi(t)\,x(0) + \int_0^t \phi(t-\tau)\,Bu(\tau)\,d\tau$$

at $$t = t_0$$

$$x(t_0) = \phi(t_0)\,x(0) + \int_0^{t_0} \phi(t_0-\tau)\,Bu(\tau)\,d\tau \quad ...(6.24)$$

Multiply both the sides by $\phi^{-1}(t_0)$

$$\phi^{-1}(t_0)\,x(t_0) = x(0) + \int_0^{t_0} \phi(-t_0)\,\phi(t_0-\tau)\,Bu(\tau)\,d\tau = x(0) + \int_0^{t_0} \phi(-t_0+t_0-\tau)\,Bu(\tau)\,d\tau$$

$$\phi^{-1}(t_0)\,x(t_0) = x(0) + \int_0^{t_0} \phi(-\tau)\,Bu(\tau)\,d\tau$$

or $$x(0) = \phi(-t_0)\,x(t_0) - \int_0^{t_0} \phi(-\tau)\,Bu(\tau)\,d\tau \quad ...(6.25)$$

Put the value of $x(0)$ from equation (6.25) in equation (6.23)

$$x(t) = \left[\phi(-t_0)\,x(t_0) - \int_0^{t_0} \phi(-\tau)\,Bu(\tau)\,d\tau\right]\phi(t) + \int_0^t \phi(t-\tau)\,Bu(\tau)\,d\tau$$

or, $$x(t) = \phi(t-t_0)x(t_0) + \int_{t_0}^t \phi(t-\tau)\,Bu(\tau)\,d\tau \quad ...(6.26)$$

Equation (6.26) can also be written as

$$x(t) = e^{A(t-t_0)}x(t_0) + \int_{t_0}^t e^{A(t-\tau)}\,Bu(\tau)\,d\tau \quad ...(6.27)$$

EXAMPLE 6.5. Find the time response of the system described by the equation

$$\dot{x}(t) = \begin{bmatrix} -1 & 1 \\ 0 & -2 \end{bmatrix} x(t) + \begin{bmatrix} 0 \\ 1 \end{bmatrix} u(t)$$

$$x(0) = \begin{bmatrix} -1 \\ 0 \end{bmatrix},\; u(t) = 1,\; t > 0$$

Solution : $$x(t) = \phi(t)\,x(0) + \int_0^t \phi(t-\tau)\,Bu(\tau)\,d\tau \quad ...(6.28)$$

Step 1 : Calculation of $\phi(t)$:

$$\phi(t) = \mathcal{L}^{-1}\phi(s) = \mathcal{L}^{-1}(sI - A)^{-1}$$

$$\phi(s) = [sI - A]^{-1}$$

$$sI - A = \begin{bmatrix} s & 0 \\ 0 & s \end{bmatrix} - \begin{bmatrix} -1 & 1 \\ 0 & -2 \end{bmatrix} = \begin{bmatrix} s+1 & -1 \\ 0 & s+2 \end{bmatrix}$$

$$[sI - A]^{-1} = \begin{bmatrix} \frac{s+2}{s^2+3s+2} & \frac{1}{s^2+3s+2} \\ 0 & \frac{s+1}{s^2+3s+2} \end{bmatrix}$$

$$\therefore \qquad \phi(t) = £^{-1}\,\phi(s) = \begin{bmatrix} £^{-1}\frac{s+2}{s^2+3s+2} & £^{-1}\frac{1}{s^2+3s+2} \\ 0 & £^{-1}\frac{s+1}{s^2+3s+2} \end{bmatrix}$$

$$\therefore \qquad \phi(t) = \begin{bmatrix} e^{-t} & e^{-t} - e^{-2t} \\ 0 & e^{-2t} \end{bmatrix}$$

Put all the values in equation (6.28)

$$x(t) = \begin{bmatrix} e^{-t} & e^{-t} - e^{-2t} \\ 0 & e^{-2t} \end{bmatrix} \begin{bmatrix} -1 \\ 0 \end{bmatrix} + \int_0^t \begin{bmatrix} e^{-(t-\tau)} & e^{-(t-\tau)} - e^{-2(t-\tau)} \\ 0 & e^{-2(t-\tau)} \end{bmatrix} \begin{bmatrix} 0 \\ 1 \end{bmatrix} 1\, d\tau$$

$$= \begin{bmatrix} -e^{-t} \\ 0 \end{bmatrix} + \int_0^t \begin{bmatrix} e^{-(t-\tau)} - e^{-2(t-\tau)} \\ e^{-2(t-\tau)} \end{bmatrix} d\tau$$

$$x_1(t) = -e^{-t} + \int_0^t e^{-(t-\tau)} - e^{-2(t-\tau)}\, d\tau$$

$$x_2(t) = \int_0^t e^{-2(t-\tau)}\, d\tau$$

or, $\quad x_1(t) = \frac{1}{2} - 2e^{-t} + \frac{1}{2}e^{-2t}$ **Ans.** ...(6.29)

$$x_2(t) = \frac{1}{2} - \frac{1}{2}e^{-2t}$$

if $\quad y = Cx \qquad C = [1\ 0]$

$$y = [1\ 0] \begin{bmatrix} \frac{1}{2} - 2e^{-t} + \frac{1}{2}e^{2t} \\ \frac{1}{2} - \frac{1}{2}e^{-2t} \end{bmatrix} = \frac{1}{2} - 2e^{-t} + \frac{1}{2}e^{-2t} \qquad ...(6.30)$$

Equations 6.29 and 6.30 are the required answers.

6.6. TRANSFER MATRIX

We know that

$$\dot{x}(t) = Ax(t) + Bu(t) \qquad ...(6.31)$$

$$y(t) = Cx(t) \qquad ...(6.32)$$

Take the laplace transform, assuming initial conditions zero*

$$s\,X(s) = AX(s) + Bu(s) \qquad ...(6.33)$$

$$Y(s) = CX(s) \qquad ...(6.34)$$

Consider the equation 6.33, I is the identity martix

$$sI\,X(s) = AX(s) + Bu(s)$$

or,
$$X(s)\,[sI - A] = Bu(s)$$

Multiplying both the side by $[sI - A]^{-1}$ we get

$$X(s) = u(s)[sI - A]^{-1}B \qquad \text{...(6.35)}$$

Put the values of $X(s)$ from equation (6.35) in equation (6.34)

$$Y(s) = C[sI - A]^{-1}\,Bu(s) \qquad \text{...(6.36)}$$

We know that

$$Y(s) = G(s)\,u(s) \qquad \text{...(6.37)}$$

Compare the equation 6.36 and equation 6.37 we get

$$G(s) = C[sI - A]^{-1}\,B \qquad \text{...(6.38)}$$

where $G(s)$ = Transfer matrix

The transfer matrix given by the equation 6.38 is for MIMO system. For single input single output system, the transfer matrix is given by

$$G(s) = C^T[sI - A]^{-1}\,b \qquad \text{...(6.39)}$$

Also, $[sI - A]^{-1} = \phi(s)$ = Resolvant matrix, Then equation (6.39) becomes

$$G(s) = C\phi(s)\,B \qquad \text{...(6.40)} \qquad \text{For MIMO system}$$

$$G(s) = C^T\,\phi(s)\,b \qquad \text{...(6.41)} \qquad \text{For SISO system}$$

EXAMPLE 6.6. The system equations are given by

$$\dot{x}(t) = \begin{bmatrix} 0 & 1 \\ -2 & -3 \end{bmatrix} x(t) + \begin{bmatrix} 0 \\ 1 \end{bmatrix} u(t)$$

$$y(t) = [1 \quad 0]\,x(t)$$

Find the transfer function of the system.

Solution : Given that

$$A = \begin{bmatrix} 0 & 1 \\ -2 & -3 \end{bmatrix}, B = \begin{bmatrix} 0 \\ 1 \end{bmatrix}, C^T = [1 \quad 0]$$

$$I = \begin{bmatrix} 1 & 0 \\ 0 & 1 \end{bmatrix}$$

$$sI - A = \begin{bmatrix} s & -1 \\ 2 & s+3 \end{bmatrix} \quad \therefore \quad [sI - A]^{-1} = \begin{bmatrix} \dfrac{s+3}{s^2+3s+2} & \dfrac{1}{s^2+3s+2} \\ -\dfrac{2}{s^2+3s+2} & \dfrac{s}{s^2+3s+2} \end{bmatrix}$$

$$C^T[sI - A]^{-1}.\,b = [1 \quad 0] \begin{bmatrix} \dfrac{s+3}{s^2+3s+2} & \dfrac{1}{s^2+3s+2} \\ \dfrac{-2}{s^2+3s+2} & \dfrac{s}{s^2+3s+2} \end{bmatrix} \begin{bmatrix} 0 \\ 1 \end{bmatrix} = \frac{1}{s^2+3s+2}$$

Hence, required transfer function $= \dfrac{1}{s^2+3s+2}$ **Ans.**

6.7. COMPUTATION OF STATE TRANSITION MATRIX : e^{At}

6.7.1. Laplace Transform Method

We know that

$$\dot{x}(t) = Ax + Bu(t)$$

For unforced system $u(t) = 0$

$$\dot{x}(t) = Ax$$

Take the laplace transform

$$s\,X(s) - x(0) = A\,X(s)$$

or $$[sI - A]\,X(s) = x(0)$$

Multiply both the sides by $[sI - A]^{-1}$

$$X(s) = [sI - A]^{-1}\,x(0)$$

$[sI - A]^{-1} = \phi(s) =$ Resolvant matrix

$$\phi(t) = e^{At} = \pounds^{-1}\,\phi(s) = \pounds^{-1}[sI - A]^{-1}$$

where, $\phi(t) = e^{At} =$ state transition matrix

EXAMPLE 6.7. Compute the STM when

$$A = \begin{bmatrix} -1 & 1 \\ 0 & 2 \end{bmatrix}$$

Solution : $$sI - A = \begin{bmatrix} s & 0 \\ 0 & s \end{bmatrix} - \begin{bmatrix} -1 & 1 \\ 0 & 2 \end{bmatrix} = \begin{bmatrix} s+1 & -1 \\ 0 & s-2 \end{bmatrix}$$

$$[sI - A]^{-1} = \begin{bmatrix} \dfrac{s-2}{s^2+3s+2} & \dfrac{1}{s^2+3s+2} \\ 0 & \dfrac{s+1}{s^2+3s+2} \end{bmatrix}$$

$$\phi(t) = \pounds^{-1}[sI - A]^{-1} = \begin{bmatrix} \pounds^{-1}\dfrac{s-2}{s^2+3s+2} & \pounds^{-1}\dfrac{1}{s^2+3s+2} \\ 0 & \pounds^{-1}\dfrac{s+1}{s^2+3s+2} \end{bmatrix}$$

$$\phi(t) = \begin{bmatrix} e^{-t} & e^{-t} - e^{-2t} \\ 0 & e^{-2t} \end{bmatrix}$$

6.7.2. Series Summation Method

$$\phi(t) = e^{At} = I + At + \frac{A^2t^2}{2!} + ...$$

EXAMPLE. 6.8. Evaluate the STM by series summation method

$$\mathbf{A} = \begin{bmatrix} 2 & -2 & 3 \\ 1 & 1 & 1 \\ 1 & 3 & -1 \end{bmatrix}$$

Solution : Given that

$$A = \begin{bmatrix} 2 & -2 & 3 \\ 1 & 1 & 1 \\ 1 & 3 & -1 \end{bmatrix}$$

$$\therefore \quad A^2 = \begin{bmatrix} 5 & 3 & 1 \\ 4 & 2 & 3 \\ 4 & -2 & 7 \end{bmatrix}, A^3 = \begin{bmatrix} 14 & -4 & 17 \\ 13 & 3 & 11 \\ 13 & 11 & 3 \end{bmatrix} \text{ and so on.}$$

* By the definition of transfer function.

$$e^{At} = \begin{bmatrix} 1 & 0 & 0 \\ 0 & 1 & 0 \\ 0 & 0 & 1 \end{bmatrix} + \begin{bmatrix} 2t & -2t & 3t \\ t & t & t \\ t & 3t & -t \end{bmatrix} + \frac{t^2}{2}\begin{bmatrix} 5 & 3 & 1 \\ 4 & 2 & 3 \\ 4 & -2 & 7 \end{bmatrix} + \frac{t^3}{6}\begin{bmatrix} 14 & -4 & 17 \\ 13 & 3 & 11 \\ 13 & 11 & 3 \end{bmatrix} + ..$$

$$= \begin{bmatrix} 1+2t+\frac{5}{2}t^2+\frac{14}{6}t^3+... & -2t+\frac{3t^2}{2}-\frac{2t^2}{3}+... & 3t+\frac{t^2}{2}+\frac{17}{6}t^3+... \\ t+2t^2+\frac{13}{6}t^3+... & 1+t+t^2+... & t+\frac{3t^2}{2}+\frac{11t^3}{6}+... \\ t+2t^2+\frac{13}{6}t^3+... & 3t-t^2+\frac{11t^3}{6}+... & 1-t+\frac{7t^2}{2}+\frac{3t^3}{6}+... \end{bmatrix}$$

6.7.3 Eigen Values

The roots of the characteristic equation are known as eigen values of matrix.

6.7.4. Eigen Vectors

Any non-zero vector P_i that satisfies the matrix equation $(\lambda_i I - A)\ P_i = 0$ where λ_i is i[th] eigen value of A is known as eigen vector of A associated with the eigen value λ_i.

6.7.5. Use of Diagonal

$$\text{let } A = \begin{bmatrix} \lambda_1 & 0 & 0 \\ 0 & \lambda_2 & 0 \\ 0 & 0 & \lambda_3 \end{bmatrix}$$

$$e^{At} = I + At + \frac{A^2t^2}{2!} + ...$$

$$e^{At} = \begin{bmatrix} 1 & 0 & 0 \\ 0 & 1 & 0 \\ 0 & 0 & 1 \end{bmatrix} + \begin{bmatrix} \lambda_1 t & 0 & 0 \\ 0 & \lambda_2 t & 0 \\ 0 & 0 & \lambda_3 t \end{bmatrix} + ... = \begin{bmatrix} e^{\lambda_1 t} & 0 & 0 \\ 0 & e^{\lambda_2 t} & 0 \\ 0 & 0 & e^{\lambda_3 t} \end{bmatrix}$$

let λ_i is the eigen value of A and P_i is the eigen vector, then for each of the n eigen values of A

$$AP_1 = \lambda_1 P_1$$
$$AP_2 = \lambda_2 P_2$$
$$: :$$
$$AP_n = \lambda_n P_n$$

or $$AM = M\wedge \qquad ...(6.42)$$

where $\wedge$ = diagonal matrix with eigen values on its main diagonal

M = modal matrix

$$\wedge = M^{-1}AM$$

or $$e^{\wedge t} = M^{-1}e^{At}M$$

or $$e^{At} = M\, e^{\wedge t}\, M^{-1}$$

EXAMPLE 6.9. Compute e^{At} when

$$\mathbf{A} = \begin{bmatrix} 0 & 1 \\ -3 & -4 \end{bmatrix}$$

Solution : $$\lambda I - A = \lambda\begin{bmatrix} 1 & 0 \\ 0 & 1 \end{bmatrix} - \begin{bmatrix} 0 & 1 \\ -3 & -4 \end{bmatrix} = \begin{bmatrix} \lambda & -1 \\ 3 & \lambda+4 \end{bmatrix} \qquad ...(6.43)$$

$$|\lambda I - A| = \lambda^2 + 4\lambda + 3 = 0$$

or,
$$\lambda_1 = -1$$
$$\lambda_2 = -3$$

let the eigen vector

$$P_1 = \begin{bmatrix} p_{11} \\ p_{21} \end{bmatrix} \text{ and } P_2 = \begin{bmatrix} p_{12} \\ p_{22} \end{bmatrix}$$

Put $\lambda = -1$ in equation (6.43)

$$\begin{bmatrix} -1 & -1 \\ 3 & 3 \end{bmatrix}\begin{bmatrix} p_{11} \\ p_{21} \end{bmatrix} = \begin{bmatrix} 0 \\ 0 \end{bmatrix}$$

$$-p_{11} - p_{21} = 0$$
$$3p_{11} + 3p_{21} = 0$$

let $p_{11} = 1$ then $p_{21} = -1$

$$\therefore \quad P_1 = \begin{bmatrix} 1 \\ -1 \end{bmatrix}$$

Put $\lambda = -3$

$$\begin{bmatrix} -3 & -1 \\ 3 & 1 \end{bmatrix}\begin{bmatrix} p_{12} \\ p_{22} \end{bmatrix} = \begin{bmatrix} 0 \\ 0 \end{bmatrix}$$

$$-3p_{12} - p_{22} = 0$$
$$3p_{12} + p_{22} = 0$$

let $p_{12} = 1$ then $p_{22} = -3$

$$p_2 = \begin{bmatrix} 1 \\ -3 \end{bmatrix}$$

$$\therefore \quad M = [P_1 \; P_2] = \begin{bmatrix} 1 & 1 \\ -1 & -3 \end{bmatrix}$$

$$M^{-1} = \begin{bmatrix} 1.5 & 0.5 \\ -0.5 & -0.5 \end{bmatrix}$$

$$\wedge = \begin{bmatrix} -1 & 0 \\ 0 & -3 \end{bmatrix}$$

$$\therefore \quad e^{At} = M e^{\wedge t} M^{-1} = \begin{bmatrix} 1 & 1 \\ -1 & -3 \end{bmatrix}\begin{bmatrix} e^{-t} & 0 \\ 0 & e^{-3t} \end{bmatrix}\begin{bmatrix} 1.5 & 0.5 \\ -0.5 & -0.5 \end{bmatrix}$$

$$e^{At} = \begin{bmatrix} 1.5\,e^{-t} - 0.5\,e^{-3t} & 0.5e^{-t} - 0.5e^{-3t} \\ 1.5e^{-3t} - 1.5e^{-t} & 1.5e^{-3t} - 0.5e^{-t} \end{bmatrix} \quad \textbf{Ans.}$$

6.8. BLOCK DIAGRAM OF A LINEAR SYSTEM IN STATE VARIABLE FORM

The block diagram or state diagram is the interconnection of there units : (*i*) Scalars, (*ii*) Adders, (*iii*) Integrators. Scalars are amplifiers having required gain.

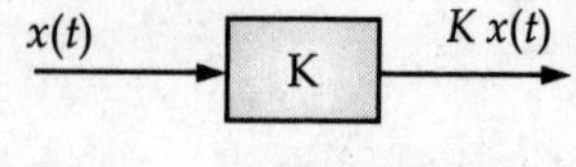

(i) Scalar

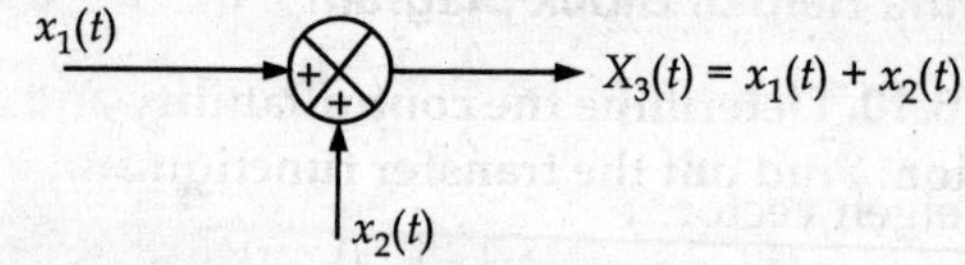

(ii) Adder

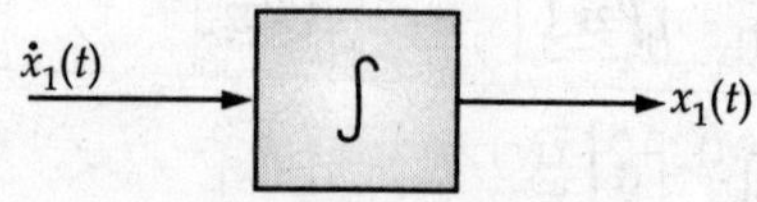

Integrator

6.8.1 MIMO System

The system is represented by the following equations

$$\dot{x} = Ax + Bu$$

$$y = Cx + Du$$

The block diagram is shown in the Fig. 6.5.

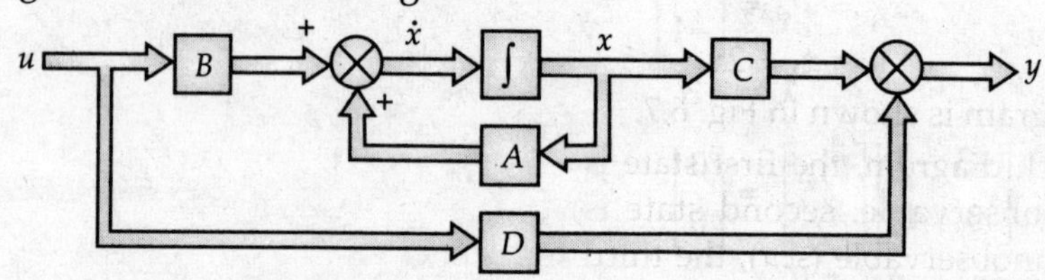

Fig. 6.5.

6.8.2. Single Input Single Output System

$$\dot{x}(t) = Ax + bu$$

$$y(t) = C^T x + du$$

The block diagram is shown in Fig. 6.6.

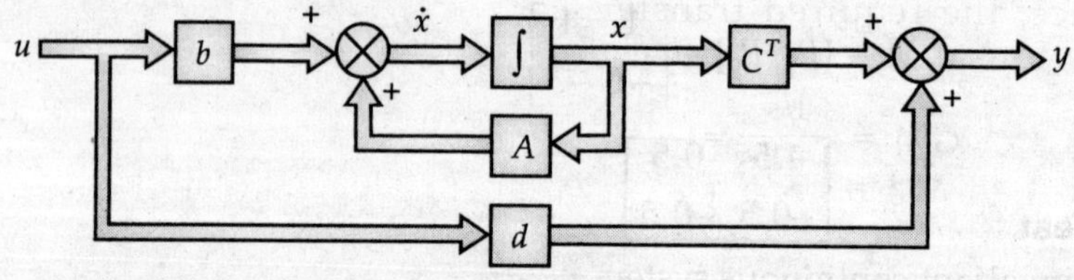

Fig. 6.6.

6.9. CONTROLLABILITY AND OBSERVABILITY

A system is said to be controllable if any initial state $x(t_0)$ or x_0 can be transferred to any final state $x(t_f)$ in a finite time interval $(t_f - t_0)$, $t \geq 0$ by some control u.

A system is said to be observable if every state x_0 can be exactly determined from the measurement of the output 'y' over a finite interval of time $0 \leq t \leq t_f$.

The concept of controllability and observability were introduced by Kalman and they play an important role in the design of control system in state space.

There are four possible states for the system.

(i) The system is controllable and observable

(ii) The system is controllable and unobservable

(iii) The system is uncontrollable and observable

(iv) The system is uncontrollable and unobservable

Consider the following example to determine the controllability and observability.

6.9.1. With the Help of Block Diagram

EXAMPLE 6.10. Determine the controllability and observability of the system described by the state equation. Find out the transfer function and draw the block diagram.

$$\dot{x}(t) = \begin{bmatrix} -1 & 0 & 0 \\ 0 & -2 & 0 \\ 0 & 0 & -3 \end{bmatrix} x(t) + \begin{bmatrix} 1 \\ 1 \\ 0 \end{bmatrix} u(t)$$

$$y(t) = \begin{bmatrix} 1 & 0 & +2 \end{bmatrix} x(t)$$

Solution :

$$\dot{x}(t) = \begin{bmatrix} -1 & 0 & 0 \\ 0 & -2 & 0 \\ 0 & 0 & -3 \end{bmatrix} \begin{bmatrix} x_1 \\ x_2 \\ x_3 \end{bmatrix} + \begin{bmatrix} 1 \\ 1 \\ 0 \end{bmatrix} u$$

$$\dot{x}_1 = -x_1 + u$$

$$\dot{x}_2 = -2x_2 + u$$

$$\dot{x}_3 = -3x_3$$

$$y = x_1 + 2x_3$$

The block diagram is shown in Fig. 6.7.

From the block diagram, the first state is controllable and observable, second state is controllable and unobservable (*scu*), the third state is uncontrollable and observable (*suo*). Since, all the states are not controllable and obserbvable the whole system is uncontrollable and unobservable.

The transfer function is available only for the state which is both controllable and observable. Hence, the required transfer function is

$$G(s) = \frac{1/s}{1+1/s} = \frac{1}{s+1}$$

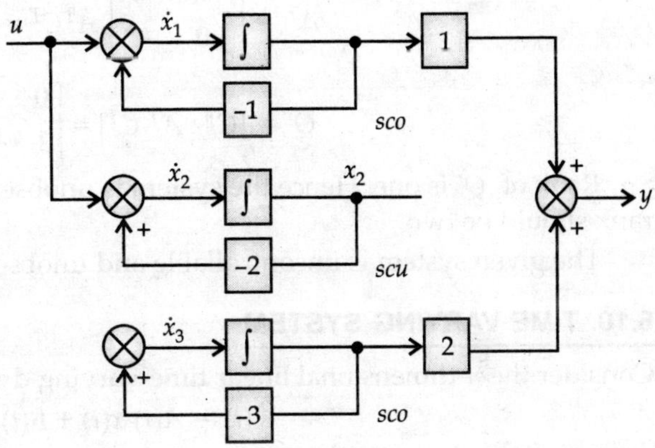

Fig. 6.7.

6.9.2. Kalman's Test

A linear time invariant continuous system described by the state equation

$$\dot{x} = Ax + Bu \qquad \text{...(6.44)}$$

$$y = Cx$$

is completely controllable if and only if the rank of the controllability matrix is defined as

$$Q = [B : AB : A^2 B : ....A^{n-1}B]$$

is equal to rank 'n'.

The system represented by the equation (6.44) is observable if and only if, the $n \times n_m$ composite matrix is obtained as

$$Q = [C^T : A^T C^T\ (A^T)^2 C^T: .....A^{T(n-1)} C^T]$$

EXAMPLE 6.11. Consider the following system

$$\begin{bmatrix} \dot{x}_1 \\ \dot{x}_2 \end{bmatrix} = \begin{bmatrix} -0.5 & 0 \\ 0 & -2 \end{bmatrix} \begin{bmatrix} x_1 \\ x_2 \end{bmatrix} + \begin{bmatrix} 0 \\ 1 \end{bmatrix} u(t)$$

$$y(t) = \begin{bmatrix} 0 & 1 \end{bmatrix} \begin{bmatrix} x_1 \\ x_2 \end{bmatrix}$$

Test for controllability and observability.

Solution :

$$Q = [B : AB]$$

Given that $A = \begin{bmatrix} -0.5 & 0 \\ 0 & -2 \end{bmatrix}, B = \begin{bmatrix} 0 \\ 1 \end{bmatrix}$

$$AB = \begin{bmatrix} -0.5 & 0 \\ 0 & -2 \end{bmatrix} \begin{bmatrix} 0 \\ 1 \end{bmatrix} = \begin{bmatrix} 0 \\ -2 \end{bmatrix}$$

$\therefore$ $Q = \begin{bmatrix} 0 & 0 \\ 1 & -2 \end{bmatrix}$

Rank of 'Q' is one. Hence, system is uncontrollable.*

$$C = \begin{bmatrix} 0 & 1 \end{bmatrix} \therefore C^T = \begin{bmatrix} 0 \\ 1 \end{bmatrix}$$

$$A^T = \begin{bmatrix} -0.5 & 0 \\ 0 & -2 \end{bmatrix} A^T C^T = \begin{bmatrix} -0.5 & 0 \\ 0 & -2 \end{bmatrix} \begin{bmatrix} 0 \\ 1 \end{bmatrix} = \begin{bmatrix} 0 \\ -2 \end{bmatrix}$$

$$Q' = [C^T : A^T C^T] = \begin{bmatrix} 0 & 0 \\ 1 & -2 \end{bmatrix}$$

Rank of 'Q' is one. Hence the system is unobservable. For controllability and observability the rank should be two.

The given system is uncontrollable and unobservable.

6.10. TIME VARYING SYSTEM

Consider the n-dimensional linear time-varying dynamical equation

$$\begin{aligned} \dot{x}(t) &= A(t)\, x(t) + B(t)\, u(t) \\ y(t) &= C(t)\, x(t) + D(t)\, u(t) \end{aligned} \quad ...(6.45)$$

where,

$x(t)$ is an n column vector, $A(t)$ and $B(t)$ are respectively $n \times n$ matrix and $n \times m$ matrix whose elements are function of time, $u(t)$ is m column vector of forcing function.

The solution of equation (6.45) with $x(t_0) = x_0$ is given by

$$x(t) = \phi(t, t_0)\, x_0 + \int_{t_0}^{t} \phi(t, \tau)\, B(\tau)\, u(\tau)\, d\tau \quad ...(6.46)$$

$$x(t) = \phi(t, t_0) \left[x_0 + \int_{t_0}^{t} \phi(t_0, \tau)\, B(\tau)\, u(\tau)\, d\tau \right] \quad ...(6.47)$$

where $\phi(t, \tau)$ is the state transition matrix of $\dot{x} = A(t)\, x$, or, equivalently, the unique solution of

$$\frac{\partial}{\partial t} \phi(t, \tau) = A(t)\, \phi(t, \tau) \qquad \phi(\tau, \tau) = I \quad ...(6.48)$$

Equation (6.47) is obtained from the equation (6.46) by using

*For controllability and observability the rank should be two.

$$\phi(t, \tau) = \phi(t, t_0)\, \phi(t_0, \tau) \qquad ...(6.49)$$

$\phi(t, t_0)$ is given by the series

$$\phi(t, t_0) = I + \int_0^t A(\tau_1)\, d\tau_1 + \int_{t_0}^t A(\tau_1)\left[\int_{t_0}^{\tau_1} A(\tau_2)\, d\tau_2\right] d\tau_1 + ...$$

EXAMPLE 6.12. Determine the state transition matrix $\phi(t, 0)$ for the system

$$\dot{x} = \begin{bmatrix} 0 & 1 \\ t & 0 \end{bmatrix}\begin{bmatrix} x_1 \\ x_2 \end{bmatrix}$$

Solution : $t_0 = 0$

$$\int_0^t A(\tau)\, d\tau = \int_0^t \begin{bmatrix} 0 & 1 \\ \tau & 0 \end{bmatrix} d\tau = \begin{bmatrix} 0 & t \\ t^2/2 & 0 \end{bmatrix}$$

$$\int_0^t A(\tau_1)\left[\int_0^{\tau_1} A(\tau_2)\, d\tau_2\right] d\tau_1 = \int_0^t \begin{bmatrix} 0 & 1 \\ \tau_1 & 0 \end{bmatrix}\begin{bmatrix} 0 & \tau_1 \\ \tau_1^2/2 & 0 \end{bmatrix} d\tau_1 = \begin{bmatrix} t^3/6 & 0 \\ 0 & t^3/3 \end{bmatrix}$$

$$\int_0^t \begin{bmatrix} 0 & 1 \\ \tau_1 & 0 \end{bmatrix}\begin{bmatrix} \tau_1^3/6 & 0 \\ 0 & \tau_1^3/3 \end{bmatrix} d\tau_1 = \begin{bmatrix} 0 & t^4/12 \\ t^5/30 & 0 \end{bmatrix}$$

$$\therefore \quad \phi(t, 0) = \begin{bmatrix} 1 + \frac{t^3}{6} + ... & t + \frac{t^4}{12} + ... \\ \frac{t^2}{2} + \frac{t^5}{30} + ... & 1 + \frac{t^3}{3} + ... \end{bmatrix}$$

6.11. CONTROLLABILITY AND OBSERVABILITY TIME VARIANT SYSTEMS

Consider n-dimensional state equation

$$\dot{x} = A(t)\, x(t) + B(t)\, u(t)$$

Assume that the matrices A and B in the above state equation are $n - 1$ times continuously differentiable. Then the state equation is controllable at t_0 if there exists a finite $t_1 > t_0$ such that

$$P[M_0(t_1) : M_1(t_1) : ..... : M_{n-1}(t_1)] = n \qquad ...(6.50)$$

$$M_{K+1}(t) = -A(t)\, M_K(t) + \frac{d}{dt} M_K(t) \qquad ...(6.51)$$

$K = 0, 1, 2 .....n - 1$

$$M_0(t) = B(t) \qquad ...(6.52)$$

EXAMPLE 6.13. Consider

$$\begin{bmatrix} \dot{x}_1 \\ \dot{x}_2 \\ \dot{x}_3 \end{bmatrix} = \begin{bmatrix} t & 1 & 0 \\ 0 & t & 0 \\ 0 & 0 & t^2 \end{bmatrix}\begin{bmatrix} x_1 \\ x_2 \\ x_3 \end{bmatrix} + \begin{bmatrix} 0 \\ 1 \\ 1 \end{bmatrix} u$$

From equation 6.52

$$M_0(t) = \begin{bmatrix} 0 \\ 1 \\ 1 \end{bmatrix}$$

$$M_1(t) = -A(t)\,M_0(t) + \frac{d}{dt}M_0(t) = -\begin{bmatrix} t & 1 & 0 \\ 0 & t & 0 \\ 0 & 0 & t^2 \end{bmatrix}\begin{bmatrix} 0 \\ 1 \\ 1 \end{bmatrix} + \frac{d}{dt}\begin{bmatrix} 0 \\ 1 \\ 1 \end{bmatrix} = \begin{bmatrix} -1 \\ -t \\ -t^2 \end{bmatrix}$$

$$M_2(t) = -A(t)\,M_1(t) + \frac{d}{dt}M_1(t) = -\begin{bmatrix} t & 1 & 0 \\ 0 & t & 0 \\ 0 & 0 & t^2 \end{bmatrix}\begin{bmatrix} -1 \\ -t \\ -t^2 \end{bmatrix} + \frac{d}{dt}\begin{bmatrix} -1 \\ -t \\ -t^2 \end{bmatrix}$$

$$M_2(t) = \begin{bmatrix} 2t \\ t^2 - 1 \\ t^4 - 2t \end{bmatrix}$$

$$[M_0(t) : M_1(t) : M_2(t)] = \begin{bmatrix} 0 & -1 & 2t \\ 1 & -t & t^2 - 1 \\ 1 & -t^2 & t^4 - 1 \end{bmatrix}$$

The above matrix has rank 3 for all $t \neq 0$, hence the system is controllable at every time 't'.

OBSERVABILITY : Consider the state equation

$$\dot{x} = A(t)\,x(t) + B(t)\,u(t)$$
$$\dot{y} = C(t)\,x(t) + D(t)\,u(t)$$

assume that the matrices A and C in the above state equation are n–1 times continuously differentiable. The state equation is observable at t_0 if there exists a finite $t_1 > 0$ such that

$$P\begin{bmatrix} N_0(t_1) \\ N_1(t_1) \\ \vdots \\ N_{n-1}(t_1) \end{bmatrix} = n \qquad \text{...(6.53)}$$

where

$$N_{K+1}(t) = N_K(t)\,A(t) + \frac{d}{dt}N_K(t) \qquad \text{...(6.54)}$$
$$K = 0, 1, 2, \ldots, n-1$$
$$N_0(t) = C(t) \qquad \text{...(6.55)}$$

ILLUSTRATIVE EXAMPLES

EXAMPLE 6.14. A system is given below, find the state and output equation of the system.

(R.M.L. University Faizabad, 2001)

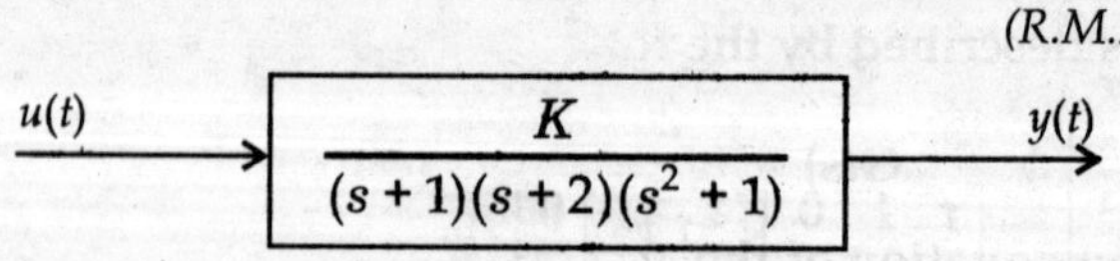

Solution :

$$\frac{y(s)}{u(s)} = G(s) = \frac{K}{(s+1)\,(s+2)\,(s^2+1)}$$

$$\frac{y(s)}{u(s)} = \frac{K}{s^4 + 3s^3 + 3s^2 + 3s + 2}$$

$$y(s)[s^4 + 3s^3 + 3s^2 + 3s + 2] = Ku(s)$$

$$s^4 y(s) + 3s^3 y(s) + 3s^2 y(s) + 3sy(s) + 2y(s) = Ku(s)$$

Taking inverse laplace

$$\overset{....}{y}(t) + 3\dddot{y}(t) + 3\ddot{y}(t) + 3\dot{y}(t) + 2y(t) = Ku(t)$$

$$y(t) = x_1$$

$$\dot{y}(t) = \dot{x}_1 = x_2$$

$$\ddot{y}(t) = \dot{x}_2 = x_3$$

$$\dddot{y}(t) = \dot{x}_3 = x_4$$

$$\overset{....}{y}(t) = \dot{x}_4$$

$$\therefore \overset{....}{y}(t) = \dot{x}_4 = Ku(t) - 3x_4 - 3x_3 - 3x_2 - 2x_1$$

Rewriting

$$\dot{x}_1 = x_2$$

$$\dot{x}_2 = x_3$$

$$\dot{x}_3 = x_4$$

$$\dot{x}_4 = Ku(t) - 3x_4 - 3x_3 - 3x_2 - 2x_1$$

$$\begin{bmatrix} \dot{x}_1 \\ \dot{x}_2 \\ \dot{x}_3 \\ \dot{x}_4 \end{bmatrix} = \begin{bmatrix} 0 & 1 & 0 & 0 \\ 0 & 0 & 1 & 0 \\ 0 & 0 & 0 & 1 \\ -2 & -3 & -3 & -3 \end{bmatrix} \begin{bmatrix} x_1 \\ x_2 \\ x_3 \\ x_4 \end{bmatrix} + \begin{bmatrix} 0 \\ 0 \\ 0 \\ K \end{bmatrix} u(t)$$

$$y(t) = x_1$$

$$y(t) = \begin{bmatrix} 1 & 0 & 0 & 0 \end{bmatrix} x(t)$$

Block diagram :

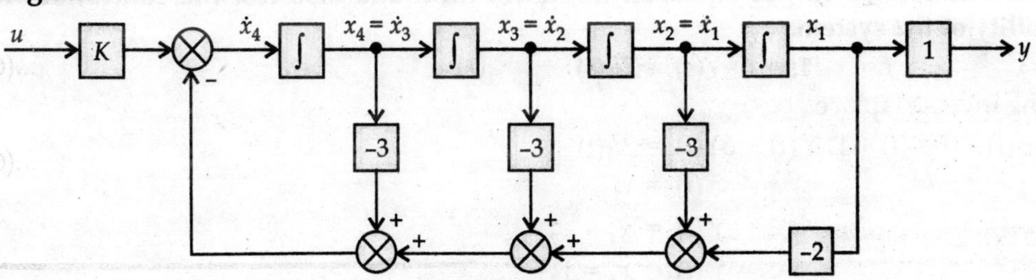

Fig. 6.8.

EXAMPLE 6.15. A system is described by the following transfer function

$$\mathbf{G(s)} = \frac{\mathbf{20\,(10s+1)}}{\mathbf{s^3+3s^2+2s+1}}$$

Find the state and output equation of the system. *(A.M.U., Aligarh 1989-90)*

Solution :

$$G(s) = \frac{y(s)}{u(s)} = \frac{20\,(10s+1)}{s^3+3s^2+2s+1}$$

$$\frac{y(s)}{u(s)} = \frac{X_1(s)}{u(s)} \cdot \frac{y(s)}{X_1(s)} = \frac{20}{s^3+3s^2+2s+1} \cdot (10s+1)$$

Consider

$$\frac{X_1(s)}{u(s)} = \frac{20}{s^3+3s^2+2s+1}$$

$$X_1(s)\,[s^3 + 3s^2 + 2s + 1] = 20\,u(s)$$

Taking inverse laplace

$$\dddot{X}_1(t) + 3\ddot{X}_1(t) + 2\dot{X}_1(t) + X_1(t) = 20\,u(t)$$

Select
$$\dot{X}_1 = X_2$$
$$\ddot{X}_1 = \dot{X}_2 = X_3$$
$$\dddot{X}_1 = \dot{X}_3 = X_4.$$
$$\dddot{X}_1 = 20u - 3\ddot{X}_1 - 2\dot{X}_1 - X_1 = 20u - 3X_3 - 2X_2 - X_1$$

Rewriting
$$\dot{X}_1 = X_2$$
$$\dot{X}_2 = X_3$$
$$\dot{X}_3 = 20u - 3X_3 - 2X_2 - X_1$$

$$\begin{bmatrix}\dot{X}_1\\ \dot{X}_2\\ \dot{X}_3\end{bmatrix} = \begin{bmatrix}0 & 1 & 0\\ 0 & 0 & 1\\ -1 & -2 & -3\end{bmatrix}\begin{bmatrix}X_1\\ X_2\\ X_3\end{bmatrix} + \begin{bmatrix}0\\ 0\\ 20\end{bmatrix}u$$

$$\frac{y(s)}{X_1(s)} = 10s + 1$$
$$y(s) = (10s + 1)\,X_1(s)$$

or,
$$y(t) = X_1 + 10\,X_2$$
$$y(t) = \begin{bmatrix}1 & 10 & 0\end{bmatrix} X(t)$$

EXAMPLE 6.16. A system characterised by the transfer function

$$\frac{Y(s)}{u(s)} = \frac{2}{s^3 + 6s^2 + 11s + 6}$$

Find the state and output equation in matrix form and also test the controllability and observability of the system.

Solution : $(s^3 + 6s^2 + 11s + 6)\,Y(s) = 2u(s)$.

Taking inverse laplace
$$\dddot{Y}(t) + 6\ddot{Y}(t) + 11\dot{Y}(t) + 6Y(t) = 2u(t).$$

let
$$Y(t) = x_1$$
$$\dot{Y}(t) = \dot{x}_1 = x_2$$
$$\ddot{Y}(t) = \dot{x}_2 = x_3$$
$$\dddot{Y}(t) = \dot{x}_3 = x_4$$

$\therefore$
$$\dddot{Y}(t) = x_4 = \dot{x}_3 = 2u(t) - 6x_3 - 11x_2 - 6x_1$$

Rewriting
$$\dot{x}_1 = x_2$$
$$\dot{x}_2 = x_3$$
$$\dot{x}_3 = 2u(t) - 6x_3 - 11x_2 - 6x_1$$

$$\begin{bmatrix}\dot{x}_1\\ \dot{x}_2\\ \dot{x}_3\end{bmatrix} = \begin{bmatrix}0 & 1 & 0\\ 0 & 0 & 1\\ -6 & -11 & -6\end{bmatrix}\begin{bmatrix}x_1\\ x_2\\ x_3\end{bmatrix} + \begin{bmatrix}0\\ 0\\ 2\end{bmatrix}u(t)$$

$$y(t) = \begin{bmatrix}1 & 0 & 0\end{bmatrix} x(t)$$

For controllability

$$Q = [b : Ab : A^2 b]$$

$$A = \begin{bmatrix} 0 & 1 & 0 \\ 0 & 0 & 1 \\ -6 & -11 & -6 \end{bmatrix}, b = \begin{bmatrix} 0 \\ 0 \\ 2 \end{bmatrix}$$

$$Ab = \begin{bmatrix} 0 & 1 & 0 \\ 0 & 0 & 1 \\ -6 & -11 & -6 \end{bmatrix}\begin{bmatrix} 0 \\ 0 \\ 2 \end{bmatrix} = \begin{bmatrix} 0 \\ 2 \\ -12 \end{bmatrix}$$

$$A^2b = \begin{bmatrix} 2 \\ -12 \\ 50 \end{bmatrix}$$

$$Q = \begin{bmatrix} 0 & 0 & 2 \\ 0 & 2 & -12 \\ 2 & -12 & 50 \end{bmatrix}$$

$$|Q| \neq 0$$

It is full of rank, hence the system is controllable

For observability :

$$Q' = \left[C^T : A^T C^T : A^{T^2} C^T\right]$$

$$C^T = \begin{bmatrix} 1 \\ 0 \\ 0 \end{bmatrix}, A^T = \begin{bmatrix} 0 & 0 & -6 \\ 1 & 0 & -11 \\ 0 & 1 & -6 \end{bmatrix}$$

$\therefore$
$$A^T C^T = \begin{bmatrix} 0 \\ 1 \\ 0 \end{bmatrix}, \quad A^{T^2} C^T = \begin{bmatrix} 0 \\ 0 \\ 1 \end{bmatrix}$$

$\therefore$
$$Q' = \begin{bmatrix} 1 & 0 & 0 \\ 0 & 1 & 0 \\ 0 & 0 & 1 \end{bmatrix}$$

Rank of the matrix = 3, Hence the system observable.

EXAMPLE 6.17. A system is described by the following equations

$$\dot{x}(t) = \begin{bmatrix} -1 & 1 \\ 0 & -2 \end{bmatrix} x(t) + \begin{bmatrix} 1 & 0 & 1 \\ 0 & 1 & 1 \end{bmatrix} u(t)$$

$$y(t) = \begin{bmatrix} 1 & 2 \\ 1 & 0 \\ 1 & 1 \end{bmatrix} x(t)$$

Find the transfer function of the system. *(B.E., A.M.U., Aligarh 89-90)* *(R.M.L. University, Faizabad, 2003)*

Solution : The given system is MIMO system.

$\therefore$
$$G(s) = C[sI - A]^{-1} B$$

$$B = \begin{bmatrix} 1 & 0 & 1 \\ 0 & 1 & 1 \end{bmatrix}, C = \begin{bmatrix} 1 & 2 \\ 1 & 0 \\ 1 & 1 \end{bmatrix}$$

$$sI - A = \begin{bmatrix} s & 0 \\ 0 & s \end{bmatrix} - \begin{bmatrix} -1 & 1 \\ 0 & -2 \end{bmatrix} = \begin{bmatrix} s+1 & -1 \\ 0 & s+2 \end{bmatrix}$$

$$[sI - A]^{-1} = \begin{bmatrix} s+2 & 1 \\ 0 & s+1 \end{bmatrix} \frac{1}{s^2+3s+2}$$

$$C[sI - A]^{-1} = \begin{bmatrix} 1 & 2 \\ 1 & 0 \\ 1 & 1 \end{bmatrix} \begin{bmatrix} s+2 & 1 \\ 0 & s+1 \end{bmatrix} \frac{1}{s^2+3s+2} = \begin{bmatrix} s+2 & 2s+3 \\ s+2 & 1 \\ s+2 & s+2 \end{bmatrix} \frac{1}{s^2+3s+2}$$

$$C[sI - A]^{-1} B = \frac{1}{s^2+3s+2} \begin{bmatrix} s+2 & 2s+3 \\ s+2 & 1 \\ s+2 & s+2 \end{bmatrix} \begin{bmatrix} 1 & 0 & 1 \\ 0 & 1 & 1 \end{bmatrix} = \frac{s^3-3s^2-9s+2}{s^2+3s+2}$$

$\therefore \quad G(s) = \dfrac{s^3-3s^2-9s+2}{s^2+3s+2}$ is the required transfer function.

EXAMPLE 6.18. A system is characterised by the equation

$$\frac{y(s)}{u(s)} = \frac{20\,(4s+2)}{s^3+5s^2+8s+2}$$

Find its state and output equation and express in matrix form. *(B.E., A.M.U., Aligarh 86–87)*

Solution :
$$\frac{y(s)}{u(s)} = \frac{X_1(s)}{u(s)} \cdot \frac{y(s)}{X_1(s)} = \frac{20}{s^3+5s^2+8s+2} \cdot (4s+2)$$

$$\frac{X_1(s)}{u(s)} = \frac{20}{s^3+5s^2+8s+2}$$

$$(s^3 + 5s^2 + 8s + 2)\, X_1(s) = 20\, u(s)$$

Taking inverse laplace

$$\dddot{x}_1 + 5\ddot{x}_1 + 8\dot{x}_1 + 2x_1 = 20u$$

Select

$$\dot{x}_1 = x_2$$
$$\ddot{x}_1 = \dot{x}_2 = x_3$$
$$\dddot{x}_1 = \dot{x}_3 = x_4$$
$$\dddot{x}_1 = \dot{x}_3 = -5x_3 - 8x_2 - 2x_1 + 20u$$

Rewriting

$$\dot{x}_1 = x_2$$
$$\dot{x}_2 = x_3$$
$$\dot{x}_3 = -5x_3 - 8x_2 - 2x_1 + 20u$$

$$\begin{bmatrix} \dot{x}_1 \\ \dot{x}_2 \\ \dot{x}_3 \end{bmatrix} = \begin{bmatrix} 0 & 1 & 0 \\ 0 & 0 & 1 \\ -2 & -8 & -5 \end{bmatrix} \begin{bmatrix} x_1 \\ x_2 \\ x_3 \end{bmatrix} + \begin{bmatrix} 0 \\ 0 \\ 20 \end{bmatrix} u$$

$$\frac{y(s)}{X_1(s)} = 4s + 2$$

$$y(s) = (4s + 2)\, X_1(s)$$

$$y(t) = 4\dot{x}_1 + 2x_1 = 4x_2 + 2x_1$$

$$y(t) = \begin{bmatrix} 2 & 4 & 0 \end{bmatrix} x(t)$$

EXAMPLE 6.19. A system is described by the following transfer function

$$G(s) = \frac{s+2}{s^3+3s^2+2s+10}$$

Find the values of *A*, *b* and C^T.

Solution :

$$\frac{Y(s)}{u(s)} = \frac{s+2}{s^3+3s^2+2s+10}$$

$$\frac{Y(s)}{u(s)} = \frac{X_1(s)}{u(s)} \cdot \frac{y(s)}{X_1(s)} = \frac{1}{s^3+3s^2+2s+10}(s+2)$$

$$\frac{X_1(s)}{u(s)} = \frac{1}{s^3+3s^2+2s+10}$$

or, $(s^3 + 3s^2 + 2s+10)\, X_1(s) = u(s)$

Taking inverse laplace

$$\dddot{X}_1 + 3\ddot{X}_1 + 2\dot{X}_1 + 10\,X_1 = u$$

$$\dddot{X}_1 = u - 3\ddot{X}_1 - 2\dot{X}_1 - 10X_1$$

Select

$$\dot{X}_1 = x_2$$

$$\ddot{X}_1 = \dot{X}_2 = x_3$$

$$\dddot{X}_1 = \dot{X}_3$$

Rewriting

$$\dot{X}_1 = x_2$$

$$\dot{X}_2 = x_3$$

$$\dot{X}_3 = -3x_3 - 2x_2 - 10x_1 + u$$

$$\begin{bmatrix} \dot{X}_1 \\ \dot{X}_2 \\ \dot{X}_3 \end{bmatrix} = \begin{bmatrix} 0 & 1 & 0 \\ 0 & 0 & 1 \\ -10 & -2 & -3 \end{bmatrix} \begin{bmatrix} x_1 \\ x_2 \\ x_3 \end{bmatrix} + \begin{bmatrix} 0 \\ 0 \\ 1 \end{bmatrix} u$$

$\therefore$

$$A = \begin{bmatrix} 0 & 1 & 0 \\ 0 & 0 & 1 \\ -10 & -2 & -3 \end{bmatrix} \text{ and } b = \begin{bmatrix} 0 \\ 0 \\ 1 \end{bmatrix}$$

$$\frac{y(s)}{X_1(s)} = s + 2$$

$$y(s) = (s + 2)\, X_1(s)$$

$$= \dot{X}_1 + 2X_1$$

$$y(t) = x_2 + 2x_1$$

$$y(t) = \begin{bmatrix} 2 & 1 & 0 \end{bmatrix} x(t) \;\therefore\; C^T = \begin{bmatrix} 2 & 1 & 0 \end{bmatrix}$$

EXAMPLE 6.20. A single input single output system is given as

$$\dot{x}(t) = \begin{bmatrix} -1 & 0 & 0 \\ 0 & -2 & 0 \\ 0 & 0 & -3 \end{bmatrix} x(t) + \begin{bmatrix} 1 \\ 1 \\ 0 \end{bmatrix} u$$

$$y = \begin{bmatrix}1 & 0 & 2\end{bmatrix} x(t)$$

Test for controllability and observability. *(R.M.L. University Faizabad 2001)*

Solution :

$$A = \begin{bmatrix} -1 & 0 & 0 \\ 0 & -2 & 0 \\ 0 & 0 & -3 \end{bmatrix}, B = \begin{bmatrix} 1 \\ 1 \\ 0 \end{bmatrix}, C = \begin{bmatrix}1 & 0 & 2\end{bmatrix}$$

$$AB = \begin{bmatrix} -1 & 0 & 0 \\ 0 & -2 & 0 \\ 0 & 0 & -3 \end{bmatrix}\begin{bmatrix} 1 \\ 1 \\ 0 \end{bmatrix} = \begin{bmatrix} -1 \\ -2 \\ 0 \end{bmatrix}$$

$$A^2B = \begin{bmatrix} 1 \\ 4 \\ 0 \end{bmatrix}$$

$$Q = [B : AB : A^2B] = \begin{bmatrix} 1 & -1 & 1 \\ 1 & -2 & 4 \\ 0 & 0 & 0 \end{bmatrix}$$

Rank of the matrix = 2

Hence, the system is uncontrollable

For observability

$$Q' = \left[C^T : A^TC^T : A^{T^2}C^T\right]$$

$$A^TC^T = \begin{bmatrix} -1 & 0 & 0 \\ 0 & -2 & 0 \\ 0 & 0 & -3 \end{bmatrix}\begin{bmatrix} 1 \\ 0 \\ 2 \end{bmatrix} = \begin{bmatrix} -1 \\ 0 \\ -6 \end{bmatrix}$$

$$A^{T^2}C^T = \begin{bmatrix} 1 \\ 0 \\ 18 \end{bmatrix}$$

$$Q' = \begin{bmatrix} 1 & -1 & 1 \\ 0 & 0 & 0 \\ 2 & -6 & 18 \end{bmatrix}$$

Rank of matrix = 2

Hence, system is unobservable

$\therefore$ The given system is uncontrollable and unobservable.

EXAMPLE 6.21. Consider the system shown in Fig. 6.9 and investigate whether it is observable or not.

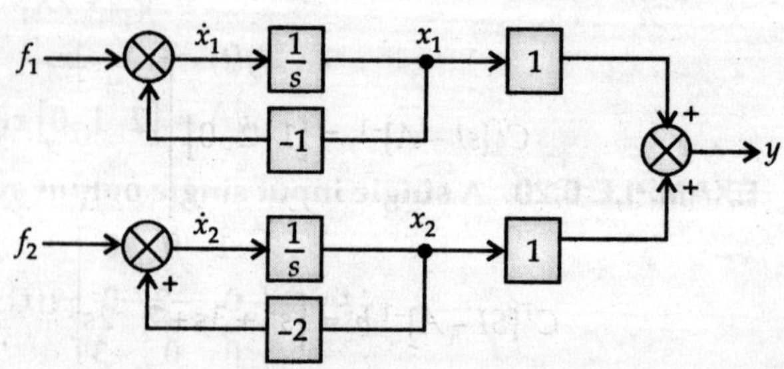

Fig. 6.9.

Solution :

$$\dot{X}_1 = -X_1 + f_1$$
$$\dot{X}_2 = -2X_2 + f_2$$

$$\begin{bmatrix} \dot{X}_1 \\ \dot{X}_2 \end{bmatrix} = \begin{bmatrix} -1 & 0 \\ 0 & -2 \end{bmatrix} \begin{bmatrix} X_1 \\ X_2 \end{bmatrix} + \begin{bmatrix} f_1 \\ f_2 \end{bmatrix}$$

$$y = \begin{bmatrix} 1 & 1 \end{bmatrix} \begin{bmatrix} x_1 \\ x_2 \end{bmatrix}$$

For observability

$$Q' = [C^T : A^T C^T]$$

$$A = \begin{bmatrix} -1 & 0 \\ 0 & -2 \end{bmatrix} \therefore A^T = \begin{bmatrix} -1 & 0 \\ 0 & -2 \end{bmatrix}$$

$$C = [1 \ 1] \quad C^T = \begin{bmatrix} 1 \\ 1 \end{bmatrix}$$

$$A^T C^T = \begin{bmatrix} -1 & 0 \\ 0 & -2 \end{bmatrix} \begin{bmatrix} 1 \\ 1 \end{bmatrix} = \begin{bmatrix} -1 \\ -2 \end{bmatrix}$$

$$Q' = \begin{bmatrix} 1 & -1 \\ 1 & -2 \end{bmatrix}$$

Rank of the matrix = 2

The system is observable.

EXAMPLE 6.22. A system is described by the matrices

$$A = \begin{bmatrix} 0 & 1 & 0 \\ 0 & 0 & 1 \\ 0 & -2 & -3 \end{bmatrix}, b = \begin{bmatrix} 0 \\ 0 \\ 1 \end{bmatrix}, C = [1 \ 2 \ 0]$$

Determine the transfer function.

Solution :

$$sI - A = \begin{bmatrix} s & 0 & 0 \\ 0 & s & 0 \\ 0 & 0 & s \end{bmatrix} - \begin{bmatrix} 0 & 1 & 0 \\ 0 & 0 & 1 \\ 0 & -2 & -3 \end{bmatrix} = \begin{bmatrix} s & -1 & 0 \\ 0 & s & -1 \\ 0 & 2 & s+3 \end{bmatrix}$$

$$[sI - A]^{-1} = \frac{1}{s^3 + 3s^2 + 2s} \begin{bmatrix} s^3 + 3s + 2 & s+3 & 1 \\ 0 & s^2 + 3s & s \\ 0 & -2s & s^2 \end{bmatrix}$$

$$C^T[sI - A]^{-1} = [1 \ 2 \ 0] \begin{bmatrix} s^3 + 3s + 2 & s+3 & 1 \\ 0 & s^2 + 3s & s \\ 0 & -2s & s^2 \end{bmatrix} \frac{1}{s^3 + 3s^2 + 2s}$$

$$C^T[SI - A]^{-1} b = \begin{bmatrix} s^2 + 3s + 2 & 2s^2 + 7s + 3 & 2s + 1 \end{bmatrix} \begin{bmatrix} 0 \\ 0 \\ 1 \end{bmatrix} \frac{1}{s^3 + 3s^2 + 2s} = \frac{2s+1}{s^3 + 3s^2 + 2s}$$

Hence required transfer function = $\dfrac{2s+1}{s^3+3s^2+2s}$

EXAMPLE 6.23. Obtain the transfer function of the system.

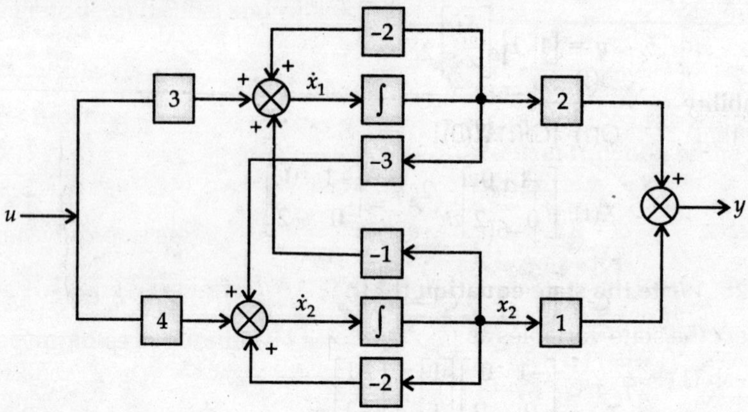

Fig. 6.10.

Solution : From Fig. 6.10.

$$\dot{x}_1 = -2x_1 - x_2 + 3u$$
$$\dot{x}_2 = -3x_1 - 2x_2 + 4u$$

Output equation:

$$y = 2x_1 + x_2$$

$$\begin{bmatrix} \dot{x}_1 \\ \dot{x}_2 \end{bmatrix} = \begin{bmatrix} -2 & -1 \\ -3 & -2 \end{bmatrix}\begin{bmatrix} x_1 \\ x_2 \end{bmatrix} + \begin{bmatrix} 3 \\ 4 \end{bmatrix} u$$

$$y = \begin{bmatrix} 2 & 1 \end{bmatrix} x(t)$$

$$[sI - A]^{-1} = \frac{1}{s^2+4s+1}\begin{bmatrix} s+2 & -1 \\ -3 & s+2 \end{bmatrix}$$

$$C^T[sI - A]^{-1} b = \frac{7s+3}{s^2+4s+1}$$

Hence, required transfer function = $\dfrac{7s+3}{s^2+4s+1}$

EXAMPLE 6.24. Consider the following matrix

$$\dot{x} = \begin{bmatrix} 0 & 1 \\ -6 & -5 \end{bmatrix} \quad x(0) = \begin{bmatrix} 1 \\ 0 \end{bmatrix}$$

Find the state transition matrix, also determine $x(t)$.

Solution :

Given $$A = \begin{bmatrix} 0 & 1 \\ -6 & -5 \end{bmatrix}$$

$$\phi(s) = [sI - A]^{-1} = \frac{1}{s^2+5s+6}\begin{bmatrix} s+5 & 1 \\ -6 & s \end{bmatrix}$$

$$\phi(t) = \pounds^{-1}\phi(s) = \pounds^{-1}\begin{bmatrix} \dfrac{s+5}{s^2+5s+6} & \dfrac{1}{s^2+5s+6} \\ \dfrac{-6}{s^2+5s+6} & \dfrac{s}{s^2+5s+6} \end{bmatrix}$$

$$\phi(t) = \begin{bmatrix} 3e^{-2t}-2e^{-3t} & e^{-2t}-e^{-3t} \\ -6(e^{-2t}-e^{-3t}) & 3e^{-3t}-e^{2t} \end{bmatrix}$$

We know that $x(t) = \phi(t)\, X(0)$

$$X(t) = \begin{bmatrix} 3e^{-2t}-2e^{-3t} & e^{-2t}-e^{-3t} \\ -6(e^{-2t}-e^{-3t}) & 3e^{-3t}-e^{-2t} \end{bmatrix}\begin{bmatrix} 1 \\ 0 \end{bmatrix} = \begin{bmatrix} 3e^{-2t}-2e^{-3t} \\ -6(e^{-2t}-e^{-3t}) \end{bmatrix}$$ **Ans.**

EXAMPLE 6.25. Write the state equation for the circuit shown in Fig. 6.11.

Solution : Select the state variables as i_{L_1} , i_{L_2} and V_c

KVL in mesh (1) :

$$\frac{-2di_{L_1}}{dt} - 1.i_{L_1} - V_1 + V_c = 0$$

$$\frac{di_{L_1}}{dt} = -\frac{1}{2}i_{L_1} - \frac{1}{2}V_1 + \frac{1}{2}V_c$$

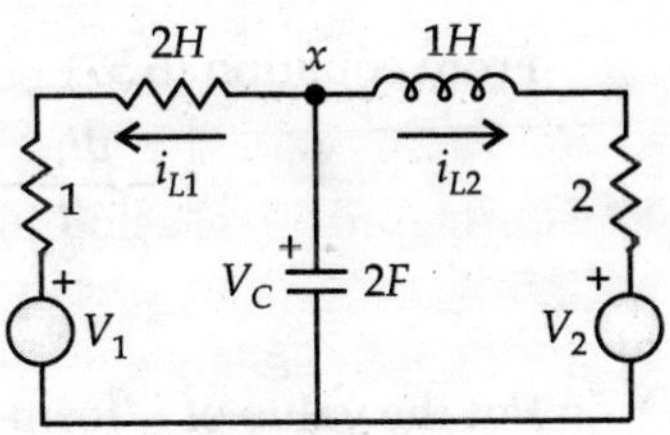

Fig. 6.11.

KVL in mesh (2) :

$$-1\frac{di_{L_2}}{dt} - 2i_{L_2} - V_2 + V_c = 0$$

$$\frac{di_{L_2}}{dt} = -2i_{L_2} - V_2 + V_c$$

KCL at node x :

$$i_{L_1} + i_{L_2} + 2\frac{dV_c}{dt} = 0$$

$$\frac{dV_c}{dt} = -\frac{1}{2}i_{L_1} - \frac{1}{2}i_{L_2}$$

$$\begin{bmatrix} \dot{i}_{L_1} \\ \dot{i}_{L_2} \\ \dot{V}_c \end{bmatrix} = \begin{bmatrix} -1/2 & 0 & 1/2 \\ 0 & -2 & 1 \\ -1/2 & -1/2 & 0 \end{bmatrix}\begin{bmatrix} i_{L_1} \\ i_{L_2} \\ V_c \end{bmatrix} + \begin{bmatrix} -1/2 & 0 \\ 0 & -1 \\ 0 & 0 \end{bmatrix}\begin{bmatrix} V_1 \\ V_2 \end{bmatrix}$$

EXAMPLE 6.26. Write the state equation for the circuit shown in Fig. 6.12.

Solution : Select the state variable as V_{c1}, V_{c2} and i_L

KVL in mesh (1) :

$$V_s - 1.i_1 - V_{c1} = 0 \qquad ...(6.56)$$

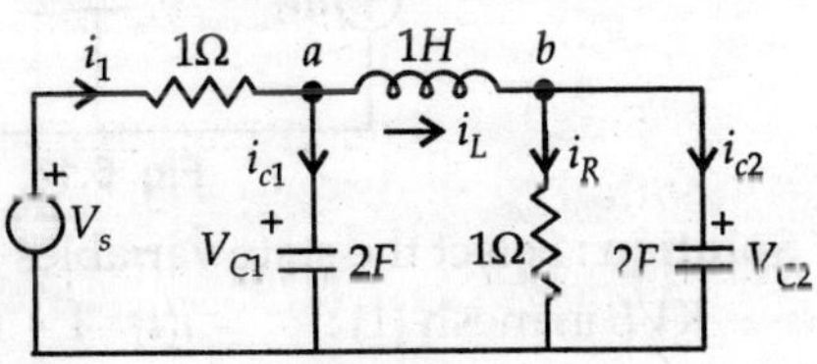

Fig. 6.12.

KVL in mesh (2) :

$$V_{C_1} - 1\cdot\frac{di_L}{dt} - 1.i_R = 0 \qquad ...(6.57)$$

KVL in mesh (3) :

$$1.\, i_R - V_{C2} = 0 \qquad ...(6.58)$$

KCL at node (*a*) :

$$i_1 = i_{C1} + i_L$$

or

$$i_1 = 2\frac{dVc_1}{dt} + i_L \qquad \text{...(6.59)}$$

KCL at node (*b*) :

$$i_L = i_R + i_{C2}$$

$$i_R = i_L - i_{C2} = i_L - 2\frac{dv_{c2}}{dt} \qquad \text{...(6.60)}$$

From equation (6.58) and (6.60)

$$i_L - 2\frac{dv_{c2}}{dt} - V_{c2} = 0$$

$$\frac{dv_{c_2}}{dt} = \frac{1}{2}i_L - \frac{1}{2}V_{c_2} \qquad \text{...(6.61)}$$

$$i_R = \frac{V_{C_2} - 0}{1} = V_{C_2}$$

From equation (6.57)

$$V_{C_1} - \frac{di_L}{dt} - V_{C_2} = 0$$

$$\frac{di_L}{dt} = V_{C_1} - V_{C_2} \qquad \text{...(6.62)}$$

Put the value of i_1 form 6.59 in equation 6.56

$$-V_{C_1} - i_L - 2\frac{dV_{C_1}}{dt} + V_s = 0$$

$$\frac{dV_{C_1}}{dt} = \frac{1}{2}V_s - \frac{1}{2}i_L - \frac{1}{2}V_{C_1}$$

$$\begin{bmatrix} \dot{V}_{C_1} \\ \dot{V}_{C_2} \\ \dot{i}_L \end{bmatrix} = \begin{bmatrix} -1/2 & 0 & -1/2 \\ 0 & -1/2 & 1/2 \\ 1 & -1 & 0 \end{bmatrix} \begin{bmatrix} V_{C_1} \\ V_{C_2} \\ i_L \end{bmatrix} + \begin{bmatrix} 1/2 \\ 0 \\ 0 \end{bmatrix} [V_s]$$

EXAMPLE 6.27. Write the state equation for the circuit shown in Fig. 6.13.

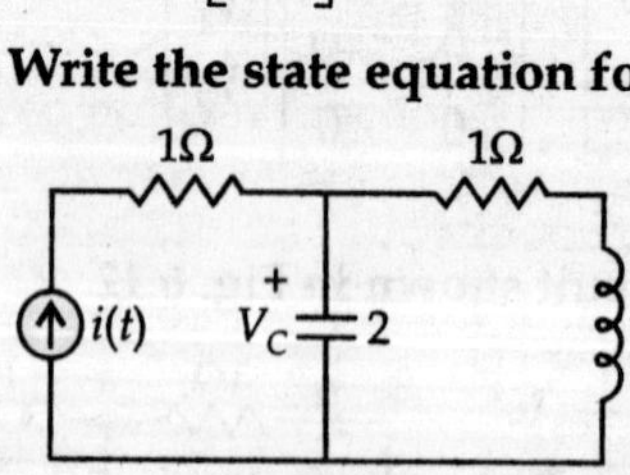

Fig. 6.13.

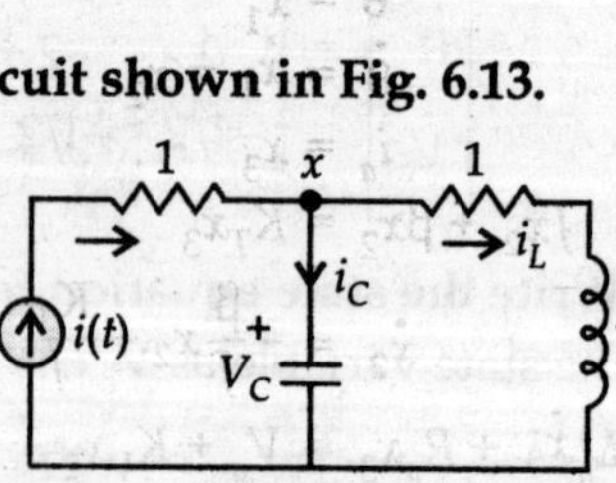

Fig. 6.14.

Solution : Select the state variables as V_C and i_L.

KVL in mesh (1) : $\quad i(t) \cdot 1 - V_C = 0$

KVL in mesh (2) : $-2i_L - 1\frac{di_L}{dt} + V_C = 0$

$$\frac{di_L}{dt} = -2i_L + V_C$$

KCL at node x :
$$i(t) = 2\frac{dV_C}{dt} + i_L$$
$$\frac{dV_C}{dt} = \frac{1}{2}i(t) - \frac{1}{2}i_L$$
$$\begin{bmatrix} \dot{V}_C \\ \dot{i}_L \end{bmatrix} = \begin{bmatrix} 0 & -1/2 \\ 1 & -2 \end{bmatrix}\begin{bmatrix} V_C \\ i_L \end{bmatrix} + \begin{bmatrix} 1/2 \\ 0 \end{bmatrix} i(t) \quad \textbf{Ans.}$$

EXAMPLE 6.28. Express the system performance by a suitable state space representation of the system shown in Fig. 6.15.

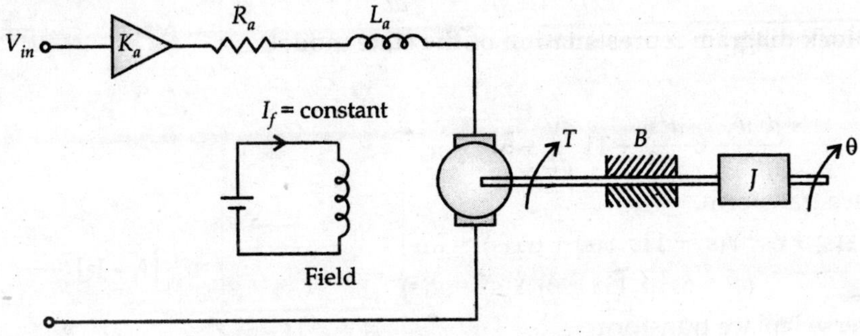

Fig. 6.15. Field control d.c. servometer

Solution : System equations
$$J\ddot{\theta} + \beta\dot{\theta} = K_T i_a \qquad ...(6.63)$$
$$L_a \dot{i}_a + R_a i_a = V_a + K_V\dot{\theta} \qquad ...(6.64)$$

where β = Coefficient of viscous friction

K_T = Torque constant of the motor

K_V = Back *e.m.f* constant of the motor

$V_a = K_a V_{in}$

$V_{in} = u$

$V_a = K_a u$

Select state variables as
$$\theta = x_1$$
$$\dot{\theta} = \dot{x}_1 = x_2$$
$$i_a = x_3$$
$$J\dot{x}_2 + \beta x_2 = K_T x_3 \qquad ...(6.65)$$

or,
$$\dot{x}_2 = -\frac{\beta}{J}x_2 + \frac{K_T}{J}x_3 \qquad ...(6.66)$$
$$L_a\dot{x}_3 + R_a x_3 = V_a + K_V x_2$$
$$\dot{x}_3 = -\frac{R_a}{L_a}x_3 + \frac{K_V}{L_a}x_2 + \frac{K_u}{L_a}u$$

Rewriting
$$\dot{x}_1 = x_2$$
$$\dot{x}_2 = -\frac{\beta}{J}x_2 + \frac{K_T}{J}x_3$$
$$\dot{x}_3 = -\frac{R_a}{L_a}x_3 + \frac{K_V}{L_a}x_2 + \frac{K_a}{L_a}u$$

$$\begin{bmatrix} \dot{x}_1 \\ \dot{x}_2 \\ \dot{x}_3 \end{bmatrix} = \begin{bmatrix} 0 & 1 & 0 \\ 0 & -\beta/j & K_T/J \\ 0 & K_V/L_a & -R_a/L_a \end{bmatrix} \begin{bmatrix} x_1 \\ x_2 \\ x_3 \end{bmatrix} + \begin{bmatrix} 0 \\ 0 \\ K_a/L_a \end{bmatrix} u$$

$$y = \begin{bmatrix} 1 & 0 & 0 \end{bmatrix} x$$

EXAMPLE 6.29. Construct the state model of a system characterized by the differential equation.

$$\frac{d^3y}{dt^3} + 6\frac{d^2y}{dt^2} + 11\frac{dy}{dt} + 6y = u$$

Give the block diagram representation of the state model.

Solution :

$$\frac{d^3y}{dt^3} + 6\frac{d^2y}{dt^2} + 11\frac{dy}{dt} + 6y = u$$

Take Laplace transform

$$s^3 Y(s) + 6s^2 Y(s) + 11s\, Y(s) + 6Y(s) = u(s)$$

$$(s^3 + 6s^2 + 11s + 6)\, Y(s) = u(s)$$

Taking inverse laplace transform

$$\dddot{Y}(t) + 6\ddot{Y}(t) + 11\dot{Y}(t) + 6\,Y(t) = u(t)$$

let

$$Y(t) = x_1$$

$$\dot{Y}(t) = \dot{x}_1 = x_2$$

$$\ddot{Y}(t) = \dot{x}_2 = x_3$$

$$\dddot{Y}(t) = \dot{x}_3$$

$\therefore$

$$\dddot{Y}(t) = \dot{x}_3 = 2u(t) - 6x_3 - 11x_2 - 6x_1$$

Rewriting

$$\dot{x}_1 = x_2$$

$$\dot{x}_2 = x_3$$

$$\dot{x}_3 = 2u(t) - 6x_1 - 11x_2 - 6x_3$$

$$\begin{bmatrix} \dot{x}_1 \\ \dot{x}_2 \\ \dot{x}_3 \end{bmatrix} = \begin{bmatrix} 0 & 1 & 0 \\ 0 & 0 & 1 \\ -6 & -11 & -6 \end{bmatrix} \begin{bmatrix} x_1 \\ x_2 \\ x_3 \end{bmatrix} + \begin{bmatrix} 0 \\ 0 \\ 1 \end{bmatrix} u(t)$$

$$y(t) = \begin{bmatrix} 1 & 0 & 0 \end{bmatrix} x(t)$$

Block diagram shown in Fig. 6.16.

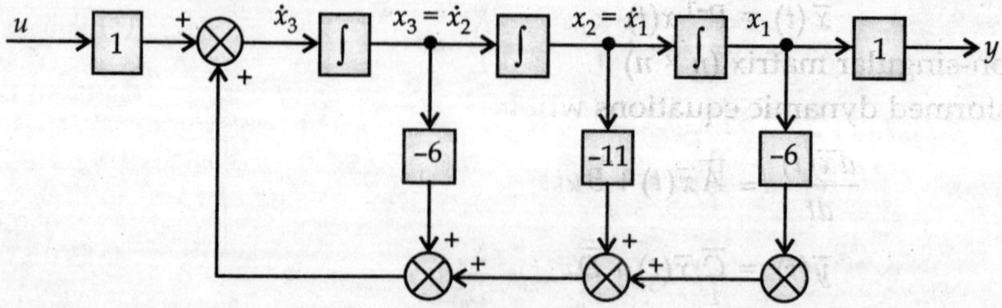

Fig. 6.16.

EXAMPLE 6.30. **A linear time invariant system is characterised by the homogeneous state equation**

$$\begin{bmatrix} \dot{x}_1 \\ \dot{x}_2 \end{bmatrix} = \begin{bmatrix} 1 & 0 \\ 1 & 1 \end{bmatrix} \begin{bmatrix} x_1 \\ x_2 \end{bmatrix}$$

Compute the solution of homogeneous equation, assuming the initial state vector

$$x(0) = \begin{bmatrix} 1 \\ 0 \end{bmatrix}$$

Solution : Given

$$A = \begin{bmatrix} 1 & 0 \\ 1 & 1 \end{bmatrix}$$

$$sI - A = \begin{bmatrix} s & 0 \\ 0 & s \end{bmatrix} - \begin{bmatrix} 1 & 0 \\ 1 & 1 \end{bmatrix} = \begin{bmatrix} s-1 & 0 \\ -1 & s-1 \end{bmatrix}$$

$$[sI - A]^{-1} = \begin{bmatrix} \dfrac{1}{s-1} & 0 \\ \dfrac{1}{(s-1)^2} & \dfrac{1}{s-1} \end{bmatrix}$$

$$\phi(t) = \mathcal{L}^{-1}[sI - A]^{-1} = \begin{bmatrix} e^t & 0 \\ te^t & e^t \end{bmatrix}$$

$$x(t) = \phi(t)\, x(0) = \begin{bmatrix} e^t & 0 \\ te^t & e^t \end{bmatrix} \begin{bmatrix} 1 \\ 0 \end{bmatrix} = \begin{bmatrix} e^t \\ te^t \end{bmatrix} \quad \textbf{Ans.}$$

6.12. SIMILARITY TRANSFORMATION

Similarity transformation means transformation of one set of dynamic equations to another set of dynamic equations. Consider the dynamic equation of a single input single output system

$$\frac{dx(t)}{dt} = Ax(t) + Bu(t) \quad \text{... (6.67)}$$

$$y(t) = Cx(t) + Du(t) \quad \text{...(6.68)}$$

where $x(t)$ = state vector ($n \times 1$)

$u(t)$ = input vector

$y(t)$ = output vector

Let the dynamic equations (6.67) and (6.68) be transformed into another set of equations.

let $$x(t) = P\,\bar{x}(t) \quad \text{...(6.69)}$$

or, $$\bar{x}(t) = P^{-1}x(t) \quad \text{...(6.70)}$$

Where, P = non-singular matrix ($n \times n$)

then the transformed dynamic equations will be

$$\frac{d\bar{x}(t)}{dt} = \bar{A}\,\bar{x}(t) + \bar{B}u(t) \quad \text{...(6.71)}$$

$$\bar{y}(t) = \bar{C}\,\bar{x}(t) + \bar{D}u(t) \quad \text{...(6.72)}$$

Take the derivative of equation (6.70)

$$\frac{d\bar{x}(t)}{dt} = P^{-1}\frac{dx(t)}{dt} \quad ...(6.73)$$

Put the value of $\frac{dx(t)}{dt}$ from equation (6.67) in equation (6.73)

$$\frac{d\bar{x}(t)}{dt} = P^{-1}Ax(t) + P^{-1}B\,u(t) \quad ...(6.74)$$

Put the value of $x(t)$ from (6.69) in equation (6.74)

$$\frac{d\bar{x}(t)}{dt} = P^{-1}A\,Px(t) + P^{-1}B\,u(t) \quad ...(6.75)$$

Compare equation (6.71) with (6.75)

$$\boxed{\bar{A} = P^{-1}AP} \quad \text{and} \quad \boxed{\bar{B} = P^{-1}B}$$

Put the value of $x(t)$ from equation (6.69) in equation (6.68).

$$y(t) = CP\bar{x}(t) + Du(t) \quad ...(6.76)$$

compare equation (6.76) with equation (6.72)

$$\boxed{\bar{C} = CP} \quad \text{and} \quad \boxed{\bar{D} = D}$$

By transformation, there is no change in characteristic equation, eigen values and transfer function.

6.13. STATE SPACE REPRESENTATION OF TRANSFER FUNCTION SYSTEMS

6.13.1. State Space Representation in Canonical Form

Consider the transfer function

$$\frac{Y(s)}{u(s)} = \frac{b_o s^n + b_1 s^{n-1} + \dots\dots\dots + b_{n-1}s + bn}{s^n + a_1 s^{n-1} + \dots\dots + a_{n-1}s + a_n} \quad ...(6.77)$$

6.13.1.1. Controllable Canonical Form

Equation (6.77) can be written as

$$\frac{Y(s)}{u(s)} = b_0 + \frac{(b_1 - a_1b_0)s^{n-1} + \dots\dots + (b_{n-1} - a_{n-1}b_0)s + (b_n - a_nb_0)}{s^n + a_1s^{n-1} + \dots\dots + a_{n-1}s + a_n} \quad ...(6.78)$$

Multiply both sides by $u(s)$

$$Y(s) = b_ou(s) + \left[\frac{(b_1 - a_1b_o)s^{n-1} + \dots\dots + (b_{n-1} - a_{n-1}b_o)s + (b_n - a_nb_o)}{s^n + a_1s^{n-1} + \dots\dots a_{n-1}s + a_n}\right]u(s) \quad ...(6.79)$$

or, $$Y(s) = b_ou(s) + \overset{*}{Y}(s) \quad ...(6.80)$$

where $$\overset{*}{Y}(s) = \left[\frac{(b_1 - a_1b_0)s^{n-1} + \dots\dots\dots + (b_{n-1} - a_{n-1}b_0)s + (b_n - a_nb_o)}{s^n + a_1s^{n-1} + \dots\dots\dots + a_{n-1}\,s + a_n}\right]u(s) \quad ...(6.81)$$

Equation (6.81) can be written as

$$= \frac{\overset{*}{Y}(s)}{(b_1 - a_1b_0)s^{n-1} + \dots\dots + (b_{n-1} - a_{n-1}b_0) + (b_n - a_nb_0)}$$

$$= \frac{u(s)}{s^n + a_1 s^{n-1} + ... + a_{n-1}s + a_n} = Q(s) \qquad ...(6.82)$$

Where $Q(s)$ is a dummy variable

Equate L.H.S of equation (6.82) to $Q(s)$

$$\overset{*}{Y}(s) = Q(s)\,[(b_1 - a_1 b_0)\, s^{n-1} + ---- + (b_{n-1} - a_{n-1}\, b_0) + (b_n - a_n b_0)]$$

or $$\overset{*}{Y}(s) = (b_1 - a_1 b_0)\, S^{n-1}\, Q(s) + ---- + (b_{n-1} - a_{n-1}\, b_0)\, Q(s) + (b_n - a_n b_o)\, Q(s) \qquad ...(6.83)$$

Equate R.H.S of equation (6.82) to $Q(s)$

$$u(s) = Q(s)[s^n + a_1 s^{n-1} + ---- + a_{n-1}\, s + a_n]$$

or, $$u(s) = s^n\, Q(s) + a_1 s^{n-1}\, Q(s) + ---- + a_{n-1}\, s\, Q(s) + a_n\, Q(s))$$

or, $$S^n Q(s) = u(s) + [-a_1 s^{n-1}\, Q(s) + ----- a_{n-1}\, s\, Q(s) - a_n\, Q(s)] \qquad ...(6.84)$$

Select the state variables

$$\left.\begin{array}{l} X_1(s) = Q(s) \\ X_2(s) = sQ(s) \\ X_3(s) = s^2 Q(s) \\ \vdots \\ X_{n-1}(s) = s^{n-2} Q(s) \\ X_n(s) = s^{n-1} Q(s) \end{array}\right\} \qquad ...(6.85)$$

Multiplying both sides of first equation of equation (6.85) by s

$$s\, X_1(s) = sQ(s) = X_2(s)$$

i.e., $$sX_1(s) = X_2(s)$$

similarly $$s^2 X_1(s) = s^2 Q(s) = X_3(s)$$

i.e., $$\left.\begin{array}{l} sX_1(s) = X_2(s) \\ sX_2(s) = X_3(s) \\ \vdots \quad \vdots \\ sX_{n-1}(s) = x_n(s) \end{array}\right] \qquad ...(6.86)$$

Equation (6.86) can be written as

$$\left.\begin{array}{l} \mathring{x}_1 = x_2 \\ \mathring{x}_2 = x_3 \\ \vdots \\ \mathring{x}_{n-1} - x_n \end{array}\right] \qquad ...(6.87)$$

Also, $$S^n\, Q(s) = s\, X_n\, (s)$$

Equation (6.84) can be written as (with the help of equation 6.85)

$$s\, X_n\,(s) = u(s) + [-a_1\, X_n\,(s)\, ... - a_{n-1}\, X_2(s) - a_n\, X_1(s)]$$

or, $$\mathring{x}_n = u(s) + [-a_1 x_1\, ... - a_{n\ 1}\, x_2\, ... - a_1\, x_n]$$

$$\mathring{x}_n = -a_n x_1\, ... - a_{n-1}\, x_2\, ...\, a_1 x_n + 4 \qquad ....(6.88)$$

Equation (6.87) and (6.88) can be written in matrix form

$$\begin{bmatrix} \mathring{x}_1 \\ \mathring{x}_2 \\ \vdots \\ \mathring{x}_{n-1} \\ \mathring{x}_n \end{bmatrix} = \begin{bmatrix} 0 & 1 & 0 & \cdots & 0 \\ 0 & 0 & 1 & \cdots & 0 \\ \vdots & \vdots & \vdots & & \vdots \\ 0 & 0 & 0 & \cdots & 1 \\ -a_n & -a_{n-1} & -a_{n-2} & \cdots & -a_1 \end{bmatrix} \begin{bmatrix} x_1 \\ x_2 \\ \vdots \\ \vdots \\ x_n \end{bmatrix} + \begin{bmatrix} 0 \\ 0 \\ \vdots \\ 0 \\ 1 \end{bmatrix} u \qquad ...(6.89)$$

From equation (6.80) and (6.83)

$$Y(s) = b_o\, U(s) + (b_1 - a_1 b_o)\, s^{n-1} Q(s) + ... + (b_{n-1} - a_{n-1} b_o)\, sQ(s) + (b_n - a_n b_o)\, Q(s)$$

from equation (6.85)

$$Y(s) = b_o\, u(s) + (b_1 - a_1 b_o)\, x_n + ... + (b_{n-1} - a_{n-1} b_o)\, X_2(s) + (b_n - a_n b_o)\, X_1(s)$$

Taking inverse Laplace

$$y = b_o u + (b_1 - a_1 b_o) x_n + ... + (b_{n-1} - a_{n-1} b_o)\, x_2 + (b_n - a_n b_o)\, x_1$$

or,

$$y = \left[(b_n - a_n b_o), (b_{n-1} - a_{n-1} b_o), ..., (b_1 - a_1 b_o)\right] \begin{bmatrix} x_1 \\ x_2 \\ \vdots \\ x_n \end{bmatrix} + b_o u \qquad ...(6.90)$$

Equation (6.89) and (6.90) are said to be in the controllable canonical form (CCF)

Now compare the equation (6.89) with (6.71)

$$\overline{A} = \begin{bmatrix} 0 & 1 & 0 & \cdots & 0 \\ 0 & 0 & 1 & \cdots & 0 \\ \vdots & \vdots & \vdots & & \\ -a_n & -a_{n-1} & -a_{n-2} & \cdots & -a_1 \end{bmatrix} = P^{-1}AP \qquad ...(6.91)$$

$$\overline{B} = P^{-1}B = \begin{bmatrix} 0 \\ 0 \\ \vdots \\ 1 \end{bmatrix} \qquad ...(6.92)$$

where $\quad P = SM \qquad ...(6.93)$

and $\quad S = \left[B\ AB\ A^2B \dots A^{n-1}B\right] \qquad ...(6.94)$

$$M = \begin{bmatrix} a_{n-1} & a_{n-2} & \cdots & a_1 & 1 \\ a_{n-2} & \vdots & & 1 & \vdots \\ \vdots & \vdots & & \vdots & \vdots \\ a_1 & 1 & \cdots & 0 & 0 \\ 1 & 0 & \cdots & 0 & 0 \end{bmatrix} \qquad ...(6.95)$$

∵ The characteristic equation is

$$|sI - A| = s^n + a_1 s^{n-1} + ... + a_{n-1}\, s + a_n$$

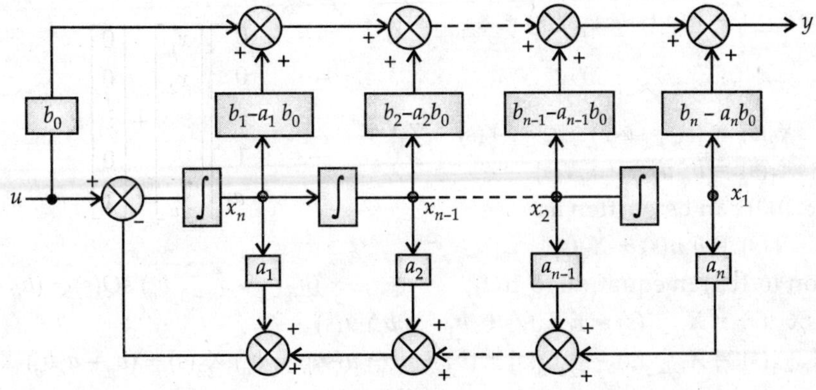

Fig. 6.17.

6.13.1.2. Observable Canonical Form

Consider the transfer function given in equation (6.77) Equation (6.77) can be written as

$$Y(s)\,[s^n + a_1 s^{n-1} + \dots\, a_2 s^{n-2} + \dots\, a_{n-1}\, s + a_n] = u(s)\,[b_o s^n + b_1 s^{n-1} + b_2 s^{n-2} + \dots + b_{n-1}\, s + b_n] \quad \text{...(6.96)}$$

or,

$$s^n\, Y(s) + a_1 s^{n-1}\, Y(s) + a_2 s^{n-2}\, Y(s) + \dots\, a_{n-1}\, sY(s) + a_n Y(s) = b_o s^n u(s) + b_1 s^{n-1}\, u(s) + b_2 s^{n-2}\, u(s) + \dots\, b_{n-1}\, su(s) + b_n u(s) \quad \text{...(6.97)}$$

Rearrange equation (6.97) by transferring RHS to LHS

$$s^n Y(s) - b_o s^n u(s) + a_1 s^{n-1}\, Y(s) - b_1 s^{n-1}\, u(s) + a_2 s^{n-2}\, Y(s) - b_2 s^{n-2}\, u(s) + \dots\, a_{n-1}\, sY(s) - b_{n-1}\, su(s) + a_n Y(s) - b_n u(s) = 0$$

or,

$$s^n\, [Y(s) - b_o u(s)] + s^{n-1}\, [a_1 Y(s) - b_1 u(s)] + s^{n-2}\, [a_2 Y(s) - b_2 u(s)] + \dots + s[a_{n-1}\, Y(s) - b_{n-1}\, u(s)] + a_n\, Y(s) - b_n\, u(s) = 0 \quad \text{...(6.98)}$$

Divide the equation (6.98) by s_n

$$[Y(s) - b_o u(s)] + s^{-1}\, [a_1 Y(s) - b_1 u(s)] + s^{-2}\, [a_2 Y(s) - b_2 u(s)] + \dots + s^{1-n}\, [a_{n-1}\, Y(s) - b_{n-1}\, u(s)] + a_n s^{-n}\, Y(s) - b_n s^{-n}\, u(s) = 0 \quad \text{... (6.99)}$$

Rearrange equation (6.99)

$$Y(s) = b_o u(s) + \frac{1}{s}[b_1 u(s) - a_1 Y(s)] + \frac{1}{s^2}[b_2 u(s) - a_2 Y(s)] + \dots\dots + \frac{1}{s^{n-1}}[b_{n-1} Y(s) - a_{n-1} Y(s)] + \frac{1}{s^n}[b_n u(s) - a_n Y(s)] = 0 \quad \text{...(6.100)}$$

Select the state variable as

$$\left.\begin{aligned}
X_n(s) &= \frac{1}{s}[b_1 u(s) - a_1 Y(s) + X_{n-1}(s)] \\
X_{n-1}(s) &= \frac{1}{s}[b_2 u(s) - a_2 Y(s) + X_{n-2}(s)] \\
&\;\vdots \\
X_2(s) &= \frac{1}{s}[b_{n-1} u(s) - a_{n-1} Y(s) + X_1(s)] \\
X_1(s) &= \frac{1}{s}[b_n u(s) - a_n Y(s)]
\end{aligned}\right\} \quad \text{...(6.101)}$$

Equation (6.101) can be written as

$$sX_{n-1} = [b_1 u(s) - a_1 Y(s) + X_{n-1}(s)] \quad \text{...(6.102)}$$

$$sX_{n-2} = [b_2u(s) - a_2\,Y(s) + X_{n-2}(s)]$$
$$\vdots$$
$$sX_2(s) = [b_{n-1}\,u(s) - a_{n-1}\,Y(s) + X_1(s)]$$
$$sX_1(s) = b_n\,u(s) - a_nY(s)$$

Equation (6.100) can be written as

$$Y(s) = b_ou(s) + X_n(s) \quad ...(6.103)$$

Put equation (6.103) in equation (6.102)

$$sX_n\,(s) = X_{n-1}(s) - a_1X_n(s) + (b_1 - a_1b_o)\,u(s)$$
$$sX_{n-1}(s) = X_{n-2}(s) - a_2X_n(s) + (b_2 - a_2b_o)\,u(s)$$
$$\vdots$$
$$sX_2 = X_1(s) - a_{n-1}\,X_n\,(s) + (b_{n-1} - a_{n-1}\,b_o)\,u(s) \quad ...(6.104)$$
$$sX_1 = -\,a_nX_n + (b_n - a_nb_o)u(s)$$

Taking inverse laplace of equation (6.104)

$$\dot{x}_n = x_{n-1} - a_1x_n + (b_1 - a_1b_0)\,u$$
$$\dot{x}_{n-1} = x_{n-2} - a_2x_n + (b_2 - a_2b_0)\,u$$
$$\vdots \quad ...(6.105)$$
$$\dot{x}_2 = x_1 - a_{n-1}\,x_n + (b_{n-1} - a_{n-1}\,b_o)\,u$$
$$\dot{x}_1 = -\,a_n\,x_n + (b_n - a_nb_o)\,u$$

Rearrange the equation (6.105)

$$\dot{x}_1 = -\,a_nx_n + (b_n - a_nb_o)\,u$$
$$\dot{x}_2 = x_1 - a_{n-1}\,x_n + (b_{n-1} - a_{n-1}\,b_o)\,u$$
$$\vdots \quad ...(6.106)$$
$$\dot{x}_{n-1} = x_{n-2}\,a_2x_n + (b_2 - a_2b_o)\,u$$
$$\dot{x}_n = x_{n-1} - a_1\,x_n + (b_1 - a_1b_o)\,u$$

Inverse laplace of equation (6.103)

$$y = b_o\,u(s) + x_n$$
$$y = x_n + b_o\,u(s) \quad ...(6.107)$$

equation (6.106) and (6.107) can be written in matrix form as

$$\begin{bmatrix} \dot{x}_1 \\ \dot{x}_2 \\ \vdots \\ \\ x_n \end{bmatrix} = \begin{bmatrix} 0 & 0 & \cdots & 0 & -a_n \\ 1 & 0 & \cdots & 0 & -a_{n-1} \\ 0 & 1 & \cdots & 0 & \vdots \\ \vdots & \vdots & & \vdots & \vdots \\ 0 & 0 & \cdots & 1 & a_n \end{bmatrix} \begin{bmatrix} x_1 \\ x_2 \\ \vdots \\ \vdots \\ a_n \end{bmatrix} + \begin{bmatrix} b_n - a_nb_o \\ b_{n-1} - a_{n-1}\,b_o \\ \vdots \\ \vdots \\ b_1 - a_1b_0 \end{bmatrix} u \quad ...(6.108)$$

$$y = [0\ 0\ \cdots\ 0\ 1]\begin{bmatrix} x_1 \\ x_2 \\ \vdots \\ x_{n-1} \\ x_n \end{bmatrix} + b_o u \qquad ...(6.109)$$

Equations (6.108) and (6.109) are said to be in the observable canonical form (OCF)
Compare equation (6.108) with equation (6.71)

$$\overline{A} = \begin{bmatrix} 0 & 0 & \cdots & 0 & -a_n \\ 1 & 0 & \cdots & 0 & -a_{n-1} \\ 0 & 1 & \cdots & \cdots & -a_{n-2} \\ \vdots & \vdots & & & \vdots \\ 0 & 0 & \cdots & 1 & a_n \end{bmatrix} \qquad ...(6.110)$$

$$\overline{B} = \begin{bmatrix} b_n - a_n b_o \\ b_{n-1} - a_{n-1} b_0 \\ \vdots \\ b_1 - a_1 b_o \end{bmatrix} \qquad ...(6.111)$$

$$\overline{C} = [0\,0 \ldots 0\,1] \qquad ...(6.112)$$

Here, we use Q instead of P as used in controllable canonical form.

$$\overline{A} = Q^{-1} AQ \qquad ...(6.113)$$

$$\overline{C} = CQ \qquad ...(6.114)$$

where $\quad Q = [MV]^{-1} \qquad ...(6.115)$

The value of M wil be same as given in equation (6.95) and V is

$$V = \begin{bmatrix} C \\ CA \\ CA^2 \\ \vdots \end{bmatrix} \qquad ...(6.116)$$

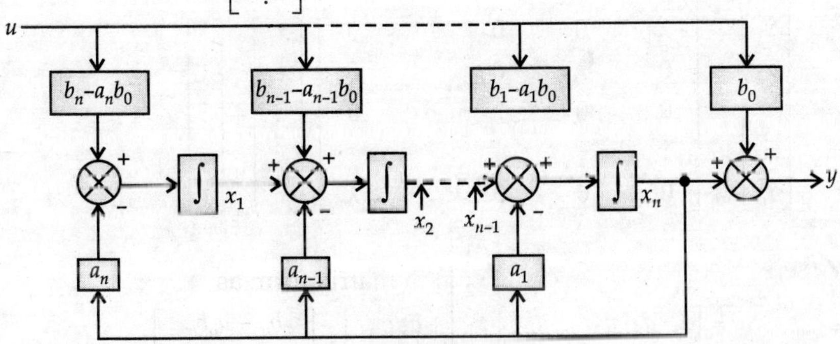

Fig. 6.18.

6.13.1.3. Diagonal Canonical Form

Let the equation (6.77) is written as

$$\frac{Y(s)}{u(s)} = b_o + \frac{d_1}{s + \lambda_1} + \frac{d_2}{s + \lambda_2} + \ldots + \frac{d_n}{s + \lambda_n} \qquad ...(6.117.0)$$

or, $$Y(s) = b_o u(s) + \frac{d_1}{s+\lambda_1}u(s) + \frac{d_2}{s+\lambda_2}u(s) + ... + \frac{d_n}{s+\lambda_n}u(s) \quad ...(6.177.1)$$

Define the state variable as

$$X_1(s) = \frac{1}{s+\lambda_1}u(s)$$
$$X_2(s) = \frac{1}{s+\lambda_2}u(s)$$
$$\vdots$$
$$X_n(s) = \frac{1}{s+\lambda_n}u(s) \quad ...(6.117.2)$$

Above equations can be written as

$$sX_1(s) = -\lambda_1 X_1(s) + u(s)$$
$$sX_2(s) = -\lambda_2 X_2(s) + u(s)$$
$$\vdots$$
$$sX_n(s) = -\lambda_n X_n(s) + u(s) \quad ...(6.117.3)$$

e.g. $X_1(s)\,[s+\lambda_1] = u(s)$
$sX_1(s) + X_1(s)\,\lambda_1 = u(s)$
or $sX_1(s) = -\lambda_1 X_1(s) + u(s)$

The inverse laplace of (6.117.3)

$$\dot{x}_1 = -\lambda_1 x_1 + u$$
$$\dot{x}_2 = -\lambda_2 x_2 + u$$
$$\vdots$$
$$\dot{x}_n = -\lambda_n x_n + u \quad ...(6.117.4)$$

put the equation (6.117.2) in (6.117.1)

$$Y(s) = b_o\, u(s) + d_1 X_1(s) + d_2 X_2(s) + ... + d_n X_n(s) \quad ...(6.117.5)$$

Inverse laplace of equation (6.117.5)

$$y = b_o u + d_1 x_1 + d_2 x_2 + ... + d_n x_n \quad ...(6.117.6)$$

equation (6.117.6) is the output equation.

The equations (6.117.4) and (6.117.6) can be written in matrix form as

$$\begin{bmatrix}\dot{x}_1\\ \dot{x}_2\\ \vdots\\ \vdots\\ \dot{x}_n\end{bmatrix} = \begin{bmatrix}-\lambda_1 & 0 & 0 & \cdots & 0\\ 0 & -\lambda_2 & 0 & \cdots & 0\\ 0 & 0 & -\lambda_3 & \cdots & 0\\ \vdots & \vdots & \vdots & & \\ 0 & 0 & 0 & & -\lambda_n\end{bmatrix}\begin{bmatrix}x_1\\ x_2\\ \vdots\\ \vdots\\ x_n\end{bmatrix} + \begin{bmatrix}1\\ 1\\ \vdots\\ \vdots\\ 1\end{bmatrix}u \quad ...(6.117.7)$$

$$y = \begin{bmatrix}d_1 & d_2 & d_3 & \ldots d_n\end{bmatrix}\begin{bmatrix}x_1\\ x_2\\ x_3\\ \vdots\\ x_n\end{bmatrix} + b_o u \quad ...(6.117.8)$$

Equations (6.117.7) and (6.117.8) are said to be in diagonal canonical form.

[***Note*** : λ_1, λ_2, ... roots of characteristic equation or eigen value of matrix A.
B = only 1, 1, 1, ... no other values (means it is controllable)
If any of d_1, d_2, ... is absent then system is unobservable]

We can also select the state variables as

$$X_1(s) = \frac{d_1}{s+\lambda_1} u(s)$$
$$X_2(s) = \frac{d_2}{s+\lambda_2} u(s)$$
$$\vdots \qquad ...(6.117.9)$$
$$Xn(s) = \frac{d_n}{s+\lambda_n} u(s)$$

equation (6.117.9) can be written as

$$sX_1(s) = -\lambda_1 X_1(s) + d_1 u(s)$$
$$sX_2(s) = -\lambda_2 X_2(s) + d_2 u(s)$$
$$\vdots \qquad ...(6.117.10)$$
$$sX_n(s) = -\lambda_n X_n(s) + d_n u(s)$$

Laplace transform of (6.117.10)

$$\dot{x}_1 = -\lambda x_1 + d_1 u$$
$$\dot{x}_2 = -\lambda_2 x_2 + d_2 u$$
$$\vdots \qquad ...(6.117.11)$$
$$\dot{x}_n = -\lambda_n x_n + d_n u$$

from equation (6.117.1)

$$Y(s) = b_o u(s) + X_1(s) + X_2(s) + X_3(s) + ... X_n \qquad ...(6.117.12)$$

Inverse Laplace of (6.117.12)

$$y = b_o u + x_1 + x_2 + x_3 + ... x_n \qquad ...(6.117.13)$$

Equation (6.117.12) and (6.117.13) can be written in matrix form

$$\begin{bmatrix} \dot{x}_1 \\ \dot{x}_2 \\ \dot{x}_3 \\ \vdots \\ \dot{x}_n \end{bmatrix} = \begin{bmatrix} -\lambda_1 & \cdots & 0 & \cdots & 0 \\ 0 & -\lambda_2 & 0 & \cdots & 0 \\ 0 & 0 & -\lambda_3 & \cdots & 0 \\ \vdots & \vdots & \vdots & & \vdots \\ 0 & 0 & 0 & & -\lambda_n \end{bmatrix} \begin{bmatrix} x_1 \\ x_2 \\ x_3 \\ \vdots \\ x_n \end{bmatrix} + \begin{bmatrix} d_1 \\ d_2 \\ d_3 \\ \vdots \\ d_n \end{bmatrix} u \qquad ...(6.117.14)$$

$$y = [1 \;\; 1 \;\; 1 \; \ldots \; 1] \begin{bmatrix} x_1 \\ x_2 \\ x_3 \\ \vdots \\ x_n \end{bmatrix} + b_o u \qquad ...(6.117.15)$$

Equation (6.117.14) and (6.117.15) are also called diagonal canonical form. This is an alternative approach.

let $\qquad x(t) = T\,\overline{x}(t) \qquad ...(6.117.16)$

or, $\qquad \overline{x}(t) = T^{-1} x(t) \qquad ...(6.117.17)$

derivative of equation (6.117.17)

$$\frac{d\overline{x}(t)}{dt} = T^{-1} \frac{dx(t)}{dt} \qquad ...(6.117.18)$$

Put the value of $\frac{dx(t)}{dt}$ from equation (6.67)

$$\frac{d\bar{x}(t)}{dt} = T^{-1}Ax(t) + T^{-1}B\,u(t)$$

or,
$$\frac{d\bar{x}(t)}{dt} = T^{-1}AT\,\bar{x}(t) + T^{-1}B\,u(t) \qquad ...(6.117.19)$$

compare equation (6.177.19) with (6.71)

$$\boxed{\bar{A} = T^{-1}AT}\ , \qquad \boxed{\bar{B} = T^{-1}B} \qquad ...(6.177.20)$$

from equation (6.68) and (6.72)

$$y(t) = CT\,\bar{x}(t) + Du(t) \qquad ...(6.117.21)$$

Compare equation (6.117.21) with (6.72)

$$\boxed{\bar{C} = CT} \quad \text{and} \quad \boxed{\bar{D} = D}$$

T matrix can be formed by the use of eigen vectors of A.

$$T = [p_1 \quad p_2 \quad p_3 \quad \cdots \quad p_n]$$

If the matrix A is of CCF and A has distinct eigen values, the DCF transformation matrix is the vandermonde matrix, then

$$T = \begin{bmatrix} 1 & 1 & \cdots & 1 \\ \lambda_1 & \lambda_2 & \cdots & \lambda_n \\ \lambda_1^2 & \lambda_2^2 & \cdots & \lambda_n^2 \\ \vdots & & & \\ \lambda_1^{n-1} & \lambda_2^{n-1} & \cdots & \lambda_n^{n-1} \end{bmatrix} \qquad ...(6.117.22)$$

where, $\lambda_1, \lambda_2, \lambda_3 \ldots \lambda_n$ are the eigen values of A.

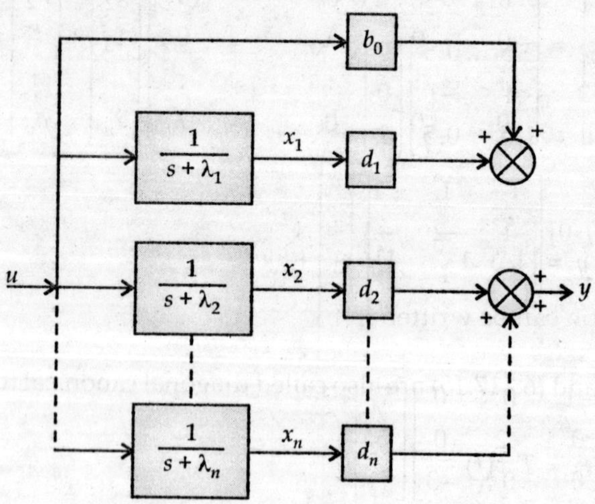

Fig. 6.19.

EXAMPLE 6.31. Consider the state space representation given in example 6.16, change it into diagonal canonical form (DCF).

Solution : Compare state equation (6.67) and (6.68)

$$A = \begin{bmatrix} 0 & 1 & 0 \\ 0 & 0 & 1 \\ -6 & -11 & -6 \end{bmatrix} \quad B = \begin{bmatrix} 0 \\ 0 \\ 2 \end{bmatrix} \quad C = [1\ 0\ 0]$$

First calculate eigen vector of A.

$$|\lambda I - A| = \lambda \begin{bmatrix} 1 & 0 & 0 \\ 0 & 1 & 0 \\ 0 & 0 & 1 \end{bmatrix} - \begin{bmatrix} 0 & 1 & 0 \\ 0 & 0 & 1 \\ -6 & -11 & -6 \end{bmatrix} = \lambda^3 + 6\lambda^2 + 11\lambda + 6$$

$$|\lambda I - A| = \lambda^3 + 6\lambda^2 + 11\lambda + 6 = 0$$

or
$$(\lambda + 1)(\lambda + 2)(\lambda + 3) = 0$$

$$\lambda_1 = -1, \lambda_2 = -2, \lambda_3 = -3$$

Now, three eigenvalues are distinct, then we calculate $\overline{A}, \overline{B}$ by the use of T, where

$$T = \begin{bmatrix} 1 & 1 & 1 \\ \lambda_1 & \lambda_2 & \lambda_3 \\ \lambda_1^2 & \lambda_2^2 & \lambda_3^2 \end{bmatrix} = \begin{bmatrix} 1 & 1 & 1 \\ -1 & -2 & -3 \\ 1 & 4 & 9 \end{bmatrix}$$

$$T^{-1} = \begin{bmatrix} 3 & 2.5 & 0.5 \\ -3 & -4 & -1 \\ 1 & 1.5 & 0.5 \end{bmatrix}$$

Note : Students are advised to calculate T^{-1} themselves

$$\overline{A} = T^{-1}AT$$

$$= \begin{bmatrix} 3 & 2.5 & 0.5 \\ -3 & -4 & -1 \\ 1 & 1.5 & 0.5 \end{bmatrix} \begin{bmatrix} 0 & 1 & 0 \\ 0 & 0 & 1 \\ -6 & -11 & -6 \end{bmatrix} \begin{bmatrix} 1 & 1 & 1 \\ -1 & -2 & -3 \\ 1 & 4 & 9 \end{bmatrix} = \begin{bmatrix} -1 & 0 & 0 \\ 0 & -2 & 0 \\ 0 & 0 & -3 \end{bmatrix}$$

$$\overline{B} = T^{-1}B = \begin{bmatrix} 3 & 2.5 & 0.5 \\ -3 & -4 & -1 \\ 1 & 1.5 & 0.5 \end{bmatrix} \begin{bmatrix} 0 \\ 0 \\ 2 \end{bmatrix} = \begin{bmatrix} 1 \\ -2 \\ 1 \end{bmatrix}$$

$$\overline{C} = CT = [1\ 0\ 0] \begin{bmatrix} 1 & 1 & 1 \\ -1 & -2 & -3 \\ 1 & 4 & 9 \end{bmatrix} = [1\ 1\ 1]$$

Now the state equation can be written in transform form (DCF) as with the help of equation (6.117.19) and (6.117.21)

$$\begin{bmatrix} \dot{\overline{x}}_1 \\ \dot{\overline{x}}_2 \\ \dot{\overline{x}}_3 \end{bmatrix} = \begin{bmatrix} -1 & 0 & 0 \\ 0 & -2 & 0 \\ 0 & 0 & -3 \end{bmatrix} \begin{bmatrix} \overline{x}_1 \\ \overline{x}_2 \\ \overline{x}_3 \end{bmatrix} + \begin{bmatrix} 1 \\ -2 \\ 1 \end{bmatrix} u \qquad \text{...(6.117.23)}$$

$$y = [1\ 1\ 1] \begin{bmatrix} \overline{x}_1 \\ \overline{x}_2 \\ \overline{x}_3 \end{bmatrix} \qquad \text{...(6.117.24)}$$

From equations (6.117.23) and (6.117.24) it is clear that transformation matrix T modifies the coefficient matrix into diagonal matrix. The diagonal elements of equation (6.117.23) are same as eigenvalues of A.

6.14. DECOMPOSITION TRANSFER FUNCTION

The process by which transfer function changes to state diagram or state equations is called decomposition of the transfer function. The transfer function can be decomposed by three methods known as direct decomposition, cascade decomposition and parallel decomposition.

6.14.1. Direct Decomposition

6.14.1.1 Direct decomposition to controllable canonical form (CCF)

Direct decomposition method is used when the transfer function of a system is not in factored form

let the transfer function is

$$\frac{C(s)}{R(s)} = \frac{b_o s^n + b_1 s^{n-1} + \ldots + b_{n-1}s + b_n}{s^n + a_1 s^{n-1} + \ldots + a_{n-1}s + a_n} \quad \ldots(6.118)$$

The following steps are involved :

Step 1 : Express the numerator and denominator in the form of rational polynomials.

Step 2 : Divide the numerator and denominator by the higest power of s in the denominator or the numerator.

Step 3 : Multiply the numerator and denominator by $Q(s)$, where $Q(s)$ is a dummy variable.

Step 4 : Express the numerator and denominator separately and use the denominator equation to write an expression for $Q(s)$.

Step 5 : Draw the state variable diagram (SVD) for $Q(s)$ obtained in step 4.

Step 6 : Draw SVD for output.

Step 7 : Combine SVD obtained in step 5 and step 6.

Step 8 : Write expressions for the derivatives of state variables.

Step 9 : Feedback paths are determined by system poles and forward paths by system zero.

EXAMPLE 6.32. Consider the following transfer function

$$\frac{C(s)}{R(s)} = \frac{s+6}{s^2+5s+6}$$

obtain the state space representation.

Solution :

$$\frac{C(s)}{R(s)} = \frac{s+6}{s^2+5s+6} \quad \ldots(6.119\ (a))$$

Step 1 : Divide numerator and denominator by s^2

$$\frac{C(s)}{R(s)} = \frac{s^{-1}+6s^{-2}}{1+5s^{-1}+6s^{-2}}$$

Step 2 : Multiply numerator and denominator by $Q(s)$

$$\frac{C(s)}{R(s)} = \frac{s^{-1}+6s^{-2}}{1+5s^{-1}+6s^{-2}} \cdot \frac{Q(s)}{Q(s)} \quad \ldots(6.119\ (b))$$

Step 3 : Express numerator and denominator separately.

$$C(s) = Q(s)\,[s^{-1} + 6s^{-2}] \quad \ldots(6.119\ (c))$$

$$R(s) = Q(s)\,[1 + 5s^{-1} + 6s^{-2}] \quad \ldots(6.119\ (c))$$

Equation (6.32.3) can be written as

$$R(s) = Q(s) + Q(s)\,[5s^{-1} + 6s^{-2}] \qquad ...(6.119\,(d))$$

or,

$$Q(s) = R(s) - Q(s)\,[5s^{-1} + 6s^{-2}] \qquad ...(6.119\,(e))$$

Step 4 : Draw SVD for equation (6.119 (*e*))

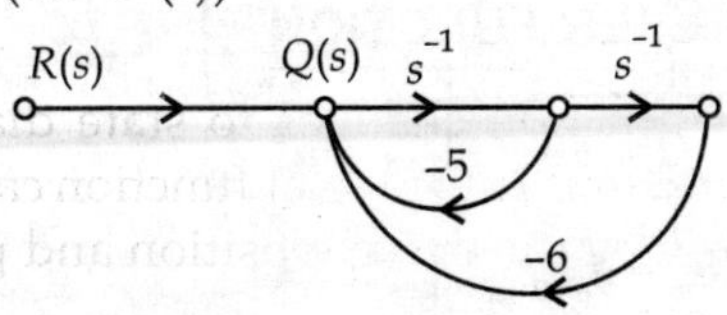

Step 5 : Draw SVD for equation (6.119 (*c*)).

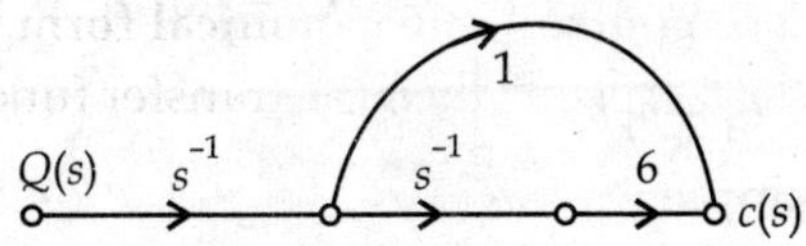

Step 6 : Combine SVD obtained in step (4) and (5)

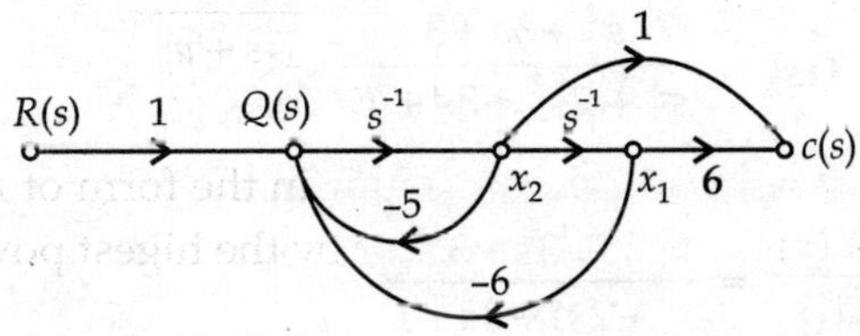

Fig. 6.20. CCF state diagram of transfer function

From diagram

$$\dot{x}_1 = x_2$$

$$\dot{x}_2 = r(t) - 6x_1 - 5x_2$$

Therefore

$$\begin{bmatrix} \dot{x}_1 \\ \dot{x}_2 \end{bmatrix} = \begin{bmatrix} 0 & 1 \\ -6 & -5 \end{bmatrix} \begin{bmatrix} x_1 \\ x_2 \end{bmatrix} + \begin{bmatrix} 0 \\ 1 \end{bmatrix} r(t) \qquad ...(6.119\,(f))$$

$$c(t) = 6x_1 + x_2$$

$$c(t) = [6 \;\; 1] \begin{bmatrix} x_1 \\ x_2 \end{bmatrix} \qquad ...(6.119\,(g))$$

Equations (6.119 (*f*)) and (6.119 (*g*)) are required state space representation. The above equations are controllable canonical form (CCF).

6.14.1.2. Direct Decomposition to Observable Canonical Form (OCF)

Multiply numerator and denominator by s^{-n} of equation (6.77)

$$\frac{Y(s)}{U(s)} = \frac{b_o + b_1 s^{-1} + ... + b_{n-1} s^{1-n} + b_n s^{-n}}{1 + a_1 s^{-1} + ... + a_{n-1} s^{1-n} + a_n s^{-n}}$$

$$(1 + a_1 s^{-1} + ... + a_{n-1} s^{1-n} + a_n s^{-n})\,Y(s) = (b_o + b_1 s^{-1} + ... + b_{n-1} s^{1-n} + b_n s^{-n})\,u(s)$$

or,

$$Y(s) = (b_o + b_1 s^{-1} + ... b_{n-1} s^{1-n} + b_n s^{-n})\,u(s) - Y(s)\,[a_1 s^{-1} + ... + a_{n-1} s^{1-n} + a_n s^{-n}] \qquad ...(6.120)$$

The output of integrators as designated as the state variables. The state variables are assigned in descending order from right to left. Apply SFG gain formula to state variable diagram and write the dynamic equation.

$$\overline{A} = \begin{bmatrix} 0 & 0 & \cdots & 0 & -a_n \\ 1 & 0 & \cdots & 0 & -a_{n-1} \\ 0 & 1 & & \vdots & \vdots \\ \vdots & \vdots & & \vdots & \vdots \\ 0 & 0 & \cdots & 1 & a_n \end{bmatrix}$$

$$\overline{B} = \begin{bmatrix} b_n - a_n b_0 \\ b_{n-1} - a_{n-1} b_0 - \\ \vdots \\ b_1 - a_1 b_0 \end{bmatrix}, \quad \overline{C} = [\,0 \;\; 0 \;\; 0 \; \ldots \; 1]$$

Consider the following example

EXAMPLE 6.33. Obtain the state model of the system whose transfer function is given by

$$T(s) = \frac{s^2 + 3s + 3}{s^3 + 2s^2 + 3s + 1} \qquad \text{...(6.121)}$$

(UPTU 2003-2004)

Solution : $$\frac{Y(s)}{U(s)} = \frac{s^2 + 3s + 3}{s^3 + 2s^2 + 3s + 1}$$

divide numerator and denominator by $s^n = s^3$

$$\frac{Y(s)}{U(s)} = \frac{s^{-1} + 3s^{-2} + 3s^{-3}}{1 + 2s^{-1} + 3s^{-2} + s^{-3}}$$

or, $$(s^{-1} + 3s^{-2} + 3s^{-3})\, u(s) = (1 + 2s^{-1} + 3s^{-2} + s^{-3})\, Y(s)$$

or, $$Y(s) = (s^{-1} + 3s^{-2} + 3s^{-3})\, u(s) - (2s^{-1} + 3s^{-2} + s^{-3})\, Y(s) \qquad \text{...(6.122)}$$

with the help of equation (6.122) draw state variable diagram (SVD)

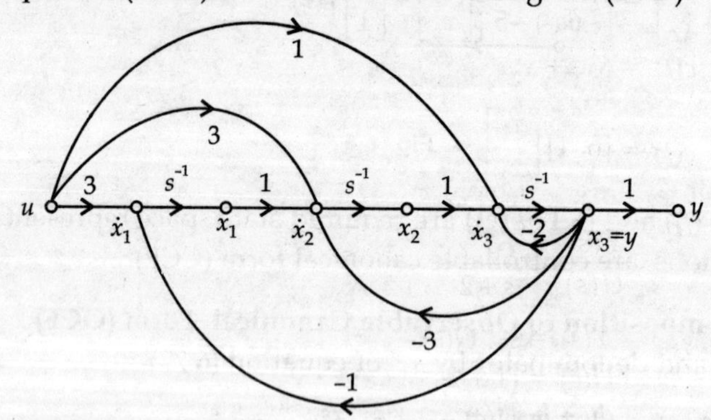

Fig. 6.21. OCF state diagram of the transfer function

From OCF diagram shown in Fig. 6.22.

$$y = x_3$$

$$\dot{x}_3 = u + x_2 - 2x_3 = Ox_1 + x_2 - 2x_3 + u$$

$$\dot{x}_2 = 3u - 3x_3 + x_1 = 1x_1 + Ox_2 - 3x_3 + 3u$$

$$\dot{x}_1 = 3u + x_3 = Ox_1 + Ox_2 + x_3 + 3u \qquad \text{...(6.123)}$$

state equations are

$$\begin{bmatrix} \dot{x}_1 \\ \dot{x}_2 \\ \dot{x}_3 \end{bmatrix} = \begin{bmatrix} 0 & 0 & 1 \\ 1 & 0 & -3 \\ 0 & 1 & -2 \end{bmatrix} \begin{bmatrix} x_1 \\ x_2 \\ x_3 \end{bmatrix} + \begin{bmatrix} 3 \\ 3 \\ 1 \end{bmatrix} u \qquad y = [0\ 0\ 1] \begin{bmatrix} x_1 \\ x_2 \\ x_3 \end{bmatrix}$$

The above equations are required answer.

6.14.2. Cascade Decomposition

When the transfer function is in factored form then cascade decomposition is used. First make the transfer function in factored form, then draw the transfer function of each part and combined all parts by unity transmittances inserted between each SVD. Consider the following example.

Example 6.34. Consider the transfer function given in example 6.32 draw SVD and obtain state space model by series (cascade) decomposition.

Solution : The given transfer function can be written in factored form

$$\frac{Y(s)}{U(s)} = \frac{s+6}{(s+3)(s+2)}$$

let, $$G_1(s) = \frac{s+6}{s+3} \text{ and } G_2 = \frac{1}{s+2}$$

First draw SVD for G_1

$$\frac{Y(s)}{U(s)} = \frac{1+6s^{-1}}{1+3s^{-1}} \cdot \frac{Q(s)}{Q(s)}$$

$$Y(s) = (1 + 6s^{-1})\, Q(s)$$

$$U(s) = (1 + 3s^{-1})\, Q(s)$$

or, $$Q(s) = u(s) - 3s^{-1}\, Q(s)$$

SVD for G_1

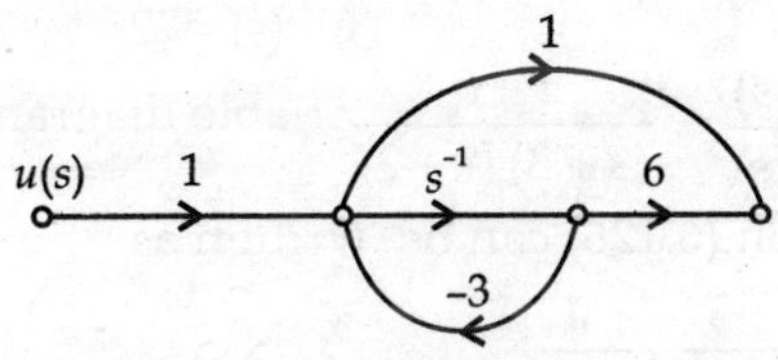

Fig. 6.22 (a)

Now draw SVD for G_2

$$\frac{Y(s)}{U(s)} = \frac{1}{s+2} = \frac{s^{-1}}{1+2s^{-1}} \cdot \frac{Q(s)}{Q(s)}$$

$$Y(s) = s^{-1}\, Q(s)$$

$$U(s) = (1 + 2s^{-1})\, Q(s)$$

or, $$Q(s) = U(s) - 2s^{-1}\, Q(s)$$

Draw SVD for G_2

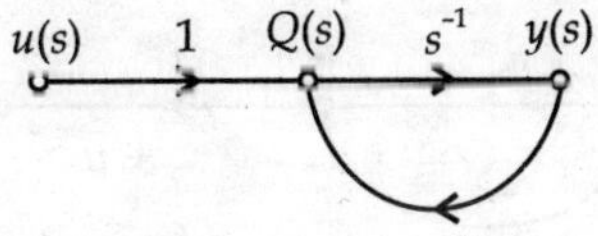

Fig. 6.22 (b)

Combine SVD for G_1 and G_2 with unity transmittance, we get complete SVD

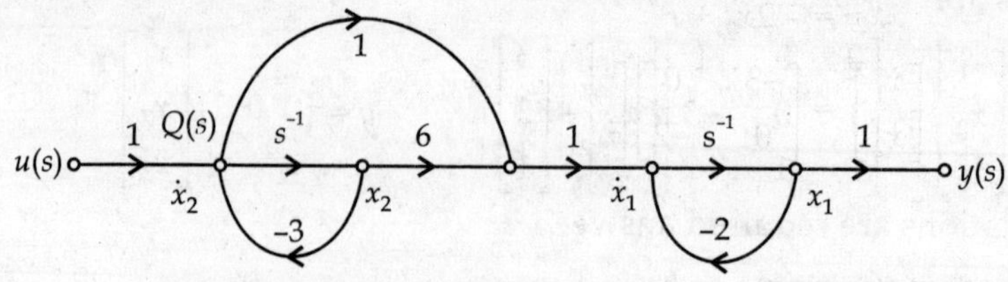

Fig. 6.22 (c). Complete SVD cascode decomposition

From SVD :

$$\dot{x}_1 = -2x_1 + \dot{x}_2 + 6x_2 \quad ...(6.124)$$

$$\dot{x}_2 = -3x_2 + u \quad ...(6.125)$$

Put the value of $\dot{x}_2$ from (6.125) in (6.124), and simplified

$$\dot{x}_1 = -2x_1 + 3x_2 + u$$

$$\dot{x}_2 = -3x_2 + u$$

In matrix form $\begin{bmatrix}\dot{x}_1\\ \dot{x}_2\end{bmatrix} = \begin{bmatrix}-2 & 3\\ 0 & -3\end{bmatrix}\begin{bmatrix}x_1\\ x_2\end{bmatrix} + \begin{bmatrix}1\\ 1\end{bmatrix}u \qquad y = \begin{bmatrix}1 & 0\end{bmatrix}\begin{bmatrix}x_1\\ x_2\end{bmatrix}$

The above equations are the required state space equations.

6.14.3. Parallel Composition

This method can be used when denominator of transfer function is in factrored form and numerator may or may not be factored form. Transfer function is expanded by partial fraction and draw SVD for each part, connect them in parallel.

EXAMPLE 6.35. Obtain the state space representation of the transfer function given in example 6.32.

Solution : $$\frac{C(s)}{R(s)} = \frac{s+6}{(s+3)(s+2)} \quad ...(6.126)$$

After partial fraction equation (6.126) can be written as

$$\frac{C(s)}{R(s)} = \frac{-3}{s+3} + \frac{4}{s+2} \quad ...(6.127)$$

(students are advised do the partial fraction)

Now draw SVD for each term and then connect in parallel. The resultant SVD is shown in figure 6.23.

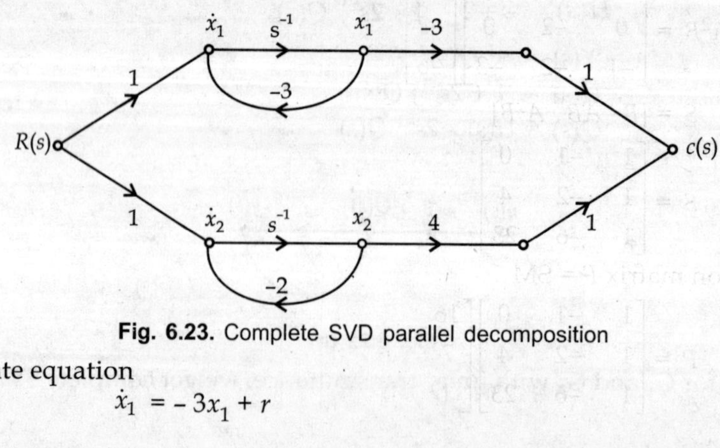

Fig. 6.23. Complete SVD parallel decomposition

The set of state equation

$$\dot{x}_1 = -3x_1 + r$$

$$\dot{x}_2 = -2x_2 + r$$

$$\begin{bmatrix} \dot{x}_1 \\ \dot{x}_2 \end{bmatrix} = \begin{bmatrix} -3 & 0 \\ 0 & -2 \end{bmatrix}\begin{bmatrix} x_1 \\ x_2 \end{bmatrix} + \begin{bmatrix} 1 \\ 1 \end{bmatrix} r$$

output equation

$$c(t) = -3x_1 + 4x_2 = \begin{bmatrix} -3 & 4 \end{bmatrix}\begin{bmatrix} x_1 \\ x_2 \end{bmatrix}$$

EXAMPLE 6.36. Given the dynamic equation

$$\frac{dx(t)}{dt} = Ax(t) + Bu(t)$$

$$y(t) = Cx(t)$$

where, $A = \begin{bmatrix} -2 & +1 & 0 \\ 0 & -2 & 0 \\ -1 & -2 & -3 \end{bmatrix}, B = \begin{bmatrix} 1 \\ 1 \\ 1 \end{bmatrix}, C = [1\ 0\ 0]$

Find the transformation matrix $x(t) = P\overline{x}(t)$.

Solution : First find out the characteristic equation

$$SI - A = s\begin{bmatrix} 1 & 0 & 0 \\ 0 & 1 & 0 \\ 0 & 0 & 1 \end{bmatrix} - \begin{bmatrix} -2 & 1 & 0 \\ 0 & -2 & 0 \\ -1 & -2 & -3 \end{bmatrix} = \begin{bmatrix} s+2 & -1 & 0 \\ 0 & s+2 & 0 \\ 1 & 2 & s+3 \end{bmatrix}$$

$$|SI - A| = s^3 + 7s^2 + 16s + 12 = 0$$

Here, $a_0 = 12,\ a_1 = 16,\ a_2 = 7,\ a_3 = 1$

$$M = \begin{bmatrix} a_1 & a_2 & 1 \\ a_2 & a_3 & 0 \\ 1 & 0 & 0 \end{bmatrix} = \begin{bmatrix} 16 & 7 & 1 \\ 7 & 1 & 0 \\ 1 & 0 & 0 \end{bmatrix}$$

$$AB = \begin{bmatrix} -2 & 1 & 0 \\ 0 & -2 & 0 \\ -1 & -2 & -3 \end{bmatrix}\begin{bmatrix} 1 \\ 1 \\ 1 \end{bmatrix} = \begin{bmatrix} -1 \\ -2 \\ -6 \end{bmatrix}$$

$$A^2B = \begin{bmatrix} -2 & 1 & 0 \\ 0 & -2 & 0 \\ -1 & -2 & -3 \end{bmatrix}\begin{bmatrix} -1 \\ -2 \\ -6 \end{bmatrix} = \begin{bmatrix} 0 \\ 4 \\ 23 \end{bmatrix}$$

$$S = [B\ \ AB\ \ A^2B]$$

$$\therefore \quad S = \begin{bmatrix} 1 & -1 & 0 \\ 1 & -2 & 4 \\ 1 & -6 & 23 \end{bmatrix}$$

$\therefore$ Transformation matrix $P = SM$

$$P = \begin{bmatrix} 1 & -1 & 0 \\ 1 & -2 & 4 \\ 1 & -6 & 23 \end{bmatrix}\begin{bmatrix} 16 & 7 & 1 \\ 7 & 1 & 0 \\ 1 & 0 & 0 \end{bmatrix} = \begin{bmatrix} 9 & 6 & 1 \\ 6 & 5 & 1 \\ -3 & 1 & 1 \end{bmatrix}$$ **Ans.**

EXAMPLE 6.37. Draw state variable diagram (SVD) for the given transfer function

$$\frac{C(s)}{R(s)} = \frac{5s}{3s^2+3s+1}$$

Solution : Multiply numerator and denominator by $\overline{s}^2$

$$\frac{C(s)}{R(s)} = \frac{5s}{3+3s^{-1}+s^{-2}} \cdot \frac{Q(s)}{Q(s)}$$

$$C(s) = 5s^{-1}\, Q(s)$$

$$R(s) = (3 + 3s^{-1} + s^{-2})\, Q(s)$$

$$R(s) = 3Q(s) + 3s^{-1}\, Q(s) + s^{-2}\, Q(s)$$

$$3Q(s) = R(s) - 3s^{-1}\, Q(s) - s^{-2}\, Q(s)$$

$$Q(s) = \frac{1}{3}R(s) - s^{-1}Q(s) - \frac{1}{3}s^{-2}Q(s)$$

$$Q(s) = \frac{1}{3}R(s) - Q(s)\left[s^{-1} + \frac{1}{3}s^{-2}\right]$$

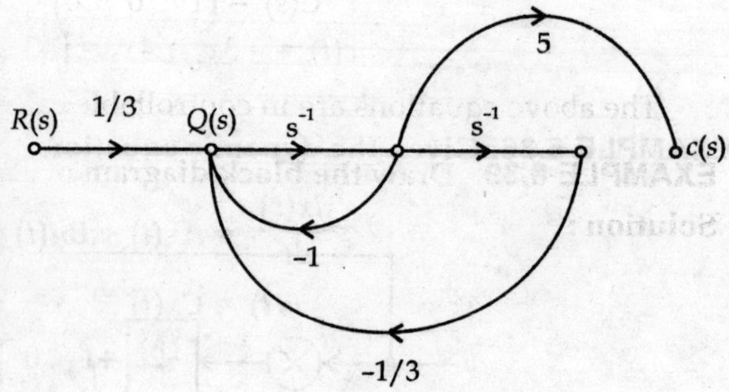

Fig. 6.24.

EXAMPLE 6.38. Write the state equations

$$\frac{d^3c}{dt^3} + 6\frac{dc}{dt} + 5c(t) = r(t)$$

Also draw the state variable diagram

Solution : Take Laplace transform of given equation

$$s^3C(s) + 6sC(s) + 5C(s) = R(s)$$

$$\frac{C(s)}{R(s)} = \frac{1}{s^3+6s+5}$$

Divide numerator and denominator by s^3

$$\frac{C(s)}{R(s)} = \frac{s^{-3}}{1+6s^{-2}+5s^{-3}} = \frac{s^{-3}}{1+6s^{-2}+5s^{-3}}\,\frac{Q(s)}{Q(s)}$$

$$C(s) = s^{-3}\, Q(s)$$

$$R(s) = Q(s)\,[1 + 6s^{-2} + 5s^{-3}]$$

$$R(s) = Q(s) + Q(s)\,[6s^{-2} + 5s^{-3}]$$

$$Q(s) = R(s) - Q(s)\,[6s^{-2} + 5s^{-3}]$$

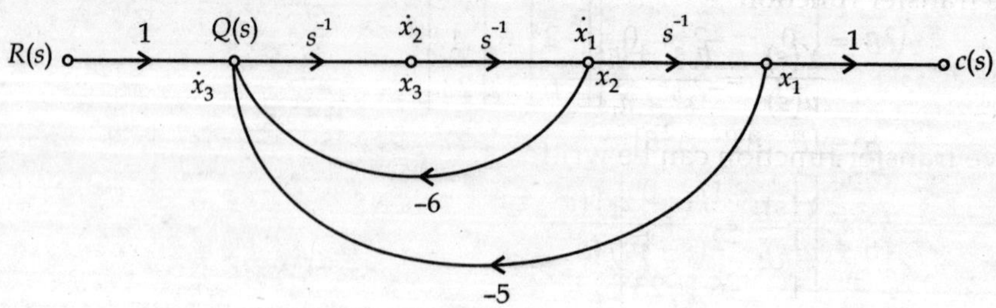

Fig. 6.25.

From SVD $\quad C(s) = x_1 \quad \dot{x}_1 = x_2$

$\dot{x}_2 = x_3 \quad \dot{x}_3 = -6x_2 - 5x_1 + r$

Above equation can be written in matrix form

$$\begin{bmatrix} \dot{x}_1 \\ \dot{x}_2 \\ \dot{x}_3 \end{bmatrix} = \begin{bmatrix} 0 & 1 & 0 \\ 0 & 0 & 1 \\ -5 & -6 & 0 \end{bmatrix} \begin{bmatrix} x_1 \\ x_2 \\ x_3 \end{bmatrix} + \begin{bmatrix} 0 \\ 0 \\ 1 \end{bmatrix} r$$

$$C(s) = \begin{bmatrix} 1 & 0 & 0 \end{bmatrix} \begin{bmatrix} x_1 \\ x_2 \\ x_3 \end{bmatrix}$$

The above equations are in controllable canonical form (CCF).

EXAMPLE 6.39. Draw the block diagram of Fig. 6.22 (c)

Solution :

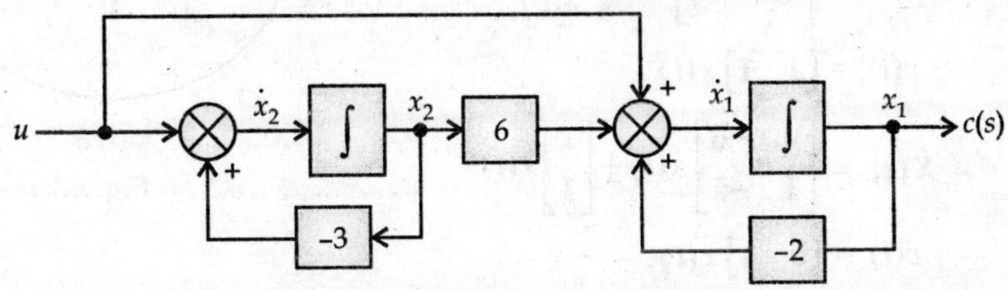

Fig. 6.26.

EXAMPLE 6.40. Draw the block diagram of example 6.35.

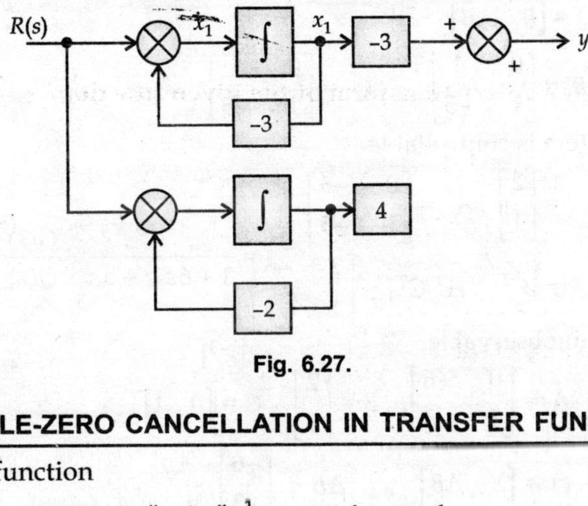

Fig. 6.27.

6.15. EFFECT OF POLE-ZERO CANCELLATION IN TRANSFER FUNCTION

Consider the transfer function

$$\frac{Y(s)}{u(s)} = \frac{b_o s^n + b_1 s^{n-1} + \ldots.. + b_{n-1} s + b_n}{s^n + a_1 s^{n-1} + \ldots. + a_{n-1} s + a_n}$$

The above transfer function can be written in factored form

$$\frac{Y(s)}{u(s)} = \frac{k(s - z_1)(s - z_2) \ldots. (s - z_n)}{(s - p_1)(s - p_2) \ldots. (s - p_n)}$$

$$= \sum_{k=1}^{k} \frac{d_k}{(s - p_k)}, \text{ where } d_k \text{ are residues of poles at } s = p_k$$

Suppose, transfer function has identical pole and zero at $z = p$ thus $d_i = 0$. The effect of this cancellation on controllability and observability depends on how the state variables are defined.

For example, if the state variables are defined by equation (6.117.7) and (6.117.8) the d_i will be zero in output vector c and if the state variables are defined by equation (6.117.14 and 6.117.15) then d_i will appear in control vector B. Thus, if pole zero cancellation occurs in transfer function, then the controllability and ovservability vary, depending how the state variables are choosen. So, for completely state controllable and observable the transfer function must not have any pole zero cancellation.

EXAMPLE 6.41. Consider the following state equations and check controllability and observability of the given systems and explain why there is a difference in controllability and observability of the same system.

(i)
$$\dot{X}(t) = \begin{bmatrix} 0 & 1 \\ -6 & -5 \end{bmatrix} x(t) + \begin{bmatrix} 0 \\ 1 \end{bmatrix} r(t)$$
$$c(t) = \begin{bmatrix} 2 & 1 \end{bmatrix} x(t)$$

(ii)
$$\dot{X}(t) = \begin{bmatrix} 0 & -6 \\ 1 & -5 \end{bmatrix} x(t) + \begin{bmatrix} 2 \\ 1 \end{bmatrix} r(t)$$
$$c(t) = \begin{bmatrix} 0 & 1 \end{bmatrix} x(t).$$

Solution : (*i*)
$$A = \begin{bmatrix} 0 & 1 \\ -6 & -5 \end{bmatrix}, \; B = \begin{bmatrix} 0 \\ 1 \end{bmatrix}, \; C = [2 \quad 1]$$

For controllability $Q = \begin{bmatrix} B & AB \end{bmatrix}$
$$AB = \begin{bmatrix} 0 & 1 \\ 1 & -5 \end{bmatrix} \; ; \; Q = \begin{bmatrix} 0 & 1 \\ 1 & -5 \end{bmatrix}$$

$Q \neq 0$, Rank = 2, system is controllable
$$C^T = \begin{bmatrix} 2 \\ 1 \end{bmatrix}, A^T = \begin{bmatrix} 0 & -6 \\ 1 & -5 \end{bmatrix}$$
$$Q' = [C^T \quad A^T C^T] = \begin{bmatrix} 2 & -6 \\ 1 & -3 \end{bmatrix}$$

Rank = 1, system is unobservable.

(*ii*)
$$A = \begin{bmatrix} 0 & -6 \\ 1 & -5 \end{bmatrix}, \; B = \begin{bmatrix} 2 \\ 1 \end{bmatrix}, \; C = \begin{bmatrix} 0 & 1 \end{bmatrix}$$

For controllability $Q = \begin{bmatrix} B & AB \end{bmatrix}$; $AB = \begin{bmatrix} -6 \\ -3 \end{bmatrix}$
$$Q = \begin{bmatrix} 2 & -6 \\ 1 & -3 \end{bmatrix}$$

Rank = 1, the system in uncontrollable.

For observability $Q' = [C^T \quad A^T C^T]$
$$A^T C^T = \begin{bmatrix} 1 \\ -5 \end{bmatrix} \; ; \; Q' = \begin{bmatrix} 0 & 1 \\ 1 & -5 \end{bmatrix}$$

Rank = 2, the system is observable.

Now find the transfer function
$$SI - A = \begin{bmatrix} s & 0 \\ 0 & s \end{bmatrix} - \begin{bmatrix} 0 & -6 \\ 1 & -5 \end{bmatrix} = \begin{bmatrix} s & 6 \\ -1 & s+5 \end{bmatrix}$$

$$|SI - A| = s^2 + 5s + 6 = (s + 2)(s + 3)$$

$$[SI - A]^{-1} = \begin{bmatrix} \frac{s+5}{s^2+5s+6} & \frac{-6}{s^2+5s+6} \\ \frac{1}{s^2+5s+6} & \frac{s}{s^2+5s+6} \end{bmatrix}$$

Transfer function is given by

$$C^T [SI - A]^{-1} B = \begin{bmatrix} 0 & 1 \end{bmatrix} \begin{bmatrix} \frac{s+5}{s^2+5s+6} & \frac{-6}{s^2+5s+6} \\ \frac{1}{s^2+5s+6} & \frac{s}{s^2+5s+6} \end{bmatrix} \begin{bmatrix} 2 \\ 1 \end{bmatrix}$$

$$\text{Transfer function} = \frac{s+2}{(s+2)(s+3)}$$

From the transfer function it is clear that there is a pole at s = – 2 and a zero at s = – 2, there is a pole zero cancellation. Therefore for completely controllable and observable the transfer function must not have any pole zero cancellation.

SUMMARY

Advantages of State Space Techniques

1. This method can be applied to linear or non-linear, time variant or time invariant system.
2. It is easy to apply where laplace transform cannot applied.
3. It is time domain approach.
4. This method is suitable for digital computer computation.
5. This method can be designed for optimal conditions.

A state vector is a vector which determines uniquely the system state $x(t)$ for any time $t \geq t_0$.

$$f(t) = \text{State transition matrix} = e^{AT}$$

$$= I + At + \frac{A^2 t^2}{21} + \ldots$$

Properties of state transition matrix (STM)

1. $\phi(0) = e^{A(0)} = I$
2. $\phi(t) = [\phi(-t)]^{-1}$
3. $\phi^{-1}(t) = \phi(-t)$
4. $[\phi(t)]^n = \phi(nt)$
5. $\phi(t_1 + t_2) = \phi(t_2)\,\phi(t_1)$

$$\phi(t) = \mathcal{L}^{-1}\,\phi(s) - \mathcal{L}^{-1}(SI - A)^{-1}$$

$$\phi(s) = \text{resolvent matrix}$$

Transfer matrix

$G(s) = C\phi(s)B$ for MIMO system

$G(s) = C^T\phi(s)b$ for SISO system

The roots of the characteristic equation are known as eigen values of matrix.

Any non-zero vector P_i that satisfies the matrix equation $(\lambda_i I - A) P_i = 0$.

λ_i is the eigen value of A

A system is said to be controllable if any initial state $x(t_0)$ or x_0 can be transferred to any final state $x(t_f)$ in a finite time interval $(t_f - t_0)$, $t_0 \geq 0$ by control U.

A system is said to be observable if every state x_0 can be exactly determined from the measurement of the output y over a finite interval of time $0 < t < t_f$

There are four types of states :

1. The system is controllable and observable. (SCO)
2. The system is controllable and unobservable. (SCU)
3. The system is uncontrollable and observable (SUO)
4. The system is uncontrollable and unobservable (SUU)

KALMAN'S TEST

A linear time invariant continuous system is completely controllable if and only if the rank of the controllability matrix is defined as

$$Q = [B: AB: A^2 B; .... A^{n-1} B]$$

is equal to rank 'n'.

EXERCISE

6.1. The state equation of a linear time invariant system is given by

$$\begin{bmatrix} \dot{x}_1 \\ \dot{x}_2 \end{bmatrix} = \begin{bmatrix} 0 & 1 \\ -1 & -2 \end{bmatrix} \begin{bmatrix} x_1 \\ x_2 \end{bmatrix} + \begin{bmatrix} 0 \\ 1 \end{bmatrix} u(t)$$

Find the state transition matrix and the characteristic equation.

6.2. Find the state transition matrix for

$$\dot{x}_1 = x_2$$

$$\dot{x}_2 = -6x_1 - 5x_2$$

6.3. The state space of a system is represented by the following equations

$$\begin{bmatrix} \dot{x}_1 \\ \dot{x}_2 \end{bmatrix} = \begin{bmatrix} -3 & 1 \\ -2 & 0 \end{bmatrix} \begin{bmatrix} x_1 \\ x_2 \end{bmatrix} + \begin{bmatrix} 0 \\ 1 \end{bmatrix} u \quad ; \; t > 0$$

$$y = \begin{bmatrix} 1 & 0 \end{bmatrix} \begin{bmatrix} x_1 \\ x_2 \end{bmatrix}$$

Find the transfer function of the system.

6.4. Obtain A, B, C and D matrices of the circuit

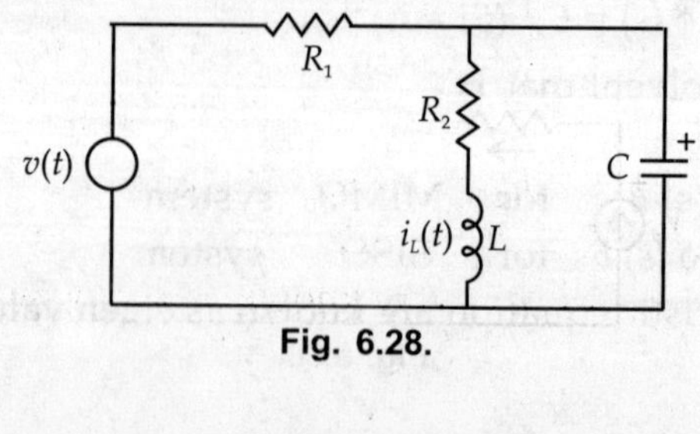

Fig. 6.28.

6.5. The state space representation of a system is

$$\begin{bmatrix} \dot{x}_1 \\ \dot{x}_2 \\ \dot{x}_3 \end{bmatrix} = \begin{bmatrix} -5 & 2 & 1 \\ 0 & 0 & 1 \\ -1 & -4 & -3 \end{bmatrix} \begin{bmatrix} x_1 \\ x_2 \\ x_3 \end{bmatrix} + \begin{bmatrix} 0 \\ 0 \\ 1 \end{bmatrix} u$$

output : $y = x_1$

Is the system controllable ?

6.6. Obtain a state space representation for the system

$$\frac{d^3y}{dt^3} + 5\frac{dy^2}{dt^2} + 6\frac{dy}{dt} + 3y = 6u(t)$$

6.7. For question no. (6.35) determine the transfer function.

6.8. Obtain the transfer function representation for a system represented by matrices

$$A = \begin{bmatrix} 0 & 1 & 0 \\ 0 & 0 & 1 \\ -10 & -2 & -3 \end{bmatrix}; \; B = \begin{bmatrix} 0 \\ 0 \\ 1 \end{bmatrix}; \; C = \begin{bmatrix} 2 & 1 & 0 \end{bmatrix}$$

6.9. The transfer function of a system is $G(s) = \dfrac{2}{(s+1)(s+2)}$. Obtain a state variable representation.

6.10. Obtain the state space representation of the following control system. *(UPSC 1975)*

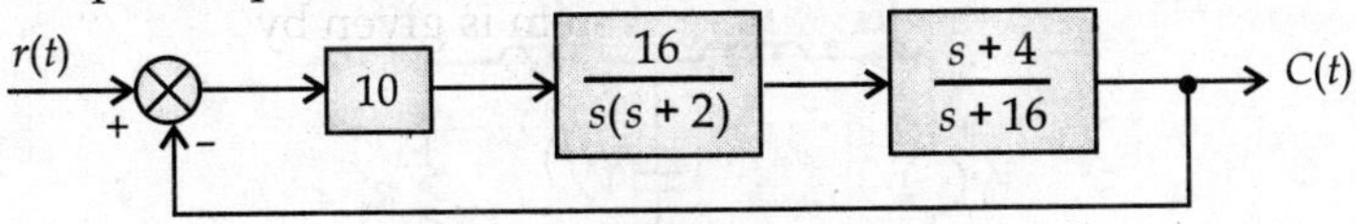

Fig. 6.29.

6.11. Determine the controllability of the equation

$$\dot{x} = \begin{bmatrix} 1 & 1 & 0 \\ 0 & 1 & 0 \\ 0 & 1 & 1 \end{bmatrix} x + \begin{bmatrix} 0 & 1 \\ 1 & 0 \\ 0 & 1 \end{bmatrix} u$$

$$y = \begin{bmatrix} 1 & 1 & 1 \end{bmatrix} x$$

6.12. Check the contrllability and observability of the equation

$$\dot{x} = \begin{bmatrix} 0 & 1 & 0 \\ 0 & 0 & 1 \\ -2 & -4 & -3 \end{bmatrix} x + \begin{bmatrix} 1 & 0 \\ 0 & 1 \\ -1 & 1 \end{bmatrix} u$$

$$y = \begin{bmatrix} 0 & 1 & -1 \\ 1 & 2 & 1 \end{bmatrix} x$$

6.13. Determine the state model for the electrical circuit shown in Fig. 6.30 select i_{L1} and i_{L2} as state variables.

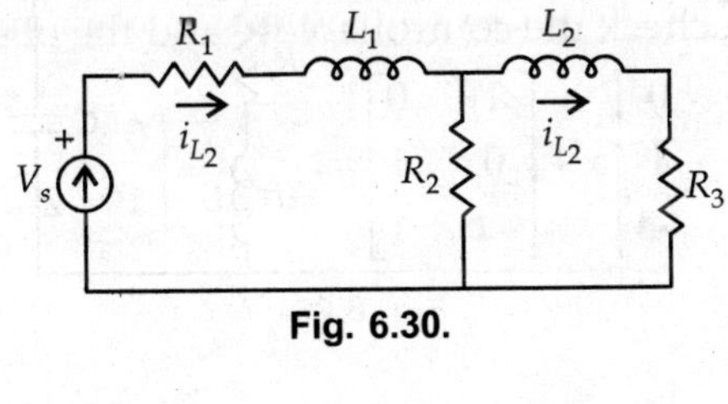

Fig. 6.30.

6.14. Obtained the state model for the given electrical circuit. Select the state variables as V_c, i_{L1} and i_{L2}

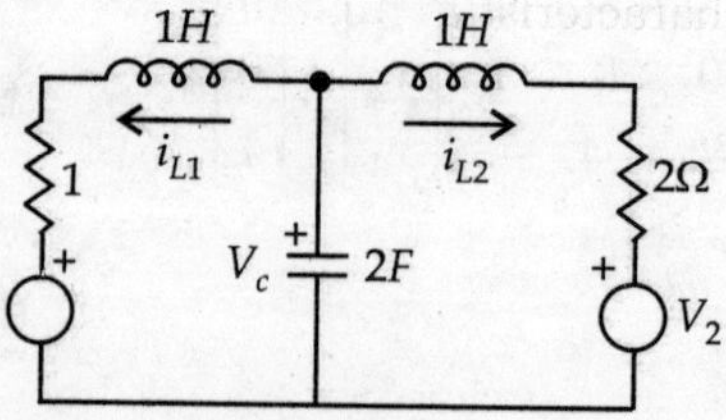

Fig. 6.31.

6.15. In question number 6.52 check the observability.

6.16. The state equation of linear time invariant system, *i.e.*,

$$\begin{bmatrix} x_1 \\ x_2 \end{bmatrix} = \begin{bmatrix} 0 & 1 \\ -1 & -1 \end{bmatrix}\begin{bmatrix} x_1 \\ x_2 \end{bmatrix} + \begin{bmatrix} 0 \\ 1 \end{bmatrix} u(t)$$

Find the stable transition matrix *(UPTU Question Bank)*

6.17. Give the advantage of state space technique & obtain the state space representation of the system

$$\frac{d^3y}{dt^3} + 5\frac{d^2y}{dt^2} + 10\frac{dy}{dt} + 20y = 8u(t)$$ *(UPTU Question Bank)*

6.18. Write the state space representation taking i, i_{L_1}, i_{L_2} and V_c as the variables.

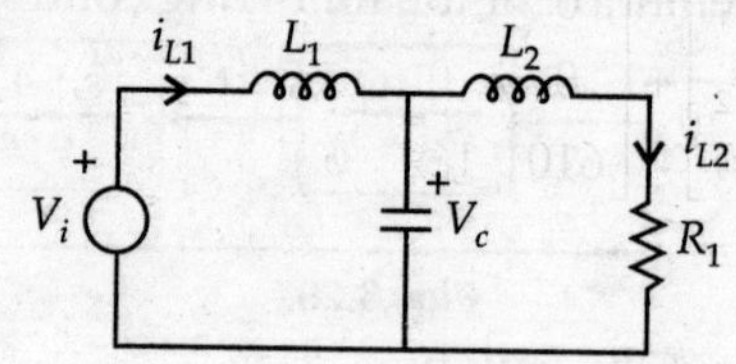

Fig. 6.32.

6.19. (*a*) Wite the properties of transition matrix.

(*b*) If $A = \begin{bmatrix} 0 & 1 \\ -6 & -5 \end{bmatrix}$; $B = \begin{bmatrix} 0 \\ 1 \end{bmatrix}$; $C = [0\ 1]$

Determine $\phi(t)$. *(UPTU Question Bank)*

6.20. (*a*) For the given transfer function, obtain the state model

$$G(s) = \frac{k}{s^3 + a_3s^2 + a_2s + a_1}$$

OR

consider a control system with the state model given by

$$\begin{bmatrix} \dot{X}_1 \\ \dot{X}_2 \end{bmatrix} = \begin{bmatrix} 0 & 1 \\ -2 & -3 \end{bmatrix}\begin{bmatrix} x_1 \\ x_2 \end{bmatrix} + \begin{bmatrix} 0 \\ 2 \end{bmatrix}[u]$$

$X_1(0)$ and $X_2(0) = 0$ and u = unit step function. Compute the state transition matrix and hence find state response. *(UPTU Question Bank)*

(*b*) For the given system check the controllability and the observability of the system

$$\begin{bmatrix} \dot{x}_1 \\ \dot{x}_2 \\ \dot{x}_3 \end{bmatrix} = \begin{bmatrix} 0 & 1 & 0 \\ 0 & 0 & 1 \\ -2 & -4 & -3 \end{bmatrix} x + \begin{bmatrix} 1 & 0 \\ 0 & 1 \\ -1 & 1 \end{bmatrix} u \ ; \ Y = \begin{bmatrix} 0 & 1 & -1 \\ 1 & 2 & 1 \end{bmatrix} x$$

(UPTU Question Bank)

ANSWERS

6.1. $\begin{bmatrix} te^{-t}+e^{-t} & te^{-t} \\ -te^{-t} & -te^{-t}+e^{-t} \end{bmatrix}$, characteristic equation = $s^2 + 2s + 1 = 0$

6.2. $\begin{bmatrix} 3e^{-2t}+2e^{-3t} & e^{-2t}-e^{-3t} \\ -6e^{-2t}+6e^{-3t} & -2e^{-2t}+3e^{-3t} \end{bmatrix}$

6.3. $\dfrac{1}{(s+1)(s+2)}$

6.4. $A = \begin{bmatrix} -\dfrac{1}{R_1C} & -\dfrac{1}{C} \\ \dfrac{1}{L} & -\dfrac{R_2}{L} \end{bmatrix}$, $B = \begin{bmatrix} \dfrac{1}{R_1C} \\ 0 \end{bmatrix}$, $C = [1 \quad 0]$, $D = [0]$

6.5. $\dfrac{s+2}{s^3+3s^2+2s+10}$

6.6. $\begin{bmatrix} \dot{x}_1 \\ \dot{x}_2 \end{bmatrix} = \begin{bmatrix} 0 & 1 \\ -2 & -3 \end{bmatrix}\begin{bmatrix} x_1 \\ x_2 \end{bmatrix} + \begin{bmatrix} 0 \\ 2 \end{bmatrix} u$ and $y(t) = [1 \quad 0]\, x(t)$

6.7. $\begin{bmatrix} \dot{x}_1 \\ \dot{x}_2 \\ \dot{x}_3 \end{bmatrix} = \begin{bmatrix} 0 & 1 & 1 \\ 0 & 0 & 1 \\ -640 & -172 & -8 \end{bmatrix}\begin{bmatrix} x_1 \\ x_2 \\ x_3 \end{bmatrix} + \begin{bmatrix} 0 & 0 & 0 \\ 0 & 0 & 0 \\ 610 & 160 & 0 \end{bmatrix}\begin{bmatrix} u_1 \\ u_2 \\ u_3 \end{bmatrix}$

6.8. not contrabble

6.9. $\begin{bmatrix} \dot{i}_{L1} \\ \dot{i}_{L2} \end{bmatrix} = \begin{bmatrix} -\dfrac{1}{L_1}(R_1+R_2) & \dfrac{R_2}{L_1} \\ \dfrac{R_2}{L_2} & -\dfrac{1}{L_2}(R_2+R_3) \end{bmatrix}\begin{bmatrix} i_{L_1} \\ i_{L_2} \end{bmatrix} + \begin{bmatrix} \dfrac{1}{L_1} \\ 0 \end{bmatrix}[V_s]$

6.10. $\begin{bmatrix} \dot{V}_c \\ \dot{i}_{L1} \\ \dot{i}_{L2} \end{bmatrix} = \begin{bmatrix} 0 & -0.5 & -0.5 \\ 0.5 & -0.5 & 0 \\ 1 & 0 & -2 \end{bmatrix}\begin{bmatrix} V_c \\ i_{L1} \\ i_{L2} \end{bmatrix} + \begin{bmatrix} 0 & 0 \\ -0.5 & 0 \\ 0 & -1 \end{bmatrix}\begin{bmatrix} V_1 \\ V_2 \end{bmatrix}$

SEMI-OBJECTIVE TYPE QUESTIONS

(*i*) What are the advantages of state space techniques. (Article 6.2)

(*ii*) Define state, state variables, state vector and state space. (Article 6.3)

(*iii*) What is state transition matrix? (Article 6.6)

(*iv*) Define Eigen values and Eigen vectors. (Articles 6.7.3 and 6.7.4.)

(*v*) What are the properties of STM. (Article 6.5.2)

(*vi*) Define controllability and observability. (Article 6.9)

(*vii*) Write a short note on Kalman's Test. (Article 6.9.2)

(*viii*) Write a short note limitations of state variable approach.

□□□

7 THE LAPLACE TRANSFORMS

7.1. INTRODUCTION

In laplace transform method, the differential equation in the time domain is transformed into s-plane. The complex frequency variable is given by $S = \sigma + j\omega$, where σ and ω are both real quantities. σ is the real part and ω is the imaginary part. The advantage of Laplace transform is that this method gives total solution. For example, when we solve the differential equation by Laplace transform, both transient and steady state component of the solution can be obtained simultaneously.

7.2. DEFINITION OF LAPLACE TRANSFORM

One-sided Laplace transform of a function of time $f(t)$ is given by

$$£[f(t)] = \int_0^\infty f(t)\cdot e^{-st}dt = £(s) = \int_0^\infty f(t)\cdot e^{-st}dt \qquad ...(7.1)$$

where $f(t)$ = is time function defined in the interval $0 \le t \le \infty$

s = complex variable

$F(s)$ = Laplace transform of $f(t)$

7.3. LAPLACE TRANSFORMS OF SOME TIME FUNCTIONS

7.3.1. Exponential Function

Consider the exponential function

$$f(t) = 0 \quad \text{for} \quad t < 0$$
$$= Ke^{-at} \quad \text{for} \quad t \ge 0$$

where 'K' and 'a' are constants.

$$£[K e^{-at}] = \int_0^\infty K e^{-at}\cdot e^{-st}dt = K\int_0^\infty e^{-(a+s)t}\,dt = K\left[-\frac{1}{a+s}e^{-(a+s)t}\right]_0^\infty$$

$$£[K e^{-at}] = -\frac{K}{s+a}[e^{-\infty} - e^{o}] = \frac{K}{s+a}$$

7.3.2. Step Function

Consider the step function

$$f(t) = 0 \quad \text{for} \quad t < 0$$
$$f(t) = A \quad \text{for} \quad t > 0$$

where 'A' is a constant

$$\pounds f(t) = \int_0^\infty Ae^{-st}\,dt = A\int_0^\infty e^{-st}\,dt = A\left[\frac{e^{-st}}{-s}\right]_0^\infty = \frac{A}{s}$$

7.3.3. Unit Step Function

Consider the unit step function

$$f(t) = 0 \quad \text{for} \quad t < 0$$
$$1 \quad \text{for} \quad t > 0$$

$$\pounds f(t) = \int_0^\infty 1.e^{-st}\,dt = \left[\frac{e^{-st}}{-s}\right]_0^\infty = \frac{1}{s}$$

7.3.4. Ramp Function

Consider the ramp function

$$f(t) = 0 \quad \text{for} \quad t < 0$$
$$= At \quad \text{for} \quad t \geq 0$$

$$\pounds\,[At] = \int_0^\infty At.e^{-st}\,dt = \frac{A}{s^2}$$

7.3.5. Sinusoidal Function

Consider the sinusoidal function

$$f(t) = 0 \quad \text{for} \quad t < 0$$
$$= A\sin\omega t \quad \text{for} \quad t \geq 0$$
$$\pounds f(t) = \pounds\,(A\sin\omega t) = A\,\pounds(\sin\omega t)$$
$$= A\,\pounds\frac{e^{j\omega t} - e^{-j\omega t}}{2j} = \frac{A}{2j}\left[\pounds\,e^{j\omega t} - \pounds\,e^{-j w t}\right] = \frac{A}{2j}\left[\frac{1}{s-j\omega} - \frac{1}{s+j\omega}\right] = \frac{A\omega}{s^2+\omega^2}$$

$$\pounds\,(A\sin\omega t) = \frac{A\omega}{s^2+\omega^2}$$

For $A = 1$

$$\pounds\sin\omega t = \frac{\omega}{s^2+\omega^2}$$

7.3.6. Laplace Transform of t^n

$$\pounds\,t^n = \int_0^\infty t^n e^{-st}\,dt$$

let $st = x$

$$dt = \frac{1}{s}dx$$

$\therefore$ $t^n = \left(\frac{x}{s}\right)^n$

When $t = 0, x = 0$

$t = \infty, x = \infty$

$$\pounds\,t^n = \int_0^\infty \left(\frac{x}{s}\right)^n e^{-x}.\frac{1}{s}dx = \frac{1}{s^{n+1}}\int_0^\infty x^n e^{-x}dx$$

Since $\overline{|n+1} \triangleq \int_0^\infty x^n e^{-x} dx = \overline{|n} = \underline{|n}$

$\therefore$ $$£\, t^n = \frac{1}{s^{n+1}}\overline{|n+1} = \frac{\underline{|n}}{s^{n+1}}$$

$$\boxed{£\, t^n = \frac{\underline{|n}}{s^{n+1}}}$$

7.3.7. Hyperbolic Sine and Cosine Functions

$$f(t) = \cosh\beta t$$

$$£f(t) = £\cosh\beta t$$

$$= £\left[\frac{e^{\beta t}+e^{-\beta t}}{2}\right] \qquad \because \cosh\beta t = \frac{e^{\beta t}+e^{-\beta t}}{2}$$

$$= \frac{1}{2}\left[£\, e^{\beta t} + £\, e^{-\beta t}\right] = \frac{1}{2}\left[\frac{1}{s-\beta}+\frac{1}{s+\beta}\right] \qquad \because £\, e^{\beta t} = \frac{1}{s-\beta}\; £\, e^{-\beta t} = \frac{1}{s+\beta}$$

$$£\cosh\beta t = \frac{s}{s^2-\beta^2}$$

$\therefore$ $$£\cosh\beta t = \frac{s}{s^2-\beta^2}$$

Similarly, $$£\sinh\alpha t = £\,\frac{e^{\alpha t}-e^{-\alpha t}}{2} = \frac{1}{2}[£\, e^{\alpha t} - £\, e^{-\alpha t}] = \frac{1}{2}\left[\frac{1}{s-\alpha}-\frac{1}{s+\alpha}\right] = \frac{\alpha}{s^2-\alpha^2}$$

$$\boxed{£\sinh\alpha t = \frac{\alpha}{s^2-\alpha^2}}$$

7.3.8. Damped Sine and Cosine Functions

$$f(t) = e^{-\alpha t}\sin\beta t$$

$$f(t) = e^{-\alpha t}\left[\frac{e^{j\beta t}-e^{-j\beta t}}{2j}\right] = \frac{1}{2j}\left[£\, e^{j\beta t}\, e^{-\alpha t} - £\, e^{-j\beta t} e^{-\alpha t}\right]$$

$$= \frac{1}{2j}\left[\int_0^\infty e^{j\beta t} e^{-\alpha t} e^{-st}\, dt - \int_0^\infty e^{-j\beta} e^{-\alpha t}\, e^{-st}\, dt\right]$$

$$= \frac{1}{2j}\left[\int_0^\infty e^{-t(s+\alpha-j\beta)} dt - \int_0^\infty e^{-t(s+\alpha+j\beta)}\right]$$

$$= \frac{1}{2j}\left[\frac{1}{s+\alpha-j\beta}-\frac{1}{s+\alpha+j\beta}\right] = \frac{\beta}{(s+\alpha)^2+\beta^2}$$

$$\boxed{£\, e^{-\alpha t}\sin\beta t = \frac{\beta}{(s+\alpha)^2+\beta^2}}$$

$$f(t) = e^{-\alpha t}\cos\beta t$$

$$\pounds f(t) = F(s) = \pounds\, e^{-\alpha t} \cos \beta t = \pounds\, e^{-\alpha t}\, \frac{e^{j\beta t} + e^{-j\beta t}}{2}$$

$$= \frac{1}{2}\left[\int_0^\infty e^{-t(s+\alpha-j\beta)}\, dt + \int_0^\infty e^{-t(s+\alpha+j\beta)}\, dt\right]$$

$$= \frac{1}{2}\left[\frac{1}{s+\alpha-j\beta} + \frac{1}{s+\alpha+j\beta}\right] = \frac{s+\alpha}{(s+\alpha)^2+\beta^2}$$

$$\boxed{\pounds\, e^{-\alpha t} \cos \beta t = \frac{s+\alpha}{(s+\alpha)^2+\beta^2}}$$

EXAMPLE 7.1. Find the Laplace transform of

$$f(t) = e^{at} \sin \omega t$$

Solution :

$$\pounds\, e^{at} \sin\omega t = \frac{1}{2j}[\pounds\, e^{j\omega t} e^{-at} - \pounds\, e^{-j\omega t} e^{at}]$$

$$\because \quad \sin\theta = \frac{e^{j\theta} - e^{-j\theta}}{2j} = \frac{1}{2j}\left[\int_0^\infty e^{j\omega t} e^{at} e^{-st} dt - \int_0^\infty e^{-j\omega t} e^{-at} e^{-st} dt\right]$$

$$= \frac{1}{2j}\left[\int_0^\infty e^{-t(s-a-j\omega)} dt + \int_0^\infty e^{-t(s-a+j\omega)}\, dt\right]$$

$$= \frac{1}{2j}\left[\frac{1}{s-a-j\omega} - \frac{1}{s-a+j\omega}\right] = \frac{\omega}{(s-a)^2+\omega^2}$$

$$\boxed{\pounds\, e^{at} \sin \omega t = \frac{\omega}{(s-a)^2+\omega^2}}$$

7.4. PROPERTIES OF LAPLACE TRANSFORMS

1. If 'a' is a constant and $F(s)$ is the Laplace transform of $f(t)$ then
$$\pounds\, af(t) = a\,\pounds f(t) = aF(s)$$
2. If $f_1(t)$ and $f_2(t)$ are the time functions and $F_1(s)$, $F_2(s)$ are their Laplace transform then
$$\pounds\, [f_1(t) \pm f_2(t)] = \pounds f_1(t) \pm \pounds f_2(t) = F_1(s) + F_2(s)$$
3. Time shift property : If 'a' is a positive real number and $F(s)$ is the Laplace transform of $f(t)$ then
$$\pounds\, [f(t-a)] = e^{-as} F(s)$$
Thus, if the original time function is shifted to the right by 'a' on time axis, its Laplace transform is equal to the original laplace transform multiplied by e^{-as}. This is the time shifting theorem.
4. If $F(s)$ is the Laplace transform of $f(t)$ then
$$\pounds\, e^{at} f(t) = F(s-a)$$
$$\pounds\, e^{-at} f(t) = F(s+a)$$

e.g., $\pounds\, e^{-at} \sin \omega t$

since $\pounds \sin \omega t = \dfrac{\omega}{s^2 + \omega^2}$

$\therefore$ $\pounds\, e^{-at} \sin\omega t = F(s + a) = \dfrac{\omega}{(s+a)^2 + \omega^2}$

5. **Real differentiation theorem:**
The Laplace transform of the derivative of a function $f(t)$ is given by

$$\pounds\left[\frac{d}{dt}f(t)\right] = sF(s) - f(o)$$

where $f(o)$ is the initial value of $f(t)$ at $t = 0$

6. **Laplace transform of higher derivative**

$$\pounds\left[\frac{d^n f(t)}{dt^n}\right] = S^n F(s) - s^{n-1} f(o) \;........ - \frac{d^{n-1} f(o)}{dt^{n-1}}$$

where $f(o)$ is the value of $f(t)$ at $t = 0$

$\dfrac{d^{n-1} f(o)}{dt^{n-1}}$ **is the value of** $\dfrac{d^{n-1} f(t)}{dt^{n-1}}$ **at $t = 0$ and so on.**

7. **Laplace transform of integral of $f(t)$**

$$\pounds\left[\int f(t)\, dt\right] = \frac{F(s)}{s} + \frac{f^{-1}(o)}{s}$$

where $F(s)$ is the laplace transform of $f(t)$ and

$$f^{-1}(o) = \int f(t)\, dt \quad at\ t = 0$$

8. **Complex differentiation theorem :**

$$\pounds\,[t^n f(t)] = (-1)^n \frac{d^n}{ds^n} F(s)$$

e.g., **Find the Laplace transform of t sinh at**

$$\pounds \sinh at = \frac{a}{s^2 - a^2}$$

$$\therefore \quad \pounds\,[\,t \sin h\, at] = -\frac{d}{ds}\left[\frac{a}{s^2 - a^2}\right]$$

$$\pounds\,[\,t \sinh at] = \frac{2as}{(s^2 - a^2)^2}$$

9. **Complex integration :**
If $F(s)$ is the Laplace transform of $f(t)$ then

$$\pounds\left[\frac{f(t)}{t}\right] = \int_0^\infty F(s)\, ds$$

e.g., **find the Laplace transform of** $\dfrac{\sin 2t}{t}$

$$\pounds \sin 2t = \frac{2}{s^2 + 4}$$

$$£\frac{\sin 2t}{t} = \int_s^{\infty}\frac{2}{s^2+4}ds = 2\cdot\frac{1}{2}\left[\tan^{-1}\frac{s}{2}\right]_S^{\infty} = \left[\tan^{-1}\infty - \tan^{-1}\frac{s}{2}\right] \quad \because \int\frac{1}{x^2+a^2} = \frac{1}{a}\tan^{-1}\frac{x}{a}$$

$$= \frac{\pi}{2} - \tan^{-1}\frac{s}{2}$$

10. **Initial value theorem:**

If initial value of the function $f(t)$ is required, then it can be calculated by initial value theorem which states

$$\lim_{t\to 0} f(t) = \lim_{s\to\infty} sF(s)$$

e.g., **Find the initial value of e^{-at}**

$$f(t) = e^{-at}$$

$$\therefore \quad F(s) = £f(t) = \frac{1}{s+a}$$

$$\lim_{t\to 0} f(t) = \lim_{s\to\infty} s\cdot\frac{1}{s+a} = \lim_{s\to\infty}\frac{1}{1+\frac{a}{s}} = 1$$

11. Final value theorem :

If final value of $f(t)$ is required then it can be calculated by final value theorem, which states that

$$\lim_{s\to\infty} f(t) = \lim_{s\to 0} sF(s)$$

e.g., find the final value of e^{-at}

$$f(t) = e^{-at}$$

$$F(s) = £f(t) = \frac{1}{s+a}$$

$$\lim_{t\to\infty} f(t) = \lim_{s\to 0} s.\frac{1}{s+a} = 0$$

7.5. INVERSE LAPLACE TRANSFORM

If $F(s)$ is the Laplace transform of $f(t)$ then

$$£^{-1}F(s) = f(t) = \frac{1}{2\pi j}\int_{c-j\omega}^{c+j\infty} F(s)\, e^{st} ds \qquad ...(7.2)$$

But this method is not generally used. In practice if F(s) is a rational function obtained the partial fraction and take the inverse Laplace with the help of standard table of transform pairs.

EXAMPLE 7.2. Find the inverse Laplace transform of $\frac{2}{s(s+2)}$.

Solution : Given: $\quad F(s) = \frac{2}{s(s+2)}$ ...(7.3)

The partial fraction of $F(s)$ is

$$F(s) = \frac{1}{s} - \frac{1}{s+2}$$

$$f(t) = £^{-1}F(s) = £^{-1}\frac{1}{s} - £^{-1}\frac{1}{s+2} = 1 - e^{-2t}$$ **Ans.**

7.6. APPLICATION OF LAPLACE TRANSFORM IN CONTROL SYSTEM

In control systems the system is described by the integrodifferential equations. It is very difficult to solve the such equations in time domain. The Laplace transforms of such equations converts them into simple algebaric equations. Then by using the inverse Laplace time domain function can be obtained.

EXAMPLE 7.3. Find i_2 by using Laplace transformation.

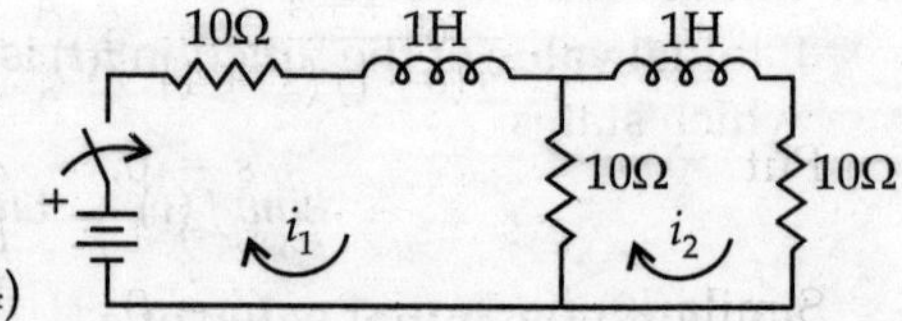

Solution : For mesh 1:

$$\frac{di_1}{dt} + 20i_1 - 10i_2 = 100\,u(t) \quad ...(7.4)$$

For mesh 2 :

$$\frac{di_2}{dt} + 20i_2 - 10i_1 = 0 \quad ...(7.5)$$

Laplace transform of (7.4) and (7.5)

$$SI_1(s) + 20I_1(s) - 10\,I_2(s) = \frac{100}{s} \quad ...(7.6)$$

$$-10\,I_1(s) + sI_2(s) + 20\,I_2(s) = 0 \quad ...(7.7)$$

equation (7.6) and (7.7) can be written as

$$I_1(s)\,[s + 20] - 10\,I_2(s) = \frac{100}{S} \quad ...(7.8)$$

$$-I_1(s)\,10 + (s + 20)\,I_2(s) = 0 \quad ...(7.9)$$

Solve (7.8) and (7.9)

$$I_2(s) = \frac{1000}{s(s^2+40s+300)} \quad ...(7.10)$$

or,

$$I_2(s) = \frac{1000}{s(s^2+40s+300)} = \frac{3.33}{s} - \frac{5}{s+10} + \frac{1.67}{s+30} \quad ...(7.11)$$

Inverse Laplace of 7.11

$$£^{-1}\,I_2(s) = i_2(t) = £^{-1}\frac{3.33}{s} - £^{-1}\frac{5}{s+10} + £^{-1}\frac{1.67}{s+30}$$

$$= 3.33 - 5\,e^{-10t} + 1.67e^{-30t}$$ **Ans.**

EXAMPLE 7.4. Obtian the solution of differential equation given below

$$\frac{d^2y(t)}{dt^2} + 3\frac{dy(t)}{dt} + 2y(t) = 5$$ **assuming zero initial conditions.**

Solution : $s^2\,y(s) + 3sy(s) + 2y(s) = \frac{5}{s}$

$$y(s) = \frac{5}{s\,(s^2+3s+2)} = \frac{5}{s\,(s+1)\,(s+2)} = \frac{2.5}{s} - \frac{5}{s+1} - \frac{2.5}{s+2}$$

$$£^{-1}\,y\,(s) = y(t) = £^{-1}\frac{2.5}{s} - £^{-1}\frac{5}{s+1} - £^{-1}\frac{2.5}{s+2}$$

$$y(t) = 2.5\,u(t) - 5\,e^{-t} - 2.5\,e^{-2t}$$ **Ans.**

EXAMPLE 7.5. Find the inverse Laplace transform of

$$\frac{s+4}{s(s-1)(s^2+4)}$$

Solution : Resolve the given function into partial fraction

$$\frac{s+4}{s(s-1)(s^2+4)} = \frac{A}{s}+\frac{B}{s-1}+\frac{Cs+D}{s^2+4}$$

Put $s = 0, \quad A = -1$

$s = 1, \quad B = 1$

Similarly, $C = 0, \quad D = -1$

$$\frac{s+4}{s(s-1)(s^2+4)} = \frac{-1}{s}+\frac{1}{s-1}-\frac{1}{s^2+4}$$

$$£^{-1}\left[\frac{s+4}{s(s-1)(s^2+4)}\right] = £^{-1}(-1/s)+£^{-1}\left(\frac{1}{s-1}\right)-£^{-1}\frac{1}{s^2+4} = -1+e^t-\frac{1}{2}\sin 2t \quad \textbf{Ans.}$$

EXAMPLE 7.6. What is Laplace transform of the function *f*(*t*) shown in Fig. 7.1.

Solution : The given waveform is a straight line, so, first calculate $f(t)$ using the equation.

$$y - y_1 = \frac{y_2-y_1}{x_2-x_1}(x-x_1) \quad ...(7.12)$$

From Fig. 7.1 : $x_1 = a, \quad x_2 = a+b$

$y_1 = 0, \quad y_2 = b$

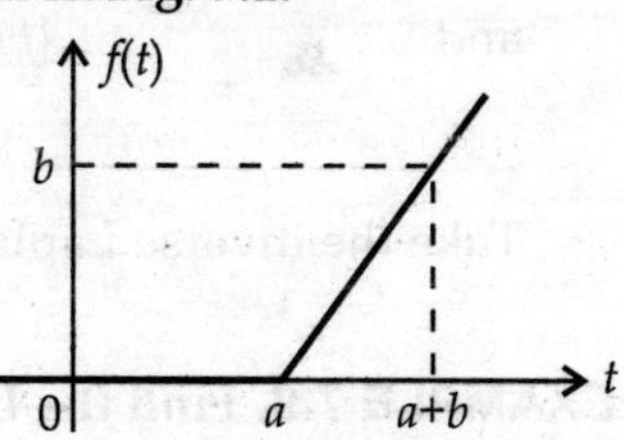

Fig. 7.1.

Put all these values in equation 7.12, we get

$$f(t) = t - a \quad ...(7.13)$$

Taking Laplace transform of equation 7.13, we get

$$F(s) = \frac{1}{s^2}-\frac{a}{s} = \frac{1-as}{s^2} \quad \textbf{Ans.}$$

EXAMPLE 7.7. Obtain the Laplace transform of the waveform shown in Fig. 7.3.

Solution: The given waveform can be divided into two parts

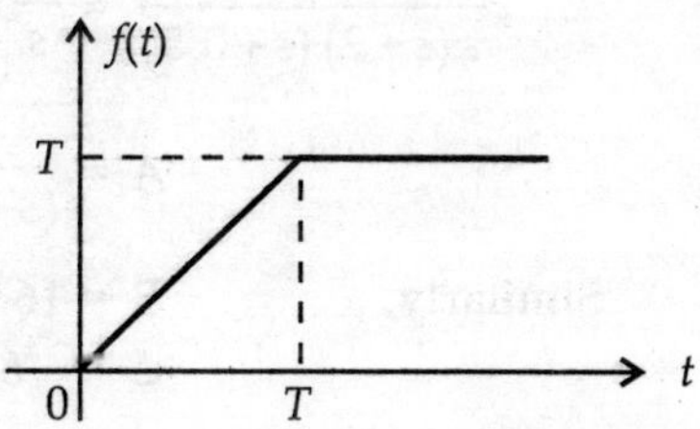

Fig. 7.2.

$$f(t) = f_1(t) + f_2(t)$$

$f_1(t)$ can be calculated with the help of equation 7.12

$\therefore \quad f_1(t) = t \quad 0 < t < T$

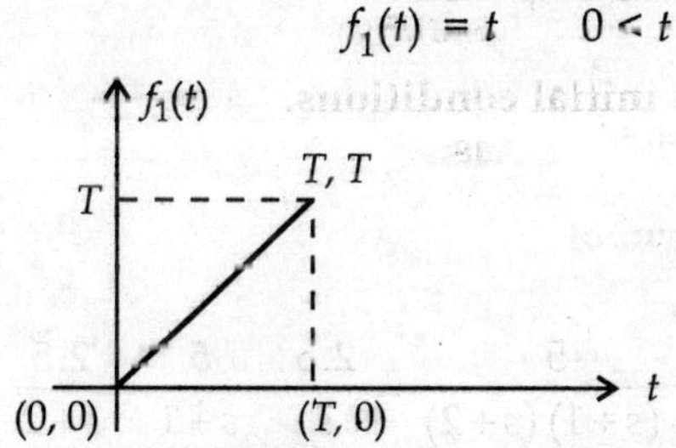

Fig. 7.2 (a)

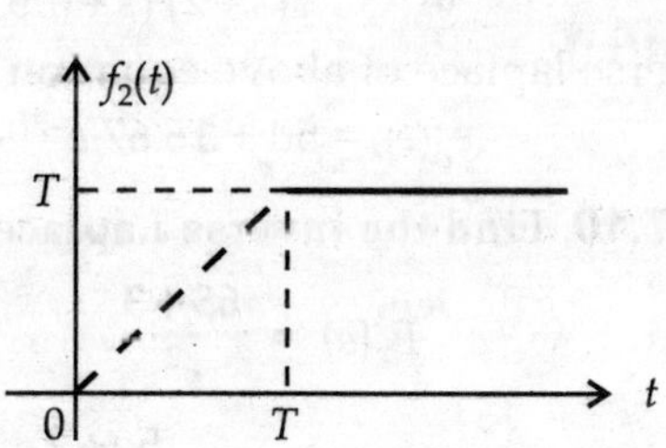

Fig. 7.2 (b)

$f_2(t)$ is shifted unit step function given by

$$f_2(t) = T\,u(t-T) \quad T < t < \infty$$

$$f(t) = \begin{cases} t & 0 < t < T \\ Tu(t-T) & t < t < \infty \end{cases} \quad ...(7.14)$$

∴ Laplace transform of 7.14 will be

$$F_1(s) = \int_0^T t\, e^{-st} dt = \frac{-1}{s} T\, e^{-sT} + \frac{1}{s^2}(1 - e^{-sT})$$

$$F^*_2(s) = £\{T\, u(t-T)\} = e^{-sT} . \frac{1}{s} . T$$

$$(s) = F_1(s) + F_2(s) = \frac{1}{s^2}(1 - e^{-sT}) \quad \textbf{Ans.}$$

EXAMPLE 7.8. Obtain the inverse Laplace transform of the following function.

$$F(s) = \frac{5e^{-s}}{s+1}$$

Solution : We know that

$$£\, e^{-at} = \frac{1}{s+a}$$

and $$£\,[f(t-a)] = e^{-as} F(s)$$

$$\therefore \quad F(s) = 5e^{-s} \frac{1}{S+1}$$

Take the inverse Laplace transform

$$f(t) = 5\, e^{-(t-1)} \quad \textbf{Ans.}$$

EXAMPLE 7.9. Find the Laplace transform of the given function

$$E_0(s) = \frac{50}{s\,(s+2)\,(s+0.5)}$$

Solution : Expand by partial fraction

$$\frac{50}{s\,(s+2)\,(s+0.5)} = \frac{A}{s} + \frac{B}{s+2} + \frac{C}{s+0.5}$$

$$A = \left.\frac{50}{(s+2)\,(s+0.5)}\right|_{s=0} \quad \therefore A = 50$$

Similarly, B = 16.67

C = – 66.67

$$E_o(s) = \frac{50}{s(s+2)\,(s+0.5)} = \frac{50}{s} + \frac{16.67}{s^2} + \frac{-66.67}{s+0.5}$$

The inverse laplace of above equation

$$e_o(t) = 50 + 16.67\, e^{-2t} - 66.67\, e^{-0.5t} \quad \textbf{Ans.}$$

EXAMPLE 7.10. Find the inverse Laplace transform of

(*a*) $$F_1(s) = \frac{6S+3}{s^2}$$

(*b*) $$F_2(s) = \frac{5s+2}{(s+1)\,(s+2)^2}$$

*Using time shift property from art. 10.4

Solution : $F_1(s) = \frac{6s+3}{s^2} = \frac{6}{s} + \frac{3}{s^2}$

Taking inverse Laplace on both sides we have

$$f_1(t) = 6u(t) + 3t \quad \text{for} \quad t \geq 0 \qquad \textbf{Ans.}$$

(*b*) $F_2(s) = \frac{5s+2}{(s+1)(s+2)^2}$

Solution : Expand by partial fraction

$$F_2(s) = \frac{5s+2}{(s+1)(s+2)^2} = \frac{A_1}{(s+2)} + \frac{A_2}{(s+2)^2} + \frac{B}{s+1}$$

$$B = (s+1)\,F_2(s)\big|_{s=-1} = -3$$

$$A_2 = (s+2)^2\,F_2(s)\big|_{s=-2} = 8$$

$$A_1 = \frac{d}{ds}\frac{5s+2}{s+1}\bigg|_{s=-2} = 3$$

$\therefore$ $$F_2(s) = \frac{3}{s+2} + \frac{8}{(s+2)^2} + \frac{-3}{s+1}$$

Inverse Laplace of above equation

$$f_2(t) = 3\,e^{-2t} + 8\,t\,e^{-2t} - 3\,e^{-t} \qquad t \geq 0 \qquad \textbf{Ans.}$$

EXAMPLE 7.11. Find the inverse Laplace transform of

$$F(s) = \frac{1}{s^2(s^2+\omega^2)}$$

Solution : Expand $F(s)$ by partial fraction

$$F(s) = \frac{1}{s^2(s^2+\omega^2)} = \frac{A}{s} + \frac{B}{s^2} + \frac{Cs+D}{s^2+\omega^2}$$

$\therefore$ $$1 = A\,s\,(s^2+\omega^2) + B\,(s^2+\omega^2) + (Cs+D)\,s^2$$

Comparing the coefficients of s^3, s^2, s, s^o we have

$$A + C = 0, \qquad \omega^2 B = 1$$

$$B + D = 0$$

$$A\omega^2 = 0$$

Thus, $A = 0$

$B = 1/\omega^2$

$C = 0$

$D = -1/\omega^2$

$\therefore$ $$F(s) = \frac{1}{\omega^2 s^2} - \frac{1}{\omega^2(s^2+\omega^2)} = \frac{1}{\omega^2 s^2} - \frac{1}{\omega^3}\cdot\frac{\omega}{s^2+\omega^2}$$

Inverse Laplace transform of above equation

$$f(t) = \frac{t}{\omega^2} - \frac{1}{\omega^3}\sin\omega t \qquad t \geq 0$$

$$f(t) = \frac{1}{\omega^2}\left(t - \frac{\sin\omega t}{\omega}\right) \qquad \textbf{Ans.}$$

EXAMPLE 7.12. Obtain the inverse Laplace transform of the given function

$$F(s) = \frac{\omega_n^2}{s(s^2+2\zeta\omega_n s+\omega_n^2)}$$

Solution: Expand the given function F(s) by partial fraction.

$$\frac{\omega_n^2}{s(s^2+2\zeta\omega_n s+\omega_n^2)} = \frac{A}{s}+\frac{Bs+C}{s^2+2\zeta\omega_n s+\omega_n^2}$$

$$\omega_n^2 = A(s^2+2\zeta\omega_n s+\omega_n^2)+(Bs+C)s$$

Comparing the coefficient, we have

$$A+B = 0$$

$$2\zeta\omega_n A+C = 0$$

$$\omega_n^2 A = \omega_n^2$$

$$\therefore \quad A = 1, B = -1, C = -2\zeta\omega_n$$

$$\therefore \quad F(s) = \frac{1}{s}-\frac{s+2\zeta\omega_n}{s^2+2\zeta\omega_n s+\omega_n^2} = \frac{1}{s}-\frac{s+\zeta\omega_n}{s^2+2\zeta\omega_n s+\omega_n^2}-\frac{\zeta\omega_n}{s^2+2\zeta\omega_n s+\omega_n^2}$$

$$= \frac{1}{s}-\frac{s+\zeta\omega_n}{(s+\zeta\omega_n)^2+\omega_n^2-\zeta^2\omega_n^2}-\frac{\zeta\omega_n}{(s+\zeta\omega_n)^2+\omega_n^2-\zeta^2\omega_n^2}$$

$$= \frac{1}{s}-\frac{s+\zeta\omega_n}{(s+\zeta\omega_n)^2+\omega_n^2(1-\zeta^2)}-\frac{\zeta\omega_n}{(s+\zeta\omega_n)^2+\omega_n^2(1-\zeta^2)}$$

Since, $1-\zeta^2$ is a positive quantity for $0<\zeta<1$

therefore $\sqrt{1-\zeta^2}$ is also a positive quantity.

$$\therefore \quad F(S) = \frac{1}{s}-\frac{s+\zeta\omega_n}{(s+\zeta w_n)^2+\left(\omega_n\sqrt{1-\zeta^2}\right)^2}-\frac{\zeta\omega_n\sqrt{1-\zeta^2}}{\left[(s+\zeta\omega_n)^2+\left(w_n\sqrt{1-\zeta^2}\right)^2\right]\sqrt{1-\zeta^2}}$$

Taking the inverse Laplace transform

$$f(t) = 1-e^{-\zeta\omega_n t}\cos\left(\omega_n\sqrt{1-\zeta^2}t\right)-\frac{\zeta e^{-\zeta\omega_n t}}{\sqrt{1-\zeta^2}}\sin\left(\omega_n\sqrt{1-\zeta^2}\right)t$$

EXAMPLE 7.13. Obtain the solution of the differential equation

$$\dot{x}+ax = A\sin\omega t,\ x(o) = b$$

Solution : The given differential equation

$$\dot{x}+ax = A\sin\omega t$$

$$sX(s)-x(0)+aX(s) = \frac{A\omega}{s^2+\omega^2}$$

given that $x(0) = b$

$$sX(s)-b+aX(s) = \frac{A\omega}{s^2+\omega^2}$$

$$\therefore \quad X(s) = \frac{A\omega}{(s^2+\omega^2)(s+a)}-\frac{(a-b)}{(s+a)} \quad ...(7.15)$$

let $\dfrac{A\omega}{(s^2+\omega^2)(s+a)} = \dfrac{K_1 s+K_2}{s^2+\omega^2}+\dfrac{K_3}{s+a}$

$$\frac{A\omega}{(s^2+\omega^2)(s+a)} = \frac{(K_1 s+K_2)(s+a)+K_3(s^2+\omega^2)}{(s^2+\omega^2)(s+a)}$$

Comparing Numerator on both sides and hence coefficients of s^2, s and s° we have

$$K_1 + K_3 = 0$$

$$aK_1 + K_2 = 0$$

$$aK_2 + K_3\,\omega^2 = A\omega$$

$$\therefore \quad K_1 = -\frac{A\omega}{a^2+\omega^2},\; K_2 = \frac{a\omega A}{a^2+\omega^2} \text{ and } K_3 = \frac{A\omega}{a^2+\omega^2}$$

Now put these values in equation (7.15), we get

$$X(s) = -\frac{A\omega}{a^2+\omega^2}\cdot\frac{s}{s^2+\omega^2}+\frac{aA}{a^2+\omega^2}\cdot\frac{\omega}{s^2+\omega^2}+\left(\frac{A\omega}{a^2+\omega^2}+b-a\right)\cdot\frac{1}{s+a}$$

Now apply inverse Laplace transform on both sides

$$x(t) = -\frac{A\omega}{a^2+\omega^2}\cos\omega t+\frac{aA}{a^2+\omega^2}\sin\omega t+\left(\frac{A\omega}{a^2+\omega^2}+b-a\right)e^{-at}$$

$$x(t) = A\left[\frac{a\sin\omega t}{a^2+\omega^2}-\frac{\omega\cos\omega t}{a^2+\omega^2}\right]+\left[\frac{A\omega}{a^2+\omega^2}+b-a\right]e^{-at} \quad \textbf{Ans.}$$

EXAMPLE 7.14. Solve the differential equation

$$\frac{d^2x(t)}{dt^2}+2\frac{d}{dt}x(t)+10\,x(t) = e^{-t}$$

when $x(0) = 0$

$$\frac{d}{dt}x(0) = 0$$

The forcing function e^{-t} is given at $t = 0$ when the system is at rest.

Solution : The given equation is

$$\frac{d^2x(t)}{dt^2}+2\frac{d}{dt}x(t)+10\,x(t) = e^{-t} \qquad \text{...(7.16)}$$

Taking Laplace transform on both sides of equation (7.16)

$$[s^2X(s) - sx'(0_+) - x(0+)] + 2[sX(s) - x(0_+)] + 10X(s) = \frac{1}{s+1}$$

Put the initial conditions

$$s^2X(s) + 2sX(s) + 10X(s) = \frac{1}{s+1}$$

$$\therefore \quad X(s) = \frac{1}{(s+1)(s^2+2s+10)} \qquad (7.17)$$

Expand the equation 7.17 by partial fraction

$$\frac{1}{(s+1)(s^2+2s+10)} = \frac{A}{s+1}+\frac{Bs+C}{s^2+2s+10} \qquad \text{...(7.18)}$$

$$1 = A(s^2+2s+10)+(Bs+C)(s+1)$$

Comparing the coefficients of s^2, s and s° we have

$$A + B = 0$$
$$2A + B + C = 0$$
$$10A + C = 1$$

$$\therefore \quad A = 1/9,\ B = -1/9 \text{ and } C = -1/9$$

put all these values in equation 7.18, we get

$$X(s) = \frac{1}{9(s+1)} - \frac{s+1}{9[(s+1)^2+9]} \qquad ...(7.19)$$

Using,

$$£\, e^{-at} = \frac{1}{s+a}$$
$$£[e^{-at} f(t)] = F(s+a)$$
$$£[e^{-at} \cos \omega t] = \frac{s+a}{(s+a)^2+\omega^2}$$

Taking inverse laplace of equation (7.19)

$$x(t) = \frac{1}{9}e^{-t} - \frac{1}{9}e^{-t}\cos 3t \qquad \textbf{Ans.}$$

EXAMPLE 7.15. For the circuit shown in Fig. 7.3, find the current at $t > 0$. The switch is closed at $t = 0$. Assume no initial charge on the capacitor.

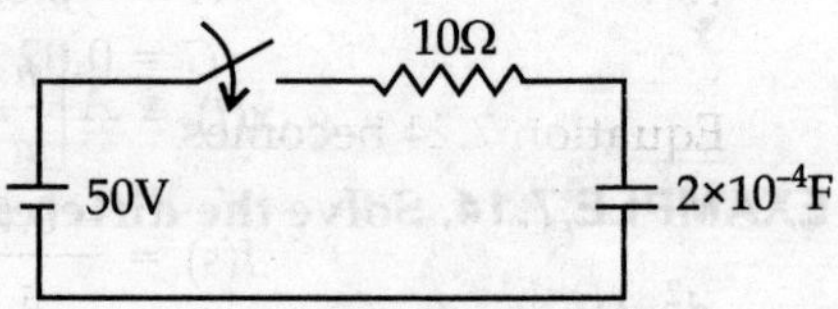

Fig. 7.3.

Solution : From the given Fig. 7.3 $Ri + \frac{1}{C}\int_{-\infty}^{t} i\,dt = 50$

$$10i + \frac{1}{2\times10^{-4}}\int_0^t i\,dt = 50 \qquad ...(7.20)$$

(because for $t < 0$ there is no charge on capacitor)

laplace transform of equation (7.20)

$$10I(s) + \frac{1}{s(2\times10^{-4})}I(s) = \frac{50}{s}$$

$$I(s)\left[\frac{s(2\times10^{-3})+1}{s(2\times10^{-4})}\right] = \frac{50}{s}$$

$$I(s) = \frac{5}{s+500} \qquad ...(7.21)$$

Inverse Laplace transform of equation 7.21

$$i(t) = 5\,e^{-500t} \qquad \textbf{Ans.}$$

EXAMPLE 7.16. In *RL* circuit the input is given by $Vs = 10t$. Determine the response of the circuit *i.e.*, $i(t)$ when $R = 50\ \Omega$ and $L = 5H$.

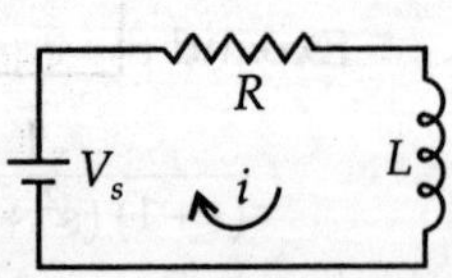

Fig. 7.4.

Solution: The differential equation of the given circuit

$$Ri + L\frac{di}{dt} = V_s$$

$$50i + 5\frac{di}{dt} = 10\,t \qquad ...(7.22)$$

Laplace transform of equation 7.22

$$50\,I(s) + 5\,I(s) = \frac{10}{s^2}$$

$$I(s) = \frac{2}{s^2(s+10)} \qquad ...(7.23)$$

Expand the equation 7.23 by partial fraction

$$I(s) = \frac{2}{s^2(s+10)} = \frac{A}{s} + \frac{B}{s^2} + \frac{C}{s+10} \qquad ...(7.24)$$

$$2 = A\,s(s+10) + B\,(s+10) + Cs^2$$

Compare the coefficients of s^3, s^2, s and s^o

$$10\,A + B = 0$$

$$A + C = 0$$

$$10\,B = 2$$

$$\therefore \quad B = 0.2$$

$$\therefore \quad A = -\,0.02$$

$$C = 0.02$$

Equation 7.24 becomes

$$I(s) = \frac{-\,0.02}{s} + \frac{0.2}{s^2} + \frac{0.02}{s+10} \qquad ...(7.25)$$

Inverse Laplace transform of 7.15

$$i\,(t) = -\,0.02 + 0.2t + 0.02\,e^{-10\,t}$$

$$i\,(t) = 0.02\,e^{-10t} + 0.2t - 0.02 \qquad \textbf{Ans.}$$

EXAMPLE 7.17. Express the pulse shown in Fig. 7.5 as a combination of step functions and obtain its Laplace transform.

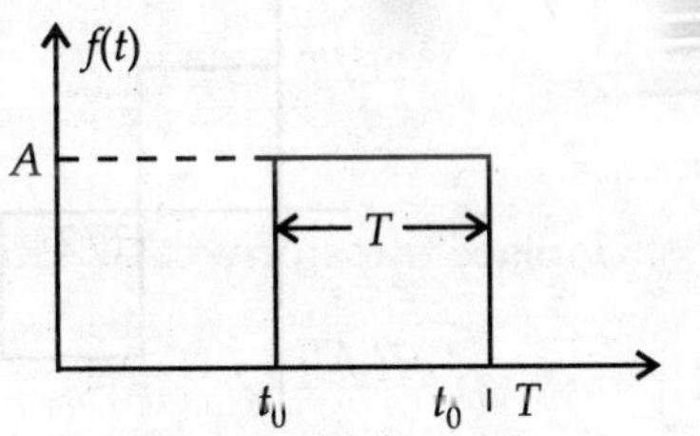

Fig. 7.5.

Solution :

$$f(t) = f_1(t) + f_2(t)$$

$$f_1(t) = Au_{-1}\,(t - t_o) \qquad \text{(shifted step function)}$$

$$f_2(t) = +\,Au_{-1}[t - (t_o + T)]$$

$$\therefore \quad f(t) = f_1(t) - f_2(t)$$

$$= Au_{-1}\,(t - t_o) - A\,u_{-1}\,[t - (t_o + T)]$$

$$\pounds f(t) = \pounds\,Au_{-1}\,(t - t_o) - \pounds Au_{-1}\,[t - (t_o + T)]$$

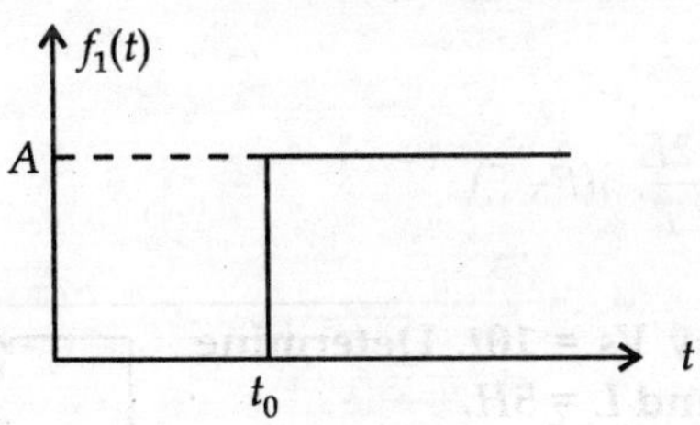

Fig. 7.5 (a)

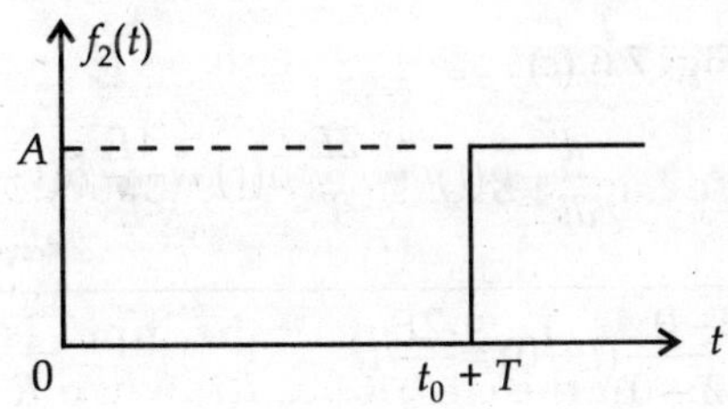

Fig. 7.5 (b)

$$F(s) = A\cdot\frac{1}{s}e^{-st_o} - A\frac{1}{s}e^{-s(t_o+T)} = \frac{A}{s}[e^{-st_o} - e^{-s(t_o+T)}] \quad \textbf{Ans.}$$

7.7. LAPLACE TRANSFORM OF PERIODIC FUNCTIONS

Theorem : The Laplace transform of a periodic function with time period T is equal to the product of $\frac{1}{1-e^{-Ts}}$ and laplace transform of the first cycle.

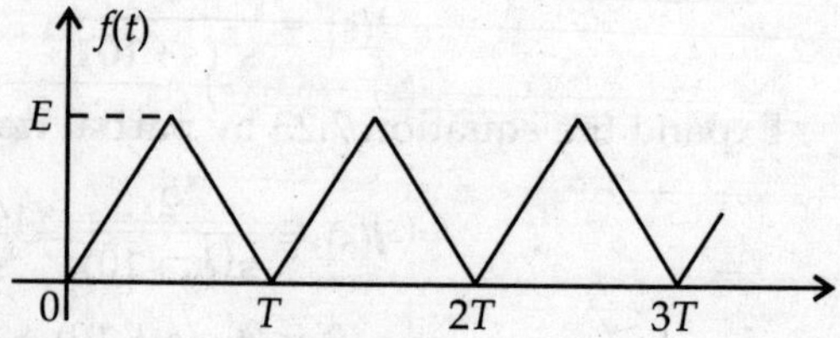

Fig. 7.6.

EXAMPLE 7.18. Find the Laplace transform of the periodic triangular wave in Fig. 7.6.

Solution : To find Laplace transform of a periodic wave, first evaluate $f(t)$ for one cycle *i.e.*, for the waveform shown in Fig. (7.6 *a*).

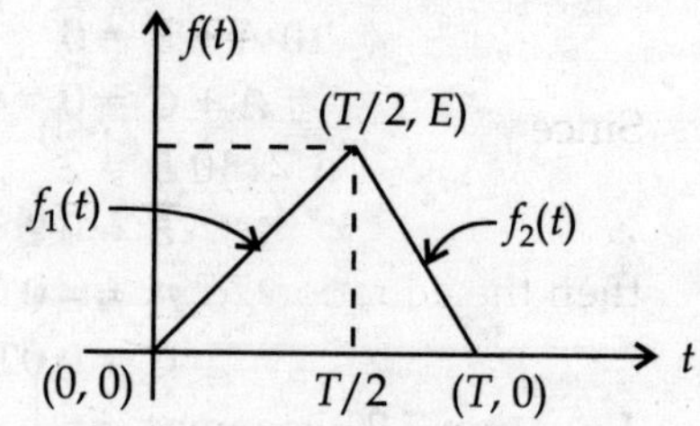

Fig. 7.6 (a).

$$g(t) = \begin{cases} \frac{2E}{T}t^* & 0 < t < T/2 \\ -\frac{2E}{T}t + 2E^* & T/2 < t < T \\ 0 & \text{otherwise} \end{cases}$$

$$\frac{d}{dt}g(t) = \begin{cases} \frac{2E}{T} & 0 < t < T/2 \\ -\frac{2E}{T} & T/2 < t < T \end{cases} \quad \text{...(7.26)}$$

The waveform of equation 7.26 is shown in Fig. 7.6 (*b*).

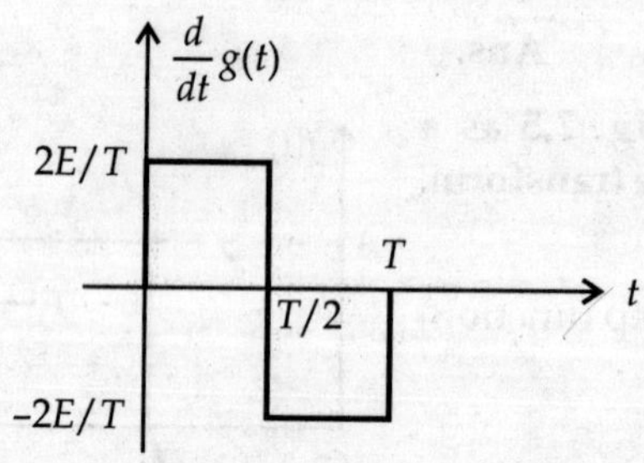

Fig. 7.6 (b).

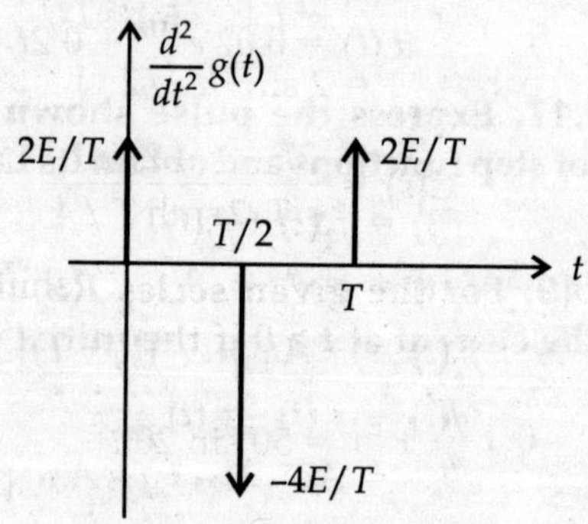

Fig. 7.6 (c).

From Fig. 7.6 (*c*)

$$\frac{d^2}{dt^2}g(t) = \frac{2E}{T}\delta(t) - \frac{4E}{T}\delta(t - T/2) + \frac{2E}{T}\delta(t - T) \quad \text{...(7.27)}$$

---

* $f_1(t) = \frac{E-0}{T/2-0}(t-0) = \frac{2E}{T}t$

$f_2(t) = \frac{E-0}{T/2-0}(t-T) = -\frac{2E}{T}t + 2E$

Take Laplace transform of equation 7.27

$$s^2 G(s) = \frac{2E}{T} \cdot - \frac{4E}{T} e^{-ST/2} + \frac{2E}{T} e^{-St} .1$$

$$G(s) = \frac{2E}{Ts^2}\left(1-2e^{-sT/2}+e^{-sT}\right) \quad ...(7.28)$$

Therefore, Laplace transform of periodic function will be

$$G(s) = G_1(s) - \frac{1}{1-e^{-sT}}$$

$$G(s) = \frac{2E}{s^2T(1-e^{-sT})}(1-2e^{-sT/2}+e^{-sT}) = \frac{2E}{s^2T}\left(\frac{1-2e^{-sT/2}+e^{-sT}}{1-e^{-sT}}\right)$$

$$= \frac{2E}{s^2T}\frac{e^{-sT/2}(e^{sT/2}-2+e^{-sT/2})}{e^{-sT/2}(e^{sT/2}-e^{-sT/2})} = \frac{2E}{s^2T}\left[\frac{e^{sT/2}+e^{-sT/2}-2}{e^{sT/2}-e^{-sT/2}}\right] \quad ...(7.29)$$

Since $\frac{e^x+e^{-x}}{2} = \cosh x$

$\therefore \quad e^x + e^{-x} = 2\cosh x$

then the equation 7.29 can be written as

$$G(s) = \frac{2E}{s^2T}\left[\frac{\frac{e^{sT/2}+e^{-sT/2}}{2}-2}{\frac{e^{sT/2}-e^{-sT/2}}{2}}\right] = \frac{2E}{s^2T}\left[\frac{2\cosh sT/2-1}{2\sinh sT/2}\right]$$

$$= \frac{2E}{s^2T}\left[\frac{\cosh sT/2-1}{\sinh sT/2}\right] \quad ...(7.30)$$

Also, $\cosh x - 1 = 2\sin^2 h\, x/2$

$$\therefore \quad G(s) = \frac{2E}{s^2T}\frac{2\sinh^2 sT/4}{2\sinh T/4\cosh sT/4} = \frac{2E}{s^2T}\tanh sT/4 \quad \textbf{Ans.}$$

EXAMPLE 7.19. For the given series *RL* circuit shown in Fig. 7.7 find the current at $t > 0$ if the initial current is zero.

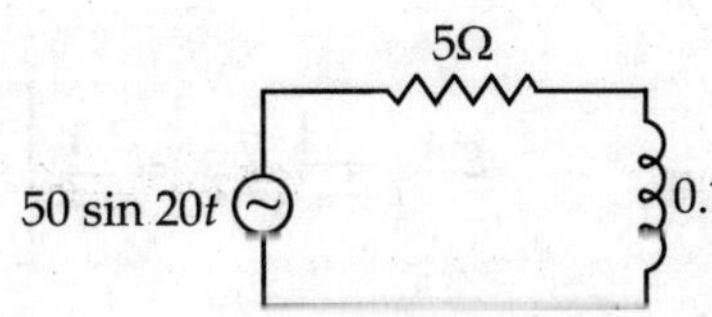

Fig. 7.7.

Solution : $0.1\frac{di}{dt} + 5i = 50\sin 20\,t$

or, $$\frac{di}{dt} + 50i = 500\sin 20\,t \quad ...(7.31)$$

Take laplace transform of equation (7.31)

$$sI(s) + 50\,I(s) = 500 \cdot \frac{20}{s^2+400} \quad ...(7.32)$$

$$I(s) = \frac{10000}{(s+50)(s^2+400)} \quad ...(7.33)$$

$$\frac{10000}{(s+50)(s^2+400)} = \frac{A}{s+50} + \frac{Bs+C}{s^2+400} \quad ...(7.34)$$

from 7.34 we get

$$A = 3.4$$
$$B = -3.4$$
$$C = 172.4$$

then the equation 7.34 becomes

$$\frac{10000}{(s+50)(s^2+400)} = \frac{3.4}{s+50} - 3.4\frac{s}{s^2+400} + \frac{172.4}{s^2+400}$$

$$= \frac{3.4}{s+50} - 3.4\frac{s}{s^2+400} + 8.62\frac{20}{s^2+400}$$

Take inverse laplace transform of above equation

$$i(t) = 3.4\, e^{-50t} - 3.4 \cos 20\, t + 8.62 \sin 20\, t \qquad \textbf{Ans.}$$

7.8. CONVOLUTION THEOREM

If $\qquad £ f_1(t) = F_1(s)$ and $£ f_2(t) = F_2(s)$ then

$$£\left[\int_0^t f_1(t-\tau)\, f_2(\tau)\, d\tau\right] = £\left[\int_0^t f_1(\tau)\, f_2(t-\tau)\, d\tau\right] = F_1(s)\, F_2(s)$$

EXAMPLE 7.20. Find $£^{-1}\dfrac{1}{(s^2+a^2)^2}$

Solution: $\qquad F_1(s) = F_2(s) = \dfrac{1}{s^2+a^2}$

$\therefore \qquad f_1(t) = f_2(t) = \dfrac{1}{a}\sin at$

$$£^{-1}\frac{1}{(s^2+a^2)^2} = f_1 * f_2 = \int_0^t f_1(t-\tau)\, f_2(\tau)\, d\tau = \int_0^t \left[\frac{1}{a}\sin a(t-\tau)\right]\left[\frac{1}{a}\sin a\tau\right] d\tau$$

$$£^{-1}\frac{1}{(s^2+a^2)^2} = \frac{1}{a^2}\left[\sin at \int_0^t \cos a\tau \sin a\tau\, d\tau - \cos at \int_0^t \sin^2 a\tau\, d\tau\right]$$

$$= \frac{1}{a^3}(\sin at - at\cos at) \qquad \textbf{Ans.}$$

EXAMPLE 7.21. Find i_2 by using Laplace transformation.

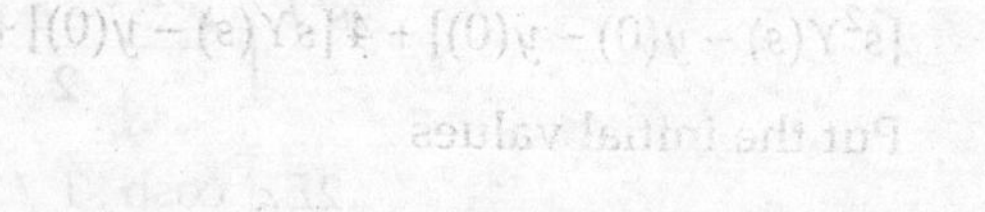

Fig. 7.8.

Solution : For mesh (1) :

$$\frac{di_1}{dt} + 20i_1 - 10i_2 = 100\, u(t) \qquad ...(7.35)$$

For mesh (2)

$$\frac{di_2}{dt} + 20i_2 - 10i_1 = 0 \qquad ...(7.36)$$

The initial values of i_1 and i_2 are zero. The Laplace transform of equation 7.35 and 7.36..

$$(s + 20)\, I_1(s) - 10 I_2(s) = \frac{100}{s} \quad ...(7.37)$$

$$-10\, I_1(s) + (s + 20)\, I_2(s) = 0 \quad ...(7.38)$$

Solve 7.37 and 7.38 for $I_2(s)$

$$I_2(s) = \frac{1000}{s(s^2 + 40s + 300)} \quad ...(7.39)$$

Expand (7.39) by partial fraction

$$I_2(s) = \frac{3.33}{s} - \frac{5}{s+10} + \frac{1.67}{s+30} \quad ...(7.40)$$

Taking inverse Laplace transform of equation 7.40

$$£^{-1}\, I_2(s) = 3.33\, £^{-1}(1/s) - 5\, £^{-1}\frac{1}{s+10} + 1.67\, £^{-1}\frac{1}{s+30}$$

$$i_2(t) = 3.33 - 5\, e^{-10t} + 1.67\, e^{-30t} \qquad \textbf{Ans.}$$

EXAMPLE 7.22. Solve the differential equation

$$\ddot{y}(t) + 4\dot{y}(t) + 3y(t) = 6$$

if $$y(0) = \dot{y}(0) = 0$$

Solution : The given equation is

$$\ddot{y}(t) + 4\dot{y}(t) + 3y(t) = 6 \quad ...(7.41)$$

laplace transform of equation 7.41

$$[s^2 Y(s) - y(0) - \dot{y}(0)] + 4\,[sY(s) - y(0)] + 3Y(s) = \frac{6}{s}$$

Put the initial values

$$Y(s)\,(s^2 + 4s + 3) = \frac{6}{S}$$

$$Y(s) = \frac{6}{s(s^2 + 4s + 3)} \quad ...(7.42)$$

Expand the equation (7.42) by partial fraction

$$Y(s) = \frac{6}{s(s^2 + 4s + 3)} = \frac{2}{s} - \frac{3}{s+1} + \frac{1}{s+3} \quad ...(7.43)$$

Take inverse laplace of equation (7.43)

$$y(t) = 2 - 3\, e^{-t} + e^{-3t} \qquad \textbf{Ans.}$$

EXAMPLE 7.23. A voltage pulse of unit height and width T is applied to the RC circuit at $t = 0$. Determine the voltage across the capacitance C as a function of time.

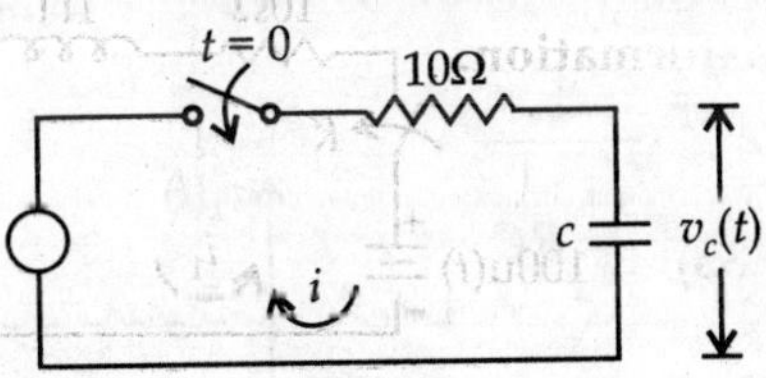

Fig. 7.9 (a).

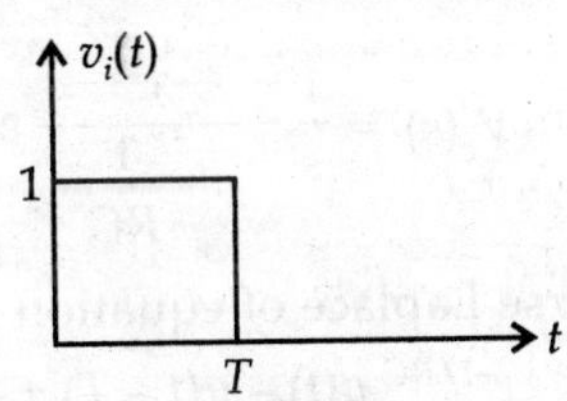

Fig. 7.9 (b).

* See the example 10.17 for $v_i(t)$, Put $t_0 = 0$ such form is called as rectangular function or gate function.

Solution : From Fig. 7.9 *(b)*

$$^*v_i(t) = u(t) - u(t-T) \qquad \text{...(7.44)}$$

Now at $t = 0$, switch will be closed, the circuit is shown in Fig. 7.9 *(c)*

Applying KVL in the circuit, we have

$$v_i(t) = Ri(t) + \frac{1}{C}\int_{-\infty}^{t} i(t)\,dt \qquad \text{...(7.45)}$$

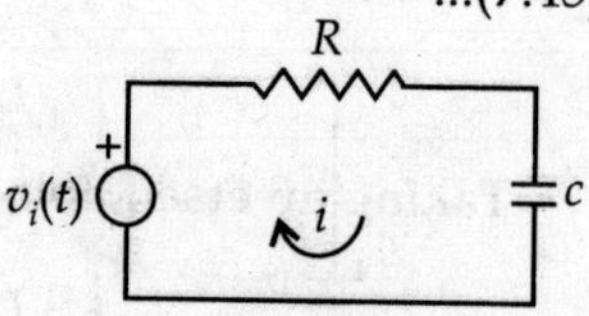

Fig. 7.9 (c)

$$\frac{1}{C}\int_{-\infty}^{t} i\,dt = \frac{q}{C}$$

$$i = \frac{dq}{dt} = C\frac{dv_c}{dt}$$

$$\therefore \quad v_i(t) = RC\frac{d}{dt}v_c(t) + v_c(t) \qquad \text{...(7.46)}$$

$$u(t) - u(t-T) = RC\frac{d}{dt}v_c(t) + v_c(t) \qquad \text{...(7.47)}$$

Taking Laplace transform on both sides

$$\frac{1}{s} - \frac{e^{-sT}}{s} = RC\,[sV_c(s) - v_c(0_+)] + V_C(s)$$

$$\frac{1}{s} - \frac{e^{-sT}}{s} = (1 + sCR)\,V_c(s)$$

$$\therefore \quad V_c(s) = \frac{1}{s(1+sCR)} - \frac{e^{-sT}}{s(1+sCR)} \qquad \text{...(7.48)}$$

let $\quad V_{c1} = \dfrac{1}{s(1+sCR)}$ and $V_{c2} = \dfrac{e^{-sT}}{s(1+sCR)}$

Expand V_{c1} and V_{c2} by partial fraction

$$V_{c1} = \frac{1}{s} - \frac{1}{s+\frac{1}{RC}} \text{ and } V_{c2} = e^{-sT}\left[\frac{1}{s} - \frac{1}{s+\frac{1}{RC}}\right]$$

$$V_c(s) = \frac{1}{s} - \frac{1}{s+\frac{1}{RC}} - e^{-sT}\cdot\frac{1}{s} + e^{-sT}\cdot\frac{1}{s+\frac{1}{RC}} \qquad \text{...(7.49)}$$

Taking inverse Laplace of equation (7.49)

$$v_c(t) = u(t) - e^{-t/RC}\,u(t) - u(t-T) + e^{-(t-T)/RC}\cdot u(t-T)$$

$$v_c(t) = (1 - e^{-t/RC})\,u(t) - [1 - e^{-(t-T)/RC}]\,u(t-T) \qquad \text{...(7.50)}$$

The graph of equation 7.50 is shown in Fig. 7.9*(d)*

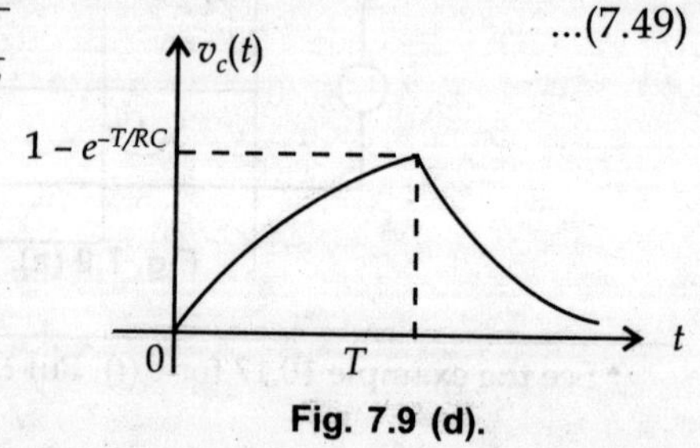

Fig. 7.9 (d).

EXAMPLE 7.24. The circuit shown in Fig. 7.10 is initially under steady-state condition. The switch is moved from position 1 to position 2 at $t = 0$. Find the current after switching.

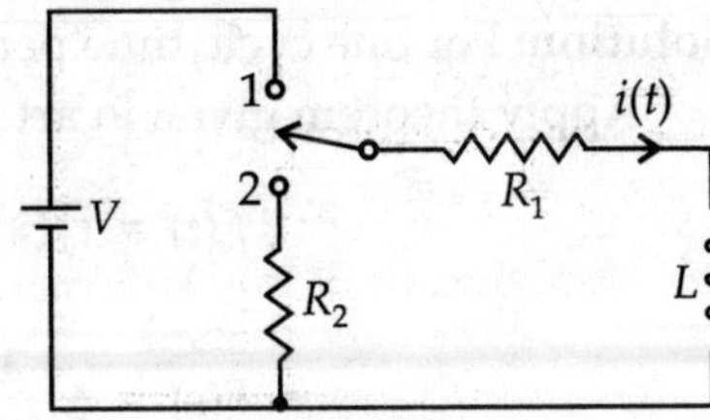

Fig. 7.10.

Solution: Consider the circuit when switch is at position 1 from Fig 7.10 (*a*).

$$i_L(0_-) = \frac{V}{R_1} \qquad ...(7.51)$$

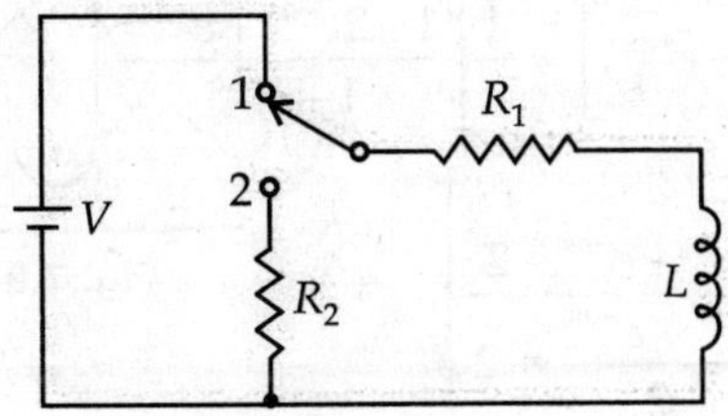

Fig. 7.10 (a)

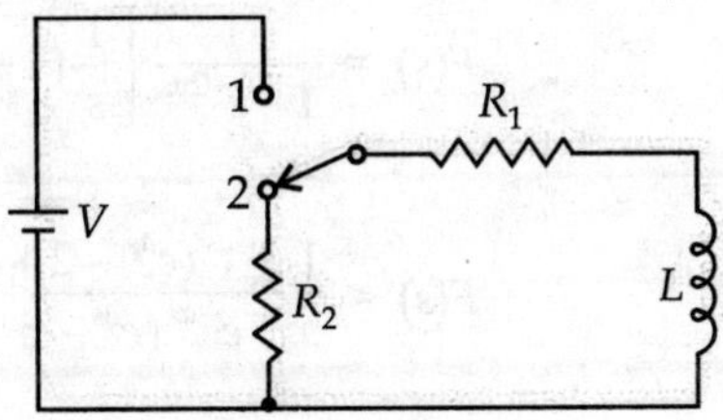

Fig. 7.10 (b)

Now, when the switch at position 2, the equivalent circuit will be

Apply KVL in Fig. 7.10 (*b*)

$$L\frac{di}{dt} + (R_1 + R_2)\,i = 0 \qquad ...(7.52)$$

Taking the inverse Laplace of equation (7.52)

$LsI(s) - i_L(O_+)\ (R_1 + R_2)\ I(s) = 0$

$$I(s) = \frac{Li_L(O_+)}{sL + R_1 + R_2} = \frac{i_L(O_+)}{s + \dfrac{R_1 + R_2}{L}} \qquad ...(7.53)$$

Taking inverse Laplace of equation 7.53

$$i(t) = i_L(O_+)\,e^{-\left(\frac{R_1+R_2}{L}\right)t} \qquad ...(7.54)$$

Put the value of $i_L(O_+)$ from equation (7.51) Hence required value of $i(t)$ will be

$$i(t) = \frac{V}{R_1}\,e^{-\left(\frac{R_1+R_2}{L}\right)t} \qquad \textbf{Ans.}$$

EXAMPLE 7.25. Find the Laplace transform of the wave shown in Fig. 7.11.

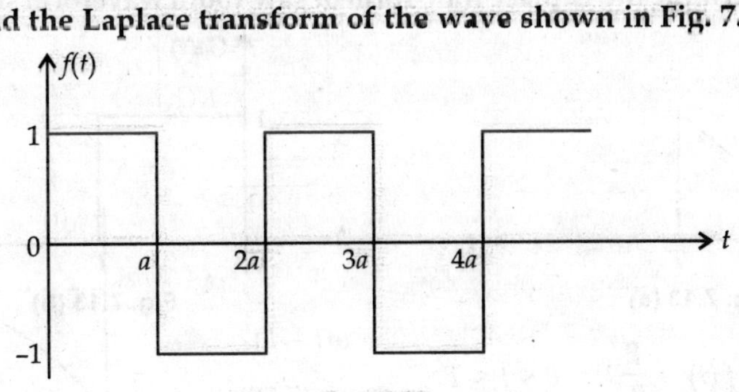

Fig. 7.11.

Solution: For one cycle time period = 2*T*

Apply theorem given in art. 7.7

$$\pounds f(t) = F_1(s)\cdot\frac{1}{1-e^{-Ts}}$$

$$\therefore \quad F(s) = \frac{1}{1-e^{-2as}}\left[\int_0^a 1.e^{-st}dt + \int_a^{2a}(-1)\,e^{-st}dt\right] \quad ...(7.55)$$

$$F(s) = \frac{1}{1-e^{-2as}}\left[\frac{1}{s}(1-e^{-as}+e^{-2as}-e^{-as})\right] = \frac{1}{s}\left[\frac{1-2e^{-as}+e^{-2as}}{1-e^{-2as}}\right] \quad ...(7.56)$$

$$F(s) = \frac{1}{s}\frac{e^{-as}(e^{as}-2+e^{-as})}{e^{-as}(e^{as}-e^{-as})} = \frac{1}{s}\left[\frac{e^{as}+e^{-as}-2}{\frac{e^{as}-e^{-as}}{2}\times 2}\right]$$

$$F(s)4 = \frac{1}{s}\left[\frac{2\cosh as - 2}{\sinh as \times 2}\right] \qquad \because \frac{e^x+e^{-x}}{2} = \cosh ax$$

$$= \frac{1}{s}\left[\frac{2\sin^2 h\frac{as}{2}}{2\sinh\frac{as}{2}.\cosh\frac{as}{2}}\right] \qquad \because \cosh x - 1 = 2\sin^2 h\, x/2$$

$$F(s) = \frac{1}{s}\tanh\frac{as}{2} \qquad \textbf{Ans.}$$

7.9. GATE FUNCTIONS

A rectangular pulse starting at $t = t_1$ and duration of pulse is T having unit height is represented as

$$G(T) = u(t-t_1) - u(t-t_1-T) \quad ...(7.57)$$

is known as gate function.

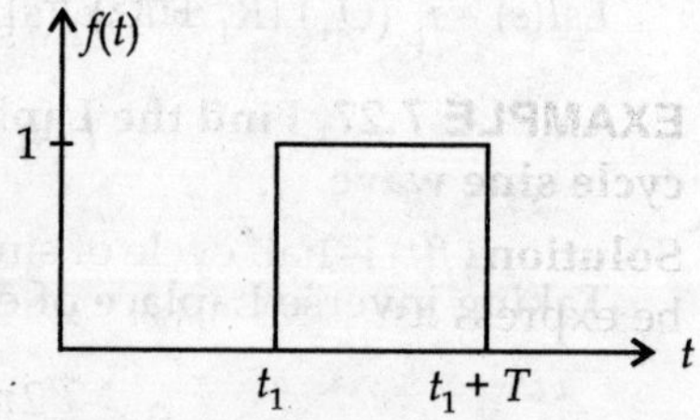

Fig. 10.12

Any function multiplied by a gate function will have zero value outside the duration of the gate $(t_1 + T) < t < t_1$ and the value of the function will be unaffected within the duration of the gate $t_1 < t < (t_1 + T)$. The gate function is also known as rectangular function.

EXMAPLE 7.26. Determine the Laplace transform of saw-tooth waveform shown in Fig. 7.13 (*a*).

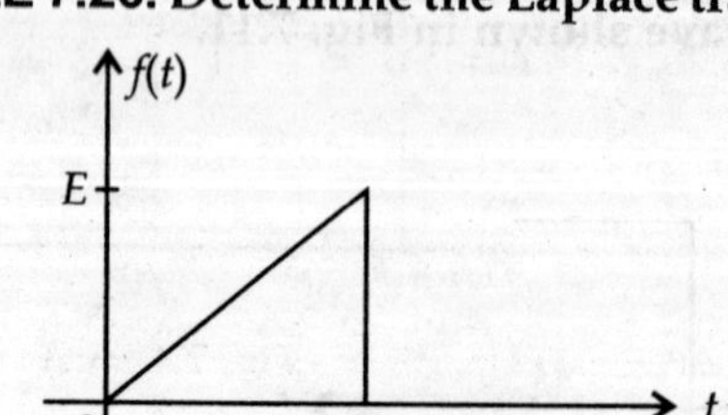

Fig. 7.13 (a)

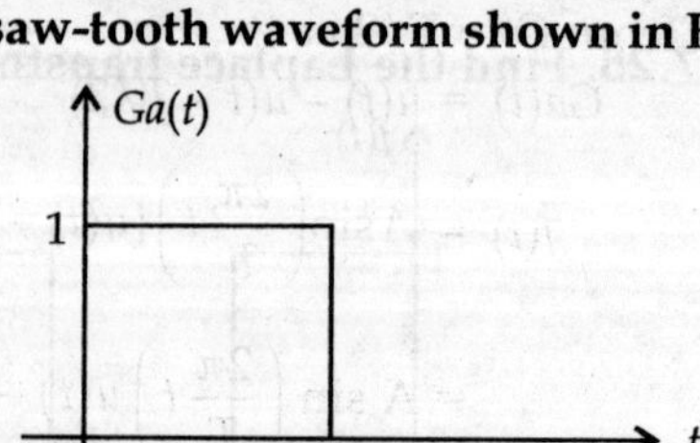

Fig. 7.13 (b)

Solution : $$f(t) = \frac{E}{T}t \quad 0 < t < T \quad ...(7.58)$$

* In the second term we cannot apply time shifting property as the whole function $tu(t-T)$ is not of form $f(t-a)$.

We can evaluate Laplace transform of $f(t)$ by replacing the time limits by gate function as shown in Fig. 7.13 (*b*)

$$Ga(t) = u(t) - u(t-T) \qquad \because t_1 = 0 \qquad \text{From eqn. (7.57)}$$

Hence, the gate function starts from $t = 0$ to $t = T$ and $f(t)$ can be represented as

$$f(t) = \frac{E}{T} t \times Ga(t)$$

$$f(t) = \frac{E}{T} t\,[u(t) - u(t-T)] = \frac{E}{T} t\,u(t) - \frac{E}{T} t\,u(t-T)^{*}$$

$$= \frac{E}{T} t\,u(t) - \frac{E}{T}(t - T + T)\,u(t-T)$$

$$= \frac{E}{T} t\,u(t) - \frac{E}{T}(t-T)\,u(t-T) - \frac{E}{T} T\,u(t-T)$$

$$f(t) = \frac{E}{T} t\,u(t)^{**} - \frac{E}{T}(t-T)\,u(t-T) - Eu(t-T) \qquad ...(7.59)$$

Taking Laplace transform of equation (7.59) we get

$$F(s) = \frac{E}{T}\frac{1}{s^2} - \frac{E}{T} e^{-sT}\frac{1}{s^2} - E\,e^{-sT}\frac{1}{s}$$

$$F(s) = \frac{E}{s^2 T}[1 - e^{-sT} - sT\,e^{-sT}] \qquad \textbf{Ans.}$$

EXAMPLE 7.27. Find the Laplace transform of following half cycle sine wave.

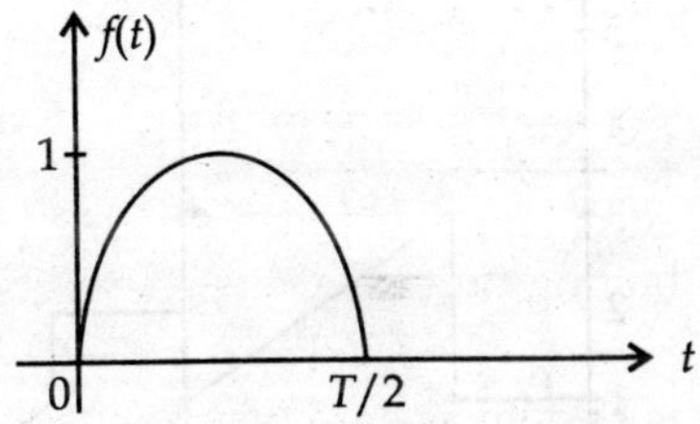

Fig. 7.14.

Solution : $f(t)$ is half cycle of sine wave (whose period was T) can be express as

$$f(t) = A\sin\left(\frac{2\pi}{T}t\right) \qquad 0 < t < T/2 \qquad ...(7.60)$$

The time limits can be replaced by the use of gate function.

$$f(t) = A\sin\left(\frac{2\pi}{T}t\right) \times Ga(t)$$

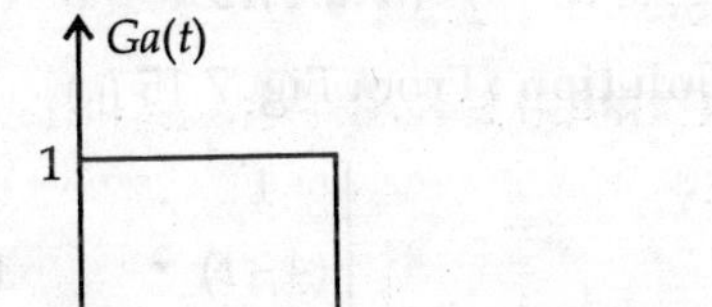

Fig. 7.14. (a)

$Ga(t)$ is from O to $T/2$ *i.e.*,

$$Ga(t) = u(t) - u(t - T/2) \qquad ...(7.61)$$

$$\therefore \quad f(t) = A\sin\left(\frac{2\pi}{T}t\right)[u(t) - u(t - T/2)]$$

$$= A\sin\left(\frac{2\pi}{T}t\right)u(t) - A\sin\left(\frac{2\pi}{T}t\right)u(t - T/2)$$

** Here, multiplying $f(t)$ by $u(t)$ makes $£f(t)\,u(t) = \int_0^\infty f(t).1.e^{-st}\,dt = £f(t)$. Here $f(t) = t$, therefore multiplication by $u(t)$ does affects its laplace transform.

$$£[t\,u(t)] = \int_0^\infty t.1.e^{-st}dt = £[t]$$

Now,
$$= A\sin\left(\frac{2\pi}{T}t\right)u(t-T/2) = A\sin\left[\frac{2\pi}{T}\left(t-\frac{T}{2}\right)+\frac{T}{2}\right]u\left(t-\frac{T}{2}\right)$$

$$= A\left[\sin\frac{2\pi}{T}\left(t-\frac{T}{2}\right)\cos\left(\frac{2\pi}{T}\cdot\frac{T}{2}\right)+\cos\frac{2\pi}{T}\left(t-\frac{T}{2}\right)\cdot\sin\frac{2\pi}{T}\cdot\frac{T}{2}\right]u\left(t-\frac{T}{2}\right)$$

$$= A\sin\left[\frac{2\pi}{T}\left(t-\frac{T}{2}\right)\right](-1)u\left(t-\frac{T}{2}\right) = -A\sin\left[\frac{2\pi}{T}\left(t-\frac{T}{2}\right)\right]u\left(t-\frac{T}{2}\right)$$

$$\therefore \quad f(t) = A\sin\left(\frac{2\pi}{T}t\right)u(t)+A\sin\left[\frac{2\pi}{T}(t-T/2)\right]u(t-T/2) \qquad ...(7.62)$$

Take the Laplace transform of 7.62.

$$F(s) = \frac{A\cdot(2\pi/T)}{s^2+(2\pi/T)^2}+\frac{e^{-sT/2}\cdot A(2\pi/T)}{s^2+(2\pi/T)^2} = \frac{2\pi A}{T\left[s^2+\left(\frac{2\pi}{T}\right)^2\right]}(1+e^{-sT/2}) \qquad \textbf{Ans.}$$

EXAMPLE 7.28. Resolve $f(t)$ in terms of step, impulse and ramp functions. Also find the laplace transform of $f(t)$. *(I.E.S 2002, Elec. & Tele. Comm.)*

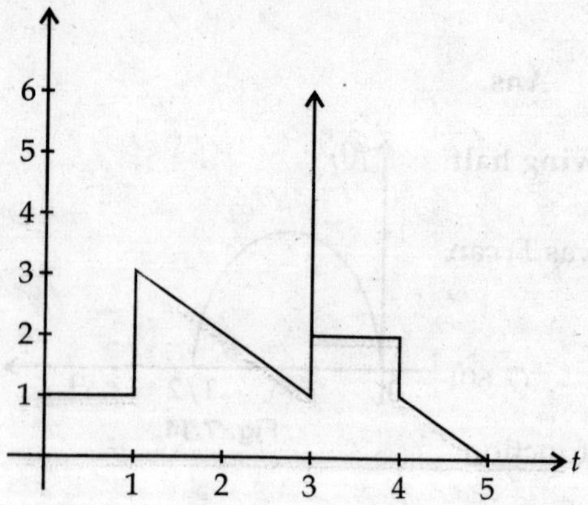

Fig. 7.15.

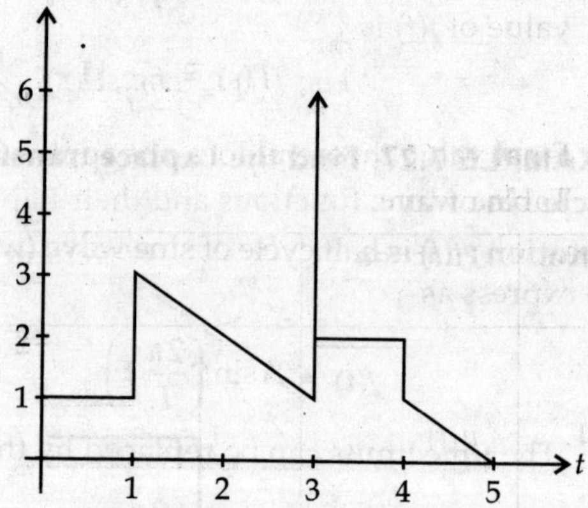

Fig. 7.15 (a)

Solution : From Fig. 7.15 (*a*)

$$f(t) = \begin{bmatrix} 1 & 0 \le t < 1 \\ (4-t) & 1 \le t < 3 \\ 6 & t = 3 \\ 2 & 3 < t \le 4 \\ (5-t) & 4 \le t \le 5 \end{bmatrix} \qquad ...(7.63)$$

$$f(t) = [u(t)-u(t-1)]+(4-t)[u(t-1)-u(t-3]+4\,\delta(t-3)+2[u(t-3)-u(t-4)] + (5-t)[u(t-4)-u(t-5)]$$

$$f(t) = u(t)-u(t-1)-(t-4)[u(t-1)-u(t-3)]+4\delta(t-3)+2u(t-3)\,2u(t-4) - (t-5)\,u(t-4)+(t-5)\,u(t-5)$$

$$= u(t)-u(t-1)-(t-1)\,u(t-1)+3\,u(t-1)-(t-3)\,u(t-3)-u(t-3)+4\delta\,(t-3) + 2u(t-3)-2u(t-4)-(t-4)\,u(t-4)+u(t-4)+(t-5)\,u(t-5)$$

$$f(t) = u(t) + 2u(t-1) - (t-1)\,u(t-1) + (t-3)\,u(t-3) + u(t-3) + 4\delta(t-3) - (t-4)\,u(t-4) - u(t-4) + (t-5)\,u(t-5) \quad ...(7.64)$$

Taking the Laplace transform of equation 7.64

$$F(s) = \frac{1}{s} + 2\frac{e^{-s}}{s} - \frac{1}{s^2}e^{-s} + \frac{1}{s^2}e^{-3s} + \frac{e^{-3s}}{s} + 4e^{-3s} - \frac{1}{s^2}e^{-4s} - \frac{e^{-4s}}{s} + \frac{e^{-5s}}{s^2}$$

$$F(s) = 4e^{-3s} + \frac{1}{s}\left[1 + 2e^{-s} + e^{-3s} - e^{-4s}\right] + \frac{1}{s^2}\left[-e^{-s} + e^{-3s} - e^{-4s} + e^{-5s}\right] \quad \textbf{Ans.}$$

SUMMARY

1. Advantages of Laplace transforms :
   (*a*) It simplifies the operations
   (*b*) By Laplace transform, the complete solution comes out.
   (*c*) Linear differential equations can be solved easily.
2. Initial value theorem. If $f(t)$ and its first derivative are Laplace transform able, then initial value of $f(t)$ is

$$f(O_+) = \lim_{t\to O_+} f(t) = \lim_{S\to\infty} sF(s)$$

3. Final value theorem : If $f(t)$ and its first derivatives are Laplace transformable then final value of $f(t)$ is

$$\lim_{t\to\infty} f(t) = \lim_{s\to 0} sF(s)$$

4. Final value theorem does not apply when $f(t)$ is a periodic function.
5. Table of time functions and their Laplace transform.

S.No.	Function $f(t)$	Waveform	Laplace transform	Pole location
1.	$u(t)$	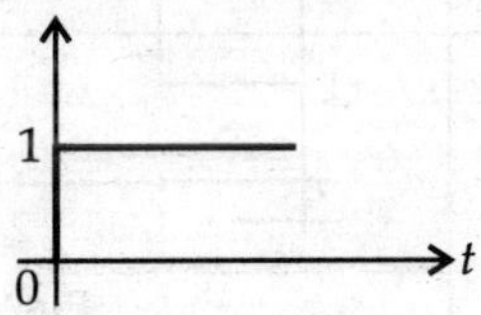	$\frac{1}{s}$	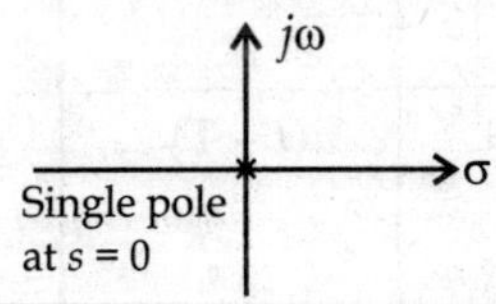
2.	t	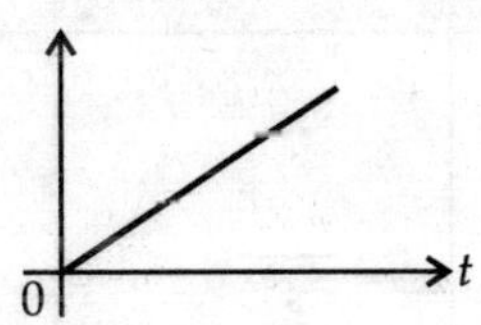	$\frac{1}{s^2}$	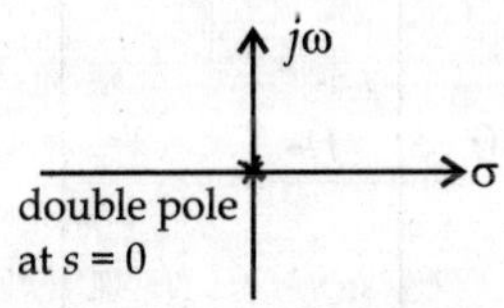
3.	t^2	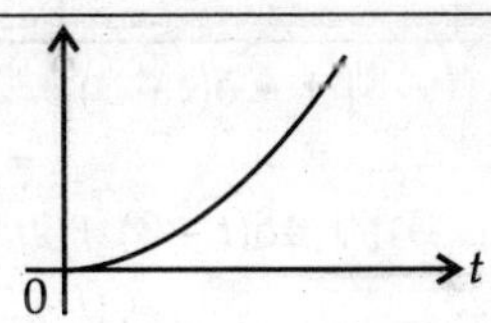	$\frac{2}{s^3}$	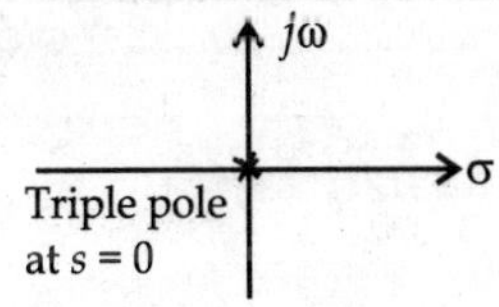

S.No.	Function $f(t)$	Waveform	Laplace	Pole location
4.	$\sin \omega t$	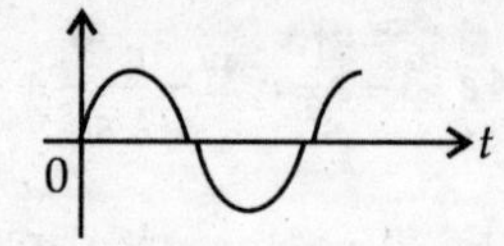	$\frac{\omega}{s^2+\omega^2}$	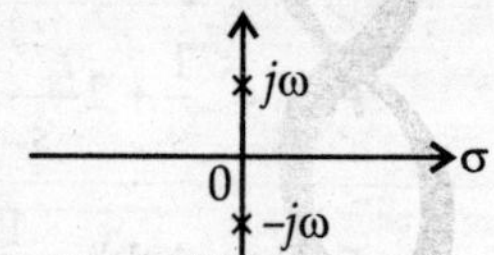
5.	$\cos \omega t$	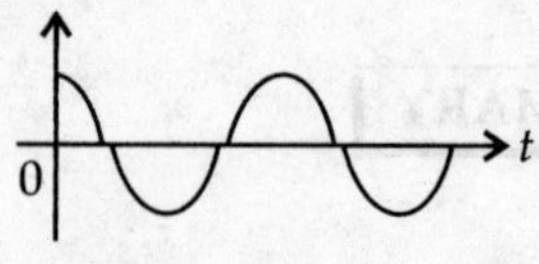	$\frac{s}{s^2+\omega^2}$	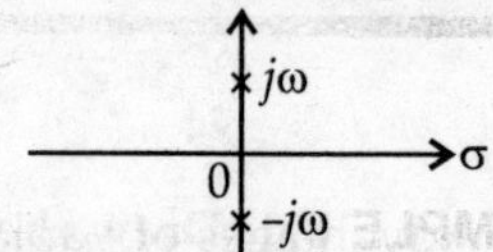
6.	e^{-at}	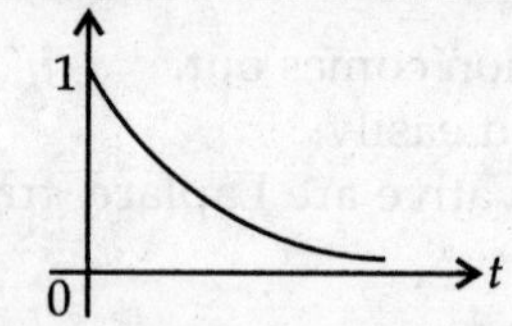	$\frac{1}{s+a}$	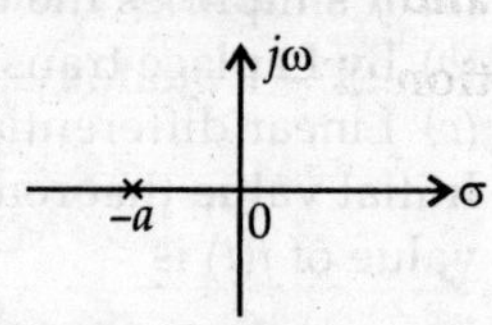
7.	$e^{-at} \sin \omega t$	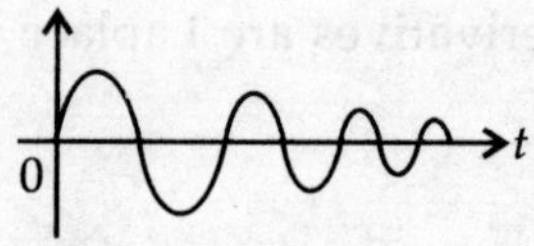	$\frac{\omega}{(s+a)^2+\omega^2}$	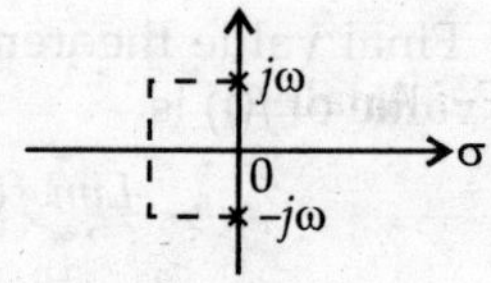
8.	$e^{-at} \cos \omega t$	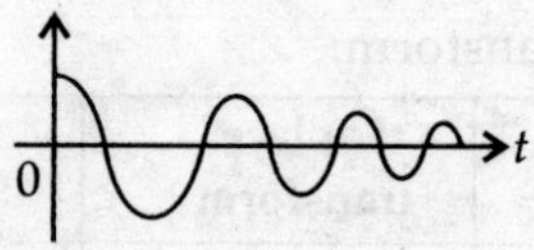	$\frac{s+a}{(s+a)^2+\omega^2}$	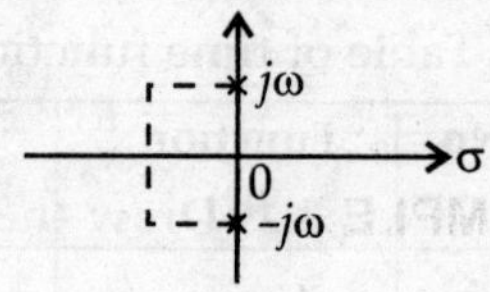
9.	$u(t-T)$	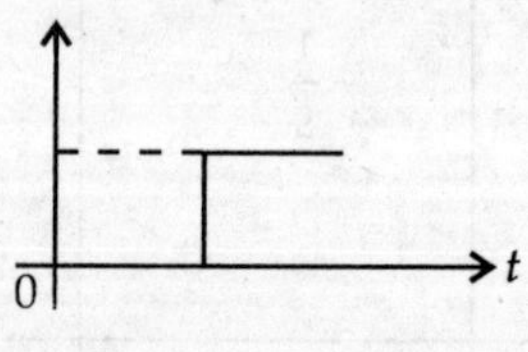	$\frac{e^{-sT}}{s}$	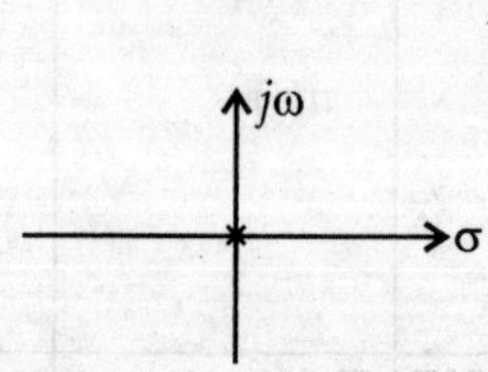
10.	$t-T$	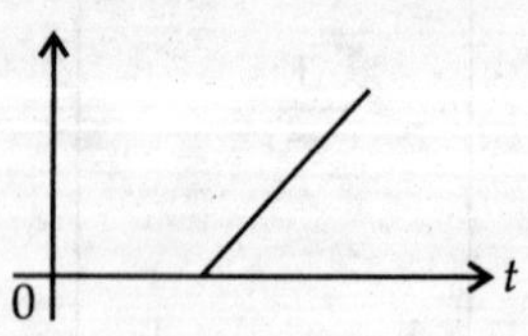	$\frac{e^{-sT}}{s^2}$	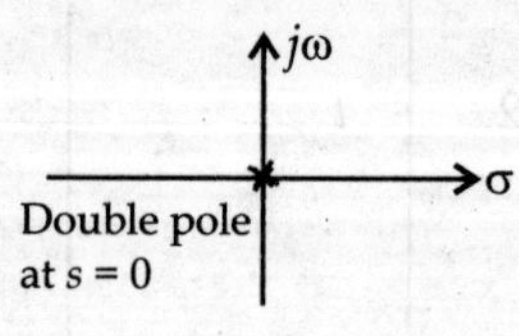

8 MISCELLANEOUS PROBLEMS

EXAMPLE 8.1. Draw the electrical analogous circuit of the system shown in Fig. 1.46 using $f - v$ and $f - i$ analogy and also write the equations.

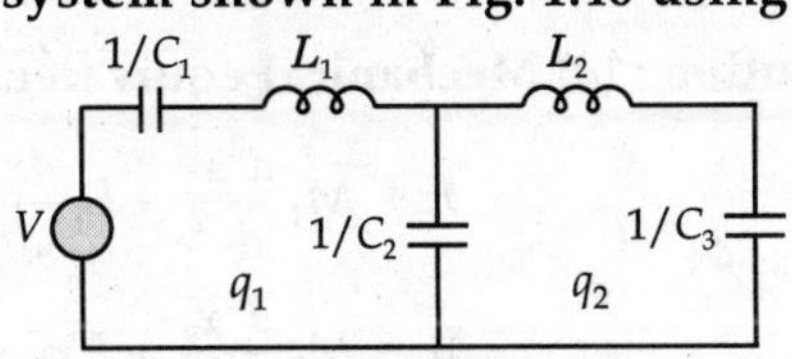

Fig 8.1

Solution : F – V analogy :

$$V(s) = q_1 \frac{1}{C_1} + s^2 L_1 q_1 + \frac{1}{C_2}(q_1 - q_2)$$

$$\frac{1}{C_2}(q_2 - q_1) + s^2 L_2 q_2 + \frac{1}{C_3} q_2 = 0$$

***F–i* Analogy**

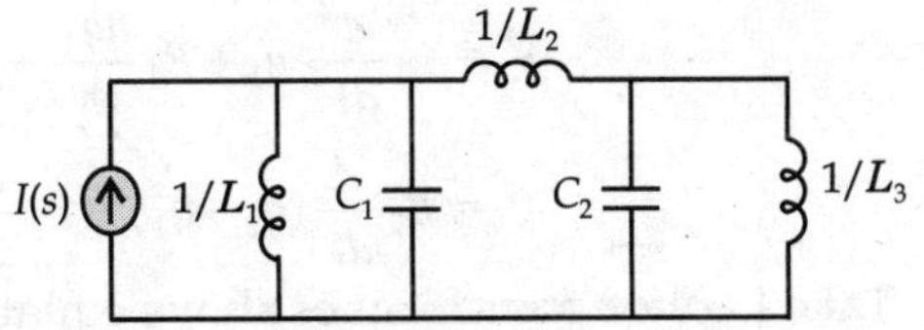

Fig. 8.2.

$$I(S) = \frac{1}{L_1}\phi_1 + s^2 C_1 \phi_1 + \frac{1}{L_2}(\phi_1 - \phi_2)$$

$$\frac{1}{L_2}(\phi_2 - \phi_1) + s^2 C_2 \phi_2 + \frac{1}{L_3}\phi_2 = 0$$

EXAMPLE 8.2. Draw the mechanical equivalent network of the system shown in Fig. 1.35.

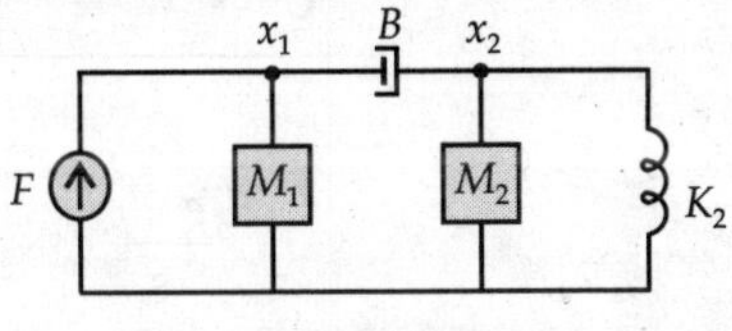

Fig 8.3.

EXAMPLE 8.3. Draw the mechanical equivalent network of the system shown in Fig. 1.53.

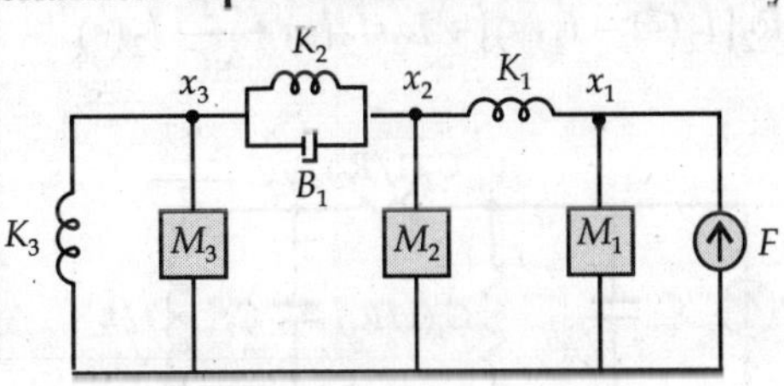

Fig. 8.4.

EXAMPLE 8.4. Draw the mechanical equivalent network of the system shown in Fig. Also draw the electrical analogous circuit and write the set of equation using *f-v* and *f-i* analogy.

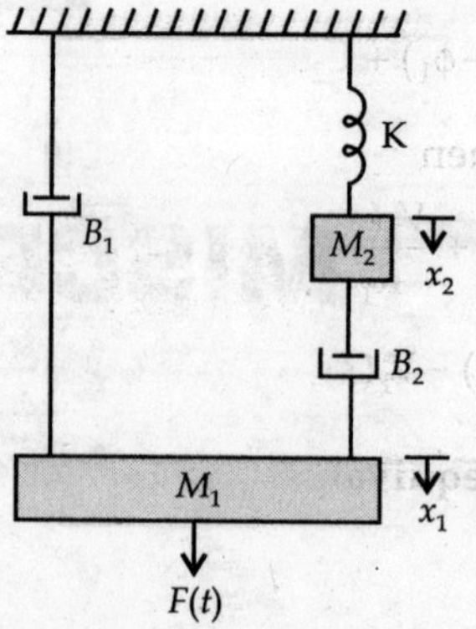

Fig. 8.5

Solution : (*a*) Mechanical equivalent network

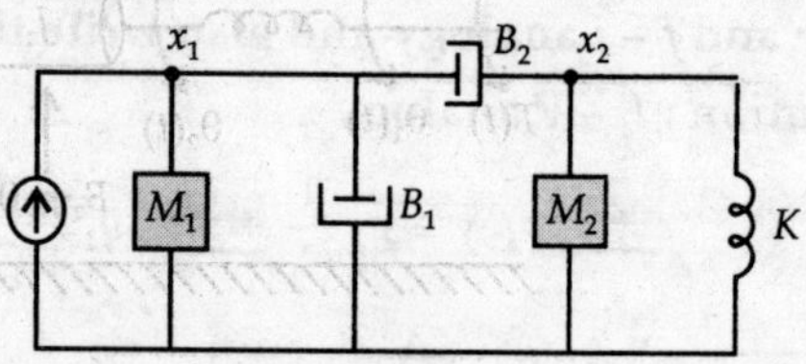

Fig. 8.6.

$$F = M_1 \frac{d^2x_1}{dt^2} + B_1 \frac{dx_1}{dt} + B_2 \frac{d}{dt}(x_1 - x_2)$$

$$0 = M_2 \frac{d^2x_2}{dt^2} + Kx_2 + B_2 \frac{d}{dt}(x_2 - x_1)$$

(*b*) *F-V* analogy

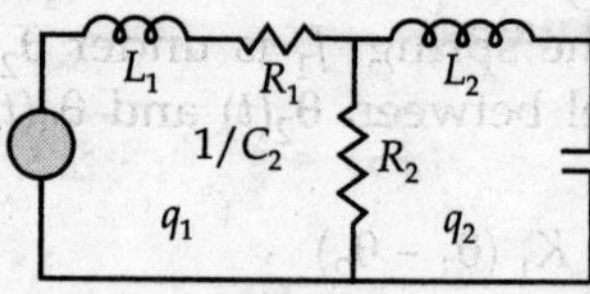

Fig. 8.7.

$$V = L_1 \frac{d^2}{dt^2} q_1 + R_1 \frac{dq_1}{dt} + R_2 \frac{d}{dt}(q_1 - q_2)$$

$$0 = R_2 \frac{d}{dt}(q_2 - q_1) + L_2 \frac{d^2}{dt^2} q_2 + \frac{1}{C} q_2$$

Take Laplace transform of above equations

$$V(s) = L_1 s^2 q_1 + R_1 sq_1 + R_2 s(q_1 - q_2)$$

$$0 = R_2 s(q_2 - q_1) + L_2 s^2 q_2 + \frac{1}{C} q_2$$

since, $I = \dfrac{dq}{dt}$

or, $I(s) = sq$

or $q = \dfrac{I(s)}{s}$

then, $V(s) = L_1 s I_1(s) + R_1 I_1(s) + R_2[I_1(s) - I_2(s)]$

$$0 = R_2[I_2(s) - I_1(s)] + L_2 s I_2(s) + \frac{1}{sC} I_2(s)$$

(*c*) *F-i* analogy :

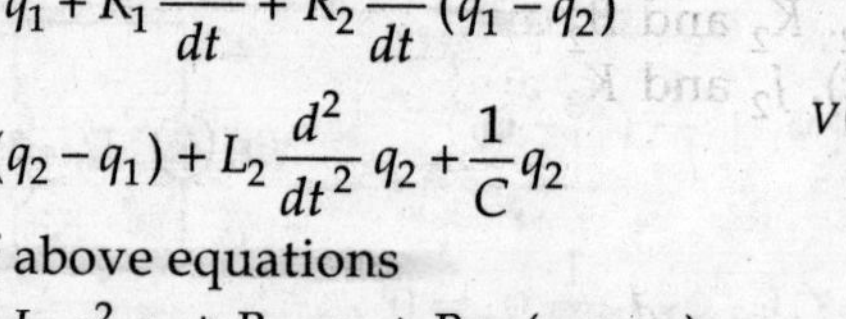

Fig. 8.8.

$$I(s) = C_1 s^2 \phi_1 + \frac{1}{R_1} s\phi_1 + \frac{1}{R_2} s(\phi_1 - \phi_2)$$

$$0 = \frac{1}{R_2} s(\phi_2 - \phi_1) + C_2 s^2 \phi_2 + \frac{1}{L}\phi_2$$

Since $V(s) = s\,\phi\,(s)$ then

$$I(s) = C_1 s V_1(s) + \frac{V_1(s)}{R_1} + \frac{1}{R_2}[V_1(s) - V_2(s)]$$

$$0 = \frac{1}{R_2}[V_2(s) - V_1(s)] + C_2 s\, V_2(s) + \frac{1}{sL} V_2(s)$$

EXAMPLE 8.5. Draw the mechanical equivalent network of the following rotational system and write the system equations.

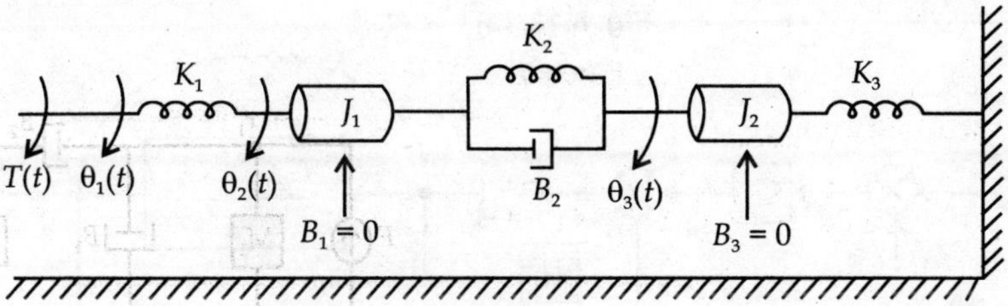

Fig. 8.9.

Solution : In the given fig the torque is applied to the spring so, there are two displacements at the ends of the spring. J_1 is under θ_2. K_2 and B_2 are in parallel between $\theta_2(t)$ and $\theta_3(t)$. J_2 and K_3 are under θ_3.

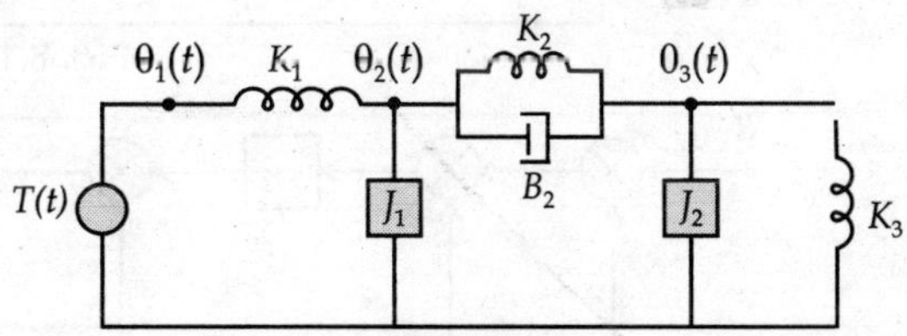

Fig. 8.10.

$$T(t) = K_1\,(\theta_1 - \theta_2)$$

$$K_1(\theta_2 - \theta_1) + J_1\frac{d^2\theta_2}{dt^2} + K_2(\theta_2 - \theta_3) + B_2\frac{d}{dt}(\theta_2 - \theta_3) = 0$$

$$J\frac{d^2\theta_3}{dt^2} + K_3\theta_3 + K_2(\theta_3 - \theta_2) + B_2\frac{d}{dt}(\theta_3 - \theta_2) = 0$$

EXAMPLE 8.6. For Example (8.5) draw the electrical analogous circuit using *T-V* analogy.

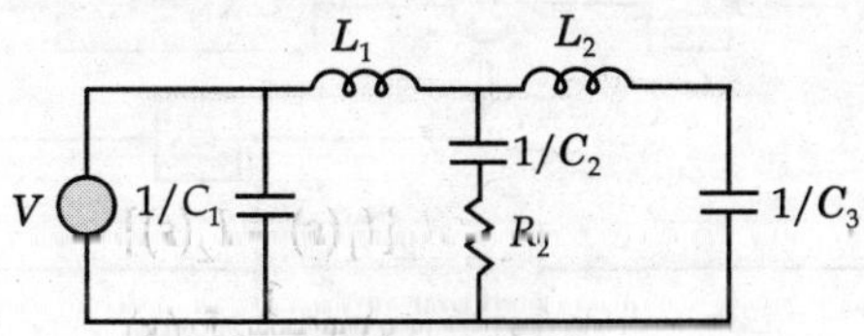

Fig. 8.11.

EXAMPLE 8.7. Find $\frac{C(s)}{R(s)}$, using block reduction method.

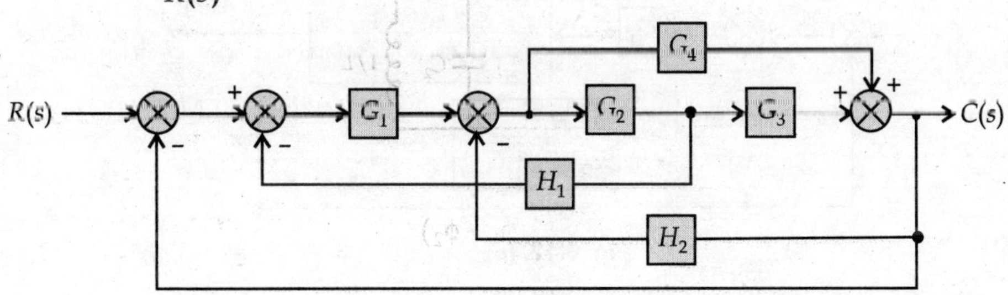

Fig. 8.12.

Solution :

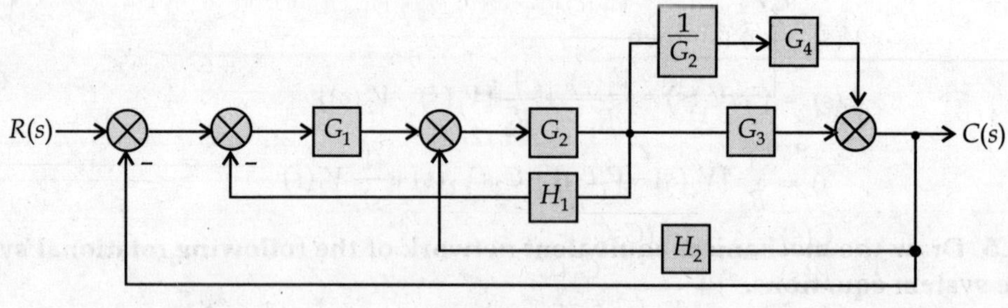

Fig. 8.12. (a)

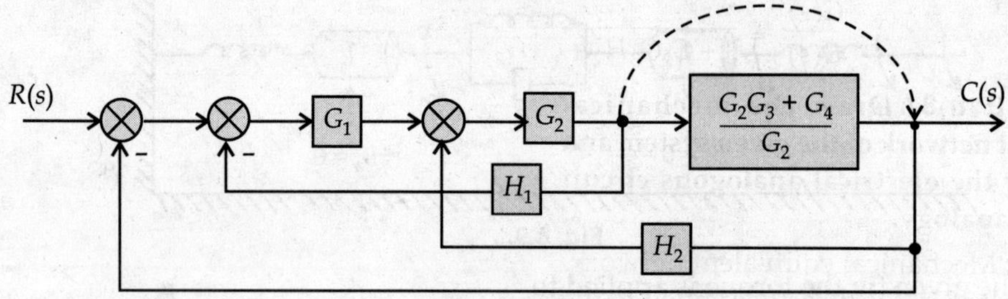

Fig. 8.12. (b)

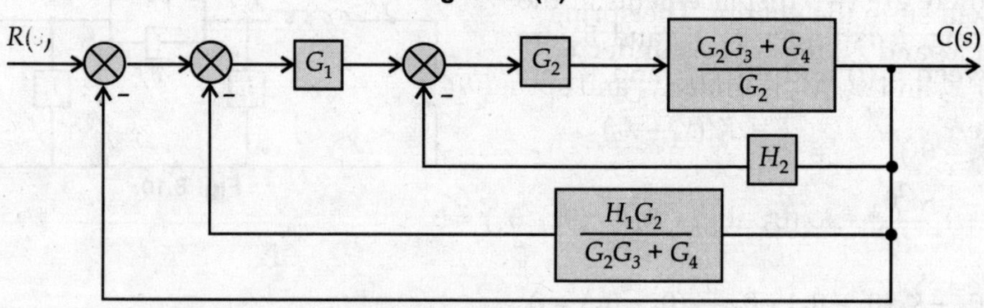

Fig. 8.12. (c)

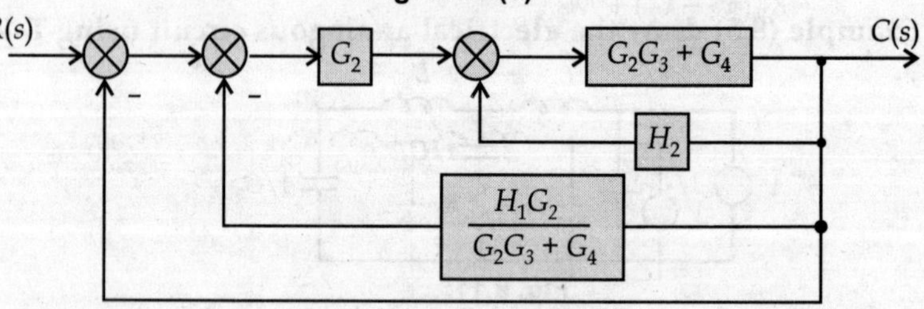

Fig. 8.12. (d)

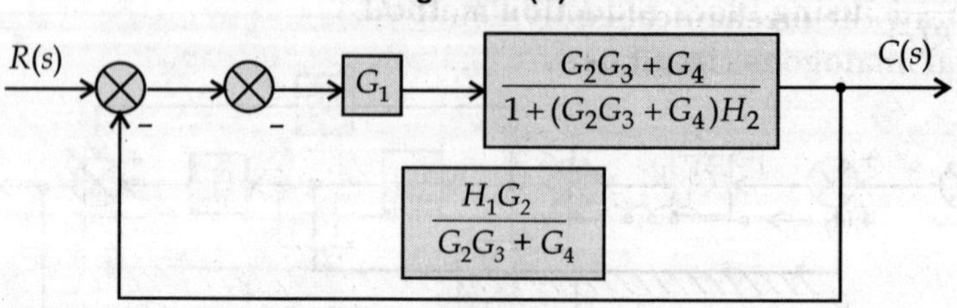

Fig. 8.12. (e)

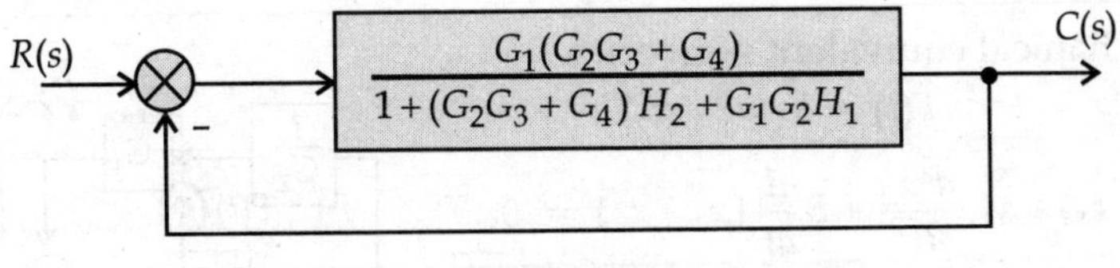

Fig. 8.12. (f)

$$\frac{C(s)}{R(s)} = \frac{\dfrac{G_1(G_2G_3+G_4)}{1+(G_2G_3+G_4)H_2+G_1G_2H_1}}{1+\dfrac{G_1(G_2G_3+G_4)}{1+(G_2G_3+G_4)H_2+G_1G_2H_1}\cdot 1}$$

$$\frac{C(s)}{R(s)} = \frac{G_1G_2G_3+G_1G_4}{1+G_2G_3H_2+G_4H_2+G_1G_2H_1+G_1G_2G_3+G_1G_4}$$

EXAMPLE 8.8. Draw the mechanical equivalent network of the given system and also draw the electrical analogous circuit using *f-v* analogy.

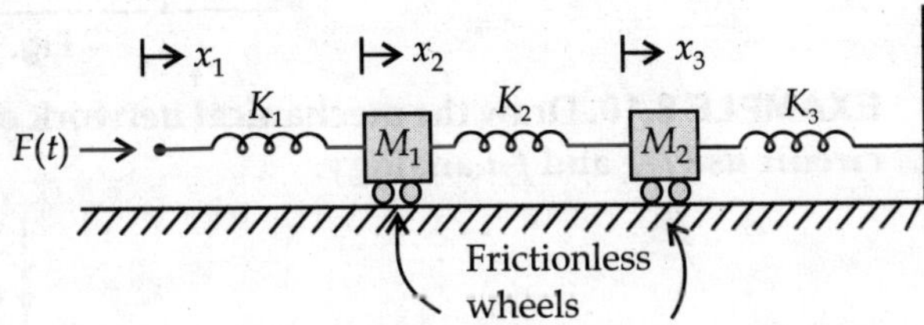

Fig. 8.13.

Solution : Mechanical equivalent network.

Since the force is directly applied to the spring, it will store the energy. The spring K_1 is under x_1 and x_2, mass M_1 is under x_2, K_2 is under x_2 and x_3, M_2 is under x_3 and spring K_3 is also under x_3.

at node x_1 : $F = K_1(x_1 - x_2)$

at node x_2 :

$$K_1(x_2-x_1)+M_1\frac{d^2x_2}{dt^2}+K_2(x_2-x_3) = 0$$

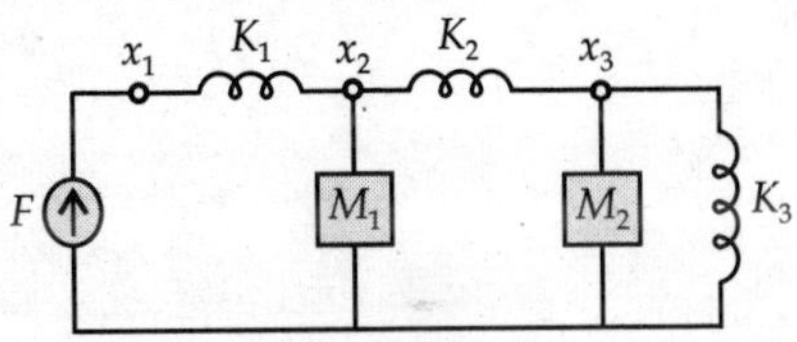

Fig. 8.14.

at node x_3 :

$$K_2(x_3-x_2)+M_2\frac{d^2x_3}{dt^2}+K_3x_3 = 0$$

***f-v* analogy :**

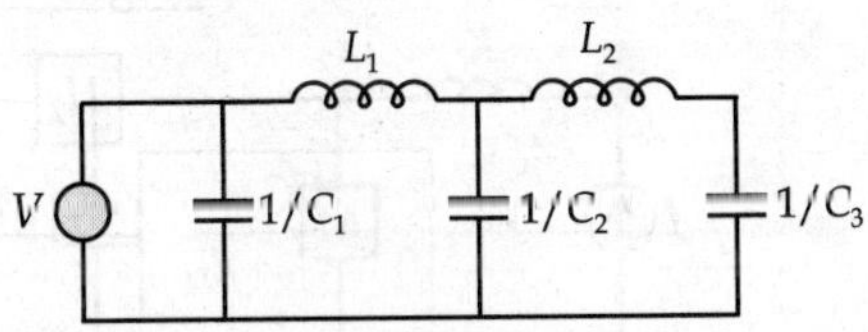

Fig. 8.15.

EXAMPLE 8.9. Draw the mechanical equivalent circuit and write the system equations. Also draw the electrical analogous circuit use *f-v* analogy.

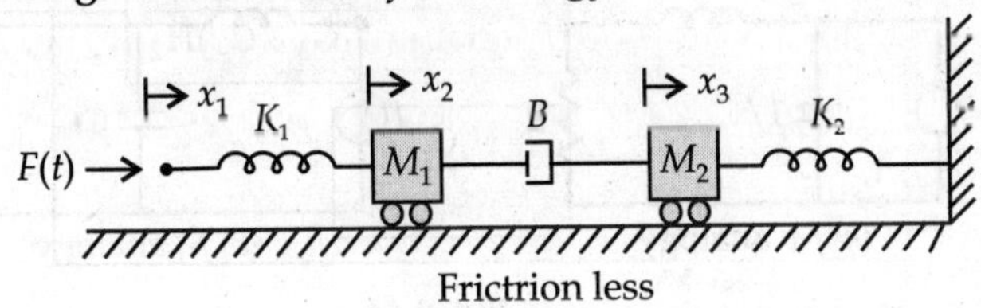

Fig. 8.16.

Solution : Mechanical equivalent network.

$$F(t) = K_1(x_1 - x_2)$$

$$K_1(x_2 - x_1) + M_1\frac{d^2x_2}{dt^2} + B\frac{d}{dt}(x_2 - x_3) = 0$$

$$B\frac{d}{dt}(x_3 - x_2) + M_2\frac{d^2x_3}{dt^2} + k_2x_3 = 0$$

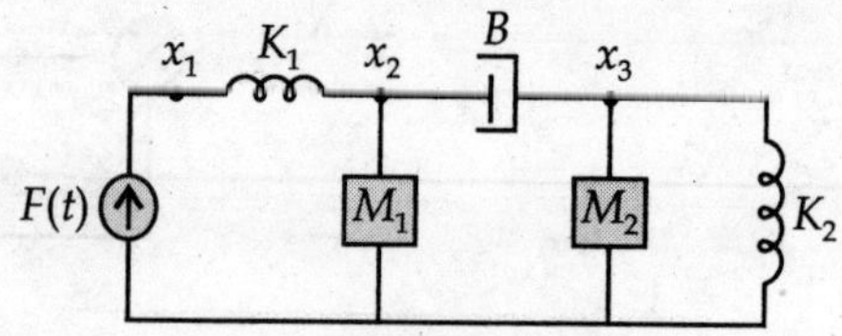

Fig. 8.17.

These are the system equations.

***f-v* analogy**

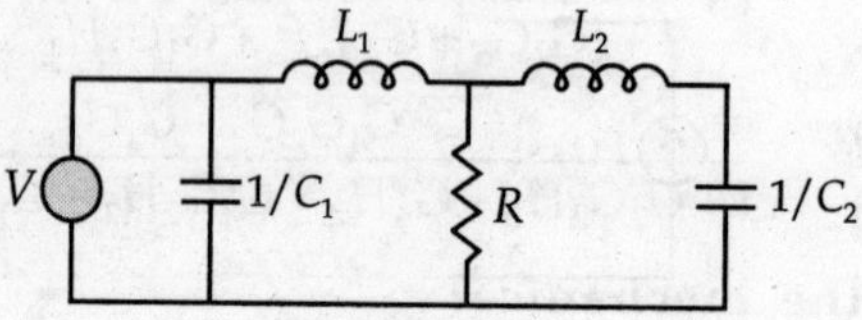

Fig. 8.18.

EXAMPLE 8.10. Draw the mechanical network of given system and draw the electrical analogous circuit use *f-v* and *f-i* analogy.

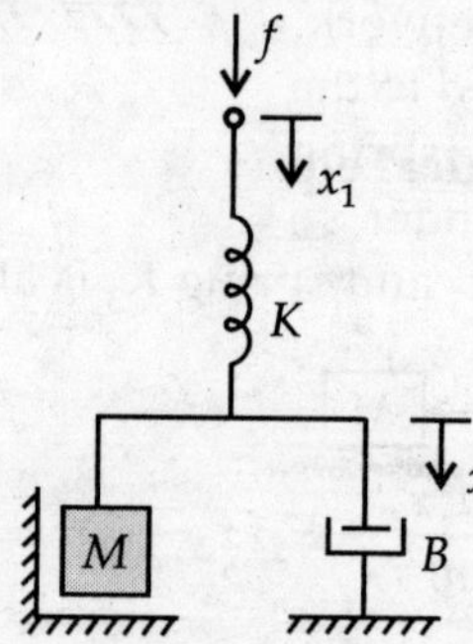

Fig. 8.19.

Solution : Mechanical equivalent network :

There are two displacements at the ends of the spring.

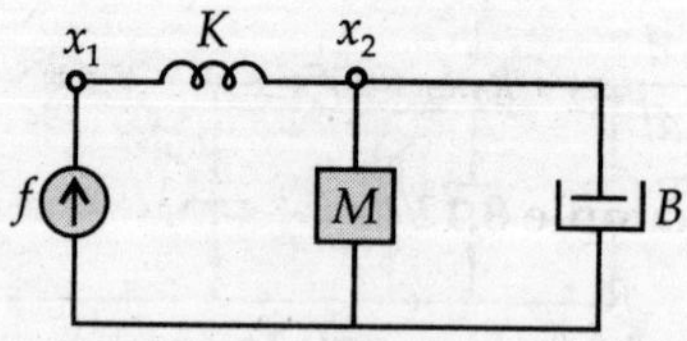

Fig. 8.20.

Electrical analogous circuits :

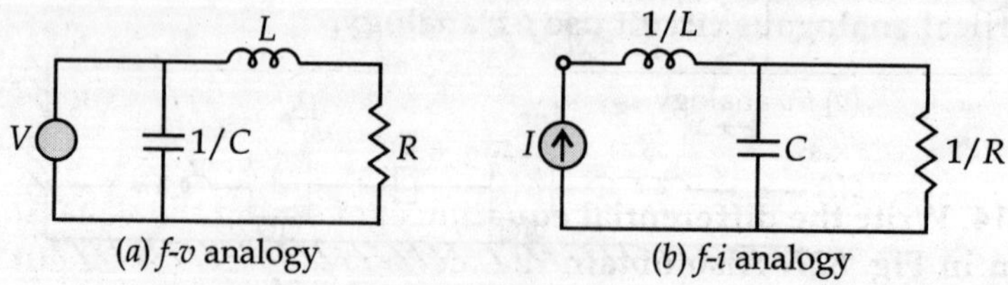

(a) *f-v* analogy (b) *f-i* analogy

Fig. 8.21.

EXAMPLE 8.11. Draw the mechanical equivalent and write the system equations.

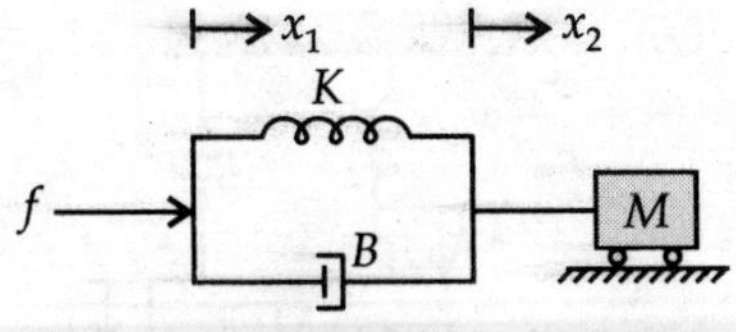

Fig. 8.22.

Solution :

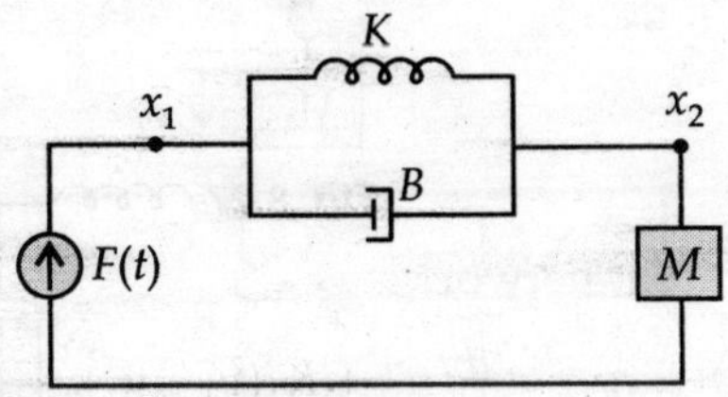

Fig. 8.23.

$$F(t) = K(x_1 - x_2) + B\frac{d}{dt}(x_1 - x_2)$$

$$K(x_2 - x_1) + B\frac{d}{dt}(x_2 - x_1) + M\frac{d^2x_2}{dt^2} = 0$$

EXAMPLE 8.12. Draw the mechanical equivalent circuit and write the system equations.

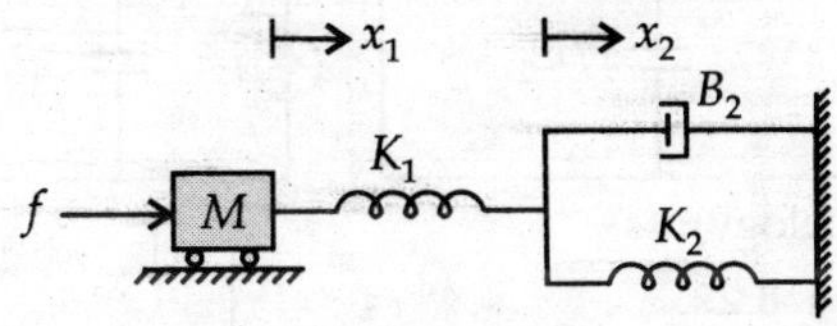

Fig. 8.24.

Solution :

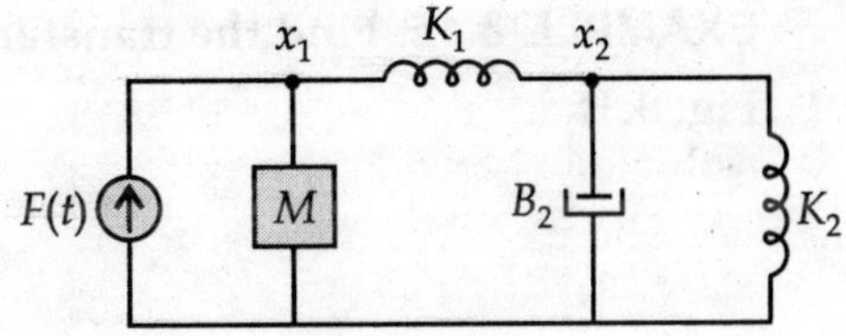

Fig. 8.25.

$$F(t) = M_1\frac{d^2x_1}{dt^2} + K_1(x_1 - x_2)$$

$$K_1(x_2 - x_1) + B_2\frac{d}{dt}x_2 + k_2x_2 = 0$$

EXAMPLE 8.13. For above example 8.12 draw *f-v* and *f-i* analogy.

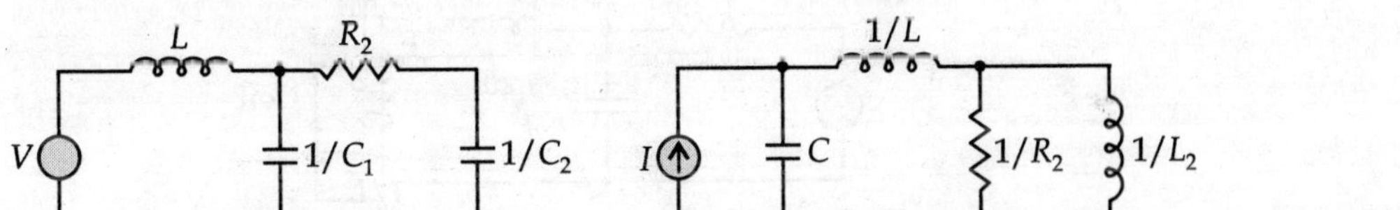

(*a*) *f-v* analogy (*b*) *f-i* analogy

Fig. 8.26.

EXAMPLE 8.14. Write the differential equations governing the behaviour of the mechanical system shown in Fig. 827. Also obtain the analogous electrical circuit based on *f-v* and *f-i* analogy.

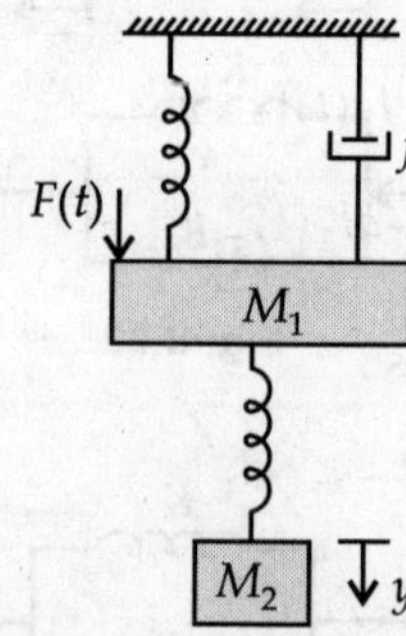

Fig. 8.27.

Solution : Mechanical equivalent network

at node y_1 : $F(t) = M_1 \frac{d^2y_1}{dt^2} + K_1 y_1 + f_1 \frac{dy_1}{dt} + K_2(y_1 - y_2)$

at node y_2 : $K_2(y_2 - y_1) + M_2 \frac{d^2y_2}{dt^2} = 0$

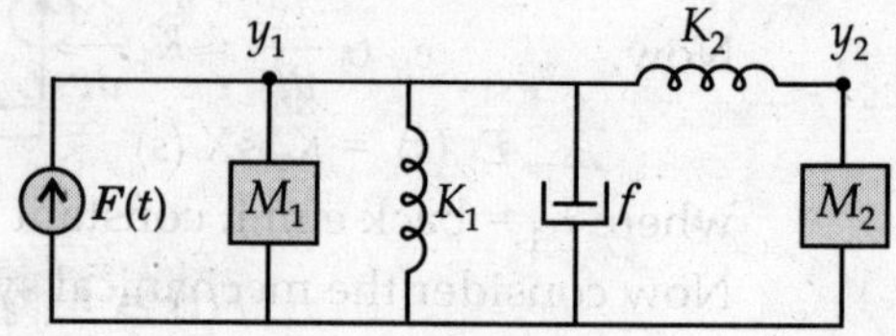

Fig. 8.28.

Analogous electrical circuits :

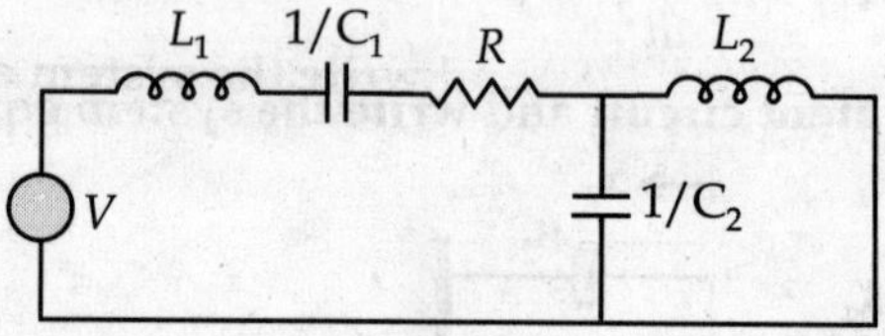

(f-v analogy)

Fig. 8.29.

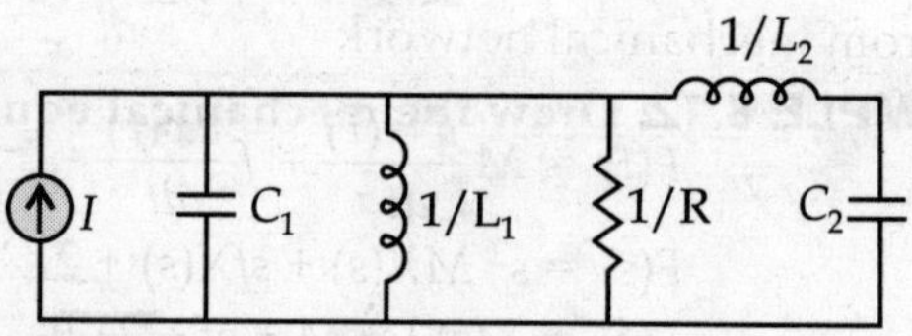

(f-i analogy)

Fig. 8.30.

EXAMPLE 8.15. Find the transfer function $\frac{X(s)}{E(s)}$ for the electromechanical system shown in Fig. 8.31.

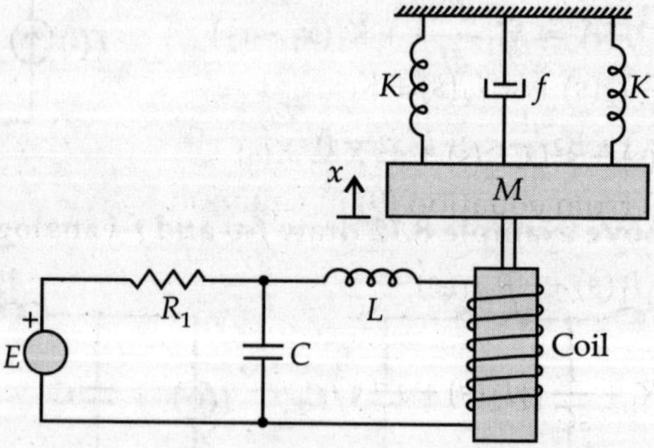

Fig. 8.31.

Solution : The given system consists of two parts:

(a) electrical, (b) mechanical

First consider the electrical system.

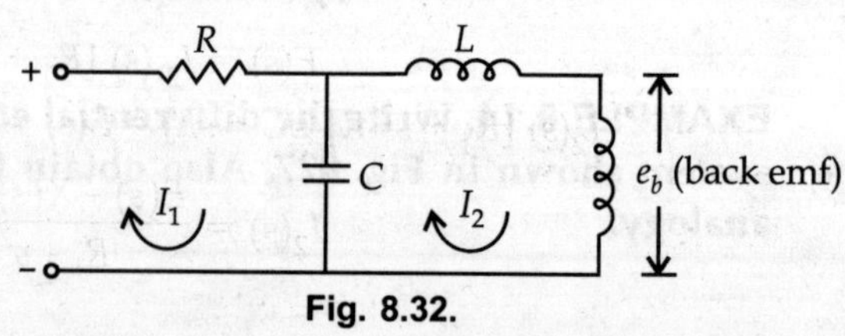

Fig. 8.32.

Apply KVL in both meshes :

$$E(t) = R_1 I_1(t) + \frac{1}{C}[I_1(t) - I_2(t)] \quad ...(1)$$

$$0 = \frac{1}{C}[I_2(t) - I_1(t)] + L\frac{dI_2(t)}{dt} + e_b \quad ...(2)$$

Laplace transform of both equations

$$E(s) = R_1 I_1(s) + \frac{1}{sC}[I_1(s) - I_2(s)] \quad ...(3)$$

$$0 = \frac{1}{sC}[I_2(s) - I_1(s)] + sL\, I_2(s) + E_b(s) \quad ...(4)$$

Now, $e_b \;\alpha\; \frac{dx}{dt} = K_1\frac{dx}{dt}$ ...(5)

$$E_b(s) = K_1\, sX(s)$$

where K_1 = back e.m.f. constant

Now consider the mechanical system :

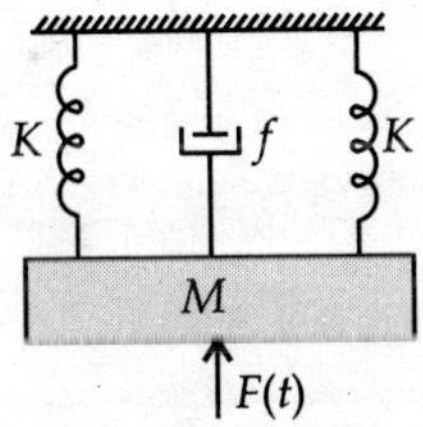

Fig. 8.33.

$$F(t) \;\alpha\; I_2(t) = K_2\, I_2(t)$$

$$F(s) = K_2\, I_2(s) \quad ...(6)$$

From mechanical network

$$F(t) = M\frac{d^2x(t)}{dt^2} + f\frac{dx(t)}{dt} + 2kx(t)$$

$$F(s) = s^2 MX(s) + sfX(s) + 2KX(s)$$

$$F(s) = X(s)\,[s^2 M + sf + 2K] \quad ...(7)$$

Put the value of $F(s)$ from (6) in (7) we get

$$K_2 I_2(s) = X(s)\,[s^2 M + sf + 2K] \quad ...(8)$$

From equation (5) put the value of $E_b(s)$ in (4)

$$0 = \frac{1}{sC}I_2(s) - \frac{1}{sC}I_1(s) + sL\,I_2(s) + K_1 s\,X(s)$$

or $$\frac{1}{sC}I_1(s) = \frac{1}{sC}I_2(s) + sL\,I_2(s) + K_1\,sX(s)$$

$$I_1(s) = I_2(s) + s^2LCI_2(s) + s^2\,XK_1\,X(s) \quad ...(9)$$

Put the value of $I_1(s)$ from equation (9) in equation (3)

$$E(s) = R_1 I_1(s) + \frac{1}{sC}I_1(s) - \frac{1}{sC}I_2(s) = \left[R_1 + \frac{1}{sC}\right]I_1(s) - \frac{1}{sC}I_2(s)$$

$$E(s) = \left[R_1 + \frac{1}{sC}\right][I_2(s) + s^2LCI_2(s) + s^2CK_1\,X(s)] - \frac{1}{sC}I_2(s)$$

$$E(s) = R_1 I_2(s) + s^2R_1LCI_2(s) + s^2R_1CK_1X(s) + \frac{1}{sC}I_2(s) + SLI_2(s) + SK_1X(s) - \frac{1}{sC}I_2(s)$$

$$E(s) = I_2(s)\,[R_1 + s^2 R_1 LC + sL] + X(s)\,sK_1\,[1 + R_1 Cs]$$

$$I_2(s)\,[R_1 + s^2 R_1 LC + sL] = E(s) - X(s)\,sK_1(1 + R_1Cs)$$

$$\therefore \quad I_2(s) = \frac{E(s) - X(s)\,sK_1(1 + R_1Cs)}{R_1 + s^2R_1LC + sL} \quad ...(10)$$

Put the value of $I_2(s)$ from equation (10) in equation (8)

$$K_2\left[\frac{E(s)-X(s)\,sK_1(1+R_1Cs)}{R_1+s^2R_1LC+sL}\right] = X(s)[s^2 M + sf + 2K]$$

$$K_2\,E(s) - X(s)\,s\,K_1\,K_2\,(1 + R_1\,Cs) = X(s)\,[R_1 + s^2\,R_1\,LC + sL]\,[s^2\,M + sf + 2K]$$

$$K_2\,E(s) - X(s)\,sK_1\,K_2\,(1 + R_1\,Cs) = X(s)\,[s^2\,R_1M + sR_1\,f + 2KR_1 + s^4\,MR_1\,LC + s^3\,R_1LC\,f + 2s^2\,R_1\,LC\,K + s^2\,ML + s^2\,fL + 2KLs]$$

$$K_2E(s) - X(s)\,s\,K_1\,K_2(1 + R_1\,Cs) = x(s)[s^4\,R_1\,LCM + s^3L\,(R_1Cf + M) + s^2\,R_1M + s^2fL + 2s^2\,R_1\,LC\,K + 2\,R_1K + sR_1f + 2KLs]$$

$$K_2\,E(s) = X(s)\,[s^4\,R_1\,LCM + s^3\,L(R_1Cf + M) + s^2\,R_1M + s^2fL + 2s^2\,R_1\,LCK + 2KLs + sR_1\,f\,2R_1\,K + s\,K_1\,K_2 + K_1\,K_2\,R_1\,Cs^2]$$

$$= X(s)\,[s^4\,R_1\,LCM + s^3L(R_1\,cf + M) + s^2(R_1\,M + fL + 2R_1\,LCK + K_1\,K_2\,R_1\,C) + 2KLS + 2R_1\,K + sR_1\,f + sK_1\,K_2]$$

$$= X(s)\,[s^4\,R_1\,LCM + s^3\,L\,(R_1cf + M) + \{R_1M + Lf + R_1C\,(2LK + K_1\,K_2)\}s^2 + s\,(R_1\,f + 2LK + K_1\,K_2) + 2R_1\,K]$$

$$\therefore \quad \frac{X(s)}{E(s)} = \frac{K_2}{[s^4R_1LCM + s^3L(R_1cf+M) + s^2\{R_1M + Lf + R_1C(2LK + K_1K_2)\} + s(R_1f + 2LK + K_1K_2) + 2R_1K]}$$

This is the required answer.

EXAMPLE 8.16. For mass-damper spring combination shown in Fig. 8.34 draw the mech. Equivalent network, write the system equations and draw *f-v*, *f-i* analog.

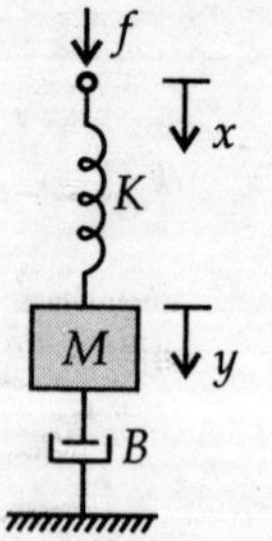

Fig. 8.34.

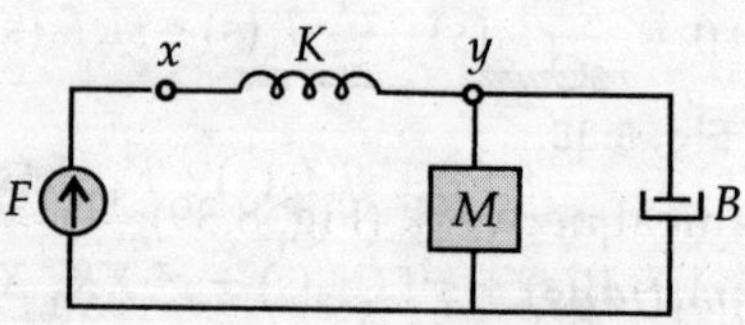

Fig. 8.35.

Solution: Mechanical network (Fig. 8.35)

at node x: $\quad F = K(x - y)$

$$0 = K(y-x) + M\frac{d^2y}{dt^2} + B\frac{dy}{dt}$$

f-v analogy

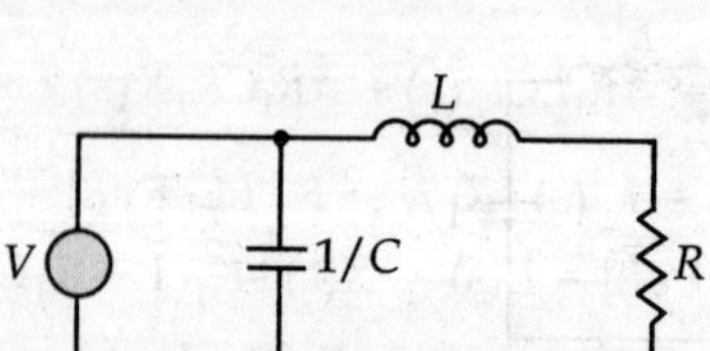

Fig. 8.36.

$$V = \frac{1}{C}(q_1 - q_2)$$

$$\frac{1}{C}(q_2 - q_1) + L\frac{d^2q_2}{dt^2} + R\frac{dq_2}{dt} = 0$$

$$I(t) = \frac{dq}{dt} \qquad \therefore I(s) = s\,q(s)$$

$$\therefore \qquad V = \frac{1}{sC}[I_1(s) - I_2(s)]$$

$$\frac{1}{sC}[I_2(s) - I_2(s)] + sLI_2(s) + RI_2(s) = 0$$

***f-i* analogy :**

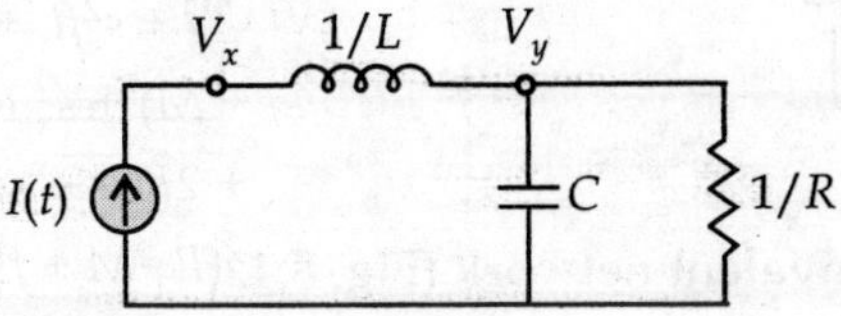

Fig. 8.37.

EXAMPLE 8.17. For the mechanical system shown in Fig. 8.38 : *(a)* draw mechanical equivalent network, *(b)* write the differential equations, *(c)* draw the electrical analogous circuit using *f-v* and *f-i* analogy.

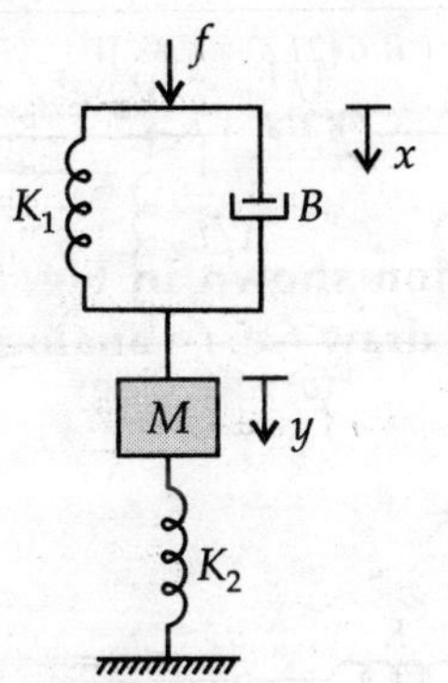

Fig. 8.38.

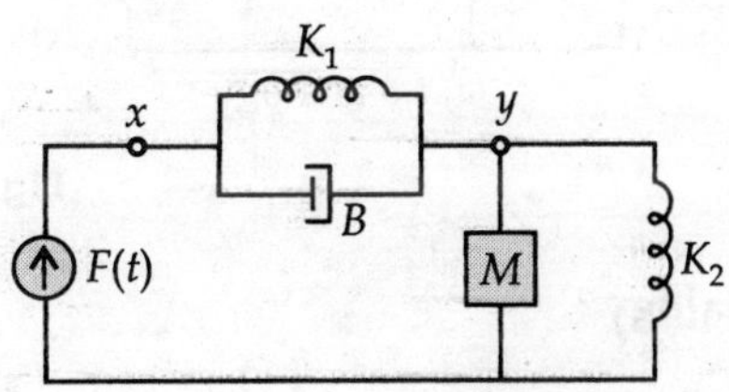

Fig. 8.39.

Solution : Mechanical network (Fig. 8.39):

(b) System equations :

$$F(t) = K_1(x - y) + B\frac{d}{dt}(x - y)$$

$$K_1(y - x) + B\frac{d}{dt}(y - x) + M\frac{d^2y}{dt^2} + K_2y = 0$$

(c) Electrical analogous circuit :

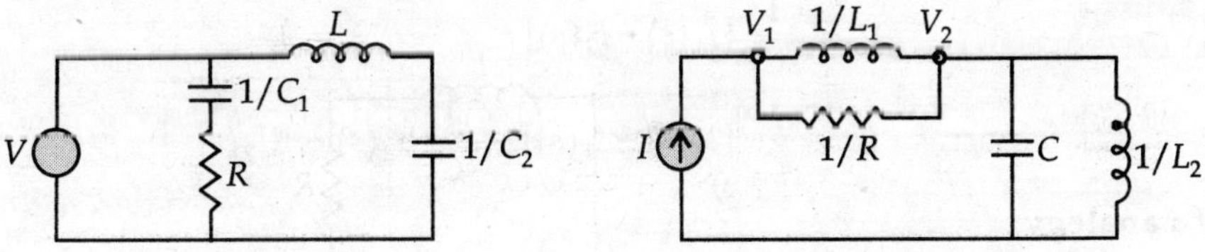

(a) f-v analogy

(b) f-i analogy (same displacement same voltage)

Fig. 8.40

EXAMPLE 8.18. For the given mechanical system draw the mechanical equivalent network and write the system equations.

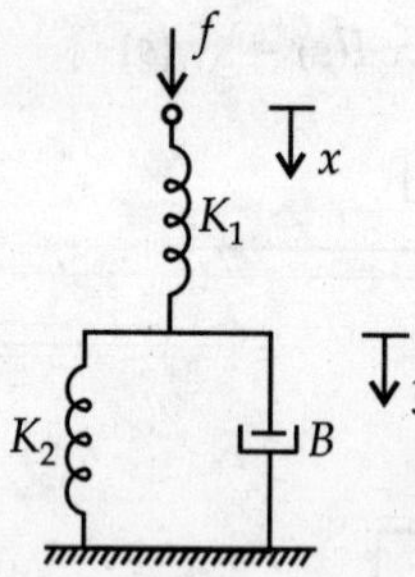

Fig. 8.41.

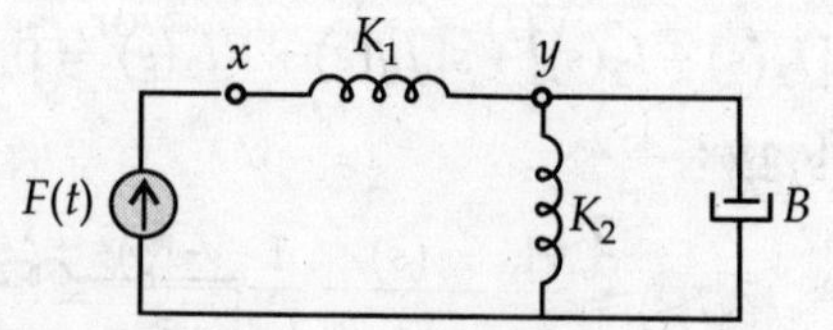

Fig. 8.42.

Solution : Mechanical equivalent network (Fig. 8.42)

at node x : $F(t) = K_1(x - y)$

at node y : $K_1(y - x) + K_2 y + B\frac{d}{dt}y = 0$

EXAMPLE 8.19. For the system shown in question (8.18) draw the electrical analogous circuit using *f-v* and *f-i* analogy.

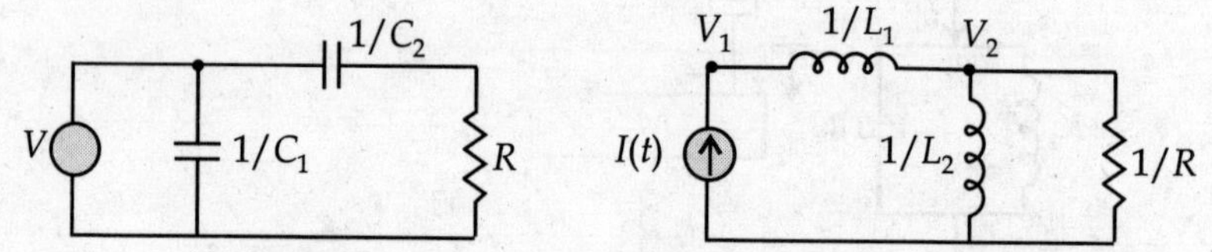

(*a*) *f-v* analogy (*b*) *f-i* analogy

Fig. 8.43.

Solution :

For *f-v* analogy :

$$V = \frac{1}{C_1}(q_1 - q_2)$$

$$0 = \frac{1}{C_1}(q_2 - q_1) + \frac{1}{C_2}q_2 + R\frac{d}{dt}q_2$$

Since $$i = \frac{dq}{dt}$$

or, $$I(s) = sq(s)$$

$$q(s) = \frac{I(s)}{s}$$

∴ $$V = \frac{1}{sC_1}[I_1(s) - I_2(s)]$$

$$0 = \frac{1}{sC_1}[I_2(s) - I_1(s)] + \frac{1}{sC_2}I_2(s) + RI_2(s)$$

For *f-i* analogy :

Laplace transform of the system equations given in question 8.18

$$F(s) = K_1 X(s) - K_1 Y(s)$$

$$K_1 Y(s) - K_1 X(s) + K_2 Y(s) + sBY(s) = 0$$

From *f-i* analogy circuit :

$$I(s) = \frac{1}{L_1}\phi_1(s) - \frac{1}{L_1}\phi_2(s)$$

$$\frac{1}{L_1}\phi_2(s) - \frac{1}{L_1}\phi_1(s) + \frac{1}{L_2}\phi_2(s) + s\cdot\frac{1}{R}\phi_2(s) = 0$$

since, $V(t) = \frac{d\phi}{dt}$ or, $V(s) = s\phi(s)$

replace $\phi(s)$ by $\frac{V(s)}{s}$

$$I(s) = \frac{1}{L_1}\frac{V_1(s)}{s} - \frac{1}{L_1}\frac{V_2(s)}{s} = \frac{1}{sL_1}[V_1(s) - V_2(s)]$$

$$\frac{1}{sL_1}[V_2(s) - V_1(s)] + \frac{1}{sL_2}V_2(s) + \frac{V_2(s)}{R} = 0$$

These are the required equations.

EXAMPLE 8.20. For the given mechanical system draw: (*a*) mechanical equivalent network, (*b*) electrical analogous circuit (*c*) write the system equations.

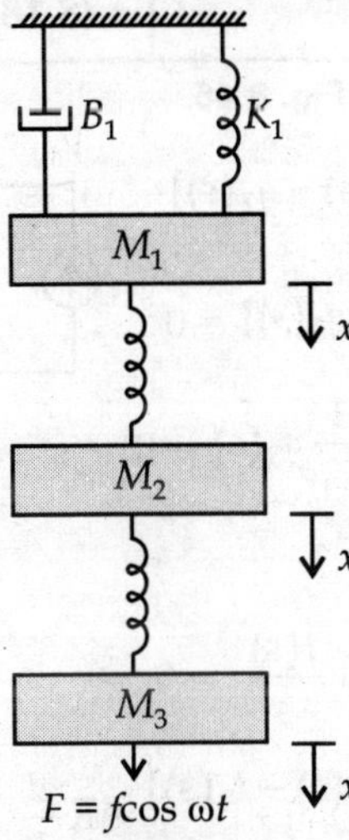

Fig. 8.44.

Solution : Mechanical equivalent network

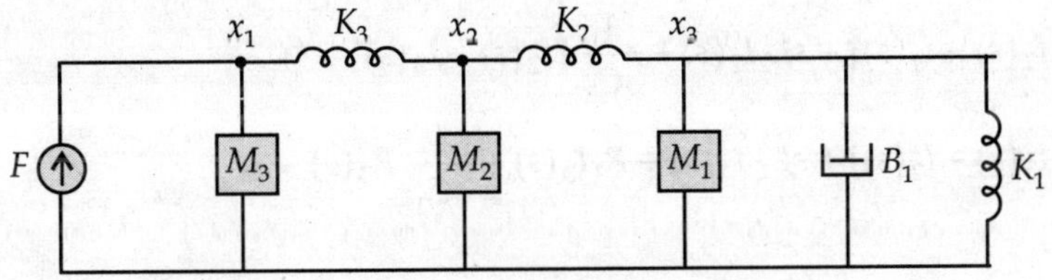

Fig. 8.45.

at node x_1 : $$F = M_3\frac{d^2x_1}{dt^2} + K_3(x_1 - x_2)$$

at node x_2 : $$K_3(x_2 - x_1) + M_2\frac{d^2x_2}{dt^2} + K_2(x_2 - x_3) = 0$$

at node x_3 : $K_2(x_3 - x_2) + M_1 \frac{d^2x_3}{dt^2} + B_1 \frac{dx_3}{dt} + K_1 x_3 = 0$

Laplace transform of above equations

$$F = f \cos \omega t \quad \text{given}$$

$$F(s) = \frac{f\omega}{s^2 + \omega^2}$$

$$\frac{f\omega}{s^2 + \omega^2} = s^2 M_3 X_1(s) + K_3 [X_1(s) - X_2(s)]$$

$$K_3[X_2(s) - X_1(s)] + s^2 M_2 X_2(s) + K_2]X_2(s) - X_3(s)] = 0$$

$$K_2[X_3(s) - X_2(s)] + s^2 M_1 X_3(s) + sB_1 X_3(s) + K_1 X_3(s) = 0$$

F-V analogy :

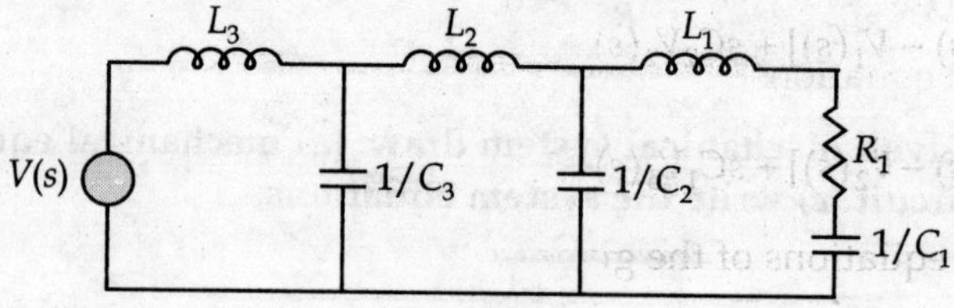

Fig. 8.46.

$$V(s) = s^2L_3q_1(s) + \frac{1}{C_3}[q_1(s) - q_2(s)]$$

$$\frac{1}{C_3}[q_2(s) - q_1(s)] + s^2L_2q_2(s) + \frac{1}{C_2}[q_2(s) - q_3(s)] = 0$$

$$\frac{1}{C_2}[q_3(s) - q_2(s)] + s^2L_1q_3(s) + sR_1q_3(s) + \frac{1}{C_1}q_3(s) = 0$$

Since $\quad i = \frac{dq}{dt}$

$$I(s) = sq(s) \quad \text{or} \quad q(s) = \frac{I(s)}{s}$$

$$V(s) = s^2L_3\frac{I_1(s)}{s} + \frac{1}{sC_3}[I_1(s) - I_2(s)]$$

$$V(s) = sL_3I_1(s) + \frac{1}{sC_3}[I_1(s) - I_2(s)]$$

$$\frac{1}{sC_3}[I_2(s) - I_1(s)] + sL_2I_2(s) + \frac{1}{sC_2}[I_2(s) - I_3(s)] = 0$$

$$\frac{1}{sC_2}[I_3(s) - I_2(s)] + sL_1 I_3(s) + R_1I_3(s) + \frac{1}{sC_1}R_3(s) = 0$$

F-i analogy :

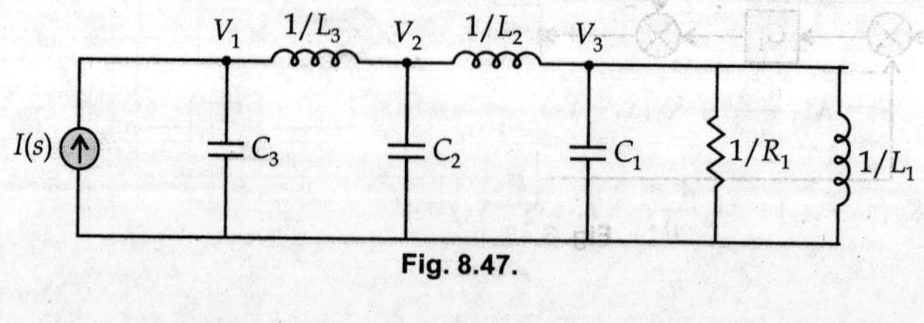

Fig. 8.47.

$$I(s) = s^2C_3\phi_1(s) + \frac{1}{L_3}[\phi_1(s) - \phi_2(s)]$$

$$\frac{1}{L_3}[\phi_2(s) - \phi_1(s)] + s^2C_2\,\phi_2(s) + \frac{1}{L_2}[\phi_2(s) - \phi_3(s)] = 0$$

$$\frac{1}{L_2}[\phi_3(s) - \phi_2(s)] + s^2c_1\phi_3(s) + s\cdot\frac{1}{R_1}\phi_3(s) + \frac{1}{L_1}\phi_3(s) = 0$$

Since $$V(t) = \frac{d\phi}{dt}$$

$$V(s) = s\phi(s) \quad \text{or,} \quad \phi(s) = \frac{V(s)}{s}$$

$$I(s) = sC_3V_1(s) + \frac{1}{sL_3}[V_1(s) - V_2(s)]$$

$$\frac{1}{sL_3}[V_2(s) - V_1(s)] + sC_2V_2(s) + \frac{1}{sL_2}[V_2(s) - V_3(s)] = 0$$

$$\frac{1}{sL_2}[V_3(s) - V_2(s)] + sC_1V_3(s) + \frac{V_3(s)}{R_1} + \frac{1}{sL_1}V_3(s) = 0$$

These are the required equations of the given system.

EXAMPLE 8.21. Using block diagram reduction technique, find closed loop transfer function of the system whose block diagram is given in Fig. 8.48 (*a*) when $R_1 = 0$ (*b*) when $R_2 = 0$.

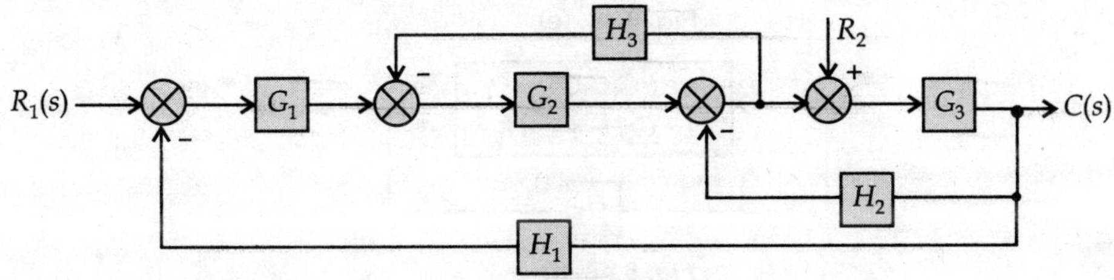

Fig. 8.48.

Solution : When $R_2 = 0$

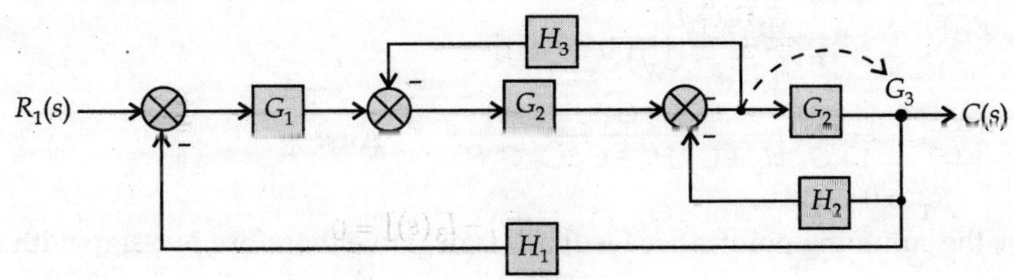

Fig. 8.48. (a)

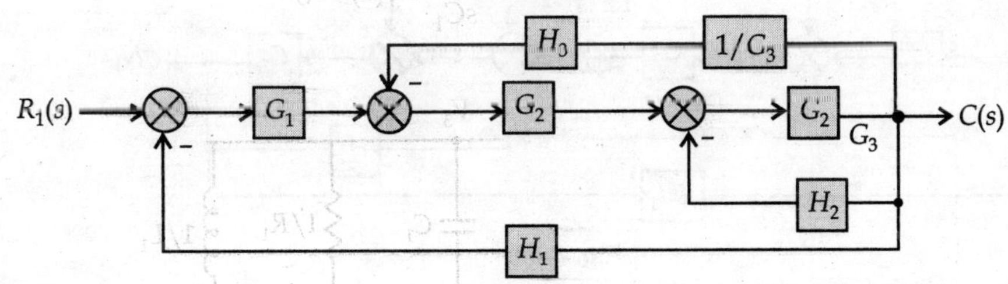

Fig. 8.48. (b)

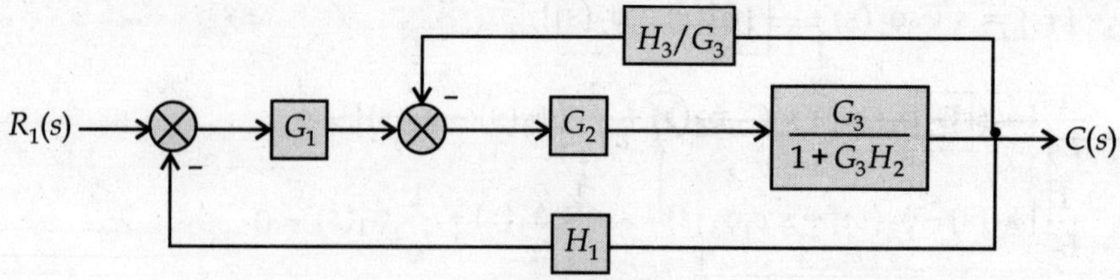

Fig. 8.48. (c)

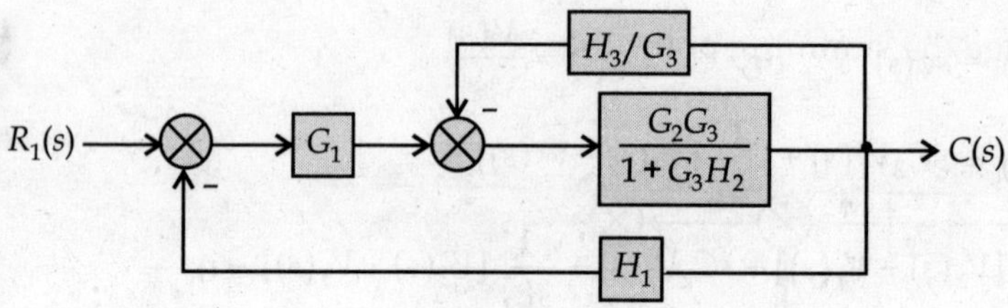

Fig. 8.48. (d)

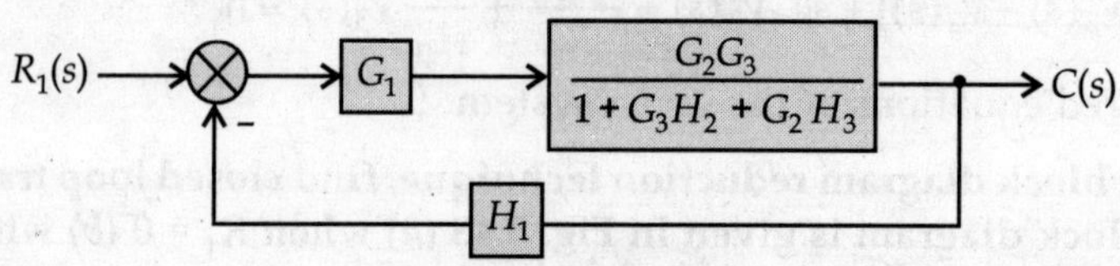

Fig. 8.48. (e)

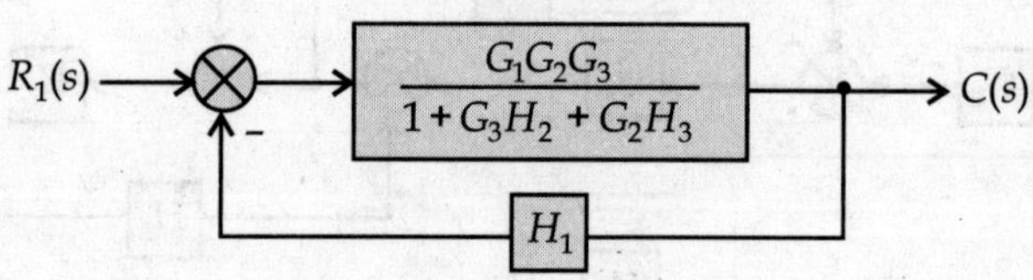

Fig. 8.48. (f)

$$\frac{c(s)}{R_1(s)} = \frac{\dfrac{G_1\,G_2\,G_3}{1+G_3H_2+G_2H_3}}{1+\dfrac{G_1G_2G_3H_1}{1+G_3H_2+G_2H_3}}$$

$$\frac{C(s)}{R_1(s)} = \frac{G_1\,G_2\,G_3}{1+G_3\,H_2+G_2\,H_3+G_1\,G_2\,G_3H_1} \quad \textbf{Ans.}$$

(*b*) when $R_1 = 0$

Now remove the summing point since feedback is negative therefore put-sign with H_1

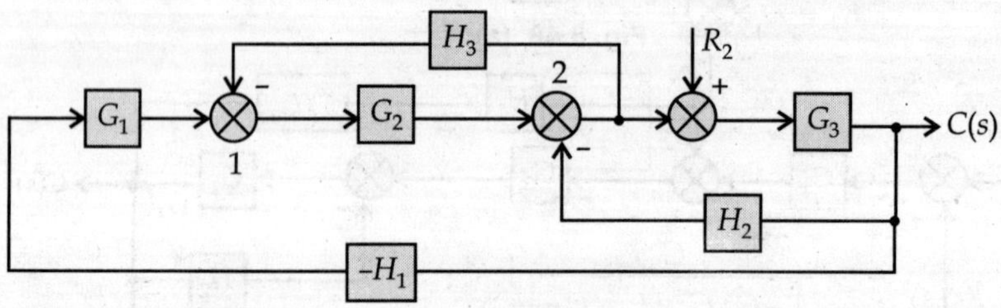

Fig. 8.49.

shift the summing point 1 after G_2 then

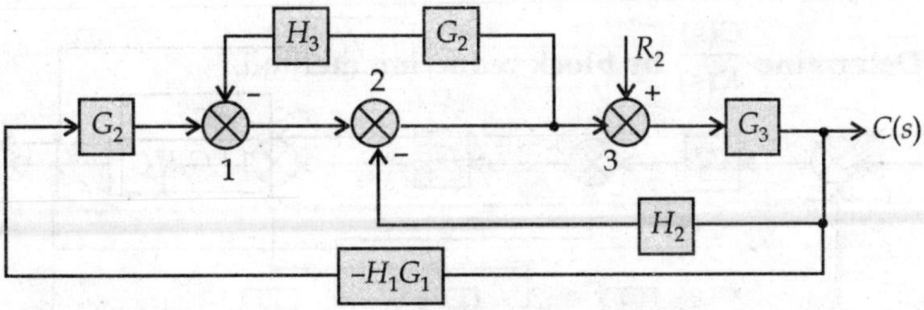

Fig. 8.49. (a)

Now interchange the summing point 1 and 2

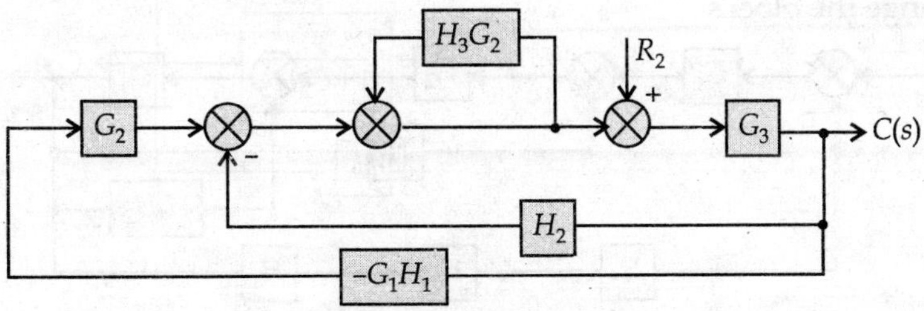

Fig. 8.49. (b)

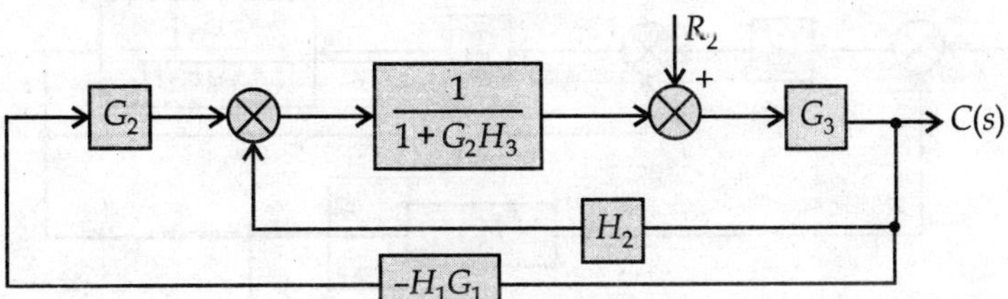

Fig. 8.49. (c)

shift the summing point after the block

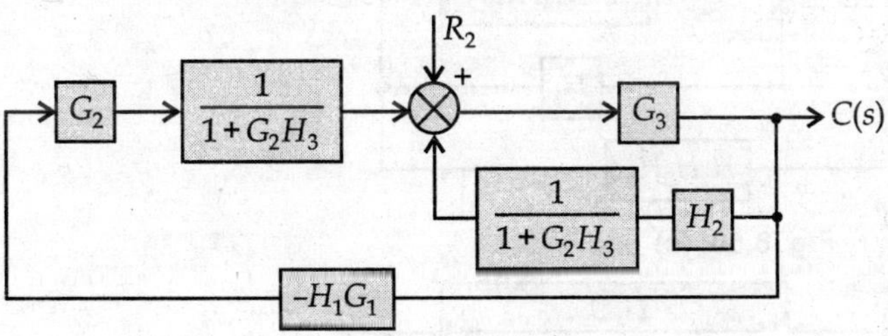

Fig. 8.49. (d)

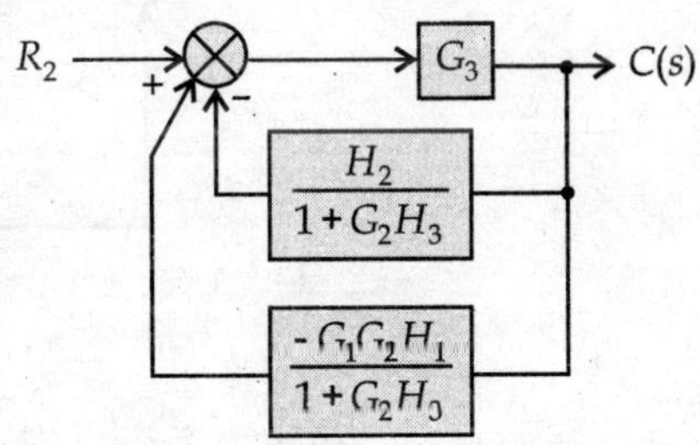

Fig. 8.49. (e)

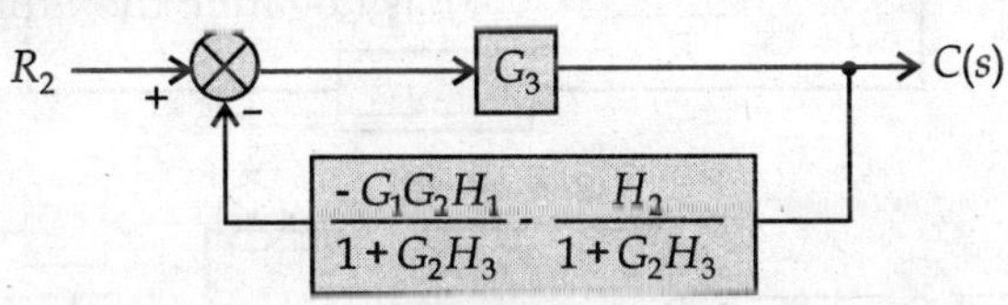

Fig. 8.49. (f)

$$\frac{C(s)}{R_2(s)} = \frac{G_3}{1 - \left[\dfrac{-G_1 G_2 H_1 - H_2}{1 + G_2 H_3}\right] G_3} = \frac{G_3(1 + G_2 H_3)}{1 + G_1 G_2 G_3 H_1 + G_3 H_2 + G_2 H_3}$$ **Ans.**

EXAMPLE 8.22. Determine $\frac{C(s)}{R(s)}$ by block reduction method.

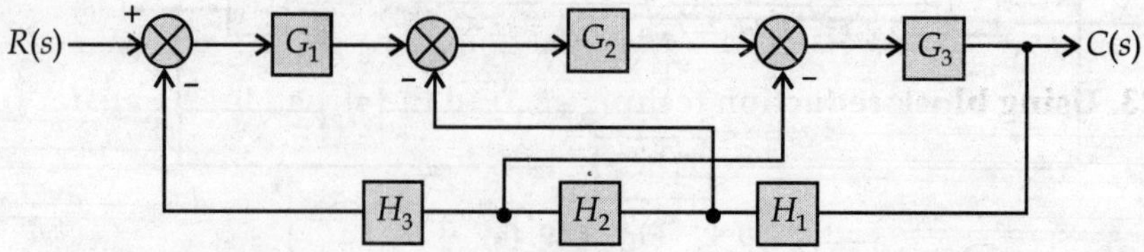

Fig. 8.50.

Solution : Rearrange the blocks

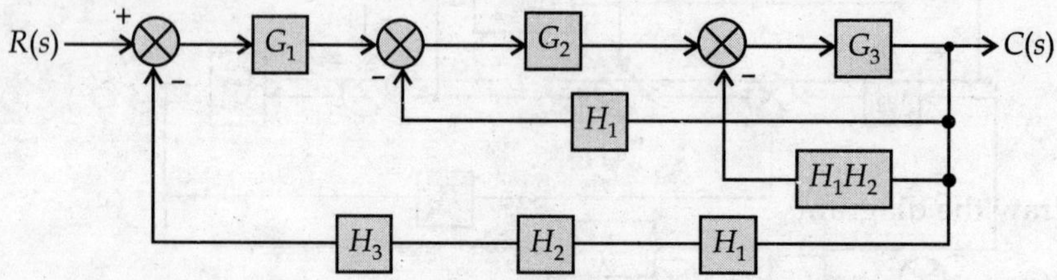

Fig. 8.50. (a)

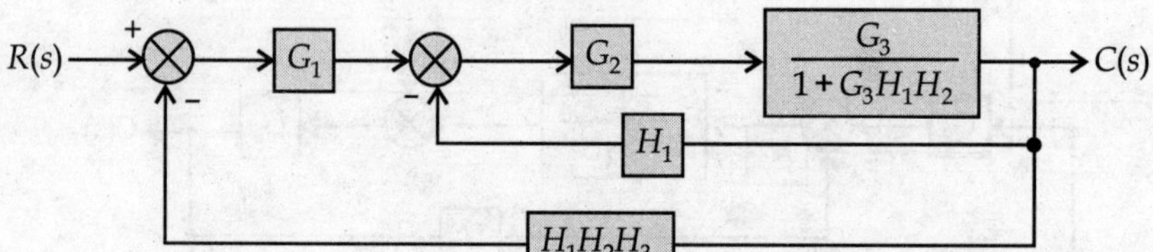

Fig. 8.50. (b)

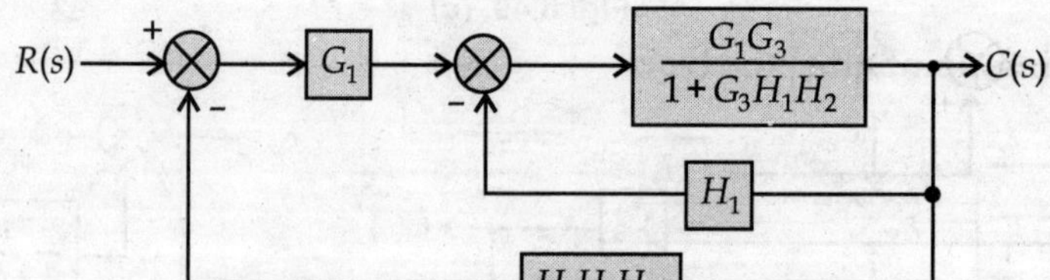

Fig. 8.50. (c)

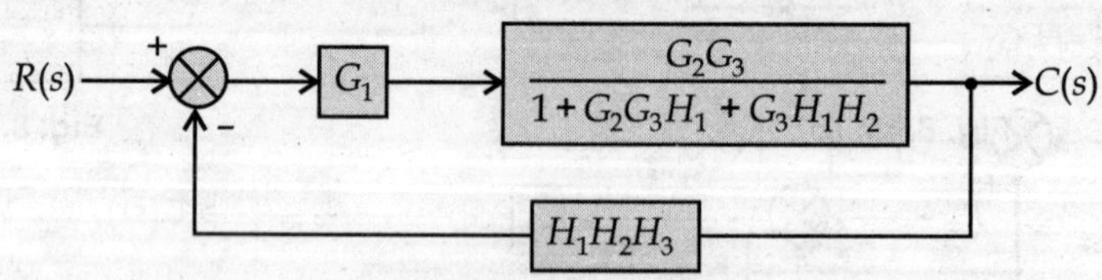

Fig. 8.50. (d)

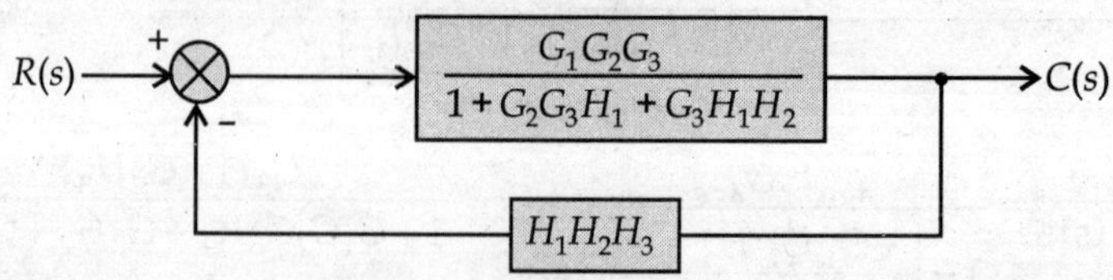

Fig. 8.50. (e)

$$\therefore \quad \frac{C(s)}{R(s)} = \frac{\dfrac{G_1\,G_2\,G_3}{1+G_3\,H_1H_2+G_2\,G_3\,H_1}}{1+\dfrac{G_1\,G_2G_3H_1H_2H_3}{1+G_3H_1H_2+G_2G_3H_1}} = \frac{G_1\,G_2\,G_3}{1+G_3H_1H_2+G_2\,G_3H_1+G_1G_2G_3H_1H_2H_3}$$

EXAMPLE 8.23. Using block reduction technique, find the closed loop transfer function.

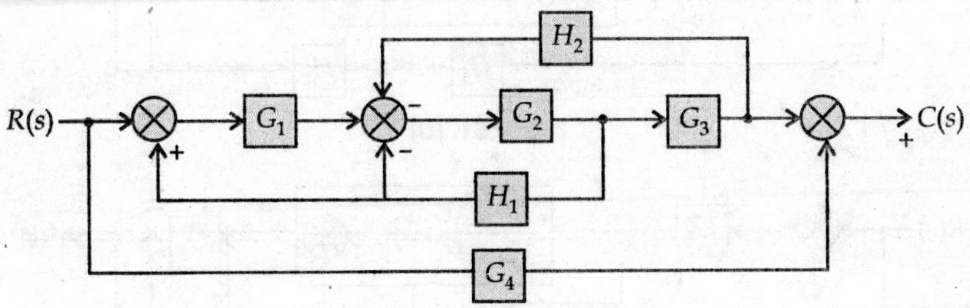

Fig. 8.51.

Solution : Redraw the diagram

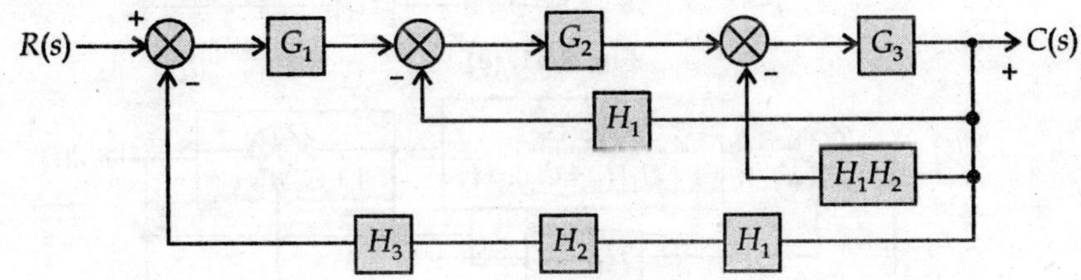

Fig. 8.51. (a)

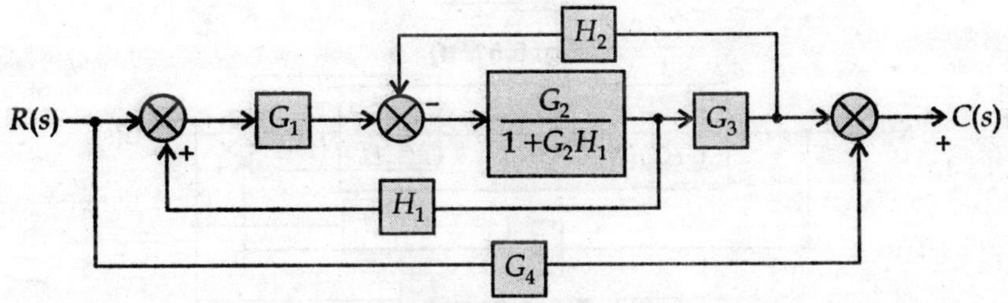

Fig. 8.51. (b)

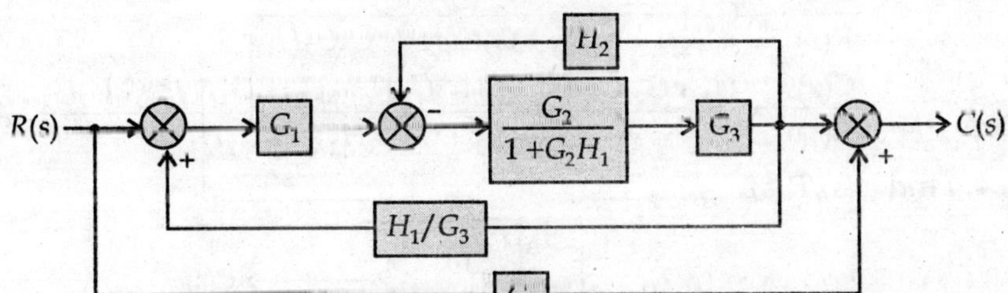

Fig. 8.51. (c)

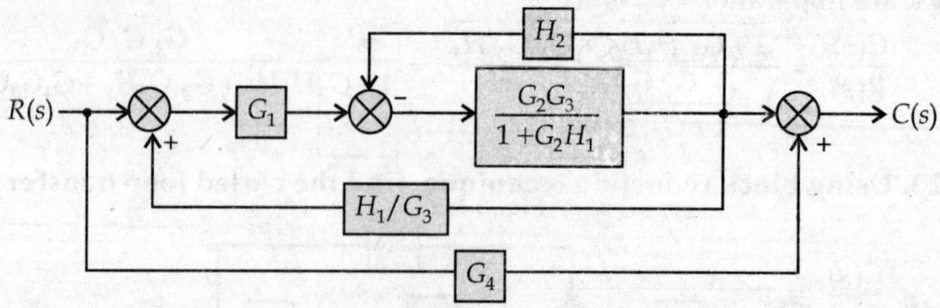

Fig. 8.51. (d)

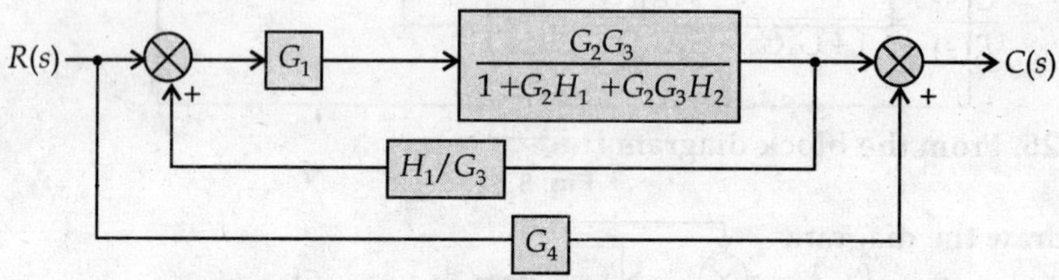

Fig. 8.51. (e)

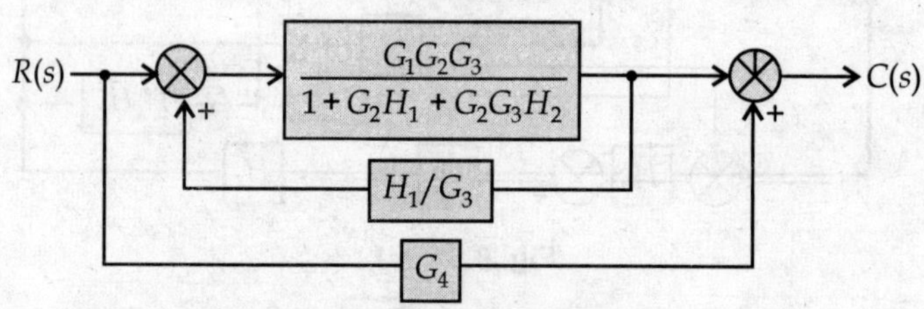

Fig. 8.51. (f)

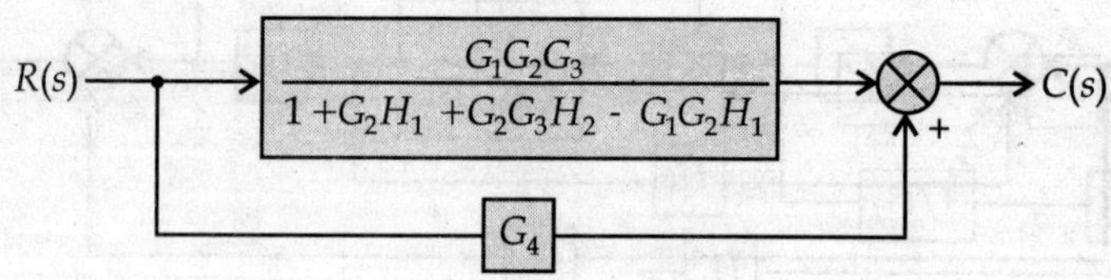

Fig. 8.51. (g)

$$\frac{C(s)}{R(s)} = G_4 + \frac{G_1\,G_2\,G_3}{1+G_2H_1+G_2G_3H_2-G_1G_2H_1}$$

$$\frac{C(s)}{R(s)} = \frac{G_4+G_2\,G_4\,H_1+G_2G_3G_4H_2-G_1G_2G_4H_1+G_1G_2G_3}{1+G_2H_1+G_2G_3H_2-G_1G_2H_1}$$ **Ans.**

EXAMPLE 8.24. Find $c(s)/R(s)$.

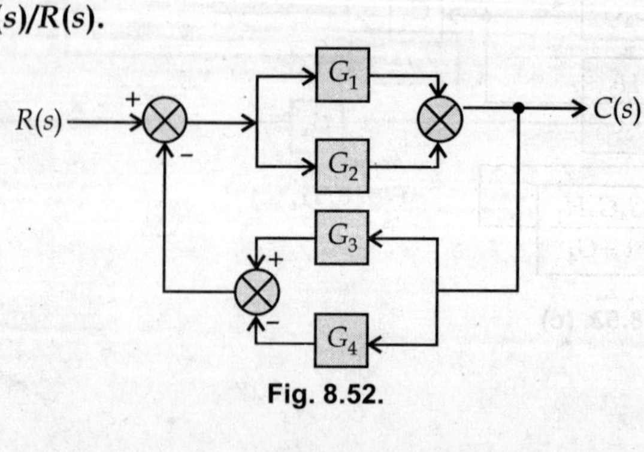

Fig. 8.52.

Solution : Block are in parallel

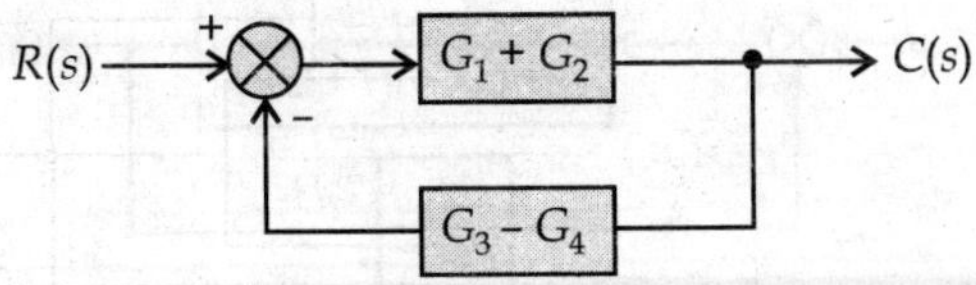

Fig. 8.52. (a)

$$\frac{C(s)}{R(s)} = \frac{G_1 + G_2}{1+(G_1 + G_2)(G_3 - G_4)}$$

$$\frac{C(s)}{R(s)} = \frac{G_1 + G_2}{1+G_1G_3 - G_1G_4 + G_2G_3 - G_2G_4}$$

EXAMPLE 8.25. From the block diagram find $\frac{C_1}{R_2}$ assuming $R_1 = 0$

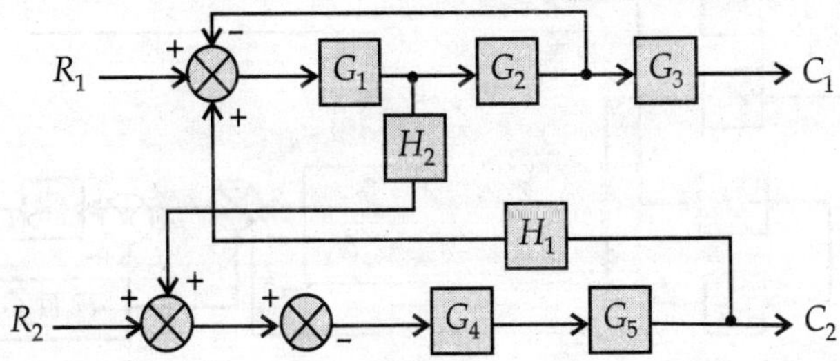

Fig. 8.53.

Solution : When $R_1 = 0$

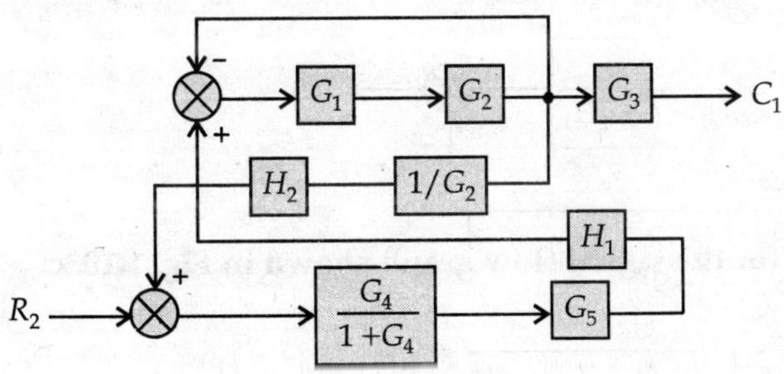

Fig. 8.53. (a)

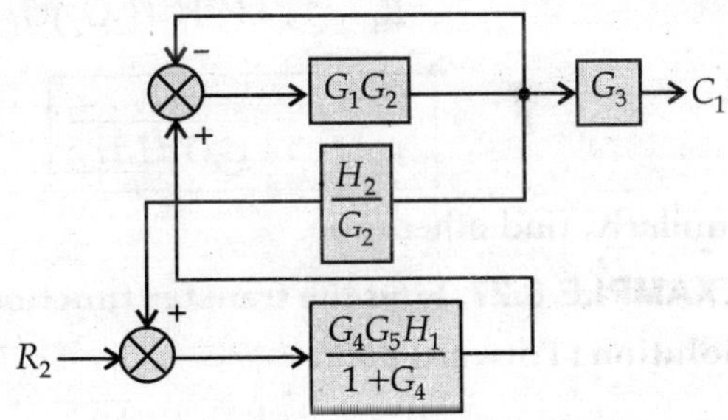

Fig. 8.53. (b)

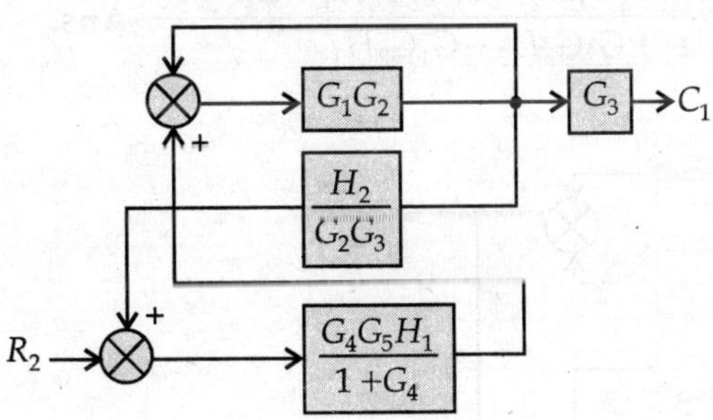

Fig. 8.53. (c)

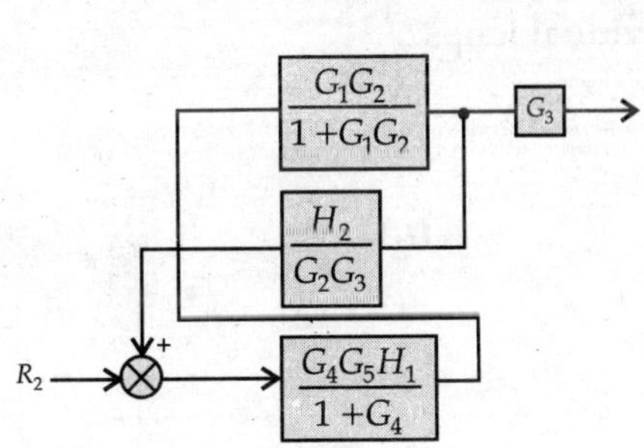

Fig. 8.5. (d)

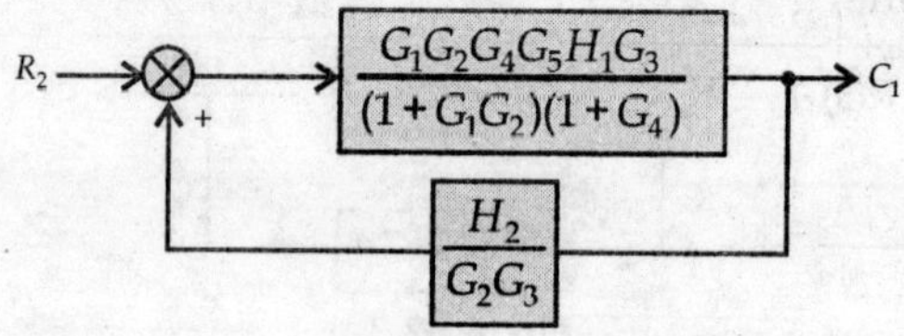

Fig. 8.53 (e)

$$\frac{C_1(s)}{R_2(s)} = \frac{G_1\, G_2\, G_3\, G_4\, G_5\, H_1}{(1+G_1\, G_2)\,(1+G_4) - G_1G_4G_5H_1H_2}$$

$$\frac{c_1(s)}{R_2(s)} = \frac{G_1G_2G_3G_4G_5H_1}{1+G_1G_2+G_4+G_1G_2G_4-G_1G_4G_5H_1H_2} \qquad \textbf{Ans.}$$

EXAMPLE 8.26. Alternative method to determine C_1/R_1 and other ratio for example 1.34.

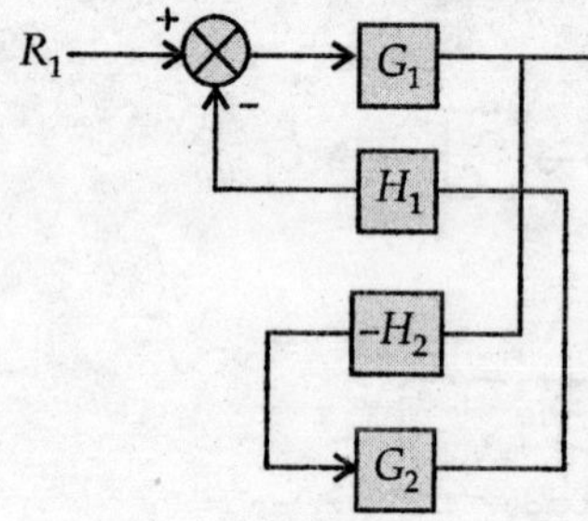

Fig. 8.54.

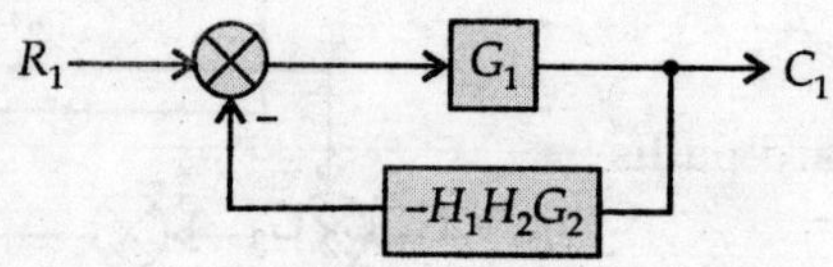

Fig. 8.54 (a)

$$\frac{C_1}{R_1} = \frac{G_1}{1+(-H_1H_2G_2)G_1}$$

$$\boxed{\frac{C_1}{R_1} = \frac{G_1}{1-G_1G_2H_1H_2}}$$

similarly, find other ratio.

EXAMPLE 8.27. Find the transfer function for the signal flow graph shown in Fig. 1.103c.

Solution : Forward path:

$$g_1 = -\frac{1}{s}\cdot\frac{1}{s}\cdot\left(-\frac{1}{s}\right) = \frac{1}{s^3} \qquad \Delta_1 = 1$$

Individual loops :

$$L_1 = -\frac{3}{s}$$

$$L_2 = \frac{-5}{s^2}$$

$$L_3 = \frac{-2}{s^2}$$

$$\Delta = 1 - [L_1 + L_2 + L_3] = 1+\left[\frac{3}{s}+\frac{5}{s^2}+\frac{2}{s^3}\right]$$

$$\Delta = 1+\left[\frac{3s^2+5s+2}{s^3}\right] = \frac{s^3+3s^2+5s+2}{s^3}$$

$$\frac{C(s)}{R(s)} = \frac{g_1\Delta_1}{\Delta} = \frac{\frac{1}{s^3}}{\frac{s^3+3s^2+5s+2}{s^3}} = \frac{1}{s^3+3s^2+5s+2} \quad \textbf{Ans.}$$

EXAMPLE 8.28. For example 8.7 draw the signal flow graph and find the transfer function.

Solution :

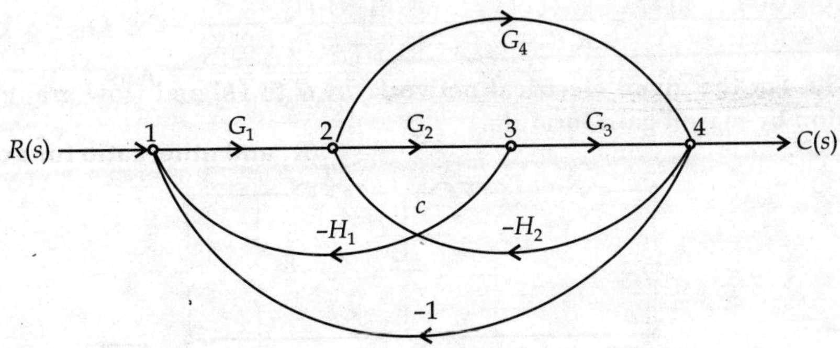

Fig. 8.55.

Forward paths :

$$g_1 = G_1\,G_2\,G_3 \qquad \Delta_1 = 1$$
$$g_2 = G_1\,G_4 \qquad \Delta_2 = 1$$

Individual loops :

$$L_1 = -\,G_1\,G_2\,H_1 \quad ; \quad L_2 = -\,G_2\,G_3\,H_2$$
$$L_3 = -\,G_1\,G_2\,G_3 \quad ; \quad L_4 = -\,G_4\,H_2$$
$$L_5 = -\,G_1\,G_4$$

Two non-touching loops : None

$$\Delta = 1 - [-\,G_1\,G_2\,H_1 - G_2\,G_3\,H_2 - G_1\,G_2\,G_3 - G_4\,H_2 - G_1\,G_4]$$
$$= 1 + G_1\,G_2\,H_1 + G_2\,G_3\,H_2 + G_1\,G_2\,G_3 + G_4\,H_2 + G_1\,G_4$$

$$\frac{C(s)}{R(s)} = \frac{g_1\Delta_1+g_2\Delta_2}{\Delta} = \frac{G_1G_2G_3+G_1G_4}{1+G_1G_2H_1+G_2G_3H_2+G_1G_2G_3+G_4H_2+G_1G_4}$$

EXAMPLE 8.29. Find the transfer function of the electrical circuit whose SFG is shown in Fig. 8.26.

Solution : Forward path :

$$g_1 = \frac{1}{R_1}\cdot R_3\cdot\frac{1}{R_2}\cdot R_4 \qquad \Delta_1 = 1$$

Individual loops :

$$L_1 = -\frac{R_3}{R_1}$$

$$L_2 = -\frac{R_3}{R_2}$$

$$L_3 = -\frac{R_4}{R_2}$$

Two nontouching loops : $L_1 L_3$

$$L_1 L_3 = \frac{R_3}{R_1} \cdot \frac{R_4}{R_2}$$

$$\Delta = 1 - [L_1 + L_2 + L_3] + [L_1 L_3] = 1 + \left[\frac{R_3}{R_1} + \frac{R_3}{R_2} + \frac{R_4}{R_2}\right] + \left[\frac{R_3 R_4}{R_1 R_2}\right]$$

$$= 1 + \frac{R_2R_3 + R_3R_1 + R_1R_4 + R_3R_4}{R_1R_2} = \frac{R_1R_2 + R_2R_3 + R_3R_1 + R_1R_4 + R_3R_4}{R_1R_2}$$

$$\frac{C(s)}{R(s)} = \frac{g_1 \Delta_1}{\Delta} = \frac{R_3R_4}{R_1R_2 + R_2R_3 + R_3R_1 + R_1R_4 + R_3R_4} \quad \textbf{Ans.}$$

EXAMPLE 8.30. For the given electrical network draw the signal flow graph and find the transfer function by mason gain formula.

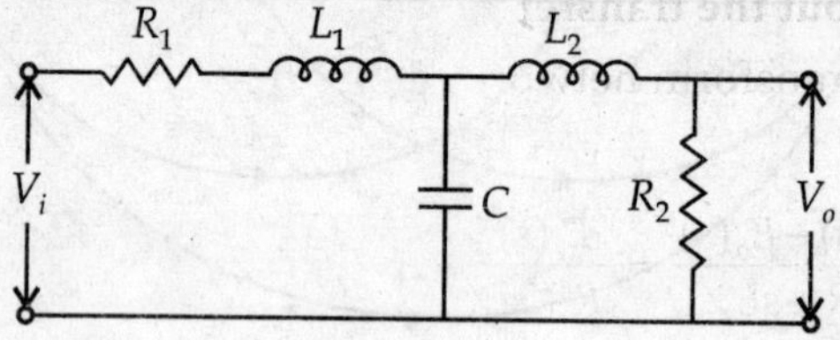

Fig. 8.56.

Solution : Draw the transform network

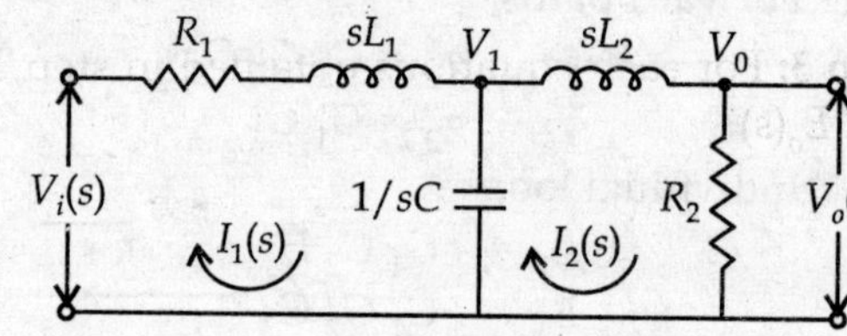

Fig. 8.56. (a)

$$I_1(s) = \frac{V_i - V_1}{R_1 + sL_1}$$

$$V_1(s) = \frac{1}{sC}[I_1(s) - I_2(s)]$$

$$I_2(s) = \frac{V_1(s) - V_o(s)}{sL_2}$$

$$V_0(s) = R_2 I_2(s)$$

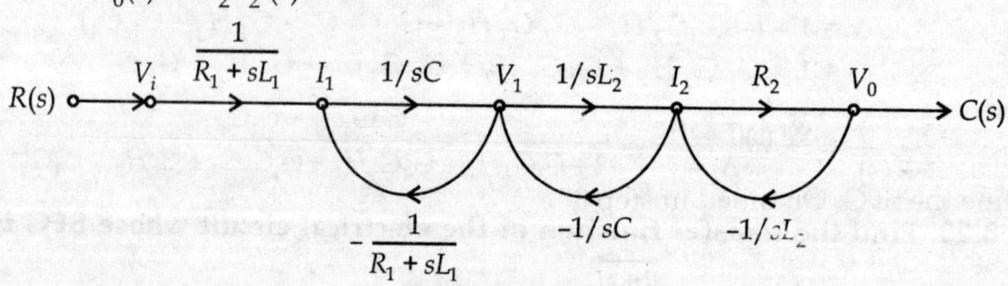

Fig. 8.56. (b)

Forward path :

$$g_1 = \frac{1}{R_1 + sL_1} \cdot \frac{1}{sC} \cdot \frac{1}{sL_2} \cdot R_2 = \frac{R_2}{s^2 L_2 C(R_1 + sL_1)}, \qquad \Delta_1 = 1 - 0 = 1$$

Individual loops :

$$L_1 = -\frac{1}{sC(R_1 + sL_1)}$$

$$L_2 = -\frac{1}{s^2 L_2 C}$$

$$L_3 = -\frac{R_2}{sL_2}$$

Two nontouching loops : $L_1\, L_3$

$$L_1\, L_3 = \frac{R_2}{s^2 L_2 C(R_1 + sL_1)}$$

$$\Delta = 1 - [L_1 + L_2 + L_3] + L_1\, L_3$$

$$\Delta = 1 + \frac{1}{sC(R_1 + sL_1)} + \frac{1}{s^2 L_2 C} + \frac{R_2}{sL_2} + \frac{R_2}{s^2 L_2 C(R_1 + sL_1)}$$

$$\frac{V_o}{V_i} = \frac{g_1 \Delta_1}{\Delta} = \frac{R_2}{s^2 L_2 C(R_1 + sL_1) + sL_2 + R_1 + sL_1 + sC(R_1 + sL_1) + R_2} \quad \textbf{Ans.}$$

EXAMPLE 8.31. Draw the signal flow graph of series RLC circuit shown in Fig. 1.10 and find out the transfer function.

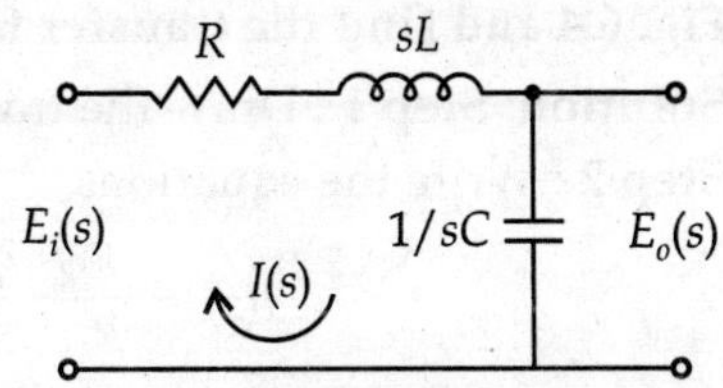

Fig. 8.57.

Solution : Step 1: Draw the transform network. (Fig. 8.57)

Step 2 : Write the equation

$$I\,(s) = \frac{E_i(s) - E_o(s)}{R + sL} = \frac{E_i(s)}{R + sL} - \frac{E_o(s)}{R + sL}$$

$$E_o(s) = I(s) \cdot \frac{1}{sC}$$

Step 3: For each equations obtained in step 2, draw SFG. Select the node variables as $I(s)$, $E_i(s)$ and $E_o(s)$.

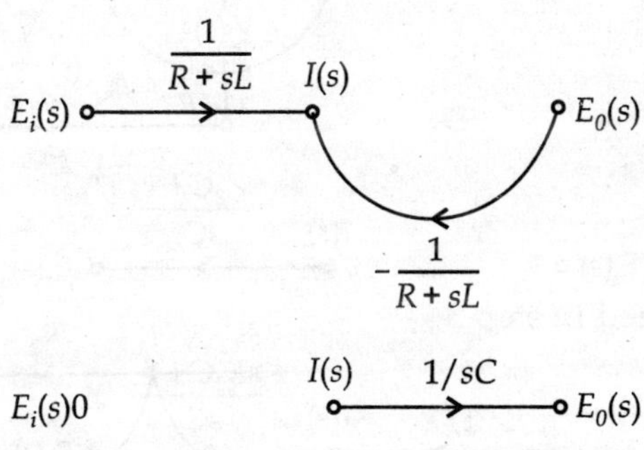

Step 4: Combine the SFG. Obtained in step 3

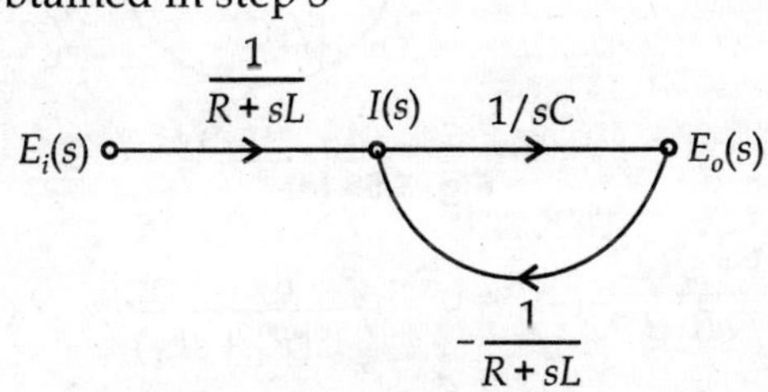

Fig. 8.57 (a)

Step 5 : Forward path

$$g_1 = \frac{1}{R + sL} \cdot \frac{1}{sC} \qquad \Delta_1 = 1$$

Individual loop $L_1 = -\dfrac{1}{(R+sL)(sC)}$

Non touching loops: None

$$\Delta = 1 - L_1 = 1 + \frac{1}{(R+sL)(sC)} = \frac{(R+sL)(sC)+1}{(R+sL)(sC)}$$

$$\frac{C(s)}{R(s)} = \frac{g_1\Delta_1}{\Delta} = \frac{\dfrac{1}{sC(R+sL)}}{\dfrac{1+(sC)(R+sL)}{sC(R+sL)}} = \frac{1}{1+RCs+s^2L_C} = \frac{1}{s^2LC+sRC+1} \quad \textbf{Ans.}$$

EXAMPLE 8.32. Draw the SFG of lag network shown in Fig. 6.4 and find the transfer function.

Solution: Step 1 : Draw the transform network (See Fig. 8.58).

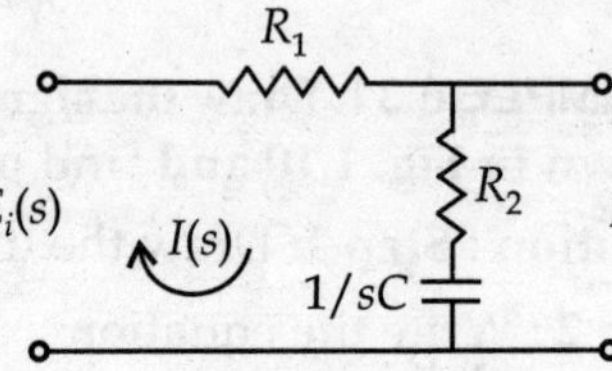

Fig. 8.58.

Step 2 : Write the equations

$$I(s) = \frac{E_i}{R_1} - \frac{E_o}{R_1}$$

$$E_o(s) = I(s)\left[R_2 + \frac{1}{sC}\right]$$

Step 3 : Draw SFG for each equation

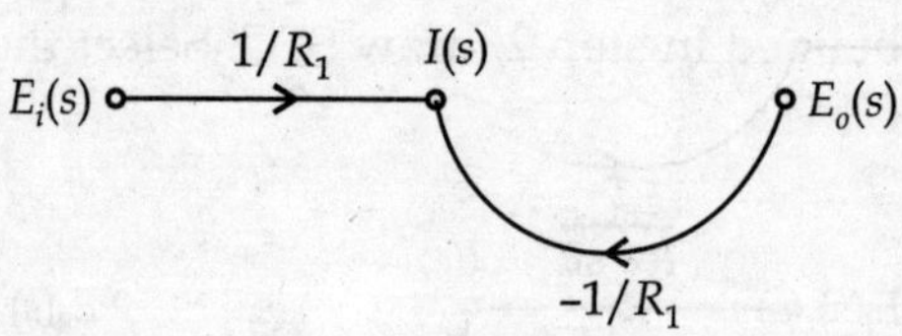

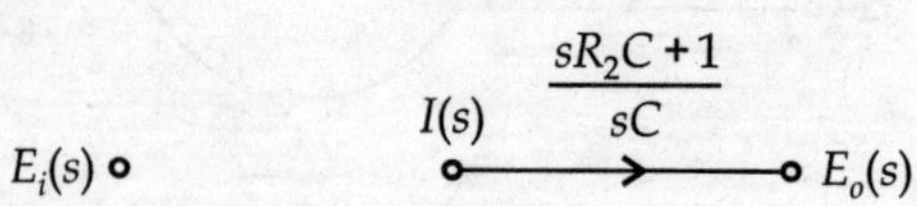

Step 4 : Combine SFG obtained in step 3

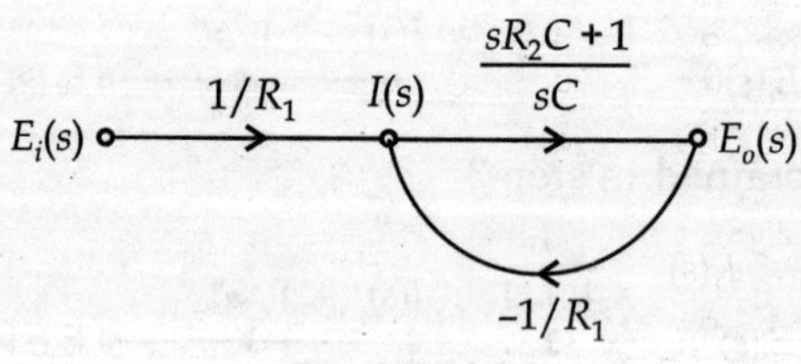

Fig. 8.58 (a)

Step 5 : Forward path $g_1 = \dfrac{1+sR_2C}{sR_1C} \qquad \Delta_1 = 0$

Individual loop $\qquad L_1 = -\dfrac{1+sR_2C}{sR_1C}$

$$\Delta = 1 - L_1 = 1 + \frac{1+sR_2C}{sR_1C}$$

$$\frac{E_o(s)}{E_i(s)} = \frac{C(s)}{R(s)} = \frac{g_1\Delta_1}{\Delta} = \frac{1+sR_2C}{1+sC(R_1+R_2)} \quad \textbf{Ans.}$$

EXAMPLE 8.33. For the given circuit draw SFG and find out the transfer function.

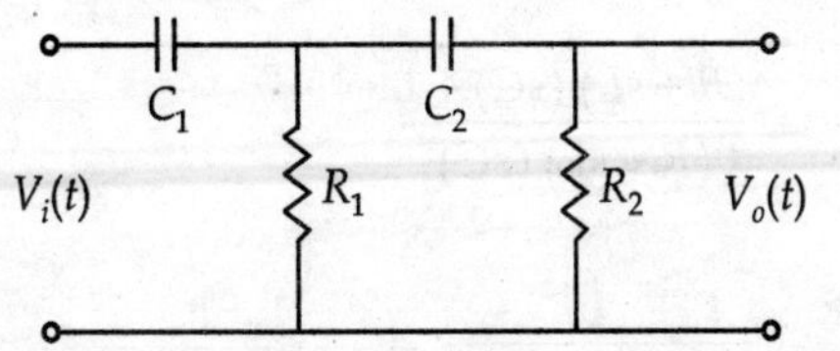

Fig. 8.59.

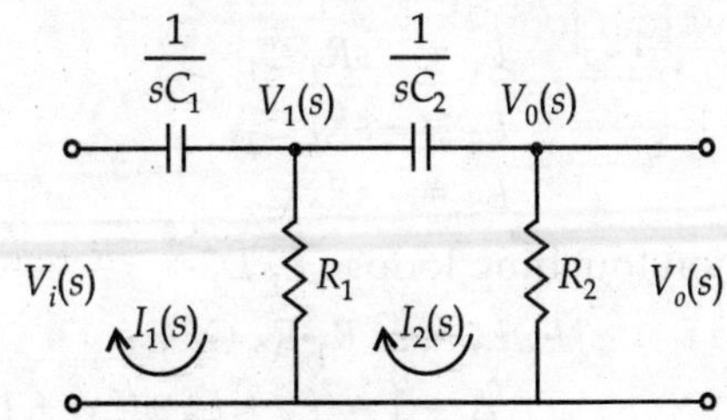

Fig. 8.60.

Solution: Draw the transform network (Fig. 8.60)

$$I_1(s) = \frac{V_i(s)}{\frac{1}{sC_1}} - \frac{V_1(s)}{\frac{1}{sC_1}} = sC_1 \, V_i(s) - sC_1 \, V_1(s)$$

$$V_1(s) = [I_1(s) - I_2(s)]$$

$$R_1 = I_1(s) \, R_1 - I_2 \, (s) \, R_1$$

$$I_2(s) = sC_2 \, V_1(s) - sC_2 V_o(s)$$

$$V_o(s) = I_2(s) \, R_2$$

Signal flow graph for each equation

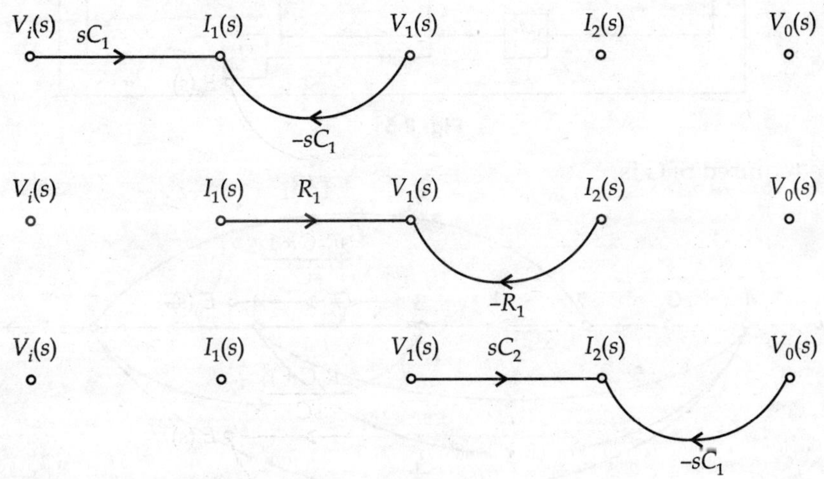

Combine all the SFG

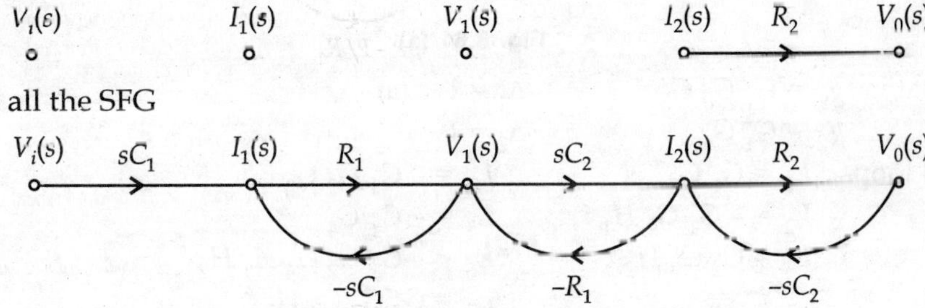

Fig 8.60 (a)

Forward path $\quad g_1 = sC_1 \, R_1 \cdot sC_2 \cdot R_2$

$$g_1 = s^2 \, R_1 R_2 \, C_1 \, C_2 \qquad \Delta_1 = 1$$

Individual loops $L_1 = -R_1 \cdot sC_1$

$$L_1 = -sR_1 C_1$$

$$L_2 = -sR_1 C_2$$

$$L_3 = -sR_2 C_2$$

Two non touching loops : $L_1 L_3$

$$L_1 L_3 = s^2 R_1 R_2 C_1 C_2$$

$$\Delta = 1 - (L_1 + L_2 + L_3) + L_1 L_3$$

$$\Delta = 1 + sR_1 C_1 + sR_1 C_2 + s R_2 C_2 + s^2 R_1 R_2 C_1 C_2$$

$$\frac{V_o(s)}{V_i(s)} = \frac{g_1 \Delta_1}{\Delta} = \frac{s^2 R_1 R_2 C_1 C_2}{1 + s^2 R_1 R_2 C_1 C_2 + s(R_1 C_1 + R_1 C_2 + R_2 C_2)}$$

EXAMPLE 8.34. For the given block diagram draw the SFG and find the ratio of $\frac{C(s)}{R(s)}$.

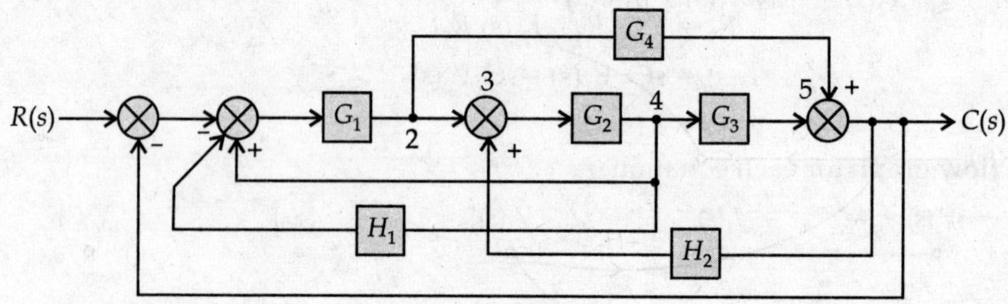

Fig. 8.61.

Solution : The required SFG is

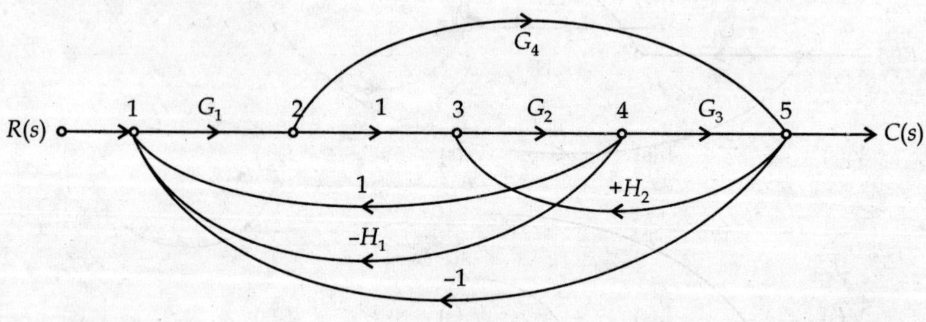

Fig. 8.61 (a)

Forward paths $g_1 = G_1 G_2 G_3$ $\quad \Delta_1 = 1$

$g_2 = G_1 G_4$ $\quad \Delta_2 = 1$

Individual loops : $L_1 = G_1 G_2$ $\quad L_4 = -G_1 G_2 G_3$

$L_2 = -G_1 G_2 H_1$ $\quad L_5 = -G_1 G_4$

$L_3 = G_2 G_3 H_2$ $\quad *L_6 = -G_1 G_2 G_4 H_1 H_2$

$*L_7 = G_1 G_2 G_4 H_2$

Two non touching loops : None

$$\Delta = 1 - (L_1 + L_2 + L_3 + L_4 + L_5 + L_6 + L_7)$$

$$= 1 - G_1 G_2 + G_1 G_2 H_1 - G_2 G_3 H_2 + G_1 G_2 G_3$$

$$+ G_1 G_4 + G_1 G_2 H_1 H_2 - G_1 G_2 G_4 H_1$$

$$\frac{C(s)}{R(s)} = \frac{g_1\Delta_1 + g_2\Delta_2}{\Delta} = \frac{G_1G_2G_3 + G_1G_4}{1 - G_1G_2 + G_1G_2H_1 - G_2G_3H_2 + G_1G_2G_3 + G_1G_4 + G_1G_2G_4H_1H_2 - G_1G_2G_4H_2}$$

EXAMPLE 8.35. For the system represented by the following equations, find the transfer function by SFG.

$$x = x_1 + \beta_3 u$$
$$\dot{x}_1 = -a_1 x_1 + x_2 + \beta_2 u$$
$$\dot{x}_2 = -a_2 x_1 + \beta_1 u$$

Solution : Let x = output

u = input

Take Laplace transform of above equations

$$X(s) = X_1(s) + \beta_3 u(s)$$
$$sX_1(s) = -a_1 X_1(s) + X_2(s) + \beta_2 u(s)$$
$$sX_2(s) = -a_2 X_1(s) + \beta_1 u(s)$$

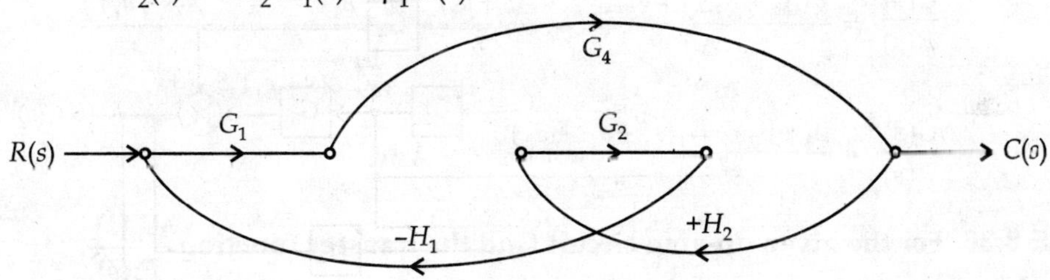

* L_6 :

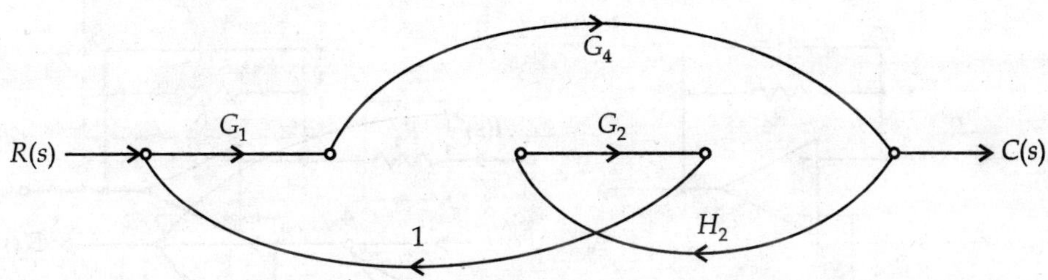

** L_7 :

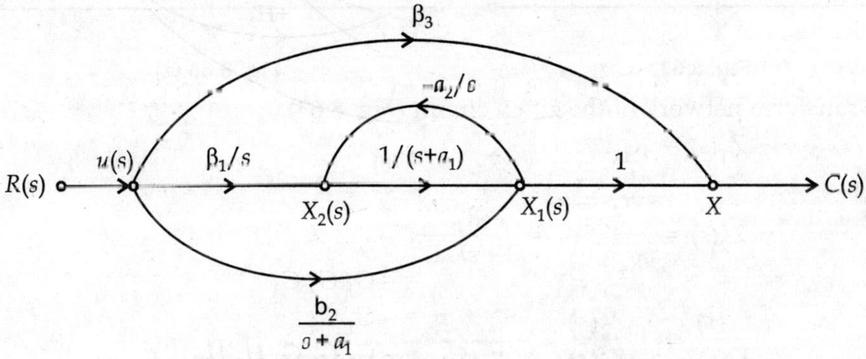

Fig. 8.62

or, $$X(s) = X_1(s) + \beta_3 u(s)$$

$$X_1(s) = \frac{1}{s+a_1} X_2(s) + \frac{\beta_1}{s+a_1} u(s)$$

$$X_2(s) = \frac{-a_2}{s} X_1(s) + \frac{\beta_1}{s} u(s)$$

Forward paths : $g_1 = \frac{\beta_1}{s} \cdot \frac{1}{s+a_1}$ $\qquad \Delta_1 = 1$

$$g_2 = \frac{\beta_2}{s+a_1} \qquad \Delta_2 = 1$$

$$g_3 = \beta_3 \qquad \Delta_3 = 1 + \frac{a_2}{s(s+a_1)}$$

Individual loops : $L_1 = -\frac{a_2}{s(s+a)}$ $\qquad \because \quad E_i(s) = Z_1(s)\, I(s)$, $\quad E_o(s) = -Z_2\, I(s)$

$$\Delta = 1 - (L_1) = 1 + \frac{a_2}{s(s+a)}$$

$$\frac{C(s)}{R(s)} = \frac{g_1\Delta_1 + g_2\Delta_2 + g_3\Delta_3}{\Delta} = \frac{\frac{\beta_1}{s(s+a_1)} + \frac{\beta_2}{(s+a_1)} + \beta_3\left(1 + \frac{a_2}{s(s+a_1)}\right)}{1 + \frac{a_2}{s(s+a_1)}}$$

$$\frac{C(s)}{R(s)} = \frac{\beta_1 + \beta_2 s + \beta_3(s^2 + a_1 s + a_2)}{s^2 + a_1 s + a_2} \quad \textbf{Ans.}$$

EXAMPLE 8.36. For the given op-amp circuit find the transfer function $\frac{E_o(s)}{E_i(s)}$.

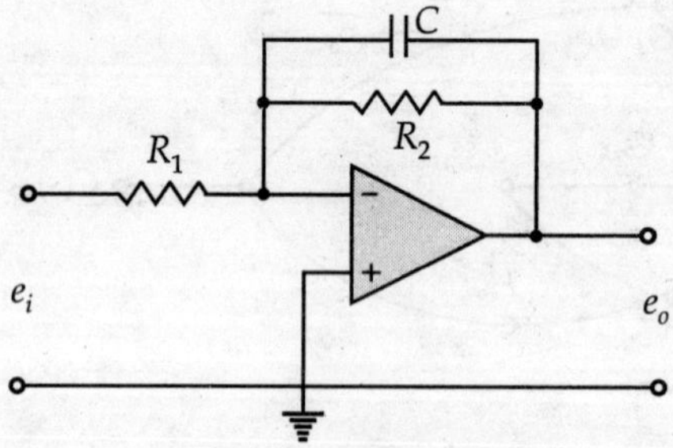

Fig. 8.63.

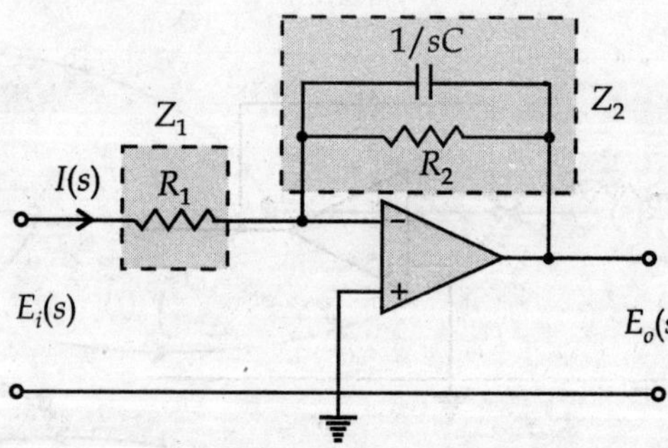

Fig. 8.63 (a)

Solution : Transform network of the given circuit (Fig. 8.63).

$$Z_1(s) = R_1$$

$$Z_2(s) = \frac{R_2 \cdot \frac{1}{sC}}{R_2 + \frac{1}{sC}} = \frac{R_2}{1 + sR_2C}$$

$$\frac{E_o(s)}{E_i(s)} = -\frac{Z_2(s)}{Z_1(s)} = -\frac{R_2}{R_1(1 + sR_2C)}$$

This is the required transfer function.

EXAMPLE 8.37. Find the transfer function of inverting amplifier as shown in Fig. 8.64.

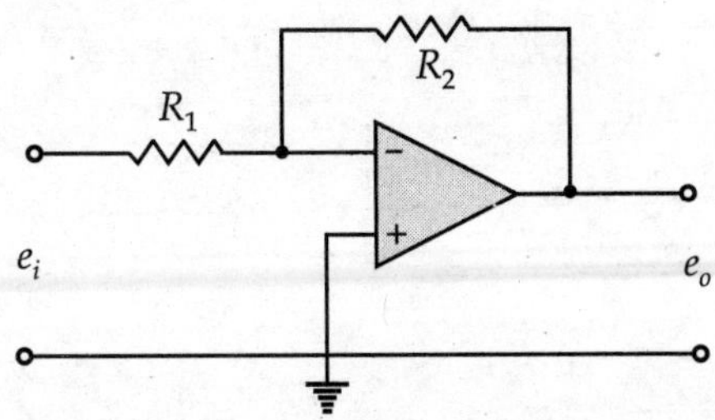

Fig. 8.64.

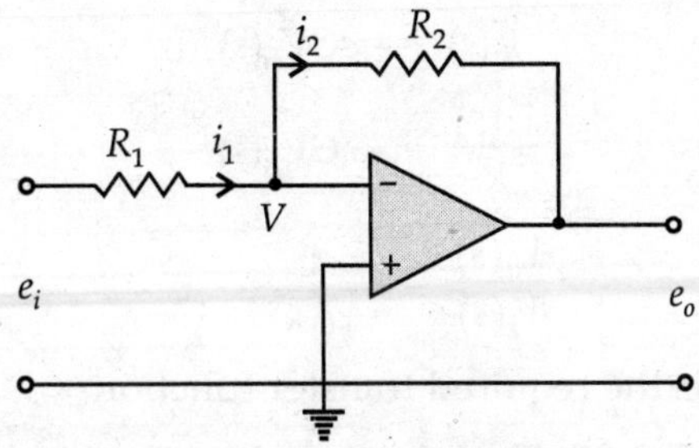

Fig. 8.64. (a)

Solution : $i_1 = \frac{e_i - V}{R_1}$ and $i_2 = \frac{V - e_o}{R_2}$

Since $V = 0 \quad \therefore i_1 = i_2$

$$\frac{e_i}{R_1} = -\frac{e_o}{R_2}$$

Laplace transform $\frac{E_i(s)}{R_1} = -\frac{E_o(s)}{R_2}$

$\frac{E_o(s)}{E_i(s)} = -\frac{R_2}{R_1}$ Required transfer function

EXAMPLE 8.38. Determine the transfer function of given circuit using op-amp as shown in Fig. 8.65.

Solution :

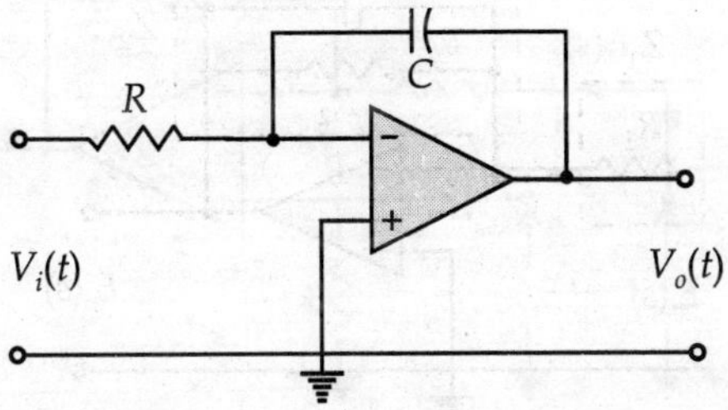

Fig. 8.65

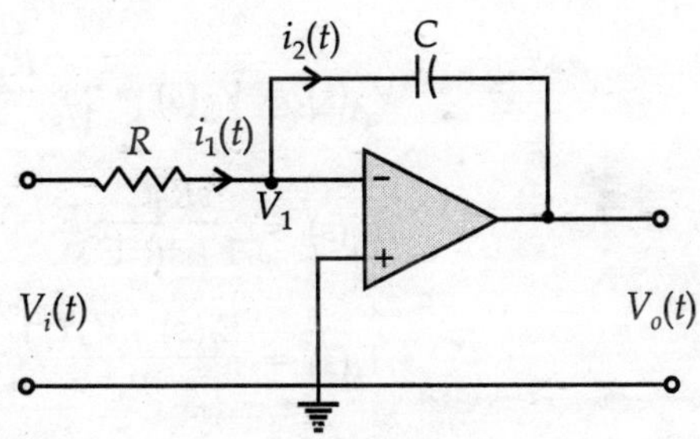

Fig. 8.65 (a)

$$i_1(t) = \frac{V_i(t) - V_1(t)}{R}$$

$$i_2(t) = C\frac{d}{dt}(V_1(t) - V_o(t))$$

Laplace transform of above equations

$$I_1(s) = \frac{1}{R}[V_i(s) - V_1(s)]$$

$$I_2(s) = sC[V_1(s) - V_o(s)]$$

Since, $V_1(s) \approx 0 \quad \therefore \quad I_1(s) = I_2(s)$

$$I_1(s) = \frac{V_i(s)}{R}$$

$$I_2(s) = -sCV_o(s)$$

$$\therefore \quad \frac{V_i(s)}{R} = -sCV_o(s)$$

$$\therefore \quad \frac{V_o(s)}{V_i(s)} = -\frac{1}{sCR}$$

This is the required transfer function

EXAMPLE 8.39. Find the transfer function of the given circuit shown in Fig. 8.66.

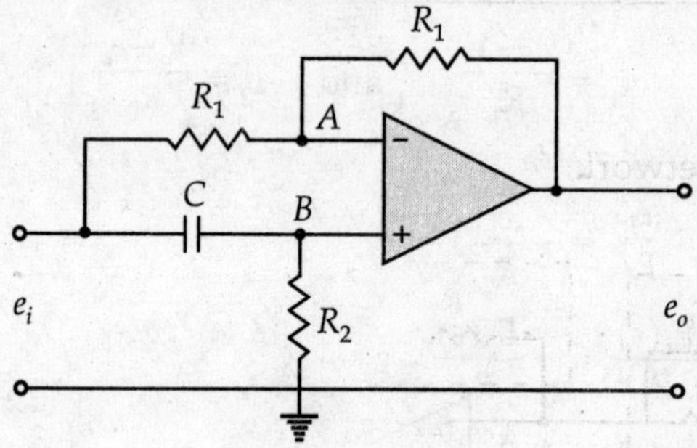

Fig. 8.66.

Solution : Redraw the given circuit as transform network.

$$V_B(s) = \frac{R_2}{R_2 + \frac{1}{sC}} \cdot E_i(s)$$

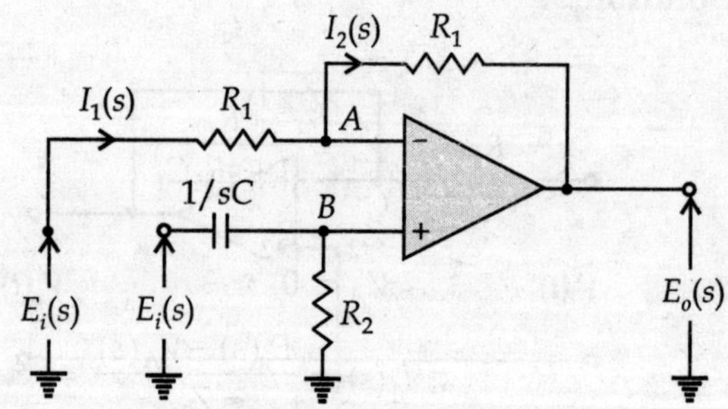

Fig. 8.66 (a)

$$V_A(s) = V_B(s) = \frac{R_2}{1/sC + R_2}$$

$$E_i(s) = \frac{sR_2C}{1 + sR_2C} E_i(s)$$

$$I_1(s) = \frac{E_i(s) - V_A(s)}{R_1}$$

$$I_2(s) = \frac{V_A(s) - E_o(s)}{R_1}$$

Since $\quad I_1(s) = I_2(s)$

$$\frac{E_i(s) - V_A(s)}{R_1} = \frac{V_A(s) - E_0(s)}{R_1}$$

or, $\quad 2V_A(s) = E_i(s) + E_o(s)$

or, $\quad E_o(s) = 2V_A(s) - E_i(s)$

$$E_o(s) = 2\left[\frac{sR_2C}{1 + sR_2C} E_i(s)\right] - E_i(s)$$

$$\frac{E_o(s)}{E_i(s)} = \frac{sCR_2 - 1}{sR_2 + 1}$$

This is the required transfer function. **Ans.**

EXAMPLE 8.40. Find the transfer function of the circuit shown in Fig. 8.67.

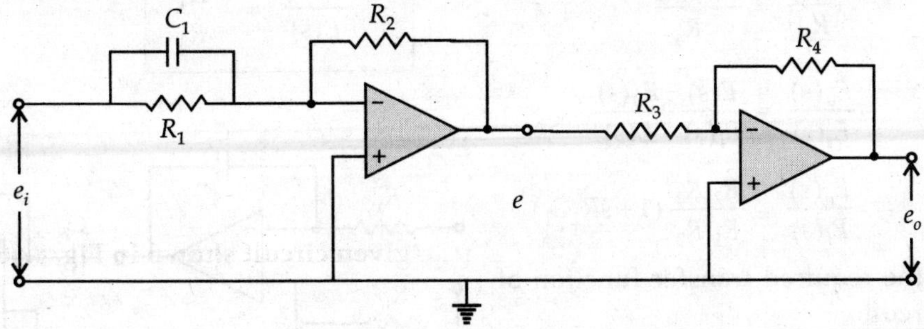

Fig. 8.67.

Solution : Transform the network

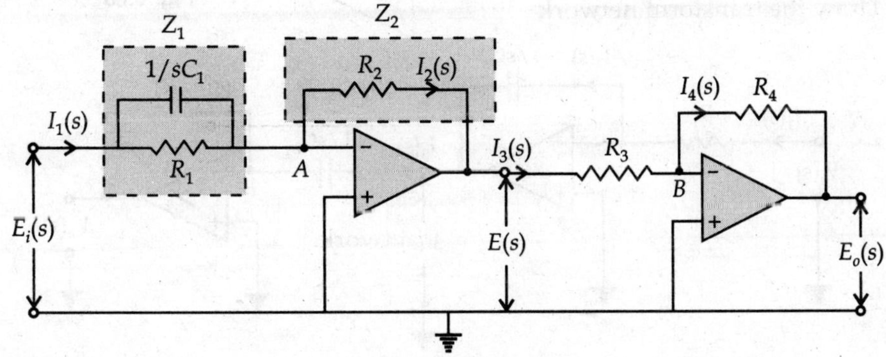

Fig 8.67 (a)

$$Z_1 = \frac{R_1}{1+sR_1C_1}$$

$$Z_2 = R_2$$

$$V_A = 0$$

$$I_1(s) = \frac{E_i(s) - V_A(s)}{Z_1} = \frac{1+sR_1C_1}{R_1} E_i(s)$$

$$I_2(s) = -\frac{E(s)}{R_2}$$

since, $I_1(s) = I_2(s)$

$$E(s) = -\frac{R_2(1+sR_1C_1)}{R_1} E_i(s) \quad \text{or,} \quad \boxed{\frac{E(s)}{E_i(s)} = -\frac{R_2(1+sR_1C_1)}{R_1}}$$

Also, $V_B = 0$

$$I_3(s) = \frac{E(s)}{R_3}$$

and $I_4 = -\dfrac{E_o(s)}{R_4}$

$$I_3(s) = I_4(s)$$

$$\frac{E(s)}{R_3} = -\frac{E_o(s)}{R_4}$$

$$\therefore \quad \frac{E_0(s)}{E(s)} = -\frac{R_4}{R_3}$$

$$\frac{E_0(s)}{E_i(s)} = \frac{E(s)}{E_i(s)} \cdot \frac{E_0(s)}{E(s)}$$

$$\therefore \quad \frac{E_o(s)}{E_i(s)} = \frac{R_2\, R_4}{R_1\, R_3}(1+sR_1C_1)$$

This is the required transfer function of the given circuit.

EXAMPLE 8.41. Find the transfer function of the given circuit shown in Fig. 8.68.

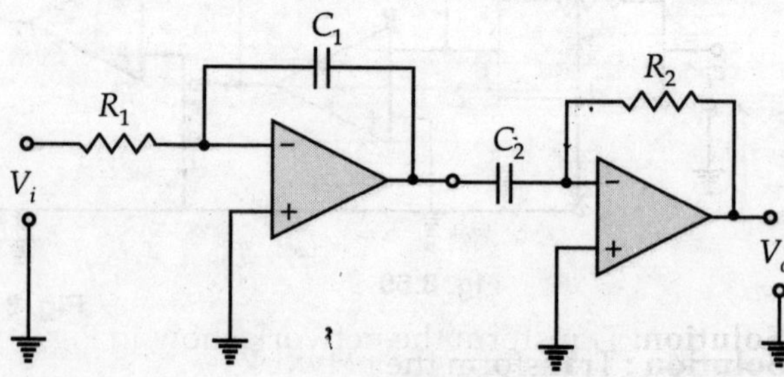

Fig. 8.68

Solution : Draw the transform network

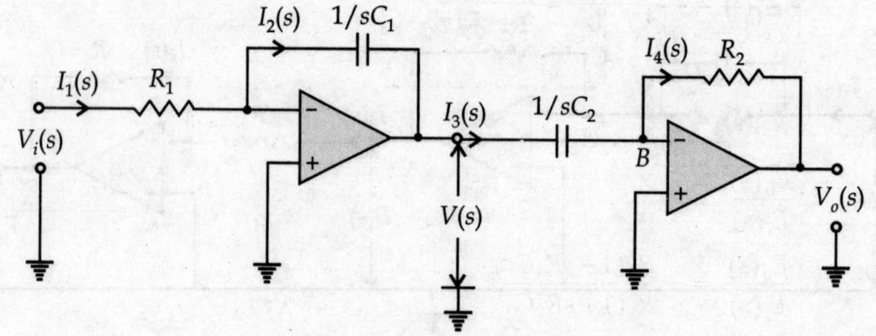

Fig. 8.68 (a)

$$I_1(s) = \frac{V_i(s)}{R_1}$$

$$I_2(s) = \frac{-V(s)}{1/sC_1} = -\,sC_1V\,(s)$$

since, $\quad I_1(s) = I_2(s)$

$$\therefore \quad \frac{V(s)}{V_i(s)} = \frac{-1}{sR_1C_1}$$

$$I_3(s) = \frac{V(s)}{1/sC_2} = sC_2V(s)$$

$$I_4(s) = -\frac{V_o(s)}{R_2}$$

$$I_3(s) = I_4(s)$$

$$\frac{V_o(s)}{V(s)} = -\,sR_2C_2$$

$$\therefore \quad \frac{V_o(s)}{V_i(s)} = \frac{V(s)}{V_i(s)} \cdot \frac{V_o(s)}{V(s)} = \frac{R_2\, C_2}{R_1\, C_1}$$

This the required transfer function.

EXAMPLE 8.42. Find the transfer function of the given circuit shown in Fig. 8.69.

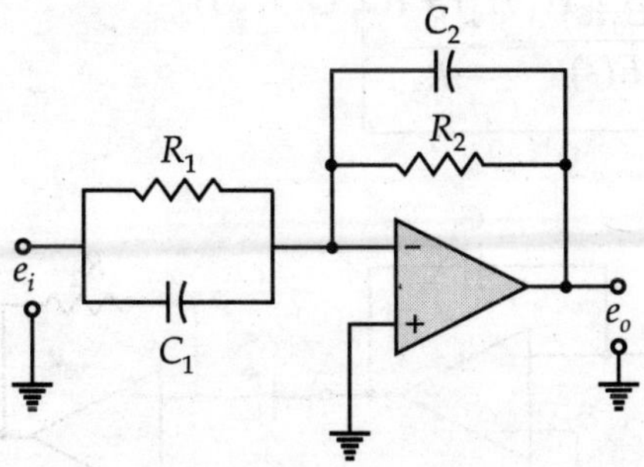

Fig. 8.69.

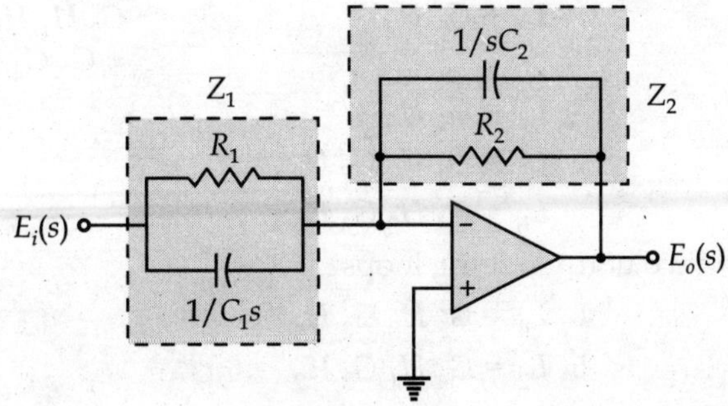

Fig. 8.69 (a)

Solution: Transform the network show in Fig. 8.69 *a*

$$Z_1(s) = \frac{R_1 \cdot \frac{1}{c_1 s}}{R_1 + \frac{1}{sC_1}} = \frac{R_1}{1 + sR_1C_1}$$

$$Z_2(s) = \frac{R_2}{1 + sR_2C_2} \qquad \because \quad E_i(s) = Z_1(s)\, I(s)$$

$$\frac{E_0(s)}{E_i(s)} = -\frac{Z_2}{Z_1} \qquad E_o(s) = -Z_2(s)\, I(s)$$

$$\frac{E_o(s)}{E_i(s)} = -\frac{R_2(1 + sR_1C_1)}{R_1(1 + sR_2C_2)}$$

EXAMPLE 8.43. For the given signal flow graph of a system find the expressions for the outputs C_1 and C_2. Also determine the condition that makes C_1 independent of R_2 and C_2 independent of R_1.

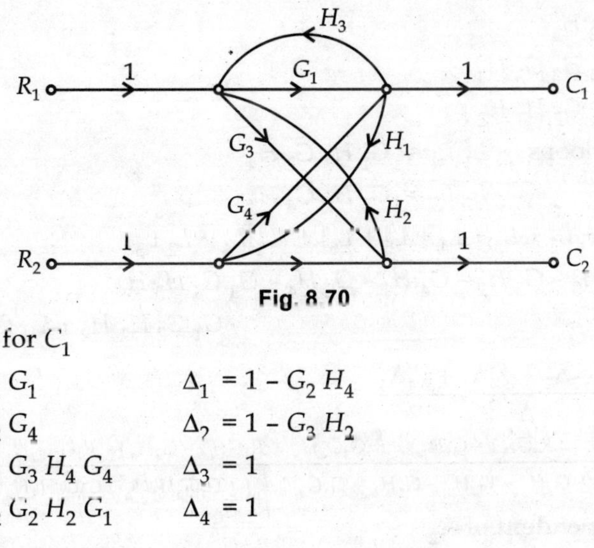

Fig. 8.70

Solution : Expression for C_1

$$g_1 = R_1\, G_1 \qquad \Delta_1 = 1 - G_2\, H_4$$

$$g_2 = R_2\, G_4 \qquad \Delta_2 = 1 - G_3\, H_2$$

$$g_3 = R_1\, G_3\, H_4\, G_4 \qquad \Delta_3 = 1$$

$$g_4 = R_2\, G_2\, H_2\, G_1 \qquad \Delta_4 = 1$$

Individual loops :

$$L_1 = G_1\, H_3 \qquad \Delta = 1 - (L_1 + L_2 + L_3 + L_4 + L_5 + L_6) + (L_1\, L_4 + L_2\, L_3)$$

$$L_2 = G_4\, H_1 \qquad \Delta = 1 - (G_1\, H_3 + G_4\, H_1 + G_3\, H_2 + G_2\, H_4 + G_1$$

$$L_3 = G_3 H_2$$
$$L_4 = G_2 H_4$$
$$L_5 = G_1 H_1 G_2 H_2$$
$$L_6 = G_4 H_3 G_3 H_4$$

$G_2 H_1 H_2 + G_3 G_4 H_3 H_4) + (G_1 G_2 H_3 H_4 + G_3 G_4 H_1 H_2)$

Two non-touching loops :

$$L_1 L_4 = G_1 H_3 G_2 H_4$$
$$L_3 L_2 = G_4 H_1 G_3 H_2$$

$$C_1 = \frac{g_1\Delta_1 + g_2\Delta_2 + g_3\Delta_3 + g_4\Delta_4}{\Delta}$$

$$C_1 = \frac{R_1G_1(1-G_2H_4) + R_2G_4(1-G_3H_2) + R_1G_3G_4H_4 + R_2G_1G_2H_2}{1-G_1H_3 - G_4H_1 - G_3H_2 - G_2H_4 - G_1G_2H_1H_2 - G_3G_4H_3H_4 + G_1G_2H_3H_4 + G_3G_4H_1H_2}$$

Expression for C_2 :

$$g_1 = R_2 G_2 \qquad \Delta_1 = 1 - G_1 H_3$$
$$g_2 = R_2 G_4 H_3 G_3 \qquad \Delta_2 = 1$$
$$g_3 = R_1 G_3 \qquad \Delta_3 = 1 - G_4 H_1$$
$$g_4 = R_1 G_1 G_2 H_1 \qquad \Delta_4 = 1$$

Individual loops :

$$L_1 = G_1 H_3$$
$$L_2 = G_3 H_2$$
$$L_3 = G_4 H_1$$
$$L_4 = G_2 H_4$$
$$L_5 = G_4 H_3 G_3 H_4$$
$$L_6 = G_1 G_2 H_1 H_2$$

Two non-touching loops $L_1 L_4 = G_1 H_3 G_2 H_4$

$$L_2 L_3 = G_3 H_2 G_4 H_1$$

$$\Delta = 1 - (L_1 + L_2 + L_3 + L_4 + L_5 + L_6) + (L_1 L_4 + L_2 L_3)$$
$$\Delta = 1 - G_1 H_3 - G_3 H_2 - G_4 H_1 - G_2 H_4 - G_3 G_4 H_3 H_4 - G_1 G_2 H_1 H_2 + G_1 G_2 H_3 H_4 + G_3 G_4 H_1 H_2$$

$$C_2 = \frac{g_1\Delta_1 + g_2\Delta_2 + g_3\Delta_3 + g_4\Delta_4}{\Delta}$$

$$C_2 = \frac{R_2G_1(1-G_1H_3) + R_2G_3G_4H_3 + R_1G_3(1-G_4H_1) + R_1G_1G_2H_1}{1-G_1H_3 + G_3H_2 + G_4H_1 + G_2H_4 + G_3G_4H_3H_4 + G_1G_2H_1H_2 - G_1G_2H_3H_4 - G_3G_4H_1H_2}$$

Condition C_1 Independent of R_2

From the expression of C_1

$$R_2G_4(1 - G_3H_2) + 2G_1G_2H_2 = 0$$
$$R_2[G_4 - G_3G_4H_2 + G_1G_2H_2] = 0$$

$$G_4 - G_3G_4H_2 + G_1G_2H_2 = 0$$

$$-G_4 = (-G_3G_4 + G_1G_2)H_2$$

$$H_2 = \frac{-G_4}{G_1G_2 - G_3G_4}$$

Similarly C_2 independent of R_1 when

$$H_1 = \frac{-G_3}{G_1G_2 - G_3G_4}$$

EXAMPLE 8.44. For the given circuit obtain the state equation. Take voltage across the capacitors $V_1(t)$ and $V_2(t)$ as two state variables and the current through the resistor as output variable.

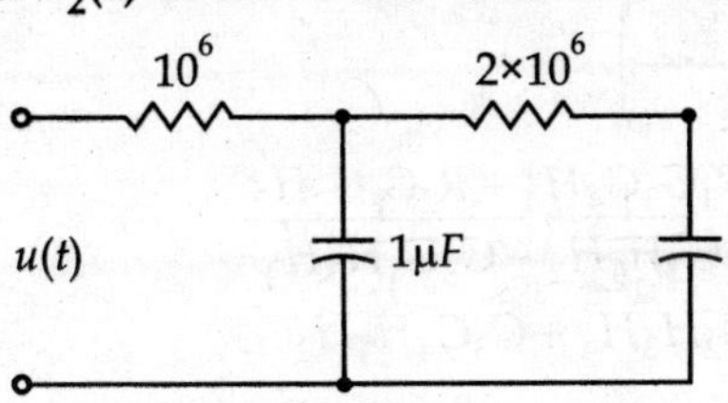

Fig. 8.71.

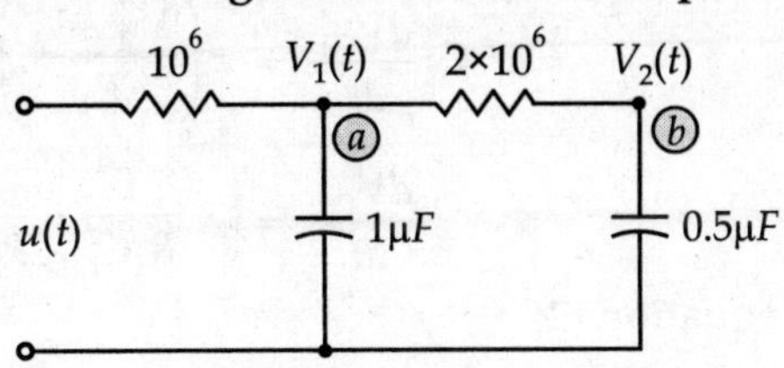

Fig. 8.71. (a)

Solution :

Apply CKL at node 'a'

$$\frac{u(t) - V_1(t)}{10^6} = 10^{-6}\frac{dV_1(t)}{dt} + \frac{V_1(t) - V_2(t)}{2 \times 10^6}$$

or,
$$\frac{dV_1(t)}{dt} = -1.5V_1(t) + 0.5V_2(t) + u(t)$$

KCL at node 'b':

$$\frac{V_1(t) - V_2(t)}{2 \times 10^6} = 0.5 \times 10^{-6}\frac{dV_2(t)}{dt}$$

or,
$$\frac{dV_2(t)}{dt} = V_1(t) - V_2(t)$$

output equation :

$$i(t) = \frac{V_1(t) - V_2(t)}{2 \times 10^6} = 0.5 \times 10^{-6}\,[V_1(t) - V_2(t)]$$

$$\begin{bmatrix} \dot{V}_1(t) \\ \dot{V}_2(t) \end{bmatrix} = \begin{bmatrix} -1.5 & 0.5 \\ 1 & -1 \end{bmatrix}\begin{bmatrix} V_1(t) \\ V_2(t) \end{bmatrix} + \begin{bmatrix} 1 \\ 0 \end{bmatrix} u(t)$$

$$i(t) = \begin{bmatrix} 0.5 \times 10^{-6} & -0.5 \times 10^{-6} \end{bmatrix}\begin{bmatrix} V_1(t) \\ V_2(t) \end{bmatrix}$$

EXAMPLE 8.45. Write the state equation for the circuit shown in Fig. 8.72.

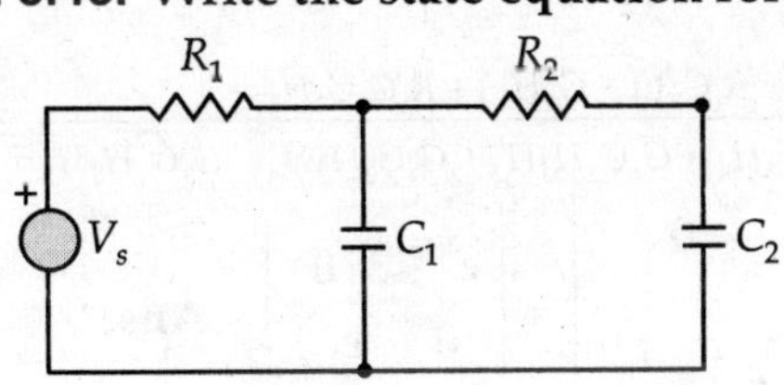

Fig. 8.72.

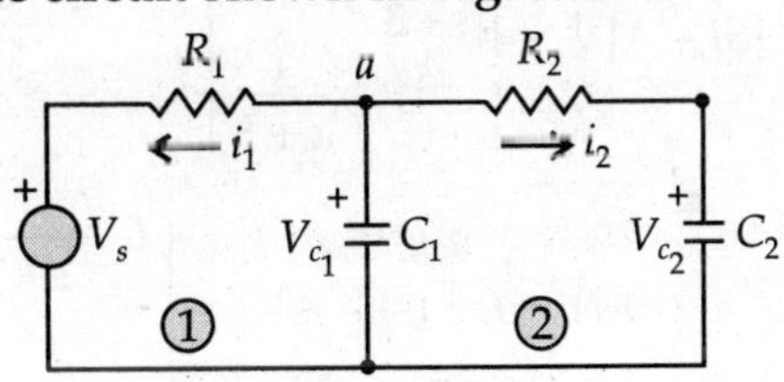

Fig. 8.72. (a)

Solution : KVL in mesh (1) :

$$-i_1 R_1 - V_s + V_{C1} = 0 \quad ...(1)$$

KVL in mech (2) :

$$-i_2 R_2 - V_{C2} + V_{C1} = 0 \quad ...(2)$$

KCL at node x :

$$i_1 + i_2 + C_1 \frac{dV_{C1}}{dt} = 0 \quad ...(3)$$

or,

$$C_1 \frac{dV_{C1}}{dt} = i_1 - i_2$$

$$C_1 \frac{dV_{C1}}{dt} = -\left(\frac{V_{c1} - V_s}{R_1}\right) - \left(\frac{V_{c1} - V_{C2}}{R_2}\right)$$

$$\frac{dV_{C1}}{dt} = \left(-\frac{1}{R_1C_1} - \frac{1}{R_2C_1}\right) V_{C1} + \frac{1}{R_2C_1} V_{C2} + \frac{1}{R_1C_1} V_s$$

From equation (2)

$$-C_2 \frac{dV_{C2}}{dt} R_2 - V_{C2} + V_{C1} = 0$$

$$\frac{dV_{C2}}{dt} = \frac{1}{R_2C_2} V_{C1} - \frac{1}{R_2C_2} V_{C2}$$

$$\begin{bmatrix} \dot{V}_{C1} \\ \dot{V}_{C2} \end{bmatrix} = \begin{bmatrix} -\frac{1}{R_1C_1} - \frac{1}{R_2C_1} & \frac{1}{R_2C_1} \\ \frac{1}{R_2\,C_2} & -\frac{1}{R_2C_2} \end{bmatrix} \begin{bmatrix} V_{C1} \\ V_{C2} \end{bmatrix} + \begin{bmatrix} \frac{1}{R_1C_1} \\ 0 \end{bmatrix} [V_s]$$

EXAMPLE 8.46. Find the state transition matrix $\phi(t)$ and the characteristic equation of A

$$A = \begin{bmatrix} -3 & 0 \\ 0 & -3 \end{bmatrix}$$

Solution : The characteristic equation is defined as

$$|SI - A| = 0$$

$$sI - A = s\begin{bmatrix} 1 & 0 \\ 0 & 1 \end{bmatrix} - \begin{bmatrix} -3 & 0 \\ 0 & -3 \end{bmatrix} = \begin{bmatrix} s+3 & 0 \\ 0 & s+3 \end{bmatrix}$$

$$|sI - A| = (s+3)(s+3) = s^2 + 6s + 9 = 0$$

$\therefore$ Required characteristic equation $\Rightarrow s^2 + 6s + 9 = 0$ **Ans.**

Calculation of $\phi(t)$:

$$[SI - A]^{-1} = \begin{bmatrix} \frac{1}{s+3} & 0 \\ 0 & \frac{1}{s+3} \end{bmatrix}$$

$$\phi(t) = \pounds^{-1}[sI - A]^{-1} = \begin{bmatrix} \pounds^{-1}\frac{1}{s+3} & 0 \\ 0 & \pounds^{-1}\frac{1}{s+3} \end{bmatrix} = \begin{bmatrix} e^{-3t} & 0 \\ 0 & e^{-3t} \end{bmatrix} \quad \textbf{Ans.}$$

EXAMPLE 8.47. Find the state transition matrix $\phi(t)$, the characteristic equation and the eigen values of A.

$$A = \begin{bmatrix} \sigma & -\omega \\ \omega & \sigma \end{bmatrix}$$

where σ and ω are real numbers.

Solution : $$sI - A = \begin{bmatrix} s & 0 \\ 0 & s \end{bmatrix} - \begin{bmatrix} \sigma & -\omega \\ \omega & \sigma \end{bmatrix} = \begin{bmatrix} s-\sigma & \omega \\ -\omega & s-\sigma \end{bmatrix}$$

characteristic equation $\Rightarrow |sI - A| = 0$

$$|sI - A| = \begin{vmatrix} s-\sigma & \omega \\ -\omega & s-\sigma \end{vmatrix} = s^2 - 2\sigma s + \sigma^2 + \omega^2$$

Required characteristic equation is

$s^2 - (2\sigma)s + (\sigma^2 + \omega^2) = 0$ **Ans.**

$$[SI - A]^{-1} = \frac{1}{(s-\sigma)^2 + \omega^2} \begin{bmatrix} s-\sigma & -\omega \\ \omega & s-\sigma \end{bmatrix}$$

$$\phi(t) = \pounds^{-1}[SI - A]^{-1} = \pounds^{-1} \begin{bmatrix} \dfrac{s-\sigma}{(s-\sigma)^2+\omega^2} & \dfrac{-\omega}{(s-\sigma)^2+\omega^2} \\ \dfrac{\omega}{(s-\sigma)^2+\omega^2} & \dfrac{s-\sigma}{(s-\sigma)^2+\omega^2} \end{bmatrix}$$

$$\phi(t) = \begin{bmatrix} e^{\sigma t}\cos\omega t & -e^{\sigma t}\sin\omega t \\ e^{\sigma t}\sin\omega t & e^{\sigma t}\cos\omega t \end{bmatrix}$$ **Ans.**

The roots of the characteristic equation are

$$s = \frac{2\sigma \pm j2\omega}{2}$$

$$s_1 = \sigma + j\omega$$

$$s_2 = \sigma - j\omega$$

$\therefore$ eigen values of A are $\sigma + j\omega$ and $\sigma - j\omega$. **Ans.**

EXAMPLE 8.48. A system is described by the following equations

$$\dot{x}(t) = \begin{bmatrix} -1 & 1 \\ 0 & -2 \end{bmatrix} x(t) + \begin{bmatrix} 1 & 0 & 1 \\ 0 & 1 & 1 \end{bmatrix} u(t)$$

$$y(t) = \begin{bmatrix} 1 & 2 \\ 1 & 0 \\ 1 & 1 \end{bmatrix} x(t)$$

Find the transfer function of the system. *(A.M.U., Aligarh B.E 89-90)*

Solution : We know that

$$G(s) = C[sI - A]^{-1} B$$

$$A = \begin{bmatrix} -1 & 1 \\ 0 & -2 \end{bmatrix}$$

$$sI - A = \begin{bmatrix} s+1 & 1 \\ 0 & s+2 \end{bmatrix}$$

$$[sI - A]^{-1} = \frac{1}{s^2+3s+2}\begin{bmatrix} s+2 & 1 \\ 0 & s+1 \end{bmatrix}$$

$$C\,[sI - A]^{-1} = \begin{bmatrix} 1 & 2 \\ 1 & 0 \\ 1 & 1 \end{bmatrix}\begin{bmatrix} s+2 & 1 \\ 0 & s+1 \end{bmatrix}\frac{1}{s^2+3s+2} = \frac{1}{s^2+3s+2}\begin{bmatrix} s+2 & 2s+3 \\ s+2 & 1 \\ s+2 & s+2 \end{bmatrix}$$

$$C\,[s\,I - A]^{-1}B = \frac{1}{s^2+3s+2}\begin{bmatrix} s+2 & 2s+3 \\ s+2 & 1 \\ s+2 & s+2 \end{bmatrix}\begin{bmatrix} 1 & 0 & 1 \\ 0 & 1 & 1 \end{bmatrix}$$

$$G(s) = \frac{s^3 - 3s^2 - 9s + 2}{s^2+3s+2}$$

This is the required transfer function **Ans.**

EXAMPLE 8.49. The mechanical system shown in Fig. 8.5 choose x_1 and x_2 of mass M_1 and M_2 as state variables as well as output obtain the state model.

Solution : The system equations are

$$F = M_1\frac{d^2x_1}{dt^2} + B_1\frac{dx_1}{dt} + B_2\frac{d}{dt}(x_1 - x_2) \quad ...(1)$$

$$0 = M_2\frac{d^2x_2}{dt^2} + Kx_2 + B_2\frac{d}{dt}(x_2 - x_1) \quad ...(2)$$

The above equations can be written as

$$F = M_1\ddot{x}_1 + B_1\,\dot{x}_1 + B_2\dot{x}_1 - B_2\dot{x}_2$$

$$F = M_1\ddot{x}_1 + (B_1 + B_2)\,\dot{x}_1 - B_2\,\dot{x}_2 \quad ...(3)$$

$$0 = M_2\ddot{x}_2 + Kx_2 + B_2\dot{x}_2 - B_2\dot{x}_1 \quad ...(4)$$

Select the state variable as

$$x_1 = Z_1 \qquad x_2 = Z_3$$

$$\dot{x}_1 = \dot{Z}_1 = Z_2 \qquad \dot{x}_2 = \dot{Z}_3 = Z_4$$

$$\ddot{x}_1 = \dot{Z}_2 \qquad \ddot{x}_2 = \dot{Z}_4$$

equation (3) and (4) can be written as

$$F = M_1\dot{Z}_2 + (B_1 + B_2)\,Z_2 - B_2\,Z_4$$

$$\dot{Z}_2 = \frac{F}{M_1} - \left(\frac{B_1 + B_2}{M_1}\right)Z_2 - \frac{B_2}{M_1}Z_4 \quad ...(5)$$

$$0 = M_2\dot{Z}_4 + KZ_3 + B_2Z_4 - B_2\,Z_2$$

$$\dot{Z}_4 = -\frac{K}{M_2}Z_3 - \frac{B_2}{M_2}Z_4 - \frac{B_2}{M_2}Z_2 \quad ...(6)$$

The state equations are

$$\dot{Z}_1 = Z_2$$

$$\dot{Z}_2 = \frac{F}{M_1} - \left(\frac{B_1 + B_2}{M_1}\right) Z_2 - \frac{B_2}{M_1} Z_4$$

$$\dot{Z}_3 = Z_4$$

$$\dot{Z}_4 = -\frac{K}{M_2} Z_3 - \frac{B_2}{M_2} Z_4 - \frac{B_2}{M_2} Z_2$$

$$\begin{bmatrix} \dot{Z}_1 \\ \dot{Z}_2 \\ \dot{Z}_3 \\ \dot{Z}_4 \end{bmatrix} = \begin{bmatrix} 0 & 1 & 0 & 0 \\ 0 & -\left(\frac{B_1 + B_2}{M_1}\right) & 0 & -\frac{B_2}{M_1} \\ 0 & 0 & 0 & 1 \\ 0 & -\frac{B_2}{M_2} & -\frac{K}{M_2} & -\frac{B_2}{M_2} \end{bmatrix} \begin{bmatrix} Z_1 \\ Z_2 \\ Z_3 \\ Z_4 \end{bmatrix} + \begin{bmatrix} 0 \\ \frac{1}{M_1} \\ 0 \\ 0 \end{bmatrix} F$$

Select displacement of mass M_1 and M_2 as output variables

$\therefore$
$$y_1 = x_1 = Z_1$$
$$y_2 = x_2 = Z_3$$

$$\begin{bmatrix} y_1 \\ y_2 \end{bmatrix} = \begin{bmatrix} 1 & 0 \\ 0 & 1 \end{bmatrix} \begin{bmatrix} Z_1 \\ Z_3 \end{bmatrix}$$

EXAMPLE 8.50. Find the characteristic equation when

$$A = \begin{bmatrix} -5 & 1 & 0 \\ 0 & -5 & 1 \\ 0 & 0 & -5 \end{bmatrix}$$

Solution : The characteristic equation $\Rightarrow |sI - A| = 0$

$$sI - A = s\begin{bmatrix} 1 & 0 & 0 \\ 0 & 1 & 0 \\ 0 & 0 & 1 \end{bmatrix} - \begin{bmatrix} -5 & 1 & 0 \\ 0 & -5 & 1 \\ 0 & 0 & -5 \end{bmatrix} = \begin{bmatrix} s+5 & -1 & 0 \\ 0 & s+5 & -1 \\ 0 & 0 & s+5 \end{bmatrix}$$

$$|sI - A| = s^3 + 15s^2 + 75s + 125$$

$\therefore$ Required characteristic equation

$$s^3 + 15s^2 + 75s + 125 = 0 \qquad \textbf{Ans.}$$

EXAMPLE 8.51. Write short note characteristic equation for differential equation, transfer function and for state equation.

Solution: Characteristic Equation from Differential Equation

Consider the differential equation of linear time invariant system

$$\frac{d^n x(t)}{dt} + a_{n-1}\frac{d^{n-1}x(t)}{dt^{n-1}} + \ldots + a_0 x(t) = b_m \frac{d^m u(t)}{dt} + \ldots + b_0 u(t)$$

laplace transform of above equation

$$(s^n + a_{n-1}\, s_{n-1} + \ldots + a_0)\, X(s) = (b_m\, s^m + \ldots\, b_0)\, u(s)$$

The characteristic equation can be obtained by setting the homogeneous part to zero. So, the characteristic equation will be

$$s^n + a_{n-1} s^{n-1} + ...+ a_0 = 0$$

For example :

The characteristic equation of

$$\frac{d^4x(t)}{dt^4} + 3\frac{d^3x(t)}{dt^3} + 2\frac{d^2x(t)}{dt^2} + \frac{dx(t)}{dt} + 5x(t) = u(t) \text{ will be}$$

$$s^4 + 3s^3 + 2s^2 + s + 5 = 0$$

Characteristic Equation from transfer function

The characteristic equation from the transfer function can be obtained by equating the denominator polynomial to zero.

For example :

$$\frac{C(s)}{R(s)} = G(s) = \frac{b_m s^m + b_{m-1}s^{m-1} + ... b_o}{a_n s^n + ... + a_o}$$

The characteristic equation is

$$a_n s^n + ... + a_0 = 0$$

If
$$G(s) = \frac{K}{s(1+.4s)(1+0.25s)}$$

$$H(s) = 1$$

$$\frac{C(s)}{R(s)} = \frac{G(s)}{1+G(s)H(s)}$$

The characteristic equation

$$1 + G(s)H(s) = 0$$

$$1 + \frac{K}{s(1+0.4s)(1+0.25s)} = 0$$

$$s(1 + 0.4s)(1 + 0.25s) + K = 0$$

is the required characteristic equation.

Characteristic Equation From State Equation :

From equation 8.38

$$G(s) = C[sI - A]^{-1} B + D = C\frac{adj\,(sI - A)}{|sI - A|}B + D$$

$$G(s) = \frac{C\,adj\,(sI - A)\,B + |sI - A|D}{|sI - A|}$$

The characteristic equation can be obtained from transfer function by equating the denominator polynomial to zero.

Hence required characteristic equation is

$$|sI - A| = 0$$

EXAMPLE 8.52. For the circuit shown in Fig. 1.114, if $R_1 = R_2 = R$ and $C_1 = C_2 = C$, show that the system is always overdamped.

Solution : Transfer function of the given circuit

$$\frac{C(s)}{R(s)} = \frac{1}{1 + R_1C_1s + R_1C_2s + R_2C_2s + s^2R_1R_2C_1C_2}$$

Put $R_1 = R_2 = R$ and $C_1 = C_2 = C$

$$\frac{C(s)}{R(s)} = \frac{1}{s^2R^2C^2 + 3RC\,s + 1} \quad \text{or,} \quad \frac{C(s)}{R(s)} = \frac{\frac{1}{R^2C^2}}{s^2 + \frac{3}{RC}s + \frac{1}{R^2C^2}}$$

compare with

$$\frac{C(s)}{R(s)} = \frac{\omega_n^2}{s^2 + 2\zeta\omega_n s + \omega_n^2}$$

$$\omega_n^2 = \frac{1}{R^2C^2} \quad \text{or} \quad \omega_n = \frac{1}{RC}$$

$$2\zeta\omega_n = \frac{3}{RC}$$

or,

$$2\zeta\cdot\frac{1}{RC} = \frac{3}{RC}$$

$\therefore$

$$\zeta = 1.5$$

Since, ζ is greater than one, hence the system is always overdamped.

EXAMPLE 8.53. The differential equation of a control system is

$$\frac{d^2Y(t)}{dt^2} + 9\frac{dY(t)}{dt} + 20\,Y(t) = 20\,X(t)$$

find the output response for unit step input $X(t)$ and $Y(t)$ are input and output respectively.

Solution : Given differential equation is

$$\frac{d^2Y(t)}{dt^2} + 9\frac{dY(t)}{dt} + 20Y(t) = 20\,X(t)$$

Laplace transform of above equation

$$s^2\,Y(s) + 9\,s\,Y(s) + 20\,Y(s) = 20\,X(s)$$

$$\frac{Y(s)}{X(s)} = \frac{20}{s^2 + 9s + 20}$$

$$X(s) = \frac{1}{s} \text{ (given)}$$

$\therefore$

$$Y(s) = \frac{20}{s\,(s^2 + 9s + 20)}$$

or,

$$Y(s) = \frac{1}{s} - \frac{5}{s+4} + \frac{4}{s+5}$$

Inverse Laplace of above equation

$$\pounds^{-1}\,Y(s) = \pounds^{-1}\frac{1}{s} - \pounds^{-1}\frac{5}{s+4} + \pounds^{-1}\frac{4}{s+5}$$

$$y(t) = 1 - 5\,e^{-4t} + 4\,e^{-5t} \quad \text{Ans.}$$

EXAMPLE 8.54. The output of a control system is

$$C(t) = 1 + 0.25\,e^{-50t} - 1.25\,e^{-10t}$$

(*i*) Obtain the expression for closed loop transfer function of the system.

(*ii*) Determine undamped natural frequency and damping ratio. Assume unit step input.

Solution : $C(t) = 1 + 0.25\, e^{-50t} - 1.25\, e^{-10t}$

Laplace transform

$$C(s) = \frac{1}{s} + \frac{0.25}{s+50} - \frac{1.25}{s+10} = \frac{500}{s(s+10)(s+50)}$$

since, $R(s) = 1/s$ given

$$\therefore \quad \frac{C(s)}{R(s)} = \frac{500}{(s+10)(s+50)}$$

$$\therefore \text{ closed loop transfer function} = \frac{500}{(s+10)(s+50)} \quad \textbf{Ans.}$$

$$\frac{C(s)}{R(s)} = \frac{500}{s^2+60s+500}$$

compare with $\dfrac{\omega_n^2}{s^2+2\zeta\omega_n\, s+\omega_n^2}$

$$\therefore \quad \omega_n^{\,2} = 500$$

or, $$\omega_n = 22.36 \text{ rad/sec.}$$

$$2\zeta\omega_n = 60$$

$$2\,\zeta\, 22.36 = 60$$

$$\therefore \quad \zeta = 1.34 \quad \textbf{Ans.}$$

EXAMPLE 8.55. Find impulse response of the given circuit. *(Mumbai university, 95)*

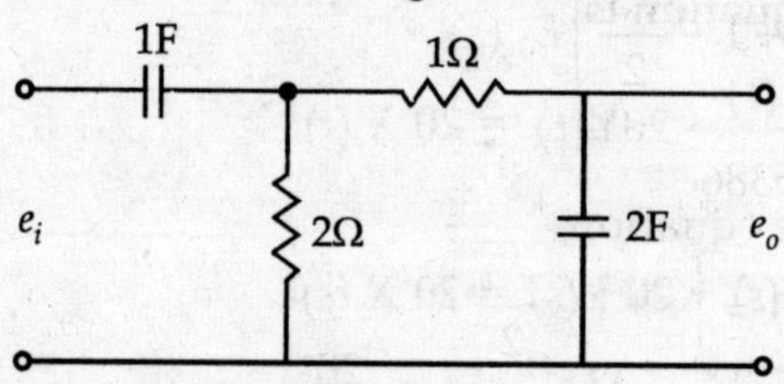

Fig. 8.73.

Solution : The transform network of the given circuit is shown in Fig. 8.73 (*a*)

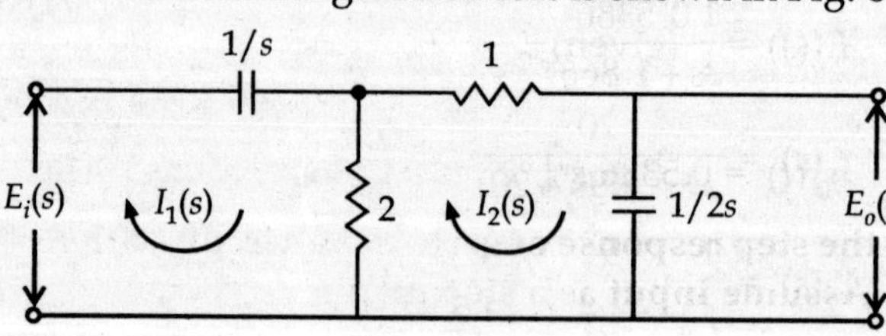

Fig. 8.73. (a)

Apply KVL in first mesh:

$$E_i(s) = I_1(s)\left[\frac{1+2s}{s}\right] - 2I_2(s) \quad \text{...(1)}$$

KVL in second mesh

$$I_2(s)\left[3+\frac{1}{2s}\right] = 2\, I_1(s) \quad \text{...(2)}$$

$$\therefore \quad I_1(s) = I_2(s)\left[\frac{6s+1}{2s}\right] \quad \text{...(3)}$$

Also, $$E_0(s) = \frac{1}{2s} I_2(s) \quad ...(4)$$

Put the value of $I_1(s)$ from equation (3) in equation (1)

$$E_i(s) = I_2(s) \frac{4s^2+8s+1}{4s^2} \quad ...(5)$$

from equation (4) and equation 5

$$\frac{E_0(s)}{E_i(s)} = \frac{2s}{4s^2+8s+1} = \frac{0.5\,s}{s^2+2s+1/4}$$

$$E_i(s) = 1 \quad \text{given}$$

$$\therefore \quad E_0(s) = \frac{0.5\,s}{s^2+2\,s+1/4} = \frac{0.5\,s}{(s+1)^2+\frac{1}{4}-1}$$

$$E_0(s) = \frac{0.5s}{(s+1)^2-\left(\sqrt{3}/2\right)^2} = \frac{0.5s}{\left(s+1+\frac{\sqrt{3}}{2}\right)\left(s+1-\frac{\sqrt{3}}{2}\right)}$$

$$E_0(s) = \frac{A}{\left(s+1+\frac{\sqrt{3}}{2}\right)} + \frac{B}{\left(s+1-\frac{\sqrt{3}}{2}\right)}$$

$$A = \left(s+1+\frac{\sqrt{3}}{2}\right) E_0(s)\Bigg|_{s=-1-\frac{\sqrt{3}}{2}}$$

$$\therefore \quad A = 0.5386$$

$$B = \left(s+1-\frac{\sqrt{3}}{2}\right) E_0(s)\Bigg|_{s=-1+\frac{\sqrt{3}}{2}}$$

$$B = -\,0.0386$$

$$E_0(s) = \frac{0.5386}{s+1.866} - \frac{0.0386}{s+0.1339}$$

Inverse Laplace

$$e_0(t) = 0.5386\, e^{-1.866\,t} - 0.0386\, e^{-0.1339\,t} \quad \textbf{Ans.}$$

EXAMPLE 8.56. Obtain the step response of series RLC circuit with $R = 1\Omega$, $L = 1\,H$ and $C = 1F$, output taken across 'R'. Assume input as a step of 10v. *(Pune University, 1994)*

Solution :

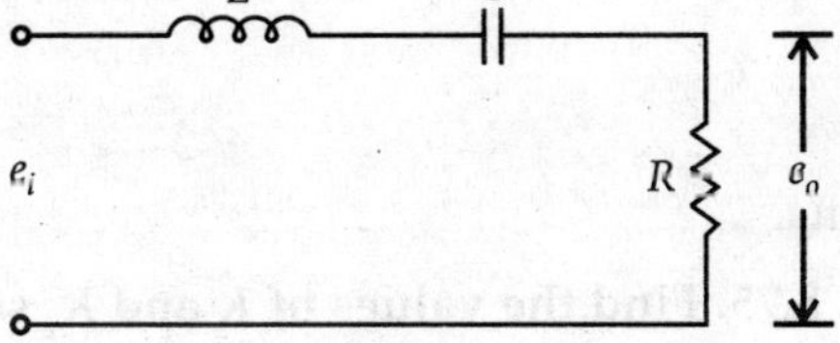

Fig. 8.74.

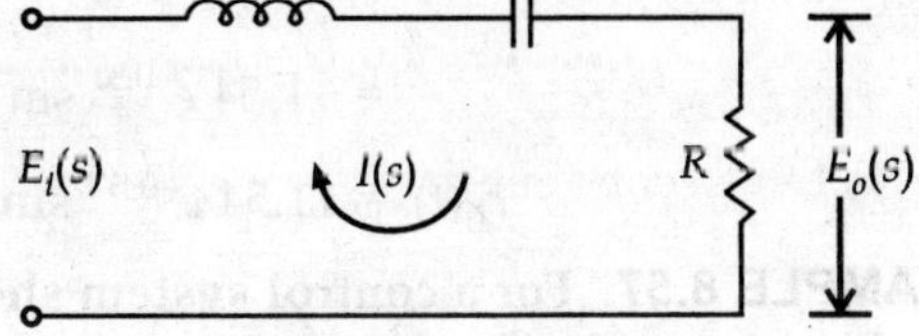

Fig. 8.74. (a)

The transform network is shown in Fig. 8.74 (*a*).

Apply KVL

$$E_i(s) = \left(R + sL + \frac{1}{sC}\right) I(s)$$

$$E_0(s) = RI(s)$$

$$\therefore \quad \frac{E_0(s)}{E_i(s)} = \frac{RCs}{s^2LC + sRC + 1}$$

$$E_i(s) = \frac{10}{s} \text{ (given) and } R = 1, L = 1, C = 1$$

$$\therefore \quad E_0(s) = \frac{10}{s^2+s+1} = \frac{10}{\left(s+\frac{1}{2}\right)^2 + \left(\frac{\sqrt{3}}{2}\right)^2} = \frac{10}{\left(s+\frac{1}{2}+j\frac{\sqrt{3}}{2}\right)\left(s+\frac{1}{2}-j\frac{\sqrt{3}}{2}\right)}$$

$$E_0(s) = \frac{A}{\left(s+\frac{1}{2}+j\frac{\sqrt{3}}{2}\right)} + \frac{B}{\left(s+\frac{1}{2}-j\frac{\sqrt{3}}{2}\right)}$$

$$A = \left(s+\frac{1}{2}+j\frac{\sqrt{3}}{2}\right) E_0(s)\Big|_{s=-\frac{1}{2}-j\frac{\sqrt{3}}{2}}$$

$$\therefore \quad A = j\frac{10}{\sqrt{3}}$$

$$B = -j\frac{10}{\sqrt{3}}$$

$$\therefore \quad E_0(s) = \frac{j10}{\sqrt{3}\left(s+\frac{1}{2}+j\frac{\sqrt{3}}{2}\right)} - \frac{j10}{\sqrt{3}\left(s+\frac{1}{2}-j\frac{\sqrt{3}}{2}\right)}$$

$$\therefore \quad e_0(t) = \frac{j10}{\sqrt{3}} e^{\left(-\frac{1}{2}-j\frac{\sqrt{3}}{2}\right)t} - \frac{j10}{\sqrt{3}} e^{\left(-\frac{1}{2}+j\frac{\sqrt{3}}{2}\right)t} = \frac{j10}{\sqrt{3}} e^{-t/2}\left[e^{-j\frac{\sqrt{3}}{2}t} - e^{+j\frac{\sqrt{3}}{2}t}\right]$$

$$= \frac{j10 \cdot 2j}{\sqrt{3}\cdot 2j} e^{-t/2}\left[e^{-j\frac{\sqrt{3}}{2}t} - e^{+j\frac{\sqrt{3}}{2}t}\right] = +\frac{20}{\sqrt{3}} e^{-t/2}\left[\frac{-e^{-j\frac{\sqrt{3}}{2}t} + e^{+j\frac{\sqrt{3}}{2}t}}{2j}\right]^*$$

$$= 11.54\, e^{-0.5t} \sin\frac{\sqrt{3}}{2}t$$

$$e_0(t) = 11.54\, e^{-0.5t} \sin 0.866\, t \quad \textbf{Ans.}$$

EXAMPLE 8.57. For a control system shown in Fig. 8.75. Find the values of K and K_t so that the damping ratio (ζ) of system is 0.6 and setting time (t_s) is 0.1sec. Use $t_s = 3.2/\zeta\omega_n$. Assume unit step input.

---

* take negative sign inside bracket

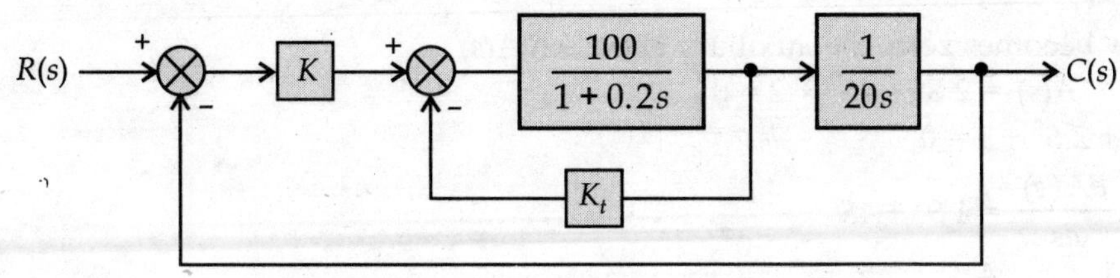

Fig. 8.75.

Solution : First solve inner closed loop.

$$\frac{\dfrac{100}{1+0.2\,s}}{1+\dfrac{100K_t}{1+0.2\,s}} = \frac{100}{1+0.2\,s+100\,K_t}$$

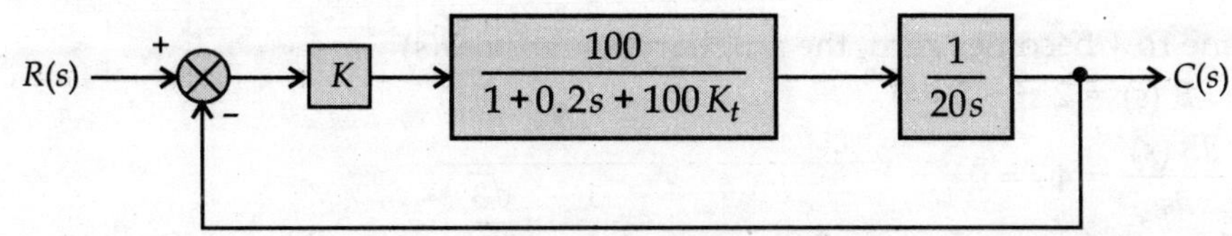

Now three blocks are in series

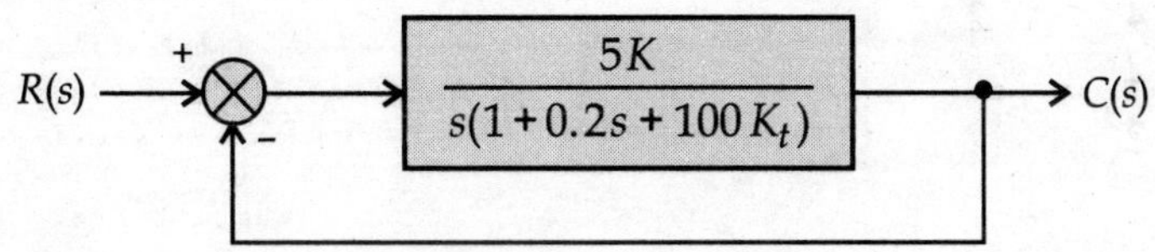

$$\therefore \quad \frac{C(s)}{R(s)} = \frac{25\,K}{s^2+5s\,(1+100\,K_t)+25K}$$

compare eq^n with $\dfrac{\omega_n^2}{s^2+2\zeta\omega_n\,s+\omega_n^2}$

$\omega_n{}^2 = 25\,k,\qquad 2\zeta\omega_n = 5\,(1+100\,K_t)$

$\therefore \quad \omega_n = 5\sqrt{K}$

Given $\quad t_s = \dfrac{3.2}{\zeta\omega_n}$

$$0.1 = \frac{3.2}{0.6\times 5\sqrt{K}}$$

$\therefore \quad K = 113.78 \qquad\qquad K = 113.78$

$\omega_n = 5\sqrt{K} = 5\sqrt{113.78} = 53.33$ rad/sec. $\qquad \omega_n = 53.33$ rad/s **Ans.**

EXAMPLE 8.58. The characteristic equation of a system is given by $S^6 + 3S^5 + 4S^4 + 6S^3 + 5S^2 + 3S + 2 = 0$. Comment on stability.

Solution :

S^6	1	4	5	2
S^5	3	6	3	–
S^4	2	4	2	
S^3	0	0	0	

one row becomes zero, the auxiliary equation $A(s)$

$$A(s) = 2S^4 + 4S^2 + 2 = 0$$

or, $S^4 + 2S^2 + 1 = 0$

$$\frac{dA(s)}{ds} = 4S^3 + 4S$$

S^6	1	4	5	2
S^5	3	6	3	
S^4	2	4	2	
S^3	4	4		
S^2	2	2		
S^1	0	0		
S^0				

Again one row becomes zero, the auxiliary equation $B(s)$

$$B(s) = 2s^2 + 2 = 0$$

$$\frac{dB(s)}{ds} = 4s = 0$$

S^6	1	4	5	2
S^5	3	6	3	
S^4	2	4	2	
S^3	4	4		
S^2	2	2		
S^1	4			
S^0	2			

All elements in first column are of same sign, therefore no root is in right half of s-plane, but there are two row of zeros, so check the system for marginally stable or unstable.

From first auxiliary equation

$$A(s) = S^4 + 2S^2 + 1$$

put $S^2 = x$

$$\therefore \quad x^2 + 2x + 1 = 0$$

$$\therefore \quad x^2 + x + x + 1 = 0$$

$$(x + 1)(x + 1) = 0$$

$x = -1, \qquad x = -1$

$S^2 = -1, \qquad S^2 = -1$

$S = \pm j, \qquad S = \pm j$

$S_1 = +j, \qquad S_3 = +j$

$S_2 = -j, \qquad S_4 = -j$

From second auxiliary equation

$$2S^2 + 2 = 0$$

$$S^2 = -1$$

$$S = \pm j$$

$$S_5 = +j \;\&\; S_6 = -j$$

On imaginary axis the roots are repeated, hence system is unstable.

EXAMPLE 8.59. Determine the stability of a system with characteristic equation.

$$S^5 + 4S^4 + 2S^3 + 8S^2 + S + 4 = 0$$

Solution :

S^5	1	2	1
S^4	4	8	4
S^3	0	0	
S^2			
S^1			
S^0			

One row becomes zero, use auxiliary equation

$$A(s) = 4s^4 + 8\,s^2 + 4 = 0 \qquad \therefore \qquad \frac{dA(s)}{ds} = 16\,s^3 + 16\,s$$

or, $S^4 + 2S^2 + 1 = 0$

put $S^2 = x$

$$x^2 + 2x + 1 = 0$$

$$(x+1)(x+1) = 0$$

$$x = -1 \quad , \quad x = -1$$

$$S^2 = -1 \quad , \quad S^2 = -1$$

$$S_1 = +j \qquad S_3 = +j$$

$$S_2 = -j \qquad S_4 = -j$$

S^5	1	2	1
S^4	4	8	4
S^3	16	16	
S^2	4	4	
S^1	0	0	
S^0			

Auxiliary equation

$$A(s) = 4\,s^2 + 4 = 0 \qquad \therefore \qquad \frac{dA(s)}{ds} = 8\,s$$

$$S^2 = -1 \;;\; S = \pm j$$

Since, the roots are multiple roots on imaginary axis at $s = \pm j$ the system is unstable.

EXAMPLE 8.60. Determine stability of a system having characteristic equation

$$S^6 + 5\,S^5 + 10\,S^4 + 24\,S^3 + 20\,S^2 + 15\,S + 10 = 0$$

Solution :

S^6	1	10	20	10
S^5	5	24	15	0
S^4	5.2	13.75	10	
S^3	10.78	–2.45		
S^2	14.93	10		
S^1	–9.67	0		
S^0	10			

No. of sign change = 2

Hence, system is unstable.

EXAMPLE 8.61. Determine number of roots with positive real parts, zero real part and negative real part for the following polynomial equation using Routh criterion.

$$Q(s) = S^4 + 2S^3 + 8S^2 + 10\,S + 15$$ *(UPTU, B.Tech, C.O. Exam. 2003, Network and Systems)*

Solution :

S^4	1	8	15
S^3	2	10	
S^2	3	15	
S^1	0		
S^0			

Auxiliary equation

$$A(s) = 3\,s^2 + 15$$

$$\frac{dA(s)}{ds} = 6\,s$$

S^4	1	8	15
S^3	2	10	
S^2	3	15	
S^1	6		
S^0	15		

Since, no sign change in first column, the system may be stable. Solve auxiliary equation and determine the location of roots for stability.

$$3\,S^2 + 15 = 0$$

$$S_{1,2} = \pm j\sqrt{5}$$

$$S^2\,(S^2 + 5) + 2s\,(s^2 + 5) + 3(s^2 + 5) = 0$$

$$(S^2 + 5)\,(S^2 + 2S + 3) = 0$$

take $\quad S^2 + 2S + 3 = 0$

$$S_{3,4} = -1 \pm j\,1.415$$

$$S_1 = 0 + j\sqrt{5}$$

$$S_2 = 0 - j\sqrt{5}$$

$$S_3 = -1 + j\,1.415$$

$$S_4 = -1 - j\,1.415$$

No. of roots with positive real part = 0

No. of roots with zero real part = 2

No. of roots with negative real part = 2

EXAMPLE 8.62. Determine the number of roots with positive real part, number of roots with zero real part and the number of roots with negative real part for the following equations, apply Routh-Hurwitz criterion.

(*a*) $2S^5 + S^4 + 6S^3 + 3S^2 + S + 1 = 0$

(*b*) $S^5 - S^4 - 2S^3 + 2S^2 - 8S + 8 = 0$

Solution : (*a*) $2S^5 + S^4 + 6\,S^3 + 3\,S^2 + S + 1 = 0$

S^5	2	6	1
S^4	1	3	1
S^3	0	− 1	
S^2			
S^1			
S^0			

Multiply the characteristic equation by $(s + 1)$

$(S + 1)(2S^5 + S^4 + 6S^3 + 3S^2 + S + 1) = 0$

$2S^6 + 3S^5 + 7S^4 + 9S^3 + 4S^2 + 2S + 1 = 0$

S^6	2	7	4	1
S^5	3	9	2	
S^4	1	2.67	1	
S^3	1	– 1		
S^2	3.67	1		
S^1	–1.27			
S^o	1			

No. of sign changes = 2

No. of roots with positive real part = 2

(*b*) $S^5 - S^4 - 2S^3 + 2S^2 - 8S + 8 = 0$

S^5	1	– 2	– 8
S^4	– 1	2	8
S^3	0	0	
S^2			
S^1			
S^o			

Auxiliary equation $A(s) = -S^4 + 2S^2 + 8 = 0$

$$\frac{dA(s)}{ds} = -4S^3 + 4s$$

S^5	1	– 2	– 8
S^4	– 1	2	8
S^3	– 4	4	
S^2	1	8	
S^1	36		
S^o	8		

No. of sign changes = 2

$\therefore$ No. of roots with positive real part = 2

solve the auxiliary equation

$$S^4 - 2S^2 - 8 = 0$$

put $S^2 = x$

$$x^2 - 2x - 8 = 0$$

$$(x + 2)(x - 4) = 0$$

when $x + 2 = 0$ and when $x - 4 = 0$

$S^2 + 2 = 0$ $\quad S^2 - 4 = 0$

$S = \pm j\sqrt{2}$ $\quad (S + 2)(S - 2) = 0$

The given characteristic equation can be written in factored form

$(S - 1)(S^2 + 2)(S + 2)(S - 2) = 0$

No. of roots with positive real part = 2

No. of roots with zero real part = 2

No. of roots with negative real part = 1

EXAMPLE 8.63. Apply Routh-Hurwitz criterion to determine: (*i*) number of roots with positive real part, (*ii*) the number of roots with zero real parts, (*iii*) the number of roots with negative real parts for the following equation.

(*a*) $S^4 + 2S^3 + 7S^2 + 10\,S + 10 = 0$

Solution :

S^4	1	7	10
S^3	2	10	
S^2	2	10	
S^1	0		
S^o			

Auxiliary equation $A(s) = 2S^2 + 10 = 0$

$$\frac{dA(s)}{ds} = 4\text{ s}$$

S^4	1	7	10
S^3	2	10	
S^2	2	10	
S^1	4		
S^o	10		

No sign change in first column, system may be stable. Solve the auxiliary equation

$$2S^2 + 10 = 0$$

$$S = \pm j\sqrt{5}$$

Hence, roots are non-repeated, the system is marginally stable. The given equation can be written in factored form as

$$(S^2 + 5)(S^2 + 2S + 2) = 0$$

for $\quad S^2 + 2S + 2 = 0$

$$S_{1,2} = -1 \pm j1$$

for $\quad S^2 + 5 = 0$

$$S_{3,4} = \pm j\sqrt{5}$$

No. of roots with positive real part = 0

No. of roots with zero real part = 2

No. of roots with negative real part = 2

$$S_1 = -1 + j1$$

$$S_2 = -1 - j$$

$$S_3 = +j\sqrt{5}$$

$$S_4 = -j\sqrt{5}$$

EXAMPLE 8.64. Find the number of roots with positive real part, zero real part and negative real parts of the following characteristic equation

$S^4 + 5S^3 + 2S + 10 = 0$

Solution :

$S^4 + 5S^3 + 0\,S^2 + 2S + 10 = 0$

S^4	1	0	10
S^3	5	2	
S^2	–2/5	10	
S^1	127		
S^o	10		

No. of sign change = 2

No. of roots with positive real part = No. of sign change = 2

The given characteristic equation in factored form can be written as :

$$(S + 5)\,(S^3 + 2)$$

or, $$(S + 5)\,(S + 1.26)\,(S^2 - 1.26\,S + 1.59)$$

$\therefore$ $$A^3 + B^3 = (A + B)\,(A^2 - AB + B^2)$$

or, $$(S + 5)\,(S + 1.26)\,(S - 0.63 - j\,1.09)\,(S - 0.63 + j\,1.09) = 0$$

$$S_1 = -5$$
$$S_2 = -1.26$$
$$S_3 = 0.63 + j\,1.09$$
$$S_4 = 0.63 - j\,1.09$$

No. of roots with positive real part = 2

No. of roots with zero real part = 0

No. of roots with negative real part = 2.

EXAMPLE 8.65. Determine the values of K for open loop transfer function

$$G(s)\,H(s) = \frac{K}{s\,(1+0.1\,s)\,(1+s)} \quad \textbf{so that}$$

(*a*) the gain margin is 15 db and

(*b*) Phase margin is 60°.

Solution : Step 1 :

$$G(s)\,H(s) = \frac{K}{s\,(1+0.1\,s)\,(1+s)}$$

corner frequencies are

$$\omega_1 = \frac{1}{0.1} = 10 \text{ rad/sec.} \quad \text{and} \quad \omega_2 = \frac{1}{1} = 1 \text{ rad/sec.}$$

Step 2 : Draw the magnitude plot without *K*. Since there is a pole at origin, the initial slope of the magnitude plot will be – 20 db/dec.

(*i*) Draw a line having slope – 20db/dec. (parallel to the I slope marker on top of left corner) passing through $\omega = 1$ upto first corner frequency 1 rad/sec.

(*ii*) At $\omega = 1$, a simple pole occurs, the resultant slope of next line will be – 20 + (– 20) = – 40 db/dec. Draw a line having slope – 40 db/dec. (parallel to II slope marker) upto II corner frequency 10 rad/sec.

(*iii*) At $\omega = 10$, a simple pole occurs, the resultant slope of next line will be – 40 + (– 20) = – 60 db/dec. Draw a line having slope – 60 db/dec. from second corner frequency towards infinity as there is no pole or zero.

Step 3 : Draw phase plot

$$\phi = -90° - \tan^{-1} 0.1\,\omega - \tan^{-1} \omega$$

ω	*Arg(jω)* ϕ_1	*Arg (1 + j 0.1 ω)* ϕ_2	*Arg (1 + jω)* ϕ_3	*Resultant* $\phi = \phi_1 + \phi_2 + \phi_3$
0.2	– 90°	– 1.14°	– 11.3°	– 102.4°
0.4	– 90°	– 2.3°	– 21.8°	–114.1°
0.6	– 90°	– 3.43°	– 30.96°	– 124.39°
0.8	– 90°	– 4.6°	– 38.65°	– 133.25°
1.0	– 90°	– 5.7°	– 45°	– 140.7°
3.0	– 90°	– 16.69°	– 71.56°	– 178.25°
4.0	– 90°	– 21.8°	– 75.96°	– 187.76°

Step 4 : From Bode plot (plot 1), gain margin available = 21 db

Required gain margin = 15 db.

For required gain margin, shift the magnitude plot upwards by 21 – 15 = 6 db. 30, shift point *A* to point *B* = + 6 db (upward shift taken as positive and downward shift negative).

$$6 \text{ db} = 20 \log_{10} K$$

$$\therefore \quad K = 1.99$$

Step 5 :

$$\text{P.M} = 180° + \angle G(j\omega)\,H(j\omega)\Big|_{\omega = \omega_{c1}}$$

$$+60° = 180° + \angle G(j\omega)\,H(j\omega)$$

$$\therefore \quad \angle G(j\omega)\,H(j\omega) = 60° - 180 = -120°$$

Find this point on phase plot i.e., locate *D* at – 120°. Extend a line from *D* on magnitude plot this intersect at *E*, but this point should be at *F* for P.M (+ 60°). So shift the magnitude plot downwards.

i.e.,

$$EF = -6$$

$$-6 = 20 \log_{10} K$$

$$\therefore \quad K = 0.501$$

Step 6 :

Value of *K* for gain margin 15 db = 1.99.

Value of *K* for phase margin 60° = 0.501.

EXAMPLE 8.66. Determine the value of K for open loop transfer function

$$G(j\omega)\,H(j\omega) = \frac{K}{j\omega\,(1 + j0.1\omega)\,(1 + j0.05\omega)}$$

such that

(*i*) the gain margin is 20 db

(*ii*) the phase margin is 30°.

Solution : Corner frequencies are

$$\omega_1 = \frac{1}{0.1} = 10 \text{ rad/sec.}$$

$$\omega_2 = \frac{1}{0.05} = 20 \text{ rad/sec.}$$

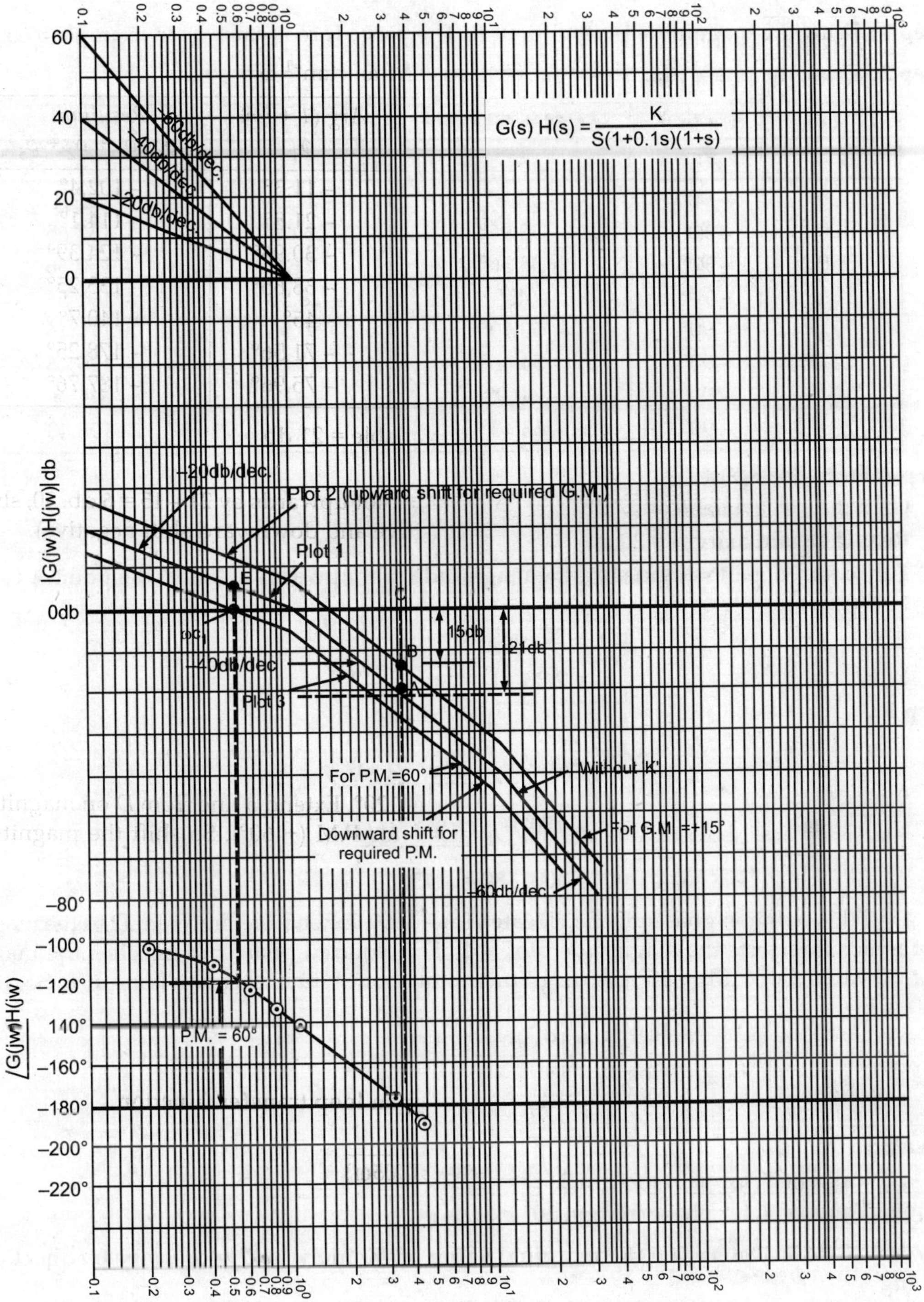

$G(s)\,H(s) = \dfrac{K}{S(1+0.1s)(1+s)}$
–60db/dec.
–40db/dec.
–20db/dec.
–20db/dec.
Plot 2 (upward shift for required G.M.)
Plot 1
E
C
0db
ωc_1
15db
21db
B
A
–40db/dec
Plot 3
For P.M.=60°
Without 'K'
Downward shift for required P.M.
For G.M. = +15°
–60db/dec
|G(jw)H(jw)|db
∠G(jw)H(jw)
P.M. = 60°
60
40
20
0
–80°
–100°
–120°
140°
–160°
–180°
–200°
–220°

Fig. 8.76.

Step 1 : Draw the magnitude plot without *K*. The procedure is same as in example 8.65.

Step 2 : Draw the phase plot

ω	*Arg* $(j\omega)$ ϕ_1	*Arg*$(1 + j\,0.1\,\omega)$ ϕ_2	*Arg* $(1 + j\,0.05\,\omega)$ ϕ_3	*Resultant* $\phi_1 + \phi_2 + \phi_3$
4.0	– 90°	– 21.8°	– 11.3	– 123.1
6.0	– 90°	– 30.96°	– 16.69°	– 137.65
8.0	– 90°	– 38.56°	– 21.8	– 150.36
10.0	– 90°	– 45°	– 26.56	– 161.56
12.0	– 90°	– 50.19°	– 30.96	– 171.46
14.0	– 90°	– 54.46°	– 35	– 179.46
18.0	– 90°	– 60.9°	– 42	– 192.9
20.0	– 90°	– 63.43°	– 45	– 198.43

Step 4: From Bode plot

Gain margin available = 27 db

Required gain margin = 20 db

For required gain margin shift the magnitude plot upwards *i.e.*, shift the point *A* to *B* = + 7 db.

$$7 \text{ db} = 20 \log_{10} K$$

$$\therefore \quad \boxed{K = 2.23}$$

Step 5 :

$$\text{P.M.} = 180° + \angle G(j\omega)\,H(j\omega)\Big|_{\omega = \omega_{c_1}}$$

$$30° = 180° + \angle G(j\omega)\,H(j\omega)$$

$$\therefore \quad \angle G(j\omega)\,H(j\omega) = 30 - 180 = -150°$$

Find this point on phase plot *i.e.*, locate *D* at – 150°. Extend the line from *D* to the magnitude plot which intersects the magnitude plot at *E*. For required phase margin shift the magnitude plot upwards. For P.M. = 30° the magnitude plot should intersects 0db lione at *F*. So, the total shift is EF

$$EF = 19 \text{ db}$$

$$20 \log K = 19$$

$$\therefore \quad \boxed{K = 8.91}$$

Result :

(*i*) Value of K for gain margin 20 db = 2.23

(*ii*) Value of K for phase margin 30° = 8.91

EXAMPLE 8.67. Determine the marginal value of K analytically for unity feedback system having

$$G(s)\,H(s) = \frac{K}{s\,(s+5)\,(s+10)}$$ **, so that the system will marginally stable.**

Solution : For marginally stable system

Phase margin = Gain margin = 0

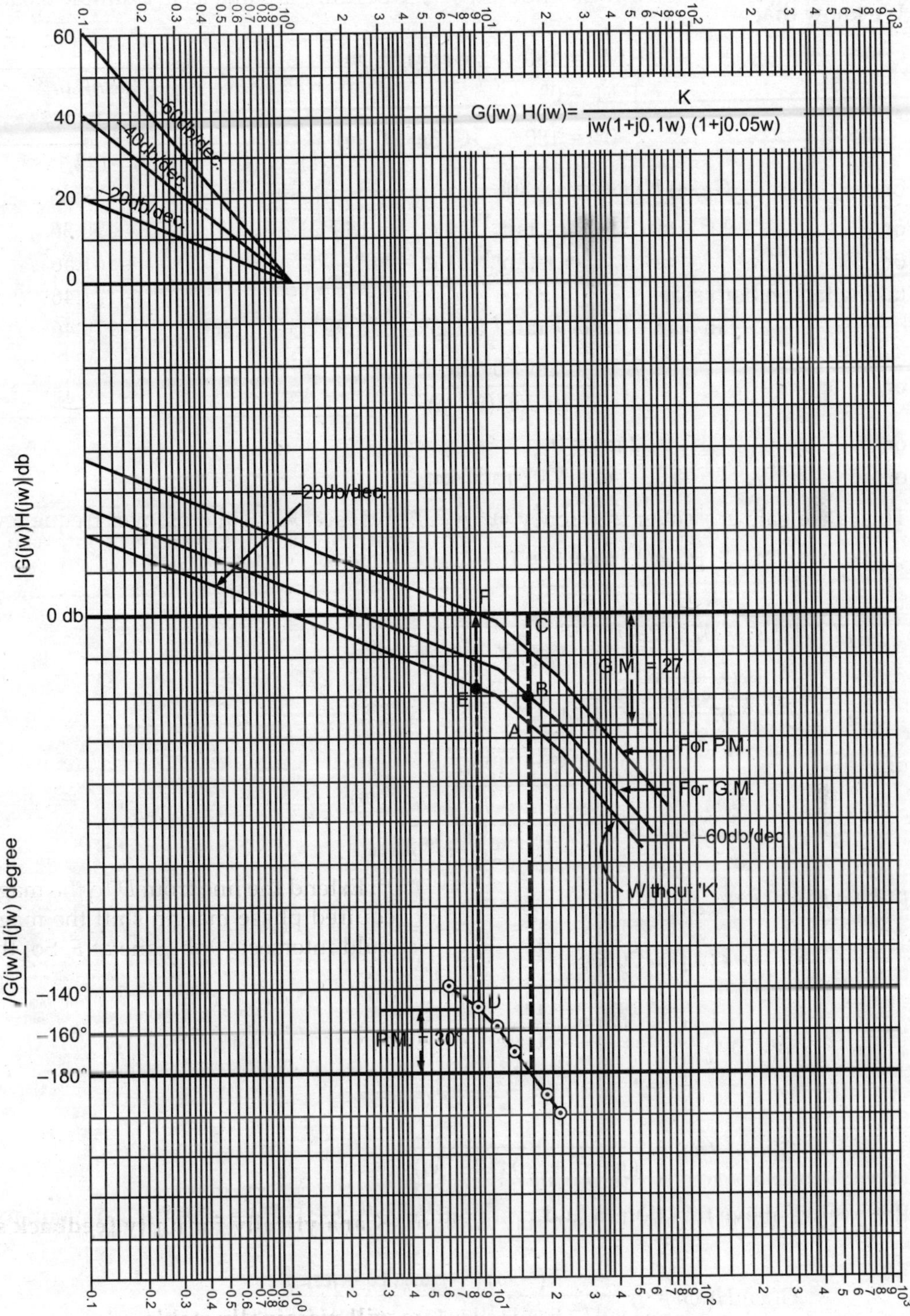

G(jw) H(jw)= K / jw(1+j0.1w) (1+j0.05w)
–60db/dec.
–40db/dec.
–20db/dec.
–20db/dec.
|G(jw)H(jw)| db
/G(jw)H(jw) degree
60
40
20
0
0 db
–140°
–160°
–180°
F
C
E
B
A
D
G.M. = 27
For P.M.
For G.M.
–60db/dec
Without 'K'
P.M. = 30°

Fig. 8.77.

Phase cross-over frequency = Gain crossover frequency.
We know that

$$\text{P.M.} = 180° + \angle G(j\omega)\,H(j\omega)$$

$$\text{P.M} = 0°$$

$$\therefore \quad 0° = 180° + \angle G(j\omega)\,H(j\omega)$$

or, $$\angle G(j\omega)\,H(j\omega) = -180°$$

or, $$-90° - \tan^{-1} 0.2\,\omega - \tan^{-1} 0.1\,\omega = -180°$$

or $$\tan^{-1} 0.2\omega + \tan^{-1} 0.1\,\omega = 90°$$

taking tan on both sides

$$\tan(\tan^{-1} 0.2\,\omega + \tan^{-1} 0.1\,\omega) = \tan 90°$$

or, $$\frac{0.2\omega + 0.1\,\omega}{1-(0.2\omega)\,(0.1\omega)} = \infty$$

or, $$1 - 0.02\,\omega^2 = 0$$

or, $$\omega = 7.07 \text{ rad/sec.}$$

This is the gain crossover frequency $\omega_{c_1} = 7.07$ rad/sec. At gain crossover frequency

$$|G(j\omega)\,H(j\omega)| = 1$$

$$\frac{K}{|j7.07|\,|5+j7.07|\,|10+j\,7.07|} = 1$$

or, $$\frac{K}{7.07 \times 8.66 \times 12.24} = 1$$

or, $$\boxed{K = 749.82}$$

$$\text{G.M} = 20\log_{10} \frac{1}{|G(j\omega)\,H(j\omega)|_{\omega=\omega_{c_2}}}$$

put $|G(j\omega)\,H(j\omega)|_{\omega=\omega_{c_2}} = x$

$$\text{G.M.} = 20\log_{10}\frac{1}{x}$$

$$0 = 20\log_{10}\frac{1}{x}$$

$$\therefore \quad \frac{1}{x} = 1 \quad \text{or,} \quad x = 1$$

$$G(j\omega)\,H(j\omega) = \frac{K}{(j\omega)\,(5+j\omega)\,(10+j\omega)}$$

Rationalize above function to find x, at point x $\angle G(j\omega)\,H(j\omega) = -180°$. *i.e.*, imaginary part is zero.

$$G(j\omega)\,H(j\omega) = \frac{K(-j\omega)\,(5-j\omega)\,(10-j\omega)}{(j\omega)\,(5+j\omega)\,(10+j\omega)\,(5-j\omega)\,(10-j\omega)}$$

$$= \frac{15K}{(25+\omega^2)\,(100+\omega^2)} - j\,\frac{K\omega\,(50-\omega^2)}{\omega^2(25+\omega^2)\,(100+\omega^2)}$$

Equate the imaginary part to zero

$$50 - \omega^2 = 0$$

or,
$$\omega^2 = 50$$
$$\omega = 7.07 \text{ rad/sec.}$$

This is the phase crossover frequency $\omega_{c_2} = 7.07$ rad/sec. Put the value of ω_{c_2} in real part, the point x is

$$x = \frac{15K}{(25+50)(100+50)}$$

$$1 = \frac{15K}{11250}$$

$$\boxed{K = 750}$$

$\therefore$ For marginally stable system $K = 750$ **Ans.**

EXAMPLE 8.68. The open loop transfer function with unity feedback is

$$G(s) = \frac{K}{s(s+2)(s+20)}.$$

Find analytically the value of 'K', for phase margin = 50°.

Solution :
$$G(s)\,H(s) = \frac{K}{s(s+2)(s+20)} = \frac{K/40}{s(1+0.5s)(1+0.05s)}$$

put $s = j\omega$

$$G(j\omega)\,H(j\omega) = \frac{K/40}{j\omega(1+j0.5\omega)(1+j0.05\omega)}$$

We know that

$$\text{P.M} = 180° + \angle G(j\omega)\,H(j\omega)$$

$$\angle G(j\omega)\,H(j\omega) = -90° - \tan^{-1} 0.5\,\omega - \tan^{-1} 0.05\,\omega$$

$\therefore$
$$50° = 180° + (-90° - \tan^{-1} 0.5\,\omega - \tan^{-1} 0.05\omega)$$

or,
$$40° = \tan^{-1} 0.5\,\omega + \tan^{-1} 0.05\,\omega$$

taking tan on both sides

$$\tan 40° = \tan(\tan^{-1} 0.5\,\omega + \tan^{-1} 0.05\,\omega)$$

or,
$$0.839 = \frac{0.55\,\omega}{1-0.025\,\omega^2}$$

or, $0.020975\,\omega^2 + 0.55\omega - 0.839 = 0$

or,
$$\omega = 1.445 \text{ rad/sec.}$$

This is the gain cross-over frequency

$$\omega_g = 1.445 \text{ rad/sec.}$$

At gain cross over frequency $|G(j\omega)\,H(j\omega)| = 1$

$$\frac{K/40}{|j1.445|\,|1+j0.7225|\,|1+j0.07225|} = 1$$

$$K/40 = 1.7873$$

or
$$\boxed{K = 71.49}$$

Required value of $K = 71.49$ **Ans.**

EXAMPLE 8.69. The open loop transfer function with unity feedback system is

$$G(s) = \frac{25}{s(s+1)(s+10)}$$

Determine phase margin and gain margin analytically.

Solution :

Step 1 : Calculation of phase crossover frequency

$$G(j\omega)\,H(j\omega) = \frac{2.5}{j\omega\,(1+j\omega)\,(1+j0.1\omega)}$$

At phase cross-over frequency the angle is – 180°

$$\therefore \quad -180° = -90° - \tan^{-1} 0.1\omega - \tan^{-1}\omega$$

$$\text{or,} \quad 90° = \tan^{-1} 0.1\omega + \tan^{-1} 1\omega$$

Take tan on both sides

$$\tan 90° = \tan(\tan^{-1} 0.1\,\omega + \tan^{-1}\omega)$$

$$\infty = \frac{0.1\omega + \omega}{1 - 0.1\omega^2}$$

$$\text{or,} \quad 1 - 0.1\,\omega^2 = 0$$

$$\omega = 3.16 \text{ rad/sec.}$$

This is the phase crossover frequency ω_{c_2}

$$|G(j\omega)\,H(j\omega)|_{\omega_{c_2}} = \frac{2.5}{|j3.16|\,|1+j3.16|\,|1+j\,0.316|} = 0.2279$$

$$\therefore \quad \text{G.M} = -20\log_{10} 0.2279 = 12.84 \text{ db} \quad \textbf{Ans.}$$

At gain crossover frequency $|G(j\omega)\,H(j\omega)| = 1$

$$\frac{2.5}{\omega\sqrt{1+\omega^2}\,\sqrt{1+0.01\omega^2}} = 1$$

$$\text{or,} \quad \omega = 1.425 \text{ rad/sec.}$$

This is the gain crossover frequency ω_{c_1}

$$\angle G(j\omega)\,H(j\omega)\Big|_{\omega=\omega c_1} = \frac{2.5}{j1.425\,(1+j1.425)\,(1+j0.1425)} = -153°$$

$$\therefore \quad \text{P.M} = 180° + (-153°)$$

$$\text{P.M.} = 27° \quad \textbf{Ans.}$$

EXAMPLE 8.70. Draw Bode plot of transfer function given in example 8.69 and find P.M. and G.M.

Solution : $$G(j\omega)\,H(j\omega) = \frac{2.5}{j\omega\,(1+j\omega)\,(1+j0.1\omega)}$$

Step 1 : Draw the magnitude plot

(*i*) Draw a line having slope – 20 db/dec. upto 2.5 and mark I corner frequency 1 rad/sec. as there is one pole at origin.

(*ii*) At ω = 1 rad/sec, there is a simple pole, so next slope will be – 20 + (– 20) = – 40 db/sec. Draw a line having slope – 40 db/dec. from I corner frequency to II corner frequency 10 rad/sec.

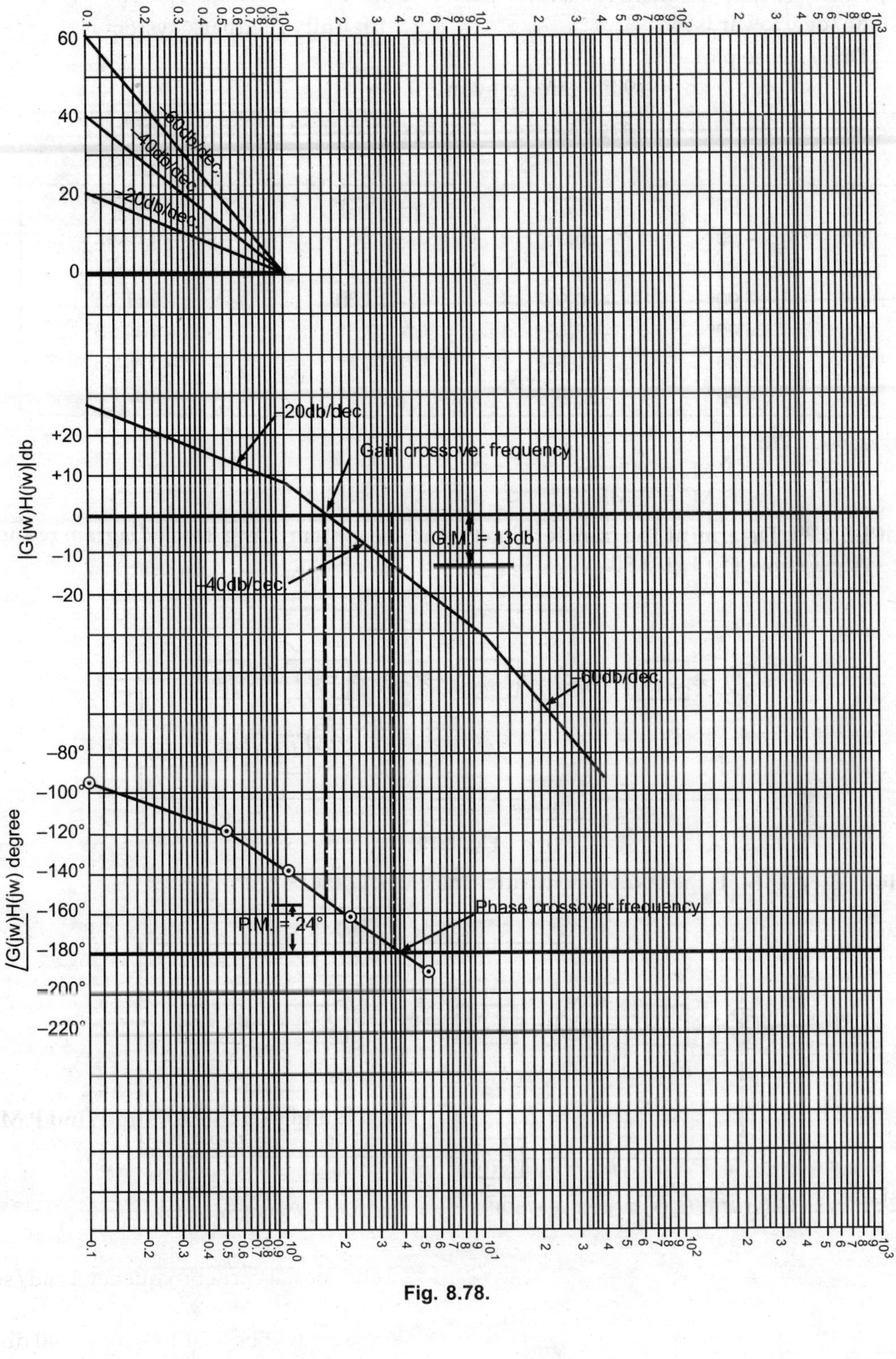

–60db/dec.
–40db/dec.
–20db/dec.
–20db/dec
Gain crossover frequency
G.M. = 13db
–40db/dec.
–60db/dec.
P.M. = 24°
Phase crossover frequency
|G(jw)H(jw)|db
/G(jw)H(jw) degree

Fig. 8.78.

(*iii*) At $\omega = 10$ rad/sec. there is a simple pole, so draw a line having slope $-20 + (-40) = -60$ db/dec. towards ∞.

Step 2 : Draw phase plot

$$\phi = -90° - \tan^{-1}\omega - \tan^{-1} 0.1\,\omega$$

ω	*Arg (jω)* ϕ_1	*Arg (1 + jω)* ϕ_2	*Arg (1 + j 0.1 ω)* ϕ_3	*Resultant* $\phi = \phi_1 + \phi_2 + \phi_3$
0.1	– 90°	– 5.71°	– 0.573°	– 96.28°
0.5	– 90°	– 26.56°	– 2.86°	– 119.42°
1.0	– 90°	– 45°	– 5.71°	– 140.71°
1.8	– 90°	– 60.95°	– 10.20	– 161.15°
2.0	– 90°	– 63.43°	– 11.3°	– 164.73°
3.0	– 90°	– 71.56°	– 16.7°	– 178.26°
4.0	– 90°	– 75.96°	– 21.8°	– 187.76°

From Bode Plot

P.M. = 24°

G.M. = 13 db

EXAMPLE 8.71. Determine the transfer function of the system using block diagram reduction of the system shown in Fig. 8.80.

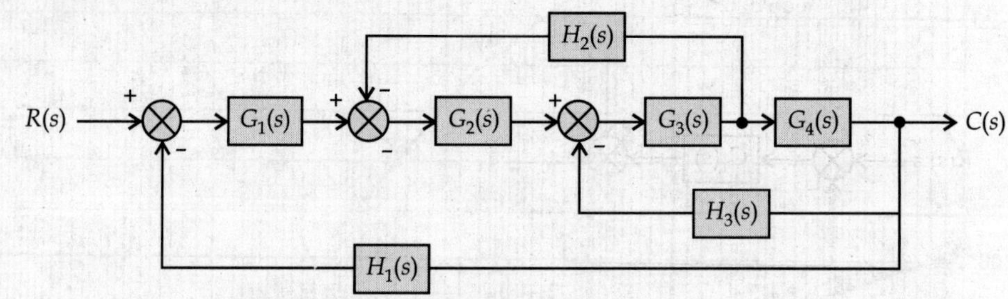

Fig. 8.79.

Solution : Step 1 : Shift the takeoff point after the block $G_4(s)$

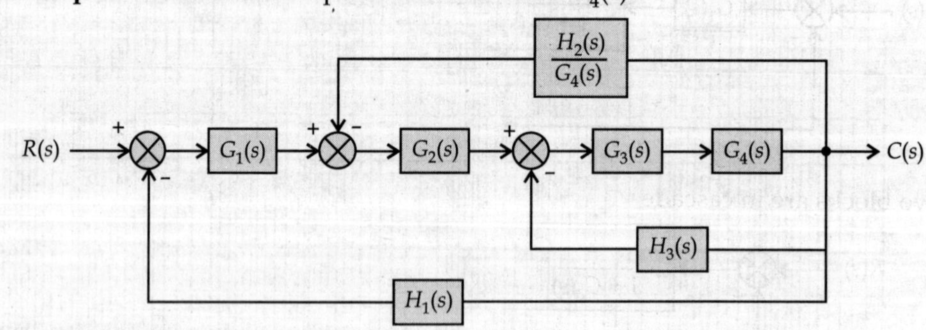

Step 2 : Blocks $G_3(s)$ and $G_4(s)$ are in cascade.

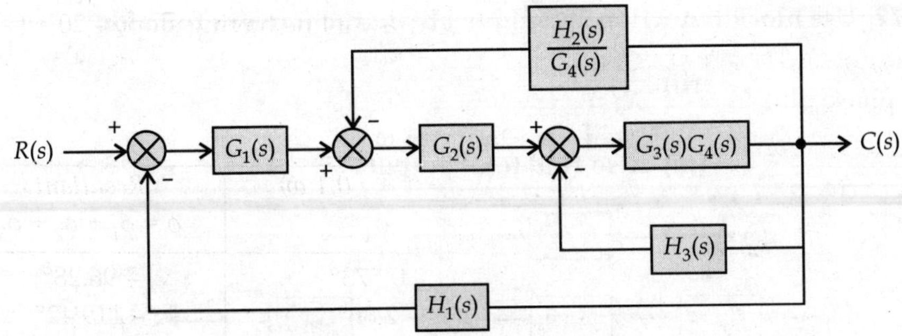

Step 3 : Solve the inner closed loop

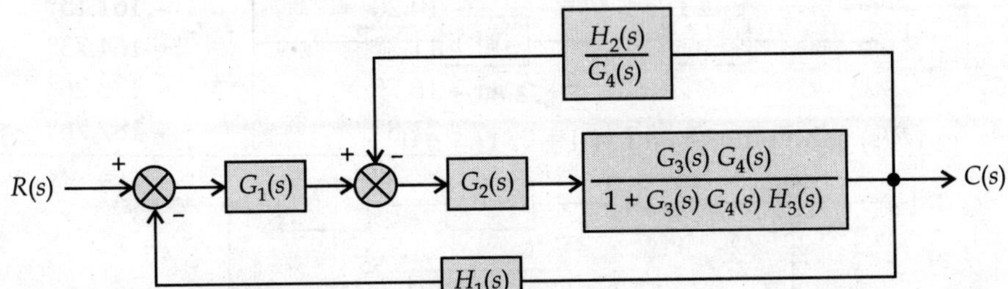

Step 4 : Two blocks are in cascade

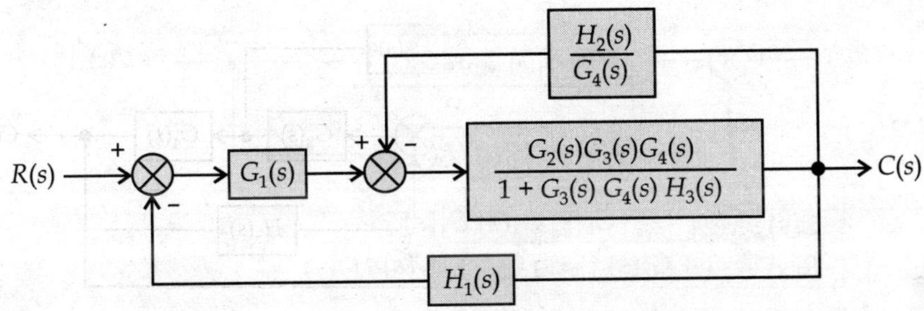

Step 5 : Solve the closed loop

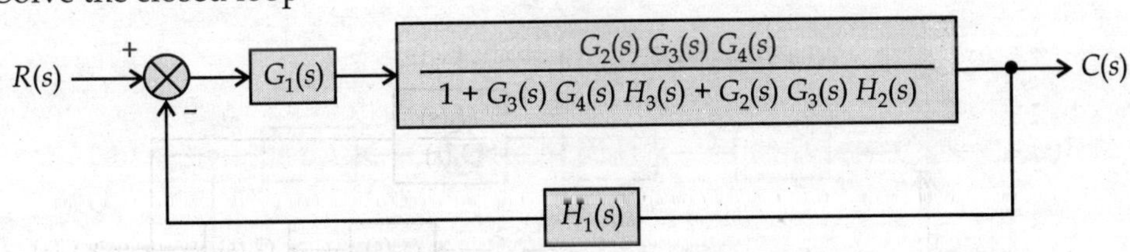

Step 6 : Two blocks are in cascade

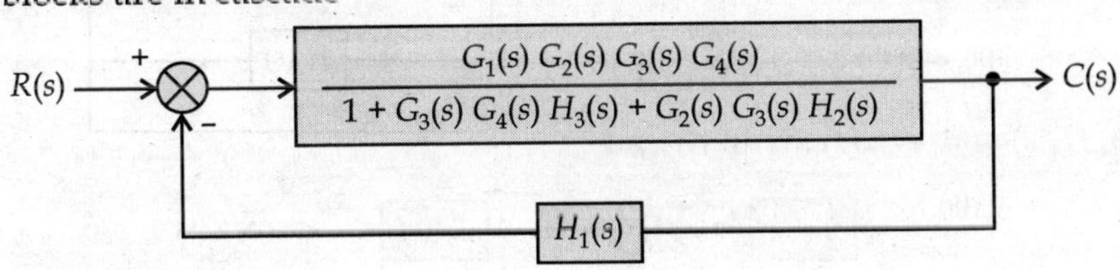

Step 7 : Find $C(s)/R(s)$

$$\frac{C(s)}{R(s)} = \frac{G_1(s)\,G_2(s)\,G_3(s)\,G_4(s)}{1 + G_3(s)\,G_4(s)\,H_3(s) + G_2(s)\,G_3(s)\,H_2(s) + G_1(s)G_2(s)G_3(s)G_4(s)H_1(s)}$$

Ans.

EXAMPLE 8.72. Use block reduction technique obtain the transfer function

(i) $\dfrac{C(s)}{R(s)}$ **(ii)** $\dfrac{C(s)}{X(s)}$

(iii) $\dfrac{C(s)}{Y(s)}$ **(iv) Also find total output**

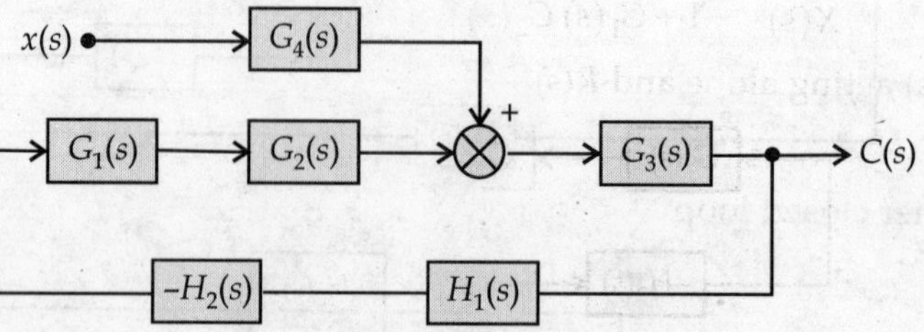

Fig. 8.80.

Solution: (*i*) When $R(s)$ acting alone and $x(s) = 0$, $Y(s) = 0$

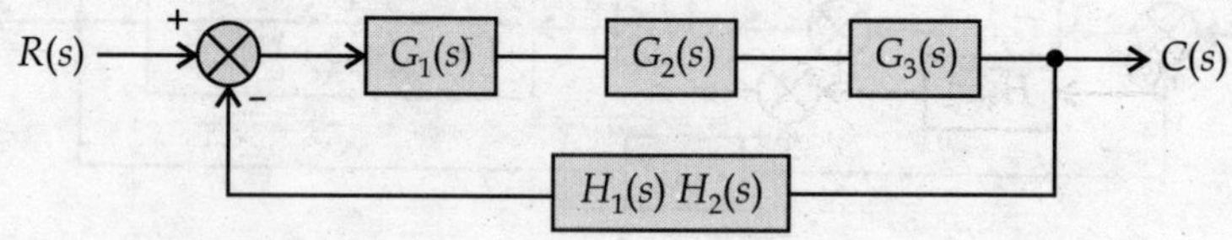

$G_1(s)$, $G_2(s)$, $G_3(s)$ are in cascade

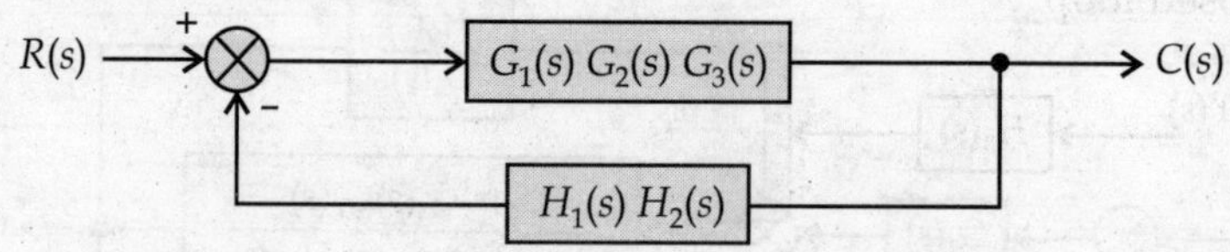

$$\frac{C(s)}{R(s)} = \frac{G_1(s)\,G_2(s)\,G_3(s)}{1 + G_1(s)\,G_2(s)\,G_3(s)\,H_1(s)\,H_2(s)} \quad \textbf{Ans.} \qquad ...(i)$$

(*ii*) When $X(s)$ acting alone and $R(s) = 0$, $Y(s) = 0$

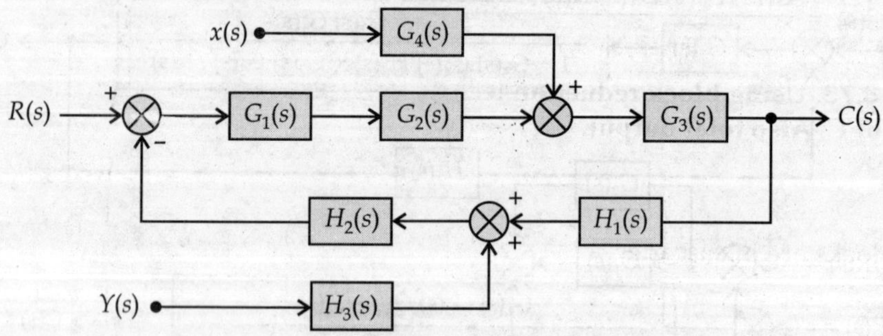

$G_1(s)$, $G_2(s)$, $H_1(s)$ and $-H_2(s)$ are in cascade

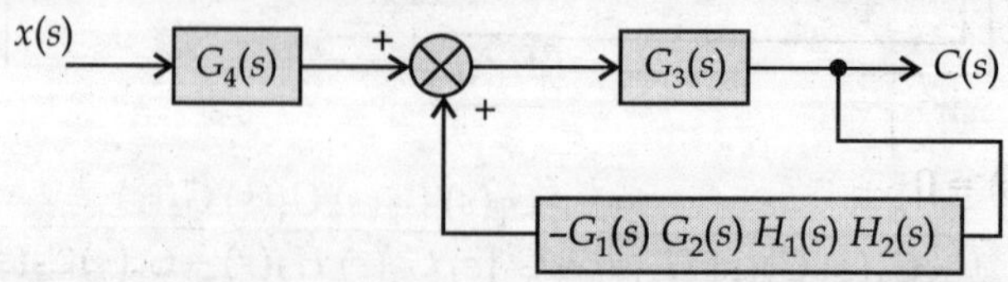

Solve the closed loop

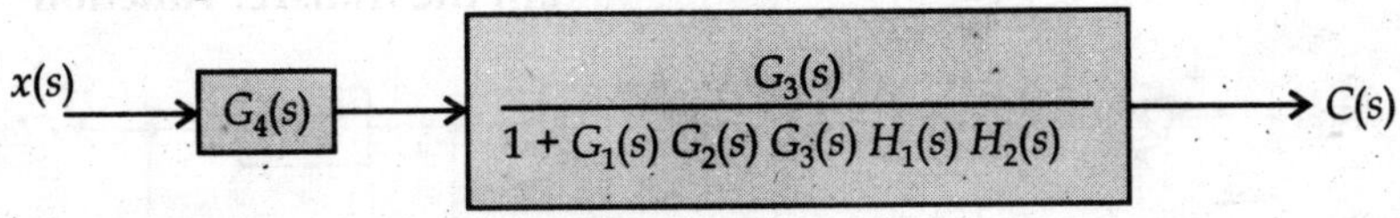

$$\frac{C(s)}{X(s)} = \frac{G_3(s)\, G_4(s)}{1 + G_1(s)\, G_2(s)\, G_3(s)\, H_1(s)\, H_2(s)} \quad \textbf{Ans.} \qquad ...(ii)$$

(*iii*) When Y(s) acting alone and $R(s) = 0$, $X(s) = 0$

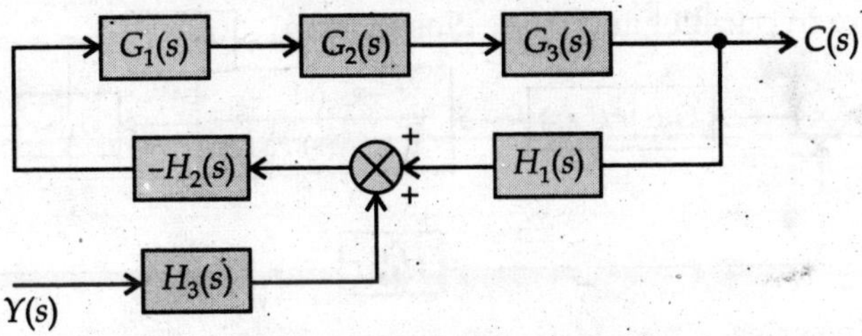

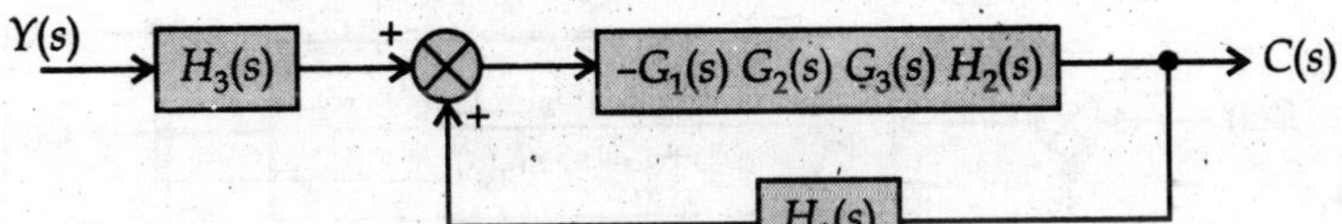

Solve the closed loop

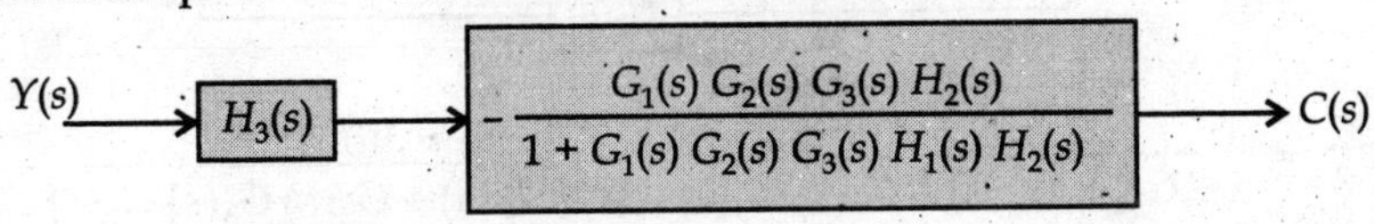

$$\frac{C(s)}{Y(s)} = -\frac{G_1(s)\, G_2(s)\, G_3(s)\, H_2(s)\, H_3(s)}{1 + G_1(s)\, G_2(s)\, G_3(s)\, H_1(s)\, H_2(s)} \quad \textbf{Ans.} \qquad ...(iii)$$

(*iv*) total output : from equtions (*i*), (*ii*), (*iii*)

$$C(s) = \frac{G_1(s)\, G_2(s)\, G_3(s)\, R(s) + G_3(s)\, G_4(s)\, X(s) - G_1(s)\, G_2(s)\, G_3(s)\, H_2(s)\, H_3(s)\, Y(s)}{1 + G_1(s)\, G_2(s)\, G_3(s)\, H_1(s)\, H_2(s)} \quad \textbf{Ans.}$$

EXAMPLE 8.73. Using block reduction technique, find the transfer function from each input to the output C. Also total output.

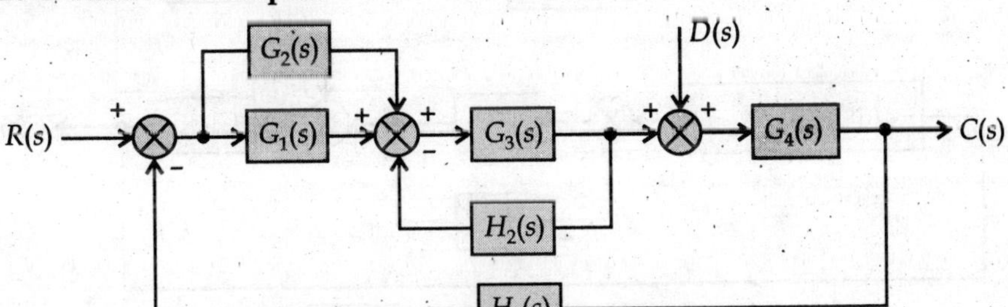

Fig. 8.81.

Solution : When $D(s) = 0$

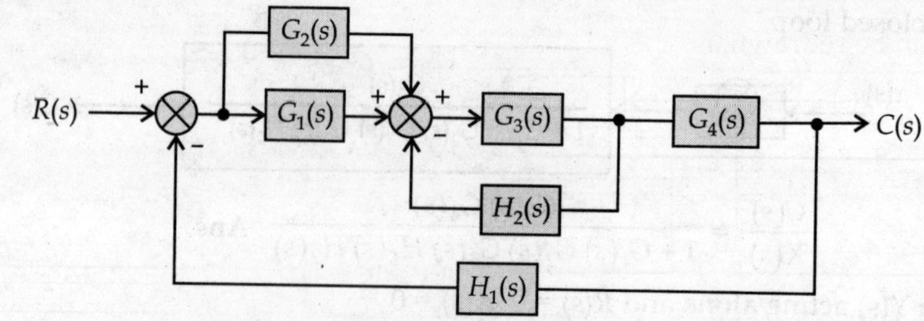

$G_1(s)$ and $G_2(s)$ are in parallel and solve closed loop

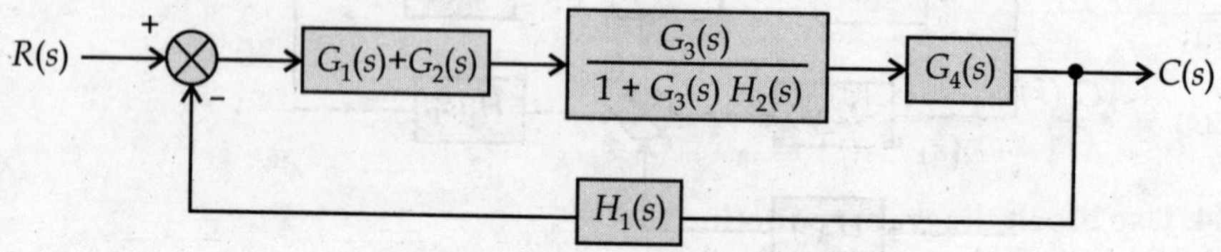

three blocks are in series

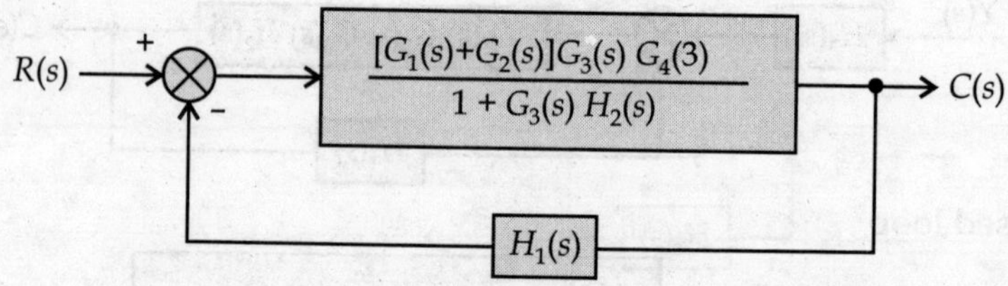

$$\frac{C(s)}{R(s)} = \frac{G_1(s)\,G_3(s)\,G_4(s) + G_2(s)\,G_3(s)\,G_4(s)}{1 + G_3(s)\,H_2(s) + G_1(s)\,G_3(s)\,G_4(s)\,H_1(s) + G_2(s)\,G_3(s)\,G_4(s)\,H_1(s)}$$ **Ans.**

when $R(s) = 0$

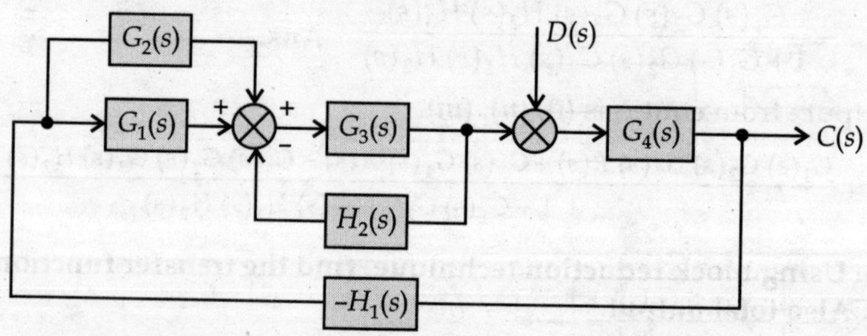

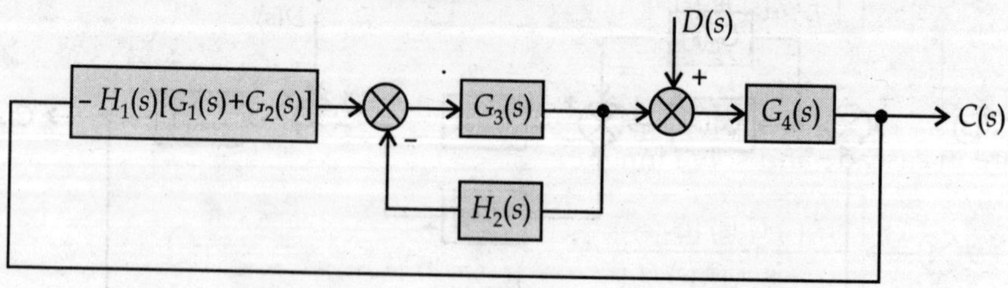

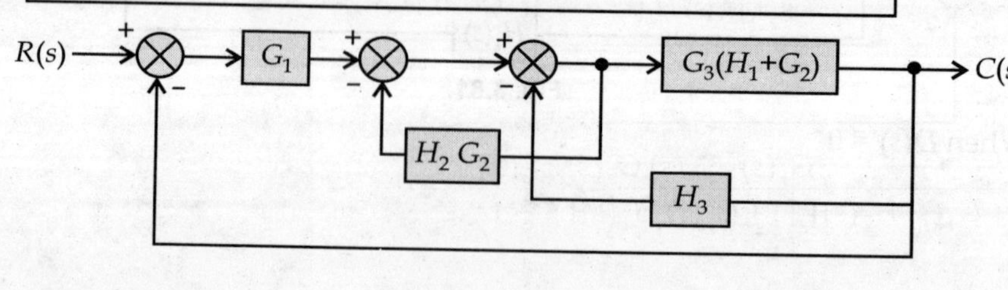

Above Fig. can be redrawn as

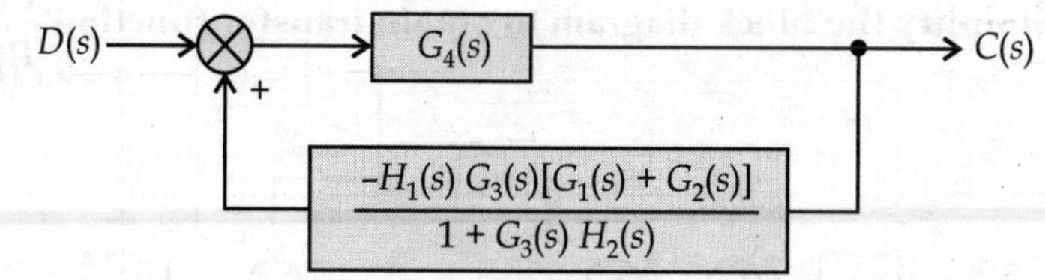

$$\frac{C(s)}{D(s)} = \frac{G_4(s)\,[1 + G_3(s)\,H_2(s)]}{1 + G_3(s)\,H_2(s) + G_3(s)\,G_4(s)\,G_1(s)\,H_1(s) + G_2(s)\,G_3(s)\,G_4(s)\,H_1(s)} \quad \textbf{Ans.}$$

Total output

$$C(s) = \frac{[G_1(s)\,G_3(s)\,G_4(s) + G_2(s)\,G_3(s)\,G_4(s)]\,R(s) + G_4(s)\,[1 + G_3(s)\,H_2(s)]\,D(s)}{1 + G_3(s)\,H_2(s) + G_1(s)\,G_3(s)\,G_4(s)\,H_1(s) + G_2(s)\,G_3(s)\,G_4(s)\,H_1(s)} \quad \textbf{Ans.}$$

EXAMPLE 8.74. Use block diagram reduction methods to obtain the equivlent tranfer function from R to C.

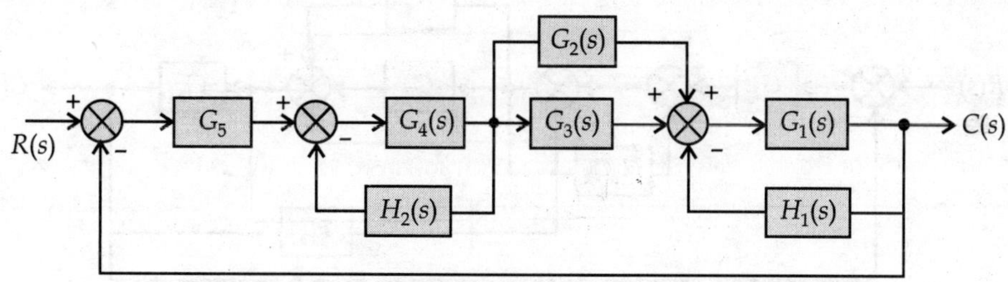

Fig. 8.82.

Solution : Solve the closed loop of $G_4(s)$ and $H_2(s)$

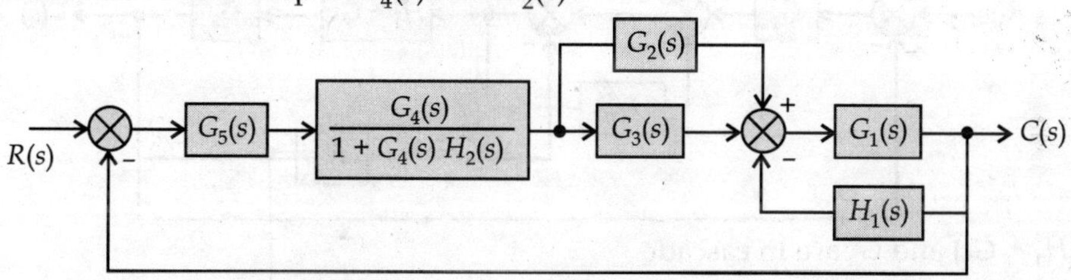

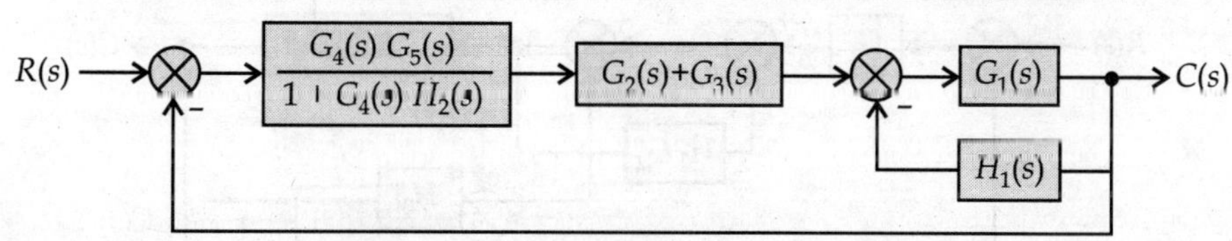

Two bloks are in cascade and solve inner closed loop

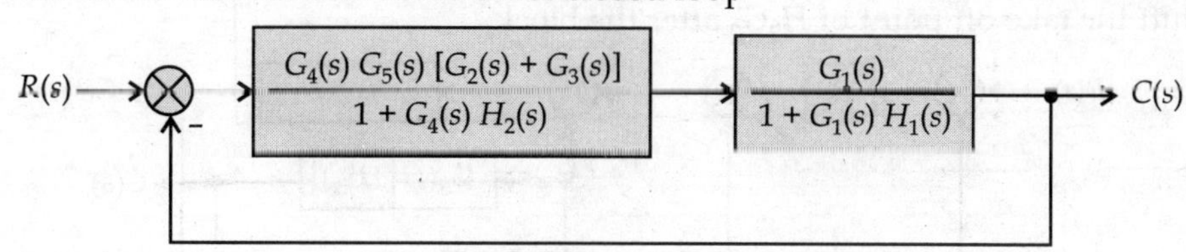

$$\frac{C(s)}{R(s)} = \frac{G_1(s)\;G_4(s)\,G_5(s)[G_2(s) +\;G_3(s)]}{[1 + G_4(s)\,H_2(s)]\;[1 + G_1(s)\,H_1(s)] + G_1(s)\,G_4(s)\,G_5(s)\,[G_2(s)\,G_3(s)]}$$

EXAMPLE 8.75. Simplify the block diagram to obtain transfer function $\dfrac{C(s)}{R(s)}$.

(UPTU 2003-2004)

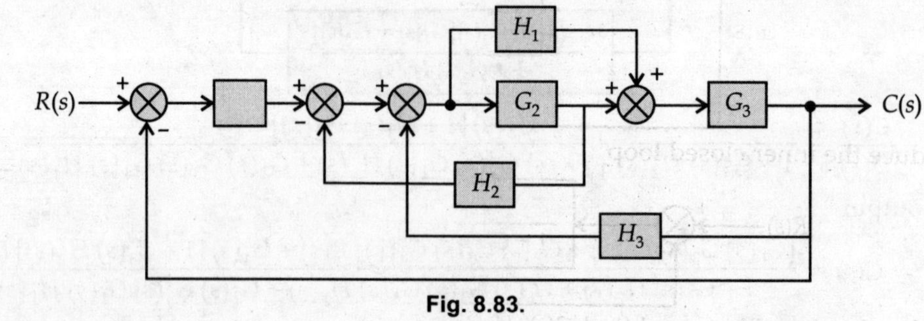

Fig. 8.83.

Step 1 : Shift the take off point of H_2 ahead of G_2.

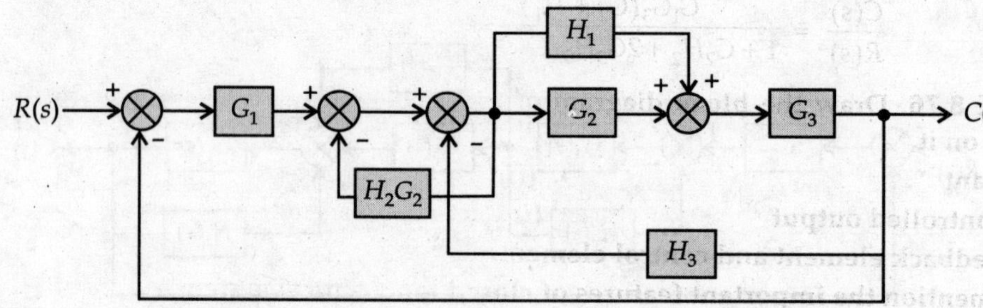

Step 2 : H_1 and G_2 are in parallel

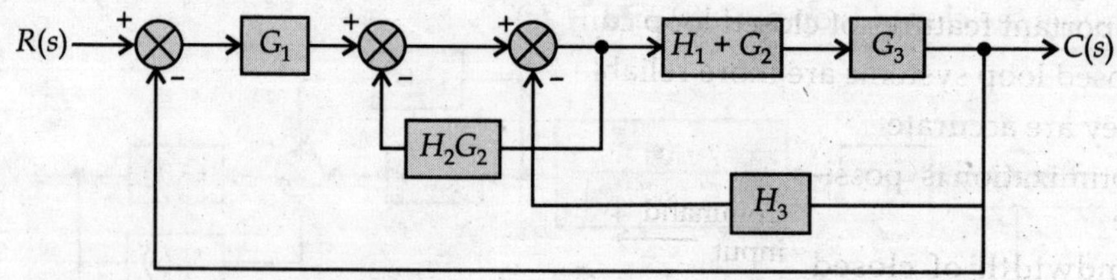

Step 3 : $(H_1 + G_2)$ and G_3 are in cascade

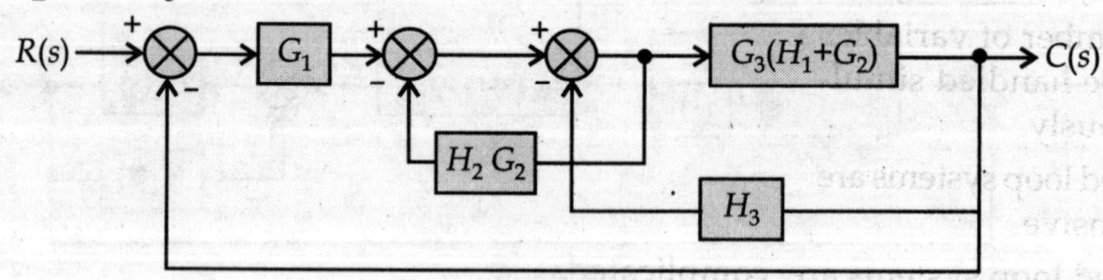

Step 4 : Shift the take off point of H_2G_2 after the block

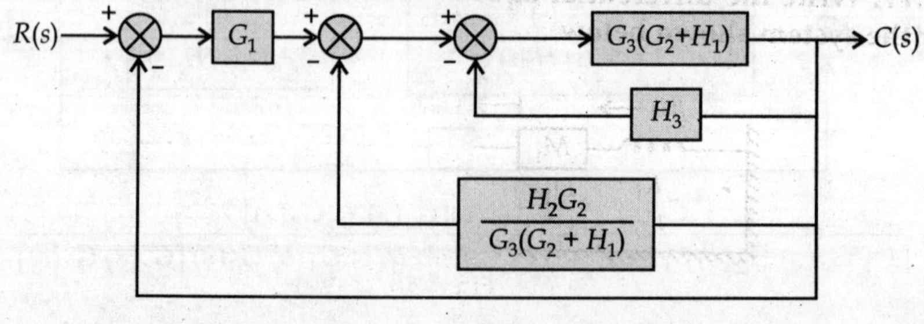

Step 5 : Solve the innermost closed loop

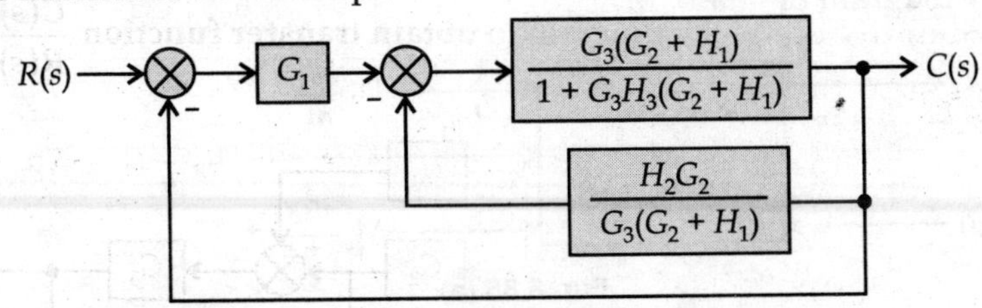

Step 6 : reduce the inner closed loop

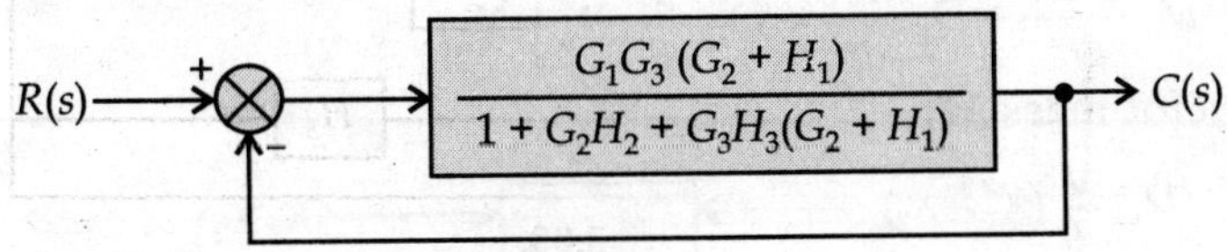

Step 7 : Solve the closed loop and find $C(s)/R(s)$

$$\frac{C(s)}{R(s)} = \frac{G_1G_3(G_2+H_1)}{1+G_2H_2+2G_3H_3(G_2+H_1)} \quad \textbf{Ans.}$$

EXAMPLE 8.76. Draw the block diagram of a closed loop control system and indicate the following on it

(*i*) Plant **(*ii*) Command input**

(*iii*) Controlled output **(*iv*) actuating signal**

(*v*) feedback element and control element.

Also mention the important features of closed loop control system. *(UPTU 2003-2004)*

Solution : Fig. 8.84 shows a closed loop control system with all given elements.

The important features of closed loop control systems are

1. Closed loop systems are more reliable
2. They are accurate
3. Optimization is possible
4. Bandwidth of closed loop system is high.
5. A number of variables can be handled simultaneously.
6. Closed loop systems are expensive.
7. Closed loop systems are complicated.

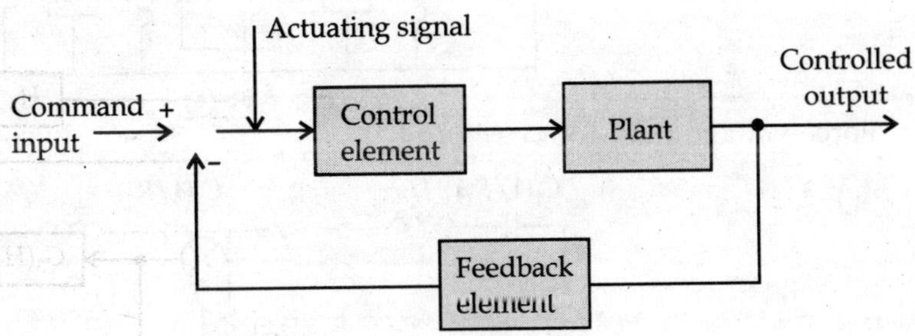

Fig. 8.84.

EXAMPLE 8.77. Write the differential equations and obtain (f – v) force-voltage analogous networks for the system shown below. *(UPTU 2003-2004)*

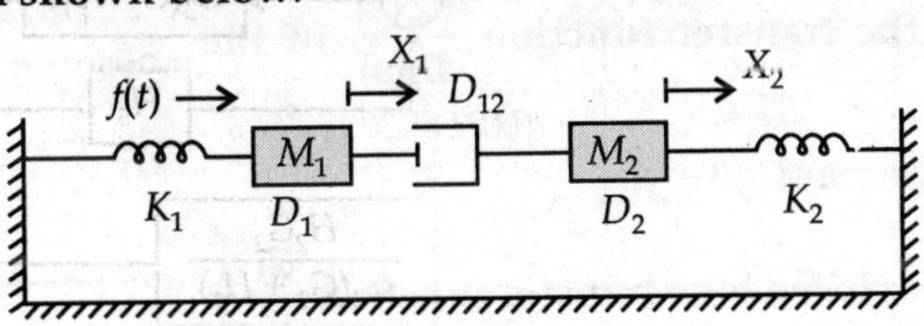

Fig. 8.85.

Solution : Freebody diagram of mass M_1

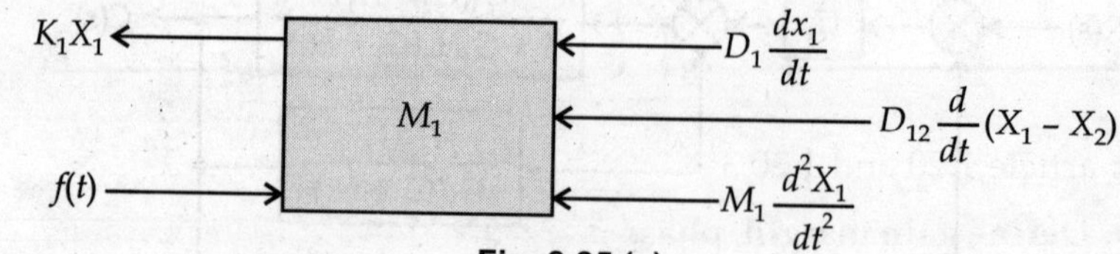

Fig. 8.85 (a)

$$\therefore \quad f(t) = M_1 \frac{d^2X_1}{dt^2} + D_1 \frac{dX_1}{dt} + D_{12} \frac{d}{dt}(X_1 - X_2) + k_1X_1 \qquad ...(i)$$

Freebody diagram for mass M_2

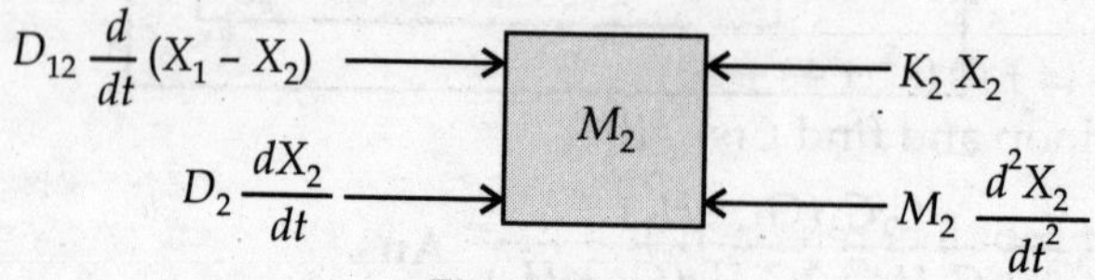

Fig. 8.85 (b)

$$\therefore \quad D_{12} \frac{d}{dt}(X_1 - X_2) = M_2 \frac{d^2X_2}{dt^2} + D_2 \frac{dX_2}{dt} + k_2X_2 \qquad ...(ii)$$

Equations (*i*) and (*ii*) are the required differential equations.
Mechanical equivalent diagram of Fig. 8.85.

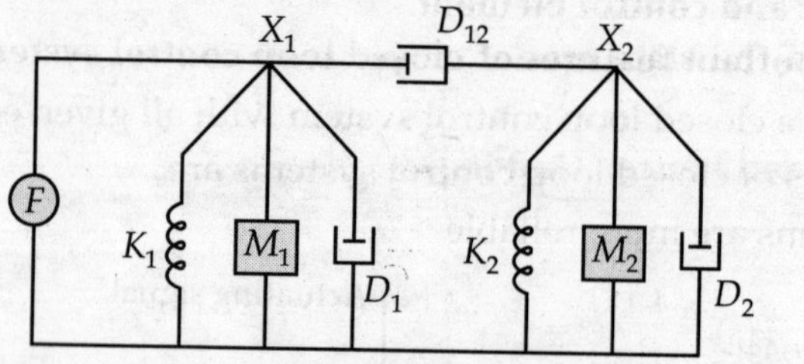

Fig. 8.85 (c)

Force-voltage analogous circuit

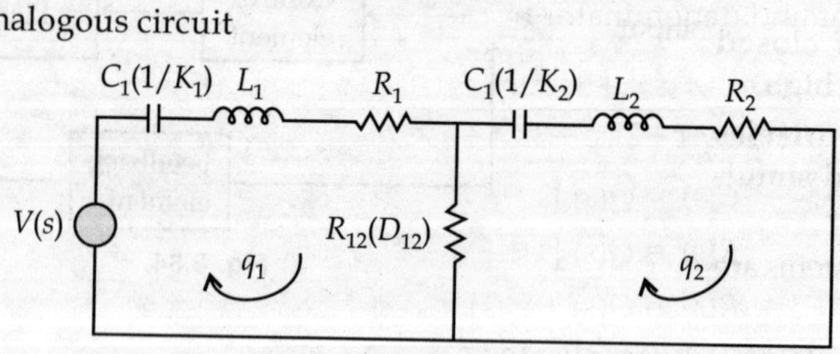

Fig. 8.85 (d)

Fig. 8.86 (*d*) is the required *f* – *v* analogous circuit.

EXAMPLE 8.78. Derive the transfer function $\frac{E_o(s)}{E_i(s)}$ of the network shown below. *(UPTU 2002-2003)*

Solution : Refer solved exampled 1.3

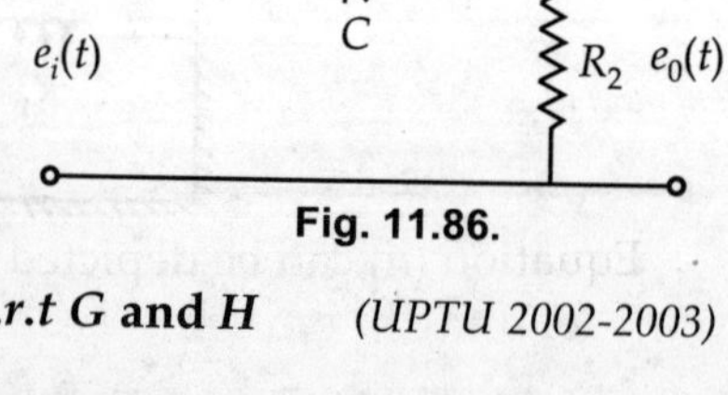

Fig. 11.86.

EXAMPLE 8.79. Derive the closed loop transfer function $\frac{C(s)}{R(s)}$ = $M(s)$ for the system shown below and find its sensitivity *w.r.t* G and H *(UPTU 2002-2003)*

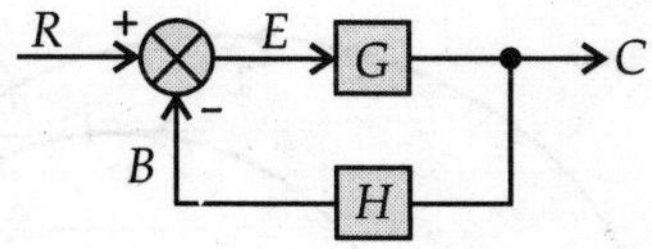

Solution : Refer article 1.20 and 1.30.3.

EXAMPLE 8.80. Define gain margin, phase margin, gain crossover frequency, phase crossover frequency in a polar plot.

Solution : Refer article 4.6 *(UPTU 2002-2003)*

EXAMPLE 8.81. State the rules for construction of root loci of $G(s)\ H(s)$. Find the breakaway points of $G(s)H(s) = \dfrac{k}{s\,(s+4)\,(s^2+4s+20)}$

Solution : *(UPTU 2002-2003)*

For construction of root loci refer article 5.8 and for breakaway point refer example 5.33.

EXAMPLE 8.82. A unity feedback system has open loop transfer function $G(s) = \dfrac{k}{s\,(0.1s+1)\,(0.2s+1)}$. **The system is required to satisfy the following performance specifications $k_v = 30$, P.M ≥ 40 bandwidth $\omega_b = 5$ rad/sec. Design a lag compensator.** *(UPTU 2003-2004)*

Solution : Refer example 6.4

EXAMPLE 8.83. Obtain the state model of the system whose transfer function is given by $T(s) = \dfrac{s^2+3s+3}{s^3+2s^2+3s+1}$ **and hence, find its state transition matrix.** *(UPTU 2003-2004)*

Solution : Let $$T(s) = \frac{C(s)}{R(s)} = \frac{s^2+3s+3}{s^3+2s^2+3s+1}$$

Divide numerator and denominator by s^3

$$\frac{C(s)}{R(s)} = \frac{s^{-1}+3s^{-2}+3s^{-3}}{1+2s^{-1}+3s^{-1}+s^{-3}} \cdot \frac{Q(s)}{Q(s)}$$

$$C(s) = Q(s)\,[s^{-1}+3s^{-2}+3s^{-3}] \quad \text{...}(i)$$

$$R(s) = Q(s)\,[1+2s^{-1}+3s^{-2}+s^{-3}] \quad \text{...}(ii)$$

or, $$R(s) = Q(s) + Q(s)\,[2s^{-1}+3s^{-2}+s^{-3}]$$

∴ $$Q(s) = R(s) - Q(s)\,[2s^{-1}+3s^{-2}+s^{-3}] \quad \text{...}(iii)$$

Equation (*iii*) can be depicted by signal flow diagram as

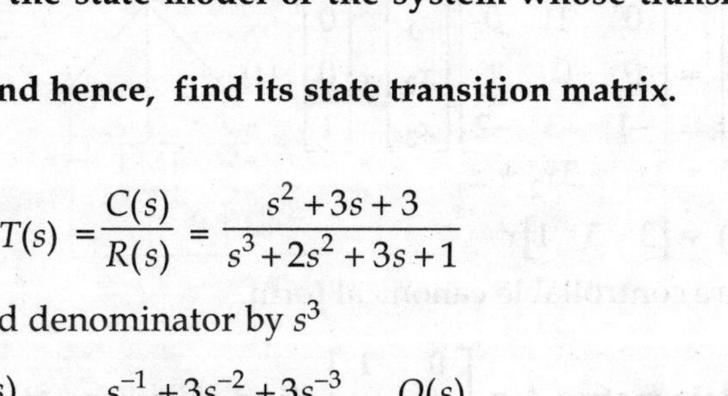

Equation (*ii*) can be depicted by SFD as

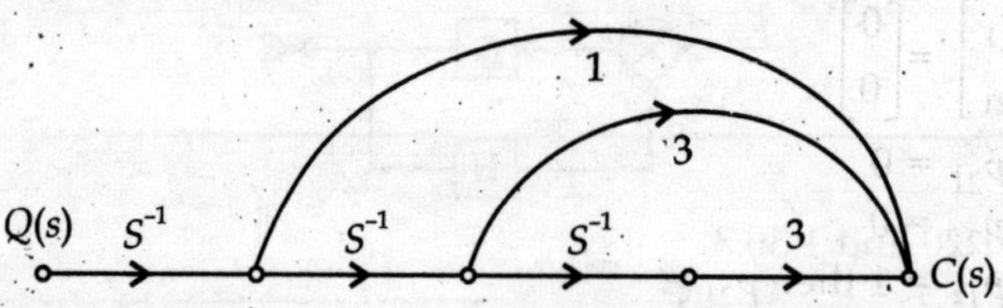

Combining the above two SFG.

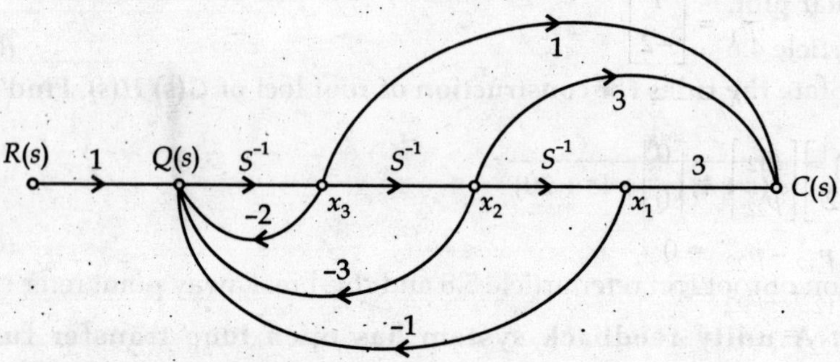

The output of each integrator is a state variable, then we have

$$\dot{x}_1 = x_2$$

$$\dot{x}_2 = x_3$$

$$\dot{x}_3 = r(t) - x_1 - 3x_2 - 2x_3$$

$$\begin{bmatrix} \dot{x}_1 \\ \dot{x}_2 \\ \dot{x}_3 \end{bmatrix} = \begin{bmatrix} 0 & 1 & 0 \\ 0 & 0 & 1 \\ -1 & -3 & -2 \end{bmatrix} \begin{bmatrix} x_1 \\ x_2 \\ x_3 \end{bmatrix} + \begin{bmatrix} 0 \\ 0 \\ 1 \end{bmatrix} r(t)$$

$$C(t) = 3x_1 + 3x_2 + x_3$$

$$C(t) = \begin{bmatrix} 3 & 3 & 1 \end{bmatrix} x$$

The above equations are controllable canonical form.

EXAMPLE 8.84. For the state matrix A = $\begin{bmatrix} 0 & 1 \\ -8 & -6 \end{bmatrix}$, find e^{At} by any two methods.

(UPTU 2002-2003)

Solution : First Method : Use of diagonal

$$A = \begin{bmatrix} 0 & 1 \\ -8 & -6 \end{bmatrix}$$

$$\lambda I - A = \lambda \begin{bmatrix} 1 & 0 \\ 0 & 1 \end{bmatrix} - \begin{bmatrix} 0 & 1 \\ -8 & -6 \end{bmatrix} = \begin{bmatrix} \lambda & -1 \\ 8 & \lambda + 6 \end{bmatrix} \quad ...(i)$$

$$|\lambda I - A| = \lambda^2 + 6\lambda + 8 = 0$$

$$(\lambda + 2)(\lambda + 4) = 0$$

$$\lambda_1 = -2, \lambda_2 = -4$$

let the eigen vector $p_1 = \begin{bmatrix} p_{11} \\ p_{21} \end{bmatrix}$ & $p_2 = \begin{bmatrix} p_{12} \\ p_{22} \end{bmatrix}$

Put $\lambda = -2$ in equation (*i*)

$$\begin{bmatrix} -2 & -1 \\ 8 & 4 \end{bmatrix} \begin{bmatrix} p_{11} \\ p_{21} \end{bmatrix} = \begin{bmatrix} 0 \\ 0 \end{bmatrix}$$

$$-2p_{11} - p_{21} = 0$$

$$8p_{11} + 4p_{21} = 0$$

let $p_{11} = 1$ then $p_{21} = -2$

$$\therefore \quad p_1 = \begin{bmatrix} 1 \\ -2 \end{bmatrix}$$

Put $\lambda = -4$

$$\begin{bmatrix} -4 & -1 \\ 8 & 2 \end{bmatrix} \begin{bmatrix} p_{12} \\ p_{22} \end{bmatrix} = \begin{bmatrix} 0 \\ 0 \end{bmatrix}$$

$$-4p_{12} - p_{22} = 0$$

$$8p_{12} + 2p_{22} = 0$$

let $p_{12} = 1$ then $p_{22} = -4$

$$p_2 = \begin{bmatrix} 1 \\ -4 \end{bmatrix}$$

$$M = [p_1\, p_2] = \begin{bmatrix} 1 & 1 \\ -2 & -4 \end{bmatrix}$$

$$\therefore \quad M^{-1} = \begin{bmatrix} 2 & 0.5 \\ -1 & -0.5 \end{bmatrix} \;\&\; \wedge = \begin{bmatrix} -2 & 0 \\ 0 & -4 \end{bmatrix}$$

$$e^{At} = Me^{\wedge t}\, M^{-1} = \begin{bmatrix} 1 & 1 \\ -2 & -4 \end{bmatrix} \begin{bmatrix} e^{-2t} & 0 \\ 0 & e^{-4t} \end{bmatrix} \begin{bmatrix} 2 & 0.5 \\ -1 & -0.5 \end{bmatrix}$$

$$\therefore \quad e^{At} = \begin{bmatrix} 2e^{-2t} - e^{-4t} & 0.5e^{-2t} - 0.5e^{-4t} \\ -4e^{-2t} + 4e^{-4t} & -e^{-2t} + 2e^{-4t} \end{bmatrix} \quad \textbf{Ans.}$$

Second Method : Series summation method.

$$A = \begin{bmatrix} 0 & 1 \\ -8 & -6 \end{bmatrix}$$

$$\therefore \quad A^2 = \begin{bmatrix} -8 & -6 \\ 48 & 28 \end{bmatrix},$$

$$A^3 = \begin{bmatrix} 48 & 28 \\ -224 & -120 \end{bmatrix},$$

$$A^4 = \begin{bmatrix} -224 & -120 \\ 960 & 496 \end{bmatrix}$$

$$e^{At} = I + At + \frac{A^2t^2}{2!} + \frac{A^3t^3}{3!} + \ldots$$

$$e^{At} = \begin{bmatrix}1 & 0\\0 & 1\end{bmatrix} + \begin{bmatrix}0 & 1\\-8 & -6\end{bmatrix}t + \frac{t^2}{2}\begin{bmatrix}-8 & -6\\48 & 28\end{bmatrix} + \frac{t^3}{6}\begin{bmatrix}48 & 28\\-224 & -120\end{bmatrix} + \frac{t^4}{24}\begin{bmatrix}-224 & -120\\960 & 496\end{bmatrix}\cdots$$

$$e^{At} = \begin{bmatrix}1 & 0\\0 & 1\end{bmatrix} + \begin{bmatrix}0 & t\\-8t & -6t\end{bmatrix} + \begin{bmatrix}-4t^2 & -3t^2\\24t^2 & 14t^2\end{bmatrix} + \begin{bmatrix}8t^3 & 4.67t^3\\-37.3t^3 & -20t^3\end{bmatrix} + \begin{bmatrix}-9.33t^4 & -5t^4\\40t^4 & 20.67t^4\end{bmatrix}\cdots$$

$$\therefore \quad e^{At} = \begin{bmatrix}t - 4t^2 + 8t^3 - 9.33t^4 \ldots\ldots & t - 3t^2 + 4.67t^3 - 5t^4 \ldots\\-8t + 24t^2 - 37.3t^3 + 40t^4 \ldots & 1 - 6t + 14t^2 - 20t^3 + 20.67t^4 \ldots\end{bmatrix} \quad \textbf{Ans.}$$

EXAMPLE 8.85. Derive the state variable model for the system shown below :

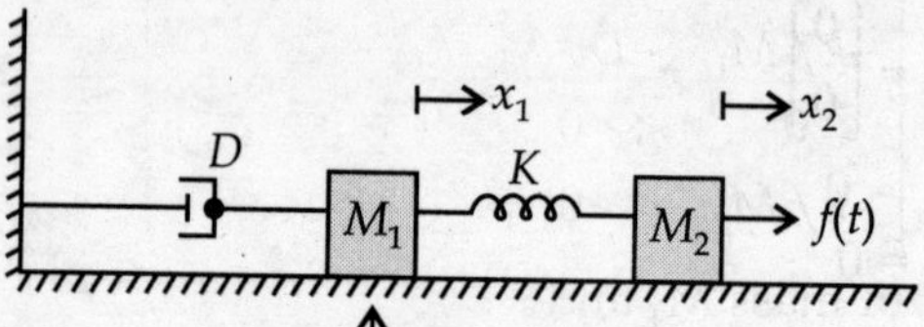

Fig. 8.87.

(UPTU 2002-2003)

Solution : Mechanical equivalent diagram

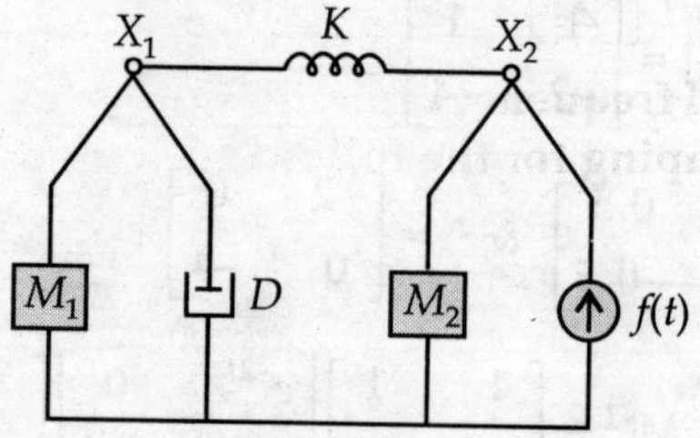

Fig. 8.87. (a)

at node (*i*)

$$M_1\ddot{x}_1 + D\dot{x}_1 + k(x_1 - x_2) = 0$$

at node (*ii*)

$$f(t) = M_2\ddot{x}_2 + k(x_2 - x_1)$$

or, $M_1\ddot{x}_1 + D\dot{x}_1 + kx_1 - kx_2 = 0$ ...(*i*)

$$f(t) = M_2\ddot{x}_2 + kx_2 - kx_1 \quad \ldots(ii)$$

Select the state variable as

$$x_1 = z_1 \qquad x_2 = z_3$$

$$\dot{x}_1 = \dot{z}_1 = z_2 \qquad \dot{x}_2 = \dot{z}_3 = z_4$$

$$\ddot{x}_1 = \dot{z}_2 \qquad \ddot{x}_2 = \dot{z}_4$$

$$f(t) = M_2\ddot{x}_2 + kx_2 - kx_1 = M_2\dot{z}_4 + kz_3 - kz_1$$

$$\therefore \quad \dot{Z}_4 = \frac{f(t)}{M_2} - \frac{k}{M_2}Z_3 - \frac{k}{M_2}Z_1 \quad \ldots(iii)$$

from equation (*i*)

$$\dot{Z}_2 = -\frac{k}{M_1}Z_1 - \frac{D}{M_1}Z_2 + \frac{k}{M_1}Z_3 \quad \ldots(iv)$$

state equations are

$$\dot{Z}_1 = Z_2$$

$$\dot{Z}_2 = -\frac{k}{M_1} Z_1 - \frac{D}{M_1} Z_2 + \frac{k}{M_1} Z_3$$

$$\dot{Z}_3 = Z_4$$

$$\dot{Z}_4 = \frac{f(t)}{M_2} - \frac{k}{M_2} Z_3 - \frac{k}{M_2} Z_1$$

$$\begin{bmatrix} \dot{Z}_1 \\ \dot{Z}_2 \\ \dot{Z}_3 \\ \dot{Z}_4 \end{bmatrix} = \begin{bmatrix} 0 & 1 & 0 & 0 \\ -k/M_1 & -D/M_1 & k/M_1 & 0 \\ 0 & 0 & 0 & 1 \\ -k/M_2 & 0 & -k/M_2 & 0 \end{bmatrix} \begin{bmatrix} Z_1 \\ Z_2 \\ Z_3 \\ Z_4 \end{bmatrix} + \begin{bmatrix} 0 \\ 0 \\ 0 \\ 1/M_2 \end{bmatrix} f(t) \quad ...(v)$$

Select the displacement of mass M_1 and M_2 as output variables.

$$y_1 = x_1 = z_1$$

$$y_2 = x_2 = z_3$$

$$\begin{bmatrix} y_1 \\ y_2 \end{bmatrix} = \begin{bmatrix} 1 & 0 \\ 0 & 1 \end{bmatrix} \begin{bmatrix} Z_1 \\ Z_3 \end{bmatrix} \quad \textbf{Ans.} \quad ...(vi)$$

EXAMPLE 8.86. Define natural frequency of oscillation ω_n and damping ratio of second order system. Find the nature of damping for the following systems. *(UPTU 2002-2003)*

(*i*) 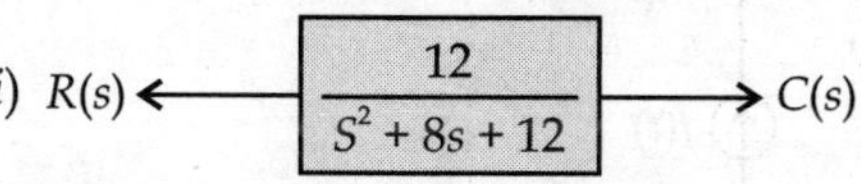

(*ii*) 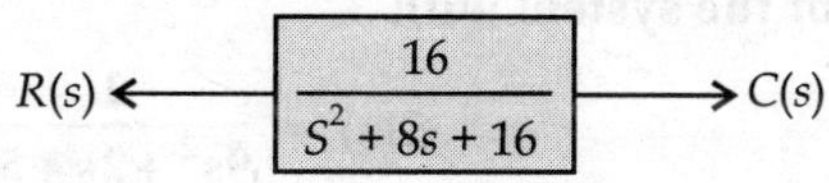

(*iii*) 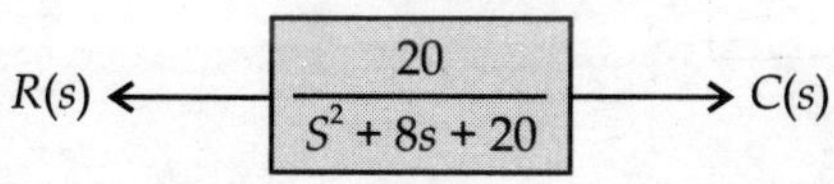

Solution : Each and every system takes finite time to reach its steady state and during this period system oscillates or increases exponentially. The type and location of closed loop poles decide the behaviour of the system. Closed loop poles are depends upon the selection of parameters of the system. Every system has a tendency to oppose the oscillatory behaviour of the system which is called as damping. This damping is measured by a factor or ratio called damping ratio. This damping ratio is denoted by ξ (zeta). In some system it is low, such system oscillates but slowly *i.e.*, with damped frequency. If it is zero, the system will oscillate with maximum frequency. As there is no opposition from system, system naturally and freely oscillates under such condition. Hence this frequency of oscillations at $\xi = 0$ called natural frequency of oscillation and denoted by ω_n red/sec.

(*i*) 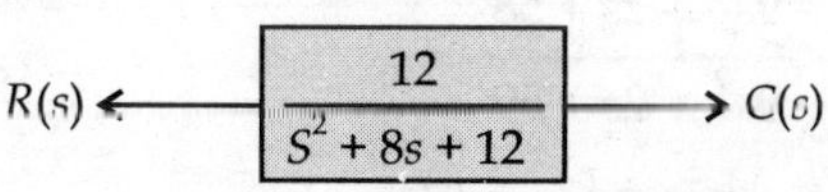

$$\frac{C(s)}{R(s)} = \frac{12}{s^2 + 8s + 12}$$

Compare with
$$\frac{C(s)}{R(s)} = \frac{\omega_n^2}{s^2 + 2\zeta\omega_n s + \omega_n^2} \quad ...(i)$$

$$\omega_n^2 = 12 \quad \therefore \omega_n = 3.464 \text{ rad/sec.}$$
$$2\zeta\omega_n = 8 \quad \therefore \zeta = 1.15$$

Since, $\zeta > 1$, the system is overdamped.

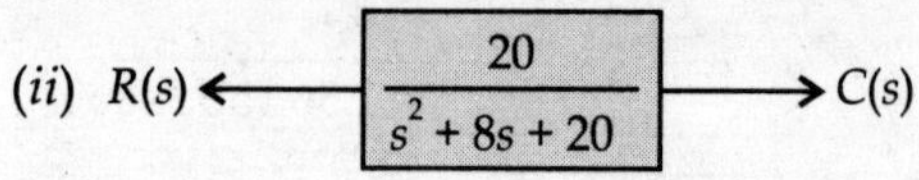

$$\frac{C(s)}{R(s)} = \frac{16}{s^2 + 8s + 16}$$

Compare with equation (*i*)

$$\omega_n^2 = 16 \qquad \omega_n = 4 \text{ rad/sec.}$$
$$2\,\zeta\omega_n = 8 \qquad \zeta = 1$$

Since, $\zeta = 1$, the system is critically damped.

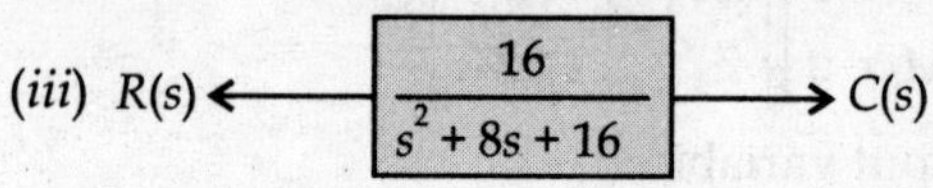

$$\frac{C(s)}{R(s)} = \frac{20}{s^2 + 8s + 20}$$

$$\omega_n^2 = 20 \qquad \omega_n = 4.472 \text{ rad/sec.}$$
$$2\zeta\omega_n = 8 \qquad \zeta = 0.894$$

Since, $\zeta < 1$, the system is underdamped.

EXAMPLE 8.87. Find the damping factor ζ, natural frequency ω_n, peak time T_p and percentage overshoot for the system with

$$G(s) = \frac{1}{9s^2 + 2s + 58}$$

Solution : Closed loop transfer function

$$\frac{C(s)}{R(s)} = \frac{G(s)}{1 + G(s)H(s)}$$

Here,

$$G(s) = \frac{1}{9s^2 + 2s + 58}, \quad H(s) = 1$$

$$\therefore \quad \frac{C(s)}{R(s)} = \frac{\dfrac{1}{9s^2 + 2s + 58}}{1 + \dfrac{1}{9s^2 + 2s + 58} \cdot 1}$$

$$= \frac{1}{\cancel{9s^2 + 2s + 58}} \cdot \frac{\cancel{9s^2 + 2s + 58}}{9s^2 + 2s + 59}$$

$$\frac{C(s)}{R(s)} = \frac{1}{9s^2 + 2s + 59}$$

or,

$$\frac{C(s)}{R(s)} = \frac{(1/9)}{s^2 + \dfrac{2}{9}s + \dfrac{59}{9}}$$

We know that

$$\frac{C(s)}{R(s)} = \frac{\omega_n^2}{s^2 + 2\zeta\omega_n s + \omega_n^2}$$

Compare the denominator, we get

$$\omega_n^2 = \frac{59}{9} \qquad \therefore\ \omega_n^2 = 6.56 \qquad \boxed{\therefore\ \omega_n = 2.56 \text{ rad/sec.}}$$

$$2\zeta\omega_n = \frac{2}{9} \qquad \text{or,} \quad 2 \times \zeta \times 2.56 = \frac{2}{9}$$

$$\therefore \qquad \zeta = 0.0434$$

$$\text{Peak Time } T_p = \frac{\pi}{\omega_n\sqrt{1-\zeta^2}}$$

$$\therefore \qquad T_p = \frac{\pi}{2.56\sqrt{1-(0.0434)^2}} = 1.23 \text{ sec} \qquad \boxed{\therefore\ T_p = 1/23 \text{ sec.}}$$

EXAMPLE 8.88. A closed loop servo system is represented by the differential equation

$$\frac{d^2C}{dt^2} + 8\frac{dC}{dt} = 64e$$

where C is the displacement of the output shaft, r is the displacement of the input shaft and $e = r - c$. Determine : (*i*) damping ratio, (*ii*) damped natural frequency, (*iii*) % M_P.

(UPTU, 2003-2004)

Solution :

$$\frac{d^2C}{dt^2} + 8\frac{dC}{dt} = 64\,(r - C)$$

$$s^2C(s) + 8sC(s) = 64\,[R(s) - C(s)]$$

$$C(s)\,[s^2 + 8s + 64] = 64\,R(s)$$

$$\frac{C(s)}{R(s)} = \frac{64}{s^2 + 8s + 64}$$

Compare with

$$\frac{C(s)}{R(s)} = \frac{\omega_n^2}{s^2 + 2\zeta\omega_n s + \omega_n^2}$$

$$\omega_n 2 = 64 \qquad \omega_n = 8 \text{ rad/sec.}$$

$$2\zeta\omega_n = 8 \qquad \therefore\ \zeta = 0.5$$

$$\%\,M_P = 100\,e^{-\frac{\pi\zeta}{\sqrt{1-\zeta^2}}} = 100\,e^{-\frac{\pi\,0.5}{\sqrt{1-0.5^2}}}$$

$$M_P = 16.3\ \% \qquad \textbf{Ans.}$$

EXAMPLE 8.89. Sketch the root locus for the system having

$$G(s) = \frac{k}{(s^2 + 2s + 2)\,(s^2 + 2s + 5)}$$

Solution :

Step 1 : Root locus is symmetrical about real axis.

Step 2 : No. of poles = 4 at $s = -1 \pm j$ and $s = -1 \pm j2$ No. of zeros = 0

Step 3 : No. of root locii = No. of poles ($\because p > z$) = 4

Step 4 : Centroid of asymptotes

$$\sigma_A = \frac{-1+j-1-j-1+j2-1-j2}{4-0}$$

$$\sigma_A = -1$$

Step 5: Angle of asymptotes

$$\phi = \frac{2k+1}{p-z} 180°$$

For $k = 0, 1, 2, 3$

$$\phi_1 = 45°, \phi_2 = 135°, \phi_3 = 225°, \phi_4 = 315°$$

Step 6: Calculation of Breakaway point

The characteristic equation : $1 + G(s)\, H(s) = 0$

$$1 + \frac{k}{(s^2+2s+2)(s^2+2s+5)} = 0$$

$$\therefore \quad k = -(s^4 + 4s^3 + 11s^2 + 14s + 10)$$

$$\frac{dk}{ds} = (4s^3 + 12s^2 + 22s + 14).$$

or,
$$\frac{dk}{ds} = 2s^3 + 6s^2 + 11s + 7$$

Note : If the poles (and zeros) are symmetical *w.r.t* some vertical line (*i.e.* s = – 1 line in this case) the equation of vertical line is one of the root of $\frac{dk}{ds} = 0$ (here s = – 1 is one of the root of $\frac{dk}{ds} = 0$)

$$\therefore \quad \frac{dk}{ds} = (s+1)(2s^2 + 4s + 7)$$

$\therefore$ Roots of $\frac{dk}{ds}$ are

$$s = -1, \quad -1 \pm j\,1.581$$

s = –1 lies on real axis, this does not belong to root loucs so this is not the breakaway point.

At $s = -1 \pm j\,1.58$ angle criterion is satisfied hence it is the breakaway point.

Step 7: Real axis is not the part of root locus.

Step 8: Point of intersection with imaginary axis.

The characteristic equation is

$$s^4 + 4s^3 + 11s^2 + 14s + (10 + k) = 0$$

s^4	1	11	$10 + k$
s^3	4	14	
s^2	7.5	$10 + k$	
s^1	$\frac{65-4k}{7.5}$	0	
s^0	$10 + k$		

For roots on imaginary axis

$$65 - 4k = 0$$

or
$$k = 16.25$$

Auxiliary equation

$$7.5s^2 + (10 + k) = 0$$

$$7.5s^2 + (10 + 16.25) = 0$$

$$\therefore \quad s = \pm\, j\,1.87$$

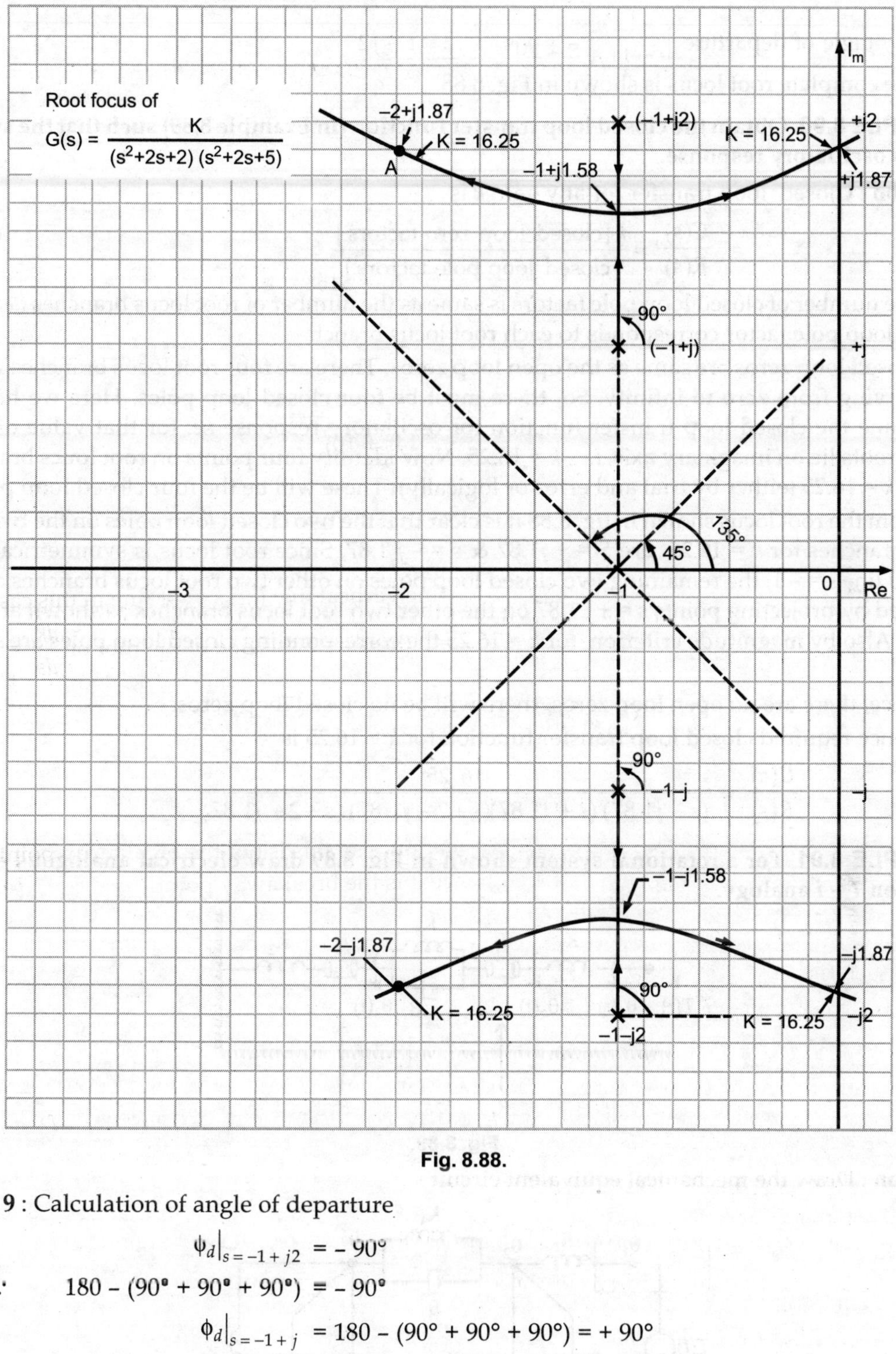

Fig. 8.88.

Step 9 : Calculation of angle of departure

$$\psi_d\big|_{s=-1+j2} = -90°$$

$$\because \quad 180 - (90° + 90° + 90°) = -90°$$

$$\phi_d\big|_{s=-1+j} = 180 - (90° + 90° + 90°) = +90°$$

$$\therefore \text{ angle of departure } \big|_{s=-1\pm j2} = \mp 90°$$

angle of departure $\left.\right|_{s=-1\pm j1} = \pm 90°$

The complete root locus is shown in Fig. 8.88.

EXAMPLE 8.90. Obtain the closed loop transfer function (in Example 8.89) such that the system exibits oscillatory response.

Solution : Closed loop transfer for any gain k is

$$\frac{C(s)}{R(s)} = \frac{k\,[\text{closed loop zero factors}]}{[\text{closed loop pole factors}]}$$

The number of closed loop pole factors is same as the number of root locus branches *i.e.*, each closed loop pole factor corresponds to each root locus branch.

Closed loop zeros are same as the open loop zeros. There are four root locus branches where k is varying from zero to infinity. So, there must be four closed loop poles. Here we have to determine the closed loop transfer function for oscillatory response *i.e.* for that value of k for which roots lie on imaginary axis, *i.e.*, $k = 16.25$. Now identify four points on root loucs branches where $k = 16.25$ (either by trial and error or logically). These will be the four closed loop poles.

From the root locus shown in fig. 8.88 it is clear that the two closed loop poles on the two root locus branches for $k = 16.25$ are $S = +j1.87$ & $s = -j\,1.87$. Since root locus is symmetrical *w.r.t* vertical line $s = -1$, the remaining two closed loop poles on other two root locus branches can be obtained by projecting points $s = \pm j\,1.87$ on the other two root locus branches as shown at point A & B. Also by magnitude criterion for $k = 16.25$ the corresponding closed loop poles are $s = -2 \pm j\,1.87$.

Since, there are no open loop zeros, there will be no closed loop zeros.

Hence required closed loop transfer function for $k = 16.25$ is

$$\frac{C(s)}{R(s)} = \frac{16.25}{(s - j1.87)\,(s + j1.87)(s + 2 - j1.87)\,(s + 2 + j1.87)}$$

EXAMPLE 8.91. For a rotational system shown in Fig. 8.89 draw electrical analogous circuit based on $T - i$ analogy.

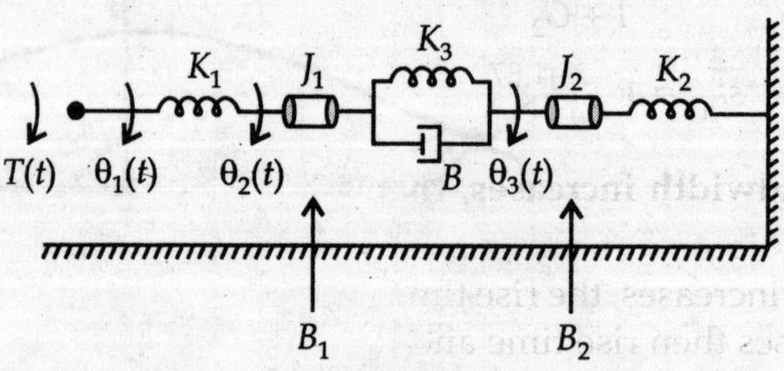

Fig. 8.89.

Solution : Draw the mechanical equivalent circuit

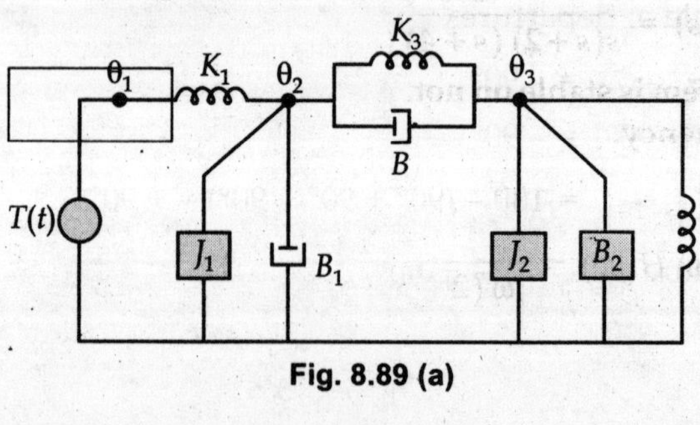

Fig. 8.89 (a)

Now electrical analogous circuit can be drawn as

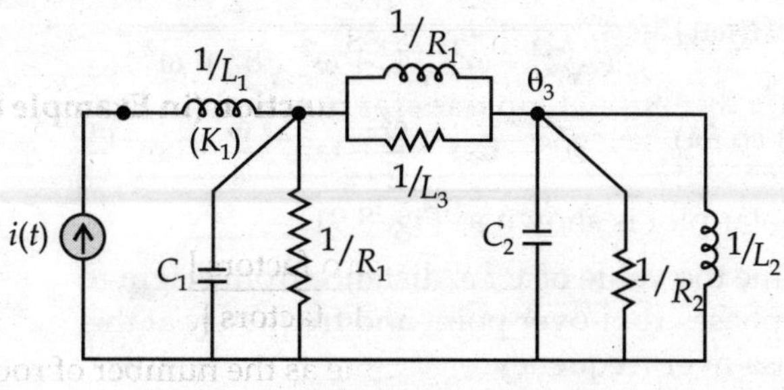

Fig. 8.89 (b)

EXAMPLE 8.92. For the system shown in Fig. 8.90. find $S_{G_1}^T$.

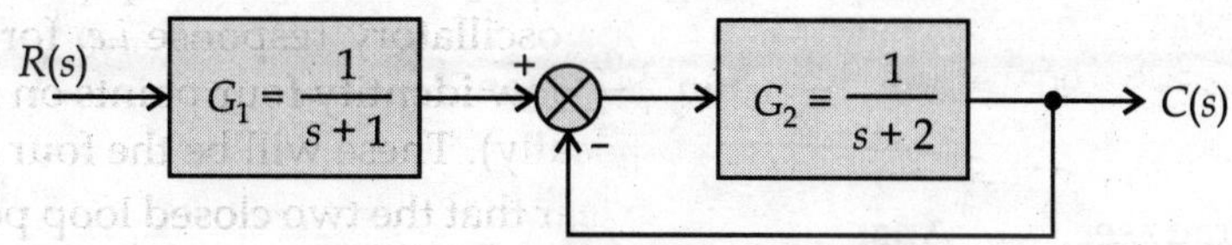

Fig. 8.90.

Solution : Overall transfer function $T(s)$ will be

$$T(s) = \frac{G_1 G_2}{1+G_2}$$

$$S_{G_1}^T = \frac{G_1}{T} \cdot \frac{\partial T(s)}{\partial G_1}$$

$$\frac{\partial T(s)}{\partial G_1} = \frac{(1+G_2)\,G_2 - G_1\,G_2\,.\,0}{(1+G_2)^2} = \frac{G_2}{1+G_2}$$

$$S_{G_1}^T = \frac{G_1}{\frac{G_1 G_2}{1+G_2}} \cdot \frac{G_2}{1+G_2}$$

$$\therefore \qquad S_{G_1}^T = 1 \qquad \textbf{Ans.}$$

EXAMPLE 8.93. If bandwidth increases, then what will be the effect on damping ratio and rise time.

Solution : If bandwidth increases, the rise time will be small, *i.e.*, the system response will be fast So, if bandwidth increases then rise time and damping ratio will decrease.

EXAMPLE 8.94. Draw the polar plot for a system given by

$$G(s)\,H(s) = \frac{100}{s(s+2)\,(s+4)\,(s+8)}$$

Find whether the system is stable on not. Also find out P.M, G.M, phase cross-over frequency and gain crossover frequency.

Solution : Put $s = j\omega$

$$G(j\omega)\,H(j\omega) = \frac{100}{j\omega\,(2+j\omega)\,(4+j\omega)\,(8+j\omega)} \qquad ...(i)$$

$$|G(j\omega)\,H(j\omega)| = \frac{100}{\omega\sqrt{2^2+\omega^2}\,\sqrt{4^2+\omega^2}\,\sqrt{8^2+\omega^2}} \quad ...(ii)$$

$$\angle G(j\omega)\,H(j\omega) = -90° - \tan^{-1}\frac{\omega}{2} - \tan^{-1}\frac{\omega}{4} - \tan^{-1}\frac{\omega}{8} \quad ...(iii)$$

A rough sketch of polar plot is shown in Fig. 8.91.

For stability determine the value of x *i.e.* distance from origin to 'a' the point 'a' is called phase cross-over point and frequency at this point is called phase cross-over frequency.

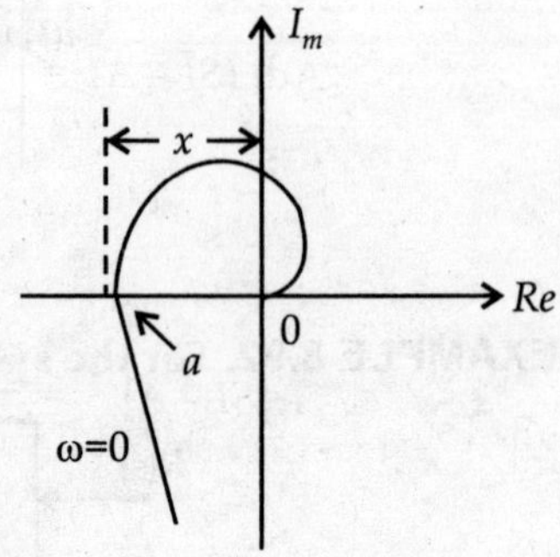

Fig. 8.91.

At point 'a' $\phi = -180°$

$$-180° = -90° - \tan^{-1}\frac{\omega}{2} - \tan^{-1}\frac{\omega}{4} - \tan^{-1}\frac{\omega}{8}$$

$$90° = \tan^{-1}\frac{\omega}{2} + \tan^{-1}\frac{\omega}{4} + \tan^{-1}\frac{\omega}{8}$$

solve for ω.

Hence $\omega \approx 2.2$ rad/sec. **Ans.**

This is the phase cross-over frequency. Now find magnitude at this frequency.

$$|G(j\omega)\,H(j\omega)|_{\omega=2.2} = \frac{100}{(2.2)^2\sqrt{2^2+(2.2)^2}\,\sqrt{4^2+2.2^2}\,\sqrt{8^2+2.2^2}} = 0.403$$

Hence, $x < 1$, the system is stable **Ans.**

$$\text{G.M.} = \frac{1}{x} = \frac{1}{0.403} = 2.47 \quad \textbf{Ans.}$$

Now calculate gain cross-over frequency. At this frequency $|GH| = 1$

$$1 = \frac{100}{\omega\sqrt{2^2+\omega^2}\,\sqrt{4^2+\omega^2}\,\sqrt{8^2+\omega^2}}$$

$\omega \approx 1.2$ rad/sec.

This is the gain cross-over frequency = 1.2 rad/sec. **Ans.**

$$\therefore \quad \phi = -90° - \tan^{-1}\left(\frac{1.2}{2}\right) - \tan^{-1}\left(\frac{1.2}{4}\right) - \tan^{-1}\left(\frac{1.2}{8}\right) = -146.18°$$

$$\text{P.M} = 180° + (-146.18°) = 33.82° \quad \textbf{Ans.}$$

EXAMPLE 8.95. Obtain the transfer function of the given state equation

$$\begin{bmatrix}\dot{x}_1\\ \dot{x}_2\\ \dot{x}_3\end{bmatrix} = \begin{bmatrix}0 & 1 & 0\\ -1 & -1 & 0\\ 1 & 0 & 0\end{bmatrix}\begin{bmatrix}x_1\\ x_2\\ x_3\end{bmatrix} + \begin{bmatrix}0\\ 1\\ 0\end{bmatrix}u$$

$$y = \begin{bmatrix}0 & 0 & 1\end{bmatrix}\begin{bmatrix}x_1\\ x_2\\ x_3\end{bmatrix}$$

Solution : Here, $A = \begin{bmatrix}0 & 1 & 0\\ -1 & -1 & 0\\ 1 & 0 & 0\end{bmatrix}$, $B = \begin{bmatrix}0\\ 1\\ 0\end{bmatrix}$, $C^T = \begin{bmatrix}0 & 0 & 1\end{bmatrix}$

$$SI - A = \begin{bmatrix} s & -1 & 0 \\ 1 & s+1 & 0 \\ -1 & 0 & s \end{bmatrix}$$

$$|SI - A| = s^3 + s + s$$

$$\text{Adj } (SI - A) = \begin{bmatrix} s^2 + s & s & 0 \\ -s & s^2 & 0 \\ s+1 & 1 & s^2+s+1 \end{bmatrix}$$

$$[SI - A]^{-1} = \frac{\text{Adj } (sI - A)}{|sI - A|} = \begin{bmatrix} \frac{s+1}{s^2+s+1} & \frac{1}{s^2+s+1} & 0 \\ \frac{1}{s^2+s+1} & \frac{s}{s^2+s+1} & 0 \\ \frac{s+1}{s(s^2+s+1)} & \frac{1}{s(s^2+s+1)} & \frac{1}{s} \end{bmatrix}$$

$$C^T[SI - A]^{-1}B = \begin{bmatrix} 0 & 0 & 1 \end{bmatrix} \begin{bmatrix} \frac{s+1}{s^2+s+1} & \frac{1}{s^2+s+1} & 0 \\ -\frac{1}{s^2+s+1} & \frac{s}{s^2+s+1} & 0 \\ -\frac{s+1}{s(s^2+s+1)} & \frac{1}{s(s^2+s+1)} & \frac{1}{s} \end{bmatrix} \begin{bmatrix} 0 \\ 1 \\ 0 \end{bmatrix} = \frac{1}{s^3+s^2+1}$$

This is the required transfer function.

EXAMPLE 8.96. Obtain state space representation of the given transfer function.

$$\frac{C(s)}{R(s)} = \frac{10.4s^2 + 47s + 160}{s^3 + 14s^2 + 56s + 160}$$

Solution :

$$\frac{C(s)}{R(s)} = \frac{10.4s^2 + 47s + 160}{s^3 + 14s^2 + 56s + 160} \cdot \frac{Q(s)}{Q(s)} = \frac{10.4s^{-1} + 47s^{-2} + 160s^{-3}}{1 + 14s^{-1} + 56s^{-2} + 160s^{-3}} \cdot \frac{Q(s)}{Q(s)}$$

$$C(s) = Q(s)\,[10.4s^{-1} + 47s^{-2} + 160s^{-3}]$$

$$R(s) = Q(s)\,[1 + 14s^{-1} + 56s^{-2} + 160s^{-3}]$$

or,

$$R(s) = Q(s) + Q(s)\,[14s^{-1} + 56s^{-1} + 160s^{-3}]$$

$$Q(s) = R(s) - Q(s)\,[14s^{-1} + 56s^{-2} + 160s^{-3}]$$

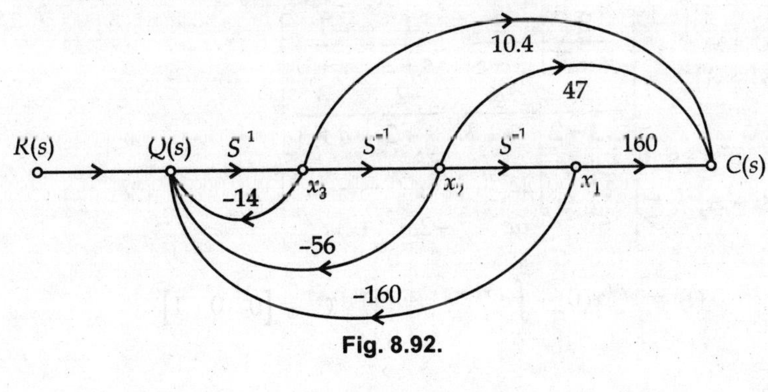

Fig. 8.92.

$$\dot{x}_1 = x_2$$
$$\dot{x}_2 = x_3$$
$$\dot{x}_3 = r(t) - 160x_1 - 56x_2 - 14x_3$$

$$\begin{bmatrix} \dot{x}_1 \\ \dot{x}_2 \\ \dot{x}_3 \end{bmatrix} = \begin{bmatrix} 0 & 1 & 0 \\ 0 & 0 & 1 \\ -160 & -56 & -14 \end{bmatrix} \begin{bmatrix} x_1 \\ x_2 \\ x_3 \end{bmatrix} + \begin{bmatrix} 0 \\ 0 \\ 1 \end{bmatrix} r(t)$$

$$C(t) = 160x_1 + 47x_2 + 10.4x_3$$

$$= \begin{bmatrix} 160 & 47 & 10.4 \end{bmatrix} \begin{bmatrix} x_1 \\ x_2 \\ x_3 \end{bmatrix}$$

The above equations are in controllable canonical form.

EXAMPLE 8.97. Obtain the time response of the following system

$$\dot{x} = \begin{bmatrix} 0 & 1 \\ -6 & -5 \end{bmatrix} \dot{x} + \begin{bmatrix} 0 \\ 1 \end{bmatrix} u$$

and $$y = \begin{bmatrix} 1 & 0 \end{bmatrix} x$$

Where $u(t)$ is a unit step input and initial conditions $x_1(0) = 0$ and $x_2(0) = 0$

(R.M.L University Faizabad 2000 Network & Systems analysis)

Solution : Here,
$$A = \begin{bmatrix} 0 & 1 \\ -6 & -5 \end{bmatrix}, B = \begin{bmatrix} 0 \\ 1 \end{bmatrix}$$

$$SI - A = \begin{bmatrix} s & 0 \\ 0 & s \end{bmatrix} = \begin{bmatrix} s & -1 \\ 6 & s+5 \end{bmatrix}$$

$$sI - A = s^2 + 5s + 6 = (s + 2)(s + 3)$$

$$\text{cofactors of } (sI - A) = \begin{bmatrix} s+5 & -6 \\ 1 & 5 \end{bmatrix}$$

$$\text{Adj of } (sI - A) = \begin{bmatrix} s+5 & 1 \\ 6 & 5 \end{bmatrix}$$

$$[sI - A]^{-1} = \begin{bmatrix} \dfrac{s+5}{(s+2)(s+3)} & \dfrac{1}{(s+2)(s+3)} \\ -\dfrac{6}{(s+2)(s+3)} & \dfrac{s}{(s+2)(s+3)} \end{bmatrix}$$

or,
$$[sI - A]^{-1} = \begin{bmatrix} \dfrac{3}{s+2} - \dfrac{2}{s+3} & \dfrac{1}{s+2} - \dfrac{1}{s+3} \\ -\dfrac{6}{s+2} + \dfrac{6}{s+3} & \dfrac{-2}{s+2} + \dfrac{3}{s+3} \end{bmatrix}$$

$$£^{-1}[sI - A]^{-1} = \begin{bmatrix} 3e^{-2t} - 2e^{-3t} & e^{-2t} - e^{-3t} \\ -6e^{-2t} + 6e^{-3t} & -2e^{-2t} + 3e^{-3t} \end{bmatrix}$$

$$x(t) = e^{At}x(0) + \int_0^t e^{A(t-\tau)} Bu(\tau)\, d\tau$$

$$x(t) = \begin{bmatrix} 3e^{-2t} - 2e^{-3t} & e^{-2t} - e^{-3t} \\ -6e^{-2t} + 6e^{-3t} & -2e^{-2t} + 3e^{-3t} \end{bmatrix} \begin{bmatrix} 0 \\ 0 \end{bmatrix}$$

$$+ \int_0^t \begin{bmatrix} 3e^{-2(t-\tau)} - 2e^{-3(t-\tau)}] & [e^{-2(t-\tau)} - e^{-3(t-\tau)} \\ -6e^{-2(t-\tau)} + 6e^{-3(t+\tau)} & -2e^{-2(t-\tau)} + 3e^{-3(t-\tau)} \end{bmatrix} \begin{bmatrix} 0 \\ 1 \end{bmatrix} 1\, d\tau$$

$$x(t) = \int_0^t \begin{matrix} e^{-2(t-\tau)} - e^{-3(t-\tau)} \\ -2e^{-2(t-\tau)} + 3e^{-3(t-\tau)} \end{matrix}\, d\tau$$

$$x_1(t) = \int_0^t e^{-2(t-\tau)} - e^{-3(t-\tau)}\, d\tau$$

put $\quad -3\,(t-\tau) = x\,, \quad \tau = 0, \quad x = -3t$

or $\quad d\tau = \frac{1}{3}dx\,, \quad \tau = t\,, \quad x = 0$

$$\frac{1}{3}\int_{-3t}^{0} e^x\, dx = \frac{1}{3} - \frac{1}{3}e^{-3t}$$

Similarly calculate $\int_0^t e^{-2(t-\tau)}\, d\tau$

$\therefore \quad x_1(t) = \frac{1}{6} - \frac{1}{2}e^{-2t} - \frac{1}{3}e^{-3t}$ **Ans.**

and $\quad x_2(t) = e^{-2t} - e^{-3t}$ **Ans.**

output response

$$y = [1 \quad 0] \begin{bmatrix} \frac{1}{6} - \frac{1}{2}e^{-2t} - \frac{1}{3}e^{-3t} \\ e^{-2t} - e^{-3t} \end{bmatrix} = \frac{1}{6} - \frac{1}{2}e^{-2t} - \frac{1}{3}e^{-3t}$$ **Ans.**

EXAMPLE 8.98. Consider the vector matrix differential equation as

$$\dot{x} = \begin{bmatrix} 0 & 1 \\ -2 & -3 \end{bmatrix} x$$

obtain the state transition matrix. *(R.M.L University Faizabad 2000 Network & Systems Analysis)*

Solution :

$$sI - A = \begin{bmatrix} s & 0 \\ 0 & s \end{bmatrix} - \begin{bmatrix} 0 & 1 \\ -2 & -3 \end{bmatrix} = \begin{bmatrix} s & -1 \\ 2 & s+3 \end{bmatrix}$$

$$|sI - A| = (s+1)(s+2)$$

$$\phi(s) = (sI - A)^{-1} = \begin{bmatrix} \frac{s+3}{(s+1)(s+2)} & \frac{1}{(s+1)(s+2)} \\ \frac{2}{(s+1)(s+2)} & \frac{s}{(s+1)(s+2)} \end{bmatrix}$$

$$\phi(t) = £^{-1}\phi(s) = £^{-1}[sI - A]^{-1} = \begin{bmatrix} 2e^{-t} - e^{-2t} & e^{-t} - e^{-2t} \\ -2e^{-t} + 2e^{-2t} & -e^{-t} + 2e^{-2t} \end{bmatrix}$$ **Ans.**

EXAMPLE 8.99. Determine the poles and zeros of the following system.

$$\dot{x} = \begin{bmatrix} 0 & 1 \\ -20 & -9 \end{bmatrix} x + \begin{bmatrix} 0 \\ 1 \end{bmatrix} r$$

$$C = \begin{bmatrix} -17 & -5 \end{bmatrix} x + [1] r$$

(IES, 1996)

Solution : The state equations are

$$\dot{x}(t) = Ax(t) + Bu(t) \quad \text{...(i)}$$

$$y(t) = Cx(t) + Du(t) \quad \text{...(ii)}$$

Laplace transform of above equations

$$sx(s) = AX(s) + Bu(s) \quad \text{...(iii)}$$

or $$[sI - A]x(s) = Bu(s) \quad \text{...(iv)}$$

$$x(s) = [SI - A]^{-1} Bu(s) \quad \text{...(v)}$$

$$Y(s) = Cx(s) + Du(s)$$

put the value of $x(s)$ from (v)

$$Y(s) = C[sI - A]^{-1} Bu(s) + Du(s)$$

or, $$\frac{Y(s)}{u(s)} = C[sI - A]^{-1} B + D = C \frac{\text{Adj}(sI - A)}{|sI - A|} B + D$$

or, $$\frac{Y(s)}{\dot{u}(s)} = \frac{C \, \text{Adj}(sI - A) B + |sI - A| D}{|sI - A|} \quad \text{...(vi)}$$

The denominator $|sI - A|$ is the characteristic equation. Hence we can calculate the poles

$$sI - A = \begin{bmatrix} s & 0 \\ 0 & s \end{bmatrix} - \begin{bmatrix} 0 & 1 \\ -20 & -9 \end{bmatrix} = \begin{bmatrix} s & -1 \\ 20 & s+9 \end{bmatrix}$$

$$|sI - A| = s(s + 9) + 20 = s^2 + 9s + 20 = 0$$

or $$(s + 5)(s + 4) = 0$$

$$s = -5, -4$$

Hence, the poles are $s = -4, -5$

$$sI - A = \begin{bmatrix} s & -1 \\ 20 & s+9 \end{bmatrix}$$

cofactors of $(sI - A)$ can be calculated as

$$(-1)^{1+1}(s + 9) = s + 9$$

$$(-1)^{1+2}(20) = -20$$

$$(-1)^{2+1}(-1) = 1$$

$$(-1)^{2+2}(s) = s$$

$\therefore$ cofactor of $$(sI - A) = \begin{bmatrix} s+9 & -20 \\ 1 & s \end{bmatrix}$$

Adj of $$(sI - A) = \begin{bmatrix} s+9 & 1 \\ -20 & s \end{bmatrix}$$

Now equate the equation (vi) to zero we get

$$C \, \text{Adj}(sI - A)B + |sI - A| \, D = 0$$

$$\begin{bmatrix}-17 & -5\end{bmatrix}\begin{bmatrix}s+9 & 1\\ -20 & s\end{bmatrix}\begin{bmatrix}0\\ 1\end{bmatrix}+(s^2+9s+20)\,1 = 0$$

$$-17 - 5s + s^2 + 9s + 20 = 0$$

$$s^2 + 4s + 3 = 0$$

$$(s + 3)(s + 1) = 0$$

$\therefore$ zeros are at $s = -1, -3$ **Ans.**

Note : In this Example value of D is given.

EXAMPLE 8.100. A close loop transfer function of a unity feedback control system is

$$\frac{C(s)}{R(s)} = \frac{20s^2}{(s+1)(s+3)(s+5)}$$

Determine the response of the system when the excitation is applied to the input terminal is

$$1 + 2t + \frac{3t^2}{2}$$ *(IES 1999)*

Solution : Given : $r(t) = 1 + 2t + \dfrac{3t^2}{2}$

Laplace transform of above equation

$$R(s) = \frac{1}{s}+\frac{2}{s^2}+\frac{3}{s^3} = \frac{s^2+2s+3}{s^3}$$

Also, given

$$\frac{C(s)}{R(s)} = \frac{20s^2}{(s+1)(s+3)(s+5)}$$

$$\therefore \quad C(s) = R(s)\frac{20s^2}{(s+1)(s+3)(s+5)}$$

$$= \frac{s^2+2s+3}{s^3}\cdot\frac{20s^2}{(s+1)(s+3)(s+5)} = \frac{20(s^2+2s+3)}{(s+1)(s+3)(s+5)}$$

After partial fraction above equation can be written as

$$C(s) = \frac{5}{s+1}-\frac{30}{s+3}+\frac{45}{s+5}$$

Inverse Laplace of above equation gives

$$C(t) = 5e^{-t} - 30e^{-3t} + 45e^{-5t}$$ **Ans.**

EXAMPLE 8.101. A unity feedback control system has a forward transfer function $\dfrac{25}{s(s+6)}$. Find the resonance peak and the corresponding frequency for the closed loop frequency response. Derive the formula used. *(IES 1997)*

Solution :

$$\frac{C(s)}{R(s)} = \frac{\dfrac{25}{s(s+6)}}{1+\dfrac{25}{s(s+6)}} = \frac{25}{s^2+6s+25}$$

Compare with $\dfrac{C(s)}{R(s)} = \dfrac{w_n^2}{s^2+2\xi w_n s+w_n^2}$

$$w_n^2 = 25 \qquad \therefore w_n = 5 \text{ rad/sec.}$$

$$2\xi w_n = 6 \qquad \therefore \xi = 0.6$$

$$w_r = w_n\sqrt{1-2\xi^2} = 5\sqrt{1-2(0.6)^2} = 2.65 \text{ rad/sec. } \textbf{Ans.}$$

$$M_r = \frac{1}{2\xi\sqrt{1-\xi^2}} = \frac{1}{2\times 0.6\sqrt{1-(0.6)^2}} = 1.042 \textbf{ Ans.}$$

For derivation refer article number 4.15.

EXAMPLE 8.102. Draw the root locus diagram for the following control system and calculate the breakin and breakaway points. *(IES 1992)*

Solution : From the Fig. 8.93

$$G(s)\,H(s) = \frac{K(s+2)(s+3)}{s(s+1)}$$

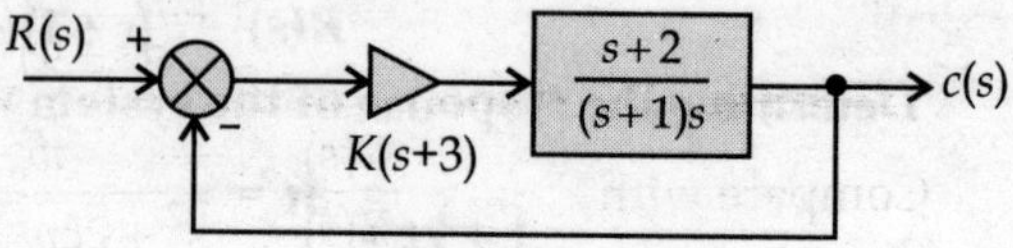

Fig. 8.93.

Step 1 : Plot the Poles and Zeros

Poles : $s_1 = 0,\ s_2 = -1$

Zeros : $s_3 = -2,\ s_4 = -3$

Step 2 : No. of poles = 2

No. of zeros = 2

Since, no. of poles = no. of zeros, therefore, there will be no asymptotes.

Step 3 : Characteristic equation $1 + G(s)\,H(s) = 0$

$$1 + \frac{k(s+2)(s+3)}{s(s+1)} = 0$$

$$\therefore \qquad k = -\frac{s^2+s}{s^2+5s+6}$$

For break points $\dfrac{dk}{ds} = 0$

$$\frac{dk}{ds} = \frac{d}{ds}\left[-\frac{s^2+s}{s^2+5s+6}\right] = 0$$

or $\qquad 2s^2 + 6s + 3 = 0$

$\therefore \qquad s = -0.634,\ \ -2.366$

As, $s = -0.634$ lies between two poles $s = 0$ and $s = -1$, therefore it is a breakaway point.

$s = -2.366$ lies between two zeros at $s = -2$ and $s = -3$ therefore it is breakin point.

Hence, root locus is a circle shown in Fig. 8.94 with centre at (– 1.5, 0) and passing through breakaway points.

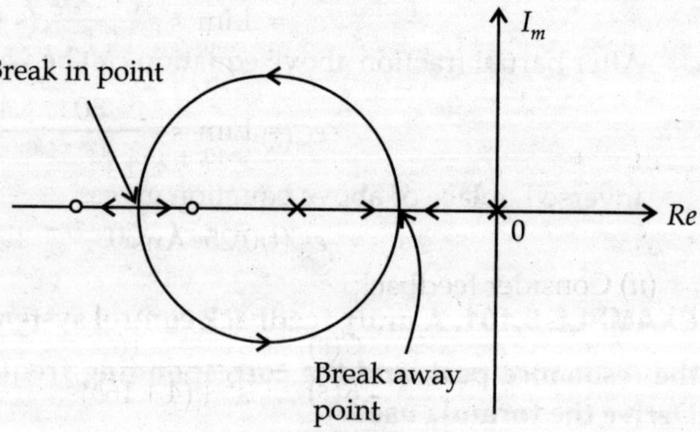

Fig. 8.94

EXAMPLE 8.103. The system shown in Fig. 8.95 consists of a unity feedback loop containing a minor rate feed-back loop

(*i*) Without any rate feedback ($b = 0$) determine the damping factor, natural resonant frequency, overshoot of the system to a unit step input and steady state error resulting from a unit ramp input.

(*ii*) Determine the rate feedback constant which will increase the equivalent damping factor of the system to 0.8. Determine the overshoot to the system in this case to a unit step input and steady state error resulting from a unit ramp input. (*IES 1994*)

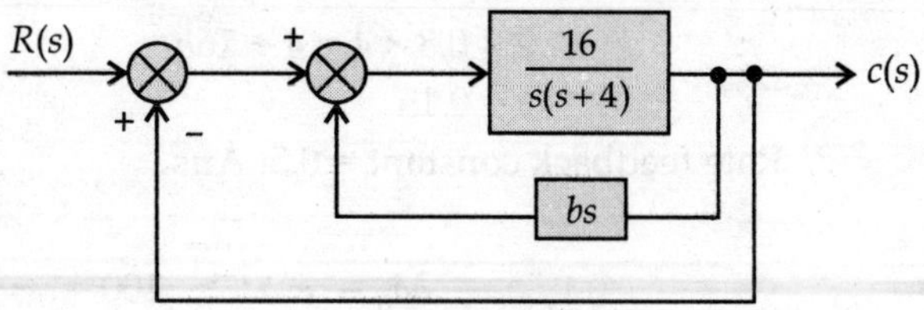

Fig. 8.95.

Solution : (*i*) When $b = 0$

$$\frac{C(s)}{R(s)} = \frac{\frac{16}{s(s+4)}}{1+\frac{16}{s(s+4)}.1} = \frac{16}{s^2+4s+16}$$

Compare with $\quad \dfrac{C(s)}{R(s)} = \dfrac{w_n^2}{s^2+2\xi w_n s + w_n^2}$

$w_n^2 = 16 \qquad \therefore\ w_n = 4 \text{ rad/sec.}$

$2\xi w_n = 4 \qquad \therefore\ \xi = 0.5$

Resonant frequency $\quad w_r = w_n\sqrt{1-2\xi^2}$

$$w_r = 4\sqrt{1-2(0.5)^2} = 2.83 \text{ rad/sec. } \textbf{Ans.}$$

$$M_p = e^{-\frac{\pi\xi}{\sqrt{1-\xi^2}}}100 = e^{-\frac{\pi 0.5}{\sqrt{1-0.5^2}}}100 = 16.32\% \ \textbf{Ans.}$$

Steady state error

$$e_{ss} = \lim_{s\to 0} s\,.\,\frac{R(s)}{1+G(s)\,H(s)}$$

Here $\quad H(s) = 1$

$$e_{ss} = \lim_{s\to 0} s\,.\,\frac{R(s)}{1+G(s)\,H(s)}$$

$$e_{ss} = \lim_{s\to 0} s\,.\,\frac{1/s^2}{1+\frac{16}{s(s+4)}} = \frac{1}{4}$$

$\therefore \quad e_{ss} = 0.25$ **Ans.**

(*ii*) Consider feedback.

$$\frac{C(s)}{R(s)} = \frac{16}{s^2+(4+16b)s+16}$$

Compare with $\quad \dfrac{C(s)}{R(s)} = \dfrac{w_n^2}{s^2+2\xi w_n s + w_n^2}$

$w_n^2 = 16 \qquad \therefore\ w_n = 4 \text{ rad/sec.}$

$2\xi w_n = 4 + 16b$

$\xi = 0.8$ given

$$\therefore \quad 2 \times 0.8 \times 4 = 4 + 16b$$

$$\therefore \quad b = 0.15$$

$\therefore$ Rate feedback constant = 0.5 **Ans.**

$$M_p = e^{-\frac{\pi\xi}{\sqrt{1-\xi^2}}} 100 = e^{-\frac{\pi 0.8}{\sqrt{1-(0.8)^2}}} 100$$

$$M_p = 1.52\% \text{ **Ans.**}$$

$$e_{ss} = \lim_{s\to 0} s \frac{1/s^2}{1+\dfrac{16}{s^2+4s+16bs}} = 0.75$$

$$\therefore \quad e_{ss} = 0.75 \text{ Ans.}$$

EXAMPLE 8.104. Find the transfer function draw the block diagram of the circuit shown in Fig. 8.96.

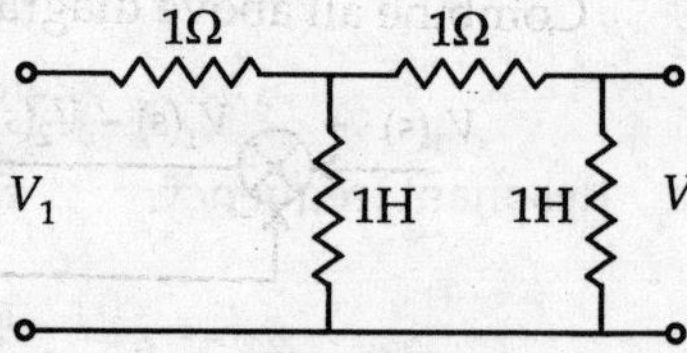

Fig. 8.96.

Solutoin : Draw the transform network

KCL at node (1)

$$\frac{V_1(s)-V(s)}{1} = \frac{V(s)}{s} + \frac{V(s)-V_2(s)}{1}$$

or

$$V(s) = \frac{s}{(2s+1)} [V_1(s) + V_2(s)] \quad ...(1)$$

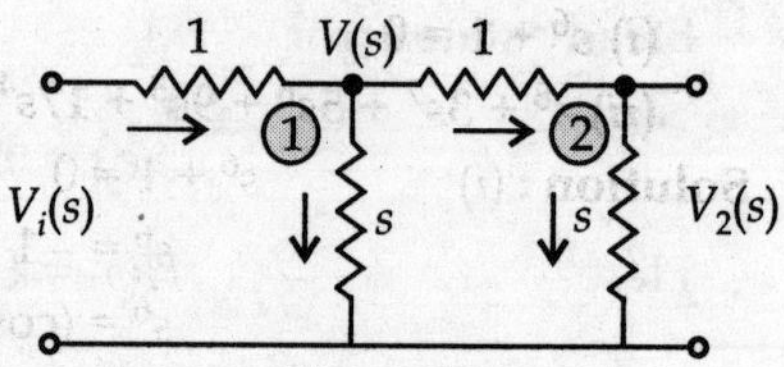

Fig. 8.96.1.

KCL at note (2)

$$\frac{V(s)-V_2(s)}{1} = \frac{V_2(s)}{s}$$

$$V(s) = V_2(s)\left[\frac{s+1}{s}\right] \quad ...(2)$$

Equate the equations (1) and (2)

$$\frac{s}{2s+1} [V_1(s) + V_2(s)] = V_2(s)\left[\frac{s+1}{s}\right]$$

$$\therefore \quad \frac{V_2(s)}{V_1(s)} = \frac{s^2}{s^2+3s+1}$$

For block diagram

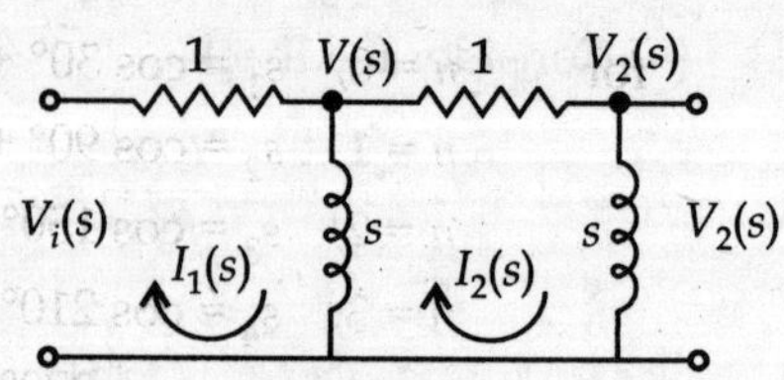

Fig. 8.96.2.

$$I_1(s) = \frac{V_1(s)-V(s)}{1}$$

$$V(s) = s\,[I_1(s) - I_2(s)]$$

$$I_2(s) = \frac{V(s)-V_2(s)}{1}$$

$$V_2(s) = s\,I_2(s)$$

Block diagram for each above equation

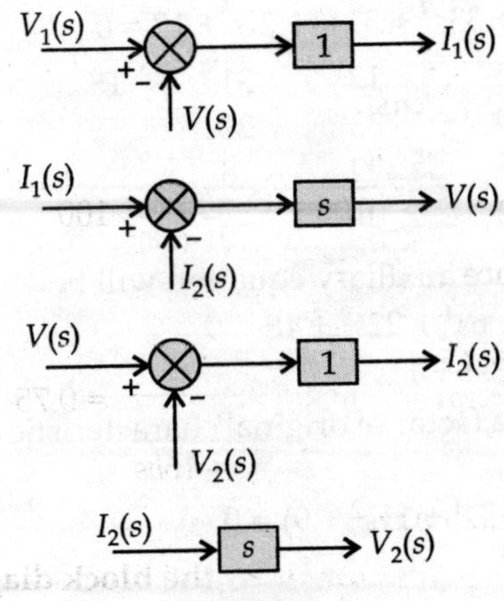

Fig. 8.96.3.

Combine all above diagram.

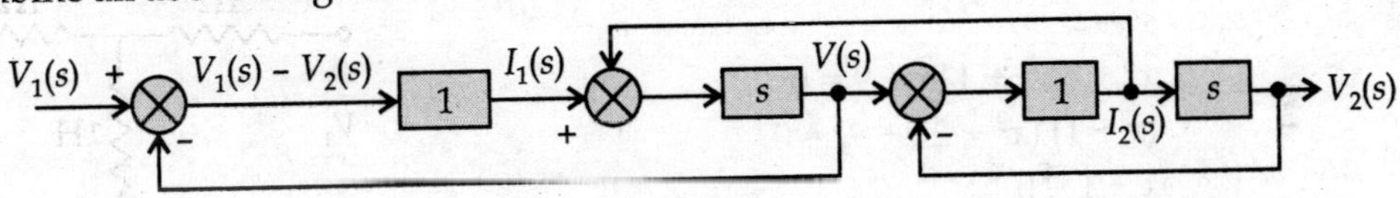

Fig. 8.97.

EXAMPLE 8.105. Find the number of roots with positive real part, negative real part and zero real part of the following equations

(i) $s^6 + 1 = 0$

(ii) $s^8 + 3s^7 + 5s^6 + 9s^5 + 17s^4 + 33s^3 + 31s^2 + 27s + 18 = 0$

Solution : (*i*)

$$s^6 + 1 = 0$$

$$s^6 = -1$$

$$s^6 = (\cos \pi + j \sin \pi) \qquad (\because -1 = \cos \pi + j \sin \pi)$$

$$s = (\cos \pi + j \sin \pi)^{1/6}$$

$$= \cos \frac{2n\pi + \pi}{6} + j \sin \frac{2n\pi + \pi}{6}$$

$$= \cos \left(\frac{2n+1}{6}\right)\pi + j \sin \left(\frac{2n+1}{6}\right)\pi$$

for $n = 0$, $s_1 = \cos 30° + j \sin 30° = +0.866 + j0.5$

$n = 1$, $s_2 = \cos 90° + j \sin 90° = 0 + j1$

$n = 2$, $s_3 = \cos 150° + j \sin 150° = -0.866 + j0.5$

$n = 3$, $s_4 = \cos 210° + j \sin 210° = -0.866 - j0.5$

$n = 4$, $s_5 = \cos 270° + j \sin 270° = 0 - j1$

$n = 5$, $s_6 = \cos 330° + j \sin 330° = +0.866 - j0.5$

$\therefore$ no. of roots with positive real part = 2

no. of roots with negative real parts = 2

no. of roots with zero real parts = 2

(ii) $s^8 + 3s^7 + 5s^6 + 9s^5 + 17s^4 + 33s^3 + 31s^2 + 27s + 18 = 0$

s^8	1	5	17	31	18
s^7	3	9	33	27	
s^6	2	6	22	18	
s^5	0	0	0		

Since, one row is zero, therefore auxiliary equation will be

$$A(s) = 2s^6 + 6s^4 + 22s^2 + 18$$

or $$A(s) = s^6 + 3s^4 + 11s^2 + 9$$

The auxiliary equation is also a factor of original characteristic equation, therefore characteristic equation can also be written as

$$(s^2 + 3s + 2)(s^6 + 3s^4 + 11s^2 + 9) = 0$$

Consider $s^2 + 3s + 2 = 0$

the roots are $s_1 = -1 + j0,\ s_2 = -2$

Now consider $s^6 + 3s^4 + 11s^2 + 9 = 0$

put $s^2 = x$

$$\therefore \quad x^3 + 3x^2 + 11x + 9 = 0$$

$$(x + 1)(x^2 + 2x + 9) = 0$$

$$x + 1 = 0$$

or $s^2 + 1 = 0$ or $s = \pm j1$

$\therefore$ $s_3 = \pm j1$ $s_4 = -j1$

$$x^2 + 2x + 9 = 0$$

or $$s^4 + 2s^2 + 9 = 0$$

or $$(s^2 + 2s + 3)(s^2 - 2s + 3) = 0$$

$s^2 + 2s + 3 = 0$, the roots will be

$$s_5 = -1 + j\sqrt{2}$$

$$s_6 = -1 - j\sqrt{2}$$

$s^2 - 2s + 3 = 0$, the roots will be

$$s_7 = 1 + j\sqrt{2}$$

$$s_8 = 1 - j\sqrt{2}$$

$\therefore$ Required roots are

$$s_1 = -1 + j0$$

$$s_2 = -2 + j0$$

$$s_3 = 0 + j1$$

$$s_4 = 0 - j1$$

$$s_5 = -1 + j\sqrt{2}$$

$$s_6 = -1 - j\sqrt{2}$$

$$s_7 = 1 + j\sqrt{2}$$

$$s_8 = 1 - j\sqrt{2}$$

$\therefore$ No. of roots with positive real part = 2

No. of roots with negative real parts = 4

No. of roots with zero real parts = 2.

EXAMPLE 8.106. For the given network, the output is V_2 and input is V_1

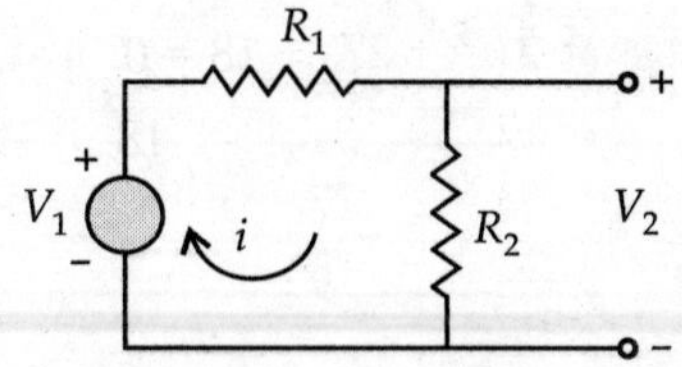

(*a*) Write an equation for V_2 in terms of V_1, R_1 and R_2 which yeilds an open loop system.

(*b*) Write an equation for V_2 which yields, a dosed loop system.

Solution: (*a*)

$$V_2 = iR_2 \quad ...(1)$$

$$i = \frac{V_1}{R_1 + R_2} \quad ...(2)$$

$$\therefore \quad V_2 = V_1 \cdot \frac{R_2}{R_1 + R_2} \quad ...(3)$$

Equation (3) is the required equation. Equation (3) can also be obtained directly by apply voltage divider rule.

(*b*)

$$i = \frac{V_1 - V_2}{R_1} \quad ...(4)$$

$$V_2 = iR_2 = \left(\frac{V_1 - V_2}{R_1}\right) R_2 = \frac{R_2}{R_1} V_1 - \frac{R_2}{R_1} V_2 \quad ...(5)$$

Equation (5) is the required equation. **Ans.**

EXAMPLE 8.107. Simplify the signal flow graph.

(*a*)

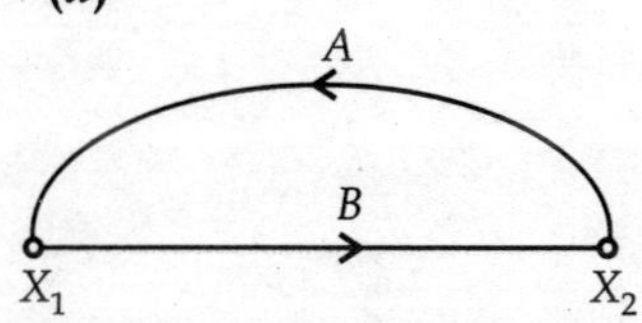

(*b*)

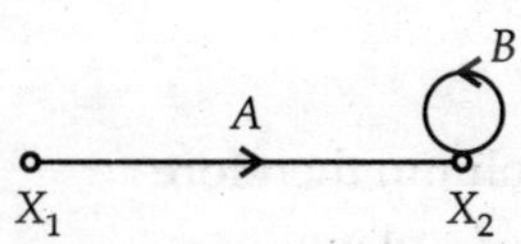

Solution : (*a*)

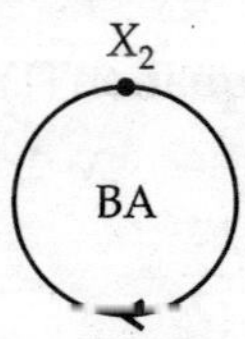

(*a*)

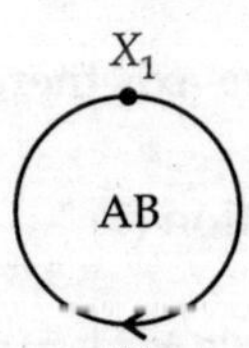

(*b*)

We have $\quad X_2 = BX_1$ and $X_1 = X_2A$

$\therefore \quad X_2 = BAX_2$ or $X_1 = ABX_1$

Yields Fig. (*a*) and (*b*).

(*b*) $\quad X_2 = X_1A + BX_2$

or $\quad X_2 = \dfrac{AX_1}{1 - B}$

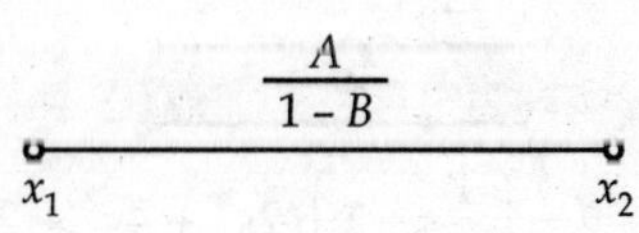

EXAMPLE 8.108. How many roots of the following polynomial are in the RHS plane and L.H.S plane and on j*w*-axis

$$P(s) = S^5 + 2S^4 + 2S^3 + 4S^2 + S + 2$$

Solution :

$$\begin{array}{llll} S^5 & 1 & 2 & 1 \\ S^4 & 2 & 4 & 2 \\ S^3 & 0 & 0 & \\ S^2 & & & \\ S^1 & & & \\ S^0 & & & \end{array}$$

Third row having all elements zero.

∴ Auxiliary equation

$$A(s) = 2s^4 + 4s^2 + 2 \qquad \text{...(1)}$$

$$\frac{dA(s)}{ds} = 8s^3 + 8s$$

$$\begin{array}{llll} S^5 & 1 & 2 & 1 \\ S^4 & 2 & 4 & 2 \\ S^3 & 8 & 8 & \\ S^2 & 2 & 2 & \\ S^1 & 0 & & \\ S^0 & & & \end{array}$$

again (V) row having all zero elements

$$A(s) = 2s^2 + 2$$

$$\frac{dA(s)}{ds} = 4s$$

$$\begin{array}{llll} S^5 & 1 & 2 & 1 \\ S^4 & 2 & 4 & 2 \\ S^3 & 8 & 8 & \\ S^2 & 2 & 2 & \\ S^1 & 4 & & \\ S^0 & 2 & & \end{array}$$

Since no sign change in first column therefore

(*a*) no roots lies on right half of *s*-plane

(*b*) since auxiliary equation contains roots on $j\omega$-axis therefore from equation (1) 4 roots lies on $j\omega$-axis

(*c*) Now divide the given polynomial by equation (1)

$$2s^4 + 4s^2 + 2 \,\big)\; s^5 + 2s^4 + 2s^3 + 4s^2 + s + 2 \;\big(\; \frac{1}{2}s + 1$$

$$\begin{array}{r} s^5 \quad + 2s^3 \quad + s \\ \hline 2s^4 + 4s^2 + 2 \\ 2s^4 + 4s^2 + 2 \\ \hline \times \\ \hline \end{array}$$

$$\therefore\; s^5 + 2s^4 + 2s^3 + 4s^2 + s + 2 = \left(\frac{1}{2}s + 1\right)(2s^4 + 4s^2 + 2)$$

put $\quad \frac{1}{2}s + 1 = 0 \;\text{ or }\; s = -\frac{1}{2}$

we can say single root lie on L.H.S.

EXAMPLE 8.109. Obtain the state model for the given transfer function.

$$\frac{y(s)}{u(s)} = \frac{s+4}{s^3+6s^2+8s+5}$$

Solution : Break the transfer function in two parts

$$\frac{y(s)}{u(s)} = \frac{x_1(s)}{u(s)} \cdot \frac{y(s)}{x_1(s)} = \frac{1}{s^3+6s^2+8s+5} . s+4$$

$$\frac{x_1(s)}{u(s)} = \frac{1}{s^3+6s^2+8s+5}$$

$$(s^3 + 6s^2 + 8s + 5)\, x_1(s) = u(s)$$

Taking inverse Laplace

$$\dddot{x}_1(t) + 6\,\ddot{x}_1(t) + 8\,\dot{x}_1(t) + 5\,x_1(t) = u(t)$$

$$\dddot{x}_1(t) = u(t) - 6\,\ddot{x}_1(t) - 8\,\dot{x}_1(t) - 5\,x_1(t)$$

Select the state variables as

$$\dot{x}_1 = x_2$$

$$\ddot{x}_1 = \dot{x}_2 = x_3$$

$$\dddot{x}_1 = \dot{x}_3$$

Rewrite

$$\dot{x}_1 = x_2$$

$$\dot{x}_2 = x_3$$

$$\dot{x}_2 = u(t) - 6x_3(t) - 8x_2 - 5x_1$$

$$\begin{bmatrix} \dot{x}_1 \\ \dot{x}_2 \\ \dot{x}_3 \end{bmatrix} = \begin{bmatrix} 0 & 1 & 0 \\ 0 & 0 & 1 \\ -5 & -8 & -6 \end{bmatrix} \begin{bmatrix} x_1 \\ x_2 \\ x_3 \end{bmatrix} + \begin{bmatrix} 0 \\ 0 \\ 1 \end{bmatrix} u(t)$$

$$\frac{y(s)}{x_1(s)} = s+4$$

$$y(s) = (s+4)\, x_1(s)$$

$$y(s) = sx_1(s) + 4x_1(s)$$

Taking inverse Laplace

$$y(t) = x_2(t) + 4\,x_1(t)$$

$$y(t) = [4 \quad 1 \quad 0] \begin{bmatrix} x_1 \\ x_2 \\ x_3 \end{bmatrix}$$

EXAMPLE 8.110. Obtain the state model for the given transfer function.

$$\frac{y(s)}{u(s)} = \frac{s+3}{(s+2)^2\,(s+1)}$$

Solution : The given transfer function can be written as (after partial fraction).

$$\frac{y(s)}{u(s)} = \frac{-1}{(s+2)^2} + \frac{-2}{(s+2)} + \frac{2}{s+1}$$

$$y(s) = \frac{-u(s)}{(s+2)^2} - \frac{2\,u(s)}{(s+2)} + \frac{2\,u(s)}{(s+1)}$$

let $$X_1(s) = \frac{u(s)}{(s+2)^2},\ X_2(s) = \frac{u(s)}{(s+2)},\ X_3(s) = \frac{u(s)}{s+1}$$

$\therefore$ $$y(s) = -X_1(s) - 2\,X_2(s) + 2\,X_3(s) \quad \text{...(1)}$$

Also, $$X_1(s) = \frac{X_2(s)}{s+2} = \frac{X_2(s)}{s+2},\ X_2(s) = \frac{u(s)}{s+2},\ X_3(s) = \frac{u(s)}{s+1} = \frac{X_2(s)}{s+2}$$

$$[s+2]\,X_1(s) = X_2(s)$$
$$sX_1(s) + 2\,X_1(s) = X_2(s)$$
$$sX_1(s) = -2X_1(s) + X_2(s)$$

From $$X_2(s) = \frac{u(s)}{s+2}$$

$$[s+2]\,X_2(s) = u(s)$$
$$sX_2(s) + 2\,X_2(s) = u(s)$$
$$sX_2(s) = -2X_2(s) + u(s)$$

From $$X_3(s) = \frac{u(s)}{s+1}$$

$$(s+1)\,X_3(s) = u(s)$$
$$sX_3(s) + X_3(s) = u(s) \quad \text{or} \quad sX_3(s) = -X_3(s) + u(s)$$

Taking inverse Laplace

$$\dot{x}_1(t) = -2\,x_1(t) + x_2$$
$$\dot{x}_2(t) = -2\,x_2(t) + u(t)$$
$$\dot{x}_3(t) = -x_3(t) + u(t)$$

In matrix form

$$\begin{bmatrix} \dot{x}_1(t) \\ \dot{x}_2(t) \\ \dot{x}_3(t) \end{bmatrix} = \begin{bmatrix} -2 & 1 & 0 \\ 0 & -2 & 0 \\ 0 & 0 & 1 \end{bmatrix} \begin{bmatrix} x_1(t) \\ x_2(t) \\ x_3(t) \end{bmatrix} + \begin{bmatrix} 0 \\ 1 \\ 1 \end{bmatrix} u(t)$$

Now take the inverse Laplace of equation (1)

$\therefore$ $$y(t) = -x_1(t) - 2\,x_2(t) + 2\,x_3(t)$$

$$y(t) = [-1 \ \ -2 \ \ 2] \begin{bmatrix} x_1(t) \\ x_2(t) \\ x_3(t) \end{bmatrix}$$

This is the canonical form of state variable model.

□□□

MATLAB PROGRAMS

MATLAB PROGRAMS

```
%EXAMPLE 01 : Obtain the unit step response of a second order system with zeta = 0.5
% and wn = 6 rad/s.
% Please note that peak time, settling time etc. can be
% obtained on clicking on the response obtained
zeta=0.5                %define zeta
wn=6                    % define wn in rad/sec.
num= [ wn*wn]           % obtain numerator of transfer function
den = [ 1 2*zeta*wn wn*wn]     % obtain coefficients of
                               % denominator of transfer function
' transfer function is'        % display label transfer function is
G= tf (num, den)               % create transfer function
                               % and display it
step (G)                       % obtain the unit step response
```

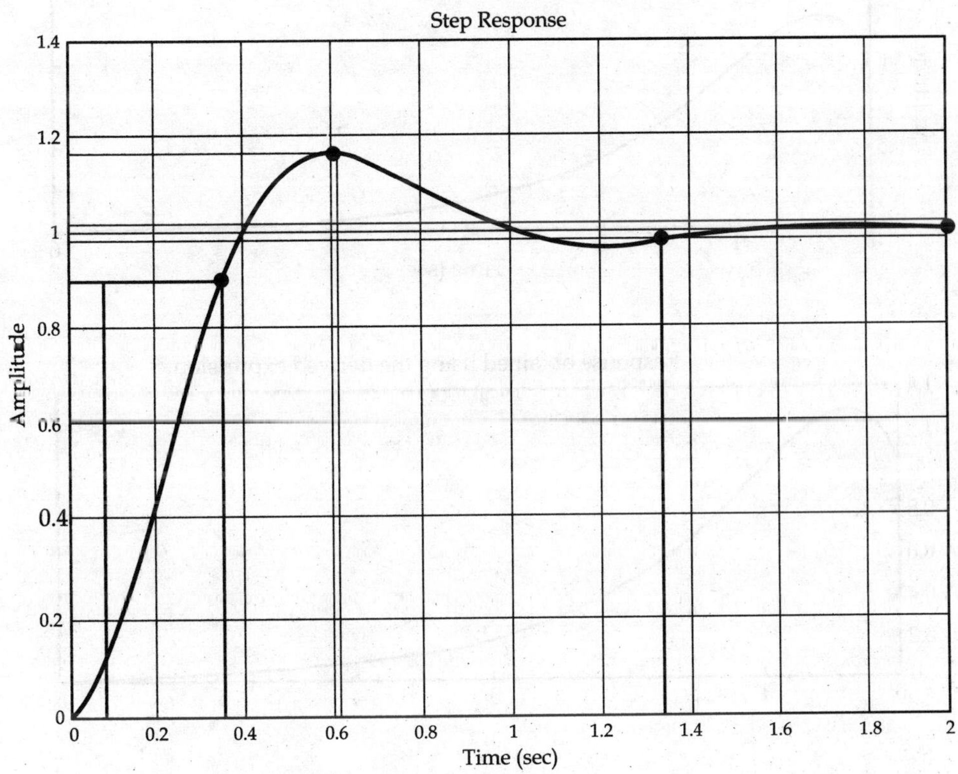

```
%EXAMPLE 02 : The closed loop transfer function of a system is
%s (s^2+9s+19) / (s^3+7s^2+14s+8). Determine the unit step response of the
% system.
clf                          % clear graph
num= [ 1 9 19 0]             % define coefficients of numerator of
                             % transfer function
den = [1 7 14 8]             % define coefficients of denominator of
                             % transfer function
G= tf ( num, den)            % obtain transfer function
subplot (2,1,1)
step (G)                     % obtain step response
title (' response obtained by using CSTB')    % CSTD= controlsystem toolbox
t=0: .01 : 6;
% obtain the step response as per expression derived
C= (11/3) *exp (-t) - (5/2) *exp (-2*t) -(1/6) *exp (-4*t) ;
subplot (2, 1, 2)
plot (t, C)
title (' response obtained using the derived expression')
```

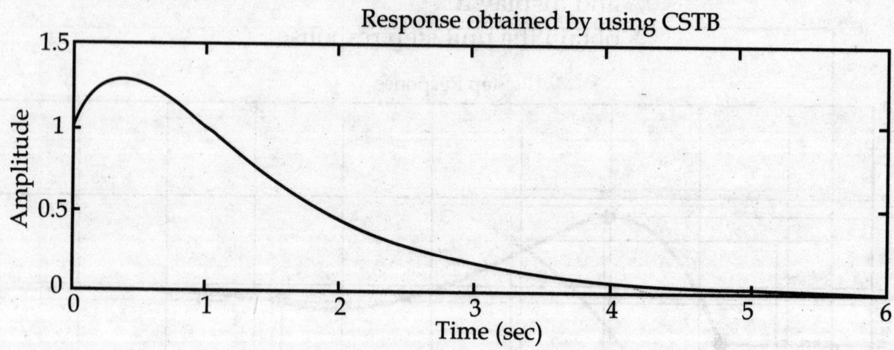

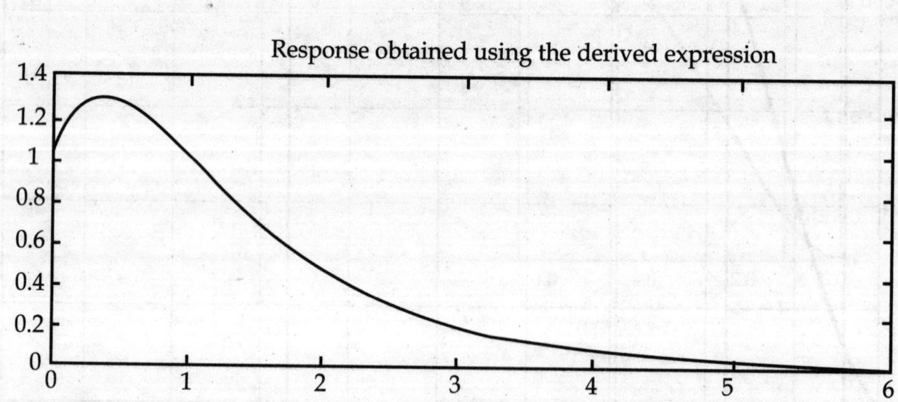

%EXAMPLE 03 : The open loop transfer function of a servo system with unity
%feedback is given by G(s) = 10/(s+2) (s+5). Determine the damping ratio,
%undamped natural frequency of oscillation. What is the percentage
% overshoot of the response to a unit step input.

```
clear
clf

num= [ 10 ];                % define the numerator of open loop tf
den=conv ([1 2] [1, 5] ) ;% define the denominator of open loop tf in
                            % the factoted form
G= tf (num, den)            % obtain open loop tf
T=feedback (G, 1)           % obtain the closed loop tf
[numt, dent]=tfdata (T, 'V') % extract the numerator and denominator
                             % of closed loop tf T
wn=sqrt (dent (3))           % find wn
zeta=dent (2) / (2*wn)       % find zeta
Mp=100*exp (-pi*zeta/sqrt (1-zeta*zeta))   % find Mp
step (T)                     % generate step response and view its characterstics
```

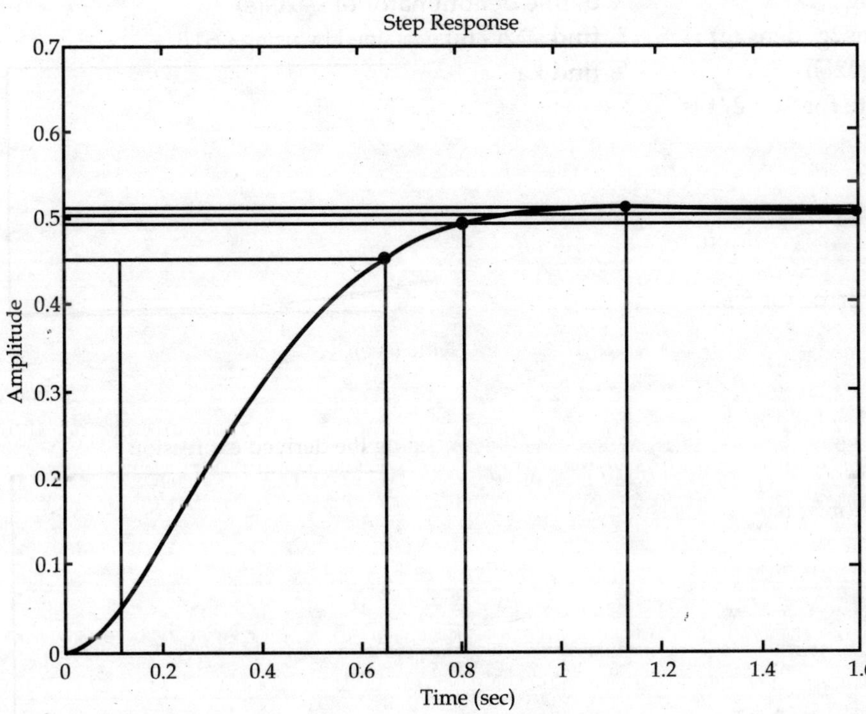

```
%EXAMPLE 04 : For a unity feedback control system the forward path function is
%given by G(s) = 20/s (s+2) (s^2+2s+20). Determine the steady state error of
%the system, when inputs are (i) 5, (ii) 5t and (iii) 3t^2/2
clear
clc
numg=[20] ;                              %define numerator of G(s)
deng= conv ( conv ( [1 0], [1 2]), [1 2 20]);    % define denominator of G(s) ;
G=tf (numg, deng) ;
kp=dcgain (G)                            % find kp
'steady state error for 5 is'
ess=5/ 1+ kp)                            % find steady state error for 5
numsg= conv ([1 0], [20] ;               % define denominator of G(s) ;
densg=deng ;                             % define denominator of sG(s)
sG= tf (numsg, densg) ;                  % find sG(s)
kv=dcgain (sG)                           % find sG (s)
kV=decgain (sG)                          % find kv
' steady state for 5t is'
ess=5 / kv                               % find, steady state error for 5t
nums2g=conv ( [1 0] , numsg) ;           % define numerator of s^2G (s)
dens2g=deng ;                            % define denominator of s^2G(s)
s2G=tf (nums2g, dens2g) ;                % find s^2G (s)
ka= dcgain (s2G)                         % find ka
' steady state for 3t ^ 2/t is'
ess=3/ka

kp =
    Inf
ans =
steady state error for 5 is
ess =
    0
kv =
    0.5000
ans =
steady state for 5t is
ess =
    10
ka =
    0
ans =
steady state for 3t^2/t is
Warning: Divide by zero.
(Type "warning off MATLAB:divideByZero" to suppress this warning.)
> In C: \MATLAB6p5\work\prob4.m at line 23
ess =
    Inf
```

%EXAMPLE 05 :

```
% sketch the polar plot of
% G(s) = 20s/ (s+1) (s+10)
w=[ [0: .1 : 5], [6:1000]] ;      % define the frequency range
n=length (w) ;                    % number of data points in the frequency range
for i=1 : n
    G_jw= (20*j*w (i))/( (j*(j*w(i) + 1)(j*w(i)+10) )*w(i)+10) ) ;
    M(i) =abs (G_jw) ;            % obtain the magnitude
    Ph (i) =angle (G_jw) ;        % obtain the phase
end
polar (Ph, M)                     % polar plot
```

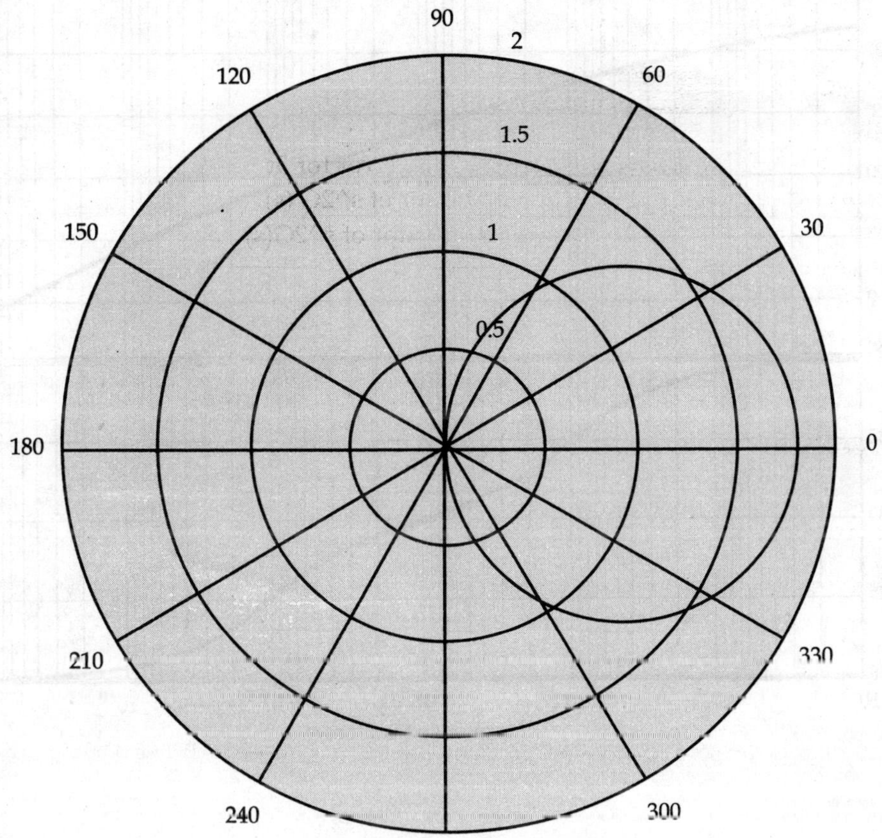

%EXAMPLE 06 :

```
% sketch the Bode plot for the transfer function
% G(s) =1000/s (1+0.1s) (1+.001s)
num= [100] ;
den1=conv ( [1 0] , [1, 0.1 ]) ;
den=conv ([den 1], [1 .001]) ;
G=t f (num, den) ;
bode (G)
grid on
[mag, phase, w] =bode(G)
```

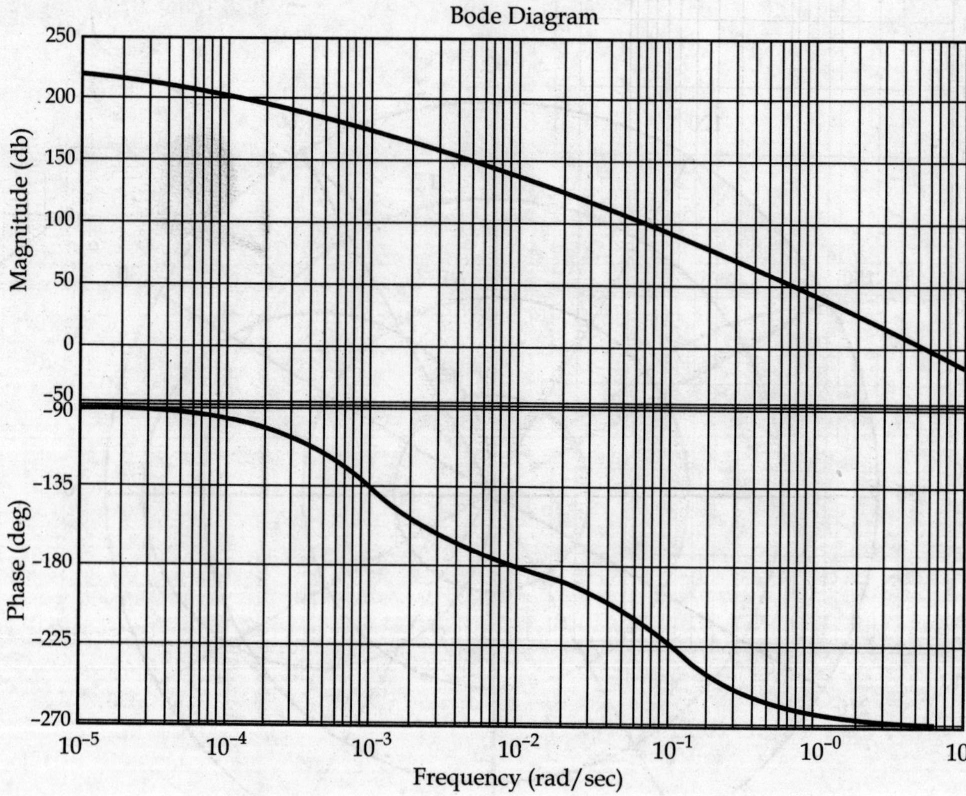

%EXAMPLE 07 : Sketch the Bode Plot for transfer function

```
% G(s) =1000/(1+0. 1s) (1+0.001s)
num=1000 ;
den=conv ([0.1 1], [.001 1] ) ;
G=t f (num, den)
margin (G)
```

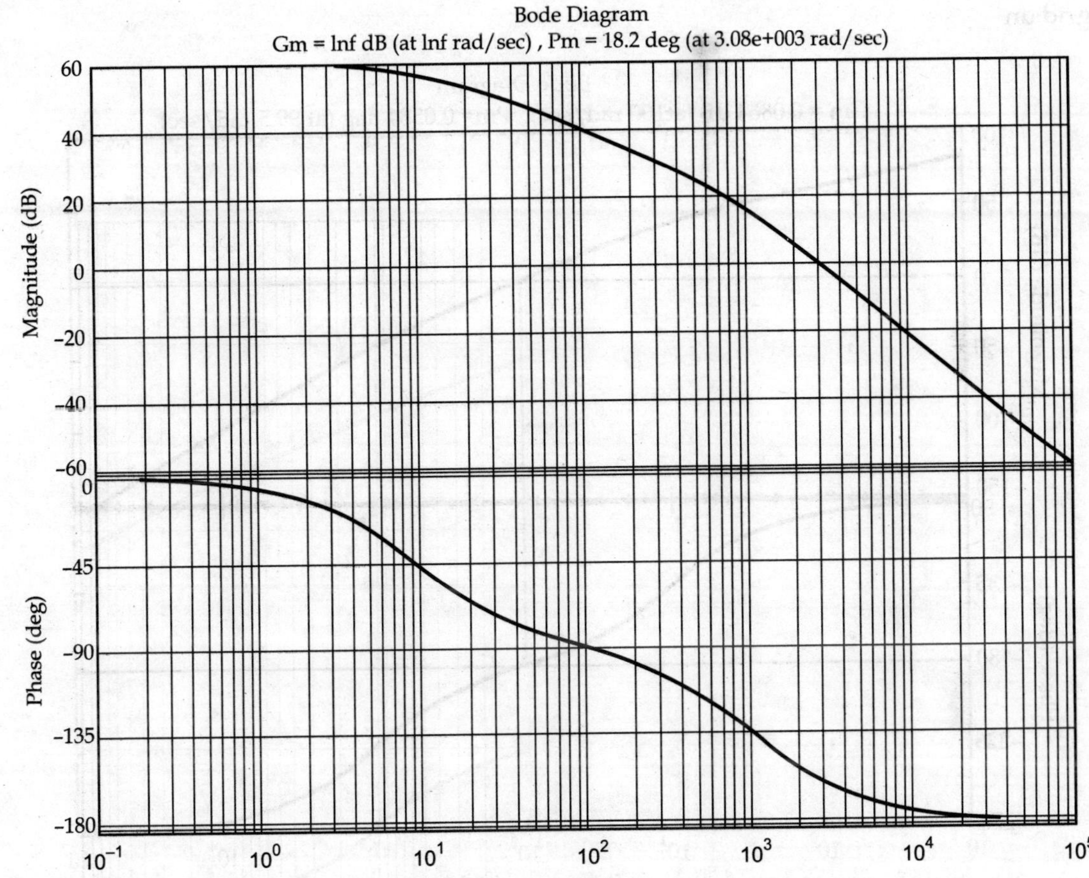

%EXAMPLE 08 : Sketch the Bode Plot for the transfer function

```
% G(s) =1000/s (1+0.1s) (1+0.001s)
num=1000;
den=conv ([1 0], conv ( [0. 1 1], [.001 1] ) );
G=t f (num,den)
margin (G)
%grid on
```

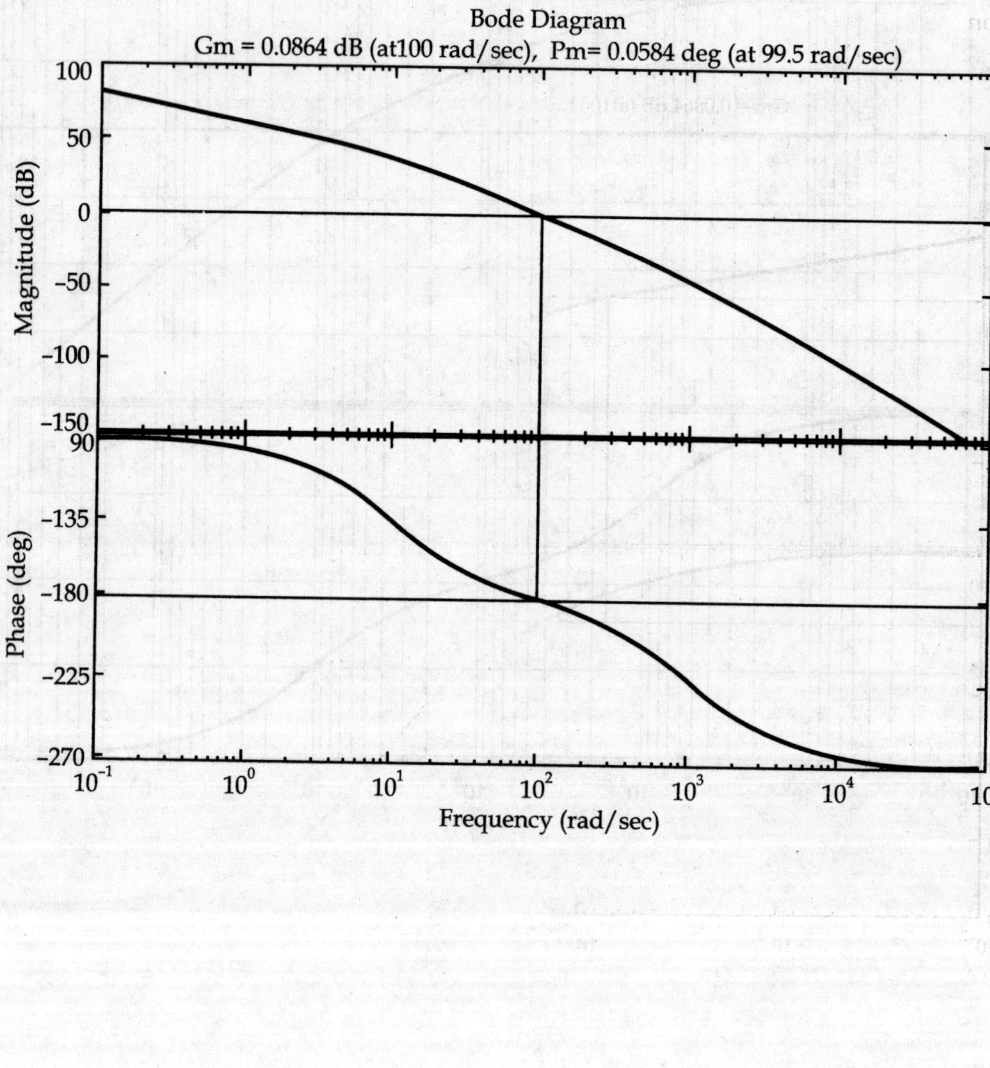

%EXAMPLE 09 :

```
% Example 09 Sketch the Bode Plot for the transfer function
% G(s) =e^-0.35/s (1+s)
den=conv ( [1 0] , [1 1] ) ;
delay=0 . 35 ;
G=t f (1, den, 'Inputadelay' , delay) ;
GG=pade (G,2) ;
margin (GG)
% grid on
```

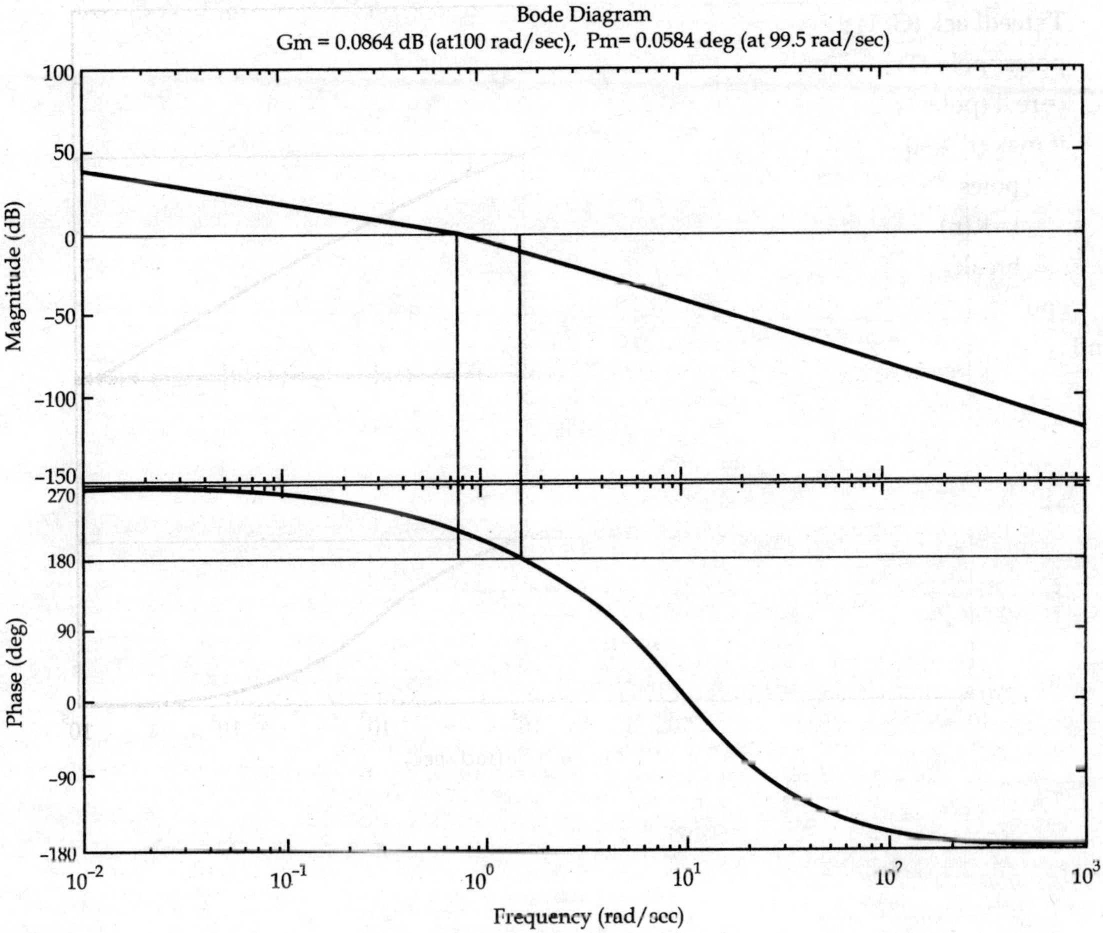

%EXAMPLE 10 : A unity feedback control system is characterized by open loop transfer

```
% function
% G(s) = k(s + 13)/s(s + 3) (s+7)
% Fidn the range of values of k for the system ot be stable
K=[ 1:100] ;
for n=1 : length (K)
    num=K (n) * [ 1  13] ;
    den=conv ( [1  0], conv ( [1 3] , [1 7] ) ) ;
    G=tf (num, den) ;
    T=feedback (G, 1) ;
    poles=pole (T) ;
    r=real (poles) ;
    if max (r) >=0
        poles
        k=K(n)
        break
    end
end
```

%EXAMPLE 11 : Sketch the root locus for

```
%G(s) = K(S+1) /s^2 (S+3.6)
clf
clc
clear
num= [1  1] ;
den=conv ( [ 1 0 0], [1 3.6] ) ;
G=tf (num, den) ;
rlocus (G) ;
```

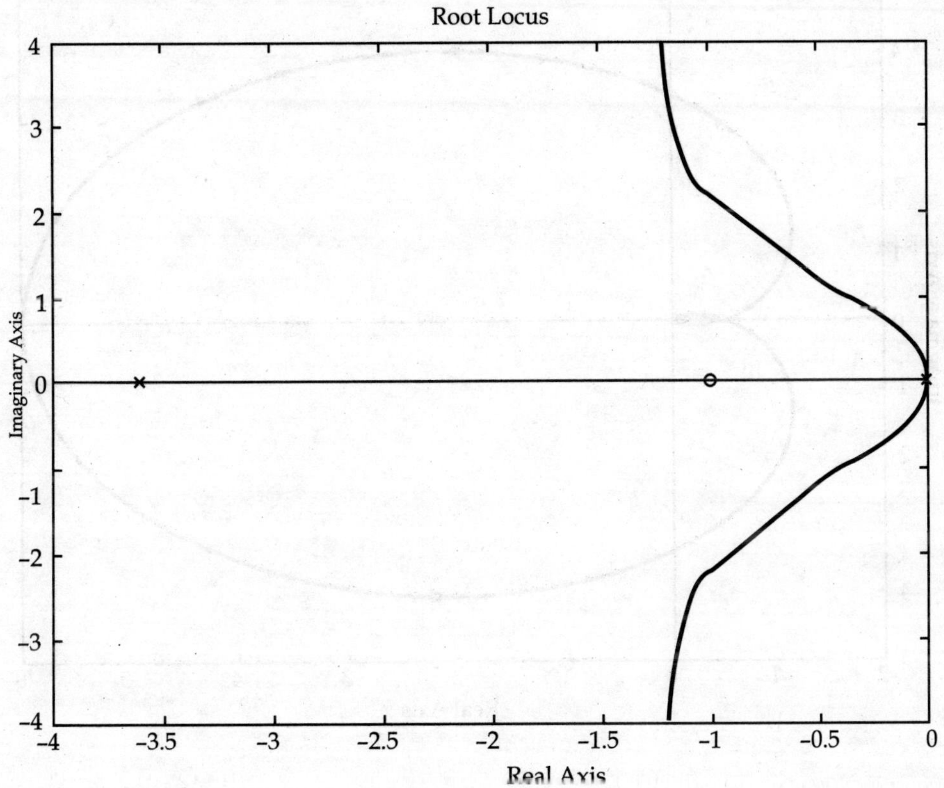

%EXAMPLE 12 : Consider the transfer function

```
% G(s) H(s) = 60/(s+1) (s+2) (s+5)
%Sketch the Nyquist plot for the system
num=60 ;
den=conv ( [1  1] , conv ( [ 1 2] , [1 5] ) ) ;
GH=tf (num, den) ;
nyquist (GH)
```

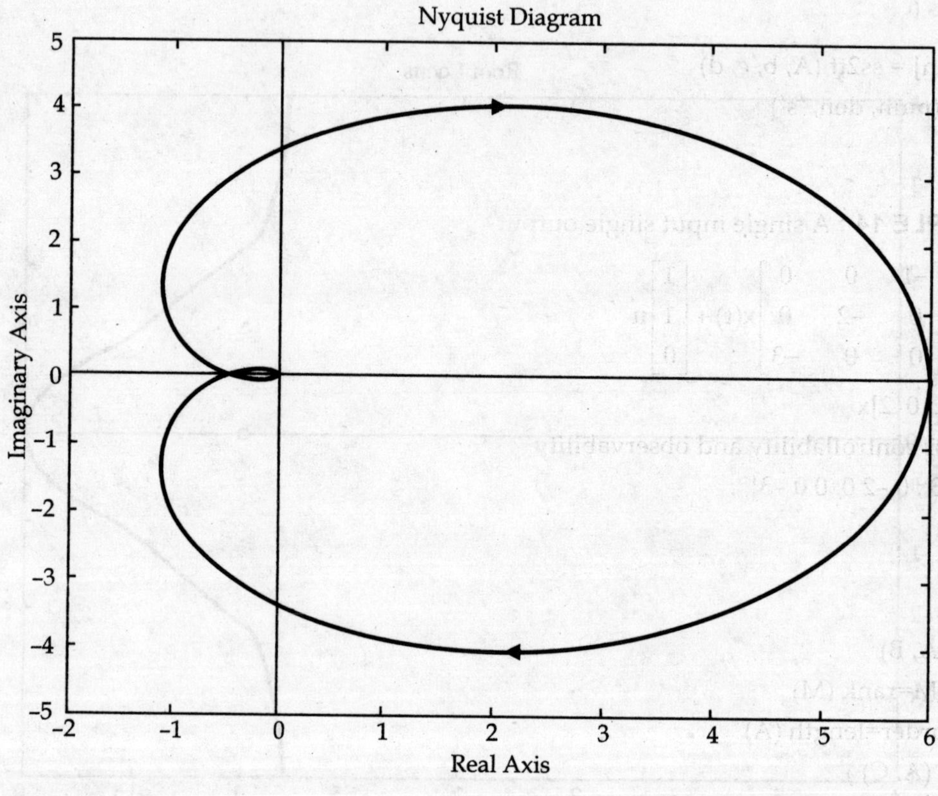

%EXAMPLE 13 : The system equations are:

$$\%xdot = \begin{bmatrix} 0 & 1 \\ -2 & -3 \end{bmatrix} x + \begin{bmatrix} 0 \\ 1 \end{bmatrix} u(t)$$

```
% y=[ 1 0] x
% find the transfer function
A = [0 1; -2 -3] ;
b=[0; 1] ;
c= [1 0];
d=[0] ;
[num, den] = ss2tf (A, b, c, d)
printsys (num, den, 's')
```

%EXAMPLE 14 : A single input single output system is given by

$$\%xdot = \begin{bmatrix} -1 & 0 & 0 \\ 0 & -2 & 0 \\ 0 & 0 & -3 \end{bmatrix} x(t) + \begin{bmatrix} 1 \\ 1 \\ 0 \end{bmatrix} u$$

```
% y = [ 1 0 2]x
% Test for controllability and observability
A=[-1 0 0 ; 0 -2 0; 0 0 -3] ;
B=[1;1;0] ;
C=[1 0 2] ;
D=[0] ;
M=ctrb (A, B)
rank_of_M=rank (M)
system_order=length (A)
N=(obsv (A, C) )'
Rank_of_N=randk (N)
```

ANSWERS

EXAMPLE 01

```
zeta =
    0.5000
wn =
    6
num =
    36
den =
    1   6   36
ans =
transfer function is
Transfer function:
        36
    ----------------------
    s^2 + 6 s + 36
```

EXAMPLE 02

```
num =
    1   9   19  0
den =
    1   7   14  8
Transfer function:
    s^3 + 9 s^2 + 19s
    ------------------------
    s^3 + 7 s^2 + 14 s + 8
```

EXAMPLE 03

```
Transfer function:
        10
    ---------------
    s^2 + 7 s + 10
numt =
    0   0   10
dent =
    1   7   20
wn =
    4.4721
zeta =
    0.7826
Mp =
    1.9260
```

EXAMPLE 07

Transfer function:

```
        1000
------------------------------
0.0001 s^2 + 0.101 s + 1
```

EXAMPLE 08

Transfer function:

```
        1000
--------------------------------
0.0001 s^3 + 0.101 s^2 + s
```

EXAMPLE 10

```
Poles =
    - 10.0000
    0.0000 + 9.5394i
    0.0000 - 9.5394i
k =
    70
```

EXAMPLE 13

```
num =
        0        0.0000        1.0000
den =
    1        3        2
```

$$\frac{8.8818e-0165+1}{s^{\wedge}2+3s+2}$$

EXAMPLE 14

```
M =
        1    -1        1
        1    -2        4
        0     0        0
rank_of_M =
        2
system_order =
        2
system_order =
        3
N =
        1    -1        1
        0     0        0
        2    -6       18
Rank_of_N
        2
```

OBJECTIVE TYPE QUESTIONS

OBJECTIVE TYPE QUESTIONS

1. Transfer function of a simple R-C integrator circuit shown in the Fig. is given by

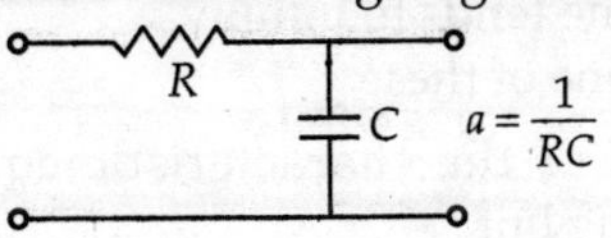

(a) $\frac{1}{s-a}$ (b) $\frac{1}{s+a}$

(c) $\frac{a}{s-a}$ (d) $\frac{a}{s+a}$

2. Transfer function of simple differentiator circuit shown in Fig. is

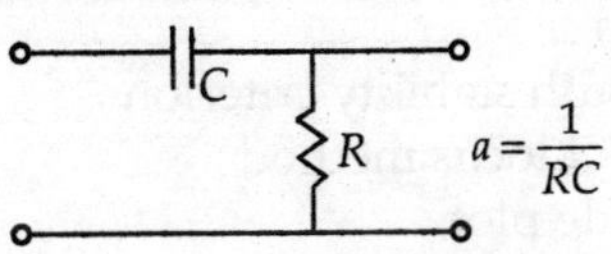

(a) $\frac{a}{s+a}$ (b) $\frac{s}{s+a}$

(c) $\frac{1}{s+a}$ (d) none of these

3. Laplace transform of t^{n+1} is

(a) $\frac{\underline{|n+1}}{s^{n+2}}$ (b) $\frac{\underline{|n}}{s^{n+1}}$

(c) $\frac{\underline{|n}}{s^{n}}$ (d) $\frac{s^{n+1}}{\underline{|n}}$

4. Laplace transform of integral of $f(t)$ is

(a) $F(s)$ (b) $sF(s) - f(o)$

(c) $sF(s)$ (d) $s^2F(s)$

5. According to Initial value theorem the initial value $f(0^+)$ of a function $f(t)$ is given by

(a) $f(0^+) = \lim_{s \to 0} sF(s)$

(b) $f(0^+) = \lim_{s \to \infty} sF(s)$

(c) $f(0^+) = \lim_{s \to \infty} F(s)$

(d) none of these

6. Laplace transform of $\frac{df}{dt}$ is

(a) $sF(s) - f(0^+)$ (b) $sF(s)$

(c) $\frac{F(s)}{s}$ (d) none

7. If $F_1(s)$ and $F_2(s)$ are the Laplace transform of $F_1(t)$ and $F_2(t)$ respectively then Laplace transform of a function $[AF_1(t) + BF_2(t)]$ will be

(a) $AF_1(s) + BF_2(s)$ (b) $\frac{F_1(s)}{A} + \frac{F_2(s)}{B}$

(c) $BF_1(s) + AF_2(s)$

(d) $\frac{AF_1(s)}{s} + \frac{BF_2(s)}{s}$

8. According to the final value theorem, the final value $f(\infty)$ of a function $f(t)$ is given by

(a) $f(\infty) = \lim_{s \to \infty} sF(s)$

(b) $f(\infty) = \lim_{s \to 0} F(s)$

(c) $f(\infty) = \lim_{s \to 0} sF(s)$

(d) none of these

9. Laplace transform $F(s)$ of any time function $f(t)$ is given by

(a) $F(s) = \int_{-\infty}^{\infty} f(t)\, e^{-st}\, dt$

(b) $F(s) = \int_{0}^{\infty} f(t)\, e^{-st}\, dt$

(c) $F(s) = \int_{0}^{\infty} f(t)\, e^{st}\, dt$ (d) none

10. The transfer function C/R for the closed loop system as shown in Fig. is

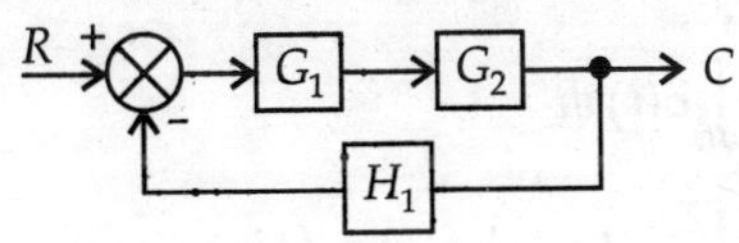

(a) $\dfrac{G_1G_2}{1+(G_1+G_2)H_1}$

(b) $\dfrac{G_1G_2}{1+G_1G_2H_1}$

(c) $\dfrac{G_1G_2}{1-G_1G_2H_1}$

(d) $\dfrac{G_1G_2H_1}{1+G_1G_2H_1}$

11. The transfer function C/R for the closed loop system is

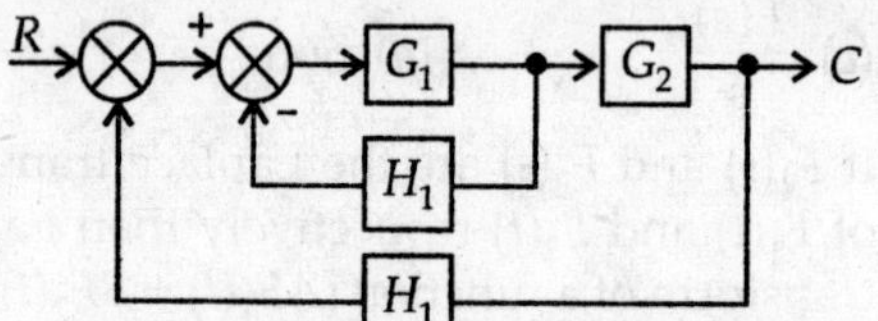

(a) $\dfrac{G_1+G_2}{1+(G_1+G_2)H_1}$

(b) $\dfrac{G_1G_2}{1+G_1G_2H_1+G_1H_1}$

(c) $\dfrac{G_1G_2H_1}{1+G_1G_2H_1}$ (d) none

12. Area under a unit impulse function is
(a) infinity (c) unity
(b) zero (d) none of these

13. Laplace transform of doublet is
(a) s (c) $1/s^2$
(b) $1/s$ (d) unity

14. If the poles of a control system lie on the imaginary axis in splane the system will be
(a) unstable (b) stable
(c) conditionally stable
(d) marginally stable

15. Some expression related to impulse functions are given below. Mark the corect expression.

(a) $\int_{-\infty}^{\infty} \delta(t)\,dt = 1$

(b) $\int_{-\infty}^{\infty} \delta(t-t_0)f(t) = f(t)$

(c) $\int_{0}^{\infty} \delta(t)\,dt = 1$

(d) $\int_{-\infty}^{\infty} \delta(t-t_0)\,f(t) = f(t_0)$

16. Which of the following system is stable
(a) $As^2 + Bs + c = 0$
(b) $As^4 + Bs^2 + c\,s + D = 0$
(c) $-As^2 + Bs - c = 0$
(d) $As^2 - Bs - c = 0$

17. A control system is stable when
(a) Roots of characteristic equation have +ve real part .
(b) Zeros of the system have negative real part
(c) Impulse response approaches zero when time tends to infinity.
(d) none of these

18. Roots of the characteristic equation are nothing but
(a) Polse of the system
(b) Zeros of the system
(c) Poles and zeros of the system
(d) none of these

19. One of the following method can be used to determine the relative stability of a control system
(a) Routh stability criterion
(b) Root locus method
(c) Bode plot
(d) none of these

20. The first column of Routh table containing the following integers 2, 3, 4, 6, 13. The system is
(a) Stable (b) unstable
(c) marginally stable
(d) none of these

21. The first colum of Routh table contains the following integers 2, 4, – 5, 3. The system is
(a) Stable
(b) unstable
(c) marginally stable
(d) none of these

22. Routh tables are separately constructed for two control systems. The first columns obtained were (1, 3, 5, 0, –1) for the system *M* and (2, 3, –4, 6,1) for system *N*. Then
(a) *M* and *N* both are stable
(b) *M* is more stable than *N*
(c) *N* is more stable than *M*
(d) Both are unstable but *N* is more unstable

23. Marginally stable systems
(*a*) are also classed as unstable system
(*b*) have one of the poles lying in R.H.S of s-plane
(*c*) equal numbers of zeros and poles
(*d*) None of these.

24. A feedback control system has transfer function given by

$$F(s) = \frac{6(s+1)(s+6)}{s^3(s+2)(s+4)}$$

It is a
(*a*) type 5 system (*b*) type 3 system
(*c*) type 2 system (*d*) type o system

25. The position error constant of a type one canonical feedback system is given by
(*a*) $\lim_{s\to 0} G(s)$ (*b*) $\lim_{s\to 0} sG(s)$
(*c*) $\lim_{s\to \infty} sG(s)$ (*d*) none of these

26. A control system is represented in a block schematic as shown in figure. What is the type of the system.

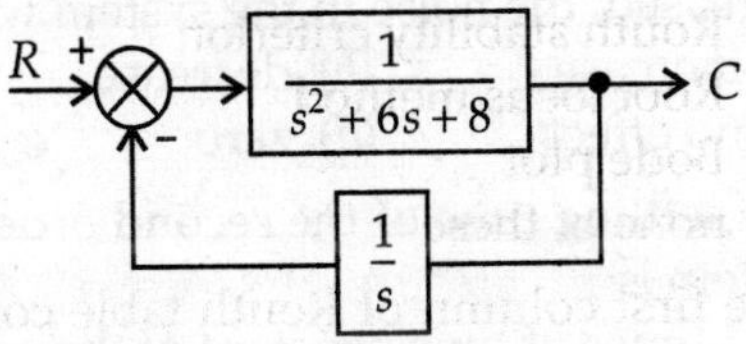

(*a*) type one (*b*) type two
(*c*) type zero (*d*) none

27. The velocity error constant of a stable canonical feedback system is given by
(*a*) $\lim_{s\to 0} G(s)$ (*b*) $\lim_{s\to 0} sG(s)$
(*c*) $\lim_{s\to 0} s^2G(s)$ (*d*) none

28. The acceleration error constant of a stable canonical feedback system is given by
(*a*) $\lim_{s\to 0} sG(s)$ (*b*) $\lim_{s\to 0} G(s)$
(*c*) $\lim_{s\to 0} s^2G(s)$ (*d*) $\lim_{s\to \infty} G(s)$

29. What is the centre of asymptotes for a open loop transfer function

$$G(s) = \frac{3(s+2)}{s^2(s+4)}$$

(*a*) –1 (*b*) –3
(*c*) – 4 (*d*) 0

30. The open loop transfer function of a control system has 5 poles and 2 zeros. The poles are located at –3, –5, –7 with remaining two at the origin. The zeros are located at –1 and –6. The centre of asymptotes is
(*a*) – 2 (*b*) – 2.666
(*c*) –3.33 (*d*) 2.666

31. Number of asymptotes in root locus for the above system is
(*a*) 2 (*b*) 5
(*c*) 1 (*d*) 3

32. What is the transfer function of a system having a gain factor of 9 and pole zero map as shown in fig.

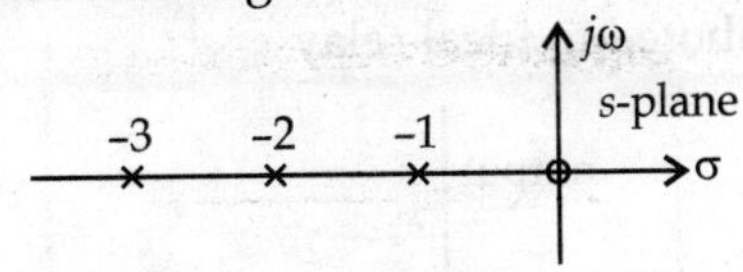

(*a*) $\frac{9(s+1)(s+2)(s+3)}{s}$

(*b*) $\frac{9(s-1)(s-2)(s-3)}{s}$

(*c*) $\frac{9s(s+1)}{(s+2)(s+3)}$

(*d*) $\frac{9s}{(s+1)(s+2)(s+3)}$

33. The transfer function of a system is $\frac{1000}{(1+0.1s)(1+0.01s)}$, the corner frequencies are
(*a*) 0.1 and 0.01 (*c*) –0.1 and –0.01
(*c*) 10 and 100 (*d*) none of these

34. A capacitance cannot be used to fabricate a lag network
(*a*) True (*b*) False

35. The inductance is not being used to fabricate a lag network because
(*a*) it has high reactance
(*b*) it produces time delay and hysteresis loss
(*c*) its size is big
(*d*) none of these

36. Which of the following is the source of non-linearity.
(*a*) backlash in gears
(*b*) saturation

(c) both (a) and (b)
(d) none of these

37. A system is critically damped. If the gain of the system is increased the system will behaves as
(a) underdamped
(b) overdamped
(c) no effect of gain
(d) none

38. Nyquist criteria gives direct value of corner frequencies.
(a) True (b) False

39. Which of the following characteristic can be attributed to ideal relay

(a)

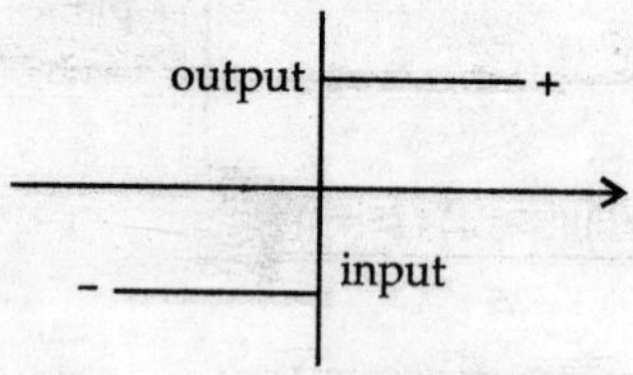

(b)

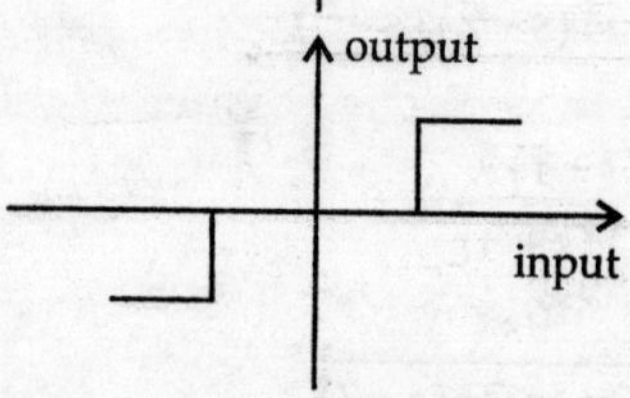

(c)

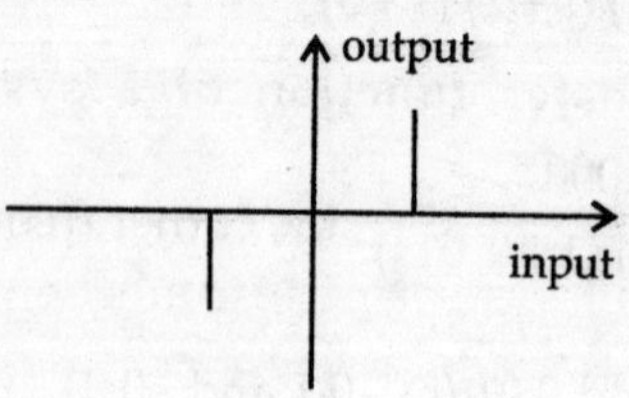

(d) none

40. The output of a feedback control system should be a function of
(a) reference and output
(b) input (c) feedback signal
(d) all of above

41. The lead-lag compensation will improve
(a) transient response
(b) transient and steady state response
(c) none of above

42. The lag compensation in the control system is achieved by
(a) adding zeros in the transfer function of the system
(b) adding poles in the transfer function of the system
(c) both (a) and (b)
(d) none

43. In frequency domain analysis both amplitude and phase varies with frequency change.
(a) True (b) False

44. The damping factor of a system is unity, the system is
(a) overdamped
(b) underdamped
(c) critical damped
(d) unstable

45. Roots of the closed loop control system can be obtained from Bode plot
(a) true (b) false

46. In open loop system, the control action depends upon the desired output
(a) true (b) false

47. If the bandwidth of a control system is increased, the noise in the system will be
(a) increase (b) decrease
(c) no effect (d) zero

48. The settling time of the second order linear system is
(a) 4 times of time constant of the system
(b) 3 times of time constant of the system
(c) 1/4 times of time constant of the system
(d) none

49. The Nyquist plot shown in the figure indicates

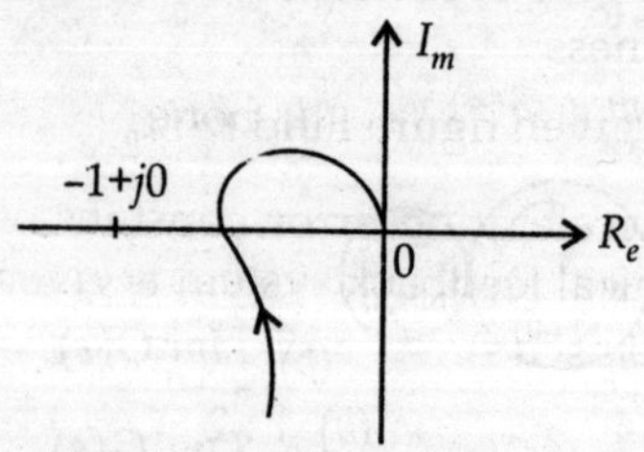

(a) marginally stable system
(b) unstable system
(c) Stable system
(d) none

50. The transient response of the system cannot be determined from the root locus analysis.
(a) true (b) false

51. The Bode plot of the transfer function $G(s) = s$ is
   (*a*) zero magnitude and zero phase shift.
   (*b*) 20 db/decade and phase shift π
   (*c*) 20 db/decade and phase shift $\pi/2$
   (*d*) –20 db/decade and phase shift $\pi/2$
52. Transfer function is the ratio of
   (*a*) Output to the input
   (*b*) Laplace transform of output to the Laplace transform of input
   (*c*) Laplace transform of input to the Laplace transform of output
   (*d*) Inverse Laplace of output to the inverse Laplace of input.
53. The electrical resistance is analogous to
   (*a*) viscous damper
   (*b*) spring (*c*) mass
   (*d*) none
54. The electrical capacitance is analogous to
   (*a*) mass (*b*) force
   (*c*) damping coefficient
   (*d*) none
55. In torque voltage analogy, inductance is analogous to
   (*a*) damping coefficient
   (*b*) moment of inertia
   (*c*) stiffness (*d*) none
56. In torque current analogy, current is analogous to
   (*a*) torque
   (*b*) moment of inertia
   (*c*) damping coefficient
   (*d*) stiffness
57. For the given figure Find $r_1 : r_2$

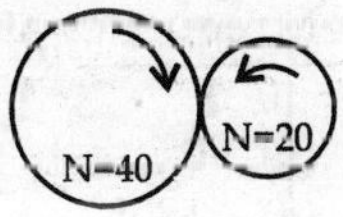

   (*a*) 20 : 4 (*b*) 1 : 2
   (*c*) 2 : 1 (*d*) 20 : 1
58. In above figure find ω_2, if ω_1 = 10 rad/sec.
   (*a*) 10 (*b*) 20
   (*c*) 40 (*d*) none
59. Find the correct one
   (*a*)

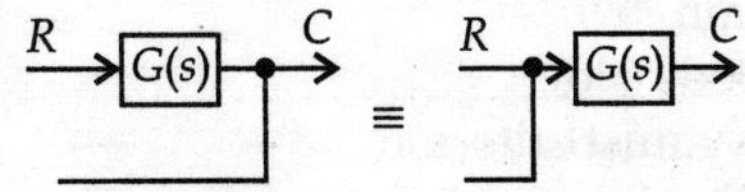

   (*b*)

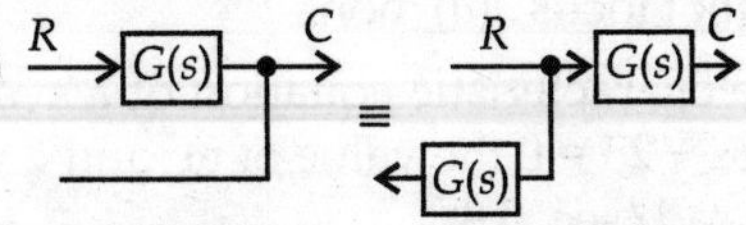

   (*c*)

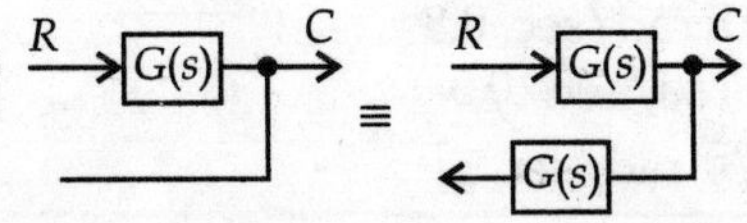

   (*d*) none
60. In general which of the following system is preferred.
   (*a*) overdamped
   (*b*) critical damped
   (*c*) underdamped
   (*d*) undamped
61. If the number of poles is '*P*' and number of zeros is '*z*' to plot a root locus, then the number of root locus segments going to infinity is
   (*a*) $P + z$ (*b*) $P - z$
   (*c*) $\frac{P-z}{P}$ (*d*) $\frac{P+z}{Pz}$
62. Gain margin is a measure of relative stablility of a system.
   (*a*) true (*b*) false
63. Laplace transform of impulse function is one
   (*a*) True (*b*) False
64. Ramp function is represented by

(*a*) 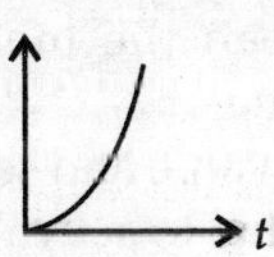

(*b*) 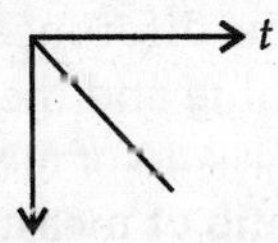

(*c*) 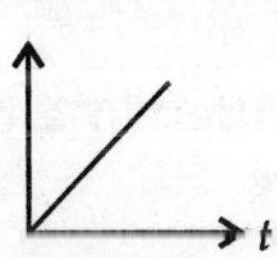

(*d*) 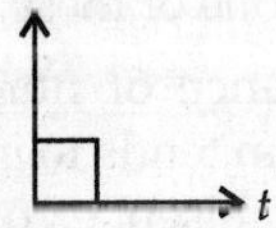

   (*a*) Figure *b* (*b*) Figure *b* and *c*
   (*c*) Figure *a* (*d*) Figure *d*

65. Which of the following is not in frequency domain.
(*a*) Bode plot
(*b*) Nyquist criterion
(*c*) Root locus (*d*) none

66. If the characteristic equation of a system is $s^2 + 8s + 25 = 0$ the value of ω_n and ξ will be
(*a*) 5 rad/sec, 0.8
(*b*) 0.5 rad/sec. 0.8
(*c*) 8 rad/sec, 0.5
(*d*) $\sqrt{8}$ rad/sec, 5

67. To define the transfer function, all initial condition should be zero.
(*a*) true (*b*) false

68. Laplace transform is not applicable to non-linear system
(*a*) true (*b*) false

69. Nyquist stablity criterion requires polar plot of
(*a*) closed loop transfer function
(*b*) open loop transfer function
(*c*) none of these

70. The phase shift of $G(s) = \frac{1}{s^2}$ is
(*a*) 90° (*b*) –90°
(*c*) 180° (*d*) – 180°

71. The transfer function is defined for
(*a*) linear and time invariant system
(*b*) linear and time variant system
(*c*) non-linear system
(*d*) none of these

72. The error signal in a control system is
(*a*) the diference between the measured value and the set value
(*b*) the sum of measured value and set value
(*c*) ratio of measured value to set value
(*d*) none of these

73. Presence of non-linearities in a control system tends to introduce
(*a*) tranisent error
(*b*) instability
(*c*) steady state error
(*d*) all of these

74. In a critically damped system, the damping factor is of the order of
(*a*) zero (*b*) less than unity
(*c*) unity (*d*) infinity

75. Damping is a function of
(*a*) gain (*b*) $\frac{1}{Gain}$
(*c*) $\sqrt{Gain}$ (*d*) $\frac{1}{\sqrt{Gain}}$

76. The response of a control system having damping factor as unity will be
(*a*) oscillatory (*b*) underdamped
(*c*) critical damped
(*d*) none

77. The rise time for type one second order system is given by

(*a*) $t_r = \dfrac{\pi - \cos^{-1}\left(\dfrac{\sqrt{1-\xi^2}}{\xi}\right)}{\tan\sqrt{1-\xi^2}}$

(*b*) $t_r = \dfrac{\pi - \sin^{-1}\left(\dfrac{\sqrt{1-\xi^2}}{\xi}\right)}{\tan^{-1}\left(\sqrt{1-\xi^2}\right)}$

(*c*) $t_r = \dfrac{\pi - \cos^{-1}\xi}{\omega_n\sqrt{1-\xi^2}}$

(*d*) $t_r = \dfrac{\tan^{-1}\left(\dfrac{\sqrt{1-\xi^2}}{\xi}\right)}{\omega_n\sqrt{1-\xi^2}}$

78. The phase crossover frequency at which the phase angle is
(*a*) – 90° (*b*) 90°
(*c*) 180° (*d*) –180°

79. The steady state error of a type one second order system to unit ramp input is
(*a*) $2\xi\omega_n$ (*b*) $\frac{2\xi}{\omega_n}$
(*c*) $\frac{\omega_n}{2\xi}$ (*d*) $\frac{2\omega_n}{\xi}$

80. The phase margin in terms of damping factor ξ is

(a) $\phi = \tan^{-1}\left[1-2\xi^2\right]^{1/2}$

(b) $\phi = \tan^{-1}\xi$

(c) $\phi = \tan^{-1}\left[2\xi\left\{\frac{1}{(4x^4+1)^{1/2}-2\xi^2}\right\}^{1/2}\right]$

(d) none of these

81. In type one, second order system, the first peak overshoot occurs at a time starting from zero to

(a) $t_p = \frac{\pi\omega_n}{\sqrt{1-\xi^2}}$

(b) $t_p = \frac{\omega_n}{\pi\sqrt{1-\xi^2}}$

(c) $t_p = \frac{\pi}{\omega_n\sqrt{1-\xi^2}}$

(d) $t_p = \frac{\omega_n}{\sqrt{1-\xi^2}}$

82. In the correlation between time and frequency response the resonant frequency is given by

(a) $\omega_r = \omega_n\sqrt{\xi^2-1}$

(b) $\omega_r = \omega_n\sqrt{1-\xi^2}$

(c) $\omega_r = \omega_d\sqrt{1-\xi^2}$

(d) $\omega_r = \omega_d\sqrt{\xi^2-1}$

83. The Bode plot is applicable to

(a) minimum phase network

(b) maximum phase network

(c) all phase network

(d) none of these

84. A mathematical model is linear if the differential equation describing it has coefficients which are

(a) zero

(b) constant

(c) Function only of the independent variable

(d) (b) or (c)

85. Knowledge a of transfer function of a system is necessary for calculation of

(a) time constant

(b) the output for a given input

(c) the order of the system

(d) none of these

86. By increasing the gain K of the system, the steady state error of the system

(a) increase

(b) decrease

(c) remains unaltered

(d) none of these

87. Zero initial condition means that the system is

(a) working with zero stored energy

(b) working with zero reference signal

(c) at rest and no energy is stored in any of its component

(d) none of these

88. The characteristic equation having more number of poles than the zeros, the number of root locus branches will be equal to

(a) no. of zeros

(b) no. of poles

(c) no. of zeros + no. of poles

(d) no. of poles – no. of zeros

89. The resonant peak is given by

(a) $M_r = \frac{1}{2\xi\sqrt{1-\xi^2}}$

(b) $M_r = \frac{1}{\xi\sqrt{1-\xi^2}}$

(c) $M_r = \frac{1}{\xi\sqrt{1+\xi^2}}$

(d) none

90. With feedback, the transient response of the system as compared to that without feedback

(a) deacay slowly

(b) rises at faster rate

(c) rises at slow rate

(d) none

91. The lag network increase the bandwidth of the system

(a) true (b) false

92. Mason's gain formula is not applicable for *SFG*.
(*a*) true (*b*) false

93. The Bode plot is a plot relation log ω with magnitude in decibel and phase angle
(*a*) true (*b*) false

94. For the transfer function

$$G(s) = \frac{K}{s(s^2+6s+10)},$$

the number of asymptotes are
(*a*) 3 (*b*) 2
(*c*) 4 (*d*) 6

95. The angle of asymptotes in question 94 are
(*a*) 60°, 180°, 120°
(*b*) 60°, 180°, 300°
(*c*) 60°, 180°, 200°
(*d*) none of these

96. Phase lag network is used to increase system stability as well as bandwidth
(*a*) true (*b*) false

97. In Nyquist criterion, the stability of the closed loop system can
(*a*) not be found
(*b*) be found by studying the behaviour of the system about a point (1, *j*0)
(*c*) be found by studying the behaviour of the system about a point (–1 + *j*0)
(*d*) none

98. In general, closed loop transfer function
(*a*) $\dfrac{\text{Direct transfer function}}{1+\text{loop transfer function}}$
(*b*) $\dfrac{\text{Direct transfer function}}{\text{Loop transfer function}}$
(*c*) $\dfrac{\text{Loop transfer function}}{\text{Direct transfer function}}$
(*d*) none

99. Closed loop systems are costly and complicated
(*a*) true (*b*) false

100. Marginally stable has some roots with real parts equal to zero, but none with positive real parts.
(*a*) true (*b*) false

Note : Q 101 to Q. 110 are from I.E.S. 2002

101.

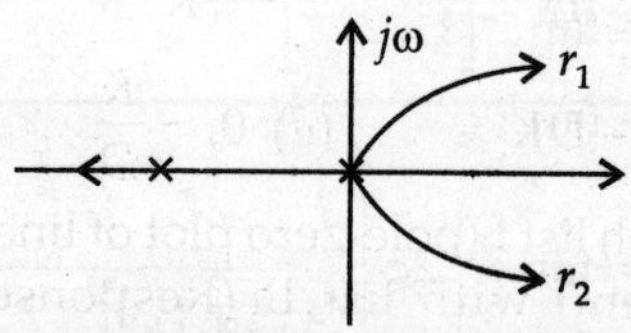

Which of the following is the open loop transfer function of the root locii shown in figure.
(*a*) $\dfrac{K}{s(1+sT_1)^2}$
(*b*) $\dfrac{K}{(s+T_1)(s+T_2)^2}$
(c) $\dfrac{K}{(s+T)^3}$
(*d*) $\dfrac{K}{s^2(1+sT_1)}$

102. A system has a single pole at orgin. Its impulse response will be
(*a*) constant (*b*) ramp
(*c*) decaying exponential
(*d*) oscillatoty

103. A control system has

$$G(s)\,H(s) = \frac{K(s+1)}{s(s+3)(s+4)}$$

Root locus of the system can lie on the real axis
(*a*) between *s* = – 1 and *s* = – 3
(*b*) between s = 0 and s = – 4
(*c*) between *s* = – 3 and *s* = – 4
(*d*) towards left of *s* = – 4

104. The mechanical system shown below has its pole(s) at

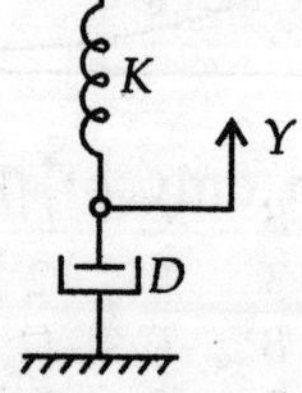

(a) $-\frac{K}{D}$ (b) $-\frac{D}{K}$

(c) $-DK$ (d) $0, -\frac{K}{D}$

105. Match list I (pole-zero plot of linear control system) with list II (Responses of the system) and select the correct answer using the codes given below the lists :

List I

Pole-zero plot of linear control system

A. 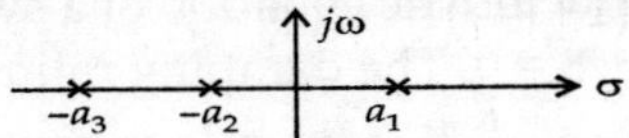

B. 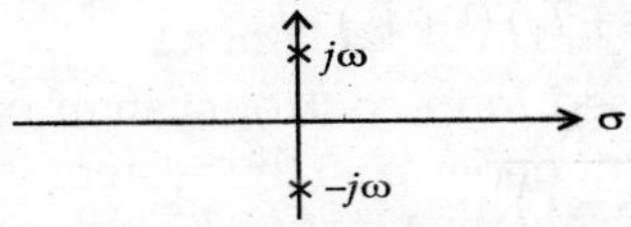

C. 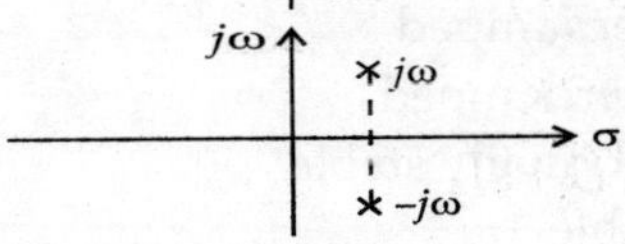

D. 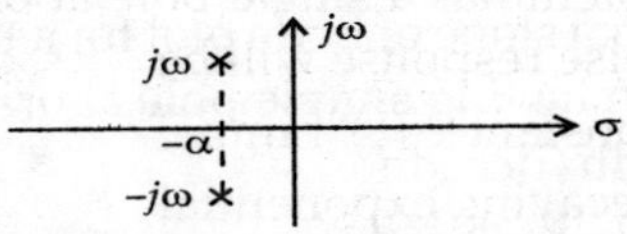

List II

Responses of the system

1. 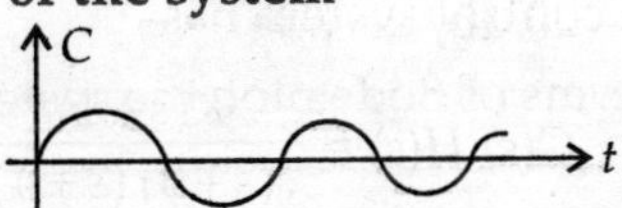

2. 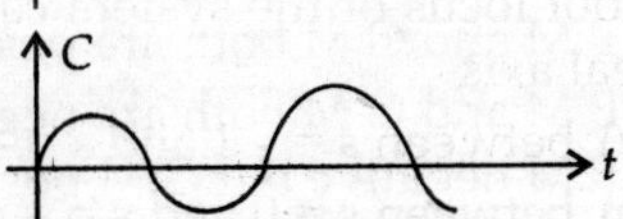

3. 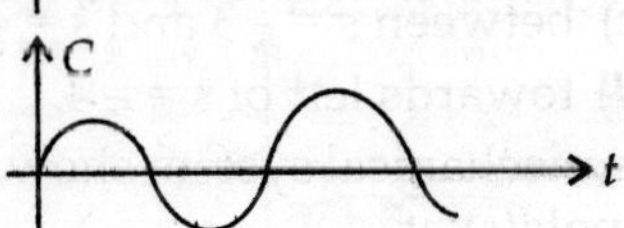

4. 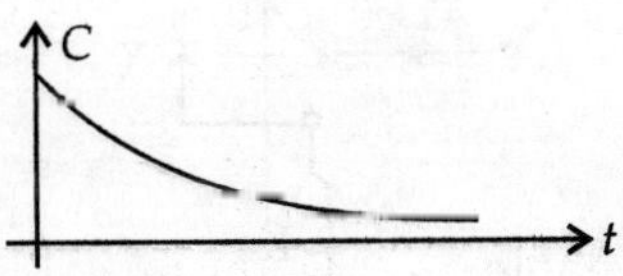

Codes :

	A	B	C	D
(a)	4	3	1	2
(b)	4	3	2	1
(c)	3	4	2	1
(d)	3	4	1	2

106. Consider the following performance characteristics :

1. Reduced velocity constant for a given relative stability.
2. Reduced gain crossover frequency.
3. Reduced bandwidth
4. Reduced resonance peak of the system.

Which of these performance characteristics are achieved with the phase lag compensation.

(a) 1 and 2 (b) 1 and 3

(c) 2, 3 and 4

(d) 1, 2, 3 and 4

107. Which of the following is the Nyquist diagram for the open loop transfer function.

$$G(s).\,H(s) = \frac{5}{s(1+0.1s)\,(1+0.01s)}$$

(a) 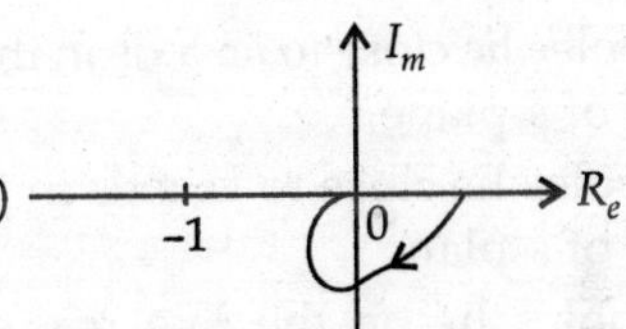

(b) 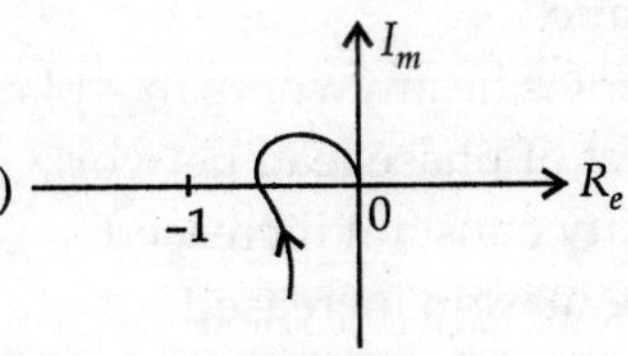

(c) 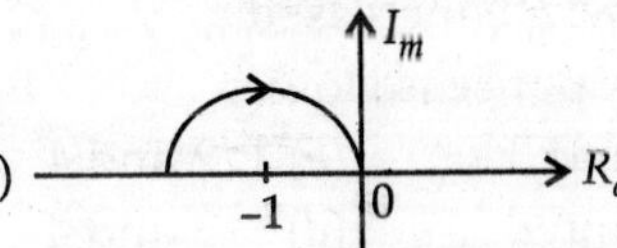

(d) 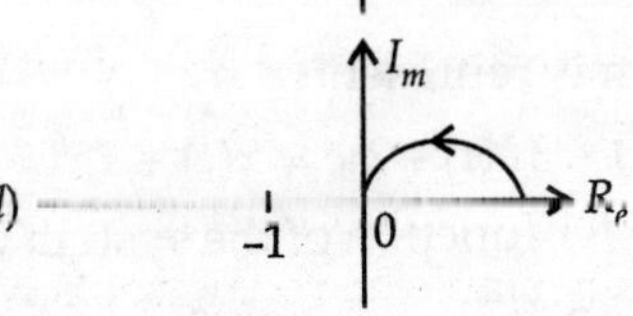

108. Match list I (Type of controller) with list II (operation) and select the correct answer using codes given below the lists

List I
(Type of controller)

A. Pneumatic controller
B. Hydraulic controller
C. Electronic controller

List II
(Operation)

1. Flexible operation
2. High torque, high speed operation
3. Fire and explosion proof operation.

Codes :

	A	B	C
(a)	1	3	2
(b)	1	2	3
(c)	3	1	2
(d)	3	2	1

109. An amplifier will generate stable sinusoidal oscillations if we provide feedback such that
(a) its poles lie close to $j\omega$-axis in the right half of s-plane.
(b) its poles lie close to $j\omega$ axis in the left half of s-plane
(c) its poles lie on the +ve real axis in s-plane.
(d) its poles lie anywhere in s-plane.

110. The effect of phase lead network
1. velocity constant increased
2. phase margin increased
3. bandwidth increased
4. the response is slower
(a) 1, 2 and 3 (b) 1, 2 and 4
(c) 2, 3 and 4 (d) 1, 3 and 4

111. A system is represented by
$$\ddot{y}(t) + 3\dot{y}(t) + 2y = \dot{x}(t) + x$$
the transfer function of the system will be
(a) $\frac{s+2}{s^2+3s+2}$ (b) $\frac{s+1}{s^2+3s+2}$
(c) $\frac{1}{s^2+3s+2}$ (d) $\frac{s}{s^2+3s+2}$

112. Forward transfer function of a closed loop control system is defined as
(a) output/input
(b) input/output
(c) output/actuating signal
(d) actuating signal/output

113. The steady error for a type 2 system subjected to unit ramp input is
(a) 2 (b) 1
(c) 0 (d) ∞

114. The characteristic equation of a system is $s^2 + 2s + 8 = 0$. The damping ratio will be
(a) < 1 (b) > 1
(c) = 1 (d) none

115. If a closed loop control system operates at a point on $j\omega$ axis, the system is
(a) overdamped
(b) underdamped
(c) marginally stable
(d) stable

116. The initial slope of Bode plot for a transfer function having simple pole at origin is
(a) 20 db/decade
(b) – 40 db/decade
(c) 40 db/decade
(d) – 20 db/decade

117. In terms of Bode plot, the system is stable if
(a) P.M = G.M
(b) P.M and G.M both are positive
(c) P.M and G.M both are negative
(d) P.M negative but G.M positive

118. For the given Nyquist plot, the system is

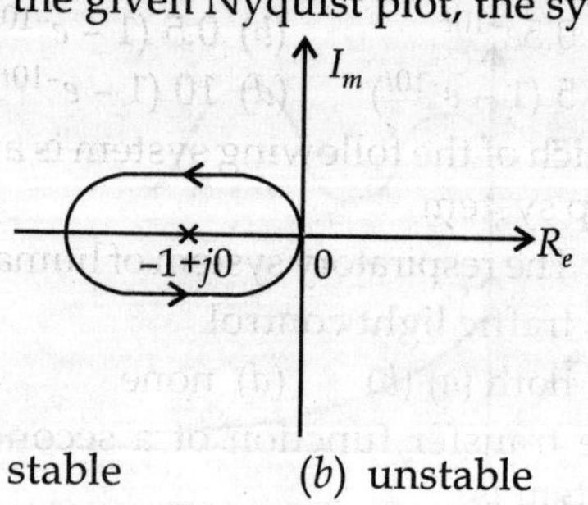

(a) stable (b) unstable
(c) marginally stable
(d) none

119. A feedback compensation is also known as
(a) cascade compensation

(b) parallel compensation
(c) cascade parallel compensation
(d) none

120. The root locii method of analysis of control system gives us
(a) transient response
(b) frequency response
(c) steady state response
(d) both (a) & (c)

121. Lyapunov's stability criterion can be used for determination of
(a) linear system
(b) non-linear system
(c) both (a) and (b)(d) none

122. By the use of PD control to the second order system the rise time
(a) decrease (b) increases
(c) remains same (d) none.

123. The given matrix is

$$\begin{bmatrix} 4 & -4 & 2 \\ -4 & 5 & -2 \\ 2 & -2 & 1 \end{bmatrix}$$

(a) positive semidefinite
(b) negative semidefinite
(c) positive definite
(d) negative definite

124. The rise time for $c(t) = 1 - e^{-t}$ is
(a) 0.2198 s (b) 2.198 s
(c) 21.98 s (d) 2.309 s

125. The impulse response of a system is $5e^{-10t}$. Its step response is equal to
(a) $0.5e^{-10t}$ (b) $0.5(1 - e^{-10t})$
(c) $5(1 - e^{-10t})$ (d) $10(1 - e^{-10t})$

126. Which of the following system is an open-loop system
(a) The respiratory system of human being
(b) traffic light control
(c) both (a) (b) (d) none

127. The transfer function of a second order system is

$$G(s) = \frac{49}{s^2 + 16s + 40},$$

input is step. The peak overshoot will be
(a) 7 (b) 4
(c) $\sqrt{7}$ (d) none.

128. The given system is

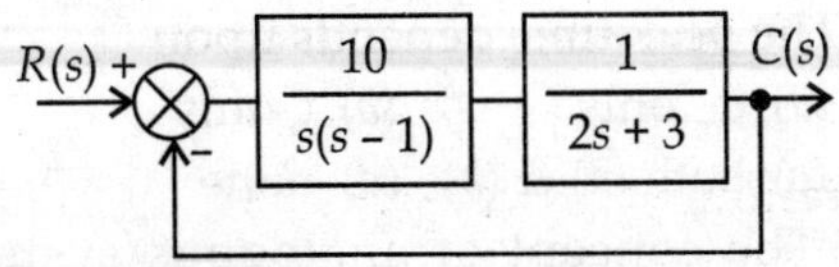

(a) stable (b) unstable
(c) marginally stable
(d) none.

129. The characteristic equation of a system is $s^4 + 2s^3 + 2s^2 + 3s + 6 = 0$. The no. of roots on RHP are
(a) 4 (b) 2
(c) 3 (d) 6

130. If $A = \begin{bmatrix} 1 & 0 \\ 1 & 1 \end{bmatrix}$ the e^{At} will be

(a) $\begin{bmatrix} e^t & 0 \\ te^t & e^t \end{bmatrix}$ (b) $\begin{bmatrix} 0 & e^t \\ e^t & te^t \end{bmatrix}$

(c) $\begin{bmatrix} e^t & 0 \\ e^t & te^t \end{bmatrix}$ (d) $\begin{bmatrix} te^t & 0 \\ e^t & e^t \end{bmatrix}$

131. If $A = \begin{bmatrix} 0.5 & 0 \\ 0 & -2 \end{bmatrix}$, $B = \begin{bmatrix} 0 \\ 1 \end{bmatrix}$, then
(a) system is controllable
(b) system is uncontrollable
(c) system is undefined
(d) none.

132. For $G(s) = \frac{s+1}{s(0.5s+1)}$. The corner frequencies are
(a) 0.5 and 1 (c) 0 and 2
(b) 2 and 1 (d) none

133. If the system has multiple poles on imaginary axis the system is
(a) stable (b) unstable
(c) marginally stable
(d) none

134. If $G(s)\,H(s) = \frac{k}{s(s^2 + 4s + 13)}$. the angle of asymptotes will be

(a) 45°, 135°, 225°
(b) 60°, 180°, 300°
(c) 90°, 180°, 270°
(d) 45°, 90°, 300°

135. The peak time depends upon
(a) ω_n only (b) ζ only
(c) both (a) & (b) (d) none.

136. The concept of analogous system is applicable to
(a) linear system only
(b) non-linear system only
(c) both (a) & (b) (d) none

137. In mechanical systems the force of friction acts
(a) in the direction of motion
(b) opposite to the direction of motion
(c) proportional to displacement
(d) none

138. In mechanical system, the spring force is proportional to
(a) motion (c) displacement
(b) acceleration
(d) mass

139. Which of the following is not a electro-mechanical system
(a) stepper motor
(b) transformer
(c) synchro (d) D.C motor

140. For the gear system shown in figure $\frac{\omega_2}{\omega_1}$ will be

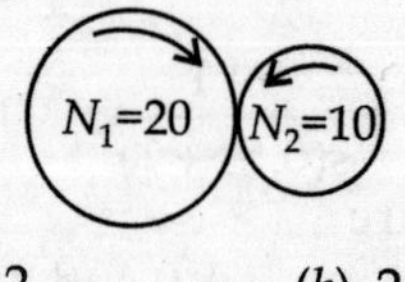

(a) 1/2 (b) 2
(c) 20 (d) 10

141. The steady state error is given by
(a) $e_{ss} = \lim_{s\to 0} sE(s)$
(c) $e_{ss} = \lim_{s\to 0} sE(s)$
(b) $e_{ss} = \lim_{s\to 0} E(s)$
(d) $e_{ss} = \lim_{s\to \infty} E(s)$

142. For given open loop transfer function the breakaway points are

$$G(s)\,H(s) = \frac{k(s+2)}{(s+1)^2}$$

(a) 1, 3 (b) −1, 3
(c) −1, −3 (d) 1, − 3

143. If $\zeta = 0.4$, then peak resonance Mr will be
(a) 1.36 (b) 2.36
(c) 13.6 (d) 0.136

144. For type 3 system, the initial slope of Bode plot will be
(a) − 20 db/decade
(b) + 60 db/decade
(c) + 20 db/decade
(d) − 60 db/decade

145. For given block diagram $C(s)/R(s)$ will be

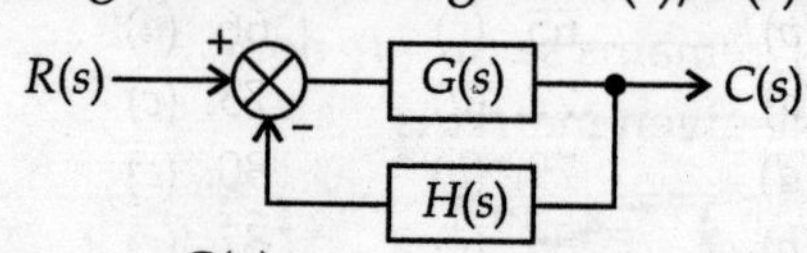

(a) $\frac{G(s)}{1-G(s)\,H(s)}$
(b) $\frac{G(s)}{1+G(s)\,H(s)}$
(c) $\frac{H(s)}{1-G(s)\,H(s)}$
(d) $\frac{H(s)}{1+G(s)\,H(s)}$

146. A unity feedback system has open loop transfer function

$$G(s) = \frac{9}{s(s+3)}$$

(a) $\omega_n = 9$ (b) $\omega_n = 5$
(c) $\zeta = 0.5$ (d) b & c

147. For closed loop transfer function with unity feedback
(a) $\frac{E(s)}{R(s)} = \frac{1}{1+G(s)}$
(b) $\frac{E(s)}{R(s)} = \frac{1}{1-G(s)}$
(c) $\frac{E(s)}{R(s)} = \frac{G(s)}{1+G(s)}$
(d) $\frac{E(s)}{R(s)} = \frac{G(s)}{1-G(s)}$

148. Synchro is a
    (*a*) electromagnetic transducer
    (*b*) electromechanical transducer
    (*c*) d.c motor (*d*) none
149. Laplace transform of $3e^{-t} - e^{-2t}$ is
    (*a*) $\dfrac{s+5}{s^2+3s+2}$ (*b*) $\dfrac{2s-5}{s^2+3s+2}$
    (*c*) $\dfrac{2s+5}{s^2-3s+2}$ (*d*) $\dfrac{2s+5}{s^2+3s+2}$

ANSWERS

1. (*a*) 2. (*d*) 3. (*a*) 4. (*a*) 5. (*b*) 6. (*a*) 7. (*a*)
8. (*c*) 9. (*b*) 10. (*b*) 11. (*b*) 12. (*c*) 13. (*a*) 14. (*d*)
15. (*a*) 16. (*a*) 17. (*d*) 18. (*a*) 19. (*b*) 20. (*a*) 21. (*b*)
22. (*d*) 23. (*a*) 24. (*b*) 25. (*a*) 26. (*b*) 27. (*b*) 28. (*c*)
29. (*a*) 30. (*b*) 31. (*d*) 32. (*d*) 33. (*c*) 34. (*b*) 35. (*b*)
36. (*c*) 37. (*a*) 38. (*b*) 39. (*a*) 40. (*a*) 41. (*b*) 42. (*a*)
43. (*a*) 44. (*c*) 45. (*b*) 46. (*b*) 47. (*a*) 48. (*a*) 49. (*c*)
50. (*b*) 51. (*c*) 52. (*b*) 53. (*a*) 54. (*b*) 55. (*b*) 56. (*a*)
57. (*c*) 58. (*b*) 59. (*b*) 60. (*c*) 61. (*a*) 62. (*u*) 63. (*b*)
64. (*b*) 65. (*c*) 66. (*a*) 67. (*a*) 68. (*a*) 69. (*b*) 70. (*d*)
71. (*a*) 72. (*a*) 73. (*c*) 74. (*c*) 75. (*d*) 76. (*c*) 77. (*c*)
78. (*d*) 79. (*b*) 80. (*c*) 81. (*c*) 82. (*b*) 83. (*a*) 84. (*d*)
85. (*b*) 86. (*c*) 87. (*c*) 88. (*b*) 89. (*a*) 90. (*b*) 91. (*b*)
92. (*b*) 93. (*a*) 94. (*a*) 95. (*b*) 96. (*b*) 97. (*c*) 98. (*a*)
99. (*a*) 100. (*a*) 101. (*d*) 102. (*a*) 103. (*c*) 104. (*a*) 105. (*b*)
106. (*c*) 107. (*b*) 108. (*d*) 109. (*b*) 110. (*a*) 111. (*b*) 112. (*c*)
113. (*c*) 114. (*a*) 115. (*c*) 116. (*d*) 117. (*c*) 118. (*a*) 119. (*b*)
120. (*d*) 121. (*c*) 122. (*a*) 123. (*a*) 124. (*b*) 125. (*b*) 126. (*a*)
127. (*d*) 128. (*b*) 129. (*b*) 130. (*a*) 131. (*b*) 132. (*d*) 133. (*b*)
134. (*b*) 135. (*c*) 136. (*a*) 137. (*b*) 138. (*c*) 139. (*b*) 140. (*b*)
141. (*a*) 142. (*c*) 143. (*a*) 144. (*d*) 145. (*b*) 146. (*d*) 147. (*a*)
148. (*b*) 149. (*d*)

❑❑❑

MODEL QUESTION PAPERS

MODEL PAPER-1

Time : 3 Hours **Total Marks – 100**

Note : *Attempt ALL the questions.*

1. Attempt any FOUR parts of the following : **(5 × 4) = 20**

(*a*) why is the stability of a linear system independent of the initial conditions and input signals?

(*b*) For the block diagram draw the signal flow graph and find $C(s)/R(s)$.

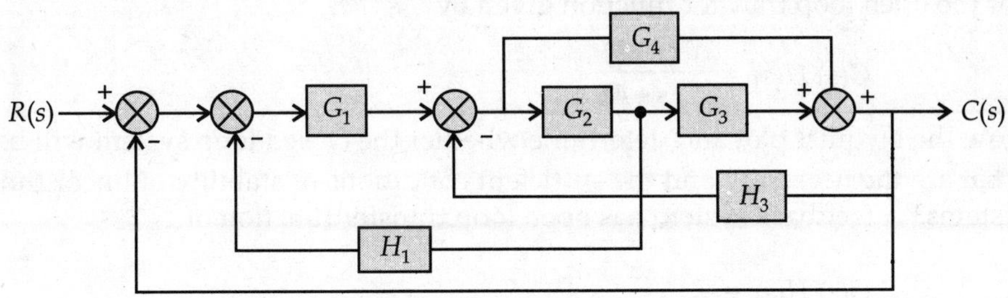

Fig. 1.

(*c*) For a unity feedback system whose open loop transfer function is $G(s) = \dfrac{50}{(1+0.1s)(1+2s)}$, find the position, velocity and acceleration error constants.

(*d*) A certain feedback system is described by the following transfer function

$$G(s) = \frac{12}{s^2+4s+16}, \quad H(s) = Ks$$

The damping factor of the system is 0.8. Determine the overshoot of the system.

(*e*) Derive the unit step response to a second order system.

2. Attempt any FOUR parts. **(5 × 4 = 20)**

(*a*) Draw the Bode plot (magnitude only) for the following transfer function and determine the gain cross-over frequency

$$G(s) = \frac{10}{s(1+0.5s)(1+0.25s)}$$

(*b*) Sketch the polar plot for the following transfer function

$$G(s) = \frac{1}{s(1+s)(1+2s)}$$

(*c*) A feedback control system is described as

$$G(s) = \frac{50}{s(s+2)(s+5)}, \; H(s) = \frac{1}{s}$$

For a unit step input, determine the steady state error constants and errors.

(*d*) Discuss the advantages and limitations of frequency response method of analysis for control systems.

(*e*) Define the following terms in reference to Bode plots for a given transfer function.

(*i*) Phase crossover frequency

(*ii*) Gain crossover frequency

(*iii*) Phase margin

(*iv*) Gain Margin

3. Attempt any TWO parts **(10 × 2 = 20)**

(*a*) Determine the range of k over which the following characteristic polynomials belong to stable systems.

(*i*) $s^3 + 6s^2 + 11s + k$

(*ii*) $s^4 + 5s^3 + 9s^2 + 20s + k$

(*b*) For the open loop transfer function given by

$$G(s)\,H(s) = \frac{10\,(s+1)}{s^2\,(s+4)}$$

draw the Nyquist plot and determine whether the closed loop system will be stable.

(*c*) What are the necessary and the sufficient conditions of stability of linear time-invariant systems? A feedback system has open loop transfer function of

$$G(s)\,H(s) = \frac{ke^{-s}}{s\,(s^2+5s+9)}$$

determine by use of Routh criterion, the maximum value of k for the closed loop system to be stable.

4. Attempt any TWO parts **(10 × 2)**

(*a*) Design a lead compensator for a plant

$$G(s) = \frac{1}{s\,(s+1)}$$

Assume that the specifications for the transient response and such that $\zeta = 0.707$ and $\omega_n = 2$ rad/sec for the desired closed loop poles will give a satisfactory response.

(*b*) Design a suitable cascade lead compensator for the system with the following open loop transfer function

$$G(s) = \frac{4}{s\,(s+2)}$$

to meet the following specifications

(*i*) $k_v = 20\ \text{sec}^{-1}$

(*ii*) P.M $\geq 50°$

(*iii*) G.M ≥ 10 db

(*c*) Explain what do you understand by cascade-lead compensation of linear control system? Give the transfer function of a typical lead compensator and explain its basic characteristics in reference to Bode plot.

5. Attempt any TWO parts **(10 × 2 = 20)**

(*a*) A system is characterised by the transfer function $\dfrac{Y(s)}{u(s)} = \dfrac{2}{s^3+6s^2+11s+6}$ obtain state space model and determine whether or not the system is controllable and observable.

(*b*) Discuss the advantages of state-space representation of systems. Point out the significance of state transition matrix is solving equations.

(c) Obtain phase variable representation for a system whose transfer function is given by

$$\frac{Y(s)}{u(s)} = \frac{6s^3 + 4s^2 + 3s + 10}{s^3 + 8s^2 + 4s + 20}$$

ANSWERS

1. (b) $\dfrac{G_1G_4 + G_1G_2G_3}{1 + G_4H_2 + G_4G_1 + G_2G_3H_2 + G_1G_2G_3 + G_1G_2H_1}$

1. (c) $k_p = 50, k_v = 0, k_a = 0$

1. (d) 1.5%

2. (c) $k_p = \infty, k_v = \infty, k_a = 5, e_{ss} = 0$

3. (a) $0 < k < 66$ and $0 < k < 20$

3. (b) stable

3. (c) $k = 7.5$

5. (a) controllable, observable

5. (c) $$\begin{bmatrix} \dot{x}_1 \\ \dot{x}_2 \\ \dot{x}_3 \end{bmatrix} = \begin{bmatrix} -8 & 1 & 0 \\ -4 & 0 & 1 \\ -20 & 0 & 0 \end{bmatrix} \begin{bmatrix} x_1 \\ x_2 \\ x_3 \end{bmatrix} + \begin{bmatrix} -44 \\ -21 \\ -110 \end{bmatrix} u$$

$$y = \begin{bmatrix} 1 & 0 & 0 \end{bmatrix} \begin{bmatrix} x_1 \\ x_2 \\ x_3 \end{bmatrix} + 6u$$

MODEL PAPER-2

Time : 3 Hours **Total Marks – 100**

Note : *Attempt ALL Five questions.*

1. Attempt all FOUR parts **(5 × 4 = 20)**

(*a*) What is the closed loop transfer function of a system with positive feedback? Explain what is the effect of positive feedback on stability.

(*b*) Determine the damping ratio, undamped natural frequency, damped natural frequency, damping coefficient and time constant for the following second order system.

$$\frac{d^2y}{dt^2}+2\frac{dy}{dt}+2y = 2x$$

(*c*) Find the transfer function of the control system shown below

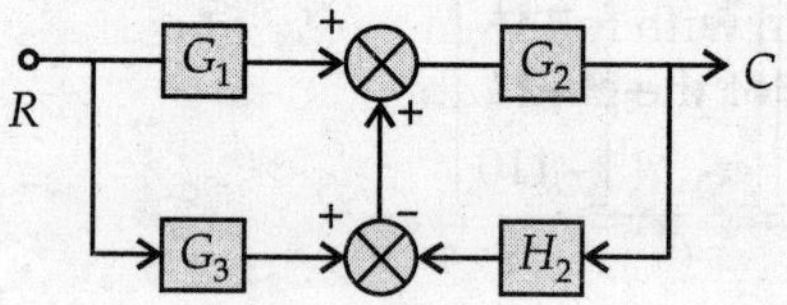

Fig. 1

(*d*) Determine the transfer function of a control system shown below and derive an expression for the output for a step input of magnitude 2 units.

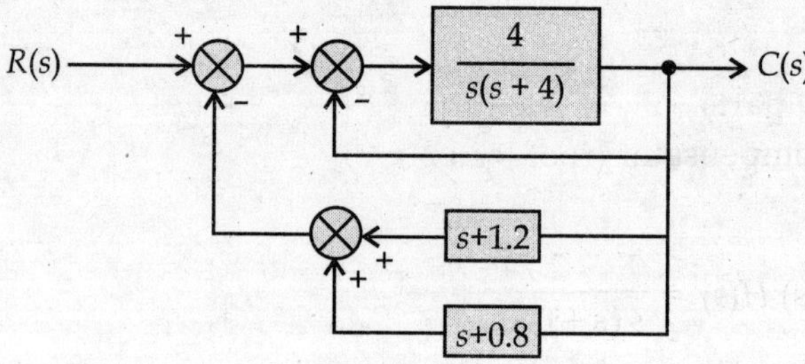

Fig. 2.

(*e*) State the D 'Alembert's principle and explain with the help of an example.

2. Attempt any TWO parts **(10 × 2 = 20)**

(*a*) The open loop transfer function of a unity feedback control system is

$$G(s) = \frac{k}{s\,(1+sT)}$$

A unit step input is applied to this system. Determine the values of k and T so that the maximum overshoot = 26% and the resonant frequency ω_r = 8 rad/sec. Calculate (*i*) resonant peak (*ii*) gain cross over frequency (*iii*) phase margin.

(*b*) Draw the asymptotic Bode plot for the transfer function

$$G(s)\,H(s) = \frac{2(s+0.25)}{s^2\,(s+1)\,(s+0.5)}$$

and determine

(*a*) static error coefficient
(*b*) phase cross-over frequency
(*c*) gain cross-over frequency
(*d*) gain margin
(*e*) phase margin
(*f*) stability of the system.

(*c*) Establish the relationship between the static error constants and the generalised error constants of a unity feedback control system.

3. Attempt any two parts **(10 × 2 = 20)**

(*a*) State Routh-Hurwitz criterion for stability and apply this criterion determine (*i*) no. of roots with positive real part (*ii*) no. of roots with zero real parts (*iii*) no. of roots with negative real part of the following equations

(*a*) $s^4 + 2s^3 + 7s^2 + 10s + 10 = 0$

(*b*) $s^5 - s^4 - 2s^3 + 2s^2 - 8s + 8 = 0$

(*b*) Is a closed loop system with following open loop transfer function and with $k = 2$ stable? Find the critical value of the gain k for stability.

$$G(s)\,H(s) = \frac{k}{s\,(s+1)\,(2s+1)}$$

(*c*) Sketch the root locus for the following given system as k varies from 0 to ∞

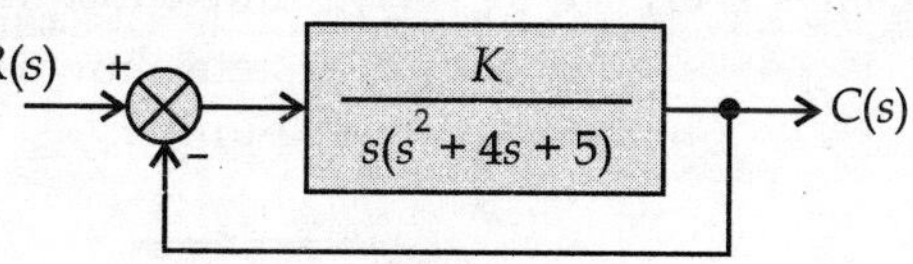

Fig. 3.

4. Attempt any TWO parts **(10 × 2 = 20)**

(*a*) Design a lag compensator (root locus) for $k_v \geq 5$, $\zeta = 0.707$ & $\omega_n = 0.81$ for

$$G(s)\,H(s) = \frac{2.74}{s\,(s+1)\,(s+4)}$$

(*b*) Design a lead compensator such that the closed loop system will have P.M of 50° and gain margin of not less than 10 db.

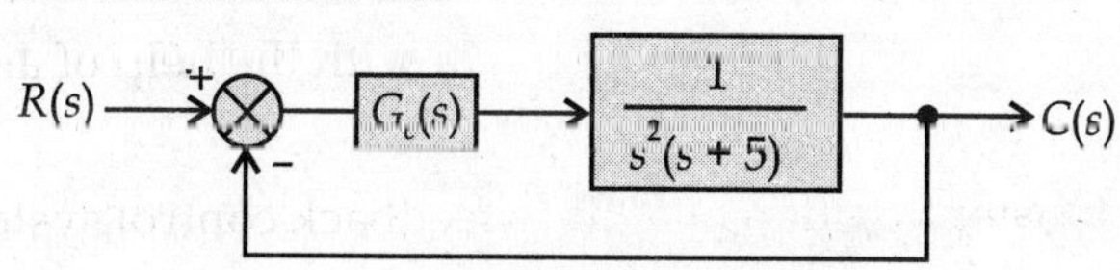

Fig. 4.

(*c*) What are the various types of compensation schemes used in control systems? Explain the concept of cascade lead compensation of control system with the help of an example.

5. Attempt any two parts. **(10 × 2 = 20)**

(*a*) Find the stability of the system whose system matrix is given by

$$\begin{bmatrix} 5 & -6 & -12 \\ -1 & 1 & 2 \\ 5 & -6 & -11 \end{bmatrix}$$

(*b*) Derive the state variable model for the system shown below

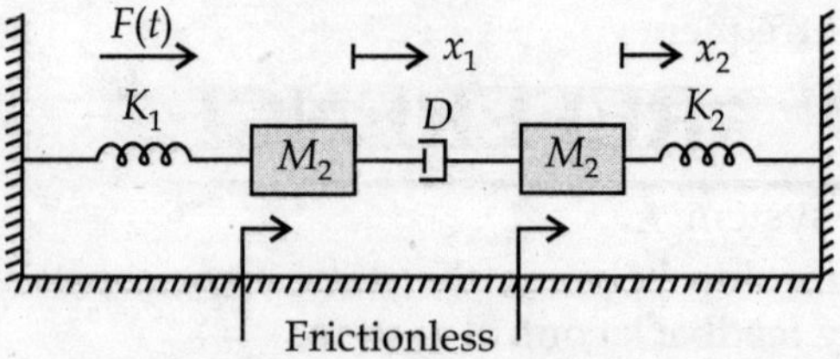

Fig. 5.

(*c*) A system is described by

$$\dot{x} = \begin{bmatrix} -1 & -4 & -1 \\ -1 & -6 & -2 \\ -1 & -2 & -3 \end{bmatrix} x + \begin{bmatrix} 0 \\ 1 \\ 1 \end{bmatrix} u \quad y = \begin{bmatrix} 1 & 1 & 1 \end{bmatrix} x$$

Find the transfer function

ANSWERS

1. (*b*) damping coefficient $\zeta\omega n = 1$, $\omega_d = 1$, $\omega_n = \sqrt{2}$, time constant $\frac{1}{\zeta\omega_n} = 1$, $\zeta = 1/\sqrt{2}$

1. (*c*) $\frac{C}{R} = \frac{G_1G_2 + G_2G_3}{1 + G_2H_2}$

1. (*d*) $\frac{C}{R} = \frac{4}{(s+2)(s+6)}$, $c(t) = \frac{2}{3} - e^{-2t} + \frac{1}{3}e^{-6t}$

2. (*a*) $k = 11.9$, $T = 0.13$, $M_{P\omega} = 1.39$, P.M = 46.5°, gain cross-over frequency = 7.47 rad/sec.

2. (*b*) 1, 0.4 rad/sec, 1.24 rad/sec, – 20db, – 35°, unstable

3. (*a*) (0, 2, 2) & (2, 2, 1)

3. (*b*) $0 < k < 1.5$, with $k = 2$ system unstable

5. (*a*) stable

5. (*c*) $\frac{2s^2 + 2s - 3}{s^3 + 10s^2 + 18s + 6}$

□□□

REFERENCES

1. CHI-TSONG CHEN, Linear system Theory and Design, Harcourt Brace College Publishers, USA.
2. **David K. CHENG.** Aanalysis of Linear Systems, Addison- wesley Publishing Company.
3. K. Ogata. Modern Control Engineering. Prentice-Hall of India Pvt. Ltd New Delhi, third edition.
4. B.C.Kuo. Automatic Control Systems. Prentice- Hall of India Pvt. Ltd. New Delhi, Seventh edition.
5. I. J. Nagrath and M. Gopal, Control System Engineering. New Age International Publishers, New Delhi. Third edition.
6. Francis H. Raven. Automatic Control Engineering, Mc Graw-Hill International Editions. Fifth edition.
7. N.K. Sinha, Control Systems. New Age International Publishers, Third edition.
8. Ashfaq Husain, Circuit Theory. CBS Publishers and Distributors, New Delhi. First edition 1992.

INDEX

H

I

K

L

M

N

O

P

R

S

Win Prizes

ATTENTION READERS

Win Prizes

We request for your frank assessment regarding various aspects of the book as given below :

CONTROL SYSTEMS-I

By: S. HASAN SAEED

Please Fill up the Sheet in Neat and Clean Writing

(i) What topics of your syllabus, important for your examination point of view, are not covered in this book ?

....

....

....

....

(ii) What are the chapters wherein the treatment of the subject matter is not systematic or organised or updated ?

....

....

....

....

(iii) Have you come across misprints/mistakes/factual inaccuracies in the book? Please specify the chapters and the page numbers.

....

....

....

....

(iv) What other books on the same subject have you found/heard better than the present book? Please specify in terms of price and quality

....

....

....

....

(v) Further suggestions and comments for the improvement of this book.

....

....

....

....

(PS: If need be please attach a separate sheet to write your views)

OTHER DETAILS

(i) Who recommended you this book ? (Please tick (✓) in the box near the option relevant to you.

☐ Teacher ☐ Friends ☐ Bookseller

....

....

....

....

(ii) Name of the recommending teacher, his designation and address :

....

....

....

....

(iii) Name and address of the bookseller you purchased the book from :

....

....

....

....

(iv) Name and address of your institution (Please mention the University or Board as the case may be)

....

....

....

....

(v) Your name and complete postal address :

....

....

....

....

(vi) Course you are enrolled in :

....

....

....

....

The best assessment will be awarded each month: The award will be in the form of our publications as decided by the Editorial Board.

Please mail the filled up coupon at your earliest to :

S.K. KATARIA & SONS®

4885/109, Prakash Mahal, Dr. Subhash Bhargav Lane,
Opposite Delhi Medical Association, New Delhi–110002
☎: +91-11-23243489, 23269384
Email : katariabooks@yahoo.com
katariabook@gmail.com